THE OXFORD HAND

ALGORITHMIC

MUSIC

Edited by

ALEX MCLEAN

and

ROGER T. DEAN

OXFORD

UNIVERSITY PRESS

OXFORD
UNIVERSITY PRESS

Oxford University Press is a department of the University of Oxford. It furthers
the University's objective of excellence in research, scholarship, and education
by publishing worldwide. Oxford is a registered trade mark of Oxford University
Press in the UK and certain other countries.

Published in the United States of America by Oxford University Press
198 Madison Avenue, New York, NY 10016, United States of America.

Library of Congress Cataloging-in-Publication Data
Names: McLean, Alex | Dean, R. T.
Title: The Oxford handbook of algorithmic music/edited by Alex McLean & Roger T. Dean.
Description: New York, NY: Oxford University Press, [2018] |
Series: Oxford handbooks |
Includes bibliographical references and index.
Identifiers: LCCN 2017013174| ISBN 9780190226992 (cloth: alk. paper) |
ISBN 9780197554364 (paper: alk. paper) |ISBN 9780190227012 (oxford handbooks online)
Subjects: LCSH: Computer composition.
Classification: LCC MT56.O94 2017 | DDC 781.3/4—dc23
LC record available at https://lccn.loc.gov/2017013174

CONTENTS

PART II WHAT CAN ALGORITHMS IN MUSIC DO?

PART III PURPOSES OF ALGORITHMS FOR THE MUSIC MAKER

PART IV   ALGORITHMIC CULTURE

Contributors

Torsten Anders, Senior Lecturer in Media Arts and Course Leader for Music Technology, University of Bedfordshire

Sarah Angliss, Visiting Research Fellow, Sound Practice Research Unit, Goldsmiths, University of London

Jan Beran, Professor, Department of Mathematics and Statistics, University of Konstanz

Oliver Bown, Senior Lecturer, Arts and Design, University of New South Wales, Sydney

Andrew Brown, Professor of Digital Arts, Queensland Conservatorium, Griffith University

Jamie Bullock, Integra Lab, Birmingham Conservatoire

Warren Burt, composer and lecturer, Box Hill Institute, Melbourne

Baptiste Caramiaux, Marie Skłodowska-Curie Research Fellow at McGill University and IRCAM

Alexandra Cárdenas, composer and improviser, Berlin

Nick Collins, Reader in Composition in the Department of Music, University of Durham

Geoff Cox, Associate Professor, Department of Aesthetics and Communication and Participatory IT Research Centre, Aarhus University

Palle Dahlstedt, Obel Professor of Art and Technology, Department of Communication and Psychology, Aalborg University

Roger T. Dean, Research Professor of Sonic Communication at the MARCS Institute for Brain, Behaviour and Development, Western Sydney University

Alice Eldridge, Research Fellow in Digital Humanities/Digital Performance, Department of Music, University of Sussex

Kristin Grace Erickson, Technical Coordinator, Digital Arts and New Media, University of California Santa Cruz

Mark Fell, independent artist, Rotherham

Rebecca Fiebrink, Lecturer, Department of Computing, Goldsmiths, University of London

Jamie Forth, Lecturer, Department of Computing, Goldsmiths, University of London

Christopher Haworth, Lecturer in Music, University of Birmingham

Mileece I'Anson, Founder, Children of Wild, Los Angeles

David Kanaga, independent composer and designer, Oakland

Yuli Levtov, founder of Reactify, London

George E. Lewis, Edwin H. Case Professor of American Music, Columbia University, New York

Thor Magnusson, Senior Lecturer in Music, Department of Music, University of Sussex

Charles Matthews, composer and researcher, London

Kaffe Matthews, founder of Bicrophonic Research Institute, London

Alex McLean, postdoctoral researcher in Weaving as a Technical Mode of Existence, Research Institute for the History of Science and Technology, Deutsches Museum, Munich

Andrew J. Milne, Australian Research Council DECRA Fellow at the MARCS Institute for Brain, Behaviour and Development, Western Sydney University

David Ogborn, Associate Professor, Communication Studies and Multimedia, McMaster University

Morten Riis, postdoctoral researcher, School of Communication and Culture, Aarhus University

Charles Roberts, Assistant Professor in the School of Interactive Games and Media at the Rochester Institute of Technology

Julian Rohrhuber, Professor for Music Informatics and Media Theory, Robert Schumann School of Music and Media, Düsseldorf

Carla Scaletti, President and Founder of Symbolic Sound Corporation

Jan C. Schacher, Research Associate, Institute for Computer Music and Sound Technology, Zürcher Hochschule der Künste

Margaret Schedel, Assistant Professor, Music Department, Stony Brook University

Mary Simoni, Dean of the School of Humanities, Arts and Social Sciences, Rensselaer Polytechnic Institute

Laurie Spiegel, composer, New York

Graham Wakefield, Assistant Professor, School of the Arts, Media, Performance and Design, York University

Renate Wieser, Researcher in Media Aesthetics, University of Paderborn

Geraint Wiggins, Professor of Computational Creativity, School of Electronic Engineering and Computer Science, Queen Mary, University of London

THE OXFORD HANDBOOK OF

ALGORITHMIC MUSIC

PART I

GROUNDING ALGORITHMIC MUSIC

....

CHAPTER 1

MUSICAL ALGORITHMS AS TOOLS, LANGUAGES, AND PARTNERS

A Perspective

....

ALEX MCLEAN AND ROGER T. DEAN

1.1 INTRODUCTION

....

IN this chapter we introduce the landscape of algorithmic music, and point to some of its burning issues and future possibilities. We also use the chapter to provide some guidance as to how we have organized the book, and where major topics are discussed: we summarize the structure in the next paragraph, and comments on individual articles in the book are made throughout this chapter. The book has been arranged to provide contrasting views on core topics, on the one hand, from theorists and analysts and, on the other, from practitioners. Happily for us, in many cases our authors pursue both types of involvement. But we have asked another group of authors (who contribute the Perspectives on Practice chapters) to foreground their own thoughts about algorithmic music and how they make it. At the same time, we have encouraged all authors to give specific musical examples of what they discuss, and to feel free to mention their own work.

Our volume brings together a diverse range of authors to explore algorithmic music in the large. We engage with meta- and post-human perspectives—pointing to the question of what new musics are now being found through algorithmic means which humans could not otherwise have made. In reciprocation with this, we also explore cultural aspects—how is algorithmic music being assimilated back into human culture, and what is its social function or meaning? Over the chapters we will gradually widen our scope, first grounding the topic and introducing its terms by exploring its artefacts, philosophies, and histories. We then survey the range (so far!) of technical approaches to composing algorithmic music, and the metaphors used that seek to install those approaches in human understanding. Then practical aspects are explored in some depth: the role of the algorithm as co-performer, and in supporting musical coordination between human performers. Finally, we explore wider cultural aspects, such as the role of algorithmic music in society, education, and commerce.

Perspectives on Practice sections (PoPs) are interspersed throughout the book as short interjections outside the main flow. But they provide prime value to the reader, connecting issues in the text with direct reflections on musical activity. PoPs provide introspection by authors on their own practice, as opposed to introductions to and analysis of the field provided by the other chapters. Authorship of chapters and PoPs divides roughly along the lines of researchers and practitioners, but not strictly; we include some practitioners who are independent researchers amongst chapter authors, and invite some respected researchers to reflect upon their practice. We have included ourselves amongst the PoPs in the form of a joint article at the end of the book: amongst other things, this serves to indicate why we came together to catalyse this volume (supported by the encouragement and enthusiasm of our OUP editor, Norm Hirschy).

This brief introduction may seem quite episodic, even sometimes temporarily taking surprising directions. But we hope it will sensitize readers to the wide range of topics to be addressed, and that after reading the book they will be left with practical understanding both of how algorithmic music is made and of what makes this activity musical.

1.2 BACKGROUND

An algorithm, essentially, is a finite sequence or structure of instructions, and we will elaborate on this terminologically in the next section. We note here that our emphasis is on algorithmic music-making, and on primarily digital computational approaches. The histories of manual and analog algorithms (mainly addressed in this chapter and by Collins in chapter 4), include experimental process music, from the 1960s in particular (by Brecht, Wolff, Glass, Stockhausen, and others), in which a piece was constructed from a precise or vague process description. But they also include musical and musicological theory (from Ancient Greece to Hiller, Ligeti, Xenakis, and beyond) and its algorithmic embodiments, including musical style modelling (Cope, Ebcioğlu, and onwards).

Leigh Landy has elaborated the often useful distinction between note-based and sound-based musics (Landy 2009). Note-based music involves a conception of discrete sequences of events, largely capable of being described in symbols (such as score notation), and most usually characterized in part by pitches: such music permits tonal, harmonic, and rhythmic hierarchies. On the other hand, sound-based music puts more emphasis on the spectral content of sounds, which may be slowly transforming, with relatively fewer discrete events and little emphasis on pitch (and usually no spectral hierarchy, but rather in-depth spectral organization). Sound-based music may also depend less on rhythmic pattern than note-based. Using this framework, we can summarize by saying that the book deals primarily with the process of creating tonal, post-tonal and sound-based music, and any other forms of music in which innovative and individual works can result from algorithmic approaches. We do not seek to provide anything more than pointers to the major algorithmic composers of the period up to about 1990, since our emphasis is on process, methods, ideas, and developments. We put little focus on work directed towards recreating earlier styles. Similarly, we try to distinguish music which is mainly an overt fulfilment of the algorithm (which might be primarily of academic interest) from that which can be musically creative. Nierhaus (2009) mentions these issues, and considers that most approaches he discerns are primarily about algorithmic 'imitation'. He covers substantially this imitative or recreative aspect and aspects concerning practical procedures. So we resurvey these aspects relatively briefly but adequately, and with a different perspective: essentially, whether and how any particular algorithmic approach might provide a path to music which is really new.

A related general question within or behind several chapters is: has there been any evaluation of the algorithm and of its products? We argue this needs to be done both by non-experts and by expert musicians, and that the field at large could usefully study features present in both the most and the least favourably evaluated works. This relates to the question of the relevance of algorithmic generation methods to perception and cognition, and the argument that syntax is either not a relevant concept in much music or it is a consistency created perhaps transiently, and in any case locally to a work.

We are now well placed in the development of algorithmic music to build upon the technical and evaluative issues to discuss cultural issues. There are a great many subcultures supporting the development of algorithmic music in diverse situations, from the close rituals of performance to the mass-market activity of smartphone and tablet 'apps'. We therefore round off the book with a section rich with viewpoints on the social function and cultural value of algorithmic music; where we use algorithms to reach for the post-human, how do cultures grow and adjust to bring the music back into the human realm?

1.3 TERMINOLOGY AND CURRENT USAGE

Defining a term such as algorithmic music is not straightforward. For one thing, it includes the word music, which stands for an unfathomable diversity of approaches, cultures, techniques, forms, and activities. Alone, the word *algorithm* can be understood as a well-defined set of operations or rules, but this definition is not of great use in understanding algorithmic music. This is because the range of structures which can be considered algorithmic is extremely broad, encompassing all computer programs (or according to some definitions, all computer programs which eventually terminate), and perhaps even all musical scores. Indeed, we could say that all music making involves exploration of rules or procedures, applied, made, or broken. Taken together, though, the words 'algorithmic music' stand for a rich field of activity, defined by the urge to explore and/or extend musical thinking through formalized abstractions. In the process of making music as (or if you prefer, via) algorithms, we express music through formal systems of notation, taking a view of music as the higher order interplay of ideas.

Over the past few decades, algorithmic music communities have formed around a number of continuing approaches, which we can arrange according to the relationship between human and algorithm. At one extreme lies the claim of the independently intelligent algorithm, in the form of computational agents which are deemed to be creative (see chapter 15 by Wiggins and Forth, for example). At the other extreme, algorithms are treated more like musical notations, which humans work with and adapt as vehicles for their own creativity. In music, the field of live algorithms (see chapter 13 by Eldridge and Bown) is situated towards the former extreme, and that of live coding (cf. chapter 16 by Roberts and Wakefield) towards the latter. The independence of live algorithms allows them to be presented as non-human musicians, often as co-performers which incorporate machine listening in order to respond to human musicians, in live interaction. On the other hand, the live coding tradition does not give an algorithm such agency, but foregrounds the human authorship of algorithms as the fundamental musical activity at play. In live coding, the algorithm may run deterministically, but this determinism is broken through live modifications by the human musician, who shapes the music through modifications to its code.

What the current traditions of live algorithms and live coding have in common is an emphasis on musical improvisation. However, it is important to note that algorithmic music practice extends far beyond improvisation and the performance of music. Indeed, throughout the early development of algorithmic music, real-time digital synthesis was challenging or even impossible, due to the lack of processing power in early computers. Accordingly, the algorithmic music heritage lies very much in music composition. Composers are often less visible than performers and improvisers, but we should bear in mind that the majority of algorithmic music making takes place in private. Indeed, algorithmic composition allows us to work with abstractions of musical time while not being subjected to the very real constraints imposed by a listening audience. So we recognize here that a core aim of many algorithmic composers is to produce works which are fixed and reproducible, but which may sometimes reach beyond the human imagination.

Chapter 34 by Levtov in this volume introduces alternative terminology from the perspective of an end-user who has purchased, or otherwise acquired, a musical algorithm to enjoy. In particular, he distinguishes between generative algorithmic music, which runs without user input; reactive algorithmic music, which responds to environmental input; and interactive algorithmic music, which end-users interact with directly to influence the music. In these terms, live coding and live algorithms are both interactive uses of algorithms in music performance, while generative and reactive forms are generally listened to in a similar way to recorded music.

A recurring theme through several of the following chapters is of the *affordance* of algorithms. This again relates to the relationship between an algorithm and its user, and the opportunities for action that the algorithm suggests or even provides. For example, affordance is core to chapter 12 by Fiebrink and Caramiaux on machine-learning algorithms, where their consideration of the musical activities suggested by machine-learning algorithms gives a practical perspective on algorithm design. Exploring the design of algorithms as a form of user interface in this way is a radical departure from the more standard purist conception of machine learning.

Issues of computational creativity and of audience perception of the source of musical ideas in a piece are discussed in some depth throughout the book. And the sociological and educational contexts in which algorithmic music is considered, for itself or incidentally, are also evaluated in the book: a diversity of attitudes continues to be present.

1.4 Origins

Ideas about what we now call 'algorithms' can be found at least as early as 900 AD, and in many different cultures, from Arabic and Greek to Indian. Clearly the word has relationships to algebra, and there is a sense in which a contemporary piece of algorithmic music has access to the whole codification of mathematics, as well as programming languages. Nierhaus (2009) has again provided many useful perspectives on these technical aspects and their application.

We set the stage of algorithmic music with this chapter. There is then a fascinating description (Collins, chapter 4) of a series of machines that link the early ideas to contemporary algorithmic thought, by way of a range of automata. One of the striking things discussed in that chapter is an analysis by Riley of the eighteenth- and nineteenth-century 'overestimation' of the novelty of automata, notably musical ones (Riley 2009). It seems that automata have been treated in some cultures and periods with a kind of reverence, in others with the demonization we tend to associate with Frankenstein's monster, and in yet others with virtual indifference. Nevertheless, in the early days of computing in the middle 1950s, toy computers such as the Geniac were sold (as the name suggests) as almost magical machines capable of making music, however simple they really were musically, or in retrospect. Even when the Geniac's producers parted company amicably, one went on to make a similar machine sold as the Brainiac. Edwards (2011) puts this machine nicely into context in his article on the development of algorithmic music. Collins points out in his chapter the salutary concern that we may still be prone to aggrandize the potential of algorithmic music.

Most of the chapters in the book take off where Collins leaves the history, and elaborate on personal compositional and communal research trends. One aspect that is worth flagging here, also mentioned by Simoni (chapter 30), is that many works that will be discussed are not necessarily or generally appreciated as being algorithmic. This, counterbalancing the risk of exaggerating the influence of algorithmic thinking in music, points out that in some respects we may tend to underestimate it.

1.5 EARLY ALGORITHMIC MUSIC

We had intended that a chapter (to come immediately after this introduction) be devoted to introducing and discussing some 'canonical' algorithmic music, such as Hiller's work, some minimal music of Reich, and some music by Ligeti; however, contractual difficulties precluded this. Consequently, we offer a brief general summary here, to link Collins's chapter to the more contemporary aspects of algorithmic music, reflected both in the personal PoP sections and in the research discussion chapters. Note again that the book is not intended to catalogue composers/improvisers and their musical outputs, but to address the context of ideas, processes, and developments. Indeed, reflection on this earlier idea of a 'canon' of algorithmic music made us question whether we are yet in a position to delineate such a canon, and even whether the idea itself is relevant. For example, the diffusion or, more importantly, understanding and analysis of algorithmic music from Africa is slight. Similarly, the role of women in algorithmic music is certainly underestimated, under researched, and possibly also under developed, an issue raised in Simoni's chapter in the preceding *Oxford Handbook of Computer Music*. As an example, consider some of the fascinating electronic and electroacoustic work by Daphne Oram or Éliane Radigue: we know all too little of their use of algorithmic processes at present.

Several authors, including Collins, point out that there is an extremely long history attached to algorithmic music. Probably the earliest parts of this used algorithms which terminated (that is, that were fulfilled): if there was a goal, it could be achieved. Here we want to link this via the practices of the 1950s on to the present. The idea of an algorithm which can terminate remained relevant in some parts of the 1960s' artistic turmoil (Banes 1993), as in some of the 'process' works, found in music, dance, theatre, and text writing. Consider George Brecht, several of whose texts describing algorithms for (sound) events are entirely feasible, if sometimes exceedingly long. For example, his Drip Music (for single or multiple performance) requires that 'A source of dripping water and an empty vessel are arranged so that the water falls into the vessel.' A single performance with a finite source would clearly terminate; multiple performances or a performance with a natural source might continue to infinity. Likewise, Jackson Mac Low, a key interface between sound and text, has produced many significant text works using such algorithmic processes, and also provided process scores for music performers. Several of his *diastics* are intended for improvisatory interpretation by both musicians and text performers, usually with finite duration. Similarly, LaMonte Young's *X for Henry Flynt* is a piece whose process is simple to initiate, but which may proceed for a finite time or essentially to infinity: a chosen event is to be repeated x times.

This raises the contrast between those composers who wished to initiate a process with no explicit termination condition and those expecting completion of their process. Stockhausen's *Setz die Segel zur Sonn* (Set sail for the sun) is a text composition amongst the set *Aus den sieben Tagen* (From the seven days): each musician is instructed to play 'a tone for so long until you hear its individual vibrations', and then after listening to the others, 'slowly move your tone' until there is 'complete harmony' and all the sounds become 'pure, gently shimmering fire'. It is difficult to achieve this. Compare this with the procedures of Xenakis, whose computational algorithmic control of elements or complete works was more pragmatic, open to termination, and applicable to the generation of both electroacoustic music in the studio and scores for instrumental performers (Harley 2004; Xenakis 1971). On the other hand, Xenakis also routinely transformed some aspects of the outputs of his algorithms, both in the electroacoustic and instrumental domain, while still intending that their product be finite in duration.

Minimal music from the 1960s onwards, in the sense of the rhythmically repetitive work of Reich, Glass, Riley, and others, is normally clearly algorithmic, though mostly manually composed and often providing a process whose completion can be identified: for example, the progressive deviation and final return of two patterns which start in unison, progressively deviate, and finally return to the original state (as in *Clapping Music, Piano Phase*). Similarly, from the pioneering works of US algorithmic music by Hiller, through those involving Laurie Spiegel, Max Mathews, James Tenney, and Larry Polansky, there were often clear target states which terminated each algorithmic section. Particularly also where improvisers were involved, as sometimes with Spiegel, there were mechanisms for them to initiate an almost Schenkerian 'prolongation' of the algorithm and process. Later, postminimalist ideas (such as those of William Duckworth) extended the range of applications of minimalist procedures, for example transforming pitch structures rather more, and they have often been used computationally in more recent times. Ligeti, on the other hand, without using a computer, employed particularly rigorous algorithmic procedures on the pitch and rhythmic structures in some of his instrumental works, such as *Continuum* for harpsichord (1968) and *Désordre* for piano (1985), permitting complex rhythmic juxtapositions and transformations. Michael Edwards provides an appealing and accessible introduction to the ideas of algorithmic music through the past, the works of Xenakis and Ligeti, and the computational approaches of Hiller, Koenig, and later composers (Edwards 2011).

Much of the music mentioned so far has been intended for performance by instrumentalists (and vocalists in the case of Jackson Mac Low and some others). Some was still note-based, but exploited synthetic sounds. Xenakis is probably the central figure in triggering the application of algorithmic processes to sound-based music, through the painstaking work in his crucial body of electro-acoustic works (Hoffmann 2002), and through the rigorous yet metaphoric and stimulating expositions in his book *Formalized Music* (Xenakis 1971). Most subsequent algorithmic musicians recognize a significant debt to him, and his work has been duly assessed in depth in several previous books.

After Xenakis, we should mention the US League of Automatic Composers, and their outgrowths such as the network ensemble The Hub, and Voyager, George Lewis's fascinating co-improviser set of algorithms. One of the key ideas of the Hub was that their sonic material (and sometimes musical process per se) should be passed around their networked computers, sometimes emerging as a more complex shared process (Brown, Bischoff, and Perkis 1996). Lewis, on the other hand, was primarily concerned with his Voyager software as a partner for improvisers, single or multiple (Lewis 2009). Lewis foregrounded what we would now call 'machine listening', and the use by the algorithms of the attended information, and these topics emerge continually in the present book. Lewis himself also provides some typically sophisticated and challenging perspectives in his contribution here (chapter 9).

1.6 Issues in Algorithmic Music

Throughout the book, there is discussion of the utility of algorithmic processes both for offline composition (as mentioned, by this we mean composition in private) and for live performance (with an audience, in public). Amongst the key issues that unite these two aspects are: (1) what benefit the algorithm, especially when computational (deterministic or stochastic), provides to the music creator; (2) whether the algorithm can become a genuine partner in the creative process; (3) whether it can make contributions which are equivalent in utility to those potentially emanating from another (human) music creator; and finally, (4) whether it can provide meta-human outputs, which we ourselves currently could not achieve but which may in the short to medium time frame become just as accessible cognitively and socially as musical outputs that we can make now. Later, it may perhaps also become possible for a human creator to produce such currently meta-human music (and hence it would no longer be meta-human). Acceptability and utility (expressed in musical and social terms) are mutable aspects of any music genre, and of any innovation or retrospection, algorithmic music included. A key question is how to endow the algorithmic creation with the humanoid power of self-evaluation, and ultimately self-evaluation that can change in nature with time (see in particular chapter 15 by Wiggins and Forth).

At the process and functional levels of making music, there are other layers of issues. For example, consider electroacoustic music since about 1950, and particularly acousmatic music (in which there are no live acoustic instrument performers involved in the presentation of a piece, and usually not in its realization and recording either). Here the previously quite distinct roles of composer and performer have been largely fused, and the level of control the initiating music creator can achieve is enhanced because it is not necessary to allocate elements of control to a separate performer. By extension, we can also observe algorithmic computational mechanisms taking over functions such as mixing, sound projection, sound spatialization, all discussed to various degrees later in the book (particularly by Schacher). In other words, algorithms in principle may contribute to all stages of music making, to what historically has been a highly differentiated series of activities: composition, performance, acoustic spatialization, recording, editing, mixing, mastering. As colleague Greg White puts it (White 2015), the overall process may tend towards 'maximal convergence' in the locale of control of the separable activities. Sometimes improvisation takes the place of the first two of that series.

We do not imply that algorithms should take over any of these steps, rather that they may do so or may contribute. To the degree that this occurs, the algorithms may be manually or automatically driven. Perhaps the strongest appeal of the automatic application of such algorithms from a creative perspective is as part of an algorithmic collaborator, with whom a human creator performs (or composes). But of course from a commercial and practical perspective for example in relation to film and tv music, the use of those automatic processes may provide more economical and efficient outputs than manual application, and hence contribute to commercial value as well as to the accessibility of creative play in music.

This brings us to a critical question: how is algorithmic music appreciated and diffused? Simoni's chapter illustrates some of the main features of the uptake and perception of such music. For example, many listeners, even with musical training, are not particularly aware of the algorithmic contribution. This is probably encouraging from the perspective that is often raised, that without physical gesture on the part of a performer that is tied to sound generation, a musical event is lacking (even boring) and requires supplement. The supplement may be a display of live code (essentially, a display of the algorithm) or a complementary dynamic visual imagery sequence. Here the editors support a diversity of views that range from a purist stance, that the code and algorithm are secondary and need not be overt in any way, to the view that we may celebrate the algorithm by making it in some way apparent, through to the idea that the algorithm in some sense is the work and should be appreciated in itself.

It seems that the main communities of algorithmic music are centred on the creators, such as the Live Algorithms in Music grouping, the Live Coding field, and its established TOPLAP community and new conference (initiated in 2015 by editor AM with Thor Magnusson), and several antecedent groups. Groupings of consumers of algorithmic music are sparse, with the possible exception of those who regularly participate in Algorave events (electronic dance music created by live coding and other algorithmic means) and precursors. The penetration to audiences of other varieties of algorithmic music, as illustrated by Simoni and in other chapters, seems to be largely as a subcomponent (overt or not) of composition and computer-interactive improvisation.

As Wiggins and Forth argue, we do not want algorithmic music to be evaluated by something akin to a Turing test, which simply asks whether an algorithmic piece seems to be plausibly a reasonable competent human creation. Rather, we want to allow for meta-human outputs and for systems which develop their own evaluation frameworks, potentially novel. The listener may or may not transform their perceptions of such music into a cognitive framework that corresponds to the algorithm's own methods, but in either case they may gradually assimilate the music into a meaningful whole. Similarly, a (human) co-performer working with a live algorithm can transform the prospective and retrospective meaning of a piece as a result of what they choose to play: the ultimate evaluation of such an algorithmic co-performer will always involve factors beyond those the algorithm itself uses. So it is perhaps fortuitously positive that much algorithmic music is simply assimilated in a context in which its nature and bounds are not transparent to most listeners (and sometimes, not to the creators either).

Nevertheless, we hope that educationalists' involvement in live music making including algorithmic music will increase alongside the desirable (indeed inevitable) rise in computational literary throughout the world. The advent of the low-cost Raspberry Pi computer has stimulated the diffusion of cheap and accessible computing machinery more widely around the world, and if this (ideally) eventually elicits a virtually universal basic literacy in programming, then algorithmic music can become accessible to almost everyone as both producer and consumer, since there will be minimal cost or cultural barriers.

1.7 CONTEMPORARY DIRECTIONS

We have taken care to ground this book in historical perspectives, particularly through Collins's chapter on the origins of algorithmic thinking in music. This grounding provides sure knowledge that there is nothing fundamentally new in the basic conception of algorithmic music, as we have known it for hundreds of years. However, in terms of the activity of algorithmic music, everything is new: the speed of modern computation allowed by microprocessors, their plummeting cost, their proliferation in handheld devices, and social shifts too; free/open source culture and online social spaces; and in much of the world, an increasingly computer-literate populus. All this means that algorithmic music is now transforming from a niche activity, shared in fringe festivals and academic conferences, into a more inclusive music culture sometimes finding large audiences, end-users, and communities of practice.

Linguists and computer scientists keenly point out that programming languages and natural languages are very different categories. Nonetheless, programming languages have always been designed for human use, and are now increasingly designed for human expression, supporting the rise of creative coding as an actual career choice for many working in art and design fields. There are now many programming languages and environments designed specifically for the expression of algorithmic music and/ or visual art, with the classic Music-N (e.g. C-Sound), Lisp (e.g. Symbolic Composer), and Patcher languages (e.g. Max/MSP, PureData) joined now by SuperCollider 3, Extempore, Gibber, Sonic Pi, Tidal, and many more. Where refinements to programming language environments are designed for human expression, we argue that they become more like the written form of natural languages. In the following, we pick out a few directions in which this new expressivity is taking us.

The foray of algorithmic music into music education is well signposted by Andrew Brown (chapter 32). As he relates, there is a long history of bringing computational media into education, but it feels as though there is a surge of interest in creative computing that can bring all this research into new fruition. The recent success of the Sonic Pi environment, designed for teaching both music and computer science, as well as supporting music practice, is particularly encouraging. While the push for computer science education in schools may at times be motivated by economic and business interests, it is also an exciting cultural experiment. What cultural shifts will algorithmic music take when our young programmers grow up, and computational literacy really takes hold? We are already seeing the growth of chiptune, a 'retro' digital music community celebrating early 8-bit computer sounds; is algorithmic folk music next?

The phrase 'paradigm shift' has certainly been overused, but some do argue that major changes to how we think about computation and human creativity are about to take place. Bret Victor, responsible for the early user interface design of the iPad, now rejects contemporary notions of 'touch' interfaces, and even the notion of technology and design, instead reaching for computational media as a means to 'think the unthinkable' (Victor 2013). Victor urges us to look beyond current practices of computer programming, towards a way of using computational representations to think through and communicate ideas. This will be familiar to many algorithmic musicians who compose music through a creative process of exploration through code, but Victor advocates finding far better representations for thinking about systems. This echoes the long-expressed motivations of the visual programming community (Blackwell 1996), and indeed Victor draws much from the unconstrained early work from the 1960s and 1970s.

A recent development in algorithmic music has found large audiences at electronic music festivals largely outside the academic context of computer music. This could be attributed to 'post-club' electronic music, which may literally be listened to after attending a nightclub, therefore taking the repetitive, timbre-focussed structures of dance music as a starting point for experiment. Autechre is a key example, generating its alien rhythms and sounds from procedures defined in software such as Max, with fans struggling to recreate patches from screenshots found in magazines. Another key example from the United Kingdom is Leafcutter John, leading club culture into unfamiliar territory through automatic remix tools and live algorithms. Elsewhere in Europe, the Viennese scene and in particular record label Mego became a strong centre for algorithmic noise and glitch, including the prolific audiovisual collective Farmers Manual (FM), which has released DVDs containing several days' worth of recordings from live performances with its handmade software. In Denmark, Goodiepal has worked more explicitly in opposition to academia, developing a post-human approach of *Radical Computer Music*. More recently several artists have grouped together under the Death of Rave label, taking an often heavily process-based approach to taking apart dance music and amplifying its structure and sound to an extreme degree. The focus for all this work is the release and performance of music, and the production methods are rarely discussed, and often form only part of a range of techniques. However, Mark Fell's work connects with this contemporary context in a multitude of ways, and he describes his approach to algorithmic music in generous detail in this volume (chapter 18).

For now though, and paradoxically, the excitement around algorithmic music is in how it is becoming everyday. Observing children build musical systems inside the hugely popular game Minecraft, and seeing the increasingly enthusiastic audience response to Algorave events, are amongst many experiences that indicate algorithmic music is beginning to enrich our lives in a multitude of possibly surprising, but fundamentally human ways. As often happens with technological shifts, from the invention of the piano to the harsh noises of the industrial revolution, the human response to mechanization is to embrace it, as a jumping-off point for creating new means of human expression. Just as the often oppressive forces of industrialization provided cultural ground and source material for astonishing new musics, the perceived threat of software automation gives way to musical compositions which reach beyond what we could do with pen and paper (or even tape and scalpel) alone.

We should be careful however not to be seduced by the idea that the future of algorithmic music is in unimaginable complexity. Algorithms also afford simplicity, and current developments lead towards new algorithmic composition environments which are accessible to anyone with sufficient curiosity. More than anything, now is the time for algorithmic music to break from perceptions of difficulty. Yes, it gives access to a rich, unfathomable creative space, but the means of access—the composition of words into code—should be thrown open to all.

1.8 Conclusion

We have minimized the discussion of algorithmic techniques per se in this introduction. They will be detailed when appropriate later in the book, though the core concern is the ideas and musical achievements of the field. For the reader interested in details of some of the principal historic techniques, Nierhaus's (2009) book is valuable. For a thorough survey and typology of techniques, with a particular emphasis on artificial intelligence, there is an extensive review (Fernández and Vico 2013). These two sources can be placed in a broad perspective by inspection of the timeline of computer music history developed by Paul Doornbusch (2009). This timeline is maintained online at http://www.doornbusch.net/chronology/.

What we have tried to do here is to point to the many flavours of algorithmic music and its wide-ranging potential. We hope that the reader will find what follows illuminates these rather deeply.

BIBLIOGRAPHY

Banes, S. 1993. *Greenwich Village, 1963*. Durham, NC: Duke University Press.

Blackwell, A. F. 'Metacognitive Theories of Visual Programming: What Do We Think We Are Doing?' Paper read at IEEE Symposium on Visual Languages, 1996.

Boden, M. A., and Edmonds, E. A. 'What Is Generative Art?' *Digital Creativity* 20, nos. 1–2 (2009): 21–46.

Brown, C., Bischoff, J., and Perkis, T. 'Bringing Digital Music to Life'. *Computer Music Journal* 20, no. 2 (1996): 28–32.

Dean, R. T. *Hyperimprovisation: Computer Interactive Sound Improvisation*. Madison, WI: A-R Editions, 2003.

Doornbusch, P. 'A Chronology of Computer Music and Related Events'. In *The Oxford Handbook of Computer Music*, edited by R. T. Dean, 557–584. Oxford: Oxford University Press, 2009.

Edwards, M. 'Algorithmic Composition: Computational Thinking in Music'. *Communications of the ACM* 54, no. 7 (2011): 58–67.

Fernández, J. D., and Vico, F. 'AI Methods in Algorithmic Composition: A Comprehensive Survey'. *Journal of Artificial Intelligence Research* 48 (2013): 513–582.

Harley, J. *Xenakis: His Life in Music*. London: Routledge, 2004.

Hoffmann, P. 'Towards an "Automated Art": Algorithmic Processes in Xenakis' Compositions'. *Contemporary Music Review* 21, nos. 2–3 (2002): 121–131.

Ingold, T. 'The Textility of Making'. *Cambridge Journal of Economics* 34, no. 1 (2010): 91–102.

Landy, L. 'Sound-Based Music 4 All'. In *The Oxford Handbook of Computer Music*, edited by R. T. Dean, 518–535. Oxford: Oxford University Press, 2009.

Lewis, G. E. 'Interactivity and Improvisation'. In *The Oxford Handbook of Computer Music*, edited by R. T. Dean, 457–466. New York: Oxford University Press, 2009.

Nierhaus, G. *Algorithmic Composition: Paradigms of Automated Music Generation*. New York: Springer, 2009.

Pressing, J. 'Improvisation: Methods and Models'. In *Generative Processes in Music: The Psychology of Performance, Improvisation and Composition*, edited by J. Sloboda, 298–345. Oxford: Clarendon Press, 1988.

Pressing, J. *Compositions for Improvisers: An Australian Perspective*. Melbourne: La Trobe University Press, 1994.

Riley, T. 'Composing for the Machine'. *European Romantic Review* 20, no. 3 (2009): 367–379.

Smith, H., and Dean, R. T. 'Practice-Led Research, Research-Led Practice: Towards the Iterative Cyclic Web'. In *Practice-Led Research, Research-Led Practice in the Creative Arts*, edited by H. Smith and R. T. Dean, 1–38. Edinburgh: Edinburgh University Press, 2009.

Victor, B. 'Media for Thinking the Unthinkable: Designing a New Medium for Science and Engineering'. *Beast of Burden* 4 (2013). http://worrydream.com/MediaForThinkingTheUnthinkable.

White, G. *Towards Maximal Convergence: The Relationship between Composition, Performance, and Production in Realtime Software Environments*. PhD dissertation, University of Newcastle, Australia, 2015.

Xenakis, I. *Formalized Music*. Bloomington: Indiana University Press, 1971.

....

CHAPTER 2

ALGORITHMIC MUSIC AND THE PHILOSOPHY OF TIME

....

JULIAN ROHRHUBER

2.1 INTRODUCTION

....

ALGORITHMS are of a liminal character. In this, they much resemble the natural numbers, for which it can be hard to tell whether they come into existence as we count, or whether we are able to count only because they have existed in the first place. An algorithm is on the *verge of time*: on the one hand, it is strictly structural—a formal, unchanging entity. On the other hand, it is not only a formula, but a formula that prescribes steps to be made one after another, depending on one another. It is a formula that exists in order to unfold, in the form of a process, in time and over time, and dependent on its past inputs.

Yet, it is not clear what kind of time algorithms involve. They describe the course of events in a way that may motivate very different of models of temporality, be they cyclic, path-like, or multidimensional, be they continuous or discontinuous. One could say that here, time and causation are entangled most inseparably. As a mathematical entity, an algorithm not only inhabits a different time from its temporal unfolding, its unfolding happens in a logical time disconnected from immediate experience, or even from its mechanical realization. Therefore, we can understand programs only if we understand how algorithms constitute temporal experience from without. But just as well one can say that an algorithmic process, and algorithmic music in particular, is always already enclosed within the same time that accommodates experience, and can be understood only on this level. That is why the distinction between a program, which unfolds its consequences mechanically, and an algorithm, which unfolds them logically, is not at all trivial.

Thus, algorithms are on the verge of time, in so far as they are on the verge between constancy and change, on the one hand, and between concrete and abstract temporality, on the other. A possible first step for a philosophical consideration of algorithms is to better understand what this liminal position means; the consideration of music is likely to help us here. Eventually, algorithmic music will turn out to be not only affected by how we understand temporality, but also it will turn out to be a possible method to constitute and convey the peculiar existence of time.

2.2 CLOCK WORKS

J'ai connu quelqu'un qui en s'endormant avait entendu, un jour, sonner quatre heures, et avait fait ainsi le compte : une, une, une, une ; et devant l'absurdité de sa conception, il s'était mis à crier : "Voilà l'horloge qui est folle : elle a compté quatre fois une heure!"[1] (Pierre Bourdin, the Jesuit mathematician, in reply to René Descartes [1642])

One of the earliest designs of automatic calculating machines was a device conceived by the seventeenth-century scholar Wilhelm Schickard, in order to, as he wrote in a letter to the astronomer Johannes Kepler, reckon up numbers 'immediately and automatically'. The skills that were necessary for such a construction, and which Kepler described as those of a practical philosopher (literally an 'ambidextrous philosopher'), came from the art of clockmaking—which might have been one reason why the machine wasn't called a calculating machine, but a calculating *clock* (*Rechenuhr*). The transmission between the digits of the numbers involved is strictly analogous to the transmission between hours, minutes, and seconds.[2] Mechanical clocks and mechanical calculators are also similar in so far as both depend on the blind constancy of some force, such as a crank or a spring, which they unfold into interlocking oscillatory movements, movements which end up being measured against the backdrop of a spatial layout or map, such as the clock face or the system of digits and numerals. Clocks serve their purpose best when set rarely, but at a precise moment. Subsequently, they should display the current state at all times without intervention, open-endedly. Reckoning machines, by contrast, can be activated at any moment. They swallow all their inputs and intermediate states and make them vanish, only in order to present, after a predictable amount of time that ideally shrinks to an instant, an adequate final result. To this end, much unlike clocks, their inner workings should not be contingent on the time at which the calculation happens. Ideally, the display of clocks depends only on time, that of reckoners only on their input. It is in the state of dysfunction, construction, and invention, as opened black boxes, that they seem to be almost identical; in their functioning, they represent the two ends of a spectrum between following and concluding, between the perfect presentation of time and its perfect vanishing.

What can we learn about time from this opposition? Can we say that a clock conveys time, like a thermometer conveys temperature? That a calculator displays the result of a term, like a telescope displays a star? Or are we obliged to say: clocks produce time, like calculators produce results? At least one thing can be noted: timekeeping and reckoning, independent of whether they involve continuous or discrete movements, are both exceptional practices insofar as they oblige us to strictly maintain an internal ratio between otherwise disconnected parts. They are both 'rational' in this sense. While other machines, like pulleys and mills, transmit movements, states, and forces from a source to achieve an effect, in the case of clocks and calculators, it is not easy to tell what it exactly is that they transmit through the movement of their internal parts. For the moment, it seems that, properly speaking, they are best considered as *media*, halfway between machines, that exert influence, and instruments, that receive it.

This ambiguity only deepens when we take into account the *sounds* that clocks produce to signal the current time. Whether it is a church bell that assembles the congregation or the alarm clock that pulls the employees out of their sleep is only a minor shift: the clock mechanism effectively exerts social power, and its internal contiguity is an agent of social unity. Because sounds from different sources diffuse in space and overlap with each other, a lack of time discipline is instantly noticed as a lack of synchronicity. By contrast, the synchronization between events, enabled by the dispersion of sound, is both possible and necessary only because events do indeed happen at their own time, take their own time, and in some respect remain completely unaffected by other events. Hence, time signals are not merely an expression of social power. They also mark moments and synchronize otherwise disconnected activities, conveying a specific moment in an internal state of their machinery that coincides with other events, such as planetary movements. No less than chronography, chronophony deals with synopses:[3] for many centuries, the sound of clocks has essentially been a signal of synchronicity or synchronization, a signal that combined the unity of the space in which it was heard with the unity of time at which it happened.[4] Sound here appears as a symptom of time. In the ambiguity between active and passive aspects of timing, it functions somewhere halfway between actively drawing together and identifying disconnected moments and passively signalling their coincidence.

2.3 SONIFICATION OF ALGORITHMS

In his transcendental aesthetics, Immanuel Kant argues that one should not try to understand time as a phenomenon or an object. Time is already an essential part of the means which our understanding of phenomena or objects requires in the first place: whenever we refer to such things, we refer to them as occurring in time and space, as existing in spatial or temporal form, as coexisting or subsequent difference. In any endeavour to understand it, one has to keep in mind that time will always have been a precondition already, and therefore a horizon for this very understanding. According to Kant, despite this predicament, understanding is still possible through the practice of mathematics (of space by geometry, of time by arithmetic), the laws of which vouch for their respective universality and unity. But in the late nineteenth- and early twentieth-century philosophy, psychology, and physics, the oppositions between empiricist and rationalist tendencies became ever more irreconcilable.[5] This led to a revision of many of the unquestioned intuitions about the nature of time and concerned basic concepts like the continuous and the instantaneous.

The foundational discourses of mathematics from that era onwards, in search of a foundation of science in the apparent certainty of the series of natural numbers (or the inexhaustibility of the continuum), have repeatedly returned to the unifying role of time as a condition of experience. But it has also become increasingly questionable whether the inner gaze of pure intuition could be trusted as an orienting limit for the under-standing of mathematics and time. Indirect means of knowledge, which do not proceed from immediate evidence, may simply be unavoidable—not only for scientific under-standing of events that happen in time, even for time itself.

But also the reconceptualization of causation by relativistic physics has had a lasting impact on the intuition of time and its apparent clarity. From the Kantian perspective, which gave us a provisional starting point, time could have been a formal condition only for the possible content of experience. But relativistic physics has broached the question again about how spacetime is related to the matter and energy that inhabits it.[6] What's more, because in spacetime simultaneity depends on relative location, the dialectics between form and content are changed: space *passes* no less than time and we *move through* time no less than through space. While relativistic effects proper are encoun-tered at a very large scale only, we do get an adequate analogue when we try to synchro-nize clocks without knowing the speed at which information travels in a medium—this issue is not only relevant in the cognition of time but has also been both a stumbling block and an inspiration for algorithmic music, network music in particular.[7] The con-frontation with situations in which time and space are a combined constraint should thus not be considered a technical detail, but an adequate, even if indirect, confronta-tion with the structure of spacetime.

The break with simultaneity in physics was closely followed by another foundational break at the interstices between mathematics and logic. It concerns the unity of arith-metic, as well as of the century-long search for a unifying formal system that could serve as an indubitable foundation of all scientific knowledge. The conceptual and technical developments in formal logic and computing have weakened, if not ended, the ambi-tious and often emancipatory hope that a single system could hold all parts together, and account for all logical contexts at once. Rather than being clear and unambiguous specifications of a schedule to be fulfilled, formalizations—and programming languages in particular—in fact have turned out to function as media that allow us to navigate and describe the interrelations between plan, process, and result.

As it seems, the ambiguous spectrum between persistent following in physical time and instantaneous conclusion in a formal leap, which we found in the technology of clocks and calculators respectively, has lost nothing of its relevance. For the invention of the computer an elaborate combination of both clock and calculator functions was necessary, and even today, *Rechenuhr* is not at all an inappropriate name for it. The combination of these two different functions fundamentally changes their very charac-ter, however. Where calculations become events that are scheduled by clocks, they are dependent on temporal coincidence with other events. Where in turn the mechanism of clocks depends on calculations, the flow of time becomes contingent, and in so far as temporal marks are set algorithmically, the meaning of duration may shift dependent on events that happen. The fact that calculations need to happen at a specific moment, but also take time for their unfolding, can make systems very hard to reason about. But even in the simpler cases, where one central clock pulls instructions and data through the bottleneck of an instantaneous memory, the order of events, rather than becoming clear and definite, is still infected by the structural incompleteness and possible inconsistency of any algorithmic formalization.

So while clocks were built to illustrate time, and reckoners to speed up a comprehensible process, programming confronted a serious problem of how to still make sense of the logic of computational processes.[8] Different ways of retaining understanding the ongoing computations have come in and out of use, notably among them the early applications of sonification in the design of computers, which deserve a brief consideration. Many of the postwar computers had a loudspeaker that was connected to a memory location in the processing unit, which thereby made audible its internal state changes during calculation.[9] This continued a practice of operators who acoustically monitored relay systems, such as those of large telephone switching stations. Just like one would listen to the ticking of a clock, skilled programmers were able to follow a part of the algorithmic process in the background. Sometimes, this feature was appropriated for playing musical tunes using the side effects of different divisions of clock times, notably for demonstrating the surprising powers of computers at public occasions. But really, calculations had to be continuously monitored because they were inherently uncertain: a process could always enter a state where it would fail to conclude and instead continue in an endless loop. In such a case, it had to be stopped, as quickly as possible, to avoid overheating—something that not only meant that one had to waste precious hours while waiting for the computer to cool down, but also that one had to face the embarrassing consequences of the failure and inform the lab engineer, who was the only person who had the privilege to break the loop externally. Background listening allowed the operator to respond to this exceptional moment before it was too late.[10]

Even if not all calculations convey time, they certainly take time. This time is structured by the material and logical resources that are needed to make the calculations happen. So it seems almost as if all the strange uncertainties of the algorithmic, and its tension between logical and physical time, finally could disappear when actualized in the apparently hard reality of 'time proper', at the point where the algorithmic law is unrolled in the form of a series of events. Time might be imagined as a dense substrate in which all logical and causal relations appear as secondary consequences of an algorithmic setup, which they can be finally moored in. Sound would be the realm of such a final order, a ground level of immediacy. Such an idea of a grounding presence, however, is misleading. Even in the optimal case, the sonification of the progress of internal state changes gives access to only one specific aspect of algorithmic time—here, all that we know about what happens is when it happens and in what order. The real challenge is a different one. The apparently immediately present and momentary events in fact happen in their specific way only because they are densely coordinated between each other, across past and future. Sometimes, this logic can be made directly audible, but in most cases, it can be comprehended only by taking into account the algorithmic system as a whole, including the different levels of formalism and their semantics. Because the coordination of events is itself subject to changes over time, the causal relations themselves cannot be reflected in their immediate consequences alone. There is, finally, no ground level of immediacy: much less than as a single continuum, time comes into existence in the form of an irreducible diffraction. Both sound and formalization are possible responses to this abstractness of time, whose understanding must, as it seems, accept an essential diversification of its modes of access.

Historically, the sonification of the time structure of algorithmic processes was motivated by the difficulty of conveying how a computation proceeds over time, a difficulty it could not resolve. The behaviour of a process is a shadow of its tacit and untamed laws, laws which are selected, encountered, or even constituted only in the course of events. On a certain level, algorithmic computer music, and also algorithmic music by other means, reflects this paradoxical situation.

2.4 HOLDING THE ERROR IN SUSPENSE

It is instructive to consider here for a moment one of the surprising and notorious results of mathematical logic: there is no formal procedure that could tell us whether an arbitrary calculation will continue forever or eventually return a result. This undecidability, usually referred to as the 'halting problem', has an obvious temporal side to it. First, it implies there is no explicit method once and for all that can *predict* whether the calculation will be conclusive. This means that possibly a procedure must be unfolded in order to reveal certain properties it implies, and often it is undecidable if such unfolding will even work at all.[11] Second, the decidability of such problems itself is posed in formal terms of finality (termination) or recurrence (infinite loop). The minimal, most elementary characteristic of finality is the absence of contradiction—and contradiction, ultimately, is interpreted as an occurrence of an endless and timeless loop, devastating and monstrous because of its unlimited potential of infecting every part of the given system. The determinate rigidity of a formal system thus amplifies its unpredictability. One can say the following: while perfect repetition is the most fundamental elementary operation of computation, reaching a state of perfect repetition globally is a sign of its failure. Repetition is the *real* of computation.[12]

The undecidability of the halting problem is the tip of the iceberg. Indeed, the causal relation between descriptions and processes involves a much broader, and perhaps an even more differentiated semantic uncertainty. Algorithms are not always total functions: just as they may fail to complete their task in some situations, they may unexpectedly fulfil a different one than specified. This unfolding of a causal relation sometimes is best understood as a mathematical transformation, for instance when we reason about algorithmic complexity, and it happens in logical time. Or it can be understood as a physical process, a runtime behaviour of a machine. In any way, in the process of investigation, debugging has hitherto proven ineluctable. At the expense of some generality, it is possible to introduce static analysis so that certain forms of error are being contained within the development process and only conceptual issues may infect the final runtime program. At some level, however, programming retains an experimental character, a dialectical oscillation between the numerous facets of failure and success.

Hence, relying on deterministic means makes music in no way more predictable, but rather less so. The difficulty actually doubles: even where it strikes us as intuitively clear what sonic character is desired, it is particularly hard to specify; so it seems almost better to avoid formalizations altogether in favour of known instruments or sound recordings. Even so, for most sound qualities, we have not yet acquired an intuitive or verbal concept, so that formal specification may be the only possible way even to approach them. In algorithmic music, the uncertainty in the formal methods, which assist such a specification, thus finally amalgamate with the uncertainty of sound description itself. What Curry and Feys wrote in the 1950s remains valid today, not only for formal logic but also for algorithmic music, namely that the '[r]esults of Gödel and the incompleteness and inconsistency theorems make it seem likely that we shall not have other criteria than the empirical ones for the most interesting systems of mathematical logic.'[13] Even if music were not already in itself an experimental field, algorithmic music would necessarily have to be.

At first sight, time seems here to take effect as an open future: programming follows an errant path, not only through one specified procedure, but through multiple alternative trajectories of causality. But rather than a global horizon of absolute uncertainty, it operates as an inner distance which, instead of forever banning inconsistency, includes the error by keeping it in suspense. It would be too rash to take this embrace of uncertainty as an obscurantist abdication of rationality; much more it is an experimental formalization, a contingent search for rationality with the means of rationality.[14] The concept of contradiction illustrates this. Two contradictory statements are not problematic by their mere juxtaposition; they become problematic where they both are consequences of general rules, rules that are thought of as synchronously valid, and thus internally connecting a given world. Hence, to diagnose a contradiction is to assume a specific temporality: something is and is not *at the same time*. Undecidability, on the other hand, is a precise notion for the lack of a formal procedure for determining whether such two paths can coexist.

On the level of a program, this structural potential of incompatibility is necessarily reflected in time. It is usually a variant of the above-mentioned halting problem: some initially given final condition is never met. It might be of more than simply technical interest that many sound algorithms place timing mechanisms in the very same place that is opened by the possibility of contradiction. By inserting the timing where loops and recursions happen, the relation between input and output (or no output at all) is converted into a relation between the input and the computational process itself (finite or infinite). In computer languages, this temporalization typically shows two perspectives: the algorithm has to determine whether a given operation on the current data should be repeated (the imperative definition) or whether a function should call itself again from the current context (the recursive one). The first case is that of an external observer who decides whether and when to make the next step; it corresponds, if you want, to a clock externally attached to a reckoner, driving its calculation. The second is its inverse, like an internal observer who decides about how to continue. Rather than a clock driving a reckoner, it is a reckoner that commands a clock.

Indeed, these are the two simplest ways to compose a computational process from the two functions of following (clock) and concluding (reckoner). Because the measure of *waiting* originates at their intersection, iterative and recursive time[15] are in many ways equivalent: arguably, what can be expressed iteratively can also be expressed recursively. However, they are radically distinct in the way time is conceptualized. As an iterative process that advances *in time*, operating on the past like a given thing, which is (partly) lost wherever it is changed. As a recursive process that advances *in its own past*, leaving it untouched, and leaving behind ever new versions of pastness, that (partly) obscure each other temporarily. Here, the end of an unfolding consists in a jump to a specific level of the past, a recapitulation, a recovery of a beginning.[16] In the imperative conception, by contrast, the end is simply the final present condition. All that was to be done has been inscribed in it already. The two modes are two ways to understand a conclusion from an ongoing process, a final break: one as a return to its beginning, one as a liberation from its future.

Perhaps these observations not only concern events that happen in time and over time but also indirectly give us an impression of the affordances that are characteristic of time itself. After all, repetition, waiting, and undecidability are irreducibly temporal phenomena. Algorithmic time might not be adequately described as a property of events in time (we have collected some evidence for this already), but the form of events, conversely, may prove to be the characteristic way of time appearing in a given world. In any case, to depict the medium of time as a continuous and homogeneous aether of presence is likely to be misleading. The suspension of error is a suspension in a situation of the possible mutual contradiction of laws, and thus a nonhomogeneous temporality. If sound should be an expression of the existence of time, time will have to appear within it in such a form.

2.5 The Self-Alienation of Time

Algorithmic music differs from other kinds of music only by degree. Music, perhaps 'the temporal art par excellence',[17] organizes sound according to explicit or implicit laws, and algorithms are instances of such laws. But also, algorithmic methods, computerized or not, prompt changes in our musical understanding of time in a particular way. Instead of directly laying out the course of events and specifying the musicians' roles on the dot, we just instigate a system of situation-dependent laws, which only indirectly give rise to a course of events. In a sense one could say that composing is less about specifying what happens when, but much more about specifying what happens why. As rule systems become more explicit, they require specific representations, and because these systems need not synchronize events relative to an imagined time axis, their representations make it much less obvious why an axis should be the best way to organize or manipulate time either. There is no axis of causality, after all. The mixture between repeatability and variation, characteristic of algorithms, is not organized by a linear order of time: as laws organize both, the resulting music is much more predictable in a certain sense, and far less in another. And as we have seen already, the degree of indeterminacy in algorithms makes experiments necessary despite, or in fact because of, the determinism of rules.

Computers give explicit access to movements that happen over very short time spans, so that the timbral qualities of sound become part of the same temporal organization as the whole composition. Sound synthesis and algorithmic composition may thereby require different strategies, because their respective frequency ranges concern different perceptual levels. The algorithms involved can nevertheless fill the whole space from sample rate and microsound to long-duration pieces.[18] Such practices of generalization and reconfiguration of temporality have ignited fierce discussion about the nature of music, with regard to the perception as well as to the politics of time. Karlheinz Stockhausen's text '... wie die Zeit vergeht ...',[19] for instance, which was inspired by the idea of an absolute uniformity of temporality, irrespective of apparent differences between rhythmic and timbral qualities, provoked Gérard Grisey to deplore such unifications as a confusion of 'the map with the lie of the land', suggesting instead a 'rhythm of our lives' as a backdrop, against which experience of time ought to be brought.[20] It is not obvious, however, whether the limits of human experience should coincide with the limits of musical time, and whether they can be clearly delimited at all. If there is a unity of time that can only be known indirectly but not heard directly, does it exist nevertheless? Even under the assumption that time is only a universal form of intuition, there would be no reason why this form itself and its structural effects should be limited to what can be immediately intuited. Grisey's originally rhetoric question '*Who* perceives them?' should therefore perhaps be understood literally, namely as a search for an unspecified or alien audience.

The proliferation of algorithmic music to a broader public has refreshingly banalized the experimentation with the sublime and the subliminal: where we delegate decisions and actions to algorithmic processes, we are faced with phenomena that are not tuned to the human perceptual apparatus or to social conventions. Not least, this is a political issue, because the knowledge about the potential manipulation of perception goes hand in hand with the liberating, truly aesthetic, potential of practices that are able to change or protect social and cognitive categories. The everyday experiences of the more or less subtle alienation through algorithmic phenomena make it more plausible that individual and collective learning is able to change cognition and that it may challenge foundational assumptions about what it means to organize time. In such a way, alienation should perhaps be considered deplorable only where it is the symptom of exploitation.

As we have seen, algorithmic methods suggest a break with the idea of time as an immediate grounding; brushing against the fur of intuition, their effects sometimes become a stumbling block, which forces us, within experience, to reconsider the limits of possible experience. Electronic composers know very well that, while it is easy to mark certain physiological limits of sense experience, the limits of perception remain largely underdetermined. It is therefore one of the aesthetic challenges of algorithmic music to find new ways of navigating paths across a space that is not a space of human experience as such, paths that trace out the alienating potential of abstract consequences as a whole. What has this to do with time? Certainly this: every delegation of an action is an alienating loss of immediacy that can only be sustained by an unconditional trust combined with an equally unconditional responsibility. And perhaps at the core of this loss is the alienation of temporality itself, an alienation that counters the structural violence of the idea that everything that appears must appear in the form of presence.[21]

Arguably one of the strongest philosophical breaks with conventional wisdom concerning time is the dissolution of the identity between what is immediately given and what can be indirectly conveyed as existing. This break should not be understood as the disclosure of a manipulation, an extrinsic division forced upon the unity of time by technology, but as one of the central characteristics of time itself, which is 'out of joint',[22] and that haunts technology from within. Hence, algorithmic sound lets us think of this dissolution as a *self-alienation of time*. Rather than alienating us from some assumed original immediacy, it allows us to inhabit the resulting zone of provisional existence.[23] In the following, we substantiate this thesis, and discuss a few implications. Algorithmic sound thereby helps understand the complementary and incommensurability between time understood as *passage* and time understood as *encounter*, which is an opposition that informs philosophical discourse of time still today.

2.6 A Split between Presence and Existence

A: I'd prefer to transform a value which is my composition, rather than arrange the side effects that will result in my composition.
B: What if the value undergoing transformation is a process?
A: Since the value is a music composition then the value is a process by definition.
B: Perhaps the process that yields the side effects is the composition.
A: Yes—as an Arrow.[24]

Most noncomputer music is divided into two clear realms: the sonic properties of the instruments, and the sonic events caused by playing them. This is essentially a temporal organization. Its fascinating tension is constituted by the ambiguity between the two aspects: hearing a sudden event means encountering the *existence* of an instrument; but it also means, 'at the same time', witnessing a *change* to which it is subject and which is mediated through it. An orchestra is like a screen that displays the changes in light, and yet it is also like the changing light itself, more or less well rendered on the screen of its stage.

In computer music, this specific arrangement between instrument and performance grows brittle. Just like in the sound of wind, where the moving air and the rigid bodies become indistinguishable,[25] here, timbral and performative qualities are events as much as anything else. The solid character of objects, understood as a static background condition that mediates the interventions of the players, turns into a variable attribute that can occur at any time scale of a given piece, and anywhere in the hierarchy of concepts in a composition or improvisation. And also for events this is the case: they may happen at any level, and may have effects anywhere else. Generally speaking, existence and change are only aggregate states of any part of the system. By consequence, an algorithmic composition is an entity under transformation, just as much as the sound that results from it; it effectively represents a rich spectrum of layers that gradually mediate between one and the other.

Somewhat unfortunately, this lack of a clear distinction between object and event leads to a loss of a fundamental tension between instrument and player. Where there is no clear separation, there also seems to be missing the fascinating, infinitely fine boundary between them, a distinct connector between existence and change. But luckily, as we saw, rather than a complete dissolution, the boundary is allowed to disseminate to any part whatsoever of the whole system. Hence, what really comes to the fore as *inexistence*—as, in lieu of Jacques Derrida,[26] Alan Badiou has named this vanishing point of touch—is the *question* of causality. Rather than taking for granted what follows from what, inexistence prompts us to reconsider the conditions of causality, through which it yields agency in the most general sense, and indifferently spans conscious and unconscious, automaton and automatism,[27] human and nonhuman. If we have assumed so far that we know what it means to 'be an event', it was only as a manner of speaking. Instead of being resolved, the inner tension of time infects any part of the algorithmic system with a distance to itself.

Rather than being clear from the beginning, the programmer encounters these distinctions in the process, as obstacles and surprises.[28] In Jean-François Lyotard's introduction to his article 'Time Today', the event is the agent of such an alienation or disappropriation: 'Because it is absolute, the presenting present cannot be grasped: it is *not yet* or *no longer* present. It is always too soon or too late to grasp presentation itself and present it. Such is the specific and paradoxical constitution of the event. That something happens, the occurrence, means that the mind is disappropriated.'[29] As we shall see, this constitutive split in time is articulated in sound as soon as we ask: what is it that we hear? The idea that 'something happens' implies that it happens at some specific moment in time. There is an irreducible point at which that which happens turns from being a future event into a past event, and, in experience, turns from expectation to memory. If we want to know more about *what* it actually is that happened at that time, we can only refer to what happened *at* that very moment, at the same time. A sonic quality, however, is an oscillatory movement, a movement that consists in nothing else but a cluster of *multiple* moments or a spectrum of frequencies, which in itself have no momentary existence: when one of them happens, the next is still to be expected and the previous has already disappeared. Each moment has no quality and quality has no moment. Sound is a case in point for an existence that consists as a split in time: *what* and *when* are mutually incommensurable.

Similar to the idea of the *perpetuum mobile*, which had to be slowly proven to be impossible, this incompatibility between frequency and time dawned only slowly upon the engineers.[30] Time really seems not to be composed of intuitive points. So if we want to consider it as being composed of another kind of elementary units (which it somehow seems to be), what would these be then? Inspired by the suggestion of Albert Landé and further results from applied mathematics,[31] in the 1940s, Dennis Gabor takes an original step: reinterpreting the Heisenberg principle in quantum mechanics, he defines what he calls an acoustic *logon*, an elementary signal that has *both* frequency and time aspects, respectively corresponding to momentum and position of a particle. A counterpart of Planck's quantum of action, this minimal 'information diagram' does not fix the two variables separately, but conjugates them, so that they lend each other the limit of possible certainty. This model tries to account for the apparently self-evident idea of a 'changing frequency', which mathematically speaking is a 'contradiction in terms',[32] at least as far as the idea of the necessarily infinite and timeless frequency spectrum of Fourier analysis is concerned. In other words, instead of deriving frequency from time coordinate, or vice versa, it treats them as two coexistent dimensions of each event. Now as much as Gabor's critique of the unreflected notion of *frequency* in physics is justified—isn't it interesting that in the formal treatment, the notion of *time* is perfectly symmetric with its alter ego? Innocuously represented as a single parameter, it certainly deserves at least as much attention: the self-evidence of a time axis may be just as deceiving as that of a frequency axis. What is called 'time' here actually means the 'instant'. And indeed, for the very idea of the event as a part of time, as it is mathematically articulated here, the two aspects are mutually irreducible, and thus are *both* objective aspects of time.

One way to deal with such a rather unsafe-looking territory is the operationalist method, a view that in quantum physics is associated with the so called Copenhagen interpretation. According to this, all such contradictory phenomena should be associated with the technical means of observation only, and in fact what is observed should better not even be mentioned in the account at all. The same interpretation of acoustics would imply that two mutually excluding perspectives of sound, its instants and its frequencies, exist *merely* for us hearing subjects, or for our measurement devices. They are distortions necessarily caused by our own limitations. In this, Gabor more or less followed the tradition of psychophysics, and continued to solidify his thesis empirically by identifying the limits of certainty with the limits of human constitution. But this is not the only possibility—also in quantum physics, the operationalist interpretation was never without alternatives.[33] It is possible to elude what Gabriel Catren called a 'narcissistic illusion of converting a conjectural limit into an absolute principle.'[34] Completely reducing a double-nature to the means necessary for its display has the disadvantage of implying an unquestioned unity somewhere else, which then tends to get tacitly associated with one of the aspects one tried to analyse in the first place. But what is incommensurable does not automatically become irreal. It is not at all unthinkable that the duality which we encounter in acoustics is, instead of merely an observation arte-fact that distorts our view, an adequate expression of the inherent necessity of the spe-cific, situated structures of temporality. Looking back, we can also say that, rather than a reflection of our own limits, it should be understood as one of several possible manifest-ations of the self-alienation of time. Finally, Gabor's acoustic uncertainty relation can be read not only as a formalization of sonic information, but also as a formalization of time itself. The problem that sonic events have partly incompatible temporal dimen-sions of 'frequency' and 'time' turns out to witness the fact that time is not reducible to the instantaneous, but split into pairwise incommensurable aspects already. The acous-tic uncertainty relation points to a possible solution to the central problem at hand—namely how something can properly be said to exist that is not present.

A split between presence and existence runs through the event. Already the para-doxes of the Eleatic School connect this insight to the mathematical navigation between continuity and discontinuity, a topic that inspired later Greek, and then early twentieth-century philosophy, right up until today.[35] The views differ in various ways, many of which matter for the questions at hand, and still need to be left out here. One central issue that informs most of them, however, is the idea that time is not reducible to imme-diate presence. Something is temporal in so far as it can exist in distance to presence. In such a way, we can take algorithms and sounds to be adequate means in the endeavour to understand time, because they share with time the common characteristic of being neither properly present nor independent of a temporal unfolding. This is one sense in which algorithmic music can be considered a philosophical practice that addresses the question of time through the experimentation with events and their causal structures.

2.7 PASSAGE OR ENCOUNTER?

To accept the split between presence and existence means to accept that there is a real contradiction, a contradiction that affects not only the understanding of what happens *in* time but also the understanding of time itself. As Gilles Deleuze showed for film, even though the thinking of movement has been an important task of cinema for a long time, it became possible through this medium to shift perspective and instead focus on the thinking of time.[36] That algorithms are one of the most complex methods to mediate between laws and their consequences not only provides us with a good means for speci-fying sound. Given the internal incommensurability of sound, algorithmic music is an appropriate way to understand the ramified consequences of the self-alienation of time.

For this, let's consider how some of the typical temporal attributes come to bear in algorithmic sound events, like temporal succession, causation, date, duration, futu-rity, presence, and pastness. Regarding the last three, futurity, presence, and pastness, one can easily see that what is irreducibly temporal about an event is its change in modality—that an event implies a *passage*, from lying ahead to becoming present, to becoming past. For example, it is certainly an essential property of a date with the dentist whether it is still scheduled or has already happened. It is an essential property of a sound to have sounded or not. It is an essential property of future that it is ahead, and of history that it has ceased to exist.[37] According to this view of time as passage, temporality corresponds to such an irreversible and inevitable shifting.

Already a brief consideration shows how different approaches to computer music orientate themselves within these three classical modalities of futurity, presence, and pastness, relating them in particular ways, and distribute them relative to a possible listener and her environment. The use of nondeterministic methods, for instance, has intensified a mode of time where the future becomes an open space of unspecified or systematically underspecified events.[38] A phenomenological perspective, deriving from Husserl's work on time consciousness, is common in electroacoustic composition: the phenomenological closure of the experience of time 'fuses' overlapping acts of expectation and memory (protension and retension) into an extended field.[39] Such an extension of the present into the nonpresent is not specific to computer music at all. But the possibilities of analogue and digital recording and reproduction technologies have enabled an understanding of the past as a condition for producing an infinitely malleable material, a phenomenological sound object (*objet sonore*), which is experienced in the multiple perspectives of random, and thus poietic, access to a time axis.[40]

These are just two most familiar cases which exemplify how *pastness* and *futurity* (rather than just time) can function as necessary conditions and themes for algorithmic music. The relevance of *presence* seems to be obvious, in so far as sound, much more than image or text, is typically associated with the momentary. As discussed already, there is indeed a particular path-specific dimension of the algorithmic, which requires an 'acting out' that cannot safely be circumvented. Within computer music, a specific understanding of presence and a corresponding desire for interactive immediacy have been predominant for many years. They have drawn their inspiration from the 1990s technological turning point where computers became powerful enough to calculate faster than the sampling rates necessary for sound output, which, after a long period of rendering static recordings, made it possible to closely interact with complicated computational processes at runtime. Algorithmic music became live music. Such focus on realtime, however, shared the spirit of the era that made programming predominantly an activity of producing a framework of parameters to be modified later when in use, in the case of computer music over the time of the performance. In this way, the requirement of realtime interaction has led to a widespread misunderstanding of the algorithm as a set-up for production, a mere precondition of events. Paradoxically, synchronicity is thereby indebted to an even more severe asynchronicity.

This implicit philosophy of time casts a very specific light on algorithmic music. It is an anticipation of presence insofar as it conceives of the event as something that happens *inside* a formal structure, rather than *to* it. This explains how it could be that through the very focus on real-time systems, programming itself remained a strictly preparational activity—programming has not much to find out about time, it is a way to preclude it.

While it may have seemed a relief that this practice finally acknowledges the performative aspect of the formal, the embodied aspect of the program, if you will, it carries with it a peculiarly unsatisfying deemphasis of the causal structures that are otherwise so central to the reasoning of programming. Even within the paradigm of time as pure passage, the algorithmic should instead be allowed to remain on the verge of time.

Now indeed, the word 'programming' shares a semantic field with 'prediction' and 'predicate', all of which mark an activity that is interested in the relations between possible moments, relations that can be qualified as causal in the broadest sense. Because from this point of view events are specified as complicated maps of causes and dates, the experience of events is that of an actualization of a hitherto implicit or virtual entity. Rather than *passing* from future to past, we *encounter* events, just like we encounter an unknown thing we stumble upon. Thus, even though this encounter may be surprising, it is nevertheless a latent reality that had been given already 'before' we encountered it. The programmatic, and usually textual representation of algorithms, like calendars and other models of time, are a way to reason about why and how something has happened, is happening, and will happen. In a world where many things happen, algorithms coordinate encounters, waiting times, postponed plans, prognoses, and retrodictions. Change becomes a function of the relations between events, as a result of the logic that underlies a certain world, a logic which is not immediately experienced, but can be conveyed only indirectly, through such media.

In algorithmic music, the distance from immediacy has several aspects. Certainly one of them is the difficulty of intuiting what it is that caused an event, where it is that it comes or followed from; it is a difficulty abundant in the logic of proofs, as we have seen, and it has found its way into the logic of music as well. In such a way, even in a completely deterministic algorithmic composition, both composer and audience may be justified in wondering why a certain sound appeared and how it came about. In the early period of computer-based algorithmic composition, composers were familiar with the immense temporal distance between a program and its rendering, which led to the understanding of musical compositions as carefully and explicitly specified workflows of some kind. Its precision notwithstanding, the final outcome was nevertheless able to surprise and often contradict original intentions and aesthetic expectations. This tension between the presentation of a plan and the presentation of its implications sheds quite a different light on time than its architectonic understanding as anticipation of presence.[41]

As we have seen, the idea of passage is essentially captured in an emphasis of a continuously changing state; the idea of *encounter*, by contrast, conceptualizes time as something like a spatial dimension, a location without place. In this sense, events are situated in time. In computer languages this style of reasoning is epitomized by the concept of a pure function. The program 'text'[42] represents general laws, whose propositions are held true over the whole of its unfolding, and whose actualization has no side effect on the laws themselves—it is the immobile relations between events only which determine the cause of events. Happening figures as an effect, not as a cause. Time is, at least in those systems where it is understood in terms of the purely functional idea, an external parameter that causes a specific state of the runtime system for each of its values. A program represents a model of past and future independent of its particular state, a time-geography.[43] So in a certain sense, with respect to algorithmic music, from this perspective programming is purely preparational as well, but now in a completely different sense: an event does not happen inside a formal system, but the event actualizes itself as the formal system itself. Programming is a throw into a future, an anticipation of certainty,[44] which may be unknown but confronts experience from the outside. One can characterize this conception as hauntological:[45] in a program we encounter a *past future* which, as we can infer from mathematical logic, can be reduced to neither its presence nor its end.

2.8 ALGORITHMIC MUSIC AS A THEORY OF TIME?

Passage or encounter: many of the disagreements in the philosophy of time can be shown to be a disagreement between these two conceptions. Following the philosopher John McTaggart, they have often been called *A-series* and *B-series*, or *tensed* and *tenseless*, respectively.[46] The notorious argument, in which he shows their paradoxical relation, is instructive: if I say every event is characterized by *passage*, I accept that the pastness, presence, or futurity of an event is prone to change over time like any other attribute. 'Having rung' is an attribute of nothing else than the alarm sound itself. For this to be possible, however, such events must have some unchanging order: for example, the alarm rang at half past seven, or I woke up after the alarm had rung. One after the other, events are *encountered*. From this perspective, the ringing is not really a proper part of time, and neither is pastness and futurity. Here, McTaggart finds a deadlock: the very order of dates that we encounter and which orders temporal attributes, can be assigned only to events that have actually been recorded, and thus happened in a moment. This, however, would have required a passage in time in the first place. Refuting the idea of 'realtime' *avant la lettre*, McTaggart thus claimed the 'unreality of time'.

Understanding time means renegotiating the frontiers between different, and sometimes contradictory, concepts. Arguably, this negotiation has no access to any ground level of immediacy, so that it can work only indirectly, that is through experimental use of those media which convey and organize time. Within algorithmic music, such a renegotiation can be found in live coding,[47] which shall serve us as a final example. It is meant to demonstrate how the specifically acoustic complementarity between frequency and instantaneous time, as established through Gabor's theory, has its counterpart in a different, namely an *algorithmic complementarity*.

In general, the practice of live coding is born out of the negation of the idea of programming as preparational activity, as 'anticipation of presence': it counters the understanding of computer language as a means to build an interface, and instead takes the rewriting of language as its very means of interaction. Its particular difficulty lies not so much in the act of writing a running program from blank slate; this is only a specific stylistic decision; the pervasive subtle challenge is the intervention in an ongoing process through the modification of its laws rather than of its immediate state. Thus, live coding becomes a particular cross section of different modes of experience and reasoning through formal writing and its computational processes.

Paradoxically, perhaps, for a form of performance, this brittle focus on the liminal aspects of algorithms deemphasizes the interactive control of a present state by parameters. It is not about real-time control. And as it turns out, this allows us to rethink the self-alienation of time in a particularly clear way. A peculiar variant of temporal incommensurability becomes manifest in live coding, which is well expressed in the vocabulary of physics (even though it applies not only to physical temporality, but also to logical time, or pure sequence). Like in sound, this incommensurability has two sides, or conjugate dimensions. But what Gabor understood as a frequency spectrum is here the law—an enduring representation of causality, like a plan, the descriptive side of an algorithm. Indeed, a frequency spectrum is also a specific form of law that tells us once and for all what is bound to happen. The program as description, which is often a text, figures here as an indispensable means for reasoning about the causation in an unfolding sound process. In a way, this aspect is a model of time as encounter, a model which can be called a *causal picture* insofar as it implies time in the form of a prediction of encounter. Live coding as a practice is possible, however, only by combining this implicit anticipation by a second aspect, namely the intervention into the program text at a specific moment of its unfolding process—that is, in fact, by changing its laws, by changing the past of what constitutes this moment and of what motivates the very change that is made. This reformulation responds to the passage of time, with its momentary changes of states—it is, in technical terms, made relative to a *state picture* of the passing current moment.[48]

There are cases where such a change to the state of affairs can be made to precisely coincide with the change of a program: think of a global variable that has one single accessible state at each point in time. A change in the description of its value has an immediate counterpart in the change of the process. But in all those cases where the value depends on an algorithmic description which is changed, the modified program text is not a valid representation any more: it fails to convey the reasons *why* things are now as they are. To retain this causal picture, it would be necessary to reset and completely unfold the program once again, this time with new premises. But this is not a general solution either: by the time we had reached the moment of intervention, its corrected state may have become completely inadequate with regard to the respective moment in time. This may be the case because time has passed, or because other states have now turned out differently as well. But the measure of what counts as 'now' has no absolute reference, which it could use as a comparison. Under these conditions, also in algorithms, causal aspect and state aspect are conjugate or, in other words, complementary perspectives: where we focus on the first, we partly lose the second, and vice versa. If laws are real and subject to change, this is the logical consequence.[49]

Concerning itself with intervention and law, live coding—independent of its being understood more as a performative or compositional practice—marks a point of greatest tension in a wide field of algorithmic methods. As we found, this can be read as a result of the characteristic of time to appear in the form of a split between presence and existence, as alienated from itself, as a tension that essentially remains irresolvable. Just as sonic complementarity, but in another way, algorithmic complementarity is a typical dilemma in the rethinking of time: unable to resolve the contradiction, it gives rise to numerous partial solutions, each of which provide a medium of time, and inform their own theory of time.

As I have tried to argue, by being mutually incommensurable, concepts or observations do not automatically become irreal. Some aspects of reality may by inner necessity require incommensurable, and seemingly 'irreal', perspectives. To still account for them, we need mediating structures and practices, which only indirectly convey them. This explains why, for example, passage and encounter do feature together in such situations. Drawing from all kinds of cultural, aesthetic, physical, or formal resources, which hardly ever match up to a unified picture, such indirect images of time often seem irreal, as if those who make and use them were acting out of an irrational ideology. But, as the social anthropologist Alfred Gell argues, the paradoxes in the models of time 'do not arise from disturbances in the logic which governs ordinary experience, including temporal experience'. Intuition is challenged, in 'moments of rapture'—necessary consequences of 'our reveries of the real, the rational, the practical, which are full of surprises'.[50] Thereby time turns out to be a category that is somehow indifferent to the distinction between cultural and physical. And yet, as we may say now, it is exposed to its own incommensurability.

The liminal nature of algorithms is central for understanding the specificity of time. It is their temporal ambiguity, being half in and half out of time, that makes these 'unfoldable formalizations' such remarkable means and subjects of investigation. From some distance it seems clear why algorithmic music is an intriguing case here: it inhabits mathematical, cultural, aesthetic, and physical spheres alike, and because it is concerned with sound, which is an ongoing affair, a consideration of time cannot be avoided as easily as in other subject matters. Implicitly or explicitly philosophically, it thinks simultaneously through programming, process, and sound. And rather than taking it for granted—as a self-evident resource, as capital perhaps, that can be invested in the production of surplus time—implicitly or not, algorithmic music reasons about time.

ACKNOWLEDGEMENTS

I would like to thank those who have taken the time to read and discuss this text in the process: Roger Dean, Alex McLean, Renate Wieser, and Michael Durnin gave suggestions for very important improvements. An essential part of my argument came up in co-authoring the article 'Algorithms Today' with Renate Wieser and Alberto de Campo. Thanks also to Maarten Bullynck and Gabriel Catren, who encouraged and inspired me very much in detailed discussions on the topic.

NOTES

1. I knew someone who one day while falling asleep heard the clock strike four, and counted up: one, one, one, one; and faced with the absurdity of his notion, he called out: "what a mad clock, it has counted four times one o'clock!" (author's translation).

2. Although only plans of it survived, Schickard's reckoning clock is a standard fitment within the furnishings of the early history of computation. See, for example, Williams 1997, 119f. The mechanism combined cogwheels and tables, and the then widely spread calculating rods called 'Napier Bones'. Note that in German it wasn't unusual to call mechanisms clocks, for instance in words like *Spieluhr* (music box); Napier Bones were also extended for various purposes, notably by the Jesuit Gaspard Scott's *music bones*.

3. For a beautifully and wildly varied collection of historical chronography, see Rosenberg and Grafton 2010.

4. Borrowing the words of Hans Reichenbach, the sound of clocks marks a superposition between two uniformities: the uniformity of *consecutive intervals in time* with the uniformity of *parallel intervals in space*. Reichenbach, 1958, 123.

5. This opposition in mutual entanglement is maybe best exemplified by the work of Helmholtz, in which, in the words of B. Erdmann, 'Kant's rationalist thoughts' are 'twisted round into their empiricist counterpart', and Husserl, who subsequently found an a priori from establishing an explicitly phenomenological and thus antipsychological basis. Cf. Helmholtz 1977, 168 (commentary by Moritz Schlick), who refers to Erdmann 1921, 27.

6. For a detailed discussion of time concepts in relativity theory, see Jammer 2006.

7. Everyday clock synchronization is historically relevant in the history of relativistic physics: Galison 2004. For a discussion in the context of algorithmic music, see e.g. Blackwell, McLean, Noble, Otto, and Rohrhuber 2014, 16f.; Rohrhuber and Campo 2005.

8. Proofs are required not only to 'work out'. Written forms of calculation have always had the *double role* of allowing an individual to perform a calculation and conveying to others an account of how it proceeded. In much the same way (and also with varying success), programming languages not only talk to humans just as to machines, but also accommodate all levels from the processing of numbers to the organization of tasks. See e.g. Mahoney 2011, 77ff.

9. As so much in the history of programming, the practice of the audification of computations was mostly forgotten, until more recently rediscovered by the Dutch historian Gerard Alberts; cf. Alberts 2000, 2005. For subsequent research, see also the work of Miyazaki 2012, 2013. See also Geoff Cox and Morten Riis, chapter 33 in this volume.

10. Gerard Alberts, personal communication. This monitoring practice was common until its usefulness diminished as the clock rates significantly exceeded the audible range, and the diversity of the translation stages between programme and process increased.

11. One should keep in mind that the undecidability results are about formal systems and may or may not be equivalent to the description of a computation in a computer language and its physical process. This text is concerned with the liminal nature of algorithms rather than their absolute categorical distinction. Independent of a decision about this relation, the undecidability of a procedure is the best description for the inherent unpredictability of computation where a total enumeration is impossible, for whatever reason.

12. Originating from the French philosopher of science Émile Meyerson, the concept of the *Real* typically refers to an absolute, existential precondition of knowledge which escapes our view like a blind spot, not, however, without giving rise to occasional, seemingly

irrational obstacles within the field it determines. For Meyerson, the real is a retroactive irrational effect of the attempt of creating identity, and thus unity. Cf. Wiener 1935. Jacques Lacan's work takes up these ideas in psychoanalysis, and proliferates the now common term. A further reading would have to include works like Deleuze 1994. See also Lacan 1998.

13. Curry and Feys 1958, 276.

14. In a more general context, the philosopher Joseph Vogl has given this tarrying suspension a formidable description in Vogl 2011.

15. Temporal recursion occurs naturally where the basic programming paradigm is recursive rather than iterative. While formally exchangeable, the two approaches differ in their specific affordances. Andrew Sorensen gives a thorough description of temporal recursion in Sorensen 2013.

16. This recursive movement itself can again take very different paths. Iterations over data structures are a good example, as they project structural into temporal order most explicitly; for instance, the movements implied by left fold and right fold in functional languages. For an accessible overview see e.g. Van Roy and Haridi 2003.

17. 'L'art du temps par excellence' ; cf. Brelet 1949, 25. Quotation after Mohr 2012.

18. Different temporal levels relevant to sound are comprehensively laid out in Roads 2004. For an approach that places more emphasis on perceptual layers, see Snyder 2001.

19. Stockhausen 1956, 13ff. ("wie die Zeit vergeht" translates to "how time passes")

20. Grisey 1982–1983. Örjan Sandred concludes: 'Grisey's criticism of some earlier examples is crushing. There has to be a connection between the composer's ideas and what the listener experiences. "They [the ideas] became ridiculous when our elders ended up *confusing the map with the lie of the land.*" He uses *Gruppen* as an example: "... the tempi have a great structural importance. *Who* perceives them?" ' (Sandred 1994, 24; emphasis added).

21. Paradoxically, compared to the idea of the living presence, the cold and dead mechanism of the algorithmic turns out to be much less dead than expected. As Derrida writes in 'Violence and Metaphysics': 'If the living present, the absolute form of the opening of time to the other in itself, is the absolute form of egological life, and if egoity is the absolute form of experience, then the present, the presence of the present and the present of presence, are all originarily and forever violent. The living present is originally marked by death. Presence as violence is the meaning of finitude, the meaning of meaning as history' (Derrida 1978, 133; quoted from Hodge 2007, 95–96).

22. Deleuze opens his book on Kant with this idea: 'Time is no longer related to the movement which it measures, but movement is related to the time which conditions it: this is the first great Kantian reversal in the Critique of Pure Reason' (Deleuze 1984, vii).

23. See e.g. Laboria Cuboniks 2016. My notion of the 'self-alienation of time' is inspired by their positive embracing of the 'self-alienation of thought'.

24. Email conversation between James McCartney (A) and Ross Bencina (B) on livecode@ toplap.org, May 2013. For some information on the concept of arrows in computer science, which is here only alluded to, see Hughes 2000.

25. Wishart 1996, 180.

26. Alain Badiou in his obituary of Jacques Derrida: Badiou 2009, 143. Following instead Marcel Duchamp, I could also have chosen the suitable term *inframince*: Becker, Cuntz, and Wetzel 2011.

27. See Renate Wieser, chapter 8 in this volume.

28. A necessity of the refactorization of code, for example, may force itself upon the program-mer, just as a factorization of a term may force itself upon a mathematician. There are reasons to believe that these are necessarily social forces mediated by human-made tech-nology, but there are also good reasons not to rely on them too much: Rohrhuber 2013.

29. Lyotard 1991, 59. The engagement with this text was also an inspiration for Rohrhuber, Campo, and Wieser 2005.

30. Gabor 1946. For a shorter version, see Gabor 1947.

31. The physicist Albert Landé, after giving up his occupation as a piano teacher in 1918, became one of the influential protagonists in quantum theory. See Landé 1930. See also Stewart 1931. For an excellent phenomenological analysis, see Palmieri 2014.

32. 'If the term "frequency" is used in the strict mathematical sense which applies only to infinite wave-trains, a "changing frequency" becomes a contradiction in terms, as it is a statement involving both time and frequency' (Gabor 1946, 431).

33. A detailed and unbiased historical treatment is given in Beller 1999, 172ff.

34. 'Indeed, the first step for pushing further the Copernican deanthropomorphisation of sci-ence is to reduce the narcissistic illusion of converting a conjectural limit into an absolute principle. We must avoid at all costs being like a congenitally deaf person trying to demon-strate the absolute impossibility of music' (Catren 2009, 466).

35. To name only two: Bergson 1991, and Whitehead 1929, 68. For a rereading of time con-ceptions in antiquity against the background of media and music technology, see e.g. Carlé 2007.

36. This work takes up the thread of the discussion of the 'Kantian turn' under completely dif-ferent conditions; Deleuze 1986, 1989.

37. The different orientations in this 'field of passage' have distinguished many philosophies of time from each other, and are often retrospectively arranged to make sense as a move-ment of progress. A detailed discussion of these positions would exceed the scope of this chapter.

38. The classic source is Xenakis 1992.

39. Husserl 1991. For a discussion of these ideas in the context of electronic music, see Mark Fell, chapter 18 in this volume.

40. Schaeffer 1966. See also Kane 2003. Following F. Kittler, media have been characterized as those technologies which allow a 'time-axis manipulation'.

41. This experience brings about an architecture beyond the architectonic as anticipation of presence. In his theory of music, Iannis Xenakis has emphasized (and formalized) a strict separation between an *algebra outside-time*, a *temporal algebra*, and their combination of an *algebra in-time*. See Xenakis 1992, 155ff., in particular 160f. See also Exarchos and Stamos 2005.

42. Considering that technically, *text* is now more broadly understood as *inscription, pro-gramme text* seems to be a good term for any initial representation of an algorithm. This prime mover may have a geometrical or even a sonic form.

43. Thrift 1977.

44. The classic discussion of the idea of a logical time as an anticipated certainty is Lacan 2006.

45. In this way, all writing is a kind of programming, whose future will have been always that of a ghost of its past; Derrida 1994. For a discussion on Derrida's and Blanchot's work on time, see Hodge 2007, 91ff. The term *hauntology* was coined by Jacques Derrida to describe a past future befalling the present as lack (Derrida 1994, 10), and has been reestablished more recently in the context of political theory and sound, especially by Mark Fisher.

46. McTaggart 1908; as well as Gale 1967.
47. See Charlie Roberts and Graham Wakefield, chapter 16 in this volume.
48. In the physics literature, these two perspectives have become associated with the protagonists in the debate in the quantum physics of the mid-1920s as the (causal) *Schrödinger picture* and (state) *Heisenberg picture*.
49. There is a sense in which it is this uncertainty that logically renders time irreversible.
Whether or not this is a valid argument at all would have to be discussed in the context of
philosophy of physics; see e.g. Savitt 1995.
50. Gell 1992, 314.

BIBLIOGRAPHY

Alberts, G. 'Rekengeluiden. De lichamelijkheid van het rekenen'. *Informatie und
Informatiebeleid* 1, no. 18 (2000): 42–47.

Alberts, G. 'Das Verschwinden der Konsole und die Vorläufer des interaktiven "User"'. In
Informatik 2005, Informatik LIVE!, edited by P. M. Armin, B. Cremers, R. Manthey, and V.
Steinhage, 205–209. Bonn: Gesellschaft für Informatik, 2005.

Badiou, A. *Pocket Pantheon: Figures of Postwar Philosophy*. London: Verso, 2009.

Becker, I. Cuntz, M., and Wetzel, M., eds. *Just Not in Time: Inframedialität und non-lineare
Zeitlichkeiten in Kunst, Film, Literatur und Philosophie*. Paderborn: Wilhelm Fink, 2011.

Beller, M. *Quantum Dialogue: The Making of a Revolution*. Chicago: Chicago University
Press, 1999.

Bergson, H. *Matter and Memory*. New York: Zone Books, 1991.

Blackwell, A., McLean, A., Noble, J., Otto, J., and Rohrhuber, J., eds. *Collaboration and Learning
through Live Coding*, Dagstuhl Seminar 13382, *Dagstuhl Reports* 3, no. 9 (2014): 130–168.

Brelet, G. *Le Temps musical: Essai d'une esthétique nouvelle de la musique*. Vol. 1, *La forme
sonore et la forme rythmique*. Paris: Presses Universitaires de France, 1949.

Carlé, M. 'Zeit des Mediums'. In *Medien vor den Medien*, edited by F. Kittler and A. Ofak, 31–60.
Munich: Willhelm Fink, 2007.

Catren, G. 'A Throw of the Quantum Dice Will Never Abolish the Copernican Revolution'.
Collapse: The Copernican Imperative 5 (2009): 453–500.

Curry, H. B., and Feys, R. *Combinatory Logic*. Vol. 1. Amsterdam: North-Holland, 1958.

Deleuze, G. *Cinema 1: The Movement-Image*. Translated by H. Tomlinson and B. Habberjam.
Minneapolis: University of Minnesota Press, 1986.

Deleuze, G. *Cinema 2: The Time-Image*. Translated by H. Tomlinson and B. Habberjam.
Minneapolis: University of Minnesota Press, 1989.

Deleuze, G. *Difference and Repetition*. Translated by P. Patton. New York: Columbia University
Press, 1994.

Deleuze, G. *Kant's Critical Philosophy*. Translated by H. Tomlinson and B. Habberjam.
London: Athlone, 1984.

Derrida, J. *Specters of Marx*. Translated by P. Kamuf. New York and London: Routledge, 1994.

Derrida, J. 'Violence and Metaphysics'. In *Writing and Difference*. Translated by A. Bass, 97–192.
London: Routledge, 1978.

Descartes, R. Septièmes Objections aux Méditations, § 2, A.T. VII, Garnier-Flammarion, t. II,
p. 654, 1642.

Erdmann, B. 'Die philosophischen Grundlagen von Helmholtz' Wahrnehmungstheorie.' *Abhandlungen der Preussischen Akademie der Wissenschaften (Philosophisch-historische Klasse)* 1 (1921): 1–45.

Exarchos, D., and Stamos, Y. 'Inside/Outside-Time: Metabolae in Xenakis's *Tetora* (1990).' In *International Symposium Iannis Xenakis: Conference Proceedings*, edited by A. Georgaki and M. Solomos, 169–176. Athens, 2005.

Gabor, D. 'Acoustical Quanta and the Theory of Hearing.' *Nature* 159 (1947): 591–594.

Gabor, D. 'Theory of Communication. Part 1: The Analysis of Information'; 'Part 2: The Analysis of Hearing'; 'Part 3: Frequency Compression and Expansion'. *Journal of the Institution of Electrical Engineers* 93, no. 6 (1946): 429–441; 442–445; 445–457

Gale, R. M. *The Philosophy of Time.* New York: Doubleday, 1967.

Galison, P. *Einstein's Clocks, Poincaré's Maps: Empires of Time.* New York: W. W. Norton, 2004.

Gell, A. *The Anthropology of Time: Cultural Constructions of Temporal Maps and Images.* Oxford and Providence, RI: Berg, 1992.

Grisey, G. 'Tempus ex machina: Reflexionen über die musikalische Zeit'. *Neuland: Ansätze zur Musik der Gegenwart* 3 (1982–1983): 190–202.

Helmholtz, H. L. F. von. *Epistemological Writings.* Edited by R. S. Cohen and Y. Elkana. Translated by M. F. Lowe. Dordrecht and Boston, MA: D. Reidel, 1977. First published in 1921.

Hodge, J. *Derrida on Time.* London and New York: Routledge, 2007.

Hughes, J. 'Generalising Monads to Arrows'. *Science of Computer Programming* 37, nos. 1–3 (2000): 67–111.

Husserl, E. *On the Phenomenology of the Consciousness of Internal Time (1893–1917).* Dordrecht: Kluwer Academic, 1991.

Jammer, M. *Concepts of Simultaneity: From Antiquity to Einstein and Beyond.* Baltimore, MD: Johns Hopkins University Press, 2006.

Kane, B. 'L'Objet Sonore Maintenant: Pierre Schaeffer, Sound Objects and the Phenomenological Reduction'. *Organised Sound* 12, no. 1 (2003): 15–24.

Laboria Cuboniks. *Xenofeminism: A Politics for Alienation*, 12 June 2015. http://www.laboria-cuboniks.net/#firstPage.

Lacan, J. *The Four Fundamental Concepts of Psychoanalysis.* New York: W. W. Norton, 1998.

Lacan, J. 'Logical Time and the Assertion of Anticipated Certainty'. In *Écrits: The First Complete Edition in English*, translated by B. Frink, 161–175. New York: W. W. Norton, 2006.

Landé, A. *Vorlesungen über Wellenmechanik.* Leipzig: Academische Verlagsgesellschaft, 1930.

Lyotard, J.-F. 'Time Today'. In *The Inhuman: Reflections on Time*, translated by G. Bennington and R. Bowlby, 58–77. Stanford, CA: Stanford University Press, 1991.

Mahoney, M. S. *Histories of Computing.* Cambridge, MA: Harvard University Press, 2011.

McTaggart, J. E. 'The Unreality of Time'. *Mind: A Quarterly Review of Psychology and Philosophy* 17 (1908): 456–473.

Miyazaki, S. 'Algorhythmics: Understanding Micro-Temporality in Computational Cultures'. *Computational Culture* 2 (2012). http://computationalculture.net/article/algorhythmics-understanding-micro-temporality-in-computational-cultures.

Miyazaki, S. *Algorhythmisiert: Eine Medienarchäologie digitaler Signale und (un)erhörter Zeiteffekte.* Berlin: Kadmos, 2013.

Mohr, G. 'Musik als erlebte Zeit'. *Philosophia Naturalis* 49, no. 2 (2012): 319–347.

Palmieri, P. '"The Postilion's Horn Sounds": A Complementarity Approach to the Phenomenology of Sound-Consciousness?'. *Husserl Studies* 30, no. 2 (2014): 129–151.

Reichenbach, H. *The Philosophy of Space and Time*. Translated by M. Reichenbach and J. Freund. New York: Dover, 1957.

Roads, C. *Microsound*. Cambridge, MA: MIT Press, 2004.

Rohrhuber, J. 'Intractable Mobiles: Patents and Algorithms between Discovery and Invention'. In *Akteur—Medien—Theorie*, edited by T. Thielmann, E. Schüttpelz, and P. Gendolla. 265–305. Bielefeld: Transcript, 2013.

Rohrhuber, J., and Campo, A. de. 'Waiting and Uncertainty in Computer Music Networks'. *Proceedings of ICMC 2004: The 30th Annual International Computer Music Conference*. San Francisco, CA: International Computer Music Association, 2004.

Rohrhuber, J., Campo, A. de, and Wieser, R. 'Algorithms Today: Notes on Language Design for Just In Time Programming'. In *Proceedings of International Computer Music Conference*, 455–458. Barcelona, 2005.

Rosenberg, D., and Grafton, A. *Cartographies of Time: A History of the Timeline*. New York: Princeton Architectural Press, 2010.

Sandred, Ö. *Temporal Structures and Time Perception in the Music of Gerard Grisey: Some Similarities and Differences to Karlheinz Stockhausen's Ideas*. Technical report, McGill University, Montréal, 1994.

Savitt, S. F., ed. *Time's Arrows Today: Recent Physical and Philosophical Work on the Direction of Time*. Cambridge: Cambridge University Press, 1995.

Schaeffer, P. *Traité des objets musicaux: Essais interdisciplines*. Paris: Seuil, 1966.

Snyder, B. *Music and Memory: An Introduction*. Cambridge, MA: MIT Press, 2001.

Sorensen, A. 'The Many Faces of a Temporal Recursion'. 2013. http://extempore.moso.com.au/temporal_%20recursion.html.

Stewart, G. W. 'Problems Suggested by an Uncertainty Principle in Acoustics'. *Journal of the Acoustical Society of America* 2, no. 3 (1931): 325–329.

Stockhausen, K. '... wie die Zeit vergeht ...'. *die Reihe* 3 (1956): 13–42.

Thrift, N. *An Introduction to Time-Geography*. Norwich: University of East Anglia, 1977.

Van Roy, P., and Haridi, S. *Concepts, Techniques, and Models of Computer Programming*. Cambridge, MA: MIT Press, 2003.

Vogl, J. *On Tarrying*. London and New York: Seagull, 2011.

Whitehead, A. N. *Process and Reality: An Essay in Cosmology*. New York: Free Press, 1929.

Wiener, P. P. 'On Émile Meyerson's Theory of Identity and the Irrational'. *Philosophical Review* 44, no. 4 (1935): 375–380.

Williams, M. R. *A History of Computing Technology*. 2nd ed. Los Alamitos, CA: IEEE Computer Society Press, 1997.

Wishart, T. *On Sonic Art*. Amsterdam: Harwood Academic, 1996.

Xenakis, I. *Formalized Music: Thought and Mathematics in Composition*. 2nd ed. Hillsdale NJ: Pendragon, 1992.

CHAPTER 3

ACTION AND PERCEPTION

Embodying Algorithms and the Extended Mind

PALLE DAHLSTEDT

AN artist friend once asked me: 'Why do they always play the musical saw in science-fiction movies?' He meant the ethereal sine waves from theremins or early synthesizers that often accompany a spaceship gliding through the void—seemingly without effort. That sound, which is similar to the sound of the musical saw, has perfect periodicity, no timbral dynamics, and is difficult to directionally locate. It is seemingly detached from the physical world, from the strife against gravity and inertia that has shaped us, our movements, and the sounds that emanate from them. Sounds from acoustic instruments strongly imply causality and agency—someone is playing on something. But the space saw is a lonely, disembodied sound.

My friend's question made me think about the physicality of music and sound, and about the strong link between music making and human effort. I became aware of my urge to 'conduct' when listening to works in progress in my studio, and of a tendency to prepare musically for sudden transitions in my music in a way that is reminiscent of physical movements. I clearly want to feel, and even anticipate, the dynamics, transitions, and rhythms in my body, and to experience the music not just intellectually, but with mind and body together. I realized that what I am trying to do in my music and research is to bring that physicality and embodiment—that I know so well from my acoustic musicianship (as a pianist)—into my electronic music making, and onto the stage.

So, there clearly is a connection between sound, mind, and body, in both directions. But what about algorithmic music? Algorithms are abstract, formal procedures. Still, they are used by humans, to produce music that may be performed by and is appreciated by humans with bodies. How does algorithmic music relate to the physical and to the body? Can a musician be a part of an algorithm? And how do we think about algorithms? In this chapter I discuss this connection in the light of cognitive science, and in the light of my own artistic practice.

3.1 Introduction

Most research about algorithmic music from a cognitive perspective concerns computational modelling of musical cognition, that is, how we can form computer models of musical thought, perception, and generation, and learn to perform musical tasks in machines, based on how humans do it. The other perspective, how humans think and work with algorithmic music, is equally important, but less often discussed. What are our cognitive processes like when we create musical algorithms, use them to compose, and perform side by side with them or in direct interaction with them?

The cognitive perspective is not complete unless the whole human being is included, both the cognitive or immaterial and the bodily or physical. Human performance on, with, and in algorithms necessarily involves embodied action and interaction, and some kind of internalization of the algorithms in the performer's mind. How does the cognitive relate to the physical here?

There is no single comprehensive take on the cognition of algorithmic music, nor on the embodiment of musical algorithms, so this chapter draws from a number of related fields, such as musical cognition, cognition, and psychology of programming, embodied performance, neurological research, as well as from personal experience as an artist working in the field. The main topics of the chapter are:

- The relationship between the cognitive and the physical, the mind–body problem, and the materiality of algorithms.
- How do we think about algorithms—what mental devices do we have to think about algorithms, and form mental models of them?
- Situated and embodied cognition—how the mind depends on context, and how action and perception are integrated in the human cognitive mechanisms.
- Agency and the listener—who is the sender, and how do we communicate what goes on in algorithmic music, and in the mind?
- Algorithm as a way to extend the mind of the composer or performer.
- Performing on, with, and in algorithms—can a human performer be an active part of the algorithm?

Where appropriate, I will describe works that illustrate these ideas.

3.2 The Mind–Body Problem

How do we bridge the gap from symbolic computation to physical actions? This question pertains to all computation and cognition, and is a modern equivalent to the old mind–body problem which has kept philosophers busy since ancient times. The musical results of algorithms have to take physical form to be perceived by listeners and human co-players. The symbolic code has to be translated to physical sound vibrations through computers, speakers, actuators, as part of a performative context that as some part of the chain includes humans as performers, co-pilots, interactors, or listeners. The words (the executable code) have to become flesh (physical reality), to paraphrase the words from the Gospel of John (1:14) in the New Testament. There is a rich cultural history behind this transformation, which spans from the classical mind–body problem of philosophy (from Plato to Descartes, and onwards), through the idea of magic as performative words, and finally to executable code in machines. Florian Cramer (2005) has given a rich and elaborate account of this cultural history in his book *Words Made Flesh*, from the perspective of what he calls 'imaginative computation'. He writes:

> From magic spells to contemporary computing, this speculative imagination has always been linked to practical—technical and artistic—experimentation with algorithms. The opposite is true as well. Speculative imagination is embedded in today's software culture. (Cramer 2005, 9)

In some cases, such as when composing batch-generated algorithmic pieces and in live coding performances (see this volume, chapter 16), the composer's bodily involvement with the creation of the music is minimal (mostly typing code), but still, an embodied perspective on what is a good result is crucial. And in a live setting, the bodily experience of the music certainly feeds back into the composer or performer's actions. This is evident, for example, in the subfield of live coding called Algorave, that is, live coding of dance music, where bodily perception and evaluation of the result are part of both the artists' creative process and the listeners' appreciation of the music.

In other cases, the bodily engagement in the creation of the music is more obvious, and absolutely crucial, for example, when performing with new instruments employing algorithms at various levels to generate the sounding results under user control. Such instruments are sometimes called hyperinstruments (Machover 1992), and can range from triggering prerecorded segments to embellishing acoustically played phrases into complex textures.

Algorithms can also be used to more directly map control input (physical gestures) to synthesized or processed sound (synthesis algorithms). Then, the algorithm is not necessarily constructed to generate musical structure, but the focus is on creating a musically meaningful relationship between the performer's physical gestures and the sound. One challenge is to not put too many layers of (temporal) processing between input and output, in order to retain the bodily presence and fingertip control over the sound.

Such an instrument may also consist of a generative algorithm that is controlled in realtime through algorithmic mapping. For example, in my piece *Circle Suite* (Dahlstedt 2014), the composer is performing physically on a generative system as a musician, reacting on a millisecond scale to the music, employing subconscious processes, learned reaction patterns, and physical gestural expressions to control the behavior of the algorithm. It is an attempt at combining gestural expressivity with algorithmic structural complexity.

3.3 MATERIALITY OF ALGORITHMS

A cognitive and cultural perspective on what algorithms can be is given by Goffey (2008) in his chapter on algorithms in the anthology *Software Studies: A Lexicon* (Fuller 2008). He says that an algorithm is an abstraction that is independent of programming languages and implementations, and its embodiment in a particular machine implementation is irrelevant. But algorithms can also have a materiality, which was already inherent in the work of the early computer scientists, Goffey suggests, supported by Rosen (1988). This is evident, for example, in the very physical implementation (although hypothetical at the time) of the Turing machine, as an embodiment of computation in a rather simple machine that we can grasp and understand. An algorithm becomes real, just as its mental counterparts (thoughts), through its consequences and actions. Also, just like the thoughts and musical intentions of a fellow musician are mental configurations, the resulting musical patterns are material, as propagating sound waves that we can perceive. So, we can relate to, play with, and react to music coming from algorithms in very much the same way as we relate to music coming from fellow musicians. Empathetically, we can follow, find the beat, find the spaces and complement the sound with our own sounds, in a game of musical dialogue.

But the process goes both ways, according to Goffey. An algorithm requires structured data to be operable, and the critical operation of translating such data might be seen as 'an incorporeal transformation, a transformation that, by recoding things, actions, or processes as information, fundamentally changes their status' (Goffey 2008, 18). An algorithm is a statement (in Foucault's meaning of the word), that operates, also indirectly, on the world around it.

Goffey concludes: 'What is an algorithm if not the conceptual embodiment of instrumental rationality within real machines?' But since formal logics are incomplete and machines break down or are hacked, algorithms are not as abstract as we want them to be.

A related argument is given by Kitchin and Dodge (2011), who describe software as both a product of the world and a producer of the world, since software mediates so many aspects of modern life. One of the founders of the field of cognitive musicology, Otto E. Laske (1988), argues that computers, by their very existence, provide a link between symbolical processing and the physical. They bridge the gulf between the mental and the physical, but also between natural science and the humanities (in this case, music studies). The prediction of this merger of computer science and humanities was quite prescient, since it has recently become prevalent under the name of Digital Humanities (Schreibman, Siemens, and Unsworth 2008).

The materiality of algorithms has been shown quite literally by Japanese composer Masahiro Miwa (Berry 2008). In a series of works based on an approach he calls *reverse simulation music*, he let a group of musicians perform simple algorithms manually. In the first piece of the series, *Matarisama* (2002), musicians are seated in a circle, each with their hands on the shoulders of the next, holding bells and castanets. In a circular sequence they perform a simple Boolean process, resulting in complex rhythmic patterns. Each musician has to remember and perform only a small part of the algorithm. The work was first prototyped in software and then turned into a physical performance.

Around the same time, I experimented with similar approaches. In one project, I translated a set of real-time algorithms from a public installation, originally implemented in digital signal processing hardware, to verbal instructions for choir singers, *The Time Document: Algorithmic Structures for Choir* (2000). This required, for example, inventing new ways of conducting with big circular movements to allow for gradually shifting phase patterns. As in Miwa's piece, each singer had to keep track of only a small component of a larger algorithm, but together they produced considerable complexity.

Algorithms can also be made concrete using electromechanical means. With my students at the Academy of Music and Drama (University of Gothenburg), I have arranged a series of projects with a modern player piano (the Yamaha Disklavier) which is a real acoustic piano capable of being controlled from a computer. The explicit purpose of the project is to render the students' algorithms and their musical results into physical action and movement of matter. The piano is always placed in the middle of the stage with all lids removed (Figure 3.1). This project would be totally meaningless to perform with a digital piano, even though it would in theory sound the

FIGURE 3.1 A photo from one of the Disklavier concerts, where my students' algorithms take physical form in a player piano. The setup in the picture also includes microphones and speakers for acoustic feedback, which was part of some of the pieces.

same. The physicality of the algorithms playing the piano has a strong impact on the students, and it sometimes inspires them to explore the physicality to the extreme, where they abuse the piano and cause it to malfunction, producing unintended (by the manufacturer) noises.

3.4 FROM ALGORITHM TO INSTRUMENT

In 1999 I developed a tool for exploring a huge synthesis parameter space using interactive evolutionary algorithms, called MutaSynth (Dahlstedt 2001a). It operates on any sound engine, your own or an existing one, with a large number of parameters, up to a few hundred. You start with a few random parameter sets, and if you like one of them, the algorithm generates a set of 'children', offspring that are slight variations of the parent. You listen to the sound of these new parameter sets and choose the one you like best, and so on. You can also mate sounds with each other to produce new sounds through recombination. Through this simple approach you can explore a huge search space guided by ear, looking for new or interesting sounds or musical material. It can be applied to almost any sound-generating algorithm having a fixed number of input variables. Later, this technique was implemented as a sound design tool in the Nord line of synthesizers, under the name PatchMutator (Dahlstedt 2007; see Eldridge and Bown, chapter 13 in this volume, for a more thorough description of this approach to sound design).

Using such a tool quickly becomes a physical act, approaching a performance situation. Each decision instantly changes the output. The continuous sound encourages precise timing of operations, and the unfolding search trajectory instinctively engages the physical musician in me. I start to embody the algorithm, and interact with it not only intellectually, but as a musician. MutaSynth was designed as a sound design and composition tool, but quickly became something like an instrument, with the limitation that each action is a stepwise move in the parameter space, due to the discrete nature of evolutionary algorithms.

The next step was to find a way to translate these stepwise explorations into smooth trajectories in timbre space, which coincided with the start of my duo pantoMorf (with Per Anders Nilsson). Our explicit research goal was to invent electronic instruments that would allow for physical improvisatory playing in a free improvisation setting. To enforce physicality and embodied playing we adopted a few dogma, inspired by acoustic instruments: If we lift our hands, the instrument goes quiet. Volume is proportional to physical effort. And every change in the sound corresponds to a physical gesture.

I developed an algorithm that allows for smooth exploration of a huge timbre space with a set of basic controllers (initially pressure sensors), mapped to a set of vectors in timbre space (Dahlstedt 2009). Through this mechanism, very complex timbral trajectories can be expressed, and if an interesting sound is found, the whole mapping mechanism can be moved to that point instantly. Algorithmically, it is still an interpolated genetic algorithm, but with musician-friendly control (Dahlstedt and Nilsson 2008). The instrument is called the exPressure Pad, and it was the first of a whole family of instruments based on the same ideas.

This search for a physical way of performing on electronics was inspired by two very different sources. First, my studies in baroque keyboard playing in my youth introduced me to the clavichord, a keyboard instrument where each finger has a direct connection to the sounding string, while maintaining the polyphony of the keyboard. Second, I wanted to achieve, in the digital domain, the same direct physicality of circuit-bending instruments—analogue circuits which played through touch, allowing your body to become part of the circuit.

To bring this process one step closer to closure, I wanted to integrate this performance paradigm with the origin of my musicianship—the piano. I designed a version of the mapping algorithm that used key velocities to control a set of processing parameters for the acoustic piano sound. This became the Foldings instrument (Dahlstedt 2015), which I have since used in a large number of performances, solo and in ensemble playing. It has also been used by pianist John Tilbury.

When I perform on any of these instruments, I execute an intricate search algorithm through my physical playing, in an embodied exploration which at the same time is music making. The trajectory through search space unfolds in front of the audience, and becomes part of the musical narrative.

3.5 Music and Cognitive Science

Cognitive science emerged as an interdisciplinary field of enquiry in the middle of the 1950s, out of computer science, cybernetics, and psychology. The actual term was coined in the early 1970s. It is concerned with the study of the mind and its processes, in particular how information is represented and processed in the brain and in machines, but also with behaviour that relates to such processes. Music was one of the first art forms to be subject of computation, together with poetry (with Hiller's and Isaacson's 1958 *Illiac Suite* and Strachey's 1954 *Love Letters* as very early examples). Both text and event-based music are easy to represent in compact symbolic form, suitable for storage and processing in early computers. Such symbolic processing was not new, but it had previously been carried out by hand (e.g. in serial composition or cut-up poetry). One might hypothesize that the hiding of the actual execution of these rather simple algorithms inside a computer made them appear more like cognitive processes instead of the mechanical processes they were. This may have triggered an interest in both understanding and emulating musical cognitive processes. In one of the founding papers of cognitive musicology, Laske (1975) described something akin to simple mental models (schemata) for musical behavior, including higher-level control structures. He discussed representations of musical knowledge, but also how such knowledge must be linked to musical behaviour and to social context.

Laske also argued that musical competence is paradoxical, since it serves as an intermediary between logical and empirical knowledge. But it is irreducible to either, since it is symbolic, while still being bound to spatiotemporal contents. Also, musical knowledge differs from linguistic competence. It deals not with the properties of sound objects, but instead with their contextual function as signifiers of something. While being symbolic and based on logical reasoning, music deals with continuous (nondiscrete) objects. It is about the design of artefacts, not (only) about communicative skills. Hence, it is fundamentally different from linguistic competence. According to Laske, musical competence is more akin to poetic competence.

Also, Laske suggests that music is 'intercompetential', since it is a composite knowledge form that relies on adapting extraneous knowledge forms to its own functioning. He also argues that music is prelinguistic, being a semiotic function to acquire and manipulate symbolic representations of its external world.

As we shall see, Laske was already in 1975 thinking in terms that later would become established in the field of cognitive science, and his thinking about social context was way ahead of its time.

3.6 Thinking about Algorithms

The early cognitive science was focussed on the idea of the mind as a computer, centred on a processing unit capable of symbolic processing, with associated memory (Pylyshyn 1984). According to this view, sometimes called 'cognitivism', our thoughts are like programs. This is closely related to good old-fashioned artificial intelligence. It is a functional theory, where the actual neural implementation is not studied. Within this view, dealing with algorithms in mental processes is not so strange, since this is the matter of thought.

Another perspective that for a long time lived in the shadow of cognitivism was connectionism, which focussed more on parallelism and the neural structure of the brain. A third perspective is interactionism, concentrating on agency and interaction between thinking agents.

Early computer music developed in parallel with the development of early computer science and artificial intelligence. At this time the computational or cognitivist perspective was predominant, and computers were slow and made for batch processing of procedural programs. This had practical and aesthetic implications for the computer music of the time.

Today, thanks to the development of sensor systems, cheap and accessible physical and gestural interfaces, and faster computers, we can run complex interactive real-time systems on stage. At the same time, the connectionist and interactionist approaches to cognitive science have developed immensely. This certainly has had aesthetic implications.

For this reason, and because some of the older theories still make sense, this chapter will include both old and new theories of human cognition.

A good overview of how one can think about algorithms is given by Ormerod (1990) in a text about human cognition and programming, written from a cognitivist viewpoint. Cognitive processes are determined by their function, not by the underlying hardware, which is also a reason for a later controversy (Searle 1980), since the brain with its parallelism and analogue design is very different from a serial computer based on the Neumann model. Ormerod mentions connectionism, but deems it not so relevant for cognitive research into programming, since it focusses on the hardware of the mind (the brain) and not on its processes, and programming languages are serial and procedural. However, this is based on the assumption that our thinking *about* programming must be of the same type as programming itself. But nothing says that the creation and design of programs, which is a high-level activity (at least higher than executing the programs) must be based on the same procedural paradigm. It might be just the opposite, that creative processes would instead require a different method, maybe situated, parallel, and nondeterministic.

In all cognitive theories, representations of knowledge and processes are crucial, and this has been one of the core problems of cognitive science. Many early theories focus on symbolic representations, but Ormerod and colleagues (1986) showed that different mental representations were used in different kinds of programming tasks, and that this had an effect on performance. In his dual coding theory, Paivio (1971) suggested that knowledge is represented in both symbolic or linguistic form and in visual form (or some other analogue perceptually based format). Both of these are needed to capture essential properties, and they are then used together, and can be reciprocally triggered.

In Ormerod's (1990) analysis of the cognition involved in programming tasks, he concentrated on the concepts of schemata and production rules for the representation of knowledge and processes. A schema is a set of propositions organized by their semantic contents, and it may be evoked or triggered by perceptual input (Bartlett 1932). Then the schema provides a relevant body of knowledge, which can guide further actions. A schema is used to represent declarative knowledge.

A production rule is a set of proposition pairs of the form condition–action, the first being the desired goal and the second being a set of actions required to reach that goal. A production rule may be triggered by perceptual input, by information from long-term memory, or by the results of another rule. Essentially, large sets of stored schemata and production rules form an extensive web of knowledge and associated actions, forming the basis for carrying out complex programming tasks.

Another important theory is Shiffrin and Schneider's (1977) theory of controlled and automated processes. Controlled processes require attentional resources and are limited in capacity, while automated processes do not require any attentional resources and are much more efficient. On the other hand, they are not modifiable. When repeating a controlled process many times, it can gradually become an automated process, increasing efficiency but reducing flexibility, for example when learning to play an instrument. In the beginning you have to think about every movement of every finger to produce a specific note sequence, but after extensive training, the task becomes automated and you can concentrate on the high-level control processes (Pressing 1988). This also applies to programming and algorithm design, so that a skilled programmer can concentrate on the higher-level features of the program, while the repetitive details are handled by automated processes. However, since automated processes are preconscious, they may be inappropriately triggered, which may be a cause of error. Then, they may have to be performed slowly as controlled processes again, to make it possible to find the mistake, something that probably many programmers recognize.

3.7 MENTAL IMAGERY AND MENTAL MODELS

As a programmer, I know well that I can perform computations and simulations of computations of limited complexity in my mind, to find out how to solve a particular programming problem. The result may sometimes be wrong (or we would have no bugs) but it is often worth the effort. But it is hard to draw any further general conclusions from such an introspective observation. How does it work, and what are we capable of in terms of mental processing, prediction, and simulation? To discuss mental models in the context of algorithmic music, we must tackle this, since these are the kinds of mental models a composer will need to work consciously with in programming generative algorithms and using algorithmic tools, and these are the models that an advanced listener may form in her mind from repeated listening.

Let us start with mental imagery. A *mental image* is a configuration of perceptual-like features, in analogy with a real-world image. It can be a memory of something previously perceived or a construction of known elements. According to anecdotal evidence from many artists and scientists, visual imagery and visual thinking play a large role in their work. In a rather thorough review of empirical research into creative and generative use of mental imagery, Finke, Ward, and Smith (1992) establish the basic characteristics of such images, and they also tell of a number of experiments in combinatorial creative tasks performed through mental images and mental models.

There seems to be a direct neural link between mental images and the actual perceptual neural circuits that deal with real perceptions. In other words, mental images can contain more than just verbal descriptions and propositions, and it has been shown that mental imagery of music is very often multimodal (Bailes 2007). It takes longer to form a mental image of a complex entity than a simple one, and it seems directly related to the number of components and their relationship (Kosslyn et al. 1983).

Still following Finke, Ward, and Smith, we are also able to perform various transformations on mental images, such as movement and rotation. Such transformations are carried out continuously and holistically, with preserved relations between parts. So, we can draw the conclusion that there is a transformational equivalence between mental images and physical configurations, and this makes it possible to anticipate changes and to predict future states of physical systems and processes.

Since computation can often be thought of as spatial configurations of various entities (with, for example, variables as slots containing figures or different amounts of something, loops as repeated actions on such objects, etc.), we may assume that we can in the same way simulate and predict future states of computational processes, within certain complexity constraints.

More interestingly, we are able to mentally synthesize and transform images so that they show unexpected properties, such as novelty, ambiguity, and emergence. For example, ambiguous images can be thought of as organized in several different ways, and emergent properties are found when the parts of an image are configured such that additional, unexpected features result, which were not intentionally created. Finke, Ward, and Smith review a series of experiments on mental image synthesis based on combinatory creative tasks, involving selection and transformation of graphical primitives, which are put together, transformed, and evaluated as mental images.

The above experiments concentrated on visual mental images, but since a departure point was that mental images can contain different kinds of perceptual-like properties, similar mental processes, including transformation and synthesis of new configurations and simulations of simple physical processes, are most likely equally possible with sonic and musical structures.

Mental images represent perceptual configurations. But how do we process symbols, relations, images, and so on in our mind? To understand this is crucial for understanding programming and working with algorithms. One theoretical construction for how we do this is the idea of a *mental model*. The term was first used by Craik (1943) and its current meaning was established by Johnson-Laird (1983) and Gentner and Stevens (1983).

While a mental image primarily consists of a configuration of perceptual entities, a mental model is a more elaborate, dynamic, and active construction used to make predictions or generate expectations. A mental model can be thought of as a simplified mental representation of a problem space, and is an analogue to the real-world representation of a problem—it has the same functional and structural nature as the system it models. It can contain propositional knowledge, relations, and sets of possible operations.

Mental models are important in, for example, creative exploration, predicting results of hypothetical solutions, examining recombinations of elements, and in considering extremes of different kinds of situations. One important property of mental models is that they provide a way to represent interactions between components from different knowledge domains. They can be tested and modified, and hence form an essential tool for learning. Through what is known as analogical reasoning (Gentner 1989), we can translate the models of computational processes (algorithms) into sound and musical events. And since imagination and creativity are structured activities influenced by existing knowledge frameworks (Finke, Ward, and Smith 1992, 114–115), learned mental models, for example in the form of internalization of new tools or theoretical concepts in the field of algorithmic music, influence how we think about new algorithms, and about new music (Dahlstedt 2012). Hence, they have a profound aesthetic influence on the music we compose. Such mechanisms are at work in our minds to make it possible to work with algorithmic music.

3.8 Situated and Embodied Cognition

Mental models as described above do not contain contextual information. This has been criticized, and already Laske (1975) recognized that musical knowledge is dependent on social context to be meaningful. To remedy this, many situated approaches to cognition have been developed. One idea is that processes that involve tools and environment are not located in our mind but dispersed between the different participating interacting entities.

For example, Greeno (1989) describes how new knowledge is generated from mental models extended with situated information. A primary problem is the insulation of symbolic knowledge. In many problem domains, symbolic notations are used to represent real objects and structural relationships, for example in maths and physics. Students who do not use semantic information in their mental models risk missing important relationships that might be necessary for solving the problems. The models are less complete, and generated knowledge is less likely to be applicable to real situations, since the students are unable to infer from them structural relationships in the real world. Since music in general, and algorithmic music in particular, also uses symbolic notations of different kinds, these results seem highly relevant, for example when translating algorithmically generated structures to musical structures or applying algorithms to solve real-world musical problems. If the composer does not know the physical behaviour of sonic material and how our aural perception works, she risks generating musical structures that do not work as intended.

A more recent (and radical) take on embodied cognition is that external objects in the immediate environment not only are part of cognitive processes—they form an extended part of our mind (Clark and Chalmers 1998; Chemero 2011). For electronic music composers, highly dependent on external tools, this is not as strange as it may initially seem. Without the tools, or with other tools, our cognitive processes would be completely different.

3.9 Action and Perception

Today, the focus is on cognitive models that connect perceptual information with actions, and a number of researchers do not want to distinguish between perception and action. They mean that they are inseparable and are represented together in the brain. Some also mean that we actively perceive through action. We also perceive our actions and their results, in a continuous feedback loop.

One essential discovery that triggered this synthesis was the discovery of mirror neurons (Rizzolatti et al. 1996). These are neural circuits, forming a part of the motor neural system, which are active in similar ways both when we perform an action and when we observe somebody else performing the same action. Their purpose is probably to enable us to empathetically perceive the intentions of others, and to enable prediction of their movement (Keysers 2011). From an evolutionary standpoint, these were necessary abilities for survival in the wild, for hunting, and to avoid predators. The mirror neuron system also enables us to learn motor patterns by imitation.

It should be noted that the existence of specific mirror neurons in humans is hard to directly demonstrate due to the kinds of experiments needed, but there is indirect evidence of mirror functionality in the form of activity in certain brain regions being triggered equally by action and observation of action. See Turella and colleagues (2009) for a review of the research in the area, and, for example, Hari and colleagues (2014) for some recent results.

This discovery resonates well with the common coding theory of action and perception (Van Der Wel, Sebanz, and Knoblich 2013). According to this, actions and perceptions are stored together in the brain in a kind of common neural 'code'. This is a functional theory, which does not concern the neural implementation of the described phenomena. There are a number of interesting results, which are relevant for the study of music making in general and interactive algorithmic music in particular, where a musician has to learn complicated reaction patterns. There are also implications for the notion of agency, which is a controversial concept in algorithmic music—who is playing that sound?

The main properties of action-perception coupling (according to Van Der Wel, Sebanz, and Knoblich) are: mirror neurons fire only when observing goal-directed actions, and action observation induces action simulation. Predictive or internal models of one's own actions can be used to simulate someone else's actions. Action simulation is better with actions one is able to perform, but impossible with unknown actions. It is possible to improve motor expertise with both visual training and motor training, which is a strong indication that they share an underlying representation. One common property shared by both action and perception is timing information, to which we are highly sensitive. Perception of movement follows human motor system constraints, so we are conditioned by our bodies also when observing others (or possibly machines).

This focus on synchrony resonates well with the composer Paulo Chagas's descriptions of embodiment in ensemble playing:

> Musical embodiment is a temporal experience that requires the synchronization of temporal objects and events. In traditional musical practices, such as the Yoruba drum ensembles or Western symphony orchestras, the presence of performers and listeners who physically share the same time and space provides the framework for the synchronization. This mode of embodiment creates the unique 'aura' of the work of music, which, according to Benjamin... , has been eliminated by mechanical reproduction. (Chagas 2006)

Embodiment relies on more than synchrony, and it should probably be regarded as a necessary rather than a sufficient condition. Chagas connects his focus on synchrony to John Cage's ideas that digital music should reestablish temporality by coupling with other media Chagas doubts the ability of electronic music to convey a sense of simultaneity and affect, saying that 'electronic and digital signals are opaque' and that they 'break the transparency of musical flow'. This rather dystopian view may apply to some old-fashioned ways of making and performing computer music, but as we have seen, there certainly are approaches that are able to achieve both synchronicity and flow. The technology is certainly not stopping you anymore.

According to a study with more neurological perspective (Pulvermüller et al. 2014), the mirror neuron system is not enough to explain the strong coupling between action and perception. Instead, the idea of action–perception circuits (APC) is proposed, or 'learned mirror circuits', which are interlinked neuronal action–perception representations, mapping learned motor patterns to perceived actions (e.g. seen or heard). APCs offer mechanisms for repetition and simulation, and when an APC is triggered by the perception of someone else performing a similar or related action, there is a kind of resonance (Zwaan and Taylor 2006), which is particularly strong when empathy and sharing of feelings are involved (Gallese, Keysers, and Rizzolatti 2004). When action repertoires are similar between individuals, perhaps due to intense shared practice, they connect similar motor and perceptual schemas. These neuronal congruencies allow for very refined shared interactions, for example in ensemble music performance (Bangert et al. 2006; Zatorre, Chen, and Penhune 2007).

3.10 AGENCY AND THE LISTENER

The concept of agency (who caused a particular action or event) is regarded as quite important in music. As an improviser, I respond to a particular musician's actions, not just to any sound. In electronic music, where the physical source of the sound is often an anonymous loudspeaker, this is crucial, and Pierre Schaeffer (1966), Michel Chion (1983), and Dennis Smalley (1992) have written extensively about acousmatic or reduced listening, where a listener tries to focus on the abstract sound object and not on its original cause or source. This is not easy.

Following Van der Wel, Sebanz, and Knoblich (2013), one proposed theory for the experience of agency (of one's own actions) is based on perceived mental causal relationship (Wenger 2002), which may lead to an illusory perception of agency under some conditions. Base on a series of studies, Van der Wel and colleagues instead argue that the shared sensorimotor cues can be used to determine agency, with some added benefits. When you perform an action, the motor commands match the sensory feedback from the movement. Also here, synchrony of time cues is crucial. Furthermore, when performing an action, we predict the sensory consequences of the action through internal models (Kawato and Wolpert 1998). Finally, an advantage with this theory is that it can be used also to explain detection of agency in others' actions, thanks to the common representation of action and perception based on the mirror neuron system.

Perception of agency is also important for embodied experience of music, and since the mirror neuron system also reacts to the sounds of actions (Keysers 2011), one might speculate that when sensory information correlates to (perceived) motor information, pseudo-agency can be experienced from pseudo-physical sonic gestures in music. This may be one explanation for the strong urge to dance to music—the dancer appropriates the music by an acquired sense of agency through synchronized movement, and thus achieves a heightened sense of musical embodiment.

The altered role of the listener in electronic music, without visible cause or source, has already been mentioned above. A similar question is the perception of agency of algorithmically generated musical structures, and how much of the underlying algorithm the audience should be able to decipher. According to Collins (2003), this question of the acousmatic sound source has been transplanted to laptop music without a solution. This is a step (or rather no step) in the wrong direction, since acousmatic compositions are often human-composed collages or sequences of sounds, while laptop music relies to a high degree on algorithms and processes, the inner workings of which are harder to grasp for the unknowing ear.

This is important in relation to the various mental models discussed in the previous sections. If the listener is supposed to form a mental model of the underlying computational processes, some information in terms of program notes or explanation may be needed, and possibly also previous knowledge. If perception based on inherent physical models of movement is desired, it is possible to aid the listener by using sonic gestural content mimicking such gestures, and visual cues and displays can also help here, as long as they are causally connected, or in synchrony with the underlying algorithms.

Collins compares an ordinary laptop performer to somebody who is checking her email, and McLean (cited in Collins 2003) has proposed, from a live coding viewpoint, to project the screens of the performers, both to reveal those who do not do much (maybe actually check their email) and to show the tools and actions used, which can be perceived as beautiful even if the exact meaning of the actions is not known by the listener. After all, a nonmusician can appreciate empathetically the actions and efforts performed by a musician, and connect it to audible results, thanks to the mirror neuron system discussed previously, and thanks to synchronicity between visual and aural perceptual cues, or zero-order transcoding from physical gesture to sound. However, the gestural synchronicity is what differentiates gesture from algorithm. An algorithm potentially introduces layers of processing, which obscures synchronicity and causality. This dichotomy between gesture and algorithm is related to Boden and Edmonds's (2009) distinction between *interaction*, actions with direct effect on the output, and *influence*, which affect only the internal states of the generative system, with potential long-term effects on the output. Judging from my own works, these are not always so well defined, since mappings from input to output can be of arbitrary complexity and consist of a combination of real-time interactions and long-term influence. Also, the mapping itself can change over time. The emphasis on effort has been defended by Joel Ryan (1992) and myself (Dahlstedt 2009), among others, as a way to aid listeners who do not know the instruments and algorithms, but also to keep the physical performative role of the musician, which is rewarding in itself (for the musician).

There are also those who think that a visual performance component is unnecessary, and that modern listeners need to get used to the idea of focussing on sound rather than action (Dean 2003). A similar viewpoint is held by Stuart (2003), who introduced the term *aural performativity* as a substitute for the 'visual spectacle and physical gesture of musical performance'. Dean's and Stuart's stance does not disregard traditional musical performance practices, but rather celebrates the new disembodied and aphysical possibilities that algorithmic music provides, which are further elaborated by d'Escriván (2006) around the idea of the effortlessness of computer music. But it does put an increased burden on listeners to reeducate themselves, or maybe this is solved by itself as new generations grow up, perfectly comfortable with the new paradigm.

A stance somewhere in between the advocates of effort-based performance and those who want us to adapt a new paradigm of electronic and algorithmic immateriality was held by David Wessel. His late experimental instruments (Wessel et al. 2007) concentrated on 'intimacy' (Wessel, Wright, and Schott 2002) with the sound, that is, minimal finger gestures and careful but rather direct mappings provided exactly this kind of intimate sonic interaction, producing rich sonic landscapes from minimal movement. And it was free from the connection to effort. Judging from performances with Wessel and my own duo pantoMorf, his instrument lent itself well to the painting of broad generative landscapes as well as to intimate control of microgestures, as contrasted with the bodily, physical, and effort-based instruments that we performed on (Dahlstedt 2009). It was also clear that the different capabilities and design approaches of our instruments made us assume different musical roles. Wessel controlled a virtual ensemble of detailed sound structures, each with only a few degrees of freedom, while we performed as solo musicians, controlling many parameters of a single sonic structure.

3.11 ALGORITHM AS AUGMENTED MIND

Formal methods, that is, algorithms, have been used by artists since antiquity (Dahlstedt 2004). One purpose for which they have been used is to go beyond the known, and help generate something that cannot be directly envisioned, thanks to our limited predictive capacity. Our mental models are not perfect, and it is easy for a composer to devise a few lines of code, without being able to predict exactly what comes out. Even simple algorithms stretch beyond our cognitive resources, because of lack of precision, lack of speed, or limited working memory. So, the only way to predict the result is to execute the algorithm.

This can be further elaborated through a spatial model of creative process (Dahlstedt 2012). If we regard a theoretical space of all possible results in the chosen medium, any tool defines pathways through this space, connecting points that differ by just one application of the tool. At any point the tool can be substituted for another, and the trajectory will follow other pathways, defined by this new tool. Hence, learning a new tool allows the artist to reach new parts of the space or to navigate with greater precision. Forming a mental model of the tool through study and training helps in predicting results and in choosing which path to take, and when using complex algorithms as tools, the steps are large and the predictive capacity is not sufficient. Hence, the algorithmic tool can be used to transcend the limits of the known part of this space.

Because of this, the material generated from an algorithm can feel alien at first, but if the composer spends enough time assimilating the material and forming mental models of the inner logic of the generated music, the new material can gradually become part of the composer's way of thinking, and she might be able to compose in the style of the algorithm, by ear (Dahlstedt 2001b). Then the mind has not only become temporarily augmented, but it has acquired new permanent abilities thanks to the use of algorithms.

The invention of writing fundamentally changed human culture, becoming an augmentation of the human mind, aiding it with memory and knowledge. Written notation, in the same way, fundamentally changed the way composers could work. Since notation allows the composer to forget without loss, she can work in a nonlinear fashion, concentrating on one small part of the work at a time. We can think of algorithms as having the same effect, but regarding process. In a way similar to the previously mentioned automated mental processes (Shiffrin and Schneider 1977), we can start an algorithm and let go of it, like spawning a new thread in a computer program. The algorithm will keep going, and we can afford to forget, instead focusing on the next thread of the process. This way of working is especially evident in the genre of live coding, where the computer becomes a real-time extension of the performer's brain, allowing for more complex and exact processes than she can perform with her brain and body.

Thus, algorithms can, in several different ways, help create an augmented mind.

From a situated cognition perspective, algorithms, when considered as tools, can also be considered as carriers of intelligent behaviour (Gregory 1981) or, in Vygotsky's terms, as mediators of culture (Vygotsky 1978). We are shaped by the tools that are given us by our culture, and we contribute by developing new tools, to be used by forthcoming generations.

Similar views are expressed by Brown (2006) and Reybrouck (2006). They view music as embodied action, and argue that when working with algorithms, the performer is part of a cybernetic control system consisting of a feedback loop of perception–cognition–action. New instruments not only augment our abilities to produce sound, but also our abilities to perceive it, since the body can be regarded as both subject and object, and perception is an action following patterns shaped by the available tools. Following this reasoning, we should continue to develop more advanced tools, tools for sense making and musical knowledge acquisition.

Since today's young grow up with a rich diversity of music apps in new formats, from programmable generative systems to gestural instruments, they are likely to form different mental models of compositional tools and the process of music making. No tool is free of stylistic implications, and the young navigate the space of the possible along paths defined by the tools they know, and if cognition is embodied and situated, and the tools form part of their extended mind, they will create very different algorithmic music.

3.12 PERFORMING ON, WITH, AND IN ALGORITHMS

Algorithms can be used in many different ways, and it is hard to generalize, since an algorithm can be almost any formally described process. Apart from Masahiro Miwa's algorithmic performances, where the algorithm is literally performed by people, as described earlier, there are three fundamentally different configurations in a live setting. You can play *on*, *with*, or *in* an algorithm. I will describe the first two, and then elaborate a little more on the last, since it constitutes a special situation with regard to action, perception, and embodiment, and it has been an important part of my own work.

If you, through some interface, interact with an algorithm, or control its parameters in some way, one can say that you perform *on* the algorithm. This is very similar to performing on the aforementioned hyperinstruments, or augmented instruments, and depending on its complexity, the underlying algorithm may include autonomous processes that evoke the feeling of extra musicians. In essence, you perform *with* these parts of the algorithms. They are not affected by your playing, but you have to learn their inner logic, that is, form a mental model of it, or develop skills in reacting to it, namely action–perception patterns that embody the desired (or necessary) interaction. To summarize, there's a gliding scale from *on* to *with*, depending on the level and nature of interaction between human and algorithm.

In a more tight-knit situation, the musician is part of a real-time algorithmic system, where every interaction has audible results and affects other parts of the system, algorithmic or human. A musician here enters in such a close synergetic relationship with an algorithm (in a performance setting or in a studio situation) that it is perceived as something you step into and become a part of, just as you do in a free improvisation setting. If performing on an algorithm involves sufficiently strong feedback connections between algorithm and human, in both directions, you become a part of a system, and essentially perform *in* the algorithm. Or rather, you become part of a larger composite algorithmic system that involves the computer algorithm, yourself (and possibly other musicians), and your internal cognitive algorithms in the form of musical preferences and behaviour patterns and the links between all these.

Such real-time algorithms do not have a beginning or end, and can be regarded more as systems you step into and form a part of, as one node in a web of interactions. For lack of a better term, I call this *systemic improvisation*, and designing such systems and performing with them has formed a major part of my artistic output the last eight years. I will give a few examples of such works.

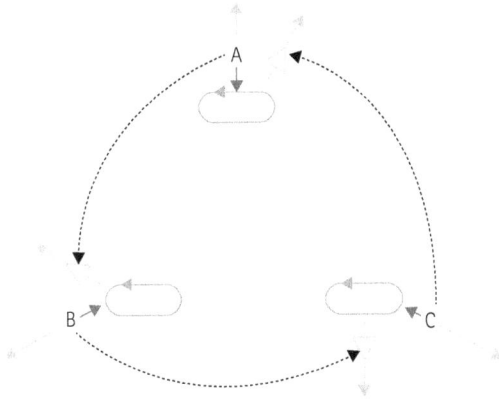

FIGURE 3.2 The connections in *Dynamic Triads* (2008). Each musician's playing is heard in the room, feeds into a circular buffer and simultaneously pulls out sound from the neighbour's buffer.

In *Dynamic Triads* (2008), three musicians are interconnected in a triangle on stage (Figure 3.2). If player A plays a sound, it has three different effects. It is heard in the room, it is stored in a small circular buffer (to be used by player C), and it brings out sounds from the buffer of player B. What happens is that as soon as you play something, it is merged with a sound from what your neighbour recently played, and since the dynamic contour of this sound is completely matched to your own sound, they merge also perceptibly, and appear as one. You play your neighbour, but you also provide sound to your other neighbour at the same time. It takes some time to get used to this unusual computer-mediated interaction model, usually a couple of days of practice. But musicians enjoy playing this system, and it has been performed in many countries with vastly different sets of musicians.

In *Brush Stroke Conversation* (2011), two to four players are connected to a central computer (Figure 3.3). They receive aural or visual cues either through individual headphones or screens. A cue gives them a temporal intensity shape contour, which they are supposed to interpret into a phrase. With two players, the contour shown to player A is generated from the previous phrase of a player B, and the phrase then played by A generates a new cue contour for player B.

In these works, there is no initial material, no prescribed timeline, only a characteristic system with a given interaction model. You enter the algorithm, and you have to survive in there by learning new interaction patterns. The result is often described as 'I have never heard these musicians play in this way'. Which was part of my goal.

A key feature is that the musicians are part of the algorithm. If they do not play, nothing happens. If they misinterpret a cue or an instruction, that is what counts. The actual sound is what goes into the system. And humans fail, often in characteristic ways, so these works rely on what I call the meaningful mistake.

This is especially clear in the work *La Chasse Évitante* (2008) for two musicians (Figure 3.4), where each tries to follow (by ear) a series of cue notes played through a

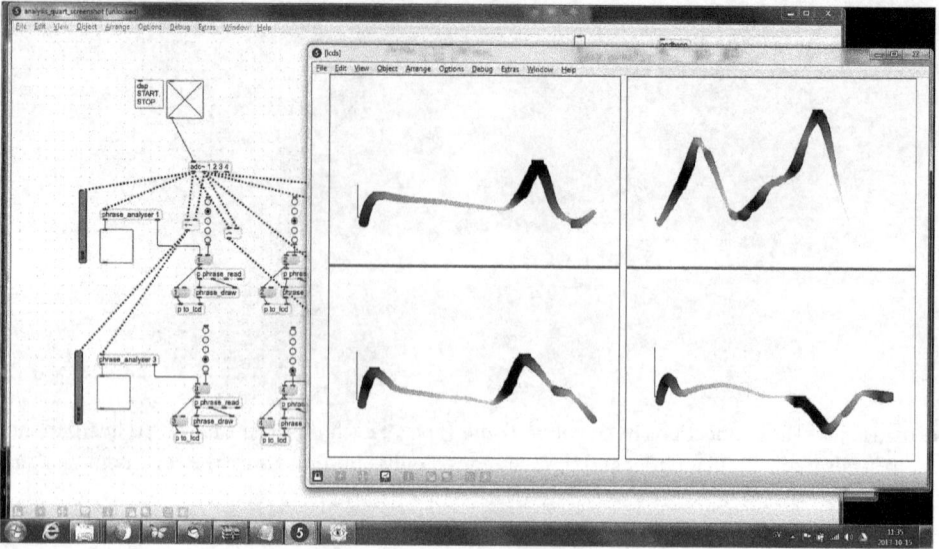

FIGURE 3.3 A screenshot from *Brush Stroke Conversation* (2011). Each musician follows one of the graphs and interprets the brush strokes as musical phrases, while new brush strokes are generated from his playing. The piece is sometimes performed with aural cues instead, in the form of time-compressed filtered noise gestures.

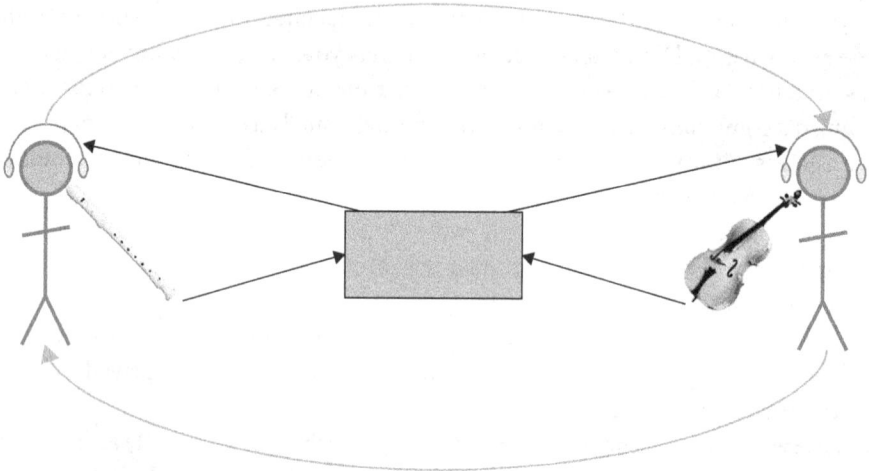

FIGURE 3.4 A schematic of the interconnections in *La Chasse Évitante* (2008). Red lines indicate connections to/from to the processor, while yellow lines indicate acoustic sound.

headphone (inaudible to the audience). They probably fail, because the cue notes move fast. The cue notes for player A are calculated from what player B plays, from a simple arithmetic formula (that can be substituted). And vice versa. The two players form essential links in a figure-8 feedback circuit. Their characteristic (in)ability to accurately follow the cue notes creates the musical content of the piece.

Performing in such systems is like being part of an ensemble of musicians, the reaction patters which you do not fully know. Because of the complexity of the situation, and the high degree of interconnectedness in the system, it is difficult to consciously calculate an appropriate musical response—the situation is unpredictable in the mathematical sense. Instead you have to practise with the system and form a mental model (however incomplete) of its behaviour. Alternatively, through practice you acquire a set of action–perception complexes characterizing the situation, allowing you to enter a flow-like state to 'become' part of the music, of the system. In this state, reaction time is minimal and you are maximally attentive, which facilitates performance.

One cognitive theory that fits playing with such a system is Rosen's (1985) anticipatory systems, where you do not react to what has happened, but to what you think will happen in the near future. The guess may be wrong, but this does not matter—the short-term predictions are still used to guide action at every instant. Even mistakes become meaningful. Essentially it is a feedback system where the feedback emanates from a predictive mental model of the system you interact with, essentially forming a hybrid feedback-feedforward system. This kind of work also relates closely to Roy Ascott's view of artworks as networks or systems (Ascott and Shanken 2003).

3.13   MUSIC AS INTELLECTUAL EFFORT

As a final perspective on the idea of perceived effort in musical performance, we will consider the perception of intellectual effort in a performance or composition. A traditional composer may try to create more and more intricate structures, with maximal mental effort pushing the limits of her own mind, and of what can be played. But such intricacies gain their meaning in relation to the musician's strife and effort to realize them and, partly, in relation to what the composer can compose. But since composition is a nonlinear process, there is no upper bound for the complexities that can be composed, or for the density of the result. The intellect, when decoupled from time and space, can prescribe anything, and without the physicality of the body it has no meaning in terms of difficulty or effort, except possibly under the constraint of limited time.

This is completely different in the act of improvisation. Here, the mind acts in realtime and the intellectual effort is directly related to the limits of the human mind, because of the constraint of linear time. The human body and mind define the time scale and the scope of gestures, phrases, and structures. It is the measure of everything in the music. The music gets its meaning in relation to the physical *and* mental constraints. If these constraints did not exist, the effort would be meaningless, since also the complexity is measured in relation to what a human can perform. In contrast, a machine can generate and perform almost anything, at any speed.

So, what happens when we listen to an algorithmic composition or performance? We may try to experience it physically, through the body, with action–perception circuits triggered by sonic hints of action. We may also reflect intellectually, trying consciously or unconsciously to form mental models of the perceived processes. Maybe the combination of these two, based on the mechanisms described earlier in this chapter, is to experience the music aesthetically.

But we can also experience the music as a human-to-human communication. Algorithms are composed, performed, and perceived. Maybe the mirror neurons also react to mediated effort—the perceived intellectual effort inherent in complex musical structures or behind elaborate studio work. We appreciate empathetically the time and skill it must have taken to realize the piece, here, again, under the constraint of time. Mental models and cognitive constraints can form a basis for such a link. When forming mental models is difficult, the observed system is perceived as complex, that is, it represents a considerable effort.

Aesthetically, this can be thought of as a cognitive and perceptual tension between human and computer: what our mind and body can create with our limited resources versus the computational sublime (McCormack and Dorin 2001), the algorithmically generated emergence that we cannot fathom. In algorithmic music, and in algorithmic performance in particular, both these perspectives have a place; a place where mind, body, and machine make music together.

BIBLIOGRAPHY

Ascott, R., and Shanken, E. A. *Telematic Embrace: Visionary Theories of Art, Technology, and Consciousness*. Berkeley: University of California Press, 2003.

Bailes, F.. 'The Prevalence and Nature of Imagined Music in the Everyday Lives of Music Students'. *Psychology of Music* 35, no. 4 (2007): 555–570.

Bangert, M., Peschel, T., Schlaug, G., Rotte, M., Drescher, D., Hinrichs, H., Heinze, H.-J., and Altenmüller, E. 'Shared Networks for Auditory and Motor Processing in Professional Pianists: Evidence from fMRI Conjunction'. *Neuroimage* 30, no. 3 (2006): 917–926.

Bartlett, F. C. *Remembering: A Study in Experimental and Social Psychology*. Cambridge: Cambridge University Press, 1932.

Berry, D. M. 'A Contribution towards a Grammar of Code'. *Fibreculture Journal* 13 (2008).

Boden, M. A., and Edmonds, E. A. 'What Is Generative Art?' *Digital Creativity* 20, nos. 1–2 (2009): 21–46.

Brown, N.. 'The Flux between Sounding and Sound: Towards a Relational Understanding of Music as Embodied Action'. *Contemporary Music Review* 25, nos. 1–2 (2006): 37–46.

Chagas, P. C.. 'The Blindness Paradigm: The Visibility and Invisibility of the Body'. *Contemporary Music Review* 25, nos. 1–2 (2006): 119–130.

Chemero, A. *Radical Embodied Cognitive Science*. Cambridge, MA: MIT Press, 2011.

Chion, M. *Guide des objets sonores : Pierre Schaeffer et la recherche musicale*. Paris: Buchet/ Chastel, 1983.

Clark, A., and Chalmers, D. 'The Extended Mind'. *Analysis* 58, no. 1 (1998): 7–19.

Collins, N. 'Generative Music and Laptop Performance'. *Contemporary Music Review* 22, no. 4 (2003): 67–79.

Craik, K. *The Nature of Explanation*. Cambridge: Cambridge University Press, 1943.

Cramer, F. *Words Made Flesh: Code, Culture, Imagination*. Rotterdam: Piet Zwart Institute, 2005.

d'Escriván, J. 'To Sing the Body Electric: Instruments and Effort in the Performance of Electronic Music'. *Contemporary Music Review* 25, no. 1 (2006): 183–191.

Dahlstedt, P. 'Creating and Exploring Huge Parameter Spaces: Interactive Evolution as a Tool for Sound Generation'. In *Proceedings of the International Computer Music Conference*, 235–242. San Francisco, CA: International Computer Music Association, 2001a.

Dahlstedt, P. 'A MutaSynth in Parameter Space: Interactive Composition through Evolution'. *Organised Sound* 6, no. 2 (2001b): 121–124.

Dahlstedt, P. *Sounds Unheard of: Evolutionary Algorithms as Creative Tools for the Contemporary Composer*. PhD dissertation, Chalmers University of Technology, Göteborg, 2004.

Dahlstedt, P. 'Evolution in Creative Sound Design'. In *Evolutionary Computer Music*, edited by E. R. Miranda and J. A. Biles, 79–99. London: Springer, 2007.

Dahlstedt, P. 'Dynamic Mapping Strategies for Expressive Synthesis Performance and Improvisation'. In *Computer Music Modeling and Retrieval: Genesis of Meaning in Sound and Music*, edited by S. Ystad, R. Kronland-Martinet, and K. Jensen, 227–242. Berlin: Springer, 2009.

Dahlstedt, P.. 'Between Material and Ideas: A Process-Based Spatial Model of Artistic Creativity'. In *Computers and Creativity*, edited by J. McCormack and M. d'Inverno, 205–233. Heidelberg: Springer, 2012.

Dahlstedt, P. 'Circle Squared and Circle Keys: Performing on and with an Unstable Live Algorithm for the Disklavier'. In *Proceedings of the International Conference on New Interfaces for Musical Expression* 114–117. London, 2014.

Dahlstedt, P. 'Mapping Strategies and Sound Engine Design for an Augmented Hybrid Piano'. In *Proceedings of the International Conference on New Interfaces for Musical Expression*, 271–276. Baton Rouge, LA, 2015.

Dahlstedt, P., and Nilsson, P. A. 'Free Flight in Parameter Space: A Dynamic Mapping Strategy for Expressive Free Impro'. In *Application of Evolutionary Computing*, 479–484. Berlin: Springer, 2008.

Dean, R. *Hyperimprovisation: Computer-Interactive Sound Improvisation*. Middleton, WI: AR Editions, 2003.

Finke, R. A., Ward, T. B., and Smith, S. M. *Creative Cognition: Theory, Research, and Applications*: Cambridge, MA: MIT Press, 1992.

Fuller, M., ed. *Software Studies: A Lexicon*. Cambridge, MA: MIT Press, 2008.

Gallese, V., Keysers, C., and Rizzolatti, G. 'A Unifying View of the Basis of Social Cognition'. *Trends in Cognitive Sciences* 8, no. 9 (2004): 396–403.

Gentner, D. 'The Mechanisms of Analogical Learning'. In *Similarity and Analogical Reasoning*, edited by S. Vosniadou and A. Ortony, 199–241. New York: Cambridge University Press, 1989.

Gentner, D., and Stevens, A. L. *Mental Models*. Hillsdale, NJ: Lawrence Erlbaum, 1983.

Goffey, A.. 'Algorithm'. In *Software Studies: A Lexicon*, edited by M. Fuller, 15–20. Cambridge, MA: MIT Press, 2008.

Greeno, J. G.. 'Situations, Mental Models, and Generative Knowledge'. In *Complex Information Processing: The Impact of Herbert A. Simon*, edited by D. Klahr and K. Kotovsky, 285–318. Hillsdale, NJ: Lawrence Erlbaum, 1989.

Gregory, R. L.. *Mind in Science*. London: Weidenfeld and Nicolson, 1981.

Hari, R., Bourguignon, M., Piitulainen, H., Smeds, E., De Tiege, X., and Jousmäki, V. 'Human Primary Motor Cortex is Both Activated and Stabilized during Observation of Other

Person's Phasic Motor Actions'. *Philosophical Transactions of the Royal Society of London B: Biological Sciences* 369, no. 1644 (2014): 2013.0171.

Hiller, L. A., Jr, and Isaacson, L. M. 'Musical Composition with a High Speed Digital Computer'. *Journal of the Audio Engineering Society* 6, no. 3 (1958): 154–160.

Johnson-Laird, P. N.. *Mental Models: Towards a Cognitive Science of Language, Inference, and Consciousness*. Cambridge, MA: Harvard University Press, 1983.

Kawato, M., and Wolpert, D. 'Internal Models for Motor Control'. In *Sensory Guidance of Movement*, 291–307. Novartis Foundation Symposium. Chichester: John Wiley, 1998.

Keysers, C.. *The Empathic Brain: How the Discovery of Mirror Neurons Changes Our Understanding of Human Nature*. Lexington, KY: Social Brain, 2011.

Kitchin, R., and Dodge, M. *Code/Space: Software and Everyday Life*. Cambridge, MA: MIT Press, 2011.

Kosslyn, S. M., Reiser, B. J., Farah, M. J., and Fliegel, S. L. 'Generating Visual Images: Units and Relations'. *Journal of Experimental Psychology: General* 112, no. 2 (1983): 278–303.

Laske, O. E.. 'Toward a Theory of Musical Cognition'. *Journal of New Music Research* 4, no. 2 (1975): 147–208.

Laske, O. E.. 'Introduction to Cognitive Musicology'. *Computer Music Journal* 12, no. 1 (1988): 43–57.

Machover, T.. *Hyperinstruments: A Progress Report, 1987–1991*. Cambridge, MA: MIT Media Laboratory, 1992.

McCormack, J., and Dorin, A. 'Art, Emergence and the Computational Sublime'. In *Proceedings of Second Iteration: A Conference on Generative Systems in the Electronic Arts*, 67–81. Melbourne: CEMA, 2001.

Ormerod, T.. 'Human Cognition and Programming'. In *Psychology of Programming*, edited by J.-M. Hoc, T. R. G. Green, R. Samurçay and D. J. Gilmore, 63–82. London: Academic, 1990.

Ormerod, T. C., Manktelow, K. I., Robson, E. H., and Steward, A. P. 'Content and Representation Effects with Reasoning Tasks in PROLOG Form'. *Behaviour and Information Technology* 5, no. 2 (1986): 157–168.

Paivio, A. *Imagery and Verbal Processes*. New York: Holt, Rinehart and Winston, 1971.

Pressing, J. 'Improvisation: Methods and Models'. In *Generative Processes in Music*, edited by J. A. Sloboda, 129–178. Oxford: Clarendon Press, 1988.

Pulvermüller, F., Moseley, R. L., Egorova, N., Shebani, Z., and Boulenger, V. 'Motor Cognition— Motor Semantics: Action Perception Theory of Cognition and Communication'. *Neuropsychologia* 55 (2014):71–84.

Pylyshyn, Z. W. *Computation and Cognition*. Cambridge: Cambridge University Press, 1984.

Reybrouck, M.. 'Music Cognition and the Bodily Approach: Musical Instruments as Tools for Musical Semantics'. *Contemporary Music Review* 25, no. 1–2 (2006): 59–68.

Rizzolatti, G., Fadiga, L., Gallese, V., and Fogassi, L. 'Premotor Cortex and the Recognition of Motor Actions'. *Cognitive Brain Research* 3, no. 2 (1996): 131–141.

Rosen, R.. *Anticipatory Systems: Philosophical, Mathematical and Methodological Foundations*. Oxford: Pergamon, 1985.

Rosen, R.. 'Effective Processes and Natural Law'. In *The Universal Turing Machine: A Half-Century Survey*, edited by R. Herken, 485–498. Oxford: Oxford University Press, 1988.

Ryan, J.. 'Effort and Expression'. In *Proceedings of the 1992 International Computer Music Conference*, 414–416. San Jose, CA, 1992.

Schaeffer, P.. *Traité des objets musicaux*. Paris: Seuil, 1966.

Schreibman, S., Siemens, R., and Unsworth, J., eds. *A Companion to Digital Humanities*. Malden, MA: Wiley Blackwell, 2008.

Searle, J. R.. 'Minds, Brains, and Programs'. *Behavioral and Brain Sciences* 3, no. 3 (1980): 417–424.

Shiffrin, R. M., and Schneider, W. 'Controlled and Automatic Human Information Processing: II. Perceptual Learning, Automatic Attending and a General Theory'. *Psychological Review* 84, no. 2 (1977): 127–190.

Smalley, D. 'The Listening Imagination: Listening in the Electroacoustic Era'. In *Companion to Contemporary Musical Thought*, edited by J. Paynter, T. Howell, R. Orton, and P. Seymour, 1:514–554. London: Routledge, 1992.

Strachey, C.. 'The "Thinking" Machine'. *Encounter* 3, no. 4 (1954): 25–31.

Stuart, C.. 'The Object of Performance: Aural Performativity in Contemporary Laptop Music'. *Contemporary Music Review* 22, no. 4 (2003): 59–65.

Turella, L., Pierno, A. C., Tubaldi, F., and Castiello, U. 'Mirror Neurons in Humans: Consisting or Confounding Evidence?' *Brain and Language* 108, no. 1 (2009): 10–21.

Van Der Wel, R. P. R. D., Sebanz, N., and Knoblich, G. 'Action Perception from a Common Coding Perspective'. In *People Watching: Social, Perceptual, and Neurophysiological Studies of Body Perception*, edited by K. L. Johnson and M. Shiffrar, 101–120. Oxford: Oxford University Press, 2013.

Wenger, D. M. *The Illusion of Conscious Will*. Cambridge, MA: MIT Press, 2002.

Wessel, D., Avizienis, R., Freed, A., and Wright, M. 'A Force Sensitive Multi-touch Array Supporting Multiple 2-D Control Structures'. In *Proceedings of the International Conference on New Interfaces for Musical Expression*, 41–45. New York, 2007.

Wessel, D., Wright, M., and Schott, J. 'Intimate Musical Control of Computers with a Variety of Controllers and Gesture Mapping Metaphors'. In *Proceedings of the International Conference on New Interfaces for Musical Expression*, 192–194. Dublin, 2002.

Vygotsky, L. S. *Mind in Society*. Cambridge, MA: Harvard University Press, 1978.

Zatorre, R. J., Chen, J. L. and Penhune, V. B.. 'When the Brain Plays Music: Auditory–Motor Interactions in Music Perception and Production'. *Nature Reviews Neuroscience* 8, no. 7 (2007): 547–558.

Zwaan, R. A., and Taylor, L. J. 'Seeing, Acting, Understanding: Motor Resonance in Language Comprehension'. *Journal of Experimental Psychology: General* 135, no. 1 (2006): 1–11.

CHAPTER 4

ORIGINS OF ALGORITHMIC THINKING IN MUSIC

NICK COLLINS

4.1 INTRODUCTION

> [T]he phono-lecturer came to the main theme of the evening—to our music as a mathematical composition (mathematics is the cause, music the effect). The phono-lecturer began the description of the recently invented musicometer.
> 'By merely rotating this handle anyone is enabled to produce about three sonatas per hour. What difficulties our predecessors had in making music! They were able to compose only by bringing themselves to attacks of inspiration, an extinct form of epilepsy.'
>
> (Zamyatin 1924, 17)

THE roots of algorithmic composition in music are elaborate, and twine through many intriguing early experiments, some well known and some more surprising. In this chapter, we consider the union of music and mathematics before exploring early algorithmic procedures for music generation, and surveying mechanical music precedents. A wider sense of music within world cultures outside the Western canon is also encountered. The whole text is intended to open the reader up to the deep-rooted foundations of more recent computer-led automatic composition and show the rich connection to general musical endeavour in human history.

4.2 MUSIC AND MATHEMATICS

The confluence of music and mathematics isn't just a case of smart people demonstrating a joint aptitude to the two subjects through self-disciplined practice, but a far-reaching historical interaction (Benson 2007; Fauvel, Flood, and Wilson 2003; Harkleroad 2006; Loy 2007; O'Keeffe 1972). From Pythagorean joint investigation of numerical and musical whole-number ratios (Crocker 1964) to the medieval quadrivium—teaching music alongside arithmetic, geometry, and astronomy—to the intensive combinatorics of more recent music theory mathematics problems (Fripertinger 1999) and computer music as a discipline (Moore 1990), the physics, technology, and theory of music making are replete with mathematical links. Coxeter (1968) even advances the theory that the times and places producing great mathematicians also tend to be those producing great musicians, though such circumstances are more likely linked to a general background of the appropriate socioeconomic circumstances for such studies.

Interaction is two directional. J. S. Bach, perhaps the most influential and respected of all composers, was highly mathematically aware, oft noted through the evidence of his contrapuntal wizardry and musical puzzles, a mind in Hofstader's well-known book on a level with arch-logician Kurt Gödel (Hofstader 1999). Scientists have dabbled with musical construction; both Kepler and Newton devised scales, Kepler from ratios of planetary orbit minimum and maximum speeds (Field 2003) and Newton with a seven-note diatonic scale analogous to the seven-colour spectrum (Bibby 2003).

The direction of influence on developments in music and mathematics goes both ways. Perhaps most often, music theory has reacted to innovations in mathematics (Nolan 2002); Catherine Nolan gives an example of great pertinence to this present volume. Mersenne, soon after the introduction of combinatorial methods in Western mathematics, calculated for his *Harmonie Universelle* (1636) the number of possible serialist melodies of from 1 to 22 notes (in his case, permutations of the order of diatonic notes up to a three octave gamut); his table lists the permutation note counts as the factorials from 1! to 22!

There are also occasions of novel mathematical results in musical theory, from the aforementioned Pythagorean investigations, which discovered the harmonic mean, to campanology's anticipation of group theory within the design of change ringing permutation chains (Roaf and White 2003), or a claim for an eleventh-century Venn diagram (Edwards 2006). The more recent theory of maximally even scales has been noted to provide a solution to a problem in Ising model spin configurations in physics (Douthett and Krantz 2008), and the related Euclidean rhythms have many linked applications in scientific disciplines (Demaine et al. 2009).

The twentieth-century preoccupation with formalist and computational method, closely detailed elsewhere in this handbook, was often carried out with an awareness of the historical precedents. Aside from Schoenberg's sense of historical inevitability for the twelve-tone method, Xenakis is the modernist composer perhaps most closely associated with cross-fertilization between the fields of mathematics and music (Harley 2004; Xenakis 1992). Yet as a Greek expatriot, he was highly aware of the ancient heritage of Greek mathematical music, and often returned to early Greek natural philosophers in his writing.

For the interested reader, Gareth Loy's two volumes of 'musimathics' (Loy 2007) provide a strong introduction to areas of mathematics with musical application, with a particular emphasis on acoustics, signal processing, and elements of tuning and algorithmic composition. David Benson's university course in music and mathematics is covered by a freely available online book and associated Cambridge University Press publication (Benson 2006), though it is perhaps less immediately accessible to the non-specialist mathematician.

4.3 PRE-COMPUTER ALGORITHMIC COMPOSITION PRECEDENTS

The standard touchpoints for algorithmic composition include the vowel-to-pitch algorithm of Guido d'Arezzo (1026) and the fad for *ars combinatoria*, the musical dice games of the later eighteenth century (Collins 2010; Loy 2007; Nierhaus 2009; Roads 1996). Hedges (1978) notes twenty examples of the latter being published following Kirnberger's brilliantly titled *The Always Ready Polonaise and Minuet Composer* (1757); Nolan (2000) makes clear that *ars combinatoria* is a more general trend within musical treatises of the time, dealing with the combinatorial possibilities of musical material within established template musical structures, as a stimulant to the work of composers.

Further procedural stimulations for composition in the eighteenth century include the use of divination, as in Vogt's 1719 casting of hobnails to furnish melodic contours (Loy 2007, 1:294), and Hayes's splattering of paint on a musical score to determine notes, as suggested in his 1751 pamphlet 'The Art of Composing Music by a Method Entirely New, Suited to the Meanest Capacity' (Hiller and Isaacson 1979, 52). These were acts more than 200 years ahead of John Cage's conceptions of chance operations, Cage also being beaten to the idea by his idol Marcel Duchamp's *Erratum Musical* (1913), where notes are drawn from a hat! Athanasius Kircher had been a century ahead of his time in the 1650 tract *Musurgia Universalis*, in which he described a music-generating machine, the 'arca musarythmica', a box full of options for different components necessary to a composition, halfway between Guido's method and the musical dice games.

Some authors have seen the mechanical potential, without building the machine. One of the most celebrated anticipations of the possibilities of computational music, more poignant for the fact that Babbage's computer was never built, is Ada Lovelace's footnote prediction on the composition of 'elaborate and scientific pieces of music of any degree of complexity' (Collins 2010). John R. Pierce, famous as director of research at Bell Laboratories, sponsor of Max Mathews's early computer music, and the coiner of the term 'transistor', had his own prescience (pun intended). Writing in November 1950 in *Astounding Science Fiction* (1968), Pierce outlined the potential of Shannon's new information theory to generate music according to Markov chains. Ahead of Hiller and Isaacson's celebrated 1956 computer-generated string quartet experiments, Pierce was carrying out experiments with human calculators on Markov music generation (Hiller and Isaacson 1979, 33). The potential for more advanced computer music yet was also being discussed; as Turing said of Shannon, reputedly during a Second World War lunch conversation, impressed by his verve in plotting a course for artificial intelligence, 'Shannon wants to feed not just data to a Brain [computer], but cultural things! He wants to play music to it!' (Hodges 2012, 251).

In as much as computer-era algorithmic composition is one manifestation of radical experimental music technique, we might go on to consider many further precedents through novel compositional ideas. Lejaren Hiller points to mappings from data to music, such as Renaissance 'eye music' experiments as early graphical scores, and Charles Ives's 1907 baseball game sonification (Hiller and Isaacson 1979, 47–48): 'notes set on paper like men on a football field' (48). Karlheinz Essl (2007) dwells on serialism in particular in his chapter survey of algorithmic composition, though he finds literary precedent in Goethe. We might also point to Earle Brown's experiments with statistical generation of graphical scores in 1952, ahead of Xenakis's own paper and pen stochastic music experiments (Xenakis gained access to a computer only in 1962) (Collins 2010). Reginald Smith Brindle (1956) highlights the new formal techniques inspiring the integral serialists and the necessity for some element of perceptual accessibility in human terms: 'A "computational" composer is in the position of a designer who uses a kaleidoscope to discover new and striking patterns. Such use is legitimate, but the patterns only captivate us if there is a certain element of familiarity inter-woven in their strangeness' (1956, 356).

One author whose conception of mathematical constructions for music composition are key here is Joseph Schillinger, who was influential on a generation of composers in the 1930s, including Gershwin, but who fell into relative obscurity in later decades (Brodsky 2003; Glinsky 2000, 131–135). Two enormous posthumously edited volumes are not enough to fully describe his composition system (Schillinger 1946); his *The Mathematical Basis of the Arts* includes strong speculation on the possibilities for automated music composition (Schillinger 1948). In a review of the latter text, whilst disparaging much of its writing style, John Myhill (1950) is impressed by the scientific approach to aesthetics, and sees no loss of human choice in the automations, or 'artomations'. Human intervention has merely moved back to the setting up of the system: 'It is in the *presetting of the controls* of the machine that the "freedom" or "individuality" of the artist expresses itself' (Myhill 1950, 113), a position much echoed in later understanding of the act of algorithmic composition. Schillinger himself writes evocatively of the potential for musical machines, the never-built Musamatons. Categorizing forms of machine creation for the arts he notes:

3. Instruments for automatic composition of music:
    a. limited to specified components, such as rhythm, melody, harmony, harmonization of melodies, counterpoint, etc.
    b. combining the above functions, and capable of composing an entire piece with variable tone qualities (choral, instrumental chamber music, symphonic and other orchestral music)

4. Instruments for automatic variation of music of the following types:
   a. quantitative reproductions and variations of existing music
   b. modernizing old music
   c. antiquating modern music
5. Instruments of groups 3 and 4, combined with sound production for the purpose of performance during the process of composition or variation.
6. Semi-automatic instruments for composing music. These instruments will be used as a hobby for everyone interested in musical composition, whether amateur or professional, and will not require any special training. The prospective name for instruments of this type will be 'Musamaton.' (Schillinger 1948, 673)

4.4 MUSICAL AUTOMATION

Automation is an essential characteristic of much algorithmic music; a process runs independent of human gestural energy, with or without higher-level human intervention. David Cope (1991) takes the aeolian harp as a paradigmatic example, but we might also dig into the more substantial engineering precedents. Self-playing mechanical musical instruments, also known as musical automata, have a long and distinguished history of over a millennium (Fowler 1967; Kapur 2005; Ord-Hume 1973). Founded on clockwork mechanisms, and making heavy use of drum-roll sequencing, early automata range from miniature music boxes through humanoid figures to larger-scale orchestrion automatic orchestras.

It is striking that in a text on mechanical precedents to the mid-twentieth-century computer, Teun Koetsier (2001) is unable to avoid multiple references to musical automata. The link of the famous eighteenth-century automaton builder Jacques de Vaucanson to the programmable loom is notable. On the back of the international fame of his automatic flute player, digesting and defecating duck, and pipe and tabor robot, Vaucanson was invited in 1740 to look into automation for looms in the French weaving industry. Though his designs were not implemented immediately, making their way to an institutional attic, they were later rediscovered. They became one of the stimulations (alongside innovations from other French engineers working earlier in the eighteenth century) to Jacquard in the creation of his eponymous loom (1801) with its punch-card control mechanism. The punch-card system would go on to inspire Babbage, and thence to twentieth-century computing.

The earliest confirmed musical automata mentioned in the Koetsier article shows the strong Islamic engineering link. Even before Al-Jazari's drinking-party robots (circa 1200), and 900 years before Vaucanson, the ninth-century Musa brothers of Baghdad deployed hydraulics in the construction of a flute-playing robot!

Nonetheless, even if indebted to Islamic science, the flourishing of mechanical music in Europe followed the thirteenth-century development of the mechanical clock. By the time of Shakespeare at the bridge between the sixteenth and seventeenth centuries, automata were well established. Adam Max Cohen notes that amongst the nobility, 'Some of the most popular automata were Christ figures on the cross, clockwork Madonnas, trumpeters, men on horses, and even men sailing ships ... clockwork beasts had eyes that shifted back and forth with the tick-tock of their verge-and-foliot escapements, and some moved or performed on the hour. ... With early clockmakers crafting increasingly elaborate and increasingly lifelike automata it was only a matter of time before a few attempted to build life-sized human and animal automata' (2012, 714).

Riley (2009) makes the interesting point that to the audiences of the early nineteenth century, mechanical music was not a novelty, but a well-known avenue, often unregarded or unsurprising, and certainly not controversial in the terms of later debates around technology threatening human jobs. Whilst certain musical engineering innovations could have a short-term run of success, the age of *ars combinatoria* was ending. Much of the more ornate work may have been fuelled by fads amongst the nobility, and it was undermined by the rise of the bourgeoisie and the industrial revolution. Illustrating this theory of public automata ennui, Riley writes of a celebrated 1814 concert containing Beethoven symphonic premieres:

> And on this spectacular and much-heralded program, following the joyful Seventh and before the concluding Battle Symphony, a ten-minute performance by Mälzel's Automatic Trumpet Player, accompanied by full orchestra, in marches by Ignace Pleyel and Jan Dussek. After the first piece, someone, presumably Mälzel, would have gone out on stage, opened the Trumpet Player's back to replace the Pleyel cylinder with the Dussek cylinder, inserted a large crank into the figure's thigh ... and wound it up for its second performance. This mechanical interlude was duly recorded in the press of the day and in the memoirs of several participants; but in none of the eyewitnesses' accounts is there any suggestion that the android's appearance on the program was strange or unexpected or interesting or uninteresting. (Riley 2009, 372).

The most fantastical plot recounted by Riley (2009) is the interaction of Johann Nepomuk Mälzel and Diederich Winkel. In 1815, Mälzel was on tour with the Panharmonicon, his second automatic orchestra. Visiting Amsterdam, he also visited the workshop of Winkel, where he discovered (i.e. stole) the idea for an accurate clockwork metronome. Whilst Mälzel grabbed the patents and the international market for the metronome, and even gave his name to the metronome's tempo markings, Winkel's revenge was to outdo the Panharmonicon. Winkel's mechanical Componium (see Figure 4.1), premiered in 1821, combined the idea of an orchestrion with the musical dice game to create a musical automata capable of playing variations on a theme. It did this via an ingenious use of two synchronized barrel rolls, alternating two-measure phrases; whilst one played back, the other was silent (having no pegs in that section), and slid horizontally on a random walk to select its next material. A revised estimate of the combinatorial capability of this fantastic device puts the number of possible variations at a modest 256 million or so, with a prediction of forty-one years of continuous play before any full sequence would be repeated (Bumgardner 2013). This combination of algorithmic composition and musical robotics is a startling early precedent, but has been

FIGURE 4.1 Crossing the mechanical musical instrument and the musical dice game: The Componium of Dietrich Winkel, photographed in the Museum of Musical Instruments, Brussels. Photo by Jbumw, used under CC BY-SA 4.0 license (original from https://commons. wikimedia.org/wiki/File:Winkel%27s_Componium_at_Brussells.jpg).

little publicized; the machine itself was a modest initial success, but with the death of its inventor, it fell into obscurity (the gutted machine survives in the Brussels Museum of Musical Instruments).

The metronome itself, perhaps more than automata, brought home the mechanization of musical process, since it was the subject of mass production and mass adoption. The changes in performance practice brought about by the arrival of the metronome as a central musical teaching aid are documented by Alexander Bonus (2014) and reveal the tensions between an absolute sense of musical time fitting the industrial-urban-scientific age and a nostalgia for the looser human time of musical tradition. The protagonists of the two sides of this debate prefigure the more twentieth-century anxieties around musical technology, whether silent film musicians losing jobs to the talkies or MIDI threatening the Musicians' Union. The heel-and-toe clog dancing of female machine operators in Lancashire dating back to the 1820s, and subsequent machine-inspired choreography, might provide a further link to the mechanization of the arts (Radcliffe and Angliss 2012); the clog link evokes sabotage by sabot, and the Luddites of a similar time period (1810s) as the metronome's arrival. Ultimately, the metronome won: metronome time guiding human musicianship permeates throughout the click tracks and computer time of modern recording process, and is indispensable to conservatoire practice.

Aside from street organs and orchestrions, musical automata had a further flowering with the fad from 1890 to 1920 for player pianos, an outgrowth of bourgeois aspiration to home pianism sneaking in before the booms in radio and higher-quality record players. The vast majority of early automata may be 'reprogrammable' in the sense of changing sequencer data by substituting a new cylinder, altering peg positions on an existing cylinder, or punching new holes in a piano roll, but they are not programmable in any more profound sense. Indeed, it is surprising how much music for mechanical musical instruments imitates human-speed music making, at least before the hypervirtuosity of Nancarrow (Gann 1995). Nancarrow began his player piano studies around the middle of the twentieth century, coincidentally mirroring the rise of the electronic computer, and perfectly anticipating the inhumanly fast sequencing capabilities so unleashed.

4.5 Ethnomusiconomy

In contrast to musicology as the humanities activity of 'speaking' (*logos*) about music, Gareth Loy suggests musiconomy as more suitable to investigation of the 'laws' (*nomos*) of music (Loy 2007). Extending this from ethnomusicology, we might consider the subdiscipline of *ethnomusiconomy* as pondering generative laws within the world's musics. Chemillier (2002) has treated the situation of 'ethnomathematics' as applied within ethnomusicology, and provides a wonderful example of Nzakara court harp music, founded on string combinatory patterns.

Mathematical thinking is hardly limited to Western music. The thousand-year-old Tibetan Buddhist tradition of *Rol Mo* is detailed by Ellingson (1979), who observes compositions based around two cymbals. The music places an emphasis on time spacing of events and the timbral effects possible by altering the location of strikes on the cymbals and the after-strike interaction of the two cymbals in proximity. *Days of the Waxing Moon*, a sequence of repetitions for events, uses an arithmetic series run from 1 to 15 by steps of 2. *Invitation to Mahākāla* uses an accelerating 'countdown', beginning by stepping through 180, 170, ... , 20, 15, then from 10 down to 1. The same piece also includes an action sequence based on the symbolism of a hexagram inscribed in a circle, where cymbal strikes draw out the geometric figure in space. This sort of gesture piece gazumps recent experiments in action composition in contemporary Western music!

Indian classical music is full of complicated temporal, pitch, and timbral structures; an introduction to rhythmic *tāla* alone might take a whole book (Clayton 2008). Perhaps the most developed research project in algorithmic modelling of North Indian (Hindustani) music is that of Kippen and Bel (1992); the associated Bol Processor software has been made available as an open-source project, and its modelling tools are generally amenable to many musics (http://bolprocessor.sourceforge.net). North Indian music is not alone in its complexity and inspiration; Virtual Gamelan Graz is a project seeking heightened understanding of compositional rules and the sound world within Indonesian gamelan (Grupe 2008). Godfried Toussaint (2013) has researched many world rhythms, especially with respect to those expressible as a subset of a cycle of isochronous pulses (see also Demaine et al. 2009); he provides an analysis of the clave timeline pattern and its historical evolution.

It may be tempting, though it is fallacious, to believe that because a music is amenable to mathematical modelling such modelling constitutes a final proof of that music's construction. This is especially dangerous to ethnomusicologists when devoid of the further cultural factors. Steven Feld has cautioned against the uncritical and often unscientific adoption of linguistic models in his field; just because a given music can be represented via a particular model does not mean that the model is thereby proved the only analytical approach, nor that the model is the original representation in culture (Feld 1974)! Nonetheless, mathematically minded music theorists and ethnomusicologists have found much of interest in the world's musics beyond Western ideas. Participants within alternative musical culture are themselves adopting and adapting formal methods to treat their own musics (see, for instance, Kippen and Bel 1994 and Sen and Haihong 1992), though the vast majority of algorithmic composition research, much like music psychology research, is limited to the Western canon.

4.6 CONCLUSIONS

Algorithms for music have rich precedents, across cultures and across eras. The expansion of scientific knowledge has often interacted with music's cultural evolution, especially where developments in technology have been necessitated by artistic concerns, such as the novelty of musical automata, or where scientific investigation of a physical phenomenon is intimately tied to aural resultant, such as the Greek study of stringed instruments. We have seen that music technology is sometimes unshocking, falling into a trend of the day such as the clockwork universe, and sometimes more threatening to older tradition (whilst attractive to others), such as the metronome's new time regulation.

The wealth of precursors may act as a check on the hyperbole of some algorithmic composers. Should we be surprised at the unsurprising nature of generative music, when only one presentation or realization can occur at a time and it is so hard in human terms to hear out the immense combinatorial space of possibility, or to program for substantial perceptual variation? Instead, one might see only the relationship to that existing generative system par excellence, human improvisation. The fate of Winkel's Componium should be borne in mind.

One unfortunate aspect of the survey carried out here is the reduced role of female composer–engineers. Although there are electronic computer-generated music pioneers such as Laurie Spiegel and the lesser-known Harriet Padberg (Ariza 2011), Lancashire clog dancers and Ada Lovelace are not enough to avoid the sense of male preserve in earlier historical time. At least the present era is replete with more equal opportunity. Confronting the challenges of algorithmic composition will without doubt be led by female composers in the coming decades.

BIBLIOGRAPHY

Ariza, C. 'Two Pioneering Projects from the Early History of Computer-Aided Algorithmic Composition'. *Computer Music Journal* 35, no. 3 (2011): 40–56.

Benson, David J. *Music: A Mathematical Offering*. Cambridge: Cambridge University Press, 2006.

Bibby, N. 'Tuning and Temperament: Closing the Spiral'. In *Music and Mathematics: From Pythagoras to Fractals*, edited by J. Fauvel, R. Flood, and R. Wilson, 13–27. New York: Oxford University Press, 2003.

Bonus, Alexander E. 'Metronome'. *Oxford Handbooks Online* (2014): 1–18. Web. 24 Apr. 2015.

Brodsky, W. 'Joseph Schillinger (1895–1943): Music Science Promethean'. *American Music* 21, no. 1 (2003): 45–73.

Bumgardner, J. 'Variations of the Componium'. Unpublished draft, 2013. https://jbum.com/papers/componium_variations.pdf.

Chemillier, M. 'Ethnomusicology, Ethnomathematics: The Logic Underlying Orally Transmitted Artistic Practices'. In *Mathematics and Music: A Diderot Mathematical Forum*, edited by G. Assayag, H. G. Feichtinger, and J. F. Rodrigues, 161–183. Berlin: Springer, 2002.

Clayton, M. *Time in Indian Music: Rhythm, Metre, and Form in North Indian Rāg Performance*. New York: Oxford University Press, 2008.

Cohen, A. M. 'Science and Technology'. In *The Oxford Handbook of Shakespeare*, edited by A. F. Kinney, 702–718. Oxford: Oxford University Press, 2012.

Collins, N. *Introduction to Computer Music*. Chichester: Wiley, 2010.

Cope, D. *Computers and Musical Style*. Oxford: Oxford University Press, 1991.

Coxeter, H. S. M. 'Music and Mathematics'. *Mathematics Teacher* 61, no. 3 (1968): 312–320.

Crocker, R. L. 'Pythagorean Mathematics and Music'. *Journal of Aesthetics and Art Criticism* 22, no. 3 (1964): 325–335.

Demaine, E. D., Gomez-Martin, F., Meijer, H., Rappaport, D., Taslakian, P., Toussaint, G. T., Winograd, T., and Wood, D. R. 'The Distance Geometry of Music'. *Computational Geometry* 42, no. 5 (2009): 429–454.

Douthett, J., and Krantz, R. 'Dinner Tables and Concentric Circles: A Harmony in Mathematics, Music, and Physics'. *College Mathematics Journal* 39 (2008): 203–211.

Edwards, A. W. F. 'An Eleventh-Century Venn Diagram'. *BSHM Bulletin* 21, no. 2 (2006): 119–121.

Ellingson, T. 'The Mathematics of Tibetan Rol Mo'. *Ethnomusicology* 23, no. 2 (1979): 225–243.

Essl, K. 'Algorithmic Composition'. In *The Cambridge Companion to Electronic Music*, edited by N. Collins and J. d'Escriván, 107–125. Cambridge: Cambridge University Press, 2007.

Fauvel, J., Flood, R., and Wilson, R. *Music and Mathematics: From Pythagoras to Fractals*. New York: Oxford University Press, 2003.

Feld, S. 'Linguistic Models in Ethnomusicology'. *Ethnomusicology* 18, no. 2 (1974): 197–217.

Field, J. V. 'Musical Cosmology: Kepler and His Readers'. In *Music and Mathematics: From Pythagoras to Fractals*, edited by J. Fauvel, R. Flood, and R. Wilson, 29–44. New York: Oxford University Press, 2003.

Fowler, C. B. 'The Museum of Music: A History of Mechanical Instruments'. *Music Educators Journal* 54, no. 2 (1967): 45–49.

Fripertinger, H. 'Enumeration and Construction in Music Theory'. In *Diderot Forum on Mathematics and Music Computational and Mathematical Methods in Music*, edited by H. G. Feichtinger and M. Dörfler, 179–204. Vienna: Österreichische Computergesellschaft, 1999.

Gann, K. *The Music of Conlon Nancarrow*. Cambridge: Cambridge University Press, 1995.

Glinsky, A. *Theremin: Ether Music and Espionage*. Urbana and Chicago: University of Illinois Press, 2000.

Grupe, G., ed. *Virtual Gamelan Graz: Rules—Grammars—Modeling*. Aachen: Shaker, 2008.

Harkleroad, L. *The Math behind the Music*. New York: Cambridge University Press, 2006.

Harley, J. *Xenakis: His Life in Music*. New York: Routledge, 2004.

Hedges, S. A. 'Dice Music in the Eighteenth Century'. *Music and Letters* 59, no. 2 (1978): 180–187.

Hiller, L. A., and Isaacson, L. M. *Experimental Music: Composition with an Electronic Computer*. 2nd ed. Westport, CT: Greenwood, 1979.

Hodges, A. *Alan Turing: The Enigma*. London: Vintage, 2012.

Hofstadter, D. *Gödel Escher Bach: An Eternal Golden Braid*. 20th anniversary ed. London: Penguin, 1999.

Kapur, A. 'A History of Robotic Musical Instruments'. In *Proceedings of the International Computer Music Conference*. Barcelona, 2005.

Kippen, J., and Bel, B. 'Modelling Music with Grammars: Formal Language Representation in the Bol Processor'. In *Computer Representations and Models in Music*, edited by A. Marsden and A. Pople, 207–238. London: Academic Press, 1992.

Kippen, J., and Bel, B. 'Computers, Composition and the Challenge of "New Music" in Modern India'. *Leonardo Music Journal* 4 (1994): 79–84.

Koetsier, T. 'On the Prehistory of Programmable Machines: Musical Automata, Looms, Calculators'. *Mechanism and Machine Theory* 36, no. 5 (2001): 589–603.

Loy, G. *Musimathics*. 2 vols. Cambridge, MA: MIT Press, 2007.

Moore, F. R. *Elements of Computer Music*. Englewood Cliffs, NJ: P. T. R. Prentice Hall, 1990.

Myhill, J. 'Review of *The Mathematical Basis of the Arts*'. *Philosophy and Phenomenological Research* 11, no. 1 (1950): 109–113.

Nierhaus, G. *Algorithmic Composition: Paradigms of Automated Music Generation*. New York and Vienna: Springer, 2009.

Nolan, C. 'On Musical Space and Combinatorics: Historical and Conceptual Perspectives in Music Theory'. In *Bridges: Mathematical Connections in Art, Music, and Science*, 201–208. Winfield, Kansas, 2000.

Nolan, C. 'Music Theory and Mathematics'. In *The Cambridge History of Western Music Theory*, edited by T. Christensen, 272–304. Cambridge: Cambridge University Press, 2002.

O'Keeffe, V. 'Mathematical-Musical Relationships: A Bibliography'. *Mathematics Teacher* 65, no. 4 (1972): 315–324.

Ord-Hume, A. W. J. G. *Clockwork Music: An Illustrated History of Mechanical Musical Instruments*. London: George Allen and Unwin, 1973.

Pierce, J. R. *Science, Art, and Communication*. New York: Clarkson N. Potter, 1968.

Radcliffe, C., and Angliss, S. 'Revolution: Challenging the Automaton: Repetitive Labour and Dance in the Industrial Workspace'. *Performance Research* 17, no. 6 (2012): 40–47.

Riley, T. 'Composing for the Machine'. *European Romantic Review* 20, no. 3 (2009): 367–379.

Roads, C. *The Computer Music Tutorial*. Cambridge, MA: MIT Press, 1996.

Roaf, D., and White, A. 'Ringing the Changes: Bells and Mathematics'. In *Music and Mathematics: From Pythagoras to Fractals*, edited by J. Fauvel, R. Flood, and R. Wilson, 113–129. New York: Oxford University Press, 2003.

Sen, W., and Haihong, Z. 'Scale-Tone Functions and Melodic Structure in Chinese Folk Music'. In *Computer Representations and Models in Music*, edited by A. Marsden and A. Pople, 111–120. London: Academic Press, 1992.

Schillinger, J. *The Schillinger System of Musical Composition*. 2 vols. New York: C. Fischer, 1946.

Schillinger, J. *The Mathematical Basis of the Arts*. New York: Philosophical Library, 1948.

Smith Brindle, R. 'The Lunatic Fringe: III, Computational Composition'. *Musical Times* 97, no. 1361 (1956): 354–356.

Toussaint, G. T. *The Geometry of Musical Rhythm: What Makes a "Good" Rhythm Good?* Boca Raton, FL: CRC Press, 2013.

Xenakis, I. *Formalized Music: Thought and Mathematics in Composition*. 2nd ed. Hillsdale, NJ: Pendragon Press, 1992.

Zamyatin, Y. *We*. Translated by G. Zilboorg. New York: E. P. Dutton, 1924.

CHAPTER 5

···

ALGORITHMIC THINKING
AND CENTRAL JAVANESE
GAMELAN

···

CHARLES MATTHEWS

PLAYING gamelan is an intrinsically communal activity. The music often typifies an aesthetic quality called *ramai*, or *ramé* in Javanese: a feeling of busyness and filling of space, as many instrumentalists create unbroken streams of sound, punctuated by a hierarchy of gong pulses (Sutton 1996, 258). Gamelan is closely associated with community events such as weddings, religious ceremonies, and all-night shadow theatre (*wayang kulit*), in which ensembles must call upon a vast repertoire of pieces on demand, interacting with other performers such as dancers and puppeteers. As a result, the playing conventions found in gamelan are flexible, providing opportunities for interaction both within and outside the ensemble that are maintained when the music is played for its own sake.

One of the reasons gamelan has attracted the interest of musicologists and composers of new music—both in Indonesia and abroad—is the way that musicians appear to be capable of working out coherent interdependent parts from a notated outline or by listening to guidance from other instruments. This applies in particular to the central Javanese style (also known as *karawitan*), in which pieces are rarely played exactly the same way twice. Such variability stems from oral tradition, and reflects other aspects of Indonesian culture, such as the variations upon classic outlines found in storytelling and *batik* patterns. Although notation and recording technology are gaining popularity, many musicians believe that traditional works for gamelan (called *gendhing*) truly exist only in the moment of performance, as a combination of all parts in the ensemble (Supanggah 1988; 2011, 181–182).

Musicians traditionally learn to play gamelan from experience, by observing, imitating, and playing from an early age, but recent attempts at formalizing the music have produced an assortment of rules abstracted from performance practice. As a result, students with limited experience can fit into an ensemble to play acceptable parts alongside expert musicians; composers who are familiar with the constraints and rules of the tradition can provide musicians with a relatively simple part from which can be generated a complex piece of music. But to what extent can the thought processes behind this rich musical tradition be described as algorithmic? How can composers and theorists draw on algorithmic thinking, not only to understand and maintain the tradition, but to build upon it?

In this chapter I discuss some rule-based approaches to learning and performing gamelan music, particularly the central Javanese classical tradition concentrated in the cities of Solo and Jogja, and explore some ways that it can hold inspiration for algorithmic thinking in musicology and new compositions. In particular, I discuss notions of outlines, generative structures, and methods of elaboration in performance practice. This discussion will be illustrated in the second part by applications in composition and musical research that specifically refer to Javanese music theory. The final part of the chapter—originally written as a *perspective on practice* contribution to the handbook—is somewhat more reflective, and explores new applications for theory from central Javanese gamelan in computer-aided composition. Examples are presented from my own practice involving gamelan and live electronics, namely Augmented Gamelan (http:// www.augmentedgamelan.com/) and Pipilan (http://www.ardisson. net/gamelan/pipilan/). In these projects I have used algorithmic processes in augmentation of traditional-style ensembles, and explored ways for audiences to join in the composition and performance processes.[1]

It is difficult to discuss gamelan music theory without becoming familiar with some Javanese terminology, since ostensibly equivalent terms in English can be misleading. Some terms will therefore be reviewed to illustrate key concepts. It is also important to consider that some generalizations overlooking a range of regional and personal variations will be inevitable in a brief chapter such as this. More comprehensive introductions to gamelan music and its cultural context may be found elsewhere (particularly Pickvance 2005; Sumarsam 1995; Supanggah 2011; and Sutton 1991; for Sundanese practice, see Cook 1992; for Balinese varieties, see Tenzer 2000).

5.1 Theoretical Foundations: Structures and Rule-Based Systems in Traditional Gamelan music

One way of approaching gamelan music is as the intersection of two systems: its cyclic gong patterns provide robust hierarchies of high and low pulses, through which are threaded rhythms and melodies of a more linear and intricate nature.[2] It is these latter elements that provide opportunities for personal expression and interaction, as they can be treated in realtime by musicians in various ways, and expanded through different time scales. Musical time in gamelan is elastic, as melodic lines and rhythms can be stretched out in the course of performance; the same essential pattern might last anywhere from a few seconds to a couple of minutes, with some players doubling up their parts to maintain a sense of flow and musical identity. This temporal expansion can be reminiscent of zooming in on a fractal, revealing hidden structures and contours that appear to refer back to the shape of the whole.

The relationship between instrumental parts in gamelan music performance is often called 'polyphonic stratification': many different melodic–rhythmic lines form distinct layers or strata of sound, each maintaining its own character in melodic contour, rhythmic idioms, and relative density (Hood 1982, 52). Attempts to describe their coherence have led to somewhat more contentious descriptions of these layers as a form of heterophony, involving many simultaneous variations of the same melody (see Perlman 2004, 62; Pickvance 2005, 22–23). The nature of shared reference points in performance can sometimes be unclear, shifting between contexts. Pieces are ostensibly defined by outlines and instructions, but closer inspection often points towards a fuzzier sort of shared abstraction, made up of various threads played aloud in the ensemble, but not always traceable to a single source.

5.1.1 Structures and Colotomy

The most fundamental characteristic shared by gamelan compositions is that of the gong cycle: a repeated pattern lasting anywhere from a few seconds to several minutes, marked by the largest and deepest hanging gong in an ensemble. This cycle is divided and punctuated by sets of other hanging and horizontally mounted gongs, to form what is known as a 'colotomic structure' (Kunst 1949). Musicians may create complexity through further subdivision of rhythmic pulses, repetition, phase offset, and the interlocking of similar parts. Other layers of more linear melodies are constructed within idiomatic constraints, relying heavily on a conceptual framework called *pathet* (a term loosely equivalent to mode if approached in terms of pitch, although it has many other connotations; see Perlman 2004, 42–43; Pickvance 2005, 52–57; Supanggah 2011, 299–313). Patterning in gamelan is generally end-weighted; parts tend to anticipate structural markers rather than responding to them. Therefore, if a pattern is expanded or subdivided, its phase is shifted backwards so that its endpoint matches the central pulse (see Figures 5.2 and 5.3).

Colotomic structures give shape and identity to traditional gamelan pieces. These formal structures form subgenres of pieces with closely associated functions in dance and drama, and their names, such as *ladrang, ketawang,* and *gendhing,* are generally included in the title of pieces, along with *pathet* and other structural information (Pickvance 2005, 37, 81–90). Gong cycles are typically repeated until cued by melodic or rhythmic leaders of the ensemble, through a change in pitch register or tempo, which can lead to other sections (often the same colotomic structure filled with different melodic information), or another piece altogether.

5.1.2 Balungan: Melodic Outlines and Notation

Beyond colotomic structures, most forms of gamelan use some sort of outline, played aloud, that can be used as a shared reference point between musicians. The melodic outline most typically found in Javanese gamelan is called a *balungan*.[3] Literally translatable as 'skeleton' or 'frame', the *balungan* in musical contexts is a metronomic part, typically written as a series of numbers (see Figure 5.1).[4] These sequences are usually split into four-note units called *gatra*, which form the basis of much performance practice and analysis of Javanese gamelan music (Pickvance 2005, 29; Supanggah 2004; 2011, 176).

The outlines presented by *balungan* can take a variety of shapes, ranging from more dense, prescriptive sequences (Figure 5.1A) to the sparse *nibani* style (Figure 5.1B). The latter type of sequence gives the most basic outline on the strongest beats, and is often played at such a slow pace that it can no longer be considered a melody, becoming a set of aural landmarks to confirm the key resting points for phrasing within the wider structure.

Alongside certain vocal parts, the *balungan* (closely matching parts played by a set of keyed instruments called *saron*) is one of the most consistent elements of a gamelan

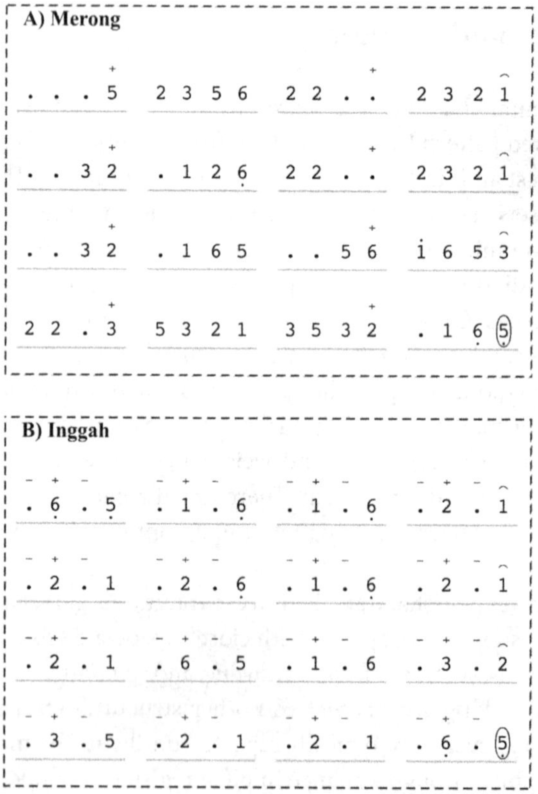

FIGURE 5.1 Two *balungan* sequences from the traditional Javanese piece *Gendhing Gambirsawit, kethuk 2 kerep minggah 4, laras slendro, pathet sanga* (see *Gendhing Jawi* 2017 for full version).

composition, and was therefore chosen for posterity when attempts to standardize nota-tion were introduced in the late nineteenth century (Sumarsam 1995, 111).[5] Although notated parts were originally intended to function as memory aids or records of aurally transmitted information, they are increasingly used prescriptively in educational insti-tutions (see Sumarsam 2004 for discussion of teaching methods in Java and abroad). Collections of *balungan* and related songs can be found at bookshops, photocopy shops, and stalls at shadow puppet performances, and more recently for free download on the Internet. For instance, the *Gendhing Jawi* collection hosts over 1600 outlined sequences that are readily usable as notation (http://www.gamelanbvg.com/gendhing/index.php).

The *balungan* is sometimes considered to be the basis of most parts in the ensem-ble; early ethnomusicological studies compared it to the cantus firmus or 'fixed mel-ody' found in other musical systems (see Perlman 2004, 123; Sumarsam 1995, 145). This idea has recently been challenged, as musicians often state that they refer to a less fixed framework, an unplayed or 'inner melody' in performance. The opinions of various theorists and musicians on this subject are collected and discussed at length by Marc Perlman (2004; also see Sumarsam 1975b; Supanggah 1988, 2011). The *balungan* might be thought of as a quantized version of this internal framework, a convenient point of reference when teaching and learning the music, and confirmation of the resting points of melodic phrasing when played out loud.

The effectiveness of transmitting traditional works, whether by succinct notation or through observation of other performers, is largely due to the overlapping of rhythmic and melodic information between compositions. Gamelan pieces are most frequently composed using recombination of existing material, a process sometimes referred to as 'centonization' (borrowed from descriptions of Gregorian chants; see Sumarsam 1995, 162; Sutton 1987). It is often assumed that composers work in the classical style by creating a *balungan* sequence, as this tends to be what is written down (Sutton 1987). However, in doing so, they may be working with placeholders for more complex melodic information associated with particular *balungan* fragments. Another point of reference at this level can be found in the form of *cengkok*, more detailed melodic formulae or stock patterns that are often given names, and which can be recognized across variations on different instru-ments. These patterns tend to fit together in certain orders, serving larger-scale idiomatic vocal contours and *pathet*, and providing familiar pathways for musicians to pass through when approaching unfamiliar pieces without the aid of notation.

5.1.3 Garap: Idiomatic Treatment of Musical Information

The reinterpretation and elaboration of existing musical material in Javanese gamelan music (including, but not limited to, the *balungan*) is known as *garap* (Perlman 2004, 60; Supanggah 2011). Since there is a great deal of potential for variability throughout the ensemble, it is common to describe this process in terms of improvisation, in contrast to more fixed sequences of other musical systems such as the Western classical tradition. However, musicians rarely have the freedom to play whatever they wish; they must refer back to the central framework, upholding the integrity, *pathet*, and *rasa* (feeling) of the piece, and conforming to playing conventions. As R. Anderson Sutton concludes in a thorough review of the subject of improvisation in gamelan, 'while very little is *entirely* fixed beforehand in a Javanese performance, a great deal is *almost* fixed (or expected to be)'; gamelan music involves improvisation, but it would not be appropriate to describe it broadly as improvisatory (Sutton 1998, 87). The type of variability in Javanese gamelan might be described as 'idiomatic improvisation', in contrast with free or 'unidiomatic' improvisation: 'improvisation serves the idiom and is the expression of that idiom' (Bailey 1992, 18).

Given its constraints, improvisation in traditional practice might be considered a surface detail filling the deeper structures of tuning, mode, and colotomy discussed thus far, in combination with more specific instrumentally oriented grammars (Hood 1972).[6] While this description might seem to undervalue individual creativity, it still allows for more spontaneous or individualistic actions, most notably through interaction. Drummers often play in unplanned interaction with a shadow puppeteer while leading the tempo of the ensemble. Forms such as *palaran* do not use an explicit *balungan part*, but rather require musicians to follow vocal patterns, adapting to the pace of the singer and mediation from the drummer (see Brinner 1995, 234–244, for a breakdown of this interactive system). Similarly, musicians might play holding patterns or adjust their playing to imitate other players if they lose their way in unfamiliar pieces, especially when playing without notation (*ngeli*—floating; 142).

Each of the instruments in the ensemble generally has a clearly defined role, and is associated with a set of playing conventions or 'instrumental idioms' (see Brinner 1995, 55–56; Matthews 2014, 13–16; Perlman 2004, 43–49). At their most basic level, these idioms provide the techniques needed for sound production, appropriate pitch range and dynamics, and stylised elements such as damping. Idiomatic conventions can also extend to the way that outlines can be treated (often through application of abstract patterns or a repertoire of stock phrases), and ways to respond to and interact with other instruments in the ensemble. Some subfamilies of instruments play together to create patterns that complement each other and interlock, such as the *imbal* patterns played on the *saron* or *bonang*, and *pinjalan* or *banyakan* played across the ensemble (see Figure 5.4; Brinner 1995, 223–226; Pickvance 2005, 170–183; Supanggah 2011, 288; Sutton 1991, 49–51). These treatments can vary between contexts; the same instruments can be associated with a range of techniques according to regional styles, or the same musician might play a given instrument differently for dance or shadow puppet theatre.

Different types of patterning are linked to the classification of gamelan instruments as either 'loud' or 'soft' style (referring to the relative volume of the instruments in an unamplified setting). 'Loud style' instruments mostly consist of thick-keyed metallophones and gong-chimes, and their associated idioms tend to make direct reference to the outline provided by the *balungan*. The 'soft style' family comprises instruments with thinner keys, stringed instruments, and xylophones, and are generally associated with vocal parts and more elaborate melodic figurations.

More specific playing techniques are associated with the physical construction and affordances of the instruments. For example, some loud-style instruments such as the keyed *saron* are restricted to a single octave, and so must employ a characteristic 'folding' of melodies back upon themselves in order to match the multiregister lines of other instruments and voices, while preserving the integrity of melodic contour (Pickvance 2005, 111–112). On other sets of gongs that might not span a complete octave, suitable intervals must be found as substitutes (209–210). Instruments are often played as fast as comfortably possible (called 'saturation density' by Hood 1982, 115), and in many cases rhythmic density is connected with the resonant decay and tessitura of the instrument; the higher the pitch range of the instrument, the faster it tends to be played.

Other conventions are dependent on the behaviour of other players—the most common example being the drummer—such as in the choice of drum (which can dictate the degree of ornamentation) and in speed changes to cue temporal expansion (Brinner 1995, 225; Pickvance 2005, 61–62; Sumarsam 1975a). Patterns played on the instruments representing the colotomic structure and *balungan* can be sped up or slowed down, for the most part retaining their sequence. Most other instrumental parts in the ensemble are expected to maintain a steady pulsation, creating an unbroken stream of melodic–rhythmic information. This situation has led to the formalization of several levels of rhythmic density (called *irama*) that can be brought into play during performance.[7]

The amount of freedom afforded to gamelan musicians typically increases with the rhythmic density of instrumental parts in relation to the central pulse and *balungan* sequence. This is observable both vertically, in instruments that are played faster according to their pitch class, and horizontally, as instrumentalists double up their rhythmic densities through temporal expansion (although some parts such as the *bonang panerus* remain dependent on the behaviour of others, as described below). As the spaces between the structural points become sparser, there is more room for interaction and melodic and rhythmic divergence. However, the integrity of the ensemble in relation to the foundations of the piece takes priority; it is extremely rare for a part's elaboration to lead the ensemble in a direction outside conventional structures.

5.1.4 Types of Melodic Patterning in Central Javanese Gamelan

Some loud-style parts can be learnt as abstract patterns to be applied to a notated *balungan* sequence, typically involving the segmentation and repetition of phrases. For example, on the set of gong-chimes called *bonang*, a pair of *balungan* notes can be repeated in alternation before being played by the rest of the ensemble. This technique is called *pipilan*, and can be used to direct other players by indicating the contour of approaching sequences or shifts in register (see Figure 5.2; Pickvance 2005, 147; Sutton 1991, 53).[8] The basic pattern is often refined by omitting the strongest final beat, and accentuating the dampening of each note as the next one is played.

This kind of activity is typically stratified across several instrumental layers. The main *bonang* has a higher-pitched counterpart (called the *bonang panerus*), which repeats the same basic pattern at double the rhythmic density. Tiered patterns such as these, with

Balungan sequence	5			3			2		1	
Abstract note order	a			b			c		d	

Bonang barung		5 a	3 b	5 a		2 c	1 d	2 c	

Bonang panerus	5 a	3 b	5 a		5 a	3 b	5 a	2 c	1 d	2 c	2 c	1 d	2 c

Black space = rest

FIGURE 5.2 Repetition and phase offset of a *balungan* sequence (5 3 2 1) using the *pipilan* technique, Solonese style.

Balungan sequence	5		3		2		1	
I: 100BPM	5	5	3	3	2	2	1	1
II: 40BPM	5 5	3 3	5 5	3 3	2 2	1 1	2 2	1 1
III: 16BPM	5 5 3 3 5 5 3 3	5 5 3 3 5 5 3 3	2 2 1 1 2 2 1 1	2 2 1 1 2 2 1 1
IV: 11BPM	55 33 55 33 55 33 55 33 55 33 55 33 55 33 55 33	22 11 22 11 22 11 22 11 22 11 22 11 22 11 22 11

FIGURE 5.3 Elaboration of a four-note sequence (5 3 2 1) using the *saron panerus* (*nacah rangkep*, Solonese style). Formal density levels shown with approximate tempo for the *balungan* sequence.

a rest on the strongest beat giving way to the note played on the next instrument in the hierarchy, create a texture characteristic of gamelan music. They reflect the subdivision found in the fundamental colotomic structures, helping to establish a sense of integrity and self-similarity in the parts played across the ensemble.

The same principle is used in temporal expansion, as note pairs from the central sequence can be repeated to fill the space provided (see Figure 5.3; in practice the player might construct a new melody in order to avoid extensive repetition). When the drummer slows the central pulse to cross into the next rhythmic density level, players of loud-style instruments may double the repetitions of each note pair. The number of strokes played is inversely proportional to the tempo of the static structure of the *balungan* and gongs.

Patterns that refer to each note of the outlined sequence in this manner are dependent on certain types of contour. More idiomatic note substitutions might also take place if the contour creates an unnatural leap (such as intervals crossing over into other octaves), or if notes are repeated in close proximity. Furthermore, sparse *balungan* sequences such as the *nibani*-style patterns shown in Figure 5.1B provide limited notes with which to work, calling for deployment of specific phrasing.

One way to bypass this problem is to use interlocking patterns, which in Javanese practice typically make direct reference to only the strongest middle and end notes in the outlined sequence. These are generally shared between two instruments, and can create predictable textures by borrowing adjacent notes in the scale (see Figure 5.4). Patterns such as these provide a simple way to support underlying melody without

| Balungan sequence | 3 | 6 | 3 | 2 |
| Abstract note order | a | b | c | d |

A) Semarang-style imbal pattern for bonang barung and bonang panerus (played in octaves)

| Bonang barung | 6 b | | 6 b | | 2 d | | 2 d | |
| Bonang panerus | 1 b+1 | 1 b+1 | 1 b+1 | 1 b+1 | 3 d+1 | 3 d+1 | 3 d+1 | 3 d+1 |

B) Generic imbal pattern shared between two saron

| Saron 1 | 6 b | 6 b | 6 b | 6 b | 3 b-2 | 6 b | 3 b-2 | 6 b | 2 d | 2 d | 2 d | 2 d | 5 d+2 | 2 d | 5 d+2 | 2 d |
| Saron 2 | 5 b-1 | 5 b-1 | 5 b-1 | 5 b-1 | 2 b-3 | 5 b-1 | 2 b-3 | 5 b-1 | 3 d+1 | 3 d+1 | 3 d+1 | 3 d+1 | 6 d+3 | 3 d+1 | 6 d+3 | 3 d+1 |

FIGURE 5.4 Two examples of interlocking patterns based on a *balungan* sequence (3 6 3 2), using strongest notes and adjacent places in the *slendro* scale.

taking contours into account, proving useful for modern compositions that make less references to established repertoire, and can also lend a lively feel to performance.

Beyond the basic loud-style conventions described thus far, playing gamelan relies on knowledge of a multitude of more concrete rhythmic and melodic idioms. The notion of the *balungan* as the basis (or at least an indicator) of all parts is put under strain when considering vocal parts and soft-style instruments (such as the *gender, rebab*, and *gambang*). Players of these instruments play somewhat more fluid and variable parts— although they may draw upon the *balungan* for guidance or constraints, they also share a repertoire of melodic patterning called *cengkok*.

Cengkok are in themselves abstract sequences that must be elaborated upon by a musician in performance.[9] While largely compatible with notated outlines, *cengkok* often follow their own melodic contours, suggesting that a different framework is being taken as a reference. Furthermore, different *cengkok* might be chosen for the same *balungan* contours to provide variety, or to respond to other parts and fit the register of the collective internal melodic movement.

These factors become particularly apparent through temporal expansion. *Cengkok* can be expanded by inserting holding patterns in their middle, or by recombining existing patterns, either as fragments or in their complete forms (Sumarsam 1975a, 164). Figure 5.5 illustrates how each change in density level places more significance on the mid-points of each phrase, until every note of the *balungan* sequence is preceded by the equivalent of a full *cengkok* in level IV. For example, the *dualolo* and *tumurun* patterns (which exist both as compressed and full-length versions) can be combined to form a whole *tumurun* pattern in the next density level. As musicians are given more freedom to interpret material in expanded sections, certain patterns may prompt a move towards appropriate midpoint notes that diverge from the outlined sequence, before rejoining the rest of the ensemble on the most important structural resting points.

Some conventions have been established for approaching these instruments in pedagogical contexts: common contours of *balungan* phrases can act as placeholders

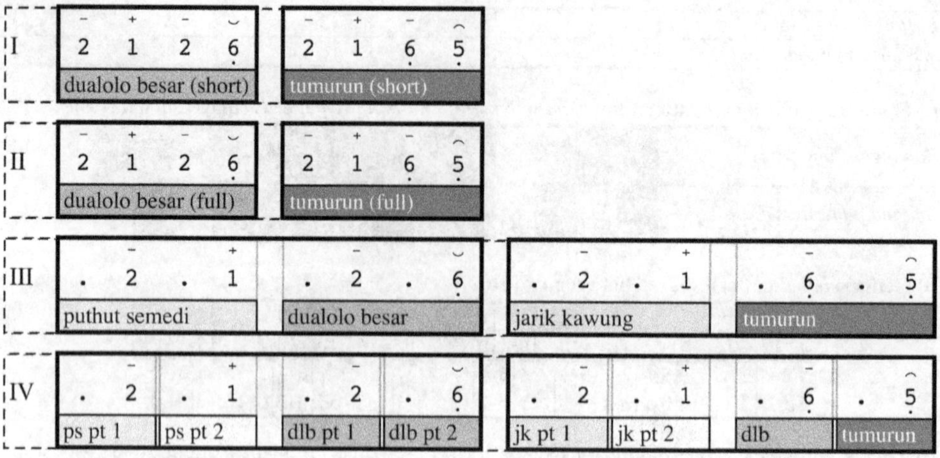

I	2	1	2	6	2	1	6	5
	dualolo besar (short)				tumurun (short)			
II	2	1	2	6	2	1	6	5
	dualolo besar (full)				tumurun (full)			
III	. 2 . 1	. 2 . 6	. 2 . 1	. 6 . 5
	puthut semedi	dualolo besar	jarik kawung	tumurun
IV	. 2 . 1	. 2 . 6	. 2 . 1	. 6 . 5
	ps pt 1 / ps pt 2	dlb pt 1 / dlb pt 2	jk pt 1 / jk pt 2	dlb / tumurun

FIGURE 5.5 Examples of named patterns for the phrase 2 1 2 6 2 1 6 5 (*pathet sanga*), expanded across four density levels (for notated *cengkok* see Martopangrawit 1973; Polansky 2005).

for more specific information, and a player may choose *cengkok* by looking at a target note from the outline and that of a preceding phrase; the main number sequence can be annotated with the names of these patterns (see Figure 5.5; Martopangrawit 1973; Polansky 2005; Sumarsam 2004, 78–79). However, these are generally considered to be 'coaching rules'—explicit information given to beginners that does not fully describe the skill to be conveyed, ideally acting as a stepping stone to implicit knowledge (Perlman 2004, 22–23).[10]

Whether filling out *pipilan*-type patterns or selecting combinations of *cengkok*, many of the rules taught to beginners work only on a relatively small selection of classic pieces before exceptions are encountered. Some have logical connections to the *balungan*, such as the replacement of notes in anticipation of stronger ones in adjacent phrases, overriding the structural points implied by the highly quantized sequence (known as *mlesed*—'slipping'—see Perlman 2004, 55; also *salah gumun*, 62; Supanggah 2004, 4).[11] Others refer to parallel phrases in sung parts and other instrumental idioms. As a result, creating a rule-based system that relies on the analysis and treatment of outlined sequences alone can become a somewhat complicated endeavour.

5.2 APPLICATIONS FOR ALGORITHMIC REPRESENTATION AND COMPOSITION OF GAMELAN MUSIC

5.2.1 Gamelan and Algorithms in Musicology

Attempts at constructing generative grammars based on melodic patterning and gong structures in gamelan have mostly been aimed at describing existing works rather than creating new ones (e.g. Becker 1980, 105–114; Becker and Becker 1979; Hughes 1988). Grammars developed by Becker and Becker were used to identify the basis of particular forms, presented as an attempt to understand and identify innovation in composition (1979, 32). Hughes's subsequent analysis emphasizes notions of deep structure and surface structure in terms of the *balungan*, suggesting that some forms are frozen versions of the temporal expansions that take place in performance practice (1988).

Grammars and related rule systems can also be useful in representing performance techniques, and the way that these structures are filled. The processes used in playing many of the loud-style instruments might be thought of as rewriting systems based on information from the central *balungan* sequence (see Milne, chapter 11 in this volume). The examples in Figures 5.2 and 5.3 illustrate letter-based systems commonly used to explain these processes (e.g. Grupe 2015, 33; Matthews 2014, 32; Perlman 2004, 57; Pickvance 2005, 147). However, these abstractions are of limited value without representation of spontaneous analysis, decision making, and interaction taking place throughout the ensemble in performance.

A practical application for systems such as these is the testing of rules by making them explicit through computer-based synthesis, and mapping out the type and degree of contextualization required for them to work. This was the impetus for the Virtual Gamelan project, which started at the Institute of Ethnomusicology in Graz, and involved the construction of a framework for the interpretation of traditional music using SuperCollider (an audio programming language). Although the majority of the project's code and audio output have yet to be made public,[12] the developers' writing on the subject presents a comprehensive discussion of the issues surrounding part selection and generation as well as data structures for representing pieces and performance conventions (Schütz and Rohrhuber 2008).

Building upon the numerical systems conventionally used to represent gamelan music, the software adopts approaches called 'literate' and 'interactive' programming, in which the code is built into the interface, and therefore becomes integrated with the notation (Schütz and Rohrhuber 2008, 132). The software is based on analysis of *balungan* sequences coupled with a rewriting system: patterns are generated by rewriting abstract note pairs unless the input sequence matches a list of phrases, which are ordered for priority (160–161). The project also presents interesting approaches to hierarchy and interaction, including creating parameters for musicians' 'empathy' and 'confidence' to emulate shifts in timing across an ensemble (179–182).

While reliable algorithmic treatment of traditional repertoire through the framework was achieved to only a limited extent within the initial timeframe of the Virtual Gamelan Graz project, an extension to the project conducted by the Institute of Ethnomusicology explored the use of fixed computer sequences to test theoretical assumptions (Grupe 2015). In lieu of an automated model, a group of expert Javanese musicians were asked to evaluate sequences created in a commercial sequencer based on rules taken from lessons and textbooks. The ensuing discussions confirmed problems with deriving formal structures from conventional learning methods, in particular reliance on the *balungan* sequence alone for synthesis, but also focussed on issues relating to phrases played by the computer exactly as they would be notated. In particular, Grupe notes that 'surface structures' in treatment, such as micro-timing and embellishment, appeared to hold as much salience as deeper structural issues (which he defines as '"correct" notes and patterns'; 2015, 41), highlighting a holistic approach taken by musicians.

5.2.2 Algorithmic Approaches in New Composition for Gamelan

Gamelan music has a complicated relationship with notions of composition, not least because of the open nature of elaboration and the frequent reuse of existing material. Although equivalent terms in Indonesian exist (such as *komponis*), many Javanese composers identify more with the role of arranger or compiler (*penyusun*; see Roth 1986). Just as traditional works are pieced together from fragments of melody or *cengkok*, innovation in contemporary practice often takes place through the combination of whole pieces into suites, and their reinterpretation with different types of treatment or vocal parts. Such larger-scale treatments are described in the same way as individual variations, called *garapan*. Where traditional structures or outlines such as *balungan* are not always present, the same essential processes are maintained: contemporary composition in Java commonly takes place through rehearsal process, with parts generated by musicians rather than being prescribed by a single person (Roth 1986).

While there are concerns that traditional practice is experiencing something of a decline, the music is continuing to evolve to fit with new technology, integrating notation and recording, and sharing the stage with synthesizers and other instruments. A popular Indonesian genre called *campur sari* (meaning a mixture of styles, often implying use of Western-style instruments) sometimes calls for traditional forms to be played alongside the auto-accompaniment functions of commercial keyboards. Applications for computers and mobile devices enable students to play in classrooms or practice at home where instruments are not available (Tempo 2009; http://www.academy.wellscathedralschool.org/free-resources). The patterns and flexibility of performance practice have attracted many composers from outside the tradition, and provided inspiration for many experimental, minimalist, and process-based approaches—Steve Reich in particular was influenced by contact with gamelan, borrowing conventions for cueing and interaction in repetitive works, and subsequently encouraged future composers to explore the structures of non-Western musics rather than simply imitating their sound (Reich 2002). Movements of composers have emerged both in Indonesia and abroad to work directly with ensembles in different ways—bringing updates to traditional styles, fusion with other musical cultures, and exploration of possibilities presented by creation of new forms of instruments (House 2014; Roth 1986; Supanggah 2011, 50).[13]

Notwithstanding problems with representing gamelan music as it is played in classically oriented groups, the application of formalized rules has provided a fertile ground for new compositions. Treatment of outlines as found in traditional practice—whether performed by musicians or automatic processes—allows for greater levels of abstraction, and thus enables concentration on different levels of detail. However, while existing structures might appear to hold convenient frameworks for composition, attempts at direct emulation without full understanding of the music's grammars and idioms (both in terms of phrasing and instrumental practice) can be restrictive. Abstraction of rules can prove difficult without tying into specific phrasing or modal frameworks, or can lead to work being received as pastiche composition if followed too rigidly. Instead it can be more beneficial for composers coming from outside the tradition to explore explicit synthesis with elements from their own background, which often entails an adaptation of performance conventions (Sorrell 2007).

For some composers, chance operations have been a sufficient means of emulating the surface complexity of gamelan music, although such approaches can also highlight expectations of idiomatic phrasing held by performers and listeners. The first recorded attempt at computer-aided composition for gamelan was conducted by a group of scholars at Gadjah Mada University in Indonesia with the aim of testing modes represented by *pathet* (Surjodiningrat, Khandelwel, and Soesianto 1977). The research centred on software written in the general-purpose programming language *Fortran*, which was used to treat melodic information as abstract sequences of numbers, identifying a library of possible phrases taken from traditional pieces. These were recombined to fit into traditional structures through processes based on randomization, creating *balungan* sequences. The resulting outlined parts generally did not conform to conventions that enabled easy instrumental treatment, and were received critically by the musical community when presented in conventional formats (Sutton 1987, 69).

These processes have been echoed in new musical contexts without traditional constraints, such as Lydia Ayer's random-part generation in her work for gamelan and tape, *Merapi* (1996), and Markov chains in Max Worgan's generative score for the shadow puppet piece *The Sound Catcher* (2009; http://www.sembler.co.uk/project_gamelan/). Patrick Hartono, an Indonesian composer who trained in electroacoustic music before working with gamelan, claims to experience patterns in gamelan music in terms of the spatial relationships between the instruments rather than their pitches (personal communication 2015). In Hartono's case, a loose imitation of the ensemble through granular synthesis—while probably unrecognizable as such to a traditionally oriented player—maintains focus on the macro structure of the work and the parameters most important to the composer.

Established Javanese performance conventions have been used to reinterpret other musical styles in seminal American gamelan compositions (Diamond 2000), as well as otherwise nonmusical information such as Lou Harrison's sonification of a social security number in his piece *Lagu Socieseknum* (Miller and Lieberman 2004, 159). While not always set out in terms of strict rule sets or instructions, pieces such as these typically use outlines inspired by traditional practice—quantized information that affords interpretation through existing instrumental idioms. The composers' knowledge of rules and conventions, whether explicit or implicit, allows them to create sequences that will be treated predictably, or generate more detailed instructions accordingly.

Modifying playing techniques from gamelan is by no means a Western innovation. Composers working on the edges of traditional idiom, such as Ki Nartosabdho (an arranger, composer, and shadow puppeteer popular in the 1960s) have introduced a range of techniques both new and adapted from older styles or regional variations. These have been borrowed by other composers and performing groups, and many have since been integrated into mainstream practice (see Pickvance 2005, 15; Sutton 1991, 220–233).

However, the creation of explicit new rules for performance, designed to be passed on and applied by other musicians, might be considered a rarity amongst contemporary Indonesian composers. Such frameworks have been explored elsewhere; for instance, the composer John Jacobs has developed what he calls *extended garap*—a set of techniques and stock phrases built upon traditional repertoire through trial and error and rehearsal process, which add to the possibilities of time signatures and polyphony within the conceptual framework of Javanese gamelan music (2013). According to Jacobs, this kind of approach is crucial to countering the mechanical feel often encountered when asking musicians to read detailed, through-composed pieces directly from notation (personal communication 2015).

Comparable processes have also been extended to computer-aided composition. The Virtual Gamelan Graz framework allows for composition involving the manipulation of rules by editing the rewriting system or applying alternative types of data input, although such development has yet to be implemented in practice (Schütz and Rohrhuber 2008). Max Worgan created an imaginary ensemble borrowing interlocking patterns from Javanese and Balinese playing styles for the *Shadow Catcher*, using real-time elaboration upon *balungan* sequences to respond to shadow puppets via video tracking (Worgan 2009). These approaches are generally rooted in contemporary teaching and learning methods that support the notion of rule-based gamelan, as will be discussed in further detail below with regards to my own practice.

5.3 CASE STUDY: COMPUTER-AIDED COMPOSITION WITH AUGMENTED GAMELAN AND PIPILAN

My first contact with gamelan came through classes at the Southbank Centre in London, as an electronic musician seeking inspiration and a theoretical framework for melody, having until that point focussed my attention on rhythm and timbre. Initially attracted to the timbre and tuning of the instruments, I felt refreshed by the logical approach taken to working out parts, and the apparent ease with which I could fit into such a large ensemble. I was also taken with the way that melodic structures could be spontaneously expanded and contracted in time, which felt reminiscent of granular audio time-stretching processes (Roads 1996, 440–446). I spent much of the ensuing ten years studying gamelan, reinforcing my learning by developing a set of patches in Max/MSP (a dataflow programming environment developed by Cycling '74; see http://cycling74.com/), and cultivating collaborations for gamelan and electronics. Although I do not consider algorithmic composition to be my primary activity, my contact with gamelan music has driven me towards a search for flexibility, challenging assumptions that introducing technology might imply a move towards fixed sequences or prerecorded material, and moving towards working on rule-based systems rather than individual pieces.

One of the principal aims of the Augmented Gamelan project is to explore the borders of tradition and the influence of physical construction of instruments on their associated idioms. The instruments are modified by attaching speakers to their bodies, playing synthesized sounds in order to stimulate metallic resonances, while projections or lights indicate each note as it is being played (see Figure 5.6). These parts can be created dynamically through a combination of sensor readings and pattern generation handled by custom software. As a result, the gestural and timbral possibilities of the instrument are expanded, and a computer and human player can share both the physical instrument and a melodic outline as foundations for part generation and interaction.

FIGURE 5.6 *Pipilan* software controlling a set of *bonang*-type instruments from the Augmented Gamelan ensemble, installed at Hackoustic Festival, Machines Room, London 2016.

For example, in *Bonang Study* (2011) for augmented gamelan instruments and computer, several sine tones are played through transducers attached to a set of gong-chimes, using patterns generated from a *balungan* sequence in synchronization with an ensemble. Matching the treatment performed by their human counterpart, these synthesized parts anticipate each note, extending the attack and decay of the acoustic instrument (Matthews 2014, 168–181; a similar effect is produced through the amplitude envelopes pictured in Figure 5.7). The resulting part provides a layer of textural elaboration interwoven with the traditional melodic parts, which can be temporally expanded and contracted in the same fashion.

Initial experiments in composing for gamelan and electronics indicated that fixed note sequences for each instrument would be impractical for these purposes. In

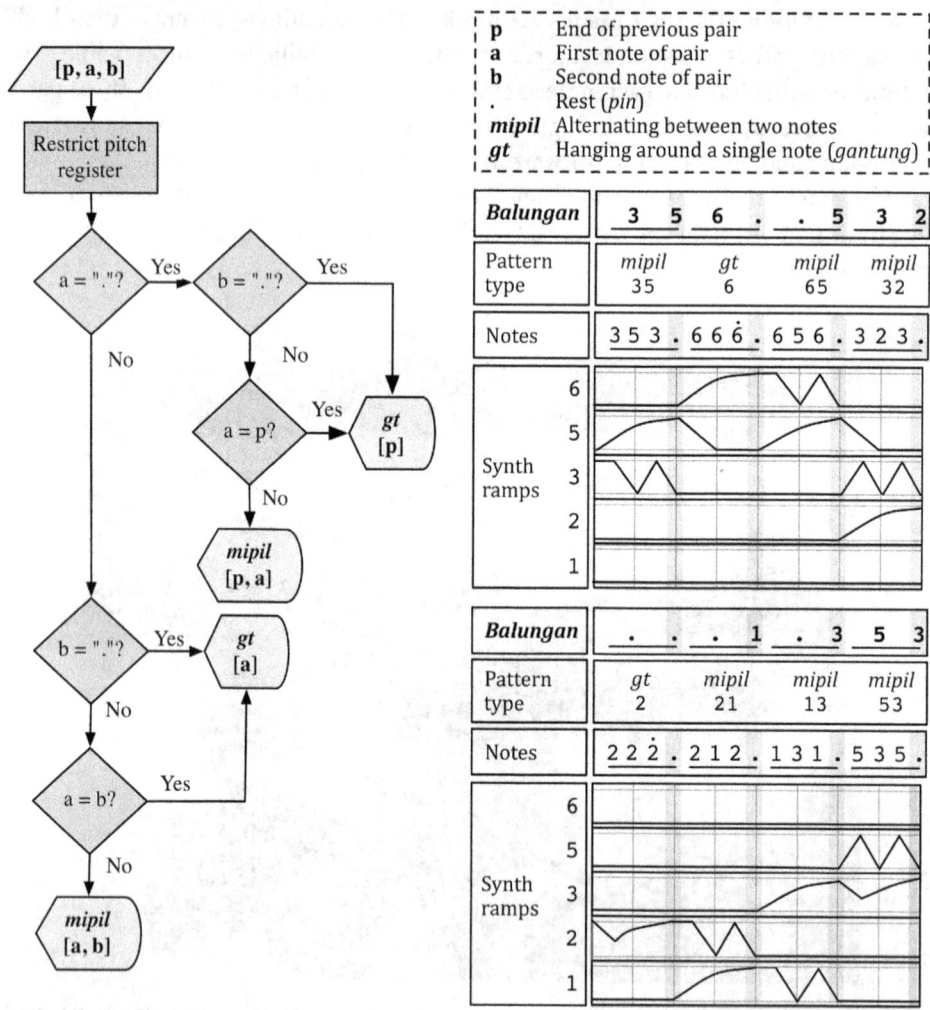

FIGURE 5.7 Flow of part selection in *Pipilan*, with example treatment shown on right.

emulating a traditional ensemble, every phrase must be prepared for several parts, duplicated and expanded across a range of rhythmic density levels. A simple change in outlines, structural information, or tempo can have dramatic consequences, necessitating substantial reworking of these sequences. Furthermore, interaction is fundamental to gamelan music, which requires flexibility. I sought to address these issues by building a framework in Max/MSP to recreate traditional music, with the intention of exploring material from outside the idiom in the future (Matthews 2014, 140–167). The resulting software—named Pipilan, after the traditional convention of segmenting and repeating notes from a sequence—has formed the foundation for performances with the Augmented Gamelan ensemble, as well as installations in which audience members are invited to interact with generative music processes.

The software is based on a set of rules for the real-time treatment of *balungan* sequences, with exceptions entered via a GUI in the manner of annotating notation. Part generation is performed with a combination of direct reference to the *balungan* and invocation of entirely fixed patterns, with a pattern matching and rewriting system similar to the Virtual Gamelan Graz framework (Schütz and Rohrhuber 2008, 160–161). By accumulating a database of variations as individual pieces are entered, a larger framework can be constructed in reflection of the learning processes that beginner musicians undergo. The instruments emulated are of the loud-style set (such as the *bonang*), as their idioms refer directly to the *balungan* sequence and are most easily represented with rules. These parts provided a natural starting point for integration with live musicians, with the understanding that they could fit into a gamelan ensemble to play basic compatible patterns, just as a beginner can employ coaching rules to join more experienced players.

The hierarchy of instrument types in the gamelan lends itself to representation with an object-oriented model. Since instrumental subgroups frequently share decisions on note selection that lead to repetition or interlocking patterns, these can form the basis for classes. For example, emulations of the *bonang barung* and *bonang panerus* parts shown in Figure 5.2 might inherit the same note selection functions, but differ in their density and therefore also their number of repetitions. Synthesized parts can then be created as children of the appropriate instrument classes, generating envelopes for timbral and spatial parameters alongside more conventional note events. This framework enables development of complementary electronic processes that can provide the foundation for an augmented ensemble to move through a piece in unison.

Using traditional repertoire as a starting point for these rule-based systems became restrictive in the context of new composition, since input sequences not fitting traditional contours in a given mode or *pathet* might either produce unusual results or call up references to arbitrary fragments of songs. As a result, the initial system was neither accurate enough to be of use to serious gamelan composers, nor flexible enough to generate interest amongst collaborators with a stronger background in electronic music. This problem stimulated the development of a system intended to establish a direct link between input and output sequences to ensure that any phrase would be playable; a foundation that could still facilitate traditional rule sets as exceptions. Consequently, the main instrumental layers in Pipilan use a set of simplified rules, largely based on a single octave range, and amalgamated from several loud-style instrumental techniques (see Figure 5.7). A collection of basic pattern types are used to generate abstract sequences to refer back to the central outline, which in turn create note events and control ramps for synthesizer parts. Depending upon the pulse of the central clock, these patterns are automatically repeated to fit different rhythmic density levels, complemented by audio time-stretching through granular synthesis techniques. The rules and reference patterns used for rewriting can also be modified as the sequence plays, allowing for auditioning of parts during composition, and a move towards rudimentary live coding of the playing styles as an element of performance (see Magnusson 2011).

While the *balungan* is retained as a model for input, unidiomatic material such as phrases from other musical systems or nonmusical data can be accommodated, provided that sequences can be broken into pairs; the software has been used effectively with random numbers, sequences input live by audience members, and even information live-tracked from synthesizers at dance-oriented Algorave events (see Roberts and Wakefield, chapter 16 in this volume). Through this somewhat more generic approach I hope to illustrate decision making in gamelan without the complications of more specific phrasing conventions in traditional practice, and thus open it up to audiences with a more casual interest in the music.

One of the benefits of using this system has been the ability to stretch compositions out from the scale of minutes to hours—an extension of traditional temporal expansion that would require impractical degrees of physical stamina and concentration for a human ensemble. It has also been useful in synchronizing players with computer parts—for example, where click tracks are required, it is possible to create individual streams of information for each musician, with a level of detail appropriate to their chosen instrument. Due to the modular nature of Max/MSP, it is relatively straightforward to build new synthesis modules or links with external software that refer back to the central sequence, adding virtual instruments as needed.

The scope for real-time integration of electronic parts with the gamelan can be expanded by building links between traditional pitch-based patterns and other domains, such as timbre and spatialization. Creation of one-to-one mappings to parameters in these areas can be problematic, as idiomatic variations and ornaments are often reduced to artefacts lost in an interesting but dislocated complexity. It is often more appropriate to create analogous processes to fit the target medium, taking information from a central outline, and creating parallel streams of events that reconverge with the note sequences at key moments.

Cases such as these can involve a mixture of cross-domain mapping that is broadly compatible with the Javanese notion of individual instrumental idioms coexisting and interacting in performance (see Zbikowski 2005). For example, a version of Pipilan adapted for Ambisonic spatialization generates parts from a central sequence in two ways: the selection and repetition of notes by the pseudo-traditional ensemble and the movement between their equivalent values in space through polar interpolation. Rather than forcing a connection to the traditional parts by linking parameters to the individual notes of the outline, the resulting movement is comparable to the convergence and divergence found in traditional Javanese music (see Perlman 2004, 62–74). Just as some instrumental parts temporarily move away from the influence of the *balungan* to find their own path to the endpoint of a phrase, these equivalent synthesizer parts take a roundabout route to navigate a spherical arrangement of speakers before reconverging with resting points in the predetermined spatial sequence.[14]

This layering of approaches was exemplified in a multichannel performance and installation at MUMUTH in Graz in 2015. Taking place in a concert hall following a traditional performance by the Southbank Gamelan Players, visitors were encouraged to collaborate in the creation of a long piece of music by writing a *balungan* sequence (see Figure 5.8). The changes in the structure were reflected in parts played by a virtual ensemble responding in realtime; certain instrumental and synthesized parts also moved through the space. Members of the gamelan ensemble were inspired to join in on instruments, adding further layers by following the same outline as the computer, either by listening for changes or reading the constantly shifting central sequence projected above the instruments. Although the algorithmically generated part was mechanical in comparison to this fluid improvisation, its robustness facilitated a unique exploratory interaction between audience and performers, simultaneously linked to spatial movement.

FIGURE 5.8 Audience members composing live using *Pipilan* in a multichannel Ambisonic installation at MUMUTH, Graz, 2015. Photo by Brad Smith.

5.4 Conclusion

The examples of rules and more general algorithmic thinking explored in this chapter are important aspects of gamelan music, but—echoing Sutton's comments on improvisation (1998)—it would be an overgeneralization to describe the tradition as predominantly or explicitly rule-based. Formal rules have been embraced in educational settings, but as the Virtual Gamelan Graz project has highlighted, in practice expert musicians tend to take a more holistic approach. Furthermore, rules and grammars might only be considered a snapshot if seeking to represent existing musical practice, particularly one of a living tradition such as gamelan.

While contemporary methods for teaching gamelan music might indicate shortcuts to part generation, few appear to translate well to computer representation. Many problems in this area stem from taking the minimal notation of the *balungan* sequence as the basis for synthesis of all parts, where in practice it often acts as a context-dependent signpost for other unnotated information. It might be more productive to find ways of analysing more complex information, with the aim of creating new forms of abstraction to build upon. A *balungan*-type sequence may then be used to form a bridge between different types of interpretation, including those of a computer and human performers, much as it mediates the loud- and soft-style instrumental approaches in traditional contexts.

Despite the apparent impracticalities of emulating whole ensembles, some of the ideas presented in this chapter might facilitate a movement towards aural approaches to learning and composition where instrumental resources, players, or recordings are not available—enabling the testing of rules and previewing outlined compositions before approaching musicians. As well as recreating existing styles, there is much to explore through the potential of interactive and generative systems in new music free from traditional obligations. While some common approaches to theory might be based on misconceptions of how gamelan music works, the imposition of Western ideals, or coaching rules usually intended to be discarded, why shouldn't the results of these be used in creating new music?

Traditional practice provides a useful model in which a structure is exposed both audibly and visually, and can be expanded and contracted. Rules designed for teaching beginners can provide interesting ways to generate coherent complexity; implemented as generative algorithms, they might be set to evolve by themselves, or enable interaction and structured improvisation with a computer. Perhaps most exciting of all is the prospect of manipulating outlines and rules as part of performance, in a feedback relationship with players or audience. Since it has been suggested that gamelan compositions exist only through the combination and interaction of many parts in realtime, it seems appropriate that gamelan-inspired composition, electronic or otherwise, might be presented the same way.

Notes

1. The approaches in my practice discussed here were developed as part of my PhD research at Middlesex University (Matthews 2014), with support from the Arts and Humanities Research Council.

2. This relationship can be found in the most common form of pieces in which the entire gamelan is played together, called *gendhing*. Other more predominantly linear forms such as *pathetan* do not rely on rigid frameworks of gongs or *balungan* described below, instead placing emphasis on interpretation, elaboration, and interaction based on a vocal or fiddle part (Brinner 1995, 245–267; Sumarsam 1975a, 164–166). The relationship between gong cycles and linear melody has been compared to broader cyclic and linear approaches to time in Javanese culture (see Hoffman 1978; McGraw 2008).

3. Some musicians do not identify these sequences as melodies, preferring the term *rangka* (framework; Martopangrawit in Perlman 2004, 103). In some varieties of gamelan found in West Java, musicians refer to structures called *patokan,* which are typically played directly by a set of pitched gongs, and can be notated to facilitate performance (Cook 1992, 18; Swindells 2004, 104–109). A similar melodic line exists in Balinese gamelan theory, referred to as the *pokok* (essence), and is played by a set of keyed bass instruments. The *pokok* sequence can be used as the basis for elaboration, providing the notes for a variety of interlocking parts called *kotekan* (Tenzer 2000, 53–54). While the *pokok* sequence's generative qualities are more consistent than its Javanese counterpart, the patterns derived from it are generally fixed in advance, at the point of composition, rather than being treated by musicians in realtime (130).

4. All examples in this chapter are presented in the standard *kepatihan* cypher-based notation system. By convention the note '4' is omitted from the five-tone *slendro* scale used here.

5. Cases of *balungan*-style outlined notation have been traced back to the fifteenth century (Sumarsam 1995, 316). Tenzer (2000, 122) notes that similar notation of *pokok* sequences were formerly used in Bali, called *grantangan* (2000, 122).

6. See Wakeling (2010) for discussion of issues around structuralist approaches to theorizing gamelan, in particular highlighting incongruity with research based on rehearsal and performance processes.

7. Five formal *irama* levels are typically recognized in Solonese style (Pickvance 2005, 60; Supanggah 2011, 292). These are commonly measured by marking the number of notes played on the *saron panerus* against the central *balungan* pulse, ranging from 1:1 to 16:1. Transitions between *irama* levels can be performed as a shared musical gesture by acceleration and deceleration of the central pulse, creating an impression of elasticity. Tempo thresholds for transitions vary between contexts; maintaining density relationships when changes might otherwise expected can be used to great effect in creating tension or space, particularly in dance accompaniment.

8. *Pipilan* (and its verb form *mipil*) can be translated as 'picking away'. Instead of notation, this instrumental technique creates parts sufficient for other players to follow aurally if they are unfamiliar with the piece in question. In some cases the *bonang* has been used to disseminate spontaneous compositions to other players (Supanggah 2011, 151–152).

9. The resulting part, which takes into account a range of parameters including performance context, personal style, and interaction with other musicians, is called *wiled* or *wiletan*

(Supanggah 2011, 286). Supanggah compares variability in the performance of *cengkok* to patterning in *batik*, in which named outlines generally remain recognizable but the manifestation varies between artists (283).

10. The rules for beginners presented in the teacher Widiyanto's 'Gambang 101' illustrate an underlying principle of many soft-style techniques: playing a pattern to 'hang' around the previous note played, followed by a bridging pattern to the next important part of the sequence (Putro 2010).

11. In cases such as these it may be more appropriate to suggest that the *balungan* diverges from the inner melody represented by the soft instruments (Perlman 2004, 147). The *balungan* is also constrained by its own idioms, such as avoiding the repetition of notes in *nibani*-type sequences, which can lead to many cases of such divergence.

12. Some source code from the framework is available online from *Virtual Gamelan*, (https://github.com/musikinformatik/VirtualGamelan/, accessed 30 June 2017).

13. Here I have chosen not to focus on the multitude of successful works that take more general influence from gamelan music or use the instruments (or samples) for other purposes. Examples of process pieces using gamelan outside its traditional context include Michael Parsons's *Changes* (written in 1981 with strong influence from change ringing; House 2014, 58), and Daniel Goode and Larry Polansky's *Eine Kleine Gamelan Computer Music*—a piece first adapted for computer realization in HMSL and subsequently ported to Max (http://eamusic.dartmouth.edu/~larry/EK/ek_readme.pdf, accessed 5 September 2017). Material from traditional gamelan has also been the subject of mutation with other styles of music in the work of Polansky (1996) and Cope (1991).

14. This mapping of melodic information to points in space was inspired by the installation piece *Framework* by Hughes and Jacobs (2012), in which spatial motion was prerecorded rather than algorithmically generated.

BIBLIOGRAPHY

Augmented Gamelan. 16 October 2016. http://www.ardisson.net/gamelan.

Ayers, L. 'Merapi: A Composition for Gamelan and Computer-Generated Tape." *Leonardo Music Journal* 6 (1996): 7–14.

Bailey, D. *Improvisation.* Cambridge, MA: Da Capo, 1992.

Becker, J. *Traditional Music in Modern Java: Gamelan in a Changing Society.* Honolulu: University Press of Hawaii, 1980.

Becker, J., and Becker, A. 'A Grammar of the Musical Genre *Srepegan*'. *Journal of Music Theory* 23, no. 1 (1979): 1–43.

Brinner, B. *Knowing Music, Making Music: Javanese Gamelan and the Theory of Musical Competence and Interaction.* Chicago: University of Chicago Press, 1995.

Cook, S. *Guide to Sundanese Music.* Bandung: Self-published, 1992. http://www.gamelan.org/library/writings/cook_sunda.pdf.

Cope, D. *Computers and Musical Style.* Middleton, WI: A-R Editions, 1991.

Diamond, J. 'In That Bright World'. *Balungan* 7–8 (2000): 101–112. http://www.gamelan.org/balungan/back_issues/balungan7-8.pdf.

Gendhing Jawi: Javanese Gamelan Notation. http://www.gamelanbvg.com/gendhing/index.php. Accessed 30 June 2017.

Grupe, G. 'From Tacit to Verbalized Knowledge. Towards a Culturally Informed Musical Analysis of Central Javanese Karawitan'. *Perifèria: Revista d'investigació i formació en Antropologia* 20, no. 2 (2015): 26–43. http://revistes.uab.cat/periferia/article/view/vol20-n2-grupe/291.

Hoffman, S. B. 'Epistemology and Music: A Javanese Example'. *Ethnomusicology* 22, no. 1 (1978): 69–88.

Hood, M. 'Music of Indonesia'. In *Handbuch Der Orientalistik: Indonesian, Malaysia und die Philippinen*. Vol. 6, *Music*, edited by M. Hood and J. Maceda, 1–28. Leiden: E. J. Brill, 1972.

Hood, M. *The Ethnomusicologist*. Kent, OH: Kent State University Press, 1982.

House, G. *Strange Flowers: Cultivating New Music for Gamelan on British Soil*. PhD dissertation, University of York, 2014. http://etheses.whiterose.ac.uk/6793.

Hughes, J, and Jacobs, J. 'Framework'. Programme notes. Roger Kirk Centre, University of York, 28 April 2012.

Hughes, D. W. 'Deep Structure and Surface Structure in Javanese Music: A Grammar of Gendhing Lampah'. *Ethnomusicology* 32, no. 1 (1988): 23–74.

Jacobs, J. *Gamelan Composition; Extended Garap*. PhD dissertation, University of York, 2013. http://etheses.whiterose.ac.uk/6214/.

Kunst, J. *Music in Java: Its History, Its Theory, and Its Technique*. The Hague: Martinus Nijhoff, 1949.

Magnusson, T. 'Algorithms as Scores: Coding Live Music.' *Leonardo Music Journal* 21 (2011): 19–23.

Martopangrawit. *Titilaras Cengkok Genderan Dengan Wiletnya* [Notation for gender patterns and their elaboration]. Solo: ASKI Surakarta, 1973.

Matthews, C. *Adapting and Applying Central Javanese Gamelan Music Theory in Electroacoustic Composition and Performance*. PhD dissertation, Middlesex University, 2014. http://eprints.mdx.ac.uk/14415/.

McGraw, A. C. 'Different Temporalities: The Time of Balinese Gamelan'. *Yearbook for Traditional Music* 40 (2008): 136–162.

Miller, L. E., and Lieberman, F. *Composing a World: Lou Harrison, Musical Wayfarer*. Urbana: University of Illinois Press, 2004.

Perlman, M. *Unplayed Melodies: Javanese Gamelan and the Genesis of Music Theory*. Berkeley: University of California Press, 2004.

Polansky, L. '*Bedhaya Guthrie/Bedhaya Sadra* for Voices, Kemanak, Melody Instruments, and Accompanimental Javanese Gamelan'. *Perspectives of New Music* 34, no. 1 (1996): 28–55.

Polansky, L. *Beginning Central Javanese Gender*. Lebanon, NH: American Gamelan Institute, 2005. http://aum.dartmouth.edu/~larry/gender_book/index.html.

Polansky, L., and Goode, D. *Eine Kleine Gamelan Computer Music*. Lebanon, NH: Frog Peak Music, 1995. http://eamusic.dartmouth.edu/~larry/EK/ek_readme.pdf

Pickvance, R. *A Gamelan Manual*. London: Jaman Mas, 2005.

Reich, S. 'Postscript to a Brief Study of Balinese and African Music'. In *Writings on Music*, edited by P. Hillier, 69–70. New York: Oxford University Press, 2002.

Putro, W. S. 'Beginning Gambang.' *Balungan* 11 (2010): 55–56. http://www.gamelan.org/balungan/current_issue/widiyanto_gambang.pdf.

Roads, C. *The Computer Music Tutorial*. Cambridge, MA: MIT Press, 1996.

Roth, A. *New Composition for Javanese Gamelan*. PhD dissertation, University of Durham, 1986. http://etheses.dur.ac.uk/1240.

Schütz, R., and Rohrhuber, J. 'Listening to Theory: An Introduction to the Virtual Gamelan Graz Framework'. In *Virtual Gamelan Graz: Rules, Grammar, Modeling*, edited by G. Grupe, 131–193. Graz: Institute of Ethnomusicology, 2008.

Sorrell, N. 'Issues of Pastiche and Illusions of Authenticity in Gamelan-Inspired Composition'. *Indonesia and the Malay World* 35, no. 101 (2007): 33–48.

Sumarsam. 'Gender Barung, Its Technique and Function in the Context of Javanese Gamelan'. *Indonesia* 20 (1975a): 161–172.

Sumarsam. 'Inner Melody in Javanese Gamelan'. Translated by M. Hatch. In *Karawitan: Source Readings in Javanese Gamelan and Vocal Music*, edited by J. Becker, 245–304. Ann Arbor: Centre for South and Southeast Asian Studies, University of Michigan. 1975b.

Sumarsam. *Gamelan: Cultural Interaction and Musical Development in Central Java*. Chicago: University of Chicago Press, 1995.

Sumarsam. 'Opportunity and Interaction: The Gamelan from Java to Wesleyan'. In *Performing Ethnomusicology: Teaching and Representation in World Music Ensembles*, edited by T. Solis, 69–92. Berkeley: University of California Press, 2004.

Supanggah, R. 'Balungan'. Translated by M. Perlman. *Balungan* 3, no. 2 (1988): 2–10. http://www.gamelan.org/balungan/back_issues/balungan3(2).pdf.

Supanggah, R. 'Gatra: A Basic Concept of Traditional Javanese Gending'. *Balungan* 9–10 (2004): 1–11. http://www.gamelan.org/balungan/back_issues/balungan(9-10)/1-Supanggah_Gatra.pdf.

Supanggah, R. *Bothekan—Garap—Karawitan: The Rich Styles of Interpretation in Javanese Gamelan Music*. Translated by J. Purwanto. Solo: ISI Press Surakarta, 2011.

Surjodiningrat, W., Khandelwel, V. J., and Soesianto, F. *Gamelan Dan Komputer: Analisa Patet Dan Komposisi Gending Jawa Laras Slendro* [Gamelan and computer: Analysis of Pathet and composition of Javanese pieces in slendro tuning]. Yogyakarta: Gadjah Mada University Press, 1977.

Sutton, R. A. 'Variation and Composition in Java.' *Yearbook for Traditional Music* 19 (1987): 65–95.

Sutton, R. A. *Traditions of Gamelan Music in Java*. Cambridge: Cambridge University Press, 1991.

Sutton, R. Anderson. 'Interpreting Electronic Sound Technology in the Contemporary Javanese Soundscape'. *Ethnomusicology* 40, no. 2 (1996): 249–268.

Sutton, R. A. 'Do Javanese Gamelan Musicians Really Improvise?' In *In the Course of Performance: Studies in the World of Musical Improvisation*, edited by B. Nettl and M. Russell, 69–92. Chicago: University of Chicago Press, 1998.

Swindells, R. *Klasik, Kawih, Kreasi: Musical Transformation and the Gamelan Degung of Bandung, West Java, Indonesia*. PhD dissertation, City University London, 2004. http://openaccess.city.ac.uk/8415/.

Tempo. 'Gamelan Virtual ala Joko Triyono (bagian ke-1)'. *Tekno*, 24 March 2009. http://tekno.tempo.co/read/news/2009/03/24/072166304/Gamelan-Virtual-ala-Joko-Triyono-bagian-ke-1.

Tenzer, M. *Gamelan Gong Kebyar: The Art of Twentieth-Century Balinese Music*. Chicago: University of Chicago Press, 2000.

Wakeling, K. *Representing Balinese Music: A Study of the Practice and Theorization of Balinese Gamelan*. PhD dissertation, School of Oriental and African Studies, London, 2010.

Worgan, M. *On Mappings, Gamelan and Shadow Puppetry*. Master's thesis, Queen Mary, University of London, 2009. http://www.sembler.co.uk/wp-content/themes/sembler/assets/papers/maw5-final-report.pdf.

Zbikowski, L. M. *Conceptualizing Music: Cognitive Structure, Theory, and Analysis*. New York: Oxford University Press, 2005.

Perspectives on Practice A

CHAPTER 6

....

THOUGHTS ON COMPOSING
WITH ALGORITHMS

....

LAURIE SPIEGEL

6.1 BACKGROUND

....

THROUGHOUT history there has been an evolution of means to define and record what we might call 'music source' (instructions which when followed produce audible music). Common music notation has been for a few hundred years a successful and useful multi-dimensional representation of actions that players can do to realize sonic compositions. For quite a long time as well, although considerably lagging behind music notation's representation of specific sonic events, there has been an evolution in the conceptualization and description of musical process, of procedures of generation of musical data, instructions that can be followed by human beings or machines.

It is not the function of this brief chapter to go into any specifics of the various attempts to compose music by predefined logical processes, either historical or personal. The evolution of today's artificial machine-executable languages has given the description of music-as-process the jump start it has for so very long needed, unleashing a vast variety of approaches. Algorithms are essentially shorthand notations for large numbers of specifics. A few operands and operators can instantiate a potentially infinite number of musical sounds. We can now relatively easily opt for a small number of relatively powerful variables instead of having to individually specify a very large number of weak variables one at a time by hand.

The distinction between 'generative music' (logic-based, algorithmically specified) and music composed by other means is vague. Composition that was to some degree rule-based was commonplace for centuries prior to computers. We composers have all studied species counterpoint, and many older 'forms' are actually process descriptions (e.g. canon and fugue) rather than abstract structures to fill in with material (e.g. sonata form, rondo, or strophic song form).

6.2 Why—Personal

Although I have always been fascinated by abstract structures and at times attempted the design of algorithmic languages for music (Spiegel 1982–1984, 1984/1985), my own use of logically defined musical processes has often begun as self-simulation (perhaps a new form of self-expression), or alternatively as a sonification of an extramusical phenomenon, or at times an exploration of curiosity or hypothesis ('what would it sound like if…?'). These attempts have run the full gamut from small logic modules that decide one specific aspect of something I am otherwise more intuitively creating (e.g. stereo placement) to standalone generative processes that, once set in motion, will go on potentially forever composing ever-changing musical material with no further human intervention.

By 'self-simulation' I mean that I discovered early in my use of computers that some of my own sonic decision making was predictable enough that I could describe it by rule. I wanted to be able to automate whatever aspects of my own musical decision making process I could delegate to logic in order to free myself to focus on and be engulfed in those aspects of the process I could not rationally explain. At times, I had the absurd fantasy of ultimately being able to automate enough of my mind's compositional processes that I could leave behind me at the end of my life an artificial simulator of my musical self that would be able to go on creating new Laurie Spiegel compositions long after I were no longer here to hear them or see the response.

Realistically, although I have been able to make logic-based musical entities that are able to play streams of ever-changing new material that embody some of my musical biases, these exercises fall far short of a true automation of my creative musical self. More often, algorithmic logic has functioned as only one part of my compositional process, combined with as-yet incomprehensible subjective components. What we might call 'musical AI' was never my intent, only a byproduct of the desire to increase my ability to manipulate and generate musical material and to interactively have more realtime musical power and control as a composer and as an improviser. No artificially constructed nonconscious logic-based entity will have the drive, passion, motivation, or inner need to express itself musically.

6.3 Why—General

To make sense of using algorithms, that is, descriptions of musical procedures encoded in logic, of process descriptions written in artificial languages as instructions that machines can follow, it would be wise to stay connected to our natural musicality and to go back to the question of why we create music at all.

Music may be background for other activities such as dance, theater, film, or our ordinary lives. For such background we may want only a texture that has a certain rhythm, mood, or quality of feeling. Music as foreground, as main focus of attention might be our personal moment-to-moment expression of emotion or of our individual sensuality, or it may attempt to capture, communicate, or express subjective experiences that manifest themselves in our emotions or imagination. We want to externalize those subtle subjective phenomena so that others can also perceive them, to make shared what has been private.

Music can also be structured to represent in an abstracted form something we experience or perceive in the world around us, a narrative drama or a data set or structure or set of relationships that can be represented in sound. Such sonic captures can run the gamut from dramatic program music to the sonification of astronomical data.

Music can also be a form of soothing for other or self, providing experiences of flow, energy, peace, physicality, emotion, or other subjective states.

6.4 COMPENSATIONS FOR LACKS

The representation of many individual sonic events as a general description of process can be a faster and more efficient method of musical fabric generation than having to specify every aspect of each individual note. This is a time- and labour-saving innovation. However, this method often constrains the music to an overall uniformity, sameness, predictability, and flatness of overall form. An algorithm, once written, is outside of oneself and in itself is invariable, not subject to a musician's momentary changes of mood, sensory responses, or ideation, such as would naturally incorporate themselves into freehand writing or spontaneous improvisation. So the choice of what to automate versus what to make malleable via means external to the algorithmic, especially by human input, is one a vitally important design consideration.

I have used various methods to attempt to overcome the dramatic flatness towards which most artificially defined generative processes tend and to impose form on algorithmically computed musical material.

6.4.1 Interactivity

One method I have often used is to make generative algorithms interactive, in effect delegating to logic only those subsets of my decision-making processes I am able to understand sufficiently to be able to encode into logic. I reserve to myself the power to specify in real-time interaction with the sound other aspects of musical creation, those that I can't automate with a sufficient sense of my own aesthetic self to be as musically satisfying as I want the output to be.

The distinction between what I delegate to automated processes and what I reserve to the less rationally comprehensible methods of more intuitive specification is perhaps the most important aspect of my algorithmic design process. Doing this has always given me an unparalleled opportunity for introspection and increased self-awareness as to how I compose, of my personal musical preferences, and of how my own creative mind works.

The variables reserved to my nonalgorithmic control may be any of many kinds, ranging from real-time interactive adjustment of variables used by an algorithm during computation of the music to ex post facto nonalgorithmic intuitive orchestration of material that was generated entirely by predefined noninteractive logic with no intervention.

6.4.2 Entropy

A second method I have frequently used is to employ the concept of informational entropy, as per Pierce's and Shannon's information theory (Pierce 1961; Shannon 1948). The informational entropy of a musical work can be varied throughout a musical composition and represented as a function of time. This curve can be designed to structure the listener's experience throughout the piece. Such a time function generates and controls the composition's emotional content in that an entropy curve represents the variation over time in the degree to which the listener can predict what will be heard in the next moment. The moment-to-moment variation in level of predictability that is embodied in an entropy curve arouses in the listener feelings of expectation, anticipation, satisfaction, disappointment, surprise, tension, frustration, and other emotions.

6.4.3 Inherent Structure

Another way to avoid overall dramatic flatness and to create form is to encode as a process description an evolutionary sequence that unfolds over time automatically until it arrives at some self-terminating culmination. Evolutionary and extrapolative processes tend to be open-ended, however, open form, not self-bounding.

Once written, an algorithm is essentially a structure external to its creator. In other words, it constitutes a new independent musical instrument or tool. I try to write such procedures in sufficiently general and adaptable form that they can be used to make a variety of different kinds of material and can be used in a variety of compositional works.

6.5 Varieties

I have previously listed some of the ways I have used algorithmic process descriptions (Spiegel 1997) and will list them again here:

(1) Allusion: to very roughly approximate or simulate a natural occurrence that appears to me somehow inherently musical, capturing more like an abstract

painting than a photograph, a perception of something's process or shape rather than an exact replication (e.g. my piece *The Expanding Universe*).

(2) Inverse analysis: simply rendering into computerized form rules based on analysis of successful music of the past (e.g. *A Harmonic Algorithm*, which resulted from analyses of Bach Chorales).

(3) Scientific modeling (designing data for the receptor): implementing, in a set of software-coded rules, generators of data designed to be cognitively meaningful or otherwise comprehensible according to perceptual or other kinds of research, such as Shannon's and Pierce's information theory that formulates how to optimally encode content for intelligible reception, used in several pieces I made at Bell Labs (Spiegel 1997).

(4) Mimicry of process: coding into a computer program the rules by which some natural phenomenon transpires, unfolds, or progresses (e.g. my realization of Kepler's *Harmonices Mundi* (Kepler 1618–1619; Sagan 1978, 154).

(5) Mimicry of process result: literal mapping of specific nonmusical data onto musical variables (e.g. my little piece *Viroid*, in which I mapped the genetic content of a simple organism to a set of pitches).

(6) Mixed (combinations of the above): the specification of one or more dimensions of a piece by one generative method while another dimension of the same piece is determined by an unrelated method (e.g. the Knowlton-Spiegel algorithm [Knowlton 1976] for an illusion of perpetual acceleration being used in the rhythmic domain at the same time as real-time interactive control of a corruption process that is being applied to the output of a data generation source in the pitch domain, in my *Orient Express*).

I choose among these approaches on the basis of the aesthetic qualities of both creative process and sonic output, on what I can learn by doing them and for their inherent fascination.

An extramusical structure can derive from an abstract emotional sequence, a dramatic or documentary scenario, a natural ongoing process or any set of sequential data, each of these as either experienced subjectively by the composer or observed as external to the self. Generative methods can therefore be viewed as having overlap with program music, which portrays an extramusical dramatic narrative, at one end of the spectrum and, at the other extreme, with scientific sonification (auditory data display), both reliant on extramusical sources of structure.

6.6 FURTHER THOUGHTS

Perhaps one of the reasons that my compositional output appears low is that the output of a generative process is potentially infinite. The overabundance of musical material that algorithmic generation can produce somehow seems to cheapen the musical result, relative to music created by intentional specification of every minute detail by more traditional means. It can feel almost deceptive to record a mere finite short run of one variant of output from within a long ongoing process that can be altered in any number of ways.

I have most often listened to the output of my logic-based generative algorithms without ever recording it, not feeling it to be 'my own music' but mere music-like texture. What do I need to do to form the material into my own personal expression, to impress upon it somehow a dramatic form that will infuse it with emotion or to invest it with my own sensuality?

To another way of thinking, that logic-based generative process, rather than any specific subset of its output, may be seen as a musical work in itself. Had computers been ubiquitous when I first did such works, instead of there being only a very few large rare computer installations owned by powerful institutions, would I have distributed the actual computer programs as musical compositions per se, for enjoyment by people at home and for them to vary and play in their own individual ways, instead of making the small number of specific finite form pieces that I did, each being only one mere single instance of a potentially infinite number of musical results that same logic could have generated? Is there any possibility or point in trying to compare the value of my program *Music Mouse* with that of the music on my *Unseen Worlds* album made almost entirely by using it? Both are musical works. They are not both music.

Instances of my individual use of my own various algorithmic logics are very much my own compositions, and they constitute musical works for which the process descriptions (computer programs) are mere compositional tools. From a more normal musical perspective, those pieces, those examples of output from individual program runs, the specific musical works themselves, were the ultimate goal: music that others could listen to that was created with the mediation of logical processes I had described to the computer. To a perhaps greater extent though than even the resultant works, I made them in order to inhabit the state of flow and concentration inherent in all forms of music making as state of mind. Simply experiencing the process of interacting with the sounds made by sonically responsive computer logic was the highest motivator. This is not really very different from how I loved playing my first guitar.

BIBLIOGRAPHY

Kepler, J. *Harmonices Mundi*. 1618–1619.

Knowlton, K. C. *A Cyclic Rhythm Perceived as 'Ever-Accelerating'*. Bell Telephone Laboratories, internal memorandum, 4 March 1976.

Pierce, J. R. *Symbols, Signals and Noise: The Nature and Process of Communications*. New York: Harper and Brothers, 1961. Reprinted unabridged and revised as *An Introduction to Information Theory: Symbols Signals and Noise*. New York: Dover, 1980.

Sagan, C. *Murmurs of Earth*. New York: Random House, 1978.

Shannon, C. E. 1948. 'A Mathematical Theory of Communication'. *Bell System Technical Journal* 27, no. 3 (1948): 379–423. Reprinted in Shannon, C. E., and Weaver, W. *Mathematical Theory of Communication*. Champaign: University of Illinois Press, 1949.

Spiegel, L. 'IMPspeak Interactive Music Processor Language'. Unpublished design specification, 1982–1984.

Spiegel, L. 'MDI Protocol Report'. June 1984 (unpublished) as referenced in and its implementation detailed in R. J. Pryor et al., *Music Description Instruction Format Preliminary Proposal: A Study to Define, Design and Develop a Prototype of a Protocol Enabling the Encoding, Storage and Transmission of Music Compatible with, and to Form a Subset of, the North American Presentation Level Protocol Syntax (NAPLPS) (TELEDON)*, Canadian Dept. of Communications Research Contract 183-00361, 1 April 1985. Vancouver: New Media Technologies, 1985.

Spiegel, L. 'Contributors' Notes: An Information Theory Based Compositional Model'. *Leonardo Music Journal* 7 (1997): 89–90.

Spiegel, L. 'Artists' Statements'. In 'Generative Music', edited by N. Collins. Special issue, *Contemporary Music Review* 28, no. 1 (2009): 127–129.

CHAPTER 7

MEXICO AND INDIA

*Diversifying and Expanding the Live
Coding Community*

ALEXANDRA CÁRDENAS

7.1 LIVE CODING IN MEXICO

LIVE coding offers a unique mixture of hacker and DIY ethics, academic intellectuality, popular and traditional music, and visual expressions that made it irresistible for a young community of artists and programmers in Mexico City. From the year 2000, composer Sergio Luque introduced SuperCollider through workshops in governmental institutions in Mexico City and for many years continued private lessons from Europe via Skype and emails. The artists he taught were mostly composers and musicians with academic training and interest in live electronics, computers, and improvisation. Luckily, many of these artists were related to the Centro Nacional de las Artes (National Centre for the Arts, or CNA), an institution of the Mexican government devoted to bringing high-quality artistic education to the young population. The CNA is characterized by its interest in the newest artistic expressions and the use of technology in the arts and is composed of different art schools as well as the Centro Multimedia (Multimedia Centre, or CMM), a place for research on the electronic arts, which is open to any person who is interested in learning, completely free of cost.[1]

Not much time passed from when Sergio Luque commenced his SuperCollider workshops until his group of pupils, led by Ernesto Romero, Eduardo Meléndez, and Ezequiel Netri, who worked at the time at the Taller de Audio (Audio Workshop) at the CMM, began to teach open-source code for sound and visuals, and attracted different artists from Mexico City and gradually from the rest of the country. The fact that the open-source workshops were given for free opened the possibility for artists from different backgrounds to learn, explore, and collaborate.

This possibility represented a radical change of vision that greatly impacted the Mexican multimedia arts scene. Live electronics and visuals were until then limited by commercial software, too expensive for most of the people who were interested in live electronics, algorithmic composition, or live visuals. Apart from the unreachable prices, the impossibility of appropriating the source code imposed a creative limitation that these Mexican artists were ready to surpass. The switch from commercial software to open-source software at the CMM gave birth to a community of musicians and visual artists who discovered the richness in sharing and exchanging. The Mexican open-source community kept in close contact with the European community through emails and forums, and some of them even travelled to Europe to experience the practice at first hand.[2] This is how live coding came to spark the curiosity in the Mexican artists who implemented it in a way that fulfilled their particular needs.

The Audio Workshop organized live coding sessions every month from December 2010, inspired by the ever-growing community of Super Collider, Fluxus, and Processing users. The first sessions consisted of duos of live coders: one for sound, the other for visuals. The rules were quite strict, asking for the code to be written from scratch and that each performance couldn't surpass the nine-minute mark. The two screens were projected side by side, and sound produced via a quality sound system.[3] The first results were mostly mere informatics exercises with not much performance potential, but the passion of the community, the free lessons at the CMM, and the monthly live coding sessions supported strong creative development of the performances over time.

By November 2012 the interest of the live coders and of the CNA had grown enough to organize the first Latin American Symposium on Music and Code, called /*vivo*/ ,[4] dedicated to live coding. During this time, the Mexican live coding community was exposed to the creators of the practice and to the pioneers of the international community. The exchange was fruitful and opened the horizons of the Mexican live coders. One of the changes this experience brought about, was the realization of the possibility to utilize pre-written code and to extend the duration of the performances. The initial nine minutes from scratch format was crucial as a means to get to know the software and to learn to think of artistic and computing solutions to create a performance. Thanks to the days the Mexican and European live coders spent together during the first /*vivo*/, the Mexican live coding community learned not only that the duration and the structures of performances are meant to be free, but that the definition of live coding is ever changing, ever expanding, and that we all are constantly creating it. This exchange was also fruitful on the other side, since European live coders were able to experience at first hand how a rich and diverse community can work.[5]

Diverse projects derived from the live coding sessions, like the HackPackt.mx project, which compiled live coded pieces from artists on a daily basis. Even though the CNA was the germinal place for live coding, the practice expanded to other venues in the city. Ensembles and collectives appeared in other cultural centres and the inclusion of acoustic instruments changed the face of the concerts.

The next step taken by the community was the incorporation of networked music. The second edition of /*vivo*/ in 2013 was based on this subject.[6] Thanks to this symposium, the Ensemble LivecodeNet was formed, as well as other live coding groups that perform in different venues for wider audiences, becoming part of the artistic life of the city. As for 2014 and 2015, universities have included live coding in their curriculum as is the case of the Escuela Superior de Física y Matemáticas (Superior School of Physics and Mathematics, ESFM), where live coders are encouraged to write their own tools in LUA. And the Universidad Nacional Autónoma de México (National Autonomous University of Mexico, UNAM) with its space SEMIMUTICAS at the applied mathematics school, where live coding is part of their research.

Other regions of Mexico have also shown interest in the practice, and workshops and concerts have taken place outside Mexico City, including in Guadalajara and Oaxaca. Nonetheless, the practice remains focused on the capital of the country. An exception to this is the recent workshop given by Sergio Luque on Sound Synthesis and Algorithmic composition using SuperCollider at the Centro Mexicano para la Música y las Artes Sonoras (Mexican Centre for Music and Sound Art, CMMAS) in Morelia during the summer of 2015. CMMAS is the most relevant institution for electroacoustic music in the country, and the fact that open source code is being included in the composers' summer course opens the possibility for the expansion of open source and live coding practice into the rest of the country.

The centralization of the practice in Mexico City may seem worrisome at first glance, but it's important to remember that the greater Mexico's capital city, with a population of about twenty-one million, is the confluence point for the economy and culture in the whole country and that people from each of the thirty-one states lives there. This, added to the fact that the country has a decentralized arts infrastructure, even though it's not always completely functional, permits one to expect that live coding practice will find a proper and healthy expansion throughout the country in the years to come.

The inherent qualities of live coding combine very well with the particular needs of the artistic community in Mexico, opening an alternative form of expression in a society where corruption and injustice suppress greatly the possibilities for diversity, inclusion, and even good education for most part of the population.

Open source and live coding are, without a doubt, more than appropriate tools for artists and young people around the country who can't possibly afford expensive computers or software, and who at the same time want to appropriate the technology to express their own voice, composed not only by the native and traditional cultures, but by the cultures that colonization brought about and that a history of mixture and changes has created: a Cyber-Mexican voice that is given the opportunity by open-source and live coding practices to grow up away from the rigidity of obsolete institutional paradigms and away from consumerism and secrecy.

A good example of this unique Mexican voice, and one which has become an important part of the international Algorave scene, has toured Europe, and is preparing to release its second album: the band Mico Rex,[7] whose music is composed by the architect

Jorge Ramírez and the composer–mathematician Ernesto Romero. Using a networked system to alter each other's code during the performance, the band composes songs inspired by Latin American popular culture, accompanied by visuals that sum up perfectly the mixture of the hacker; traditional, popular, and academic influences that are unique to the Mexican live coding community.

7.2 LIVE CODING IN INDIA

In 2014 a group of students of the College of Engineering, Guindy, Anna University in Chennai planned the yearly Kurukshetra 'the Battle of the Brains' Festival. This UNESCO-supported International Techno Management Fest is a highly regarded event in a society where technology and innovation are of very high concern.[8] The very young students were interested in bringing the latest technological and artistic trends from Europe and found live coding suitable and pertinent for their festival. This is how the first workshop on SuperCollider and live coding in India took place.

Less than a year after the Kurukshetra experience, the second workshop took place in Coimbatore, and the host was the Bannari Amman Institute of Technology. The Festival FUTURA reached its sixteenth edition and they also wanted to be part of the live coding genesis in India.[9]

In both cases the workshop was given in well-provided locations with Internet and computers for each of the more than 100 students of computer science that attended both of the workshops. The attention given to the parity of genders in these academic environments is exceptional, since both workshops were attended by groups consisting of males and females in equal numbers.

In a society so inclined to the computer sciences, bringing live coding to the academic environment seems to offer an artistic and creative input that the community has been lacking. Academic environments are highly competitive, and from all the yearly graduates very few achieve positions in technological corporations, making professional life even more competitive. Live coding offered a refreshing space where no competition is encouraged and where traditional expressions are welcome.

Most of the students are familiar with traditional and classical Indian music and it was not too difficult for them to grasp on the basic musical concepts. It was also not difficult for them to understand the improvisational nature of live coding, given the role improvisation plays in their own music. The final concert, after only three days of introduction to the concept of live coding, consisted of huge laptop orchestras playing, while singers and dancers of Carnatic music improvised.

During 2014 and 2015 the interest kept alive and the communication between the newly born community and the international community kept going through the Internet. Though there are still no cases of live coding sessions or performances, the liminal state in which the live coding community in India finds itself at the moment is remarkably promising.

The third live coding workshop took place in Kurukshetra in February 2016, where the students learnt also Tidal and Fluxus. And where the community gave one more step further in its consolidation.

7.3 CONCLUSION

Live coding is a very recent performative practice that has permeated and integrated many different cultural and philosophical currents. It has found a way to disclose what was hidden before, to transcend boundaries, and to connect the unexpected. Its philosophical concepts of diversity, sharing and transparency merges effectively with the some of the most sacred spiritual ideals of both traditional Mexican and Indian cultures, where humans can value themselves and the others by their uniqueness and not by outdated and imposed societal perceptions. It is this vision of the human being as an individual who is a part of the whole that makes live coding a sanctuary where expression and creativity are valued outside of old institutional paradigms.

Apart from this, new economies like Mexico and India share the fact that their native, precolonial cultural expressions are deeply rooted and not eradicated by the 'aggressor', therefore still existent. These cultures have managed to keep artistic expressions intact through their colonization, and have been able to sustain a very interesting, though not always healthy, balance between the Eurocentric and the native. This ambivalence creates a fertile soil for interesting and futuristic artistic expressions like live coding. Given that these cultures are more used to opening themselves, having survived colonialism while maintaining authenticity, they are more open to understanding different cultures and paradigms.

The fact that open-source and hacker philosophy are inclusive and treasure the difference and the uniqueness of expression makes live coding an ideal tool for artists of countries with new economies to express themselves in a new and unique way.

NOTES

1. Audio Workshop, CMM http://cmm.cenart.gob.mx/tallerdeaudio/.
2. Video FLOSSOFÍA, 2008, Audio Workshop, CMM, *Vimeo*, https://vimeo.com/33207474.
3. Villaseñor 2012
4. /*vivo*/ International Symposium on Music and Code, Live Coding, 2012 http://vivo2012.cenart.tv.
5. Blackwell, McLean, Noble, and Rohrhuber 2014.
6. /*vivo*/ International Symposium on Music and Code, Networking, 2013 http://toplap.org/vivo-simposio-internacional-de-musica-y-codigo-redes/.
7. Mico Rex, *Soundcloud*, https://soundcloud.com/micorex.
8. Kurukshetra Festival, 2013 Knowafest.com, 16 December 2013, http://www.knowafest.com/2013/12/kurukshetra-2014-anna-university-techno-management-festival-chennai.html.

9. Futura Festival, 2014, Knowafest.com, 21 August 2014, http://www.knowafest.com/2014/08/futura-2014-bannari-amman-institute-technology-technical-symposium-erode.html.

BIBLIOGRAPHY

Anderson, B. *Comunidades Imaginadas: Reflexiones sobre el origen y la difusión del nacionalismo*. Mexico City: Fondo de Cultura Económica, 1991.

Blackwell, A., McLean, A., Noble, J., and Rohrhuber, J. 'Collaboration and Learning through Live Coding', Dagstuhl Seminar 13382. *Dagstuhl Reports* 3, no. 9 (2014): 130–168.

Brown, J. A. *Cyborgs in Latin America*. New York: Palgrave Macmillan, 2010.

Levy, S. *Hackers: Heroes of the Computer Revolution*. Garden City, NY: Nerraw Manijaime/Doubleday, 1984.

Stallman, Richard. 'The Hacker Community and Ethics: An Interview with Richard M. Stallman'. *GNU Operating System*, 2002. http://www.gnu.org/philosophy/rms-hack.html.

Turner, V. 'Liminality and Communitas'. In *The Ritual Process: Structure and Anti-Structure*, 94–130. Chicago: Aldine, 1969.

Villaseñor, H. *Live Coding: El paradigma de la programación en vivo*. Mexico City: Centro Multimedia, 2012.

....

CHAPTER 8

DEAUTOMATIZATION OF
BREAKFAST PERCEPTIONS

....

RENATE WIESER

A while ago, a friend asked me to participate in an art project she had organized. It was concerned with shared space and individual perception. My part was to think up one in a sequel of events which were dedicated to different sense modalities. I was responsible for 'hearing' and decided to organize an 'acoustic breakfast'. There was an open invitation and anyone could come and join; the table was set, one could eat, drink coffee or tea, and at the same time, one could pick up the sounds which were created by all these actions with a self-made pickup. Two game pads controlled sound algorithms programmed in the programming language SuperCollider, allowing the guests, while having their breakfast, to record audio snippets, change, replay, and spatialize them across six speakers placed in various positions of the room.

The concept of a long breakfast was chosen, because it epitomizes an event of leisure, where there is no aim and no time pressure. For many years there had existed a broadcasting sequel on free radio called *Sunday Breakfast on Monday Mornings*, a title which I always loved for its great symbolic value. Much less subversive, the breakfast I organized was on a Sunday, and I also wanted to incite the participants to make their own music while having breakfast, to work in their time of leisure.

Of course I faced the problems usually faced in interactive art projects. My role as an organizer or artist involved far too much explaining of how the technology is meant to be used and this 'didactic' relationship didn't fit too well with the breakfast situation. In my experience, audio-based interactive works are sometimes difficult if there is no major visual layer which helps to understand what is going on. Many of the breakfast guests were completely happy to only amplify sounds and remained unexcited about all the other possibilities provided. The ideas of the project would have worked out better perhaps in a space that made it easy to listen carefully. But of course chatting is an element of the breakfast atmosphere. So it was a nice event indeed, which I remember with pleasure. But I could imagine more interesting sounds being produced. My idea was to create an environment where the participants explore the possibility by listening to the changes they are able to obtain.

My practical and also my theoretical work has been focused for several years on algorithmic processes. It is an exploration of possibilities and an investigation in related discourses. With a background in fine art, philosophy and media theory, rather than in music, I have been using programming languages for my art projects. I perform musical pieces and exhibit installations, but the ideas behind this practice are often connected to the mindset of philosophy, as well as art and media theory. As a participant and visitor, I sometimes have the impression of a curious difference between the two worlds of art and music, despite the fact that their borders have become blurred over many years. The connection between music and contemporary art has been a break with cultural conventions.

For me, art is a particular way of thinking: it implies the exploring of social and political conditions, not only with the means already provided by academic disciplines, but by finding out new ways to explore, conceptualize, and change them.

The concept of change and novelty here is not based on the idea of a world perfecting itself as time passes, an idea which one could typically find in idealistic early modern aesthetic. It is rather based on the renewal of our understanding of the world we are familiar with. An art practice of that kind is explored very compendiously by the Russian literary theorist Viktor Shklovsky in his famous 1917 paper 'Art as Technique'. Here, he coins the term *defamiliarization* (*ostranenie*) for a strategy that artists use in their work in order to bring awareness to automatized perceptions and behaviors:

> If we start to examine the general laws of perception, we see that as perception becomes habitual, it becomes automatic. Thus for example, all of our habits retreat into the area of the unconsciously automatic; if one remembers the sensation of holding a pen or of speaking in a foreign language for the first time and compares that with the feeling at performing the action for the ten thousandth time, he will agree with us.[1]

And some pages further:

> After we see an object several times, we begin to recognize it. The object is in front of us and we know about it, but we do not see it hence we can not say anything significant about it. Art removes objects from the automatism of perception in several ways.[2]

Contrary to common intuition, Shklovsky uses the term *automatism* to define routines which are neither planned nor programmed, and which largely elude conscious control.[3] The examples he refers to come from literature, but his approach is also very valuable for the understanding of art practices that involve programming. In Tolstoy's writing, Shklovsky observed techniques of bringing blanketed perceptions to awareness again. Within art he found ways not only of exploring automatisms, but also of exposing them to the readers' understanding. As a central example, Shklovsky describes how

Tolstoy writes about torture, describing the procedure as if there was no common word for it, as if he were describing something he knows nothing about. Like this, many of his examples deal with the defamiliarization of positions of power and the ways they interweave with habitualized behavior. So how does this apply to art forms that work with algorithmic procedures?

Examining the commonly used devices and artefacts helps to understand how they mediate social structures and power relations. Computers are ubiquitous in many places in the world, and most people in the richer countries are very familiar with the use of hardware and software. The multiple layers of algorithms that make them function are mostly hidden, however. Shklovsky developed his ideas with respect to literature, differentiating unambiguously between poetic and everyday language. Here, language plays an important role in the formation of automatized structures, as much as it provides the means to deautomatize them. Against this background, we can ask new questions about the relation between natural and machine language. Because a programming language is, by itself, entirely rule-based, including the source code, and taking into consideration automatisms and the artistic techniques that make them visible, we may elucidate the contrast between the purposive and the poetic, which are both equally implied in programming. Most people use software without being aware of its rules and peculiar language-like character. Because they are used to working with computers, they intuitively know the workflows and functionalities involved, but they don't mind its grammar. Here, technical necessities and wilful decision merge into an unknown terrain. Programming as a tool for artistic research, as well as for research about art, helps to illuminate and differentiate this terrain. By distinguishing between automatisms, automats, and artistic working methods, this type of critical programming becomes crucial for the understanding of changing technosocial developments. Illustrating its automatized character, as well as the impossibility of fully automating it, code shares many properties with natural language. Like literature, an algorithmic artwork can be an investigation into an ambiguous relation.

In the digital world, rules are not only blinded out by habitual behavior but also through technical standards. In IT, the word *transparency* is used in an unexpected way: it here describes the relation between 'user-friendly' graphical user interface and the hidden program code.[4] The program is transparent if it is like a clear window, which is not to be noticed, a notion which is as common as it is misleading. This strange disappearance of the work of the programmers might be an important reason for the widespread anxieties with regard to programming. Of course, many people share strong disinclinations against such strategies of disempowerment.

During the algorithmic breakfast, the computer program between pick-up device and amplified output wasn't questioned much at all. Somehow in this commonplace atmosphere of having breakfast, the computer and the set of rules governing the interactive possibilities were hardly noticeable. To me, it seems an interesting requirement that the visitor of an art space is challenged to learn about the technological aspects of an artwork. This requirement differs from the more spectacular pretension of installations which try to create illusion. I'm more inspired by artworks which reduce transparency (understood in the IT sense of the term), by making the program code traceable. The rule structure behind an audible or visible output is understandable at least in principle by reading the code, even if the reader doesn't know any programming language. Evidence from live coding is very revealing here. As described by many people who witnessed a live coding concert, this practice is very often alienating and explanatory at the same time—a combination that reminds one of the process of deautomatization. It is interesting that the mere presence of code can often cause strong reactions and become an important aesthetic component.

In my installations, the code was hidden in the black box. Thinking back, I suspect that hearing is a sense which tends to operate subconsciously. It seems to be really hard to wrest it from its automatized condition. It was part of the concept of the installations that the visitor tried to find out more about the otherwise hidden functionality through listening to the algorithmic sound. Often people were able to use the provided tools, but verbalizing or even relating to the experienced processes proved difficult. This makes explicit what can also be seen as the diagnosis of a contemporary situation. Algorithmic processes have become ubiquitous in everyday life, not only because they are unavoidable, but because their appearance is increasingly simplified and streamlined. At the same time, paradoxically, many people experience this proliferation of modes of simplicity as overburdening them with alienating routines which provokes much anxiety. This development may be an explanation for the surprising relevance of Shklovsky's work, which may inspire new possibilities of how to face algorithmic procedures.

Notes

1. V. Shklovsky, 'Art as Technique,' in *Russian Formalist Criticism. Four Essays*, edited by L. T. Lemon and M. Reis, 5–24 (Lincoln: University of Nebraska Press, 1965), at 11.
2. Shklovsky 1965, 13.
3. This definition is emphasized in: A. Brauerhoch, N. O. Eke, R. Wieser, and A. Zechner, *Entautomatisierung* (Paderborn: Willhelm Fink, 2014).
4. See I. Arns, 'Read_me, run_me, execute_me. Code as Executable Text: Software Art and its Focus on Program Code as Performative Text', *Medien Kunst Netz*, 2004, http://www.medienkunstnetz.de/themes/generative-tools/read_me/.

CHAPTER 9

··

WHY DO WE WANT OUR COMPUTERS TO IMPROVISE?

··

GEORGE E. LEWIS

I'VE been involved in live electronic music since 1975, and programming computers since around 1978. An important part of my work in improvisation and computing is rooted in practices that arose in the early 1970s, when composers used the new mini- and microcomputers to produce 'interactive' or 'computer-driven' works, which preceded the better-known new media 'interactivity' that began in the late 1980s, as well as influencing developments in 'ubiquitous computing'.[1] Composer Joel Chadabe, one of the earliest pioneers, called these machines 'interactive composing' instruments that 'made musical decisions as they responded to a performer, introducing the concept of shared symbiotic control of a musical process'.[2] This process was exemplified by the pioneering mid-1970s work of the League of Automatic Music Composers, whose members constructed networks of musical computers that interacted with each other. 'Letting the network play', with or without outside human intervention, became a central aspect of League performance practice, one that was a key influence on my own work.[3]

At some point around 1979 I started to actually put on performances with what I have come to call *creative machines*.[4] Since that time, my machines have improvised in solo and group settings, and even as a soloist with a full symphony orchestra (see Figures 9.1–9.3).[5] Of course, reception varied widely, but early on, one issue arose that I've never quite been able to put to rest was exemplified by frequent conversations that started off a bit like this: 'Why do you want to play with computers and not with people?'

Good question, and one that couldn't be dismissed by mere name calling ('You Luddite!'), since many of the people who were asking weren't in dialogue with those histories of technological scepticism anyway. In these early days of algorithmic improvisation, 'posthumanism' was well in the future, and placing computers on stage with other people seemed somehow a denigration of the latter—not least because the machines often sounded a bit clueless and really couldn't keep up with what the musicians were

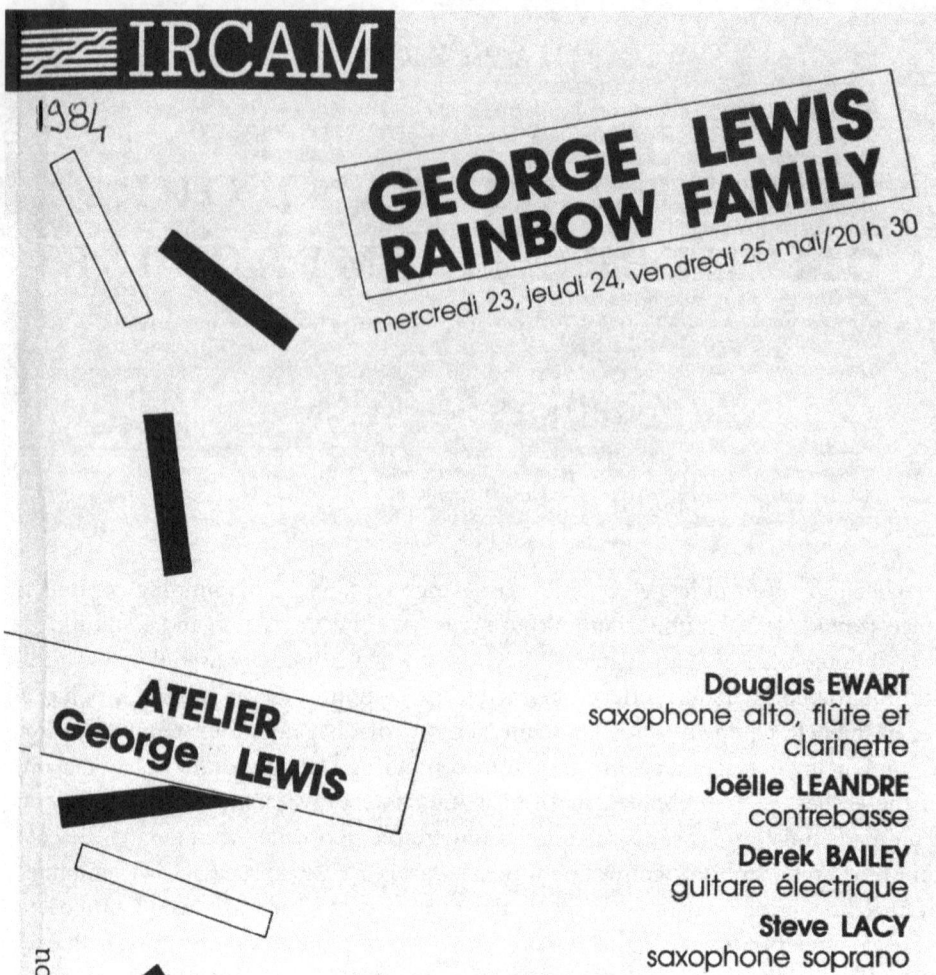

FIGURE 9.1 Poster, George Lewis, *Rainbow Family* (1984), IRCAM.

doing. One fear, expressed by many, is well articulated by anthropologist of technology Lucy Suchman in terms of a project to 'displace the biological individual with a computational one',[6] a successor to the now-ubiquitous transformation of the nature of work that began with the Industrial Revolution and gained new urgency in the late modernist technosphere of the 1950s: 'You will be replaced by a machine'.

By 2000 I was able to articulate at least one rationale for my work: 'This work deals with the nature of music and, in particular, the processes by which improvising musicians produce it. These questions can encompass not only technological or music-theoretical interests but philosophical, political, cultural and social concerns as well'.[7] By this time, however, I had got used to the fact that not everyone wanted those concerns encompassed. As one musician (and his interviewer) put it in a recent interview in the jazz magazine *Downbeat*:

```
\ phrases.fth
\ phrasing
\ formula version begun 3 16 94 george e. lewis la jolla ca
\ rev 6 3 95 gel ucsd
\ rev 6 24 95 gel vancouver

only forth also definitions internals also
decimal

: phrases ; cr .( USE PHRASES-SOLO.FTH FOR SOLO PERFORMANCE )

decimal

6 constant nreplytypes

create replytypes nreplytypes allot
|
: setreplies ( -- )
  nreplytypes 0
    do 70 %chance
      if 1        \ follow input
        else 0    \ reverse or ignore input
      then
      replytypes i + c!
    loop
;

variable voldir    \ 0 = no change 1 = down, 2 = up
variable octdir    \ 0 = no change 1 = down, 2 = up
variable keydir    \ 0 = no change 1 = down, 2 = up
variable playdir   \ 0 = no change 1 = down, 2 = up
variable legdir    \ 0 = no change 1 = down, 2 = up
variable spddir    \ 0 = no change 1 = down, 2 = up

: computer-solo? ( -- flag ) \ both solo counters have to be up
  solo1 @ 59 >
  solo2 @ 59 >
  and
;

\ *** if more than 16 players, group them according to instrumental group,
\ different groups have different speakers (virtual orchestra)

variable maxgroup
variable mingroup

: getplayers ( -- ) \ right now 64 max number of voices
\  nchans irnd startchan !
  startchan @ 8 irnd + nchans mod startchan !
  maxgroup @ irnd mingroup @ max groupsize !
\  nchans irnd startchan ! nchans groupsize !
;
```

FIGURE 9.2 1994 excerpt from Forth program code for George Lewis, *Voyager* (1987–).

George Lewis is working on another book that is about improvisation in daily life, and parsing the mental processes of which we make spontaneous decisions. It's not so mystifying, even in our dialogue. We draw upon our vocabulary, our experiences, our associations—and we spontaneously speak to each other. I think that happens in music, but there's a particular mindfulness to the craft of that in music. Maybe that's the mystery—and should remain a mystery—if it's beyond what we can say in words.[8]

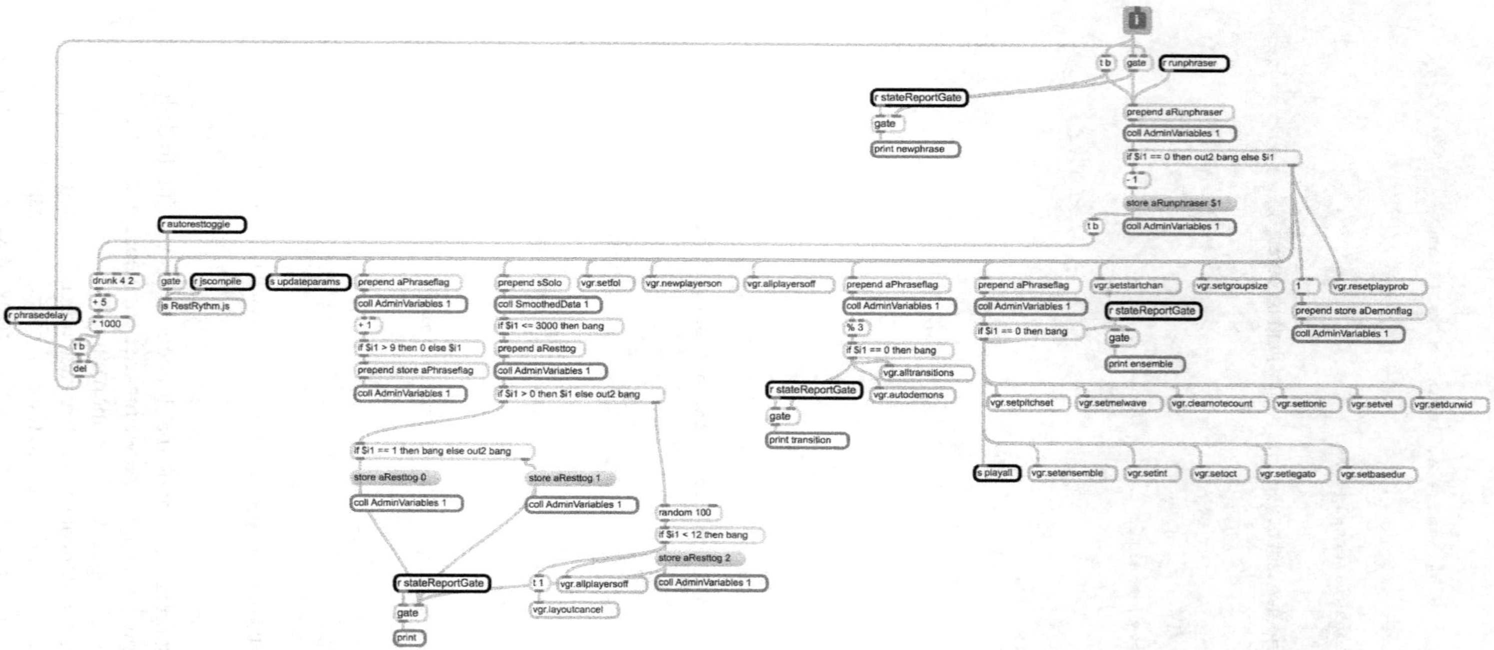

FIGURE 9.3 Excerpt from Max/MSP code for George Lewis, *Interactive Trio* (2010–); realization by Damon Holzborn.

The interviewer's response: 'I'd be disappointed if we could actually map spontaneity'.[9]

In fact, what I've come to expect from this arts-oriented notion of improvisation is an invocation of the ineffability of practice—a useful trope for artists trying to win space in an at least partially stochastic process of artworld success. In that regard, it seemed that, for some, a certain line was being crossed, and a certain transgression was in the air.

The better the machines played—and for my money, they do play quite well now—the greater the threat to the mystery, and to an artist's strategic self-fashioning as one of a select band of designated superpeople with powers and abilities far beyond those of mere mortals.

In any case, the questions weren't put to flight by the quality of the computer performances. Even as much of the music we hear in the West sounds electronic or at least electronically enhanced, and programs that play music electronically are commonplace, it seemed to many that there was just something wrong with the notion of computers as improvisers. There was something special about improvisation—something essential, fundamental to the human spirit—that one just couldn't, or shouldn't, approach with machines. Nonetheless, for the rest of this chapter I want to briefly sketch out some of the basic reasons why some perverse individuals, and indeed entire communities, persist in wanting their machines to improvise.

The practice of improvisation is thoroughly embedded within the nature of many interactive, software-driven musical works, systems, and computing platforms. For my part, I've made efforts to imbue interactive systems with values such as relative autonomy, apparent subjectivity, and musical uniqueness rather than repeatability. My musical computers were designed to stake out territory, assert identities and positions, assess and respond to conditions, and maintain relativities of distance—all elements of improvisation, artistic and otherwise. Many composers theorized relations between people and interactive systems as microcosms of the social, drawing on social aesthetics that valorized bricolage and homegrown elements. System design and real-time musical interactions with the results were marked by efforts to achieve nonhierarchical, collaborative, and conversational social spaces that were seen as manifesting resistance to institutional hegemonies—all aspects of a free improvisation ethos that had emerged in the mid-1960s.

Interactions with these systems in musical performance produce a kind of virtual sociality that both draws from and challenges traditional notions of human interactivity and sociality, making common cause with a more general production of a hybrid, cyborg sociality that has forever altered both everyday sonic life and notions of subjectivity in high technological cultures. Being present at the creation of such a new mode of everyday life is simply too interesting to pass up, so that is one reason why I want my computers to improvise.

The question of machine agency is of long standing; to offer one example (and far from the earliest), in 1869 John Stuart Mill wondered, 'Supposing it were possible to get houses built, corn grown, battles fought, causes tried, and even churches erected and prayers said by machinery—by automatons in human form.'[10] In our own time, improvisation has presaged new models of social organization that foreground agency, history, memory, identity, personality, embodiment, cultural difference, and self-determination. When we improvise, we can take part in that wide-ranging social and cultural transformation. That is because improvisation is everywhere, even if it is very hard to see—a ubiquitous practice of everyday life, fundamental to the existence and survival of every human formation, it is as close to universal as contemporary critical method could responsibly entertain. As computer scientist Philip Agre put it,

> activity in worlds of realistic complexity is inherently a matter of improvisation. By 'inherently' I mean that this is a necessary result, a property of the universe and not simply of a particular species of organism or a particular type of device. In particular, it is a computational result, one inherent in the physical realization of complex things.[11]

For the philosopher Arnold I. Davidson, 'Collective intelligibility ... unfolds in real-time when the participants in social interaction are committed to making sense of, and giving sense to, themselves and others.'[12] Thus, when our machines improvise musically, they allow us to explore how meaning is exchanged through sound. To improvise is to encounter alternative points of view and to learn from the other; improvising with computers allows us a way to look inside these and other fundamental processes of interaction. In this regard, creative machines that take part in collective musical improvisations exemplify the radical position of Lucy Suchman: 'I take the boundaries between persons and machines to be discursively and materially enacted rather than naturally effected and to be available ... for refiguring.'[13] At the moment at which musical improvisation with machines enacts this radical fluidity of identity, what we have is not a simulation of musical experience, but music making itself—a form of artificial life that produces nonartificial liveness.

In that sense, perhaps our improvising computers can teach us how to live in a world marked by agency, indeterminacy, analysis of conditions, and the apparent ineffability of choice. Through improvisation, with and without machines, and within or outside the purview of the arts, we learn to celebrate our vulnerability, as part of a continuous transformation of both Other and Self.

I'd like to conclude with Arnold Davidson's understanding:

> What we say about human/computer interaction is all too frequently dictated by an already determinate picture of the boundaries of the possible and the impossible. ... When one is pushed to go beyond already established models of intelligibility and habitual practices of the self, when one searches for new forms of self and of social intelligibility, new modes of freedom, the improvisatory way of life assumes not only all of its ethico-political force, but also all of its very real risks of unintelligibility and self-collapse.[14]

Negotiating this complex matrix is part of why many of us want our computers to improvise and why we want to improvise with them. What we learn is not about machines, but about ourselves, and our environment. In the end, I've always been in sympathy with the way in which human-computer improvisations enact, as Andrew Pickering observes of cybernetics, 'a *nonmodern* ontology in which people and things are not so different after all.'[15]

NOTES

1. For more on the anteriority of 1970s computer music interactivity, see Lewis 2003. I further develop this alternative history in an essay on the work of composer Rich Gold, who went on to become a key early figure in computer gaming and ubiquitous computing. See Lewis 2017. Also see Gold 1993; Weiser, Gold, and Brown 1999.
2. Chadabe 1997, 291.
3. See Bischoff, Gold, and Horton 1978 and Chandler and Neumark 2005. Also see Salter 2010.
4. For the earliest recorded example of my algorithmic improvised music, dating from 1979, hear George Lewis, 'The KIM and I' (Lewis 2004a).
5. See Lewis 2004b. For a documentary on my work from 1984, watch Davaud 1984. Complete performances with the *Rainbow Family* system, from 1984 at IRCAM, are available at http://medias.ircam.fr/xee3588_improvisation-george-lewis. For an example from 1994, hear George Lewis, *Voyager* (Lewis 2000). For a performance of my *Interactive Trio* piece from 2011 with pianist Geri Allen and a computer-interactive pianist, watch http://leccap.engin.umich.edu/leccap/view/7td666n16oht47labax/15205.
6. Suchman 2007, 240.
7. Lewis 2000a, 33.
8. Hwang and Staudter 2013.
9. Hwang and Staudter 2013.
10. Mill 1974, 123.
11. Agre 1997, 156.
12. Davidson 2012.
13. Suchman 2007, 12.
14. Davidson 2012.
15. Pickering 2010, 18.

BIBLIOGRAPHY

Agre, P. E. *Computation and Human Experience*. Cambridge: Cambridge University Press, 1997.

Bischoff, J., Gold, R., and Horton, J. 'Music for an Interactive Network of Microcomputers'. *Computer Music Journal* 2, no. 3 (1978): 24–29.

Chadabe, J. *Electric Sound: The Past and Promise of Electronic Music*. Upper Saddle River, NJ: Prentice Hall, 1997.

Chandler, A., and Neumark, N., eds. *At a Distance: Precursors to Art and Activism on the Internet*. Cambridge: MIT Press, 2005.

Davaud, M. 'Écoutez votre siècle 8: George Lewis, *Rainbow Family*'. *Ressources, IRCAM*, 1984. http://medias.ircam.fr/x015be3.

Davidson, A. I. 'Improvisation as a Way of Life'. Talk given at the Susan and Donald Newhouse Center for the Humanities, Wellesley College, 10 February 2012.

Gold, R. 'Art in the Age of Ubiquitous Computing'. *American Art* 7, no. 4 (1993): 2–11.

Hwang, J. K., and Staudter, T. 'Jason Kao Hwang Unleashes the Improvisors'. *Downbeat*, 11 February 2013. https://www.yumpu.com/en/document/view/12094805/dave-brubeck-downbeat/21.

Lewis, G. E. 'Too Many Notes: Computers, Complexity and Culture in *Voyager*'. *Leonardo Music Journal* 10 (2000a): 33–39.

Lewis, G. E. 'Voyager'. In *Endless Shout*. New York: Tzadik 7054, 2000b. CD.

Lewis, G. E. 'The Secret Love between Interactivity and Improvisation, or Missing in Interaction: A Prehistory of Computer Interactivity'. In *Improvisation V: 14 Beiträge*, edited by W. Fähndrich, 193–203. Winterthur: Amadeus, 2003.

Lewis, G. E. 'The KIM and I'. In *From the Kitchen Archives: New Music New York 1979*. New York: Orange Mountain Music OMM 0015, 2004a. CD.

Lewis, G. E. 'Virtual Concerto' [programme notes]. *American Composers Orchestra*, 28 April 2004b. http://www.americancomposers.org/improvise/notes20040428.htm.

Lewis, G. E. 'From Network Bands to Ubiquitous Computing: Rich Gold and the Social Aesthetics of Interactivity'. In *Improvisation and Social Aesthetics*, edited by G. Born, E. Lewis, and W. Straw, 91–109. Durham and London: Duke University Press, 2017.

Mill, J. S. *On Liberty*. Edited by G. Himmelfarb. London: Penguin, 1974.

Pickering, A. *The Cybernetic Brain: Sketches of Another Future*. Chicago: University of Chicago Press, 2010.

Salter, C. *Entangled: Technology and the Transformation of Performance*. Foreword by Peter Sellars. Cambridge: MIT Press, 2010.

Suchman, L. A. *Human-Machine Reconfigurations: Plans and Situated Actions*. 2nd ed. Cambridge: Cambridge University Press, 2007.

Weiser, M., Gold, R., and Brown, J. S. 'The Origins of Ubiquitous Computing Research at PARC in the Late 1980s'. *IBM Systems Journal* 38, no. 4 (1999): 693–696.

PART II

WHAT CAN ALGORITHMS IN MUSIC DO?

CHAPTER 10

COMPOSITIONS CREATED WITH CONSTRAINT PROGRAMMING

TORSTEN ANDERS

10.1 INTRODUCTION

THIS chapter surveys approaches that use constraint programming for algorithmically composing music. In a nutshell, constraint programming is a method of implementing compositional rules—rules as found in music theory textbooks, rules formulated by composers to model their compositional style, or piece-specific rules. Constraint solvers efficiently search for musical solutions that obey all constraints applied.

Fernández and Vico (2013) propose a taxonomy of artificial intelligence (AI) methods for algorithmic composition that distinguishes between symbolic AI, optimization techniques based on evolutionary algorithms and related methods, and machine learning. Constraint programming is a symbolic method. Other AI methods are discussed in other chapters of this book (e.g. machine learning in chapter 12, evolutionary algorithms in chapter 13, and formal grammars like rewriting systems in chapter 5).

Rule-based algorithmic composition is almost as old as computer science. The pioneering *Illiac Suite* (1956) for string quartet by Lejaren Hiller already used a generate-and-test algorithm for the composition process, where randomly generated notes were filtered in order to ensure they met constraints of different compositional styles, such as strict counterpoint for the second movement, and chromatic music for the third movement (Hiller and Isaacson 1993). Ebcioğlu (1980) proposed what is probably the first system where a systematic search algorithm was used for composing music (florid counterpoint for a given cantus firmus). Ebcioğlu later extensively modelled Bach chorales (Ebcioğlu 1987).

Algorithmic composition with constraint programming has been surveyed before. Pachet and Roy (2001) review harmonic constraint problems. Fernández and Vico

(2013) provide a comprehensive overview of music constraint problems and systems in the context of AI methods in general. Anders and Miranda (2011) survey the field in detail, and carefully compare music constraint systems. However, these surveys tend to focus on the technical side, as they are published in computer science journals.

This chapter complements these surveys by focussing on how several composers employed constraint programming for their pieces. Constraint programming systems are briefly presented to introduce techniques used for those compositions.

In this context, the composer is responsible for the final aesthetic result, and computers merely assistant in the composition process. Composers therefore cherry-pick or manually edit the musical results. To emphasize such artistic responsibility and liberty of composers, this field is often called computer-aided composition instead of algorithmic composition, but in this chapter we keep the term algorithmic composition for consistency with the rest of this book.

10.2 WHAT IS CONSTRAINT PROGRAMMING?

Constraint programming (Apt 2003) is a highly declarative programming paradigm that is well suited to automatically solve combinatorial problems, called constraint satisfaction problems. The general idea is easy to understand: constraint programs are much like a set of equations (or inequations) with variables in algebra: a solver finds values for all variables such that all equations (inequations) hold.

Each *variable* is defined with a *domain*, a set of possible values it can take in a solution. Variable domains typically contain multiple values initially, so that the variable value is unknown. In a musical application, the domain of a pitch variable may be, say, all pitches between C4 (middle C) and B4. The search process by and by reduces variable domains in order to find a solution.

Constraints restrict relations between variables. Examples include unary relations (e.g. `isOdd`), binary relations (e.g. <, =, +, −), and relations between more elements (e.g. `allDistinct`). A constraint *solver* searches for one or more *solutions*, where each variable is bound to a single value of its domain without violating any of its constraints.

The general constraint literature clearly distinguishes between variables of different domains (quasi types), such as integer variables, Boolean variables (variables of truth values), float variables, and set (of integer) variables. Modern constraint programming systems define individual constraints by algorithms that depend on such 'type' information, which reduce variable domains without search depending on the constraints applied to them (constraint propagation; Tack 2009).

By contrast, in most constraint systems developed for music composition, constraints are simply test functions returning a Boolean value (true or false). In this context we can therefore often ignore these domain distinctions. A disadvantage of not using constraint propagation is the reduced speed of the search process, but the search is nevertheless fast enough to be useful in practice. An advantage of simpler constraints is a greater flexibility. Variable domains can consist of any values (e.g. in section 10.5.1 below we discuss an orchestration example, where variable domains are symbols). Perhaps most importantly, with this simple approach virtually any function of the host programming language can be used for defining user constraints.

The existence of efficient solvers had an important impact on the success of constraint programming in general. For musicians, a major appealing factor is the relative ease with which common music theory rules can be encoded so that automatically generated music complies with them.

10.2.1 A Minimal Counterpoint Definition

The following presents a two-part counterpoint definition, to provide a practical example. The example is musically somewhat simplistic (much simpler than textbook examples, e.g. Fux 1965), but the point here is to demonstrate underlying principles of the actual implementation and not a convincing musical solution. Later, we will discuss more advanced examples, but in less detail.

For simplicity, this is a first-species counterpoint example, that is, all durations are the same and each note in one part has a simultaneous note in the other part.

The resulting music is unknown before the search: every pitch of both parts is represented by a variable with a domain that includes, say, all white keys on the keyboard between A3 and G5. It is useful to represent pitches numerically—that way concepts like intervals are easily defined. Commonly, pitches are represented as MIDI note numbers (Rothstein 1995).

Only two constraints are defined: a melodic constraint and a harmonic constraint. Melodic intervals are limited to two semitones at most (steps, or repetitions). That constraint can be defined by an inequation that first computes the interval (the absolute difference) measured in semitones between two consecutive note pitches of one part, $pitch_1$ and $pitch_2$, and then limits that interval to at most two semitones, as shown in Equation 10.1.

$$\left| pitch_2 - pitch_1 \right| \leq 2 \tag{10.1}$$

The harmonic constraint requires all simultaneous note pairs to form consonances. That constraint computes again an interval, but this time between two simultaneous pitches, p_1 and p_2, and requires that this interval is an element in a set of consonant intervals, say the intervals unison, minor third, major third, fifth, minor sixth, major sixth, and octave, all measured in semitones (Equation 10.2).

$$\left| p_2 - p_1 \right| \in \left\{0, 3, 4, 7, 8, 9, 12\right\} \tag{10.2}$$

In the music constraint programming systems introduced below, these constraints would be largely defined as above (though they will use the syntax of different programming languages—we used mathematical notation only for clarity). However, the above definitions are not complete, and it will turn out that different systems clearly differ in these missing parts.

In music constraint systems, the pitch variables of the two parts would be organized in some musical representation that defines their relations (e.g. which pitch variable belongs to which part and at what position). Different systems clearly differ in their music representations, which lead to specific capabilities and limitations.

We did not actually model above how the harmonic constraint is applied to pitches of simultaneous notes, or the melodic constraint to pitches of consecutive note pairs in the same part. Different music constraint systems implement different paradigms to control the application of a constraint to variable sets in the score.

Finally, we did not discuss how a constraint solver actually solves the above example. The search strategies of constraint solvers of different music constraint systems also differ clearly.

Figure 10.1 shows a visual example implementation. The constraint solver receives a score that specifies the rhythm of the two voices, and the pitch domains as a sequence of MIDI note numbers representing the white keys on the keyboard between A3 and G5. The constraint boxes hide the details of the constraint implementation and their application to variables, but the melodic interval threshold and the possible harmonic intervals are shown as arguments to these boxes. This implementation uses the constraint

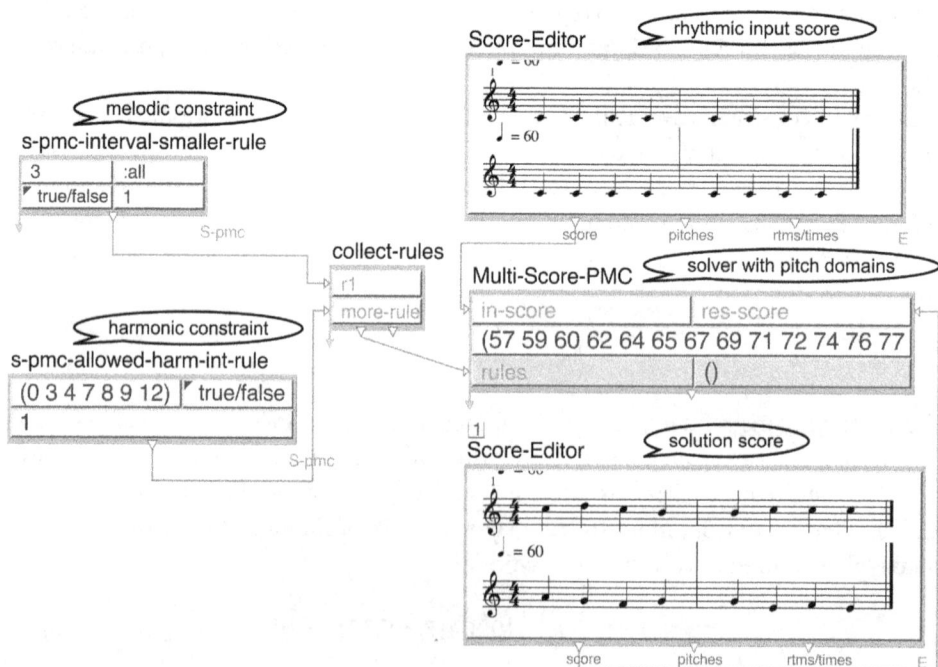

FIGURE 10.1 Implementation of a minimal counterpoint constraint problem with Score-PMC and predefined constraints by Baboni Schilingi.

system Score-PMC, and ready-made constraints from a collection by Jacopo Baboni Schilingi; both are discussed in more detail below.

10.3 CONSTRAINING A SINGLE PARAMETER

This section and the following two sections present different approaches to using constraint programming for music composition. Alternating subsections introduce music constraint systems and discuss compositions created with those systems.

Music constraint programming systems have been designed with a certain range of musical constraint problems in mind, which they can solve. The presented systems are intended to let users such as composers model their own constraints and then combine multiple constraints to their own music theories. In order to make that easier, most systems provide basically a template that conveniently solves a certain class of problems, while other problems can be solved only awkwardly or not at all.

By contrast, constraint programming systems and libraries in general (outside music) are far more flexible, and support more advanced solving strategies. Examples of widely used general constraint programming systems include several Prolog implementations, such as ECLiPSe (Apt and Wallace 2007) and SICStus (Carlsson et al. 2014), and the C++ library Gecode (Schulte, Tack, and Lagerkvist 2017). An interesting recent development is the constraint modelling language MiniZinc, which provides a high-level and relatively user-friendly front end for various state-of-the-art solvers (Nethercote et al. 2007). However, these systems are designed for experienced computer programmers, which makes them inaccessible for most users of music constraint systems.

Many music systems have been designed as libraries of the visual composition systems PWGL (Laurson, Kuuskankare, and Norilo 2009) and OpenMusic (Assayag et al. 1999). Such integration provides direct access to powerful tools such as score editors, and export functionality to commercial music notation software. It also allows for the combination of constraint programming with other algorithmic composition paradigms. Some of these libraries have already a long history that goes back to the common predecessor of PWGL and OpenMusic, PatchWork (Laurson 1996).

10.3.1 Constraining Sequences: PMC

PMC is a built-in constraint solver of PWGL for solving general constraint problems. It is inherited from PatchWork and has been partly ported to OpenMusic as the library OMCS.

PMC is particularly suitable for constraining compositional material before it is part of a score, because its music representation is always a flat sequence (list) of variables (typically with numeric domains, but any types are supported). Such representation is useful, for example, to search for twelve-tone rows, the pitches of a chord, the durations forming a rhythm, the pitch sequence of a melody, and so on.

Even though PMC is an integral part of a visual programming language, its constraints are interestingly defined by textual code in the programming language Common Lisp (the language in which PatchWork itself and its successors are defined). The melodic constraint of Equation 10.1 above can be readily translated into Lisp syntax, but it is then still incomplete.

Constraints are applied to variables in sequences by complementing them with *pattern matching* expressions (simple cousins of regular expressions): a constraint is applied to all variable sets that match its given pattern. PMC's pattern-matching language allows for wildcards: * matches any number of variables. For example, melodic constraints that restrict consecutive variable pairs can be applied with the following pattern-matching expressions: * ?1 ?2, where * matches any number of variables preceding the actual match (including none), while ?1 and ?2 denote two variables to be matched. The Lisp code part of the constraint then uses these symbols to refer to those variables. The importance of pattern matching for PMC is also reflected in its name, which stands for pattern-matching constraints.

Note that the harmonic constraint of Equation 10.2 above cannot be implemented in PMC (at least not alongside a melodic constraint): PMC's purely sequential music representation cannot express melodic and harmonic relations at the same time.

PMC solves constraint problems with the classical backtracking algorithm (Dechter 2003), which by and by completes a partial solution. The algorithm selects a variable and a value of its domain and tests whether all constraints applied to it hold. If so, this domain value is (for now) considered the solution for that variable, and the algorithm continues with the next variable. However, if any constraint fails, the algorithm by and by tries other domain values of the same variable until the algorithm finds one of them that satisfies all constraints, so it may be the solution. The algorithm then proceeds to the next variable. If no domain value of a certain variable fulfils all constraints, then the algorithm must *backtrack* by revisiting the previously visited variable to continue with its other domain values. Backtracking performs a complete search: if a solution exists, then the algorithm finds one.

An interesting feature of PMC, highly useful for musical purposes, is its support of heuristic constraints: a solution should obey *heuristic constraints* if this is (easily) possibly, otherwise such constraints can be broken. A similar concept in the general constraint literature is soft constraints (Meseguer, Rossi, and Schiex 2006). While strict constraints in PMC simply return a Boolean value, heuristic constraints return a number indicating its weight (used if multiple heuristic constraints are in conflict).

The solver favours domain values that meet heuristic constraints by trying them first: domain values are tested in the order of the numeric values returned by the heuristic constraints. While this approach does not necessarily find the best solution, it quickly finds reasonable approximations.

10.3.2 Embedded Constraint Problems: Jacopo Baboni Schilingi

Baboni Schilingi extended PMC (and Score-PMC, presented below) by a sizeable collection of ready-to-use constraints (about 120 constraints for PMC).[1] Examples include constraints disallowing various cases of repetitions, diverse constraints controlling pitches and melodic intervals (e.g. inspired by classical counterpoint), and constraints controlling short value subsequences (useful for enforcing structure, like motifs).

All constraints come as user-friendly graphical PWGL boxes (outputting Lisp code required by PMC) with various arguments to control the effect of these constraints. Also, all constraints can be easily switched to heuristic constraints, which is useful for avoiding overconstrained problems.

Baboni Schilingi likes to develop patches that invite quick editing in many ways to tweak the output or create variations.[2] The result is then output to commercial music notation software for further editing (via MusicXML). Patches typically generate musical material for a short section, where musical parameters such as the rhythm and pitches—but also music notation details like articulations, expressions, and grace notes—are given as independent sequences, written manually or created algorithmically.

Baboni Schilingi often uses constraint programming with PMC to algorithmically generate such parameter sequences. For example, *de la nature du sacre* for string quartet and computer (2012) ends with a fast section of short musical cells (motifs) with irregular accents due to constant metre changes. Material for this section was created with a patch where six individual cells were composed manually (sequences of pitches, durations, and time signatures). The patch generated a sequence that was constrained to play at least four different cells before one could be repeated.

Two interesting approaches are the constraint-based transformation (refining) of parameter sequences that have been generated with other algorithmic techniques, and heuristic profile constraints. For instance, *Aura-phoenix* for violin and computer (2015) contains long downward violin gestures that were generated in multiple steps. A rough version of the pitch sequence was created by interpolating two given pitch sequences (i.e., generating intermediate sequences—concatenated to a long sequence). PMC was then used to refine this sequence: direct repetitions were to be avoided, but otherwise the overall shape of the rough version was to be followed. A heuristic profile constraint expressed a preference to roughly follow the original pitch sequence. Profile constraints are an important means for Baboni Schilingi to control the overall development, while other constraints control local contexts (Baboni Schilingi 2009).

Note that random sequences of cells with restricted repetition as sketched above could also be obtained by other methods, for example, "shuffling" as discussed in chapter 29, or Common Music's heap pattern (Taube 2004). By contrast, sequences that comply with two or more constraints on the same values are very challenging to create with methods that do not search for a solution.

10.3.3 Constraining Pitches in a Polyphonic Score: Score-PMC

While PMC is not able to constrain musical relations in a score that go beyond mere sequential relations, its sibling Score-PMC has been designed for constraining polyphonic scores. For this purpose, Score-PMC features a music representation that supports multiple parts, where the rhythmic structure, the pitch structure, and even details like articulations are represented.

However, most of this information is static during the search process. The only variables are the pitches in the score, whose domain consists of integers representing MIDI notes.

With Score-PMC we can implement both constraints defined in Equations 10.1 and 10.2 above. Like with PMC, constraints are defined by Lisp expressions, and simple melodic constraints are largely defined as in PMC. For constraints that depend on other information (e.g. simultaneous pitches across parts or the metre), Score-PMC features an extended pattern-matching formalism (Laurson and Kuuskankare 2005).

Score-PMC also uses backtracking. For an efficient search process, it first computes a suitable order in which notes should be visited during the search process, which progresses more or less in score time. The search jumps between voices for an efficient search, instead of completing parts one after each other, because otherwise conflicts of harmonic constraints are detected too late, resulting in unnecessary work. This efficient order depends on the rhythmic structure, which is the reason why Score-PMC depends on a completed rhythmic score before the search starts.

Score-PMC can also search for rhythms, but in that mode pitches are not represented any more (Laurson and Kuuskankare 2001). While this functionality is interesting for computing certain textures (e.g. the rhythmic structure of Ligeti-like counterpoint), it is still limited to searching for only a single parameter.

10.3.4 *Engine* by Magnus Lindberg

Magnus Lindberg used Score-PMC to compose *Engine* for chamber orchestra (1996) (Laurson and Kuuskankare 2009). Lindberg was interested in working with constraints programming, because it forced him to analyse and better understand his own compositional style and also to avoid mannerisms of his style.[3]

As Score-PMC requires that the rhythmic structure is fully precomposed, Lindberg created rhythms with a self-developed library that featured a simple representation of rhythms as sequences of fractions. Such data can be manually composed, generated, or transformed in many ways.

Score-PMC was then used to compose individual sections of the piece. Constraints were assigned per section; exemplar constraints are discussed here. While the different constraints can be organized in traditional music theory categories (melodic, harmonic, and voice-leading constraints), Lindberg aimed for a clearly personal style. For example, he developed a personal collection of melodic constraints that permit certain interval successions in order to generate music with distinct characteristics. Other melodic constraints ensured that individual tones did not stand out too much: octaves should be avoided between local pitch minima and maxima that are close to each other, and repetitions between two or more consecutive notes were prohibited. Harmonic constraints required simultaneous pitches to form given pitch class sets. Further harmonic constraints refined the result (for example: no octaves). Longer chord formations tended to be more constrained than short ones. Finally, voice-leading constraints controlled the relation between adjacent parts. Example constraints forbade voice crossing, or required a minimum and maximum distance between adjacent parts.

Overall, for the composer it was important to localize the effect of constraints. Precomposed rhythmic structures already showed certain musical ideas, and single sections could display a counterpoint of different textures and characteristic gestures that called for different constraints. An easy approach is to apply constraints only to certain parts, but constraints could also depend on the rhythmic situation. For example, some phrases with longer durations (or notes interrupted by rests) allowed for large pitch skips, while some rapid phrases required smaller intervals. More generally, a melodic interval constraint could depend on its 'rhythmic interval': a short note followed by a short note could be constrained differently from a short note followed by a long note, and so on.

Results were exported into commercial music notation software (Finale, via the intermediate format Enigma), where it was edited manually. Also the orchestration was done manually in Finale.

10.4 Constraining Multiple Parameters

Score-PMC always requires a precomposed rhythmic score, because it computes the order in which all variables are visited before the search starts (*static variable ordering*). By contrast, systems presented in this section compute which variable to visit next only when this information is actually needed (*dynamic variable ordering*), which allows them to efficiently search in parallel for durations, pitches, and possibly further musical parameters of a polyphonic score.

10.4.1 PWMC

Sandred (2003) extensively studied how to model rhythm with constraints, and developed an OpenMusic library for rhythmic constraint problems (OMRC; Sandred 2000), before he started to work on a new constraint system. PWMC (Sandred 2010) is a PWGL library that solves polyphonic constraint problems, where time signatures, durations, and note pitches can be variables.

PWMC is relatively user-friendly. Constraint problems—including custom constraints—can be expressed by visual programming (experienced users can also write textual Lisp code for more concise definitions).

Users control which variables are affected by a certain constraint with special boxes for various score contexts. For example, melodic constraints are applied to consecutive pitch variables in a part with the box `access-melody`. The actual constraint is defined independently in a PWGL abstraction (a subpatch). Even inexperienced programmers can easily switch the abstraction into 'Lambda mode' so that it returns the abstraction definition wrapped in a function. As a Lisp dialect, PWGL supports functional programming, where functions themselves are values that can be passed around (Abelson, Sussman, and Sussman 1985). This function is given to a box like `access-melody` as an argument. Other arguments of constraint applicator boxes such as `access-melody` control various further details, like to which part the constraint should be applied, and whether it should hold only in certain situations. For example, a harmonic constraint may be applied to every note, or only to the first note in each bar.

Users can express groupings of durations and pitches by motifs. Interestingly, rhythmic and melodic motifs are independent and not 'synchronized', much as in isorhythm the color and talea are not synchronized. Melodic motifs can also be freely transposed for variety, but intervals between their tones will not change.

The order in which variables are visited during the search process can have a crucial impact on efficiency (Beek 2006). PWMC therefore allows users to customize the variable ordering by so-called strategy rules. For example, one strategy rule sets the search to complete multiple voices more or less in parallel. For polyphonic constraint problems, such search approach can be orders of magnitude faster than an approach that first finds all durations before searching for pitches (Anders 2011). For the actual search process, PWMC uses the solver PMC, but PMC's flat sequential representation is internally mapped to a richer music representation.

Sandred is currently developing a successor of PWMC called Cluster Engine,[4] which provides largely similar features but searches more quickly for solutions, due to a custom search algorithm. Multiple solvers search quasi in parallel for different parameter sequences (e.g. the durations, pitches, and time signatures of each part). Solvers can force each other to backtrack, resulting in a kind of back-jumping (Beek 2006): in case of a fail, the algorithm tries to analyse which variable actually caused the fail and directly jumps back to that variable. Doing so avoids redundant work at intermediate variables, as it may otherwise repeatedly run into the same failure. The library Cluster Rules[5] complements Cluster Engine by a collection of predefined constraints for various purposes comparable to Baboni Schilingi's constraint collection for PMC.

10.4.2 Constraining Both Rhythm and Pitch: Örjan Sandred

Constraint programming allowed Örjan Sandred to algorithmically compose rhythm in a way that carefully balances simplicity and complexity. *Kalejdoskop* for clarinet, viola, and piano (1999) was composed with OMRC (Sandred 2006). The composition uses uneven durations of individual notes and rhythmic motifs (e.g. a motif may last in total nine semiquavers or sixteenth notes). This leads to rhythms that do not agree with a regular meter. Sandred proposed a constraint that creates some alignment between multiple rhythms running in parallel in order to limit the overall rhythmic complexity. He used this technique in *Kalejdoskop* to rhythmically organize analytical information (i.e., information only implicitly contained in the final score, like the harmonic rhythm), but this technique can also be used for directly controlling the rhythmic relations between multiple parts.

Figure 10.2 shows the rhythm at the beginning of *Kalejdoskop*: the rhythm of the phrase level (form layer) is shown at the top with long durations; the harmonic rhythm in the middle; and the actual rhythm, which is the accumulation of all note onsets of all parts on the lowest staff. The composer constrained the rhythmic complexity by aligning the rhythms of pairs of layers: whenever an event starts at a higher level, a new event must also start at the lower level (marked by arrows in Figure 10.2). However, new events can start in the lower level between note onsets in the upper level. The result is a rhythmic texture that can freely use uneven durations and motifs, but multiple parts support that uneven rhythm in a semi-homophony. The pitch structure is also controlled with constraints in *Kalejdoskop*, but constraints for pitches and rhythm are not interdependent.

Sandred later developed PWMC (see above), which allowed him to constrain both the rhythm and pitch structure in subsequent pieces, as for example in *Whirl of Leaves* for flute and harp (2006).[6] For this piece he composed rhythmic and melodic motifs that characterize different sections, but which are not synchronized, as discussed above. Only a small subset of constraints can be discussed here, but it should be noted that the majority of constraints affect pitches (melodic and harmonic constraints). Several

FIGURE 10.2 The hierarchical structure of rhythmic information at the beginning of Örjan Sandred, *Kalejdoskop* (Sandred 2006).

constraints are inspired by conventional rules, without actually being conventional. For example, the music expresses an underlying harmony, often with a slow harmonic rhythm. A short 'seed' chord progression was composed manually, largely quartal harmony with four to six different pitch classes. The issue of complex chord progression rules is avoided by allowing this precomposed progression to traverse both forwards and backwards. The resulting harmony is represented explicitly by an extra part in the constraint problem (removed in the actual composition later), so that constraints can explicitly refer to that analytical information.

Some constraints directly link the rhythmic and pitch structure. For example, in a central section that reappears in varied form multiple times, changes of the underlying harmony are clearly marked by a short gesture in a different texture (both instruments share rhythm and pitch classes, but in random octaves, resulting in many melodic skips). This texture change is forced by certain constraints—such constraints that react to events happening only occasionally (here, harmonic changes) represent another way to control the musical form. Another constraint links metre and harmony, again inspired by traditional rules: in some sections, pitches on downbeats must exist in the underlying quartal harmony, while other pitches can come from a slightly larger pitch set of a scale associated with the harmony. Constraints can also affect certain musical characteristics. For example, in some contrasting more quiet sections, higher notes in the melody must be longer than shorter notes, and that way stand out.

10.4.3 An Extensible System: Strasheela

The main design goal of Strasheela was to allow for a wider range of musical constraint problems than previous systems (Anders and Miranda 2011). Previous systems provide a constraint problem template: a certain class of problems is defined relatively easily, but other problems are only awkwardly defined, or cannot be defined at all. Strasheela offers instead a software framework: it offers building blocks that simplify the definition of music constraint problems, but these building blocks can be extended or freely redefined.

Strasheela's music representation models musical concepts as objects (Pope 1991). Core objects are notes and temporal containers. Such containers arrange other objects sequentially (e.g. in a part), or simultaneously in time (e.g. in a chord, or multiple parallel parts) (Dannenberg 1989). Users can extend existing objects or define their own, and Strasheela itself already provides various such extensions (e.g. objects to represent harmonic information). Users create the music representation for a constraint problem by freely arranging score objects in a hierarchy.

Users apply constraints to variables in the score by functional programming techniques, much like PWMC. The differences are that variables are also directly accessible (i.e., constraints can also be applied directly), and Strasheela supports an interface for users to define their own constraint applicators from scratch, so that every score context, every possible combination of variables, can be constrained (Anders and Miranda 2010b).

Strasheela's search performs constraint propagation (depending on the constraints applied, variable domains are reduced without search), which greatly reduces the search space and thus speeds up the search process. Also, while a number of search strategies suitable for efficiently solving various problems are predefined, users can freely program dynamic variable and value orderings of score parameters with convenient building blocks (Anders 2011).

Strasheela's flexibility will be briefly demonstrated by sketching its support for motifs. Strasheela offers multiple motif models. The pattern motif model is basically a generalization of PWMC's motifs, where a sequence of parameters (e.g. the pitches of notes in a part) is constrained to consist of given subsequences. Multiple parameters can be constrained at the same time (e.g. pitches and durations, but also analytical parameters such as chord roots), and motifs of multiple parameters can be unsynchronized (as in PWMC), or synchronized. Motif definitions can contain other variables, so that the actual motifs themselves can be searched for too. How the motifs are actually mapped to variables in the score can be freely defined. For example, pitch contour motifs are possible that constrain only the direction of melodic intervals.

The variation motif model defines a music representation for motifs that explicitly represents analytical features such as the identity and variation of a motif by variables (Anders 2009). Users define what a motif identity means (e.g. which parameters are involved and how they are constrained) and which variations are possible (defined as a set of transformation functions).

In the prototype model, motifs are represented by subconstraint problems that define their own music representation and constraints. Such subproblems can then be arranged freely in higher-level temporal containers and further constrained.

All these models have their own strengths and limitations. For example, the overall number of motifs can be constrained only in the pattern motif model; only for variation motifs can the identity and variation of motifs be constrained independently, and only prototype motifs can be polyphonic.

Strasheela's implementation language, Oz (Roy and Haridi 2004), features a built-in constraint system, which was state-of-the-art at the time Strasheela development started. The nowadays widely used C++ constraint library Gecode is the successor of this constraint system. Strasheela's implementation language greatly simplified realizing much of Strasheela's strengths, such as its custom score search strategies.

Strasheela is more flexible than other music constraint systems (Anders and Miranda 2011), but that flexibility comes at a certain price. Its framework approach requires more programming experience than previously introduced systems. While it provides flexible export functionality into various formats for music notation and sound synthesis, it misses the ecosystem of a widely used composition environment (e.g. other user libraries or powerful editors). Also, its interface is a textual programming language that is rarely used for music.

10.4.4 Microtonal Music: *Tempziner Modulationen* by Torsten Anders

For the composition *Tempziner Modulationen* (2011), Torsten Anders was interested in exploring seven-limit harmony (Erlich 1998). Several seven-limit intervals sound consonant, but unusual (e.g. the harmonic seventh with frequency ratio 7:4 and the subminor third, 7:6).

Tempziner Modulationen for Fokker organ and Carrillo piano is composed in thirty-one-tone equal temperament (31-TET). 31-TET is almost identical to quarter-comma meantone (Barbour 2004), a dominant tuning system of Renaissance and early Baroque music, so all intervals of traditional music are present. Additionally, 31-TET closely approximates seven-limit intervals (Fokker 1955), for example, the harmonic seventh is only 1 cent off.

Figure 10.3 shows an excerpt. Note that 31-TET is notated with traditional accidentals, but tones that are enharmonically equivalent in 12-TET denote different pitches in 31-TET. For example, the interval C–A♯ denotes the harmonic seventh (C–B♭ is still the minor seventh), while C–D♯ is the subminor third. The subminor seventh chord in the third bar of the excerpt uses these intervals.

Instead of memorizing a large number of microtonal chords and scales with many transpositions, and then studying their relations, Anders modelled a suitable music theory in Strasheela (Anders and Miranda 2010a). This theory is based on ideas of conventional music theory, but with a twist: the resulting music is tonal in the sense of Tymoczko (2011), but it is decidedly nondiatonic.

The music theory for this composition constrains complex analytical information. For example, chords and scales represent their 31-TET pitch class set, root, transposition interval, identity and other harmonic features (see Figure 10.3). All these features are variables that can be constrained.

Various constraints control chord progressions. Only a small set of different chord types is permitted per section (harmonic consistency), the roots of consecutive chords

FIGURE 10.3 Excerpt from Torsten Anders, *Tempziner Modulationen*. The upper three staves show the actual composition and the lower two an analysis of the underlying harmony. Chord and scale tones are shown with small notes and root notes as normal notes.

must differ, and the first and last chords per section are either manually set or otherwise closely controlled. Consecutive chords are connected by a small voice-leading size, defined here as the minimal sum of the intervals by which two pitch class sets differ (typically limited to a minor third overall to connect five-tone chords). This last constraint leads to smooth chord progressions.

Further constraints may be applied to chords. In some sections, chords must fall into an underlying scale (see Figure 10.3). A few sections are concluded by a cadence—a short chord progression that sounds all pitch classes of the scale. In other sections, chords are not limited to an underlying scale, but all chords of a phrase share some centre pitch class, and between phrases this pitch class moves only by small intervals (a chromatic semitone or less).

Many more constraints where used (e.g. rhythmic constrains controlling the position of durational accents; Anders 2014), but space limitations prevent discussing these in more detail.

The relatively large pitch domains caused by the microtonal tuning system did not pose any performance problem, because constraint propagation drastically reduced these domain sizes.

10.5 Constraint-Based Orchestration

This section discusses orchestration using constraint programming.

10.5.1 *second horizon* by Johannes Kretz

The solver OMCS (an OpenMusic port of PMC, see above) allows for arbitrary data types as domain values, including symbols, such as instrument names. Johannes Kretz (2006) used this feature to automatically orchestrate sections of the composition *second horizon* for piano and orchestra (2002).

Notes of the piano part were pointillistically spread across other parts. Only few simple constraints were applied. The range of instruments had to be respected. For a better blend, only instruments that belong to predefined families could be used together in a chord. Here, the families were simply woodwind versus brass (strings were composed separately).

Also, the allocation of instruments in chords corresponded always to some standard arrangement. For example, if flute and oboe were playing together, the flute's pitch was always above the oboe's. The purpose was to rule out unusual combinations (e.g. the bassoon above the oboe).

While these constraints are obviously oversimplifying the subtleties of orchestration, they demonstrate the flexibility of PMC. To be clear, constraint problems on symbols can also be modelled with integer variables by mapping every symbol uniquely to an integer, but the resulting definition is somewhat less intuitive.

Constraint programming has also been used for other tasks in the composition process of this piece (e.g. other interesting uses of heuristic profile constraints), but space limitations make it impossible to discuss these here.

10.5.2 Searching for Orchestrations that Imitate a Target Sound: Orchidée

The orchestration environment Orchidée (Carpentier and Bresson 2010) is very different from the other systems presented here so far. It is designed to support the orchestration process only, but in contrast to the purely symbolic systems discussed so far it also depends on signal processing.

The basic idea of Orchidée is that users state an orchestra (a set of instruments), and a target sound (typically a recording, but also synthesized sounds are possible). The system then searches for orchestrations (a mix of instruments from the orchestra alongside with dynamics and playing techniques) that imitate the target sound. Orchidée is currently limited to static sounds: it can only compute individual orchestration 'time slices'.

The system depends on multiple sound features (spectral centroid, attack time, and so forth) automatically extracted from the target sound, and from a database of orchestral sound samples. It compares the perceptual similarity of multiple features of the target sound and the combined features of mixtures of orchestra samples. During a search process based on an evolutionary algorithm this similarity is maximized (Carpentier, Assayag, and Saint-James 2010).

Users can further restrict the solutions with symbolic constraints. For example, users can state the maximal number of instruments to be involved, certain pitches that should be played, and what the minimal dynamics should be.

Constraint problems in general can have multiple solutions, and it is important for users to explore different solutions. In the case of Orchidée this is particularly important: different solutions may imitate different timbral features of the target sound more closely (e.g. one solution imitates better the attack and another the spectrum), but there may be no single ideal solution. Also, Orchidée often results in highly unconventional solutions, so different solutions should be considered. The system therefore provides graphical tools for exploring the solution space.

10.5.3 *Speakings* by Jonathan Harvey

In the composition *Speakings* for large orchestra and electronics (2008), the orchestra imitates certain aspects of speech sounds, derived from, for example, baby babbles, radio interviews, and poetry readings (Nouno et al. 2009). Jonathan Harvey and his team of assistants from IRCAM (Paris) achieved such imitations on the one hand with certain orchestrations using Orchidée, and on the other hand with real-time sound-processing techniques based on analysis and resynthesis.

Orchidée was used, for example, to imitate the sound (formats) of a simple three-note mantra (sung by the composer). The mantra is repeated many times, and the orchestration progressively evolves. Over time, the resulting sound becomes louder, brighter, and also closer to the recorded mantra, the target sound. Such a progression was realized by changing symbolic constraints and by tweaking the sound features taken into account.

10.6 DISCUSSION

This chapter presents how several composers have used constraint programming for their pieces. Instrumentations ranged from chamber music to orchestral works. The constraint systems used during the composition process have been introduced first, so that their use can be discussed. The systems differ in what kind of constraints they allow for (e.g. searching for only pitches, or also rhythmic values), which is reflected in their use.

10.6.1 Similarities and Differences in Compositional Uses

While composers and pieces discussed here differ greatly, they share some common tendencies. Local pitch score contexts are commonly constrained, such as the pitches of simultaneous notes or the pitches of two or more consecutive notes in a melody. For example, melodic intervals are commonly constrained, and so are repetitions (both in various ways). Also, traditional music theory rules form an important inspiration, though the actual constraints and results commonly twist tradition.

Composers obviously aim for a personal voice, and their use of constraint programming thus clearly differs. For example, some composers aim for characteristic melodic interval successions. However, some differences are more structural.

Rhythmic constraints are less common so far, on the one hand because some systems do not support them, but possibly also because traditional music theory largely neglects rhythm. Composers who do use rhythmic constraints tend to develop their own music theories for that purpose (e.g. Sandred's metrical hierarchy and Anders's durational accents). While we can describe rhythms and their transformation, we seemingly lack widely accepted concepts for rhythmic rules.

Shaping the musical form is a central concern for composers. Diverse approaches have been proposed for controlling musical form with constraints, including heuristic profile constraints, constrained-based reactions to certain musical events, and several motif models. Most composers discussed here used motifs (i.e., groups of pitch and/or rhythmic values), but otherwise their means for controlling form clearly differ.

Note that the proposed approaches primarily address the formal development within sections. All composers discussed here controlled the global form manually, and used constraint programming only for generating sections (some composers even only very short sections, some lasting only a single bar).

10.6.2 Strengths of Constraint Programming

When compared with other algorithmic composition methods, constraint programming has particular strengths. Perhaps most importantly, constraint programming can easily control multiple contexts of a single compositional parameter. For example, it is easy to control melodic intervals within a single part and harmonic relations between parts, where both constraints affect the same pitches. With other algorithmic composition methods such control is much more difficult.

Constraints are declarative and modular, like music theory textbook rules are declarative and modular. Each constraint describes only how the result should look, not how this result is reached. Also, each constraint is independent from other constraints, and describes only one aspect of the result. Such strengths allow users to model highly complex compositional theories, including traditional theories or theories inspired by tradition, which is more difficult to do by other approaches that do not use searches.

Constraint programming also suits the mindset of certain composers well, in particular those with classical training. For them it is rather natural to think in constraints, due to the importance rules have in their training. Further, constraint programming allows for a close combination of manual and algorithmic composition, which gives composers great flexibility to shape the result, as all variables in the music representation can be either constrained or manually set (or both) (Anders and Miranda 2009).

10.6.3 Challenges

There are also noteworthy challenges connected with constraint programming. While the idea of constraints is easy to understand for composers, actually implementing new constraints can be difficult, as the desired outcome must be described exactly and in detail.

Some constraint systems (e.g. PWMC, Strasheela, and Cluster Engine) allow speeding up the search process for specific constraint problems with custom search strategies. However, programming custom search strategies requires expertise. These systems therefore offer a default search strategy, or allow users to select predefined search strategies for different categories of problems.

Constraint programs in general are hard to debug. Programming environments offer only limited help. Instead, users have to carefully analyse their programs and results. Contradictions between constraints can happen, and in case of hard constraints (e.g. no heuristics) this leads to a fail (no solution). Such contradictions can be difficult to find in a large set of possibly complex constraints. A somewhat crude but effective way around starts by disabling all constraints, and then enables them again one by one to identify the culprit.

Stochastic approaches have an important place in algorithmic composition and music theory modelling. Constraint programming is good at enforcing strict relations, but probabilities cause difficulties. Several approaches to stochastic constraint programming have been proposed in the general constraint literature (Rendl, Tack, and Stuckey 2014): these approaches distinguish between stochastic variables that follow some random distribution and decision variables that can be constrained—constraints between stochastic variables cannot be directly enforced. More promising is the approach of Sandred, Laurson, and Kuuskankare (2009), where a stochastic constraint ensures a random distributions between variables (it depends on the order in which the solver visits variables, in contrast to the previous approach). They used this approach to create new renderings of the fourth movement of the *Illiac Suite* (1956), which follows Hiller's and Isaacson's (1993) Markov chain probability tables complemented by constraints on harmony and voice leading.

The music constraint systems presented here cannot be used in realtime (or only in a limited way), but real-time support likely will come in the future. Constraint propagation (without search) has already been used for a long time in real-time applications. For example, MidiSpace (Pachet and Delerue 1998) users can control music spatialization and mixing in realtime, and constraints arrange for mixing consistency. Reasonably simple search problems can be solved within milliseconds today. Constraint problems can quasi react to user input on the fly, when the problems are cut into time slices, as a pilot project demonstrated for interactive first-species counterpoint (Anders and Miranda 2008). The speed of constraint solvers and computers has increased further since then. Missing is still the integration of high-performance solvers in real-time composition environments.

In future, constraint programming will possibly be used more often for composing directly with sound. Aucouturier and Pachet (2006) proposed a system—also running in realtime—for concatenating audio segments (samples), where constraints shape the result by restricting metadata of samples. This approach has been used, for instance, for an interactive drum machine that reacts to a MIDI live performance. Orchids (Esling and Bouchereau 2014)—the successor of Orchideé (discussed above)—meanwhile supports finding dynamically evolving sounds; besides orchestration, it is therefore also interesting for sound design and electroacoustic composition.

NOTES

1. Baboni-Schilingi's libraries are freely available at http://baboni-schilingi.com/index.php/research/, accessed 2 September 2017.
2. This section is based partly on personal communication and partly on patches created by the composer for recent compositions.

3. See the programme notes of this composition, available at http://www.musicsalesclassical. com/composer?category=Works&workid=7693, last modified 2014.
4. Cluster Engine is available at http://github.com/tanders/cluster-engine, last modified 2017.
5. Cluster Rules is available at http://github.com/tanders/cluster-rules, last modified 2017.
6. The discussion of this composition is based on personal communication and slides provided by the composer.

BIBLIOGRAPHY

Abelson, H., Sussman, G. J., and Sussman, J. *Structure and Interpretation of Computer Programs*. Cambridge, MA: MIT Press, 1985.

Anders, T. 'A Model of Musical Motifs'. In *Mathematics and Computation in Music*, edited by T. Klouche and T. Noll, 52–58. Berlin: Springer, 2009.

Anders, T. 'Variable Orderings for Solving Musical Constraint Satisfaction Problems'. In *Constraint Programming in Music*, edited by C. Truchet and G. Assayag, 25–54. London: Wiley, 2011.

Anders, T. 'Modelling Durational Accents for Computer-Aided Composition'. In *Proceedings of the 9th Conference on Interdisciplinary Musicology—CIM14*. Berlin, 2014.

Anders, T., and Miranda, E. R. 'Constraint-Based Composition in Realtime'. In *Proceedings of the 2008 International Computer Music Conference*. Belfast, 2008.

Anders, T., and Miranda, E. R. 'Interfacing Manual and Machine Composition'. *Contemporary Music Review* 28, no. 2 (2009): 133–147.

Anders, T., and Miranda, E. R. 'A Computational Model for Rule-Based Microtonal Music Theories and Composition'. *Perspectives of New Music* 48, no. 2 (2010a): 47–77.

Anders, T., and Miranda, E. R. 'Constraint Application with Higher-Order Programming for Modeling Music Theories'. *Computer Music Journal* 34, no. 2 (2010b): 25–38.

Anders, T., and Miranda, E. R. 'Constraint Programming Systems for Modeling Music Theories and Composition'. *ACM Computing Surveys* 43, no. 4 (2011): 30:1–38.

Apt, K. R. *Principles of Constraint Programming*. Cambridge: Cambridge University Press, 2003.

Apt, K. R., and Wallace, M. G. *Constraint Logic Programming Using Eclipse*. Cambridge: Cambridge University Press, 2007.

Assayag, G., Rueda, C., Laurson, M., Agon, C., and Delerue, O. 'Computer-Assisted Composition at IRCAM: From PatchWork to OpenMusic'. *Computer Music Journal* 23, no. 3 (1999): 59–72.

Aucouturier, J.-J., and Pachet, F. 'Jamming With Plunderphonics: Interactive Concatenative Synthesis of Music'. *Journal of New Music Research* 35, no. 1 (2006): 35–50.

Baboni Schilingi, J. 'Local and Global Control in Computer-Aided Composition'. *Contemporary Music Review* 28, no. 2 (2009): 181–191.

Barbour, J. M. *Tuning and Temperament*. Mineola, NY: Dover, 2004.

Beek, P. van. 'Backtracking Search Algorithms'. In *Handbook of Constraint Programming*, edited by F. Rossi, P. van Beek, and T. Walsh, 85–134. Amsterdam: Elsevier, 2006.

Carlsson, M., et al. 'SICStus Prolog User's Manual 4.3.5'. SICS Swedish ICT AB, 2014. https://sicstus.sics.se/sicstus/docs/latest4/pdf/sicstus.pdf.

Carpentier, G., Assayag, G., and Saint-James, E. 'Solving the Musical Orchestration Problem Using Multiobjective Constrained Optimization with a Genetic Local Search Approach'. *Journal of Heuristics* 16, no. 5 (2010): 681–714.

Carpentier, G., and Bresson, J. 'Interacting with Symbol, Sound, and Feature Spaces in Orchidée, a Computer-Aided Orchestration Environment'. *Computer Music Journal* 34, no. 1 (2010): 10–27.

Dannenberg, R. B. 'The Canon Score Language'. *Computer Music Journal* 13, no. 1 (1989): 47–56.

Dechter, R. *Constraint Processing*. San Francisco, CA: Morgan Kaufmann, 2003.

Ebcioğlu, K. 'Computer Counterpoint'. In *Proceedings of the International Computer Music Conference*, 534–543. San Francisco, CA: International Computer Music Association, 1980.

Ebcioğlu, K. 'Report on the CHORAL Project: An Expert System for Harmonizing Four-Part Chorales'. Report 12628. San Jose, CA: IBM, Thomas J. Watson Research Center, 1987.

Erlich, P. 'Tuning, Tonality, and Twenty-Two-Tone Temperament'. *Xenharmonikôn* 17 (1998). http://lumma.org/tuning/erlich/erlich-decatonic.pdf.

Esling, P., and Bouchereau, A. *ORCHIDS: Abstract and Temporal Orchestration Software*. Paris: IRCAM, 2014. http://repmus.ircam.fr/_media/esling/orchids-documentation.pdf.

Fernández, J. D., and Vico, F. 'AI Methods in Algorithmic Composition: A Comprehensive Survey'. *Journal of Artificial Intelligence Research* 48 (2015): 513–582.

Fokker, A. D. 'Equal Temperament and the Thirty-One-Keyed Organ'. *Scientific Monthly* 81, no. 4 (1955): 161–166.

Fux, J. J. *The Study of Counterpoint: From Johann Joseph Fux's* Gradus Ad Parnassum. New York: W.W. Norton, 1965.

Hiller, L., and Isaacson, L. 'Musical Composition with a High-Speed Digital Computer'. In *Machine Models of Music*, edited by S. M. Schwanauer and D. A. Lewitt, 9–21. Cambridge, MA: MIT Press, 1993.

Kretz, J. 'Navigation of Structured Material in Second Horizon for Piano and Orchestra'. In *The OM Composer's Book*, edited by C. Agon, G. Assayag, and J. Bresson, 1:97–114. Paris: Delatour France/Ircam, 2006.

Laurson, M. *PATCHWORK: A Visual Programming Language and Some Musical Applications*. PhD dissertation, Sibelius Academy, Helsinki, 1996.

Laurson, M., and Kuuskankare, M. 2001. 'A Constraint Based Approach to Musical Textures and Instrumental Writing'. In *Seventh International Conference on Principles and Practice of Constraint Programming, Musical Constraints Workshop*. Berlin: Springer, 2001.

Laurson, M., and Kuuskankare, M. 'Extensible Constraint Syntax through Score Accessors'. *Journées d'Informatique Musicale* (2005): 27–32.

Laurson, M., and Kuuskankare, M. 'Two Computer-Assisted Composition Case Studies'. *Contemporary Music Review* 28, no. 2 (2009): 193–203.

Laurson, M., Kuuskankare, M., and Norilo, V. 'An Overview of PWGL, a Visual Programming Environment for Music'. *Computer Music Journal* 33, no. 1 (2009): 19–31.

Meseguer, P., Rossi, F., and Schiex, T. 'Soft Constraints'. In *Handbook of Constraint Programming*, edited by F. Rossi, P. van Beek, and T. Walsh, 281–328. Amsterdam: Elsevier, 2006.

Nethercote, N., Stuckey, P. J., Becket, R., Brand, S., Duck, G. J., and Tack, G. 2007. 'MiniZinc: Towards a Standard CP Modelling Language'. In *Principles and Practice of Constraint Programming*, 529–543. Berlin: Springer, 2007.

Nouno, G., Cont, A., Carpentier, G., and Harvey, J. 'Making an Orchestra Speak'. In *Proceedings of the 6th Sound and Music Computing Conference*, 277–282. Porto, 2009.

Pachet, F., and Delerue, O. 'MidiSpace: A Temporal Constraint-Based Music Spatializer'. In *Multimedia '98: Proceedings of the Sixth ACM International Conference on Multimedia*, 351–359. New York: ACM, 1998.

Pachet, F., and Roy, P. 'Musical Harmonization with Constraints: A Survey'. *Constraints Journal* 6, no. 1 (2001): 7–19.

Pope, S. T., ed. *The Well-Tempered Object: Musical Applications of Object-Oriented Software Technology*. Cambridge, MA: MIT Press, 1991.

Rendl, A., Tack, G., and Stuckey P. J. 'Stochastic MiniZinc'. In *Principles and Practice of Constraint Programming*, edited by B. O'Sullivan, 636–645. Berlin: Springer, 2014.

Rothstein, J. *MIDI: A Comprehensive Introduction*. 2nd ed. Madison, WI: A-R Editions, 1995.

Roy, P. van, and Haridi, S. *Concepts, Techniques, and Models of Computer Programming*. Cambridge, MA: MIT Press, 2004.

Sandred, Ö. *OMRC 1.1: A Library for Controlling Rhythm by Constraints*. 2nd ed. Paris: IRCAM, 2000.

Sandred, Ö. 'Searching for a Rhythmical Language.' In *PRISMA 01*. Milan: Euresis Edizioni, 2003.

Sandred, Ö. ' "Kalejdoskop" for Clarinet, Viola and Piano'. In *The OM Composer's Book*, edited by C. Agon, G. Assayag, and J. Bresson, 1:223–235. Paris: Delatour France/Ircam, 2006.

Sandred, Ö. 'PWMC, a Constraint-Solving System for Generating Music Scores'. *Computer Music Journal* 34, no. 2 (2010): 8–24.

Sandred, Ö., Laurson, M., and Kuuskankare, M. 'Revisiting the Illiac Suite: A Rule-Based Approach to Stochastic Processes'. *Sonic Ideas/Ideas Sonicas* 2 (2009): 42–46.

Schulte, C., Tack, G., and Lagerkvist, M. Z. 'Modeling and Programming with Gecode'. Last modified 4 July 2017. http://www.gecode.org/documentation.html.

Tack, G. *Constraint Propagation: Models, Techniques, Implementation*. PhD dissertation, Saarland University, Germany, 2009.

Taube, H. *Notes from the Metalevel: An Introduction to Computer Composition*. London and New York: Taylor and Francis, 2004.

Tymoczko, D. *A Geometry of Music: Harmony and Counterpoint in the Extended Common Practice*. Oxford: Oxford University Press, 2011.

CHAPTER 11

....

LINKING SONIC AESTHETICS WITH MATHEMATICAL THEORIES

....

ANDREW J. MILNE

MATHEMATICS describes *relationships* between *objects*: collections of rules that determine how one or more object may be transformed into one or more other object, and the rich patterns that flow from these relationships. By a judicious mapping from mathematical structures to musical features, it becomes possible to imbue the latter with similarly rich structure and patterning—to link sonic aesthetics with mathematical theories.

In the following section, I explore the application of mathematical techniques to mould the raw materials of music into interesting *latent* (as yet unrealized) structures. The focus will be on musical scales and metres, both periodic and nonperiodic. In the section after that, I explore some mathematically informed procedures that can produce musical realizations of these latent structures. These include musical canons, methods for generating self-similar and fractal-like forms, and the use of the Fourier transform to dynamically change pitch, timbre, and rhythm. At the risk of perpetuating the hegemonic 'three-dimensional lattice' of discrete pitches, times, and timbres (Wishart 1983), most of the examples use discrete events; despite that, many of the techniques described here are also applicable to smooth and dynamic changes of musical variables. In the third section, prior to the conclusion, I ground some of the abstractions by discussing a real-world musical realization.

11.1 LATENT MUSICAL STRUCTURES
....

11.1.1 Periodic

Musical scales and metres are often thought of as *periodic*, which means they repeat at a regular pitch interval or time interval (nonperiodic scales and metres are considered in a subsequent section). For example, the C major scale is assumed to repeat every oct-ave, and $\frac{4}{4}$ time implies a repeating pattern of four beats—the first of which has great-est 'weight'. The most common periodicity in musical scales is at the octave. This likely arises from the (psycho)acoustical similarity of harmonic complex tones an octave apart (the harmonics in the higher tone are all in the lower tone). Many common scales also have subperiodicities that occur at subdivisions of the octave, such as the diminished or octatonic (e.g. C, D, E♭, F, F♯, G♯, A, B), which repeats every quarter octave, or the hexatonic (e.g. C, D♭, E, F, G♯, A) which repeats every third of an octave. Some scales do not exhibit periodicity at the octave: stretched octaves are commonly found in non-Western instruments (indeed intervals slightly larger than 2/1 are typically heard, cross-culturally, as corresponding to an 'octave'; Burns 1999); repetition at the *tritave* (the 3/1 frequency ratio) has been proposed for instruments with only odd harmonics (Bohlen 1978; Mathews, Roberts, and Pierce 1984); Wendy Carlos (1987) developed *alpha* and *beta* scales (every step is 78 or 63.8 cents, respectively—neither of which produces a 1200 cents octave), and Gary Morrison (1993) has used an 88 cents equal tuning. Some examples of completely nonperiodic structures are provided subsequently.

The periodicity of metres is exemplified by a straightforward Western metre that can be verbally counted as '1 & 2 & 3 & 1 & 2 & 3 & …', or by a more complex aksak metre (from the Balkan regions) such as '1 & 2 & & 3 & & 4 & 5 & & 1 & 2 & & 3 & & 4 & 5 & & …' (Fracile 2003). In distinction from scales, with metre there is no psychoacous-tic imperative for any specific period size: there is not a specific interval of time that represents a 'natural' period of equivalence; however, it is reasonable to assume that any such temporal period must be neither so short that it cannot be heard nor so long that it cannot be remembered (perceptual and categorical time limits are discussed by London 2004).

A periodic structure is naturally represented by a unit circle, which is represented in the complex plane by the formula $e^{2\pi i x}$. (The complex plane is usually visualized as hav-ing real units increasing rightwards and complex units increasing upwards.) The letter i denotes the *imaginary* unit $i = \sqrt{-1}$, which is orthogonal to the *real* unit 1. A com-plex number is a linear combination of real and imaginary units, which means it can be thought of as a vector extending from the origin to some point in this two-dimensional space. The identity $i^2 = -1$ requires a special multiplication operation between such vec-tors, which is equivalent to summing their angles (measured anticlockwise from the ascending real line) and multiplying their length. When $x = 0$, the $e^{2\pi i x}$ formula corres-ponds to a vector extending from the circle's centre to a point at 3 o'clock; as x increases, the vector's point moves anticlockwise around the circle; the circle is completed when $x = 1$; hence, this function is periodic over the unit interval in x.

In much of this section, it will be useful to think of scales or metres with K mem-bers as represented by a *scale vector* of K complex numbers (i.e., as a vector of vec-tors). Each complex number in this vector represents a specific member of the scale or metre. Its *phase*—its anticlockwise angle with respect to the horizontal axis—represents the pitch class or time of that member. In the scale vector, the complex numbers are placed in phase order (i.e., they proceed anticlockwise around the circle)—this

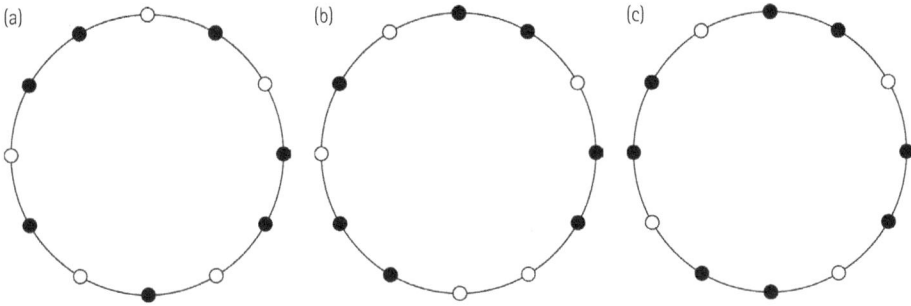

FIGURE 11.1 Three scales represented by ordered points on the unit circle in the complex plane: (a) diatonic major, (b) harmonic minor, (c) diminished.

latter requirement is necessary for some of the mathematics used later. For example, the twelve-tone equal temperament (12-TET) diatonic major scale would have the scale vector $z = \left(e^{2\pi i\, 0/12}, e^{2\pi i\, 2/12}, e^{2\pi i\, 4/12}, e^{2\pi i\, 5/12}, e^{2\pi i\, 7/12}, e^{2\pi i\, 9/12}, e^{2\pi i\, 11/12} \right)$.

Figure 11.1 shows the diatonic major, harmonic minor, and diminished scales on the unit circle. The magnitude of each complex number can additionally be used to represent that member's 'weight' (the magnitude of a complex number is changed by multiplying it by a real number). For example, the magnitude might represent loudness, or probability of occurring. In this case, some points will now lie outside the unit circle (greater weight), some will lie inside the circle (lesser weight); but, for the most part in this chapter, we will assume all members have unit weight. Under this representation, a circular shift (or rotation) represents transposition in pitch (for scales) or in time (for metres). Rotational symmetries (such as those in the diminished scale) represent *subperiodicities*. This method of representation allows for periodic structure to be mathematically analysed in a number of ways, each of which provides useful information about its musical properties, and which may guide the generation of useful periodic patterns.

11.1.1.1 *Evenness and Balance*

Two useful properties arise from the application of the discrete Fourier transform (DFT) to this circular characterization of periodic patterns in the complex plane. These are *evenness* and *balance*. Both are intuitively simple ideas that generalize aspects of many prevalent scales and metres, and can be used to guide the creation of unfamiliar scales and metres (including microtonal and nonisochronic).

A *perfectly even* rhythm or scale is one in which every step size is identical. We are likely very familiar with such musical structures; for example, any regular (isochronous) beat is perfectly even, as is the common 12-TET tuning of the chromatic scale. *Evenness* quantifies the extent to which a pattern with K elements approximates a perfectly even pattern also with K elements. A *perfectly balanced* rhythm or scale is one where the mean position, or 'centre of gravity', of all members around the circle is at the centre of the circle. *Balance* quantifies the proximity of the pattern's centre of gravity to the circle's centre.

Across the broad sweep of K-element patterns available in any given universe of N equal steps, balance and evenness are typically highly correlated (e.g. approximately .94); but there are notable examples where balance and evenness are quite different, as shown by the perfectly balanced patterns in Figure 11.2. Both balance and evenness seem plausibly important (archetypal) forms of organization, which may be relevant to musical structure. The importance of evenness in tonal and metrical structures has been extensively discussed (Clough and Douthett 1991; London 2004; Rahn 1986); prior to my own recent work, balance and its musical implications had not been explored in depth and it has often not been clearly disambiguated from evenness (Lewin 1959; Milne et al. 2015; Quinn 2004).

As mentioned above, the DFT of the scale vector provides a useful quantification of both properties. The tth coefficient ($t = 0, 1, \dots, K - 1$) of the DFT of the scale vector z (defined above) is given by:

$$\mathcal{F}z[t] = \frac{1}{K}\sum_{k=0}^{K-1} z[k]e^{-2\pi itk/K}.$$

(11.1)

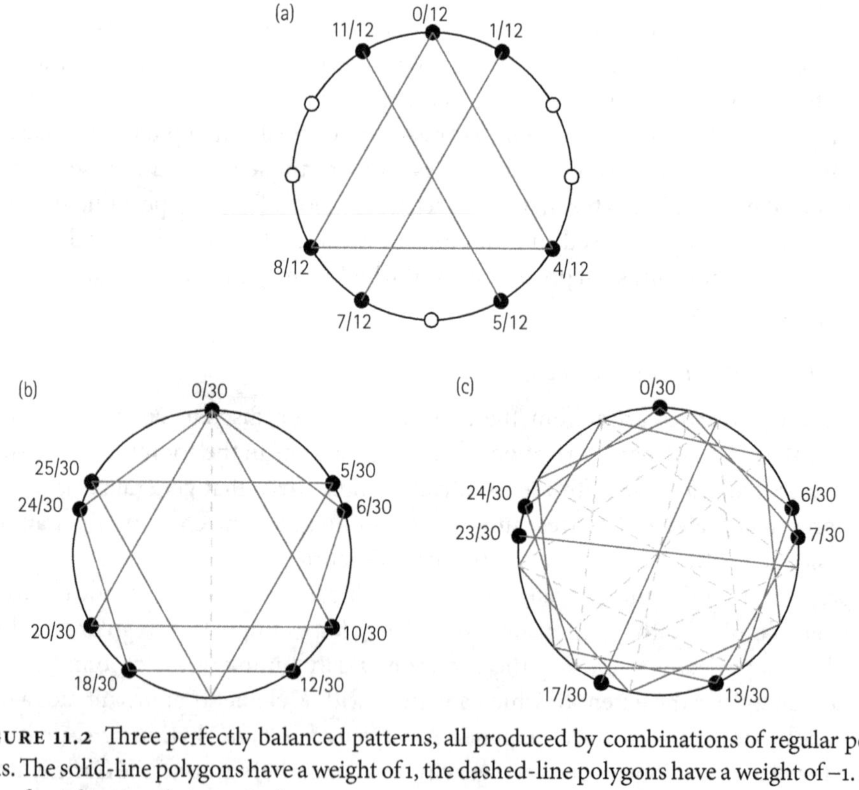

FIGURE 11.2 Three perfectly balanced patterns, all produced by combinations of regular polygons. The solid-line polygons have a weight of 1, the dashed-line polygons have a weight of −1. The sum of weights at each vertex is always 0 or 1.

As shown by Amiot (2009a), the magnitude of the first coefficient (i.e., when $t = 1$) gives the *evenness* of the pattern, which takes a value between zero and unity (the latter being perfect evenness) (Equation 11.2):

$$evenness = \left|\mathcal{F}z[1]\right| \in [0,1], \text{ where}$$
$$\mathcal{F}z[1] = \frac{1}{K}\sum_{k=0}^{K-1} z[k]e^{-2\pi itk/K}. \tag{11.2}$$

This equation can be broken down into three steps. First, it calculates the angle between each successive member of the scale vector and each successive member of a perfectly equal K-fold division of the period. Second, it calculates the circular variance of these angles (circular variance has a maximal value of unity; Fisher 1993). Finally, this circular variance is subtracted from unity. This means that if the displacements are all equal, their circular variance is zero and the pattern is *perfectly even*.

As shown by Milne and colleagues (2015), unity minus the magnitude of the zeroth coefficient (i.e., when $t = 0$) gives the *balance* of the pattern, which takes a value between zero and unity (the latter being perfect balance) (Equation 11.3):

$$balance = 1 - \left|\mathcal{F}z[0]\right| \in [0,1], \text{ where}$$
$$\mathcal{F}z[0] = \frac{1}{K}\sum_{k=0}^{K-1} z[k]. \tag{11.3}$$

This equation simply sums all the complex numbers (vectors) in the scale vector. If the result is a zero vector (all vectors have cancelled each other out), the pattern is perfectly balanced. For any given N and K, there may be many different patterns with equivalent balance values (including perfect balance). Many perfectly balanced patterns have interesting and fairly complex structures, as illustrated in Figure 11.2—in each of these patterns, the balance is achieved by combining regular polygons with coprime numbers of vertices. (Two or more numbers are *coprime* if their greatest common divisor is 1; e.g. 5 and 6 are coprime, but 4 and 6 are not because their largest common divisor is 2.) The scale in (a) is common in the Indian musical tradition (Bhairav thaat and Mayamalavagowla raga). In (b) and (c), some of the polygons have a weight of −1, some have a weight of 1, hence some of the vertices cancel out, thereby creating even more complex structures, which may provide interesting microtonal scales or nonisochronic beat rhythms. Perfectly balanced rhythms and scales such as these can be algorithmically generated by the XronoMorph application (Milne and Dean 2016; Milne et al. 2016).

Perfect evenness, on the other hand, always represents a state of extreme simplicity— such patterns have a distribution of intervals with the lowest possible entropy, and may be thought to have insufficient variation for musical purposes. An interesting solution is provided by *maximizing* evenness under an additional set of constraints. These are discussed in the following subsection.

11.1.1.2 *Well-Formedness, Christoffel Words, and Hierarchies*

There is a useful way to constrain circular patterns (scales and metres) such that when evenness is maximized, the resulting pattern is not perfectly even, but also exhibits a remarkable set of structural properties that are hierarchical, complex, and interwoven. These are *well-formed patterns*. A well-formed pattern is one that has no more than two step sizes that are arranged so as to maximize the evenness of the resulting pattern (there are many alternative definitions, all of which lead to the same end result). Well-formed patterns were first described (in a musical context) by the microtonal theorist Erv Wilson (1975) (he denoted them *moments of symmetry* or *MOS*), and they were independently rediscovered and expressed in a different mathematical framework by Carey and Clampitt (1989).

Because the definition of well-formedness is somewhat abstract—it simply requires two step sizes, but there can be any number of either step, and each step can take any precise size—they are applicable to a wide variety of both familiar Western and non-Western scales (Carey and Clampitt 1989) and rhythms (Rahn 1986), as well as to experimental microtonal musical scales and rhythms (Carey 1998). Furthermore, such rhythms or scales may or may not be embedded within an equal temperament or isochronous pulse. Indeed, when the size ratio of the large interonset interval to the small takes on certain values based on the *golden ratio* $\phi = \frac{1+\sqrt{5}}{2} \approx 1.1618$ and other so-called metallic ratios such as $\delta = \frac{2+\sqrt{8}}{2} \approx 2.414$ *(silver ratio)* and $\sigma = \frac{3+\sqrt{13}}{2} \approx 3.303$ *(bronze ratio)*, the resulting rhythms are such that there is never any faster isochronous pulse that closely approximates the rhythm (Wilson 1997). *Deeply nonisochronous* rhythms such as these can be explored with XronoMorph.

A recent conceptualization of well-formed patterns is based on word theory (Clampitt, Domínguez, and Noll 2009; Domínguez, Clampitt, and Noll 2009). A *word* is an ordered list of letters taken from a finite *alphabet*. For example, we could denote the two step sizes by the letters ℓ (for 'large') and s (for 'small'), and then any particular scale or metre is a different word made up from these letters; for example, the (well-formed) diatonic scale is the word $\ell\ell s\ell\ell\ell s$, the (non-well-formed) ascending melodic minor is $\ell s\ell\ell\ell\ell s$. The (non-well-formed) harmonic minor, for example, could not be represented by a word over a two-letter alphabet because it has three step sizes. *Well-formed words* are special types of words known as *Christoffel words* or circular rotations, thereof (i.e., moving the last symbol to the start) (Clampitt, Domínguez, and Noll 2009).

For a Christoffel word that has a total of j symbols of one type (e.g. s) and a total of k symbols of the other type (e.g. ℓ), the word is equivalent to the sequence of *cuts* a line of slope j/k makes through horizontal and vertical lines of an integer grid (Berstel et al. 2009). This is illustrated in Figure 11.3.

When the sloping line cuts through a vertical line this represents a letter ℓ, when it passes through a horizontal line this represents a letter s. When the line passes through a corner (as it does at the start and finish of the word), the first corner represents ℓ and the second represents s. This corresponds to shifting the sloping line vertically upwards an infinitesimal distance. (An alternative is for the first and second corners to represent s and ℓ respectively, but the resulting word is simply a circular rotation of the other;

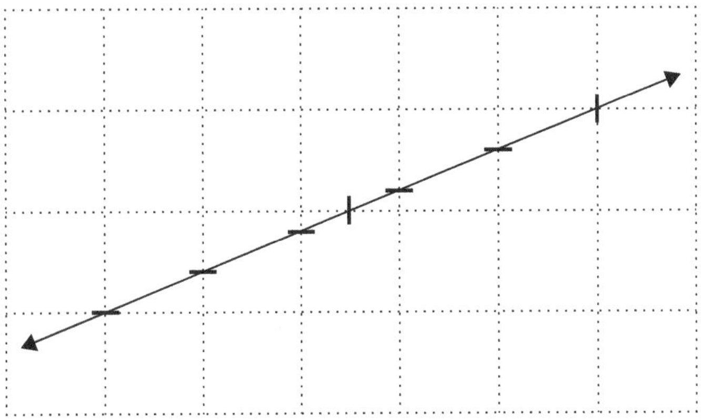

FIGURE 11.3 The cutting sequence of slope 2/5. If the horizontal and vertical cuts represent ℓ and s respectively, the well-formed diatonic word is produced (the Lydian mode, $\ell\ell\ell s\ell\ell s$). Note that the first and last cuts pass through corners, so the sequence of cuts is taken by moving the line infinitesimally upwards. Because the slope is rational, the word is periodic.

as indicated by Lemma 2.7.) Clearly, the cutting sequence is periodic over the word of length $j + k$ because the same pattern repeats every time the line passes through a corner. Furthermore, the arrangement of the two letters is as even as possible.

A remarkable feature of well-formed patterns is that they form a rich hierarchy, whereby for any given well-formed pattern there is always another *higher-level* well-formed pattern that contains all the original's events and may contain more. For example, the pentatonic scale is embedded within the diatonic scale, which is embedded within the chromatic scale—all of these scales being well-formed. Because of the generality of the well-formed definition, these hierarchical relationships also hold for unfamiliar microtonal well-formed scales, and so generalize the notion of *diatonic* and *chromatic*. For example, Figure 11.4, shows the familiar Western hierarchy of well-formed scales, as well as two alternative microtonal well-formed scale (or metrical) hierarchies; there are numerous alternatives.

Hierarchies also apply to metres, where slower beats are typically accented more than intervening pulses; for example, in rock music, the slow rhythmic level may be enunciated by a psychoacoustically prominent bass and snare, while a faster rhythmic level is enunciated by a more delicate cymbal. Well-formed hierarchies enable the creation of hierarchical metres in asymmetric time signatures (additive rhythms), as well as the deeply nonisochronous rhythms mentioned earlier.

Under the word theory representation, the generation of these higher-level rhythms can be easily understood (as discussed more extensively by Milne and Dean 2016). For example, let us take a well-formed word (e.g. $\ell\ell s\ell s$); we then define two *morphisms*. The first morphism performs the following two *rewrites*: $\ell \mapsto \ell s$ and $s \mapsto \ell$. The second morphism performs the following two *rewrites*: $\ell \mapsto \ell s$ and $s \mapsto s$. The first morphism (e.g. $\ell s\ell s\ell\ell s\ell$) corresponds to splitting the large step into a new large and small step such that the old small step becomes the new large step, the second morphism (e.g. $\ell s\ell ss\ell ss$) corresponds to splitting the large step into a new large and small step such that the old small

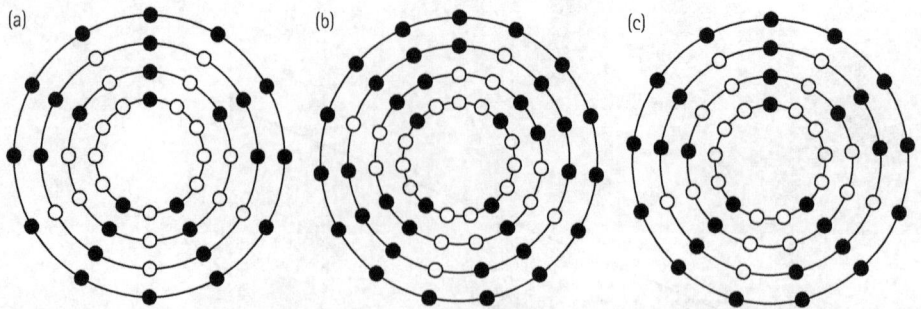

FIGURE 11.4 Three scale hierarchies: within each hierarchy, each successive level splits the large steps of the previous level. (a) chromatic; (b) scale system from *TiHYL*.

step is the same as the new small step. In this way, each well-formed pattern can generate two direct descendants (one for each morphism), hence the number of different well-formed patterns quickly proliferates as we pass down the generations. (The above two morphisms can be equivalently characterized as L-systems, which are described later.)

It is perhaps for all of these reasons that well-formed scales have been one of the most important organizational principles behind much recent work in microtonal music, as exemplified by members of microtonal internet fora and the xenharmonic wiki. In particular, there has been great interest in well-formed scales that contain numerous intervals that approximate just intonation intervals (rational frequency ratios, notably low-integer ratios like 3/2, 5/4, 7/4, etc., which are typically heard as consonant even when unfamiliar) (Erlich 2006; Milne, Sethares, and Plamondon 2008).

Such tunings of nonstandard well-formed scales are interesting because they can connect familiar major and minor chords in unfamiliar ways; for example, certain chord progressions that, in standard Western tunings, start and end with precisely the same chord (e.g. I–vi–ii–V–I) may no longer do so and, importantly, chord progressions that in Western tunings start and end on the same chord may no longer do so. An example of the latter is *TiHYL* (Milne and Rolph 2014)—the repeating eight-chord cycle in the introduction and middle section could be construed as B♭m–D♭–Dm–E11–Fm–A♭–Am–B11–Bm– ... , but this final 'Bm' chord is actually the same as the starting B♭m. In this way, a musical cycle that is impossible in the standard diatonic/chromatic system has been realized. Another example is *A Broken Stern* (Milne 2012), which is in the scale system depicted in Figure 11.4b. A wide variety of well-formed scales with flexible tunings can be explored in the MIDI sequencer Hex (Prechtl et al. 2012) and the associated software at http://www.dynamictonality.com.

11.1.2   Nonperiodic

11.1.2.1   *Fibonacci Sequences and Sturmian Words*

A Christoffel word is a finite-length word written in an alphabet of two letters, such that the two letters are as evenly distributed as possible. A *Sturmian word* is analogous except that it is nonperiodic and of infinite length. Such words can, therefore, be used to generate well-ordered patterns which, remarkably, never repeat—they are a means to pattern the infinite. Sturmian words are the *cutting sequence* made through an integer grid by a line with a slope that is *irrational* (not a ratio of two integers). This is illustrated in Figure 11.5, which shows a line with the irrational slope of $1/\phi$. When the sloping line passes through a horizontal grid-line this is denoted by a 1, when it passes through a vertical grid-line this is denoted by a 0. The sequence of 1's and 0's is then read from left to right to make the Sturmian word. (The cutting sequence for any slope and its inverse are identical, after swapping the 1's and 0's, because they are just reflections about the line $y = x$; the slope $1/\phi$ is used in Figure 11.5 rather than ϕ because this means the figure can take a landscape orientation.) Because the line has irrational slope, it never passes through a vertex after the origin (the bottom-left corner) so the cutting sequence is non-periodic. If the slope had have been rational, the sequence of cuts would become periodic and would spell out a Christoffel (well-formed) word. In this way, we can see that Sturmian words are a generalization of Christoffel words.

These are the first thirty-four members of the sequences that arise from three well-known irrational numbers; each binary symbol might represent a different timbre or a time interval, and so forth (these are Sloane's integer sequences A005614, A144610, and A144609 respectively):

$\phi \approx 1.618$ 1, 0, 1, 1, 0, 1, 0, 1, 1, 0, 1, 1, 0, 1, 0, 1, 1, 0, 1, 0, 1, 1, 0, 1, 1, 0, 1, 0, 1, 1, 0, 1, 1, 0, . . .

$e \approx 2.718$ 0, 1, 1, 0, 1, 1, 1, 0, 1, 1, 1, 0, 1, 1, 0, 1, 1, 1, 0, 1, 1, 1, 0, 1, 1, 1, 0, 1, 1, 0, 1, 1, 1, 0, . . .

$\pi \approx 3.142$ 0, 1, 1, 1, 0, 1, 1, 1, 0, 1, 1, 1, 0, 1, 1, 1, 0, 1, 1, 1, 0, 1, 1, 1, 0, 1, 1, 1, 0, 1, 1, 1, 1, 0, ...

Because such sequences never repeat they may be used to generate infinitely long metres or timbral shifts. Alternatively, arbitrarily sized segments of each such sequence

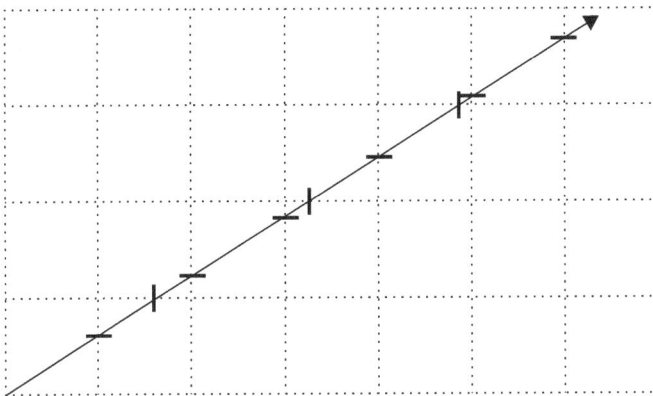

FIGURE 11.5 The cutting sequence of slope If the horizontal and vertical cuts are denoted 1 and 0, respectively, the Fibonacci Sturmian word is produced. The first cut, which passes through the origin is typically ignored. Because the slope is irrational, the word is nonperiodic.

can be extracted to form a finite structure. Sequences such as these, as generated by the golden section, have been musically utilized by Canright (1990, 2001; e.g. *Fibonacci Suite*) and in numerous pieces by Mongoven (2013).

In distinction to the two-letter Fibonacci word described above, the better-known Fibonacci sequence (1, 1, 2, 3, 5, 8, 13, 21, ...) has also been utilized to drive musical algorithms for pitch and time (White 1997). Over a longer musical time scale, Lendvai (1971) has claimed Bartók made use of Fibonacci sequences to structure his pieces (although these assertions are not without controversy; see Howat 1983).

A further generalization can be used to extend both the finite Christoffel words and the infinite Sturmian words. This is to allow for more than just two letters, but still to ensure these letters are organized in an analogous fashion. This can be done by using cutting sequences through higher-dimensional grids—where the number of dimensions is the number of letters in the alphabet in which the words are written. Such words are known as *billiard words* because they replicate the pattern of bounces a perfect billiard ball would make when propelled inside a hypercube (Bedaride and Hubert 2007; Borel 2005).

11.1.2.2 *Other Nonperiodic Structures*

An interesting method for structuring nonperiodic scales is *prime number scales* (Dean 2009), which are based on multiplying a base frequency (not log-frequency) by the first 41, 51, 61, 71, or 91 successive primes. For instance, with a base frequency of 20Hz, the resulting scale would have frequencies (in Hz) of $20 \times (2, 3, 5, 7, 11, 13, ...) = (40, 60, 100, 140, 220, 260, ...)$. The resulting frequencies are, therefore, those of a fundamental and all of its prime numbered harmonics. The resulting scale has some frequency differences that occur more than once (e.g. there are three occurrences of a 40Hz difference above) but the overall sequence is nonperiodic because the primes are themselves nonperiodic. Furthermore, there will never be any two frequency ratios (log-frequencies) that occur more than once (this is a natural consequence of the fundamental theorem of arithmetic). The same procedure could be applied to event times as well as frequencies, thereby generating infinite-length rhythms which ascend through the primes.

Also worthy of note are the pitches used by Aboriginal Australians from the western desert area, which do not exhibit periodicity, but which do seem to contain repetitions of frequency differences rather than ratios (Will and Ellis 1996).

11.2 STRUCTURED MUSICAL REALIZATIONS

Markov models, which assign probabilities to events and their transitions, and *artificial intelligence systems* (such as *Petri nets*), which are also based on transition rules, are commonly used to emulate established musical styles (Loy 2006; Nierhaus 2009). This can be achieved through statistically analysing existing musical practice and applying those insights to the probabilities and/or rules in the model. A more abstract approach is to make a natural mapping from structured latent materials to transition probabilities or rules. For instance, in a well-formed hierarchy, elements lower in the hierarchy (e.g. 'diatonic' pitches or any microtonal generalization thereof) may be assigned higher probabilities of occurring than elements higher in the hierarchy (e.g. 'chromatic' pitches or any microtonal generalization, thereof). Advanced stochastic procedures, often operating on more abstract lines, are detailed by Xenakis (1971) and are exemplified in his music (see also Dodge and Jerse 1997).

Another approach is the use of complex dynamic systems. Such systems are capable of producing complex behaviour with simple interactions. For example, multiagent systems that model flocking and swarming with simple rules can produce seemingly intelligent behaviour (think of a flock of birds where every bird suddenly changes to a new direction, apparently simultaneously). Indeed, swarming models have been used to mimic the interactions of freely improvising musicians (Blackwell and Bentley 2002; Blackwell and Young 2004; Davis and Rebelo 2005; Unemi and Bisig 2004, 2005). Other multiagent systems have simulated evolutionary processes to generate musical structures (Miranda and Biles 2007).

In the following subsections, I consider processes that are more purely mathematical (i.e., less based on models of real-world phenomena). Such procedures may be used to create additional organization in keeping with, but also beyond, that implied by the latent materials.

11.2.1 Tone Rows, Tilings, and Canons

A musical realization where all elements of a set are played just once is a principled way of realizing a nonhierarchical latent structure (such as a scale or metre whose pitches or events are represented with a simple ordered list so that none of its members is privileged). It is principled in that it does not privilege—by repetition—any one element over any other.

To achieve this, one could simply make random selections, with a uniform distribution, across the set. But, typically, composers have sought greater organization. A classic method for achieving this is Schoenberg's tone-row technique. A *tone row* is an ordered set of pitch classes, where no pitch class occurs more than once, and the tone row may be transformed, in toto, by any composition (succession) of transposition, inversion, and retrograde. If the pitch classes of the tone row are represented, in order, by a vector, these musical transformations are simple linear and invertible transforms (they can be performed by left-multiplying the tone row vector by a 12×12 matrix). Furthermore, each such row is played from start to finish. These rules ensure that, despite the large number of distinct transformations that can be made to the original tone row (forty-eight, in all), every possible member of the set of twelve pitch classes is played once and before moving onto the next iteration. It also ensures each tone row is related to every other (via transposition, inversion or retrograde), hence the rows exhibit symmetries with respect to their interval content. Related techniques can be applied to any musical dimension: indeed, the technique of *total* or *integral serialism* includes duration, loudness, timbre, articulation, and so forth (e.g. Babbitt, Nono, Boulez, Stockhausen, and other members of the Darmstadt School).

A generalization of the tone-row method is to use a *partition* of a *universal set* into *disjoint* (nonoverlapping) subsets. The universal set is one that contains all elements of interest; for example, all twelve chromatic pitch classes, or all pulses in a metre. Then we iterate through the resulting subsets (this will be exemplified in the next paragraph). This ensures that all elements of the universal set are played once, but now with a structure resulting from the partition used. The patterning resulting from any given partition may be trivial but, as with evenness, when certain additional constraints are applied, interesting results occur.

One such constraint is that each subset is identical in form to every other: they differ only by their starting pitch or start time. This is exemplified by Figure 11.6: in a metrical context, the rhythm of each subset (all sequences A, A, A, … , or B, B, B, … , etc.) is denoted the *inner rhythm* (shown by the solid lines), the rhythm made by the repetitions (all sequences A, B, C, …) is denoted the *outer rhythm* (shown by the dashed lines). Translating the inner rhythm by the outer rhythm, and summing them, results in every possible pulse being played once and only once. The inner and outer rhythms may or may not exhibit rotational symmetry (rotational symmetry means that there is a rotation by less than a full circle where the resulting pattern perfectly coincides with the unrotated version). For example, the inner rhythm in (a) has rotational symmetry because it is exactly the same whenever it is rotated by $\frac{1}{2}$ of a circle; the inner rhythm in (b) never exactly coincides when rotated by an angle less than the full circle and so does not have rotational symmetry. In both (a) and (b) the outer rhythm has rotational symmetry; in (c) it does not.

In a musical context, such rhythms mean that one voice might play a rhythmic pattern or melodic motif, then, after an appropriate delay, a second voice plays exactly the same pattern, followed, after an appropriate delay, by a third voice playing exactly the same pattern, and so on. This procedure is, therefore, intimately related to the technique of musical canons, which have formed an important part of compositional procedure, notably in the works of Josquin des Prez, Obrecht, J. S. Bach, Haydn, and Olivier Messiaen: indeed, regular tilings such as these were named *regular complementary canons* by Vuza in a series of papers (the most relevant to the present discussion being the third, Vuza 1992) that first identified them and explored their mathematical properties.

Given a universal set with a specific number of equally spaced events, there will be only a limited number of such canons (trivially, only universal sets with nonprime numbers of elements can be tiled because the number of elements in each subset must be a divisor of the number of elements in the universal set). Hall and Klingsberg (2006, Theorem 6 with $r = n$) have shown that, for periods with M equally spaced elements (e.g.

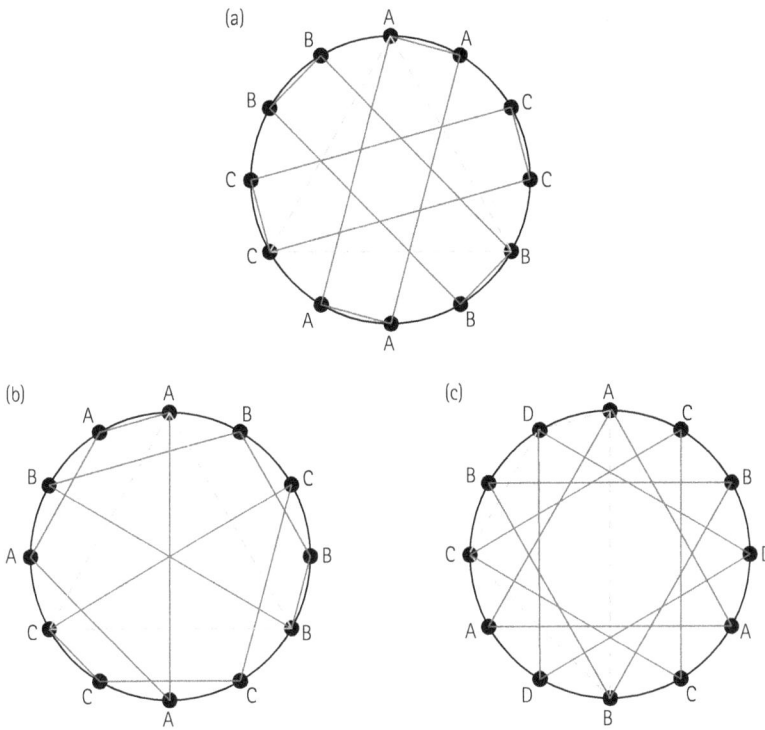

FIGURE 11.6 Three regular complementary canons. The solid lines show the inner rhythms, the dashed lines show the outer rhythms. Note that (b) and (c) are complementary—the inner and outer rhythms have swapped. Any regular canon has such a complement.

regular pulses, or equal-tempered pitches), inner rhythms with n elements where n is a divisor of M, and outer rhythms with rotational symmetry, the number of distinct regular complementary canons is as follows (Equation 11.4):

$$\frac{1}{M}\sum_{\substack{d|n \\ \gcd(d,M/n)=1}} \phi(d)\left(M/n\right)^{n/d} \tag{11.4}$$

where gcd denotes the greatest common divisor, $\phi(d)$ is the Euler totient function which gives the number of positive integers no greater than d that are coprime (defined earlier) with it, and $d|n$ are those d which are factors of n. For example, a periodic metre with twenty-four isochronous pulses (e.g. two bars of $\frac{12}{8}$) has the following numbers of distinct regular complementary canons: 6 two-pulse patterns; 22 three-pulse patterns; 54 four-pulse patterns; 172 six-pulse patterns; 278 eight-pulse patterns; 172 twelve-pulse patterns. So clearly, there is still a wealth of different possibilities available. Interestingly, canons produced by nonperiodic inner and outer rhythms form a special class known as *Vuza canons*, which are extremely rare and have been the subject of much investigation (Amiot 2009b; Vuza 1992).

Andreatta, Agon, and Amiot (2002) discuss other interesting canons where the sub-sets may be related not just by transposition but also by temporal retrogrades (or equiva-lently pitch inversion), and augmentations or diminutions. Canons such as these can be created and realized in the OpenMusic environment, which is designed for computer-aided music composition (Agon, Assayag, and Bresson 2006; Andreatta, Agon, and Chemillier 1999; Bresson, Agon, and Assayag 2011).

11.2.2 Self-Similarities and Fractals

By definition, regular canons by translation (transposition), retrograde, and inver-sion exhibit self-similarity across those three different transformations. Patterns that exhibit self-similarity at all possible *scalings* ('zooms') are known as *fractals*. Fractals are highly organized, and their similarities to natural phenomena such as coastlines, moun-tain ranges, fungi, leaves, plants, trees, and so forth, is well documented (Mandelbrot 1983). From an algorithmic point of view, a useful aspect of fractals is that they can be generated by relatively simple formulae; furthermore, these algorithms typically gen-erate *sequences* of numbers, which means each successive number can be naturally mapped to some parameter of each successive event (e.g. its pitch, duration, interonset interval, etc.).

11.2.2.1 *Fractal Sequences*

Pressing (1988) details the musical implementation of a number of different functions, each of which produces a fractal sequence of numbers. His first example—the *logistic map*—is particularly illuminating and, given its complex results, is a remarkably sim-ple algorithm. It was developed in the nineteenth century to model population dynam-ics, and generates successive values of *x*, indexed by *n*, with the following equation: $x_n + 1 = a\left(x_n\left(1 - x_n\right)\right)$, where *a* is a parameter that controls the map's behaviour. When $3 < a < 4$, successive values of *x* may oscillate between two, three, or more values but, at many *a* in the range $3.57 < a < 4$, successive *x* will oscillate *chaotically* between an infinite number of values. Pressing notes that *a*-values close to transitions between sim-ple oscillations and chaos are particularly interesting because the resulting sequence shows unpredictability but also traces of cyclic behaviour. The stream of *x* values can be straightforwardly mapped to any musical parameter of interest. For example, Pressing maps them to frequency by 2^{cx+d}, where *c* is the overall pitch range in octaves, and *d* is the lowest frequency (the resulting frequencies were not quantized to 12-TET, so the melo-dies are microtonal). The above-mentioned pseudo-cyclic behaviour of the algorithm means that the resulting melodies also exhibit some degree of correlation over time-lags; in other words, they comprise 'motifs' which approximately repeat.

In music, multiple musical variables (notably interonset interval, note duration, and pitch, for each of possibly multiple voices) must be controlled in tandem. Using differ-ent *a*-values for each variable means they lack any mutual influence, which Pressing describes as not musically desirable and resulting in too much unpredictability. In order to solve this issue, he tried fractal algorithms that generate higher-dimensional values: at each step, a pair, or more, of values is produced. These include maps using complex numbers, whose real and imaginary units naturally map into two dimensions, and quarternions, which have four independent components, and so could be used to control four musical variables at the same time. See http://www.australianmusiccentre.com.au/artist/pressing-jeff for examples of Pressing's music.

11.2.2.2 *Lindenmayer Systems*

Lindenmayer systems (also known as *L-systems*) are a way to formalize the generation of successive levels in the hierarchy of well-formed patterns. These systems were introduced by Aristid Lindenmayer (1968) to model the development of simple multicellular organisms and were later used to model the forms created by fungi, plants, and trees (Prusinkiewicz and Lindenmayer 1990). An L-system starts with a word (denoted an *axiom*) written in some alphabet. The word is then 'rewritten' using a set of rules that determine exactly how each letter is to be transformed. In this way, an L-system is closely related to a *generative grammar* (Chomsky 1963): the difference is that, in a grammar, the rewriting rule is not necessarily applied to every letter in the word; in an L-system, the rewriting is always applied to all letters (applications of generative grammars to music are not covered here; see Jones 1981 and Rohrmeier 2011). The rewriting process can be repeated indefinitely by applying the same rewriting rules to the result of the previous rewrite. This results in words exhibiting fractal self-similarities.

To give these symbolic words aesthetic value, they must be mapped to musical variables. Also, it is necessary to define an order in which a word's symbols are 'read'. Typically, they are simply read from left to right (or right to left). A common approach is to produce a visual representation of the word using *turtle drawing* (Papert 1980), exemplified below, then to map features of the resulting graphic to musical variables. For example, there may be an alphabet with three symbols F, +, and −, where F means 'draw a line of unit length', + means 'turn clockwise by 90°', − means 'turn anticlockwise by 90°'. Using this drawing scheme with the axiom $F + F + F + F$ and three rewriting rules: (1) $F \mapsto F+F−F−F\ F+F+F−F$; (2) $+ \mapsto +$; (3) $− \mapsto −$ results in the *quadratic Koch curve*—a highly patterned nonintersecting fractal comprising horizontal and vertical segments (Prusinkiewicz 1986). Prusinkiewicz suggests moving through the word from left to right and using the height of each line segment as its pitch height (alternatively, this could be scale degree), and its horizontal length as its duration or interonset interval—as if each horizontal line segment is a MIDI event in a software sequencer or piano roll.

Nelson (1996) extended this idea by using different turning angles (101° or 107° rather than 90°). The resulting graphic was then further manipulated with nonlinear stretches, twists, and warps. The height of each successive vertex (the points at which the line changes direction) was used to specify each successive pitch, and the horizontal distance between successive vertices specified the respective note's duration.

To deal with the previously mentioned problem of ensuring interrelationships between multiple simultaneous voices, Prusinkiewicz (1986) suggests the use of L-system techniques to model the *branching* found in, for example, trees where a main trunk will have branches, which themselves have subbranches (twigs), and so on. To achieve this, additional branching symbols [and] are added to the alphabet: the first marks the start of a branch, the second marks the end of the branch and instructs the drawing turtle to return to the position when [was first encountered. The use of such techniques to generate distinct musical lines has been explored by DuBois (2003), and led to his development of the [jit.linden] object in Max (https://docs.cycling74.com/max5/refpages/jit-ref/jit.linden.html). Kevin Jones (1989) discusses a number of interesting musical applications of L-systems, including multidimensional generalizations, which naturally lend themselves to multidimensional musical spaces.

The process of choosing useful L-system rewriting rules and axioms, as well as mapping the resulting symbols to musical variables, is far from trivial. DuBois (2003) suggests using statistical analyses of the symbols in the resulting word to ensure the upwards and downwards movements in pitch, duration, loudness, and so on are reasonably balanced (e.g. if a word uses U to represent a specific ascending pitch interval, and D to represent a specific descending pitch interval, and say a statistical analysis of the resulting word shows there are twice as many U symbols as there are D symbols, it might be sensible to make the ascending interval half the size of the descending so as to avoid a general upwards drift in pitch). Soddell and Soddell (2005) take a different approach in that they choose a specific biological form—the fungus Geotrichum—and use the stochastic L-system that best models its growth.

11.2.2.3 *Noise*

Random noise exhibits scaling similarity in that the general appearance of a noise waveform does not change as you zoom the time dimension; put differently, you can slow down or speed up a recording of noise and it will sound the same (assuming an adequate recording and playback system). The continuum of noise is often categorized into three types, each of which has a different high-frequency roll-off: *white noise* has constant power spectral density across frequency; *pink noise* has a power spectrum inversely proportional to frequency ($1/f$); *brown noise* (named after the related Brownian motion, which describes the random movements of particles in a fluid) has a power spectrum inversely proportional to squared frequency ($1/f^2$). In addition to their scaling self-similarities, of particular importance with respect to their musical utility are their self-similarities across time. A simple way to measure self-similarity with respect to time is to perform autocorrelation on the signal. In general, white noise has zero autocorrelation (other than at lag zero), so exhibits no similarities across time lags. Interestingly, any noise with power spectral density roll-off exhibits correlations across time lags: brown noise has strong correlations over small time scales, which gradually decreases as the time lag increases; pink noise has moderate correlation across all time scales, even across large time scales.

By mapping successive noise values to musical variables of successive events, noise can be used to generate stochastic realizations that also exhibit self-similarity across scale (augmentation/diminution), time, or pitch. For this purpose, Voss and Clarke (1978) suggested that white noise is too random, brown is too correlated, but pink is just right; they also presented empirical evidence that pink noise is prevalent in music. Nettheim (1992), however, subsequently demonstrated that, under a more precise analytical method, the pitch distributions in common-practice music (for lags of up to four bars) tend more towards $1/f^2$ than $1/f$. Recent analyses of John Coltrane's solos also show spectra closer to $1/f^2$ than to $1/f$ (Charyton 2015). Regardless of its precise specifications, if a stochastic process is required, then using noise with a downwards spectral slope is a means to produce similarities across both scaling and translation. Jones (1989) demonstrates an interesting application of brown noise to 'walk' the frequencies of sine waves across different harmonics of the same fundamental frequency, thereby producing a continuously evolving timbral texture.

11.2.3 Fourier Scratching

To end, let us return to the first-used mathematical formalization: the representation of a periodic pattern as a scale-vector in the complex domain, and the discrete Fourier transform of this vector to produce a *DFT-vector* that quantifies balance and evenness. The *Fourier scratching* technique, developed by Amiot, Noll, Andreatta, Agon, and Carlé (Amiot et al. 2006; Carlé and Noll 2010; Milne et al. 2011) reverses this process, and treats the DFT-vector as a set of user-adjustable parameters that generate—with an *inverse DFT*—a scale- or metre-like *IDFT-vector*. The coefficients of this generated pattern are cycled through, in order, and used to trigger pitches or timbres in a circular space.

More concretely, imagine we have a circular keyboard with K keys, each of which occupies a circular sector (wedge), and each different sector may subtend a different angle (the keys may have differing widths). For example, a well-formed scale with four large steps and three small could be represented by four wide sectors and three narrow arranged in an appropriate circular order. Now imagine a virtual robot with N 'fingers', such that the first finger is placed at the phase of the IDFT-vector's zeroth coefficient, the second finger is placed at the phase of the first coefficient, and so on until the positions of all N fingers are defined. A regular pulse is used to trigger finger 1, then finger 2, up to finger N, then back to finger 1, and so on. Given a circular keyboard of K keys, the locations of the N fingers around the circle will determine which of the K notes they play, and when. Furthermore, the magnitude of each coefficient determines the velocity of that finger's virtual strike. Note that smooth variations of the parameters in the DFT-vector will smoothly vary the positions and velocities of the robot's fingers, and that the robot's fingers can cross thereby changing the order in which keys are played. In this way, complex patterns of pitches and loudnesses can be quickly created and smoothly morphed between. This example has used the metaphor of a standard pitch-based keyboard, but the coefficients' phases can also be applied to any musical variable that can be made periodic (e.g. a loop 'drawn' through a multidimensional parameterization of a timbral/spectral space). In this case, the circular keyboard may be entirely dispensed with so the robot fingers directly trigger specific points within the continuum.

11.3 A MUSICAL REALIZATION

In this section, I describe an actual musical realization of some of the previously discussed ideas. As mentioned in section 11.1.1.2, I have previously used well-formedness to structure microtonal scales; in the piece I discuss here, I instead apply well-formedness to musical rhythms. I additionally apply, for the first time, the principle of perfect balance.

Babylon 19|30 alternates between two multilayered rhythmical structures. The first is a well-formed structure in a time signature of $\frac{19}{8}$, the second is perfectly balanced rhythm in $\frac{30}{8}$ (or, for comprehensibility, $\frac{15}{8}$). The harmonic contents of the loops vary but are kept relatively simple (essentially diatonic with soft 'modal' dissonances) so as not to distract from the rich and complex rhythmical structure. The rhythms were generated as live MIDI by XronoMorph and sent to synthesizers. Additional physical instrument parts were then overdubbed.

The well-formed section has five rhythmic levels, where each successive level is derived by splitting the long durations of the next lower level. The lowest level comprises just three beats, two of which are long (eight pulses), one of which is short (three pulses). The long beats are split to make the next higher-level rhythm, and this is done successively until the fifth and final level of nineteen equal-sized beats is formed. Each of the levels is characterized in Table 11.1. Interestingly, this hierarchy corresponds precisely

Table 11.1 The five rhythmic levels in the well–formed hierarchy used in *Babylon 19/30*.

Level	Morphism	Pattern	Signature	Size (ℓ)	Size (s)
1		$\ell s \ell$	$2\ell, 1s$	8	3
2	$\ell \mapsto \ell s, s \mapsto s$	$\ell s s \ell s$	$2\ell, 3s$	5	3
3	$\ell \mapsto \ell s, s \mapsto \ell$	$\ell s \ell \ell \ell s \ell$	$5\ell, 2s$	3	2
4	$\ell \mapsto \ell s, s \mapsto \ell$	$\ell s \ell \ell s \ell s \ell s \ell \ell s$	$7\ell, 5s$	2	1
5	$\ell \mapsto \ell s, s \mapsto s$	$\ell s s \ell s \ell s s \ell s s \ell s s \ell s \ell s s$	$7\ell, 12s$	1	1

to that found in the pentatonic-diatonic-chromatic scale hierarchy when a 19-TET tuning is used (I did not plan this in advance, and only noticed it while writing up this description).

A result of the well-formed method of generating successive levels is that every rhythmic event is duplicated in all higher levels. For example, all three beats in the first level are additionally played by the remaining four levels. Naturally, this gives a strong accent to low-level beats, and amplifies the inherently hierarchical nature of well-formed rhythmic structures. However, there is an interesting alternative strategy, which is to treat each successive level as the complement of all lower levels, so it plays only when no lower level is also playing. For example, say we have a lower level which, if expressed as a scale rather than as a rhythm, corresponds to the white-note diatonic scale (as in level 3 of Table 11.1), while the next higher level corresponds to a twelve-pitch chromatic scale (as in level 4 of Table 11.1). Then, instead of playing all twelve events in the latter rhythm, only those events not occurring in the lower-level pattern are played. Using the scalic analogy, this means using only the black-note pentatonic scale, which is the complement of the diatonic in a chromatic universe. Interestingly, these *complementary* well-formed rhythms are themselves well-formed, but they are displaced with respect to each other, so they never coincide. The resulting patterns are shown in Table 11.2. This creates a sparser rhythmic structure that is somewhat reminiscent of the multiple interlocking parts used by Latin or gamelan percussion ensembles—although each individual part is relatively simple, in combination, they produce a complex and interwoven totality.

The perfectly balanced section of the piece comprises six independent rhythmic components (I use the word *component* rather than *level* because, unlike well-formed rhythmic structures, there is no implied hierarchy in perfectly balanced rhythmic structures).

Table 11.2 The five complementary well-formed rhythmic levels actually used in *Babylon 19/30*. The pitches assigned to each level transition every few repetitions to give harmonic movement and melodic variety. The offset is the number of pulses after the start of the period that that level's first beat plays.

Level	Offset	Pattern	Signature	Size (ℓ)	Size (s)
1	0/19	$\ell s\ell$	$2\ell,1s$	8	3
2	5/19	ℓs	$1\ell,1s$	11	8
3	3/19	ℓs	$1\ell,1s$	11	8
4	2/19	$\ell ss\ell s$	$2\ell,3s$	5	3
5	1/19	$\ell s\ell\ell\ell s\ell$	$5\ell,2s$	3	2

Table 11.3 The six simultaneously playing balanced components in the second section of *Babylon 19/30*. The offset is the number of pulses after the start of the period that that component's first beat plays.

Balanced polygon	Offset	Pitch
regular pentadecagon	0/30	C7
regular decagon	2/30	G5
regular hexagon	4/30	G4
irregular heptagon	0/30	D3
regular triangle	2/30	F2
regular pentagon	3/30	A1

These are detailed in Table 11.3, where they are ordered by their pitch height. The irregular heptagon is that shown in Figure 11.2c. All the other polygons are regular, and have numbers of vertices (3, 5, 6, 10, and 15) that factor into 30, which is the number of underlying pulses in the period.

Both types of rhythmic pattern are illustrated as polygons in Figure 11.7 and staff notation, with pitches, in Figure 11.8. An important aspect of the algorithmic process used for this piece is that, without further user intervention, every vertex of any given polygon (both well-formed and balanced) produces the same pitch value. This introduces a rewarding compositional constraint to work with. Interestingly, it also means that when the pitch values and vertices of two polygons are close, they are likely to be perceptually streamed into a single melodic line. As the music progresses, I have successively applied different pitch values to different polygons, which results in harmonic shifts as well a variety of implied or 'half-heard' melodies emerging from the whole.

XronoMorph greatly facilitates the production of rhythms such as these. It is unlikely I would have been able to produce a piece of music like this without it; in this sense, the algorithms and the mathematical structures it embodies have expanded my creative facility. Furthermore, rhythms such as these—despite their accessible sound—can be extremely difficult to perform. In other recent live work, I have used the same algorithmic routines to generate well-formed beats on the fly with both abrupt and smoothly changing structures. Music such as this is genuinely impossible to achieve without computational means.

Using the algorithmic approach described here, intelligent compositional input is still required—not all well-formed and perfectly balanced rhythms, or transitions between

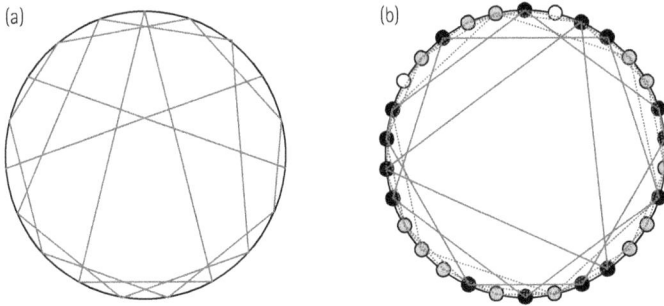

FIGURE 11.7 The two rhythmical patterns in *Babylon 19|30* visualized as polygons inscribed in a periodic circle. Imagine a playhead rotating clockwise around the circle, and outputting a MIDI note whenever it encounters a polygon vertex. The pitch, velocity, and duration of that MIDI note is a determined by the polygon to which the vertex belongs (the user inputs these values for each polygon). In the well-formed rhythm, precisely one event occurs at every nineteenth (pulse) division of the period. In the balanced rhythm, the three lowest-pitched polygons delineate a bass pattern (solid lines and dark disks), while the others have a more percussive or melodic effect (dotted lines and grey disks). At two out of the thirty (pulse) divisions of the period, no event occurs (white disks). In the piece, I have made the lengths of the periods of the well-formed and balanced rhythms take the ratio 19:30 to equalize their pulse tempos.

them, will sound appropriate. Furthermore, effective choices for pitches and durations are still required. These latter choices could feasibly be done algorithmically too—in earlier experiments we used well-formed rhythms to drive Dean's Serial Collaborator (Dean 2013) to produce rhythmically informed serial transformations of previously written tone rows.

11.4 CONCLUSION

As mentioned in the introduction, although the procedures described in this chapter have been mostly exemplified by discrete pitches and times, many of them are also applicable to sonic gestures in the continuum of musical space. For example, well-formedness, fractal structures, and (despite its name) the discrete Fourier transform, lend themselves to smooth and dynamic changes of continua of time, space, timbre, and so forth. I hope that the distinction between latent musical structures and processes by which these are realized has usefully reflected the way that compositional processes typically unfold—in both the human and the computational context. We choose our palette, we paint our picture.

FIGURE 11.8 Two loops from *Babylon 19|30* in staff notation. The one-bar loop in 19-time is well-formed. The two-bar loop notated in 15-time is perfectly balanced. In both loops, each polygon is notated with a separate staff so as not to specify where melodic streaming may arise across the levels (hocketing). For readability, all staffs are vertically arranged in order of pitch height, which means the well-formed staves are not arranged as in Table 11.2.

BIBLIOGRAPHY

Agon, C., Assayag, G., and Bresson, J., eds. *The OM Composer's Book*. Vol. 1. Paris: Delatour France/Ircam, 2006.

Amiot, E. 'Discrete Fourier transform and Bach's good temperament.' *Music Theory Online* 15, no. 2 (2009a).

Amiot, E. 'New Perspectives on Rhythmic Canons and the Spectral Conjecture.' *Journal of Mathematics and Music* 3, no. 2 (2009b): 71–84.

Amiot, E., Noll, T., Andreatta, M., and Agon, C. 'Fourier Oracles for Computer-Aided Improvisation.' In *Proceedings of the International Computer Music Conference*, 99–103. New Orleans, LA: 2006.

Andreatta, M., Agon, C., and Amiot, E. 'Tiling Problems in Music Composition: Theory and Implementation.' In *Proceedings of the International Computer Music Conference*, 156–163. Göteborg, Sweden, 2002.

Andreatta, M., Agon, C., and Chemillier, M. 'OpenMusic et le problème de la construction des canons musicaux rythmiques.' In *Actes des sixièmes Journées d'Informatique Musicale*, 179–185. Paris, 1999.

Bedaride, N., and Hubert, P. 'Billiard Complexity in the Hypercube.' *Annales de l'Institut Fourier* 57, no. 3 (2007): 719–738.

Berstel, J., Lauve, A., Reutenauer, C., and Saliola, F. V. *Combinatorics on Words: Christoffel Words and Repetitions in Words*. Providence, RI: American Mathematical Society, 2009.

Blackwell, T., and Bentley, P. 'Improvised Music with Swarms.' In *Proceedings of the 2002 Congress on Evolutionary Computation*, 2:1462–1467. Washington, DC: IEEE, 2002.

Blackwell, T., and Young, M. 'Self-Organised Music.' *Organised Sound* 9, no. 2 (2004): 123–136.

Bohlen, H. '13 Tonstufen in der Duodezime.' *Acustica* 39, no. 2 (1978): 76–86. http://www.huygens-fokker.org/bpsite/publication0178.html.

Borel, J.-P. 'A Geometrical Approach to Palindromic Factors of Standard Billiard Words.' *Discrete Mathematics and Theoretical Computer Science* 9, no. 2 (2005): 195–212.

Bresson, J., Agon, C., and Assayag, G. 'OpenMusic: Visual Programming Environment for Music Composition, Analysis and Research.' In *Proceedings of the 19th ACM International Conference on Multimedia*, 743–746. New York: ACM, 2011.

Burns, E. M. 'Intervals, Scales, and Tuning.' In *The Psychology of Music*, edited by D. Deutsch, 215–264. 2nd ed. New York: Academic Press, 1999.

Canright, D. 'Fibonacci Gamelan Rhythms.' *Journal of the Just Intonation Network* 6, no. 4 (1990): 4.

Canright, D. *Fibonacci Suite* for Retuned Piano, Seven Hands. 2001. https://sites.google.com/site/davidrcanright/compositions/fibonacci-suite, accessed 30 June 2017.

Carey, N. *Distribution Modulo One and Musical Scales*. PhD dissertation, University of Rochester, 1998.

Carey, N., and Clampitt, D. 'Aspects of Well-Formed Scales.' *Music Theory Spectrum* 11, no. 2 (1989): 187–206.

Carlé, M., and Noll, T. 'Fourier-Scratching: {SOUNDINGCODE}.' In *SuperCollider Symposium 2010*. Berlin, 2010.

Carlos, W. 'Tuning: At the Crossroads.' *Computer Music Journal* 11, no. 1 (1987): 29–43.

Charyton, C. 'The Impact of Improvisation on Creativity: A Fractal Approach.' In *Creativity and Innovation among Science and Art: A Discussion of the Two Cultures*, edited by C. Charyton, 153–178. London: Springer, 2015.

Chomsky, N. 'Formal Properties of Grammars'. In *Handbook of Mathematical Psychology*, edited by R. D. Luce, R. Bush, and E. Galanter, vol. 2:323–418. New York: Wiley, 1963.

Clampitt, D., Domínguez, M., and Noll, T. 'Plain and Twisted Adjoints of Well-Formed Words'. In *Mathematics and Computation in Music*, edited by E. Chew, A. Childs, and C.-H. Chuan, 65–80. Berlin: Springer, 2009.

Clough, J., and Douthett, J. 'Maximally Even Sets'. *Journal of Music Theory* 35, nos. 1–2 (1991): 93–173.

Davis, T., and Rebelo, P. 'Hearing Emergence: Towards Sound-Based Self-Organisation'. In *Proceedings of the International Computer Music Conference*. Barcelona, 2005.

Dean, R. T. 'Widening Unequal Tempered Microtonal Pitch Space for Metaphoric and Cognitive Purposes with New Prime Number Scales'. *Leonardo* 42, no. 1 (2009): 94–95.

Dean, R. T. 'The Serial Collaborator: A Meta-Pianist for Real-Time Tonal and Non-Tonal Music Generation'. *Leonardo* 47, no. 3 (2013): 260–261.

Dodge, C., and Jerse, T. A. *Computer Music: Synthesis, Composition, and Performance*. 2nd ed. New York: Schirmer, 1997.

Domínguez, M., Clampitt, D., and Noll, T. 'WF Scales, ME Sets, and Christoffel Words'. In *Mathematics and Computation in Music*, edited by T. Klouche and T. Noll, 477–488. Berlin: Springer, 2007.

DuBois, R. L. *Applications of Generative String-Substitution Systems in Computer Music*. PhD dissertation, Columbia University, New York, 2003.

Erlich, P. 'A Middle Path between Just Intonation and the Equal Temperaments, Part 1'. *Xenharmonikôn* 18 (2006): 159–199.

Fisher, N. I. *Statistical Analysis of Circular Data*. Cambridge: Cambridge University Press, 1993.

Fracile, N. 'The Aksak Rhythm, a Distinctive Feature of the Balkan Folklore'. *Studia Musicologica Academiae Scientiarum Hungaricae* 44, no. 1 (2003): 191–204.

Hall, R. W., and Klingsberg, P. 'Asymmetric Rhythms and Tiling Canons'. *American Mathematical Monthly* 113, no. 10 (2006): 887–896.

Howat, R. 'Bartók, Lendvai and the Principles of Proportional Analysis'. *Music Analysis* 2, no. 1 (1983): 69–95.

Jones, K. 'Applications of Stochastic Processes'. *Computer Music Journal* 5, no. 2 (1981): 45–61.

Jones, K. 'Generative Models in Computer-Assisted Musical Composition'. *Contemporary Music Review* 3 (1989): 177–196.

Lendvai, E. *Béla Bartók: An Analysis of His Music*. London: Kahn and Averill, 1971.

Lewin, D. 'Re: Intervallic Relations between Two Collections of Notes'. *Journal of Music Theory* 3, no. 2 (1959): 298–301.

Lindenmayer, A. 'Mathematical Models for Cellular Interaction in Development, Parts I and II'. *Journal of Theoretical Biology* 18 (1968): 280–315.

London, J. *Hearing in Time: Psychological Aspects of Musical Meter*. New York: Oxford University Press, 2004.

Loy, G. *Musimathics*. Vol. 1. Cambridge, MA: MIT Press, 2006.

Mandelbrot, B. B. *The Fractal Geometry of Nature*. New York: W. H. Freeman, 1983.

Mathews, M. V., Roberts, L. A., and Pierce, J. R. 'Four New Scales Based on Nonsuccessive-Integerratio Chords'. *Journal of the Acoustical Society of America* 75, S10 (1984).

Milne, A. J. 'A Broken Stern'. Stern Brocot Band (V. Angelis, S. Holland, A. J. Milne). *SoundCloud*, 2012. https://soundcloud.com/andrew-j-milne/a-broken-stern-2012.

Milne, A. J., Bulger, D., Herff, S., and Sethares, W. A. 'Perfect Balance: A Novel Organizational Principle for Musical Scales and Meters'. In *Mathematics and Computation in Music:*

Proceedings of the 5th International Conference, edited by T. Collins, D. Meredith, and A. Volk, 97–108. Berlin: Springer, 2015.

Milne, A. J., Carlé, M., Sethares, W. A., Noll, T., and Holland, S. 'Scratching the Scale Labyrinth'. In *Mathematics and Computation in Music: Proceedings of the 3rd International Conference*, edited by C. Agon, E. Amiot, M. Andreatta, G. Assayag, J. Bresson, and J. Mandereau, 180–195. Berlin: Springer, 2011.

Milne, A. J., and Dean, R. T. 'Computational Creation and Morphing of Multilevel Rhythms by Control of Evenness'. *Computer Music Journal* 40, no. 1 (2016): 35–53.

Milne, A. J., Herff, S. A., Bulger, D., Sethares, W. A., and Dean, R. 'XronoMorph: Algorithmic Generation of Perfectly Balanced and Well-Formed Rhythms'. In *Proceedings of the International Conference on New Interfaces for Musical Expression*, 388–393. Brisbane, 2016.

Milne, A. J., and Rolph, S. 'TiHYL'. Stern Brocot Band (V. Angelis, S. Holland, A. J. Milne, S. Rolph). *SoundCloud*, 2014. https://soundcloud.com/andrew-j-milne/tihyl-studio-version-2014.

Milne, A. J., Sethares, W. A., and Plamondon, J. 'Tuning Continua and Keyboard Layouts'. *Journal of Mathematics and Music* 2, no. 1 (2008): 1–19.

Miranda, E. R., and Biles, J. A., eds. *Evolutionary Computer Music*. London: Springer, 2007.

Mongoven, C. P. 'Sonification of Multiple Fibonacci-Related Sequences'. *Annales Mathematicae et Informaticae* 41 (2013): 175–192.

Morrison, G. '88 Cent Equal Temperament'. *Xenharmonikôn* 15 (1993).

Nelson, G. L. 'Real Time Transformation of Musical Material with Fractal Algorithms'. *Computers and Mathematics with Applications* 32, no. 1 (1996): 109–116.

Nettheim, N. 'On the Spectral Analysis of Melody'. *Interface: Journal of New Music Research* 21 (1992): 135–148.

Nierhaus, G. *Algorithmic Composition*. Vienna and New York: Springer, 2009.

Papert, S. *Mindstorms: Children, Computers, and Powerful Ideas*. New York: Basic Books, 1980.

Prechtl, A., Milne, A. J., Holland, S., Laney, R., and Sharp, D. B. 'A MIDI Sequencer that Widens Access to the Compositional Possibilities of Novel Tunings'. *Computer Music Journal* 36, no. 1 (2012): 42–54.

Pressing, J. 'Maps as Generators of Musical Design'. *Computer Music Journal* 12, no. 2 (1988): 35–46.

Prusinkiewicz, P. 'Score Generation with L-Systems'. In *Proceedings of the 1986 International Computer Music Conference*, 455–457. San Francisco, CA: International Computer Music Association, 1986.

Prusinkiewicz, P., and Lindenmayer, A. *The Algorithmic Beauty of Plants*. New York: Springer, 1990.

Quinn, I. *A Unified Theory of Chord Quality in Equal Temperaments*. PhD dissertation, University of Rochester, 2004.

Rahn, J. 'Asymmetrical Ostinatos in Sub-Saharan Music: Time, Pitch, and Cycles Reconsidered'. *Theory Only: Journal of the Michigan Music Theory Society* 9, no. 7 (1986): 23–27.

Rohrmeier, M. 'Towards a Generative Syntax of Tonal Harmony'. *Journal of Mathematics and Music* 5, no. 1 (2011): 25–53.

Soddell, F., and Soddell, J. 'Of Lindenmayer Systems, Fungi and Music.' In *ACMC05 Generate and Test: Proceedings of the Australasian Computer Music Conference 2005*, 143–148. Brisbane, 2005.

Unemi, T., and Bisig, D. 'Playing Music by Conducting BOID Agents: A Style of Interaction in the Life with A-Life'. In *Proceedings of A-Life IX*, 546–550. Cambridge, MA: MIT Press, 2004.

Unemi, T., and Bisig, D. 'Music by Interaction among Two Flocking Species and Human'. In *Proceedings of the Third International Conference on Generative Systems in Electronic Arts*, 171–179. Melbourne, 2005.

Voss, R. F., and Clarke, J. "1/*f* Noise' in Music: Music from 1/*f* Noise'. *Journal of the Acoustical Society of America* 63, no. 1 (1978): 258–263.

Vuza, D. T. 'Supplementary Sets and Regular Complementary Unending Canons (Part Three)'. *Perspectives of New Music* 30, no. 2 (1992): 102–124.

White, G. 'Fibonacci'. On *Present Tense*. Australysis Electroband (R. Dean, S. Evans, G. White). Tall Poppies TP109, 1997. CD.

Will, U., and Ellis, C. 1996. 'A Re-analyzed Australian Western Desert Song: Frequency Performance and Interval Structure'. *Ethnomusicology* 40, no. 2 (1996): 187–222.

Wilson, E. 'Letter to John Chalmers Pertaining to Moments-of-Symmetry/Tanabe Cycle'. 1975. http://www.anaphoria.com/mos.pdf.

Wilson, E. 'The Golden Horograms of the Scale Tree'. 1997. http://www. anaphoria.com/hrgm. PDF.

Wishart, T. 1983. *On Sonic Art*. London: Gordon and Breach.

Xenakis, I. 1971. *Formalized Music: Thought and Mathematics in Composition*. Bloomington: Indiana University Press.

....

CHAPTER 12

THE MACHINE LEARNING ALGORITHM AS CREATIVE MUSICAL TOOL

....

REBECCA FIEBRINK AND BAPTISTE CARAMIAUX

12.1 INTRODUCTION

MACHINE learning algorithms lie behind some of the most widely used and powerful technologies of the twenty-first century so far. Accurate voice recognition, robotics control, and shopping recommendations stand alongside YouTube cat recognizers (Le et al. 2012) as some of machine learning's most impressive recent achievements. Like other general-purpose computational tools, machine learning has captured the imaginations of musicians and artists since its inception. Sometimes musicians politely borrow existing machine learning algorithms and use them precisely as they were intended, providing numerous well-chosen examples of some phenomenon and then using an appropriate algorithm to accurately model or recognize this phenomenon. Other times, musicians break the rules and use existing algorithms in unexpected ways, perhaps using machine learning not to accurately model some phenomenon implicit in the data but to discover new sounds or new relationships between human performers and computer-generated processes. In still other cases, music researchers have formulated their own new definitions of what it means for a computer to learn, and new algorithms to carry out that learning, with the specific aim of creating new types of music or new musical interactions.

12.1.1 What Is This Chapter?

This chapter draws on music, machine learning, and human-computer interaction to elucidate an understanding of machine learning algorithms as creative tools for music and the sonic arts. Machine learning algorithms can be applied to achieve autonomous computer generation of musical content, a goal explored from various perspectives in other chapters of this book. Our main emphasis, however, is on how machine learning algorithms support distinct human-computer interaction paradigms for many musical activities, including composition, performance, and the design of new music-making technologies. By allowing people to influence computer behaviours by providing data instead of writing program code, machine learning allows these activities to be supported and shaped by algorithmic processes. This chapter provides new ways of thinking about machine learning in creative practice for readers who are machine learning novices, experts, or somewhere in between. We begin with a brief overview of different types of machine learning algorithms, providing a friendly introduction for readers new to machine learning and offering a complementary perspective for readers who have studied these algorithms within more conventional computer science contexts. We will then motivate a new understanding of learning algorithms as human-computer interfaces. We show that, like other interfaces, learning algorithms can be characterized by the ways their affordances intersect with goals of human users. We also argue that the nature of interaction between users and algorithms impacts the usability and usefulness of those algorithms in profound ways. This human-centred view of machine learning motivates our concluding discussion of what it means to employ machine learning as a creative tool.

12.1.2 Learning More about Machine Learning and Music

A single chapter is insufficient to properly cover the use of machine learning, even within music! We will not discuss machine learning outside the context of electronic, electroacoustic, and/or experimental music creation. Readers with a more general interest in machine learning for music recommendation and analysis should investigate the literature in music information retrieval, particularly the proceedings of the International Society for Music Information Retrieval (ISMIR) conference. Those interested in algorithmic learning and re-creation of Western classical music might find David Cope's work stimulating (e.g. Cope 1996). Related challenges include machine learning of musical accompaniment (e.g. Raphael 2001) and expressive rendering of musical scores (e.g. the Rencon workshop, Hiraga, Bresin, Hirata, and Katayose 2004). Finally, readers who are new to machine learning and interested in learning more (albeit from a conventional, not arts-centric perspective) might use textbooks by Witten and Frank (2005) for a practical introduction, or Bishop (2006) for a more thorough treatment. Such resources will be helpful in addressing practical challenges, such as understanding the differences between learning algorithms, or understanding how to improve the accuracy of a given algorithm. However, any creative practitioner will also be well served by hands-on experimentation with machine learning and healthy scepticism for any official wisdom on how learning algorithms 'should' be used.

12.2 Machine Learning as a Tool for Musical Interaction

One significant advantage of machine learning is that it allows us to tackle increasingly complex musical scenarios by leveraging advances in computation and/or data resources. In this section, we begin by describing at a very high level the utility of machine learning for these types of scenarios and by introducing the basic terms needed to describe the learning process. We then provide a practical perspective on how different families of algorithms—each with its own approach to learning from data—allow us to achieve common types of musical goals.

12.2.1 From Executing Rules to Learning Rules

Creating algorithms for music making can be thought of as defining rules that will subsequently drive the behaviour of a machine. For instance, mapping rules can be defined between input data values (e.g. sounds or gestures performed by a human musician) and output values (e.g. sounds produced by the computer). Although explicitly defining these rules provides complete control over the elements in play, there are complex situations in which execution rules cannot be defined explicitly or where defining an exhaustive set of rules would be too time-consuming.

An alternative approach is to learn these rules from examples. For instance, a gesture-to-sound mapping can be defined by providing examples of input gestures, each paired with the output sound that should be produced for that gesture. Using a learning algorithm to learn these rules has several advantages. First, it can make creation feasible when the desired application is too complex to be described by analytical formulations or manual brute force design, such as when input data are high-dimensional and noisy (as is common with audio or video inputs). Second, learning algorithms are often less brittle than manually designed rule sets; learned rules are more likely to generalize accurately to new contexts in which inputs may change (e.g. new lighting conditions for camera-based sensing, new microphone placements or acoustic environments for audio-based sensing). Third, learning rules can simply be faster than designing, writing, and debugging program code.

12.2.2 Learning from Data

A learning algorithm builds a *model* from a set of training examples (the training set). This model can be used to make predictions or decisions, or to better understand the structure of the data. Its exact nature depends on the type of learning algorithm used, as we explain below. A training dataset typically consists of many example data points, each represented as a list of numerical *features*. A feature can be thought of as a simple, informative measurement of the raw data. For example, an audio analysis system might describe each audio example using features related to its pitch, volume, and timbre. A gesture analysis system might describe each example human pose using (x, y, z) coordinates of each hand in three-dimensional space. Much research has considered the problem of choosing relevant features for modelling musical audio, gesture, and symbolic data (see, for example, the proceedings of the ISMIR and NIME conferences).

12.2.2.1 *Supervised Learning*

In supervised learning (Figure 12.1), the algorithm builds a model of the relationship between two types of data: input data (i.e., the list of features for each example) and output data (also sometimes called 'labels' or 'targets'). The training dataset for a supervised learning problem contains examples of input–output pairs. Once the model has been trained, it can compute new outputs in response to new inputs (Figure 12.2).

For example, consider a musician who would like to associate different hand positions, captured by a video camera, to different notes played by a computer. The musician can construct a training set by recording several examples of a first hand position and labelling each one with the desired note, for instance A♯. She can then record examples

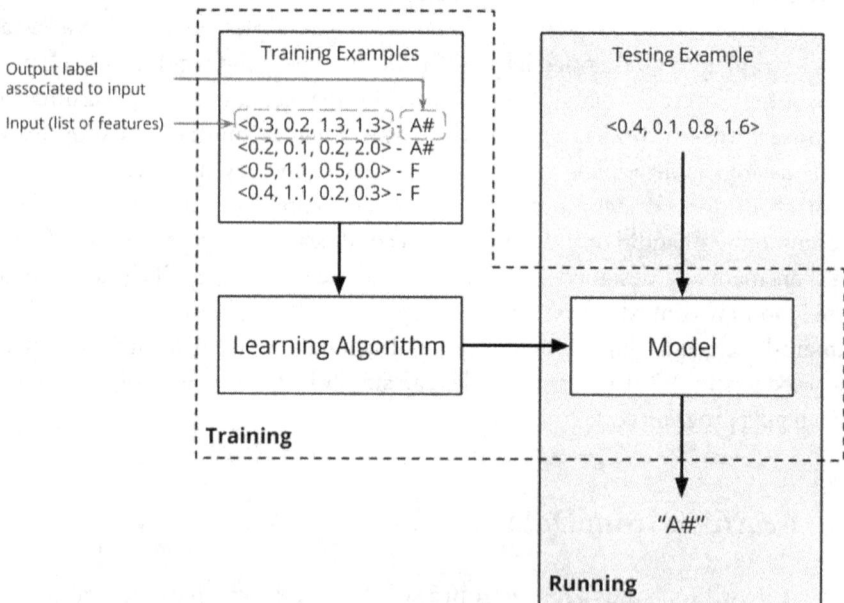

FIGURE 12.1 Supervised learning: The dashed line surrounds the training process, in which a model is built from training examples. The shaded box denotes running of the trained model, where outputs are produced in response to new inputs.

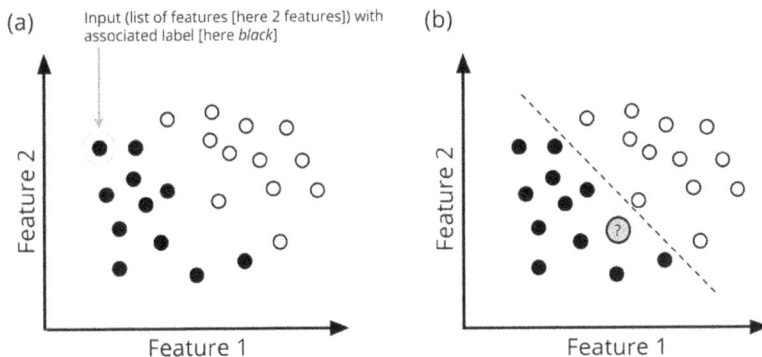

FIGURE 12.2 A classifier is trained on examples, which are each labelled with a class. (a) A train-ing dataset, where each example is a point whose position is determined by the value of its two features and whose colour is determined by its class label (black or white). (b) A classification model can be understood as a partitioning of the input space into regions corresponding to each class, separated by a decision boundary (shown here as a dashed line). When a new example is seen (corresponding to the point '?'), the classifier assigns it a label according to its position rela-tive to the decision boundary. This new point will be labelled as black.

of a second hand position, labelling each with another note, for instance F. The training process will learn what distinguishes an A♯ hand position from an F hand position. After training is complete, the musician can perform these two hand positions and use the trained model to label them as A♯ or F.

If the model outputs are categories (e.g. labels 'A♯' or 'F'), the task is typically called 'classification' or 'recognition'. If they are continuous values (e.g. if the model is to exe-cute a smooth 'glissando' from A♯ to F as the performer's hand moves from one position to the other), the task is usually called 'regression'.

12.2.2.2 *Unsupervised Learning*

In unsupervised learning (Figure 12.3), the algorithm learns the internal structure of the input data only; no corresponding output labels are provided. A musician might employ unsupervised learning simply to discover structure within the training set, for example to identify latent clusters of perceptually similar sound samples within a large sample database or to identify common chord progressions within a database of musical scores. A musician might employ this learned structure to generate new examples similar to those in the database. Or she might use this learned structure to provide better feature representations for further supervised learning or other processing. We return to this topic in section 12.2.3.4.

Consider again our musician who wants to play music by executing different hand positions in front of a video camera. However, this time she does not know beforehand how to define a suitable feature representation for hand positions. She might show the computer examples of different hand positions, without any note labels, and use an unsupervised algorithm to identify latent clusters of similar hand positions. She can

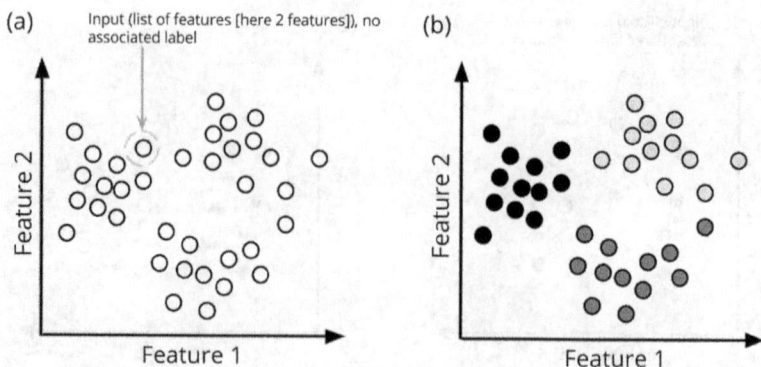

FIGURE 12.3 Unsupervised learning. (a) An unsupervised learning algorithm trains from a dataset where each example's feature values are known (here, the x and y positions of each point) but no class membership or other output information is given. (b) An unsupervised algorithm for identifying clusters of similar examples might assign these points to three clusters according to the shading here.

then compute the most likely cluster for any new hand example and use this as an input feature in her new hand-controlled instrument.

12.2.2.3 *Other Types of Learning*

Although most uses of machine learning in music employ supervised or unsupervised learning algorithms, other algorithmic families exist. For example, in semisupervised learning, some training examples include output labels but others do not. This approach is motivated by the fact that providing output labels for every input in the training set can be difficult and time consuming. Our hand-position instrument designer might create a large unlabelled example set by moving her hand in front of the camera without providing any additional information, then select a few still images from this dataset and add labels specifying what note should be played for those hand positions. She might then apply a semisupervised learning algorithm to build her hand-position classifier, with the algorithm using the labelled examples to learn how inputs should be matched to notes, but also benefitting from the numerous unlabelled examples that provide further information about the nature of inputs it is likely to encounter.

In reinforcement learning, an algorithm learns a strategy of action to maximize the value of some reward function. This reward could be an explicit value specified by a human user in response to each action of the algorithm. A simple example is a melody-generation program that could be trained to produce 'good' melodies by a human user who presses a 'thumbs up' button (positive reward value) when he likes a melody and a 'thumbs down' (penalty or negative reward value) when he dislikes the melody. The reward could alternatively be computed, for instance using an estimate of how well the melody fits with current musical material generated by human collaborators.

12.2.3 Algorithms and Musical Applications

In the previous section, we laid out the most basic ideas behind how different learning algorithms learn from data. Numerous textbooks describe how specific algorithms actually accomplish this learning, so we refer readers interested in such details to the resources mentioned in section 12.1.2. Here, though, we turn to a discussion of how these general approaches to machine learning can be matched to different types of musical goals. Specifically, we explore five goals that are relevant to many musical systems: Recognize, Map, Track, Discover New Data Representations, and Collaborate.

12.2.3.1 *Recognize*

Many types of musical systems might take advantage of a computer's ability to recognize a musician's physical gestures, audio patterns, or other relevant behaviours. Musically, recognizing such behaviours enables the triggering of new musical events or modes. To accomplish this, supervised learning algorithms can be used to perform classification.

Such interaction can be used to create new gesturally controlled musical instruments. For instance, Modler (2000) describes several hand-gesture controlled instruments that use neural networks to classify hand positions. Specific hand positions—measured using a sensor glove—were used to start and stop sound synthesis, excite a physical model, or select modes of control for a granular synthesis algorithm. Gesture recognition can also be used to augment performance on existing musical instruments. For instance, Gillian and Nicolls (2015) use an adaptive naïve Bayes classifier to recognize a set of pianist postures. Recognition of these postures during live performance allows the machine to react to gestures outside a musician's usual vocabulary.

Other types of musical systems might take advantage of real-time machine recognition of higher-level characteristics of musical audio, such as pitch, chord, tempo, structure, or genre. Implementing such recognition systems using only computer programming and signal processing can be extremely difficult, due to the complex relationships between these semantic categories and an audio signal or its spectrum. Substantial research, including much work published at the ISMIR conference, has demonstrated the potential for classification algorithms to more accurately model these relationships. By making audio understandable to machines in human-relevant terms, such classifiers can form useful building blocks for generating musical accompaniment, live visuals, or systems that otherwise respond sensitively to human musicians.

12.2.3.2 *Map*

Machine learning can also be used to map input values in one domain to output values in the same or another domain. Mapping has been widely investigated in the creative domain of gestural control of sound synthesis, where properties of a musician's gesture are measured with sensors and mapped to control sound synthesis parameters (Wanderley and Depalle 2004). Other musical applications include the generation of images from sound or vice versa (Fried and Fiebrink 2013), and the creation of sound-to-sound mappings for audio mosaicing and timbre remapping (Stowell 2010).

Designing a mapping function to generate outputs in response to inputs is a difficult task, especially for the many musical applications in which inputs and outputs are high-dimensional. The space of possible mappings is enormous, and it can be hard to know what form a mapping should take in order to satisfy the higher-level goals of the system designer, such as the creation of a musically expressive gestural controller or an aesthetically pleasing music visualization. Supervised learning algorithms are often appropriate to this type of musical challenge, since examples of inputs and corresponding outputs can be provided together. In the common case where output values are continuous, the creation of a mapping can be achieved using regression.

Supervised learning has been used to create mappings for gestural control over sound since the early work of Lee, Freed, and Wessel (1991), employing neural networks to gestural control of audio. By recording training examples in realtime as a person moves while listening to sound, training sets can be constructed to match a user's corporeal understanding of how gesture and sound should relate in the trained instrument (Fiebrink et al. 2009; Françoise, Schnell, and Bevilacqua 2013).

12.2.3.3 *Track*

Some systems that respond to human actions do more than just recognize that an action has occurred or map from a human state onto a machine behaviour. Musical applications can benefit from the machine understanding *how* an action is performed, that is, by tracking an action and its characteristics over time. After all, in many forms of musical activity, it is the dynamics of an action that communicate musical expression and expertise.

Score following is one common type of tracking problem, in which the goal is to computationally align a human's musical performance—specifically, the audio signal of this performance—to a musical score. Score following allows electronic events to be synchronized to an acoustic piece whose performance is subject to expressive changes by a human performer (e.g. Cont 2010). Synchronizing real-time sensor data to a template can also be used in the creation of new gestural controllers. For instance, work by Bevilacqua and colleagues (2010) performs real-time alignment of a gesture onto template gestures from a given vocabulary. Musically speaking, gesture alignment allows the machine to respond appropriately to the timing of a human performer, for instance playing in synchrony with the downbeat of a conductor (Wilson and Bobick 2000), or scrubbing playback position of an audio sample using the position within a gestural template (Bevilacqua et al. 2011). These types of tracking applications typically employ learning algorithms that are capable of modelling sequences of events, such as hidden Markov models or dynamic time warping. These algorithms are typically trained on user-provided examples of reference gestures or audio.

12.2.3.4 *Discover New Data Representations*

We can also employ learning algorithms to uncover structure within a collection of sound samples, musical scores, recordings of human motions, or other data. Unsupervised algorithms can uncover latent clusters of similar items, as well as re-map items into lower-dimensional spaces that preserve certain structural properties. These techniques are frequently used to facilitate human browsing and navigation of datasets. For example, self-organizing maps have been used to create two-dimensional interfaces for audio browsing (Sebastian et al. 2008) and real-time audio playback (Smith and Garnett 2012), in which perceptually similar sounds appear near each other in space. Supervised approaches such as metric learning can also be used to guide the learned representation to more closely match a user's perception of similarity between sounds or other data items (e.g. Fried., Jin, Oda, and Finkelstein 2014).

Discovering representations that succinctly account for the types of variation and structure present in a dataset can facilitate more accurate machine learning on subsequent tasks (e.g. recognition, mapping, following), when these representations are used as features for the data. For example, Fasciani and Wyse (2013) apply self-organizing maps to human gesture examples in order to establish a gesture feature representation to use in mapping within a new digital musical instrument. Fried and Fiebrink (2013) demonstrate how unsupervised deep feature learning can be applied to two domains—such as gesture and audio, or music and image—in order to subsequently build mappings between them. Such work follows a larger trend in machine learning, in which advances in machine learning of features are driving improvements in many applications involving analysis of rich media, including speech (Hinton et al. 2012), video (Mobahi, Collobert, and Weston 2009), and music (Humphrey, Bello, and LeCun 2012).

12.2.3.5 *Collaborate*

Another category of musical applications involves the computer taking on a role more similar to a human musical collaborator, imbued with 'knowledge' of musical style, structure, or other properties that may be difficult to represent using explicit rules. Building an artificial musical collaborator often requires the computer to understand the sequences of human actions taken in music performance and/or to generate appropriate sequences itself. Therefore, learning algorithms for probabilistic modelling of sequences have long been used in this context. These include Markov processes and their extensions, including hidden Markov models (HMMs) and hierarchical and variable-length Markov models (e.g. Ames 1989; Conklin 2003).

Pachet's Continuator, for example, learns musical sequence patterns from a training set using variable-order Markov models (Pachet 2003). During performance, the machine can autonomously continue sequences begun by a human musician, responding to and extending human input in a style similar to the training corpus. This type of algorithmic approach opens up several types of collaborative relationships between the algorithmic and human performers, such as mimicking the style of a famous musician or mimicking one's own style.

Assayag's Factor Oracle (Assayag and Dubnov 2004) follows a similar approach, learning patterns from a human musician's improvisation, then improvising with the musician using the same musical material and patterns. The system is not probabilistic, but based on a syntactic analysis of the music played by the improviser.

Machine learning can also explicitly train machine stand-ins for human perform-ers. Derbinsky and Essl (2012) use reinforcement learning to model rhythm sequences played by people in a collaborative digital 'drum circle'. An intelligent agent learns the drumming style of any human performer, using its degree of match with that per-son's real-time performance as the reward function to guide learning. The agent can then replace that person if they leave the drum circle. Sarkar and Vercoe (2007) use dynamic Bayesian networks (a generalization of HMMs) to model tabla sequences of human drummers to facilitate collaboration between physically distant humans performing together over the Internet. Their system is trained to recognize a tabla player's current drumming pattern. The player's machine transmits only information about this pattern (which can be thought of as an extremely compressed version of the player's audio signal) to the distant collaborator, where the pattern's audio is synthe-sized locally.

In other work, we (Fiebrink et al. 2010) have written about the role of general-purpose supervised learning algorithms as collaborative partners for exploring new musical instrument designs. Indeed, we believe that the potential for learning algorithms to sup-port human exploration, reflection, and discovery in the creation of new music and new technologies is an underrecognized strength, and we will detail this shift in perspective in the remainder of this chapter.

12.3 MACHINE LEARNING ALGORITHM AS INTERFACE

Different machine learning algorithms present different assumptions about what it means to learn and how data can be used in the learning process, and it is not always clear which algorithm is best suited to a problem. Algorithms can be stymied by noisy data, by too little data, by poor feature representations. Computational perspectives on these challenges are easy to find in machine learning textbooks, as well as in the machine learning research literature. However, purely computational perspectives on what learn-ing means and how it can be algorithmically accomplished are insufficient for under-standing machine learning's potential and consequences as a tool for music creation. In this section, we describe how applied machine learning can be understood as a type of *interface*—not a graphical user interface, but a more fundamental relationship between humans and computers, in which a user's intentions for the computer's behaviour are mediated through a learning algorithm and through the model it produces.

By understanding a human interacting with a machine learning algorithm as just another scenario in which a human interacts with a computer, we can bring concepts, methodologies, and value systems from human-computer interaction (HCI) to bear on applied machine learning. We begin with a consideration of the interactive affordances of learning algorithms in relation to people creating new musical systems.

12.3.1 Affordances of Learning Algorithms

The term 'affordance' was coined by the perceptual psychologist J. J. Gibson (1977), and it is now used in HCI to discuss the ways in which an object—such as a software program, a user-interface element, a chair—can be used by a human actor. Gaver (1991) defines affordances for an HCI readership as 'properties of the world defined with respect to people's interaction with it'. 'Most fundamentally, affordances are properties of the world that make possible some action to an organism equipped to act' (1991, 80). McGrenere and Ho (2000), writing about the historical use and adaptation of the concept within the HCI community, enumerate several 'fundamental properties' of an affordance. First among these is the fact that 'an affordance exists relative to the action capabilities of a particular actor'. That is, an affordance is not a property of an object (or a human-computer interface) in isolation; it is a property of an object in relation to a specific person using that object, with their specific abilities, goals, and context of use. Furthermore, 'the existence of an affordance is independent of the actor's ability to perceive it'.

McGrenere and Ho show how the concept of affordances can be used to frame discussion about the usefulness and usability of human-computer interfaces. They argue that the usefulness of an interface is essentially linked to the existence of necessary affordances, whereas the usability of an interface is influenced by the ease with which a user can undertake an affordance and the ease with which they can perceive it.

We draw on this concept to explore the usefulness and usability of machine learning in creative musical contexts. In such contexts, an affordance refers to the ways in which properties of a machine learning algorithm match the goals and abilities of a particular composer, performer, or musical instrument designer. The presence and nature of affordances thus help us to understand when and how machine learning can be useful to such users. Examining these affordances also allows us to compare alternative algorithms according to the degree to which they match particular users' goals (i.e., their usefulness), to consider the ways in which affordances are made understandable and accessible to users (i.e., their usability), and to envision new machine learning algorithms and tools that improve usefulness and usability by providing new or easier-to-access affordances.

In the following section, we examine affordances that are especially relevant to composers, performers, and instrument designers.

12.3.1.1 *Defining and Shaping Model Behaviour through Data*

A machine learning algorithm exposed via an appropriate software interface affords a person the ability to build a model from data, without having to describe the model explicitly in rules or code. The existence of this affordance is fundamental to the usefulness of learning algorithms for the many musical applications described above. As discussed earlier, supervised learning algorithms afford people to employ a training dataset to communicate intended relationships between different types or modalities of data, and unsupervised algorithms afford people to use data to communicate example behaviours or other properties the computer must mimic, build upon, or represent.

Most general-purpose learning algorithms employ a few basic assumptions about the training dataset, providing opportunities for users to manipulate the nature of the trained models through changes to the data. Many algorithms assume that the training set is in some sense 'representative' of data that the model will see in the future; if there are relatively more examples of a certain 'sort' in the training set, this can be interpreted as a likelihood that the model will see relatively more examples of this sort in the future. (We are glossing over all the technical details; see Bishop 2006 for a more respectable treatment.) This property can be misused in delightful ways; for example, a composer can communicate that a model's performance on some sort of input data is more important simply by providing more examples of that sort.

At the same time, one can imagine other types of musical goals that a person might communicate easily through example data, which are not taken advantage of by general-purpose algorithms. For instance, 'Don't generate anything that sounds like this!' or 'These are the body motions that are comfortable and visually evocative; make sure I can use these in musically interesting ways in my new instrument.' The design of new learning algorithms for musical applications might thus be motivated by a desire to support new useful affordances rather than only by more conventional computational goals such as accurate and efficient modelling.

Many general-purpose machine learning algorithms perform best with a lot of training data, especially when they are applied to difficult problems or to data with many features. However, it may be difficult for musical users to obtain or generate a large number of training examples for the problem of interest (e.g. building a recognizer for novel gestures). A variety of strategies have therefore emerged to afford the creation of models from small datasets. Properties of a model that would usually be tuned according to the data can instead be pre-defined, as is done in the Gesture Follower by Bevilacqua and colleagues (2010). Or, models can be trained on larger, more general-purpose datasets where available and then tuned using data from an individual user via transfer learning algorithms, as demonstrated by Pardo, Little, and Gergie (2012) in their work building personalizable audio equalizers. Other strategies include regularization and interactive machine learning, both discussed below. Alternatively, when a user's intention is not to build models that generalize but rather to use models as a basis for exploring some musical or interactive space, small datasets can suffice (Caramiaux et al. 2014; Françoise 2015).

12.3.1.2 *Exposing Relevant Parameters*

Different learning algorithms expose different control parameters—configurable properties that affect the training process and the characteristics of the trained model. Many algorithm parameters are notoriously difficult to set using human intuition alone: for example, the parameters of support vector machines (SVMs) (including kernel choice, complexity parameter, others; see Witten and Frank 2005) may substantially impact the ability to accurately model a dataset.

People using machine learning in musical contexts often care about properties of models that cannot be measured adequately with automated empirical tests, nor easily manipulated via the choice of training dataset (Fiebrink, Cook, and Trueman 2011). Sometimes, learning algorithms can expose parameters that afford users more direct control over properties they care about. Algorithms that train using iterative optimization often present a trade-off between training time and accuracy; Fiebrink et al. (2009) show how a user interface can allow musicians to exercise high-level heuristic control over this trade-off when training during live performance or similarly time-sensitive contexts. As another example, some algorithms offer *regularization* to control the degree to which a model fits a training set. In classification, this can be understood as controlling the smoothness or complexity of the decision boundary (Figure 12.4). Regularization can prevent a classifier's output from changing in unpredictable ways as a user smoothly changes its input (e.g. moving from A to B in Figure 12.4), but it can also prevent a model from accurately handling examples similar or identical to those in the training set. Françoise (2015) describes methods for regularization of probabilistic models of human movement that allow users to experimentally adjust models' fit, while also enabling models to be learned from small training sets.

Some work has explored parameterizing algorithms in ways that are specifically meaningful to musicians. For example, Morris, Simon, and Basu (2008) provide users with ability to manipulate a 'happy factor' and a 'jazz factor' in tuning a hidden Markov model chord generator. Pachet's Continuator (2003) affords tuning of the extent to which a music sequence generator is governed by patterns in the training set versus being influenced by the immediate performance context (e.g. key area or volume).

12.3.1.3 *Modelling Temporal Structure*

The organization of sound over different timescales is a defining characteristic of music. For instance, musical concepts such as phrase, form, and style can be defined only

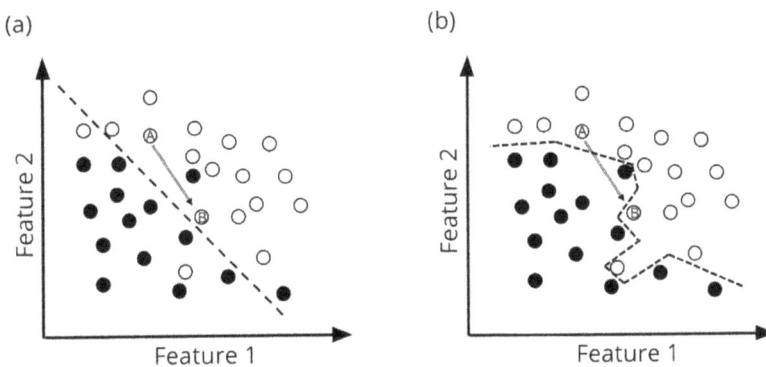

FIGURE 12.4 Regularization affects boundary complexity. (a) Greater regularization leads to smoother decision boundaries, possibly at the expense of inaccurately modelling certain training examples. (b) Less regularization leads to more jagged decision boundaries, possibly enabling greater sensitivity to the training data. In this example, moving from A to B now crosses the decision boundary twice.

with regard to patterns of sound over time. Likewise, the physical movement involved in activities such as conducting or playing an instrument entails the expressive execution of patterns of movements over time. Different learning algorithms vary in their approaches to modelling structure over time (if they model it at all); as such, models produced by different algorithms afford different types of interactions between humans, machines, and sound over time.

When machine learning is applied in a musical context, each data point often represents a single, brief instant in time. For models produced by many general-purpose algorithms, times in the past or future are irrelevant to how the model interprets data at the current time. For example, a neural network mapping that is trained to produce a given sound in response to a performer's body pose will produce the same sound whenever that pose is encountered, regardless of the way the musician is moving through that pose. Such a model affords a tight relationship between gesture and sound over time that is similar to that which occurs when a musician performs on an acoustic instrument.

On the other hand, this type of model is incapable of responding to the human as if he were a conductor. A conductor moving in front of an orchestra expects very different sounds to arise from the same movements at different times in a piece, and the dynamics of her sequence of movements may be more important than her precise pose in communicating her intention. Creating this type of interaction between human and computer requires a different type of computational understanding of movement over time. For example, human motion over time can be modelled by Markov processes (e.g. Bevilacqua et al. 2010). A first-order Markov model learns transition probabilities that describe the likelihood of moving to a particular next position (called a 'state'), given the previous position. Such a model affords interactions that rely on the computer remembering a sequence of past human actions and using them to anticipate the next actions. Musical sequences of pitches and chords, for example, can also be modelled by Markov processes. In such contexts, temporal modelling affords the creation of computational systems that learn notions of melodic or harmonic style from a set of examples.

A number of strategies have been devised to simultaneously account for both low- and high-level temporal structure, as both are important to many musical phenomena. One approach is to use hierarchies of Markov models (or their 'hidden' variants), for example devoting different levels to pitch (low-level) and phrase (higher-level) of a melody (Weiland., Smaill, and Nelson 2005). In gesture modelling, hierarchical levels can afford modelling of individual gestures as well as transitions between gestures (which also enables segmentation between gestures; see Caramiaux, Wanderley, and Bevilacqua 2012). Task-specific hierarchy definitions can also be employed; for example, Françoise, Caramiaux, and Bevilacqua (2012) decompose a physical gesture for musical control into four phases: preparation, attack, sustain, and release. A Markov model is learned for each of these phases, and the system can follow a musician as he switches between these phases in order to control phase-specific sound outputs.

General-purpose algorithms that learn temporal structure in the input data usually try to become robust to variability in the data, whether temporal, spatial, or other. Variability is usually considered as noise and modelled as such. However, such variability can also be seen as a form of intentional expression by a human user, and algorithms capable of recognizing and responding to variability afford the user possibility for new types of exploration and control. In speech, for instance, prosody is defined as the way a sentence is said or sung. Synthesis of expressive speech or singing voice exploits these potential variations in intonation. Similarly, musicians' timing, dynamics, and other characteristics vary across performances (and across musicians), and recent techniques allow for modelling the temporal structure of musical gesture while also identifying its expressive variations (Caramiaux et al. 2014).

12.3.1.4 *Running and Adapting in Realtime*

In many musical contexts, trained models have to respond in realtime to the actions of a performer. It is therefore often necessary to choose or customize machine learning algorithms so that the models they produce afford sufficient real-time responsiveness.

One unavoidable challenge is that temporal models that analyse real-time sequences of inputs may have to continually respond and adapt to a sequence before the sequence has completed. A system for real-time accompaniment or control may need to generate sound while a performer plays or moves, rather than waiting for a phrase or gesture to finish. This can necessitate a change in computational approach compared to offline contexts. For example, algorithms that analyse a sequence as it unfolds in time (such as the forward-inference algorithm for HMMs) may be less accurate than algorithms that have access to the full sequence (e.g. Viterbi for HMMs; see Rabiner and Juang 1989).

Training or adapting models to new data typically requires much more processing power than running a pre-trained model. However, some systems do manage to adapt in realtime, even during performance, through clever algorithm design or through strategically constructing real-time interactions to accommodate the time needed for learning. For instance, Assayag and colleagues (2006) describe a machine improvisation system in which the machine learns a sequence model from human musicians' playing and uses that model to generate its own sequences. They use a factor oracle (a type of variable-order Markov chain) to model sequences, structured so that the system is capable of efficiently learning and generating sequences in real-time performance.

Even when training does not occur in a real-time performance context, the time required to train a model also impacts its interactive affordances in exciting ways, as we discuss next.

12.3.2 Interactive Machine Learning

Although a user's intentions for a learning algorithm can be embedded in his or her initial choice of training data (as mentioned above), recent work shows the usefulness of enabling the user to iteratively add training examples, train the algorithm, evaluate the trained model, and edit the training examples to improve a model's performance. Such interaction is possible when training is fast enough not to disrupt a sense of interactive flow (e.g. a few seconds). Interactive machine learning is the term first used by Fails and Olsen (2003) to describe an approach in which humans can iteratively add training examples in a freeform manner until a model's quality is acceptable; it has since come to encompass a slightly broader set of techniques in which human users are engaged in a tight interaction loop of iteratively modifying data, features, or algorithm, and evaluating the resulting model (Figure 12.5). Fails and Olsen originally proposed this approach in the context of computational image analysis, but it has since been applied to a variety of other problems, such as webpage analysis (Amershi et al. 2015), social network group creation (Amershi, Fogerty, and Weld 2012) and 'debugging' personalised systems (Groce et al. 2014).

Fiebrink et al. (2009) show how interactive machine learning can be used in music composition and instrument building. They designed a machine learning toolkit, the Wekinator,[1] that allows people to create new digital musical instruments and other real-time systems using interactive supervised learning, with user-specified model inputs (e.g. gestural controllers, audio features) and outputs (e.g. controlling sound synthesis parameters, live visuals, etc.). Instrument builders, composers, and other people using Wekinator to create interactive music systems provide training examples using real-time demonstration (e.g. a body pose paired with a sound synthesis parameter vector to be triggered by that pose). General-purpose supervised learning algorithms for classification and regression can then learn the relationship between inputs and outputs. Users evaluate trained models by running them in realtime, observing system behaviour (e.g. synthesized sounds) as they generate new inputs (e.g. body movements). Users can also iteratively modify the training examples and retrain the algorithms on the modified data.

Interactive machine learning can allow users to easily fix many system mistakes via changes to the training data. For example, if the trained model outputs the wrong sound for a given gesture, the user can record additional examples of that gesture paired with the desired sound, then retrain. This also allows users to change the behaviour of the system over time, for example iteratively adding new classes of gestures until accuracy

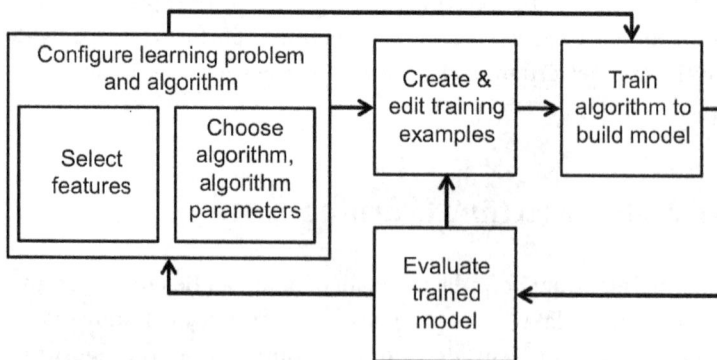

FIGURE 12.5 Interactive machine learning involves free-form iteration through different types of changes to the learning algorithm and data, followed by retraining and evaluating the modified model.

begins to suffer or there is no need for additional classes. Interactive machine learning can also allow people to build accurate models from very few training examples: by iteratively placing new training examples in areas of the input space that are most needed to improve model performance (e.g. near the desired decision boundaries between classes), users can allow complicated concepts to be learned more efficiently than if all training data were 'representative' of future data. We recommend readers see work by Fiebrink (2011, 299–303) for a more detailed discussion, and Khan, Mutlu, and Zhu (2011) for a plausible explanation combining human behavioural theory and machine learning theory.

Such interaction with algorithms can impact a musician's creative process in ways that reach beyond just producing models that are more accurate. For example, interactive machine learning using regression can become an efficient tool for exploration and for accessing unexpected relationships between human actions and machine responses. In work with composers building new gesturally controlled instruments using interactive machine learning, Fiebrink and colleagues (2010) observed a useful strategy for accessing new sounds and gesture-sound relationships, while also grounding the instrument design in the composer's own ideas: composers first decided on the 'sonic and gestural boundaries of the compositional space' (e.g. minimum and maximum [synthesis] parameter values and [gestural] controller positions), then created an initial training dataset employing a few gestures and sounds at these extremes. After training neural networks on this data, the resulting continuous regression models allowed composers to move around the gesture space, discovering new sounds in between and outside the boundaries of these 'anchor' examples.

Composers in this same study also described the value of being able to try many ideas in a short amount of time, as building instruments using machine learning was faster than writing code. Furthermore, being able to edit the training data easily meant that they could continually revise the data to reflect their own changing understanding of what the instrument should do. For example, when a composer discovered new sounds that she liked within a neural network-trained instrument, she could reinforce the presence of these sounds in the final instrument by incorporating them into new training examples.

Allowing users to evaluate trained models using exploratory real-time experimentation also allows users to judge trained models against varied and subjective criteria, such as musicality or physical comfort, and to discover information they need to improve systems' behaviour via modifications to the training examples (Fiebrink, Cook, and Trueman 2011; Zamborlin, Bevilacqua, Gillies, and d'Inverno 2014). Through iterations of modifying and evaluating models, users themselves learn how to effectively adjust the training data to steer the model behaviour in favourable ways. Also, iterative experimentation with models encourages users to reflect on the nature of the data they are providing. For instance, users building a gesture classification model can come to better understand the nature of a gesture and iteratively improve their skills in performing it, as Caramiaux, Altavilla, Pobiner, and Tanaka (2015) observed in workshops on embodied sonic interaction design.

Although conceptually attractive, involving humans in tight action-feedback loops with machine learning algorithms presents some usability challenges. Learning algorithms' inevitable mistakes can be difficult for users to understand and correct. Allowing users to act on the model itself may present opportunities for them to understand how a model works, and consequently why it sometimes fails (e.g. Kulesza et al. 2011). In a *grey-box* approach, the user has access to some parts of the internal model structure and can act on them directly. Françoise (2015) proposed a grey-box approach for creating gesture-to-sound mappings in which users can choose between models designed for gesture recognition or gesture-to-sound mapping and between instantaneous or temporal modelling. Ultimately, making machine learning more usable by novices or experts entails helping people navigate complex relationships among data, feature representations, algorithms, and models. This is a significant challenge and a topic of ongoing research across many domains (see section 12.4.2.1).

12.3.3 A Human-Centred Perspective on Machine Learning

We have presented a human-centred view of machine learning in which learning algorithms can be understood as a particular type of interface through which people can build model functions from data. We showed that these algorithms (and the models they create) provide relevant affordances for musicians, composers, and interaction designers to achieve musical goals. Different algorithms expose particular opportunities for user control, and mechanisms for users to obtain feedback to better understand the state of a model and the effects of their own actions as they build and change a model (we also refer the reader to chapter 23 by Bullock in this volume, which provides a complementary discussion on interface design in music).

A human-centred view demands that we consider the goals of the human(s) employing a learning algorithm to accomplish a particular task. While building an accurate model of a specific training dataset may be relevant to some people employing learning algorithms for music creation, it is likely that most people have other goals as well (or instead). These can include generating musically novel material within a rough space of styles, using learning algorithms to explore an unknown space of compositional possibilities, or building new instruments whose relationship between movement and sound 'feels' right to play, as the examples encode users' embodied practices better than systems designed by writing code (Fiebrink et al. 2010). Using machine learning to create interactive systems can leverage cognitive properties of musical performance, such as the role of listening in the planning and control of human actions (Caramiaux et al. 2014; Leman 2008).

There is ample room for future research to better understand the goals of people applying machine learning to musical tasks, and to develop new learning algorithms and software toolkits whose interactive affordances are better matched to these goals. Much more could be done to develop algorithms that are even easier to steer in useful directions using user-manipulatable parameters or training data, to develop mechanisms for people to debug machine learning systems when they do not work as intended, or to use algorithms that learn from data to scaffold human exploration of new ideas, sounds, and designs.

12.4 MACHINE LEARNING AS CREATIVE TOOL

Applying a human-centred perspective to the analysis of machine learning in context, as we have presented in section 12.3, shifts the focus from technical machinery to human goals and intentions. As we argue next, this shift in perspective opens up new possibilities for understanding and better supporting creative practice.

12.4.1 Roles of Machine Learning in Creative Work

Machine learning is perhaps most obviously understood as a creative tool when a model acts as a creative agent, exhibiting human-like creative behaviours. A number of works discussed above, including those by Assayag et al. (2006), Derbinsky and Essl (2012), and Pachet (2003), employ learning algorithms to generate musical material in realtime. These algorithms function as collaborators with creative agency, or even as stand-ins for other humans.

Looking more closely at such work, though, we often find that its motivations go beyond strictly replacing or augmenting human performers. Assayag and colleagues write about the desire for the machine's 'stylistic reinjection' to influence human musicians: '[A]n improvising [human] performer is informed continually by several sources, some of them involved in a complex feedback loop. ... The idea behind stylistic reinjection is to reify, using the computer as an external memory, [the] process of reinjecting musical figures from the past in a recombined fashion, providing an always similar but always innovative reconstruction of the past. To that extent, the virtual partner will look familiar as well as challenging' (2006). Pachet (2008) is motivated by a similar idea, that of 'reflexive interaction', in which the machine is trained to be an 'imperfect mirror' of the user. Through engaging with her reflections in this imperfect mirror, the user is helped to express 'hidden, ill-formulated ideas', to engage in creative expression without being stymied by limited expertise on an instrument, and to enter a state of Flow as described by Csikszentmihalyi (1991).

Learning algorithms can also be examined with regard to their affordances as *design tools*, whether they are used to create new musical material or computer behaviours during performance, composition, instrument building, or other activities.

Research in design and HCI suggests that practices such as sketching of incomplete ideas, exploration of the design space through rapid prototyping, and iterative refinement are essential for the creation of new technologies in any domain (Resnick et al. 2005). Furthermore, creators of new music technologies are often engaged with what design theorist Horst Rittel described as 'wicked' design problems: ill-defined problems wherein a problem 'definition' is found only by arriving at a solution. The 'specifications' for these new technologies—e.g. precisely what sort of music should they produce? how exactly should they interact with human performers?—are usually not known with certainty at the beginning of the design process, making prototyping and experimentation paramount in order for the designers to 'get the right design' as well as 'get the design right' (Buxton 2010).

Machine learning algorithms often naturally support these design activities (Fiebrink 2011). By enabling people to instantiate new designs from data, rather than by writing code, the creation of a working prototype can be very fast. Prototypes can be iteratively refined by adjustments to the training data and algorithm parameters. By affording people the ability to communicate their goals for the system through user-supplied data, it can be more efficient to create prototypes that satisfy design criteria that are subjective, tacit, embodied, or otherwise hard to specify in code. The data can implicitly communicate the style of a machine improviser or the feel of a digital instrument. The data can alternatively act as a rough sketch of a user's ideas, allowing instantiation of a model that allows further exploration of those ideas. Thus, machine learning can allow designers to build better prototypes, to build more of them, and to use prototypes to explore a wider space of designs than building systems by programming.

12.4.2   A Comparison with Conventional Machine Learning

We end this section by summarizing the shared aims and the divergences between a 'conventional' machine learning perspective (i.e., the perspective implicit in most machine learning textbooks) and an understanding of machine learning used as a creative tool. Both perspectives are relevant to creative practitioners who want to wield learning algorithms effectively in practice, and both can inform advances in musical machine learning research.

12.4.2.1   *Commonalities*

In both perspectives, machine learning can be seen as a powerful tool to extract information from data. Indeed, in numerous domains, machine learning algorithms are used to provide new insights into data that may otherwise be poorly understood by people. 'Big data' is driving new discoveries in scientific fields including astronomy, high-energy physics, and molecular biology (Jordan 2011), and data mining influences decision-making at companies across all sectors of the economy (Lohr 2012). As we have discussed in section 12.2.3.4, discovering latent structure in musical data can support the creation of new interfaces for human exploration of that data, although the aim of these interfaces is often to scaffold new interactions rather than to simply understand the data or make decisions from it. In music, algorithms can also lend new insight into users' own data, whether by acting as an 'imperfect mirror' that invites new types of reflection on a composer's own style (Pachet 2008), or alerting a cellist to the fact that her bowing articulation technique must be improved in order to allow an accurate classifier to be built (Fiebrink, Cook, and Trueman 2011).

Machine learning is also often used because algorithms can perform more accurately than people manually trying to build model functions or rule sets. Many learning algorithms are explicitly designed to build models that generalize well from the training data (using a rigorous definition of generalization and computational methods that can be demonstrated to achieve it). They easily outperform less theoretically rigorous human attempts to solve complex problems in domains such as computer vision, ontology creation, and robotics control. Music is full of similarly complex challenges, including making sense of musical audio, symbolic data (e.g., music scores), human motion or emotion, or any number of other problems involving semantic analysis of or control over high-dimensional, noisy, complex signals. As such, many challenges faced by musicians trying to build accurate models are similar to those faced by other machine learning practitioners. Machine learning practitioners may have to choose among many possible feature representations, learning algorithms, and parameters when building a model. A basic grasp of computational perspectives on machine learning is invaluable for choosing, implementing, and debugging machine learning techniques in any domain. Nevertheless, even expert intuition is often insufficient, and applied machine learning involves a great deal of experimentation with different features, algorithms, and so on.

Needless to say, machine learning is not magic, and users in music and beyond still encounter numerous challenges for which existing learning algorithms are just inaccurate, slow, or inapplicable. Ongoing advances in fundamental machine learning research will doubtless drive advances in musical applications as well.

12.4.2.2 *Differences*

Unlike most conventional applications, users in musical applications often have great flexibility in their choice of training data. As discussed in section 12.3.2, users can modify the training set to steer model behaviour in useful ways: fixing mistakes, adding new classes, and so on. Iterative interaction can enable learning from smaller datasets than noninteractive learning from a fixed dataset. Musicians also often adapt their own ideas for what the computer should learn based on the outcomes of their machine learning experiments. If it turns out to be too hard to teach the computer a particular concept with a limited amount of data, for example, a musician might reasonably choose to instead model a simpler concept that is almost as useful to him in practice. (Or, if it is easier than expected to teach the machine a concept, the user may choose a more difficult concept that is even more useful to him!) Whereas most machine learning practitioners might require computational techniques and software tools to efficiently compare the accuracy of models created with different algorithms and algorithm parameterizations, creative practitioners might further benefit from tools that help them diagnose how to most effectively change the training dataset, number of classes, or other characteristics of the learning problem in order for accurate modelling to take place.

When users are 'experts' in the problem being modelled (e.g. if they are the ones who will be performing with a new gestural controller), new opportunities for user evaluation of models open up. In conventional applications, model evaluation often involves running the trained model on a test dataset (separate from the training set), or partitioning the available data many times into different versions of training and test sets (i.e., 'cross-validation'). In musical applications, though, users can often generate new data on the fly and see what the trained model does with it. This type of free-form testing affords users the ability to assess models with regard to subjective criteria: for example, 'For what types of inputs is this model's behaviour most musically interesting?' or 'What are the gestures for which this classifier is likely to fail?' Fiebrink, Cook, and Trueman (2011) suggest that this free-form testing is invaluable to users in understanding how to improve models or deciding whether a model is ready to be used in performance, and that conventional metrics such as cross-validation may be poor indicators of a model's subjective quality.

Musical users' goals for learning algorithms sometimes differ from the conventional goal of creating a model that generalizes well from the given dataset. In interactive machine learning, if the user adds new training examples to steer model behaviour in a particular direction, he may prefer models that are easily influenced by new data points. This can correspond to a preference for algorithms that overfit rather than those that aim to generalize from the data (e.g. models that look like Figure 12.4b instead of Figure 12.4a)—something usually viewed as undesirable in conventional machine learning (Fiebrink, Cook, and Trueman 2011).

In music, machine learning may be also used as a way to discover new sounds or interactive relationships, and the training data may just be a way to ground this exploration in a region of the design space a user thinks is promising. In such cases, generalization may not be at all important, and learning fairly arbitrary models from just a few examples may be perfectly acceptable. When users employ a trained model for musical exploration, they may also seek out configurations of input data that look nothing like the data present in the training set. Conventional machine learning approaches tend not to be concerned with such 'long tail' configurations if the training data suggest they are not representative of the modelled population (Murphy 2012). However, from a creative perspective, such configurations may bring relevant new musical ideas; a model thus needs to take them into account as potential relevant inputs from the user, instead of treating their occurrence as an unlikely case that can be handled in a trivial manner. This brings important challenges in terms of model design, for example the need for fast adaptation to unexpected inputs from the user. Such challenges might also be relevant to advancements in machine learning concerned with a wider set of applications; in finance, for instance, rare and unanticipated events can have important consequences.

12.5 DISCUSSION

By understanding machine learning as a creative tool, used within a larger context of design practice to achieve complex and often idiosyncratic artistic goals, and in an interactive setting in which users may be able to manipulate training and testing data in dramatic ways, we can begin to imagine avenues for improving on machine learning techniques to act as better partners in creation. New algorithms and user interfaces could make it even easier to instantiate designs from data, by imposing structure on learning problems that is well-matched to the structure of particular musical tasks, as well as taking advantage of other information that users might communicate through example data. New techniques could make it even easier to explore the design space of musical technologies generated from data, to compare alternative designs, or to refine designs according to criteria that are meaningful to users.

This approach also brings scientific challenges and opportunities. A user's understanding of an algorithm's affordances can certainly be enhanced through an interactive approach to the learning phase: involving a user in teaching a model will help him to understand the model. However, the means by which a human might efficiently teach an algorithm for creative purposes remains to be explored. Moreover, the relationship between a user's perception of the quality of a model and the machine's 'perception' of its quality with regard to user-supplied inputs invites further attention. For example, Akrour and colleagues (2014) show that a learning algorithm can obtain better performance by 'trusting' the competence of a user, even when that user at first makes mistakes, because the model's performance in return impacts the user's consistency. In other words, feedback between user and computer can enable both to improve their skills.

To fully realize the potential of computers as musical tools requires taking advantage of their affordance of new interactive relationships between people and sound. Many computer music composers have written about the importance of building new human-computer relationships that transcend simple ideas of control. David Rokeby distinguishes strongly between interaction and control; his view is summarized by Lippe thus: 'if performers feel they are in control of (or are capable of controlling) an environment, then they cannot be truly interacting, since control is not interaction' (2002, 2). Robert Rowe, in his seminal book *Interactive Music Systems* (Rowe 1992), writes about the importance of feedback loops between human and machine in which each influences the other. Chadabe has proposed several metaphors for musical human-machine interaction, including systems in which interaction is 'like conversing with a clever friend' (1997, 287), or 'sailing a boat on a windy day and through stormy seas' (Drummond 2009).

Although these composers were writing about performance-time interactions between people and machines, we argue that it is productive to characterize design-time interactions in many of the same ways. It is possible to write out detailed specifications for a new musical instrument or improvisation system, implement a system to those specifications, and be done. However, how much better to be able to discover, explore, and adapt to everything that one can learn along the way! When the computer becomes a conversation partner, or a boat rocking us in unexpected directions, we may find that the technologies we build become more useful, more musical, more interesting than our original conceptions.

Machine learning allows us to forgo programming the machine using explicit rules, and instead makes it possible to create new technologies using more holistic strategies, using data to implicitly communicate goals and embodied practices. Machine learning allows us to create prototypes from half-baked ideas and discover behaviours we hadn't thought of, and to efficiently modify our designs in order to reflect our evolving understanding of what a system should do. In this, we can understand machine learning algorithms as more than a set of computational tools for efficiently creating accurate models from data. They can be wonderful stormy ships, conversation partners, imperfect mirrors, and co-designers, capable of influencing, surprising, and challenging us in many musical creation contexts.

NOTE

1. Available as open-source software at www.wekinator.org.

BIBLIOGRAPHY

Akrour, R., Schoenauer, M. Sebag, M. and Souplet, J.-C. 'Programming by Feedback'. In *Proceedings of the 31st International Conference on Machine Learning*. Beijing, 2014.

Amershi, S., Chickering, M. Drucker, S. M., Lee, B., Simard, P., and Suh, J. 'ModelTracker: Redesigning Performance Analysis Tools for Machine Learning'. In *Proceedings of the Conference on Human Factors in Computing Systems*. New York: ACM, 2015

Amershi, S., Fogarty, J., and Weld, D. 'Regroup: Interactive Machine Learning for On-Demand Group Creation in Social Networks'. In *Proceedings of the SIGCHI Conference on Human Factors in Computing Systems*, 21–30. New York: ACM, 2012.

Ames, C. 'The Markov Process as a Compositional Model: A Survey and Tutorial'. *Leonardo* 22, no. 2 (1989): 175–187.

Assayag, G., and Dubnov, S. 'Using Factor Oracles for Machine Improvisation'. *Soft Computing* 8 (2004): 1–7.

Assayag, G., Bloch, G., Chemillier, M., Cont, A., and Dubnov, S. 'OMax Brothers: A Dynamic Topology of Agents for Improvisation Learning'. In *Proceedings of the 1st ACM Workshop on Audio and Music Computing Multimedia*, 125–132. New York: ACM, 2006.

Bevilacqua, F., Zamborlin, B., Sypniewski, A., Schnell, N., Guédy, F., and Rasamimanana, N. 'Continuous Realtime Gesture Following and Recognition'. In *Gesture in Embodied Communication and Human-Computer Interaction*, edited by S. Kopp and I. Wachsmuth, 73–84. Berlin: Springer, 2010.

Bevilacqua, F., Schnell, N., Rasamimanana, N., Zamborlin, B., and Guédy, F. 'Online Gesture Analysis and Control of Audio Processing'. In *Musical Robots and Interactive Multimodal Systems*, edited by J. Solis and K. Ng, 127–142. Berlin: Springer, 2011.

Bishop, C. M. *Pattern Recognition and Machine Learning*. New York: Springer, 2006.

Buxton, B. *Sketching User Experiences: Getting the Design Right and the Right Design*. Burlington, MA: Morgan Kaufmann, 2007.

Caramiaux, B., Montecchio, N., Tanaka, A., and Bevilacqua, F. 'Adaptive Gesture Recognition with Variation Estimation for Interactive Systems'. *ACM Transactions on Interactive Intelligent Systems (TiiS)* 4, no. 4 (2014).

Caramiaux, B., Wanderley, M. M., and Bevilacqua, F. 'Segmenting and Parsing Instrumentalists' Gestures'. *Journal of New Music Research* 41, no. 1 (2012): 13–29.

Caramiaux, B., Françoise, J., Bevilacqua, F., and Schnell, N. 'Mapping Through Listening'. *Computer Music Journal* 38, no. 3 (2014): 1–19.

Caramiaux, B., Altavilla, A., Pobiner, S., and Tanaka, A. 'Form Follows Sound: Designing Interactions from Sonic Memories'. In *Proceedings of the SIGCHI Conference on Human Factors in Computing Systems*. New York: ACM, 2015

Chadabe, J. *Electric Sound: The Past and Promise of Electronic Music*. Upper Saddle River, NJ: Prentice Hall, 1997.

Conklin, D. 'Music Generation from Statistical Models'. In *Proceedings of the AISB 2003 Symposium on Artificial Intelligence and Creativity in the Arts and Sciences*, 30–35. Aberystwyth, 2003.

Cont, A. 'A Coupled Duration-Focused Architecture for Real-Time Music-to-Score Alignment', *IEEE Transactions on Pattern Analysis and Machine Intelligence* 32, no. 6 (2010): 974–987.

Cope, D. *Experiments in Musical Intelligence*. Madison, WI: AR Editions, 1996.

Csikszentmihalyi, M. *Flow: The Psychology of Optimal Experience*. New York: Harper-Perennial.

Derbinsky, N., and Essl, G. 'Exploring Reinforcement Learning for Mobile Percussive Collaboration'. In *Proceedings of New Interfaces for Musical Expression (NIME)*. Ann Arbor, MI, 2012.

Drummond, J. 'Understanding Interactive Systems'. *Organised Sound* 14, no. 2 (2009): 124–133.

Fails, J. A., and Olsen, D. R., Jr. 'Interactive Machine Learning'. In *Proceedings of the 8th International Conference on Intelligent User Interfaces*, 39–45. New York: ACM, 2003.

Fasciani, S., and Wyse, L. 'A Self-Organizing Gesture Map for a Voice-Controlled Instrument Interface'. In *Proceedings of New Interfaces for Musical Expression (NIME)*, 507–511. Daejeon, Korea, 2013.

Fiebrink, R., Cook, P. R., and Trueman, D. 'Play-Along Mapping of Musical Controllers'. In *Proceedings of the International Computer Music Conference (ICMC)*, 61–64. Montreal, 2009.

Fiebrink, R., Cook, P. R., and Trueman, D. 'Human Model Evaluation in Interactive Supervised Learning'. *Proceedings of the 2011 Annual Conference on Human Factors in Computing Systems*, 147–156. New York: ACM, 2011.

Fiebrink, R., Trueman, D., Britt, C., M. Nagai, M., Kaczmarek, K., Early, M., Daniel, M. R. Hege, A., and Cook, P. R. 'Toward Understanding Human-Computer Interaction in Composing the Instrument'. In *Proceedings of the International Computer Music Conference*. New York: International Computer Music Association, 2010.

Fiebrink, R., Trueman, D., and Cook, P. R. 'A Metainstrument for Interactive, on-the-Fly Machine Learning'. In *Proceedings of New Interfaces for Musical Expression (NIME)*. Pittsburgh, PA, 2009.

Françoise, J. *Motion-Sound Mapping by Demonstration*. PhD dissertation, Université Pierre et Marie Curie (Paris 6) and IRCAM, Paris, 2015.

Françoise, J., Caramiaux, B., and Bevilacqua, F. 'A Hierarchical Approach for the Design of Gesture-to-Sound Mappings'. In *Proceedings of the International Conference on Sound and Music Computing*, 233–240. Copenhagen, 2012

Françoise, J., Schnell, N. and Bevilacqua, F. 'A Multimodal Probabilistic Model for Gesture-Based Control of Sound Synthesis'. In *Proceedings of the 21st ACM International Conference on Multimedia*, 705–708. New York: ACM, 2013.

Fried, O., and Fiebrink, R. 'Cross-Modal Sound Mapping Using Deep Learning'. In *Proceedings of New Interfaces for Musical Expression (NIME)*, 531–534. Daejeon, Korea, 2013

Fried, O., Jin, Z., Oda, R., and Finkelstein, A. 'AudioQuilt: 2D Arrangements of Audio Samples using Metric Learning and Kernelized Sorting'. In *Proceedings of New Interfaces for Musical Expression (NIME)*, 281–286. Goldsmiths, University of London, 2014.

Gaver, W. W. 'Technology Affordances'. In *Proceedings of the SIGCHI Conference on Human Factors in Computing Systems*, 79–84. New York: ACM, 1991.

Gibson, J. J. 'The Theory of Affordances'. In *Perceiving, Acting, and Knowing: Towards an Ecological Psychology*, edited by R. Shaw and J. Bransford, 127–143. Hoboken, NJ: Lawrence Erlbaum, 1977.

Gillian, N., and Nicolls, S. 'Kinect-Ed Piano'. *Leonardo* 48, no. 3 (2015): 294–295.

Groce, A., Kulesza, T. Zhang, C. Shamasunder, S. Burnett, M. Wong, W.-K. Stumpf, S., Das, S., Shinsel, A., Bice, F., and McIntosh, K. 'You Are the Only Possible Oracle: Effective Test Selection for End Users of Interactive Machine Learning Systems'. *IEEE Transactions on Software Engineering* 40, no. 3 (2014): 307–323.

Hinton, G., Deng, L., Yu, D., Dahl, G. E., Mohamed, A.-R. Jaitly, N., Senior, A., Vanhoucke, V., Nguyen, P., Sainath, T. N., and Kingsbury, B. 'Deep Neural Networks for Acoustic Modeling in Speech Recognition: The Shared Views of Four Research Groups'. *IEEE Signal Processing Magazine* 29, no. 6 (2012): 82–97.

Hiraga, R., Bresin, R., Hirata, K., and Katayose, H. 'Rencon 2004: Turing Test for Musical Expression'. In *Proceedings of New Interfaces for Musical Expression (NIME)*, 120–123. Hamamatsu, Japan, 2004

Humphrey, E.J., Bello, J. P., and LeCun, Y. 'Moving Beyond Feature Design: Deep Architectures and Automatic Feature Learning in Music Informatics'. In *Proceedings of the International Society for Music Information Retrieval Conference (ISMIR)*, 403–408. Porto, 2012.

Jordan, M. I. 'A Message from the President: The Era of Big Data'. *ISBA Bulletin* 18, no. 2 (2011): 1–3.

Khan, F., Mutlu, B., and Zhu, X. 'How Do Humans Teach: On Curriculum Learning and Teaching Dimension'. In *Advances in Neural Information Processing Systems*, edited by J. Shawe-Taylor, R. S. Zemel, P. L. Bartlett, F. Pereira, and K. Q. Weinberger, 1449–1457. Granada, 2011.

Kulesza, T., Stumpf, S., Wong, W.-K., Burnett, M., Perona, S., Ko, A., and Oberst, I. 'Why-Oriented End-User Debugging of Naive Bayes Text Classification'. *ACM Transactions on Interactive Intelligent Systems (TiiS)* 1, no. 1 (2011): 2.

Le, Q. V., Ranzato, M. A., Monga, R., Devin, M., Chen, K., Corrado, G. S., Dean, J., and Ng, A. Y. 'Building High-Level Features Using Large Scale Unsupervised Learning'. In *Proceedings of the 29th International Conference on Machine Learning*, edited by J. Langford and J. Pineau. Edinburgh, 2012.

Lee, M., Freed, A., and Wessel, D. 'Real-Time Neural Network Processing of Gestural and Acoustic Signals'. In *Proceedings of the International Computer Music Conference (ICMC)*, 277–280. Montreal, 1991.

Leman, M. *Embodied Music Cognition and Mediation Technology*. Cambridge, MA: MIT Press, 2008.

Lippe, C. 'Real-Time Interaction among Composers, Performers, and Computer Systems'. *Information Processing Society of Japan SIG Notes* 123 (2002): 1–6.

Lohr, S. 'The Age of Big Data'. *New York Times*, 11 February 2012.

McGrenere, J., and Ho, W. 'Affordances: Clarifying and Evolving a Concept'. In *Proceedings of Graphics Interface, 2000*, 179–186. Montreal, 2000.

Mobahi, H., Collobert, R., and Weston, J. 'Deep Learning from Temporal Coherence in Video'. In *Proceedings of the 26th Annual International Conference on Machine Learning (ICML)*, 737–744. New York: ACM, 2009

Modler, P. 'Neural Networks for Mapping Hand Gestures to Sound Synthesis'. In *Trends in Gestural Control of Music*, edited by M. M. Wanderley and M. Battier, 301–314. Paris: IRCAM, 2000.

Morris, D., Simon, I., and Basu, S. 'Exposing Parameters of a Trained Dynamic Model for Interactive Music Creation'. In *Proceedings of the Twenty-Third AAAI Conference on Artificial Intelligence*, edited by A. Cohn, 2:784–791. Palo Alto, CA: AAAI, 2008

Murphy, K. P. *Machine Learning: A Probabilistic Perspective*. Cambridge, MA: MIT Press, 2012.

Pachet, F. 'The Continuator: Musical Interaction with Style'. *Journal of New Music Research* 32, no. 3 (2003): 333–341.

Pachet, F. 'The Future of Content Is in Ourselves'. *Computers in Entertainment (CIE)* 6, no. 3 (2008): article 31.

Pardo, B., Little, D., and Gergle, D. 'Building a Personalized Audio Equalizer Interface with Transfer Learning and Active Learning'. In *Proceedings of the Second International ACM Workshop on Music Information Retrieval with User-Centered and Multimodal Strategies*, 13–18. New York: ACM, 2012.

Rabiner, L. R. 'A Tutorial on Hidden Markov Models and Selected Applications in Speech Recognition'. *Proceedings of the IEEE* 77, no. 2 (1989): 257–286.

Raphael, C. 'Synthesizing Musical Accompaniments with Bayesian Belief Networks'. *Journal of New Music Research* 30, no. 1: 59–67.

Resnick, M., Myers, B., Nakakoji, K., Shneiderman, B., Pausch, R., Selker, T., and Eisenberg, M. 'Design Principles for Tools to Support Creative Thinking'. In *Proceedings of the NSF Workshop on Creativity Support Tools*, 25–36. Washington, DC, 2005.

Rowe, R. *Interactive Music Systems: Machine Listening and Composing*. Cambridge, MA: MIT Press, 1992.

Sarkar, M., and Vercoe, B. 'Recognition and Prediction in a Network Music Performance System for Indian Percussion'. In *Proceedings of the 7th International Conference on New Interfaces for Musical Expression (NIME)*, 317–320. New York: ACM, 2007.

Smith, B. D., and Garnett, G. E. 'Unsupervised Play: Machine Learning Toolkit for Max'. In *Proceedings of New Interfaces for Musical Expression (NIME)*. Ann Arbor, MI, 2012.

Stowell, D. *Making Music through Real-Time Voice Timbre Analysis: Machine Learning and Timbral Control*. PhD dissertation, School of Electronic Engineering and Computer Science, Queen Mary University of London, 2010.

Wanderley, M. M., and P. Depalle, P. 'Gestural Control of Sound Synthesis'. *Proceedings of the IEEE* 92, no. 4 (2004): 632–644.

Weiland, M., Smaill, A., and Nelson, P. 'Learning Musical Pitch Structures with Hierarchical Hidden Markov Models'. *Journées d'Informatique Musicale*, 2005.

Wilson, A. D., and Bobick, A. F. 'Realtime Online Adaptive Gesture Recognition'. In *Proceedings. International Workshop on Recognition, Analysis, and Tracking of Faces and Gestures in Real-Time Systems*, 270–275. IEEE, 2000.

Witten, I. H., and Frank, E. *Data Mining: Practical Machine Learning Tools and Techniques*. Burlington, MA: Morgan Kaufmann, 2005.

Zamborlin, B., F. Bevilacqua, F., M. Gillies, M., and d'Inverno, M. 'Fluid Gesture Interaction Design: Applications of Continuous Recognition for the Design of Modern Gestural Interfaces'. *ACM Transactions on Interactive Intelligent Systems (TiiS)* 3, no. 4 (2014): 22–50.

....

CHAPTER 13

BIOLOGICALLY INSPIRED AND AGENT-BASED ALGORITHMS FOR MUSIC

....

ALICE ELDRIDGE AND OLIVER BOWN

13.1 INTRODUCTION

....

> Whatever vibrates is a musical instrument: whatever is stable is a mechanical brain—the difficulty lies in making a particular one. (Ashby 2008, no. 158)

FOR all of humankind's creative achievements, we in turn were made by a more powerful creative force: biological evolution. Since Darwin's and Wallace's great insight (Darwin 1861), it has become widely accepted that the astonishingly beautiful and complex structure and behaviours of the living world have taken shape through a remarkable process that is mechanical, blind, and purposeless. This sublime beauty has inspired art since its beginnings, but whereas we have always incorporated natural *form* in our paintings, sculpture, and music, artists working with code now draw upon *processes* from the natural world.

The arrival of general-purpose computers in the middle of the last century transformed not only science but also compositional practice in ways that are documented throughout this book. Of importance to this chapter, it enabled us to harness *behaviours* inspired by natural systems, formalized by biologists and computer scientists into algorithms, in order to develop, perform, and compose with software instruments. We now borrow from the designs of specific biological organisms, and the properties and processes of complex natural systems, as well as from the creative mechanism of evolution itself.

In this chapter we examine a range of approaches to algorithmic music-making inspired by biological systems. In doing so we cover topics that are located at the intersection of contemporary music, computer science, and the study of creativity: optimization and problem-solving using evolutionary methods; emergence, self-organization, and complexity; adaptive behaviour; and autonomy and self-determination. Section 13.1.1 provides a brief historical context of the core intellectual, musical, and social movements which influence contemporary creative practice. Section 13.1.2 provides a primer in the concepts and tools developed for the study of systems which are foundational to the specific approaches described in the following sections.

Endeavours in this field are often hybrid and idiosyncratic, and cannot be neatly categorized. Nevertheless, we organize an overview of the key musical motivations, concepts, and computational methods of the field under four themes which map the topics above. Section 13.2, 'Evolutionary Search', outlines the application of evolutionary algorithms to design issues and opportunities associated with algorithmic music. In section 13.3, 'Multi-Agent Compositions', we look at the ways in which agent-based modelling has been used to compose emergent, self-organizing music. Section 13.4 considers how the study of adaptive behaviour has inspired the design and realization of *adaptive collaborators*—interactive software systems which begin to enable active electroacoustic partnerships. Many of these ideas come together in section 13.5, which describes the development of *creative ecosystems* inspired by ecological principles. We end by mentioning two themes that will guide future work: autonomy and agency, and the poetics of biologically inspired algorithms.

13.1.1 Intellectual Precursors

Contemporary practice in the area of biologically inspired computer music can be best understood against the backdrop of a series of interrelated intellectual, cultural, and social currents arising in the late nineteenth and early twentieth centuries. Central to these was a shift from an essentialist paradigm towards the relational thinking championed by the thinkers behind general systems theory (GST) and cybernetics. Systems thinking derives from the work of biologist Von Bertalanffy, who sought to abstract from the intractable messiness of actual biological, social, economic, chemical, and other systems in order to define *general* principles of dynamic interaction (e.g. Von Bertalanffy 1950). Similarly, cybernetics sought to understand processes of control and communication across electronic, mechanical, biological, or economic systems in terms of common principles, such as regulatory feedback. Cybernetics, expounds pioneer Ross Ashby, 'treats, not things but ways of behaving. It does not ask what is this thing? but what does it do?' (1956, 1).

These new paradigms had far-reaching influence in society and the arts as well as engineering and science, and underpin contemporary practice both conceptually and methodologically. The conceptual relationship between organisms and machines had been thoroughly explored in Western nineteenth-century thinking, typified in popular culture by publications like *Frankenstein*. Darwin's theory of natural selection (1861) sealed the direction of thinking into the twentieth century. Cybernetics made the first formal steps towards truly integrating the study of natural systems with the creation of artificial ones. Later the science of complexity and chaos—the new computational magic made famous by Mandelbrot Sets and Lorenz Attractors—in step with developments in theoretical biology, blossomed into the discipline of artificial life (ALife) (Langton 1989). In various ways, each sought to explain complex systems in terms of the interactions between the mutually inter-related parts of which they were comprised. ALife advanced the idea that complex real-world processes could be modelled computationally, imagining that we might not only recreate the phenomena of the biological world, but also uncover principles of life divorced from the biological substrate of life-as-we-know-it—'life as it could be' (Langton 1989, 1)—and in doing so, it aimed to reveal general principles of biology, both natural and artificial.

The inherently interdisciplinary and conceptually profound nature of these movements entailed a close relationship between the sciences and arts. Not only did pioneering practitioners engage across domains, but the cybernetic vision inspired a revolution throughout modern art. Recapitulating the earlier shift from essentialism in the biological sciences, Jack Burnham's 'system aesthetics' (1968) and Roy Ascott's 'behavioural tendency' (1967) drew attention from self-contained objects characteristic of modern formalism to a postmodern open-ended and immersive experimentation in which feedback loops stitched previously distinct elements of artist, artwork, and observer into an indivisible whole. This new intimacy between technology and arts was celebrated by the landmark exhibition and accompanying book, *Cybernetic Serendipity* (Reichardt 1971), which showcased computer generated work and cybernetic devices from the pioneering players: Standford Beer, Earle Brown, John Cage, Edward Ihnatowicz, Ben Lapowsky, Frieder Nake, Nam Jun Paik, Gordon Pask, Karlheinz Stockhausen, Jean Tinguely, and Iannis Xenakis.

The liberation and expansion of musical sound that was an essential part of Futurism at the start of the twentieth century had developed by midcentury into a radical rethinking of what music could become. Early electronic music collectives such as the Sonic Arts Union (Robert Ashley, David Behrman, Alvin Lucier, and Gordon Mumma) and their contemporaries, John Cage, Christian Wolff, David Tudor, and others continued to bring into question essential assumptions about music; breaking free of the Western canon, their music spoke to and drew from world music as well as architecture and science. Their early compositional explorations of principles such as chance and self-organization continue to inspire the design of digital music systems today (see section 13.3.2). These ideas were linked with social and political views that questioned the existing social order: political directions, to both the left and the right, drew in different ways on Darwinian thinking and its derivatives, through issues such as social Darwinism and sociobiology.

Evolutionary and adaptive software art and music grew into an entity in its own right in the 1990s as part of the second wave of biologically inspired systems thinking, associated with ALife. Artworks created by evolutionary computing techniques were pioneered by William Latham (Todd and Latham 1991), Karl Sims (1991), and Jon McCormack (1993), and principles of ALife were explored widely by artists in a variety of ways (see Whitelaw 2004 for a good introduction). Experiments using artificial intelligence (AI) to solve musical problems were well underway (Cope 1996; Todd and

Werner 1998), including biologically inspired cognitive systems such as artificial neural networks.

The concept of the ecosystem, which had been established in the 1930s as a critical unit of study for ecology (Tansley 1935), drew the attention of ALife arts practitioners with the promise of generative autonomy, engendered by processes of feedback, coupling, and coevolution. Ecosystem-based creative works emerged, quite literally, with work such as McCormack's *Eden* and the ecosystemic approach to composition pioneered by Agostino Di Scipio discussed in section 13.5.

The twenty-first century saw the mass adoption of computing technologies in the developed world, centred on the Internet. Increasingly productive programming languages and methods emerged, including languages and development environments specifically designed for creative coding. These factors have combined to give greater power to individual creative programmers working in music and the arts, and as the theorist DeLanda (2002) proposed, creative coders now routinely 'hack' the rules of thermodynamics, mathematics, and biology as if they were malleable artistic materials. Simple biological models may now be considered part of the creative coding canon: cellular automata, swarm models, and evolutionary algorithms are routinely included as examples in creative coding environments, such as Processing and OpenFrameworks (see e.g. Shiffman, Fry, and Marsh 2012).

With this whirlwind tour we have begun to map how the work of the pioneering thinkers of the early twentieth century has influenced how we approach biologically inspired and agent-based algorithms in music today. Our aim in this chapter is not to give a comprehensive overview of each field (pointers are given to many excellent references which provide these), but to give a flavour of the challenges and opportunities afforded by biologically inspired and agent-based computing in digital music making. We proceed with the introduction of some key concepts.

13.1.2 Core Concepts

> There exist models, principles, and laws that apply to generalized systems or their subclasses, irrespective of their particular kind, the nature of their component elements, and the relationships or 'forces' between them. It seems legitimate to ask for a theory, not of systems of a more or less special kind, but of universal principles applying to systems in general. (Von Bertalanffy 1950, 32)

The approaches to biologically inspired composition discussed in this chapter draw heavily on the conceptual and methodological tools of GST, cybernetics, chaos theory, and the study of complex systems, as well as more specific fields of ALife, connectionism and AI. Although aims and techniques vary in each specific field, some key concepts and modelling principles are shared. In this section we outline some of these foundational concepts.

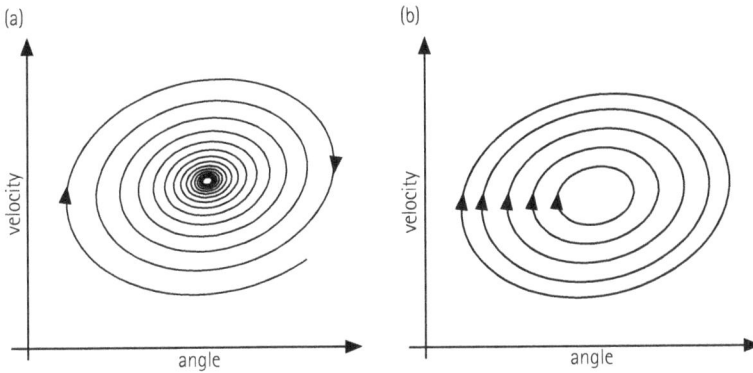

FIGURE 13.1 Trajectories showing (a) the point attractor of a damped pendulum, and (b) cyclic attractors of an undamped pendulum.

Central to modelling and understanding systems, from simple particles to complex ecosystems, is the notion of *system state*. The set of all possible states of a system is called its *state space*, and a systems approach advances by studying the *trajectory* of the system states through this space.

Take a single damped pendulum, swinging back and forth in a single plane. Its state can be defined in terms of its current position and velocity. Due to the forces of friction and gravity it swings with ever-decreasing energy, eventually coming to rest (Figure 13.1a) at its stable equilibrium point or *point attractor*. For this closed system ('closed' referring to the fact that there are no other forces acting on it), the pendulum is attracted to this point no matter where it starts in the state space. If, on the other hand, the pendulum had zero friction, then it could swing forever. In that case we would find multiple *cyclic* attractors, each described by the set of states that the pendulum passes through in its swinging cycle (Figure 13.1b).

A certain class of systems—such as a double-rod pendulum or the famous Lorenz system (Figure 13.2; Lorenz 1963)—exhibits chaotic behaviour, which is exemplified by state trajectories that are close, but never actually repeat. These paths are known as *strange attractors*. Chaotic systems are sensitive to initial conditions, meaning that similar starting states can tend towards wildly different outcomes: 'the present determines the future, but the approximate present does not approximately determine the future' (Lorenz, cited in Danforth 2013). Note that we are still talking here about closed, deterministic systems; neither external stimuli nor randomness are necessary to produce incredibly rich system behaviours, even when the system might consist of a very simple update rule. This principle has been of great interest to algorithmic musicians wishing to achieve rich, complex outcomes with algorithmic efficiency.

A state that is not an attractor state will typically lie in the *basin of attraction* of an attractor state. A helpful metaphor is to think of how rain falling on sloping ground has an attractor state depending on where it lands: rain in the Alsace region in France enters the Rhine river basin, and runs into the North Sea; rain in the Sancerre region enters the

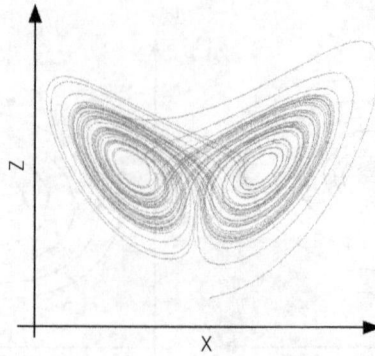

FIGURE 13.2   A 2D projection of the strange attractor of the Lorenz equation.

Loire river basin and ultimately runs into the Atlantic Ocean. The lie of the land is much like the set of state-space trajectories of a system. This landscape metaphor is similarly useful to conceptualize evolutionary search through a *fitness landscape*, which we return to in section 13.2.

Systems thinking can be applied to biological systems at a range of levels: a cell is a system of chemical and energy transfer; the heart is an open oscillatory system; an organism is a self-regulating system that operates to keep certain critical parameters within acceptable boundaries (to stave off death—the ultimate point attractor); an evolving population may arrive at an *evolutionarily stable state* or shoot off on a trajectory of runaway coevolution. Although these are very different in their details, systems thinking enables a common language for the study of system behaviours across domains.

Such generalist thinking is alluring for musicians. The language of state, trajectory, transition rules, attractors, basins of attraction, and so on resonates with sonic and musical experiences and concepts and inspires new approaches. Software models of complex, dynamic systems offer rich possibilities for musical composition and interaction, for the imitation of existing musical styles, for the creation of esoteric new forms, or as frameworks for human-computer musical interaction. In cases such as the modelling of human rhythmic perception using oscillatory models (Large and Palmer 2002), the system dynamics link explicitly with psychological theories of music perception. This state-based approach to studying systems is foundational to many of the topics discussed in this chapter.

13.2 Evolutionary Search

With the help of an electronic brain the composer turns into an astronaut pressing buttons of his musical spaceship to introduce coordinates and keep the course of his vessel on its journey through constellations and galaxies of sound, controlling from his easy-chair what the imagination

of yesteryear could have envisaged only remotely in its wildest dreams.
(Xenakis 1971, 124)

Accounts of journeys through unimagined and unchartered territories appear throughout early electronic and digital music discourse, as well as in recent generative art and
computational creativity literature (McCormack and Dorin 2001). Computational
algorithms have clear musical potential, but vast swathes of these spaces of possibility
contain nothing of interest, and efficiently navigating to the interesting areas is a nontrivial task. Searching for solutions to problems is a well-developed field in AI, and its
application in the arts also has a rich history. The search for fruitful solutions cannot
be random, and as one of a number of directed algorithmic search strategies, artificial evolution promises a powerful vehicle for discovery (see McCormack 2008; for a
detailed overview of the challenges and conceptual issues involved in algorithmic creative search).

The theory of evolution by natural selection (Darwin 1861; Wallace 1858) radically
transformed our understanding of nature by describing a seemingly simple, blind process that explains the origins and development of Earth's biological complexity. In the
reproduction of biological organisms, they observed, there is a predominant continuity of form and behaviour (heredity), but this overall continuity is corrupted by minor
random mutations (variation) that occur naturally and may accumulate over time
into radical morphological changes. Whilst the majority of these reproductive mutations are detrimental, certain variations in an organism's 'design' improve its survival
rate and reproductive success relative to its peers. Those better-off variants by definition
are prone to grow in number, whilst the weaker variants dwindle. Darwin and Wallace
posited that over time, this alone sets the sufficient conditions for new species to form
and develop sophisticated adaptations to their environments. Intense competition for
resources enhances this evolutionary effect as weaker variants are rapidly displaced by
their stronger counterparts. The combined result is *natural selection*, in which an invisible hand creates life forms intricately adapted to flourish in their present environment,
as if they were designed to do so. The discovery of the genetic system (Mendel 1866) provided the underlying mechanism for heredity, mutation, and sexual recombination that
is critical to modern evolutionary theory.

13.2.1 Genetic Algorithms

John Holland was the first to recognize that the power of this process could be harnessed
in computational models as a search tool in optimization and introduced the 'genetic
algorithm' (GA) (Holland 1975). Many variants followed, ultimately abstracted and generalized into a category known as population-based, metaheuristic optimization (see
Eiben and Smith 2003 and Mitchell 1998 for general introductions, and Burton and
Vladimirova 1999 and Husbands, Copley, Eldridge, and Mandelis 2007 for musically
oriented outlines).

To illustrate how evolutionary concepts can be adapted for optimization tasks, consider how we might go about designing a paper airplane (Figure 13.3). We could take a pile of paper and randomly fold pieces to create an *initial population*. A description of the points at which we folded the paper—e.g. *x,y* coordinates of start and end points of each fold, and so on—represent the *genotype* (roughly the underlying 'design') of each individual plane, the resultant physical form being the *phenotype* (the actual form). This population of phenotypes is then evaluated by assigning each candidate a fitness score according to how successful a solution it provides. In this case, we might cast planes across the room and measure how far they travel. This allows us to quantitatively compare the efficacy of each phenotype, that is, a *fitness function* (flight length).

In order to develop a population of planes achieving longer and longer flight lengths, we could then preferentially select those planes which flew the furthest and make modifications to the ways in which they were folded. This might include making minor random *mutations* to individual folds, or combining the folds from two winning planes. In biological reproduction, the latter is known as *crossover*; the genetic material from parents is mixed, as in sexual reproduction. We could then make a set of new planes which implemented these variations—the *replacement scheme*—either mixing them with solutions from the first round, or creating an entirely new population before launching them all in the air again.

Iterating this process can, in theory, lead to functional paper airplanes. In practice, there are many factors that affect how successful the outcome is. The encoding scheme (how we represent the phenotype as a genotype), genetic operators (mutation and reproduction schemes), and the fitness function together define what we call a *fitness landscape* across the genotype space. Fitter solutions sit at the tops of hills and the less fit solutions down in valleys. A well-designed GA acts to guide a population of candidate solutions across the fitness landscape towards higher ground, until an individual or percentage of the population reaches a prespecified value for an optimal solution.

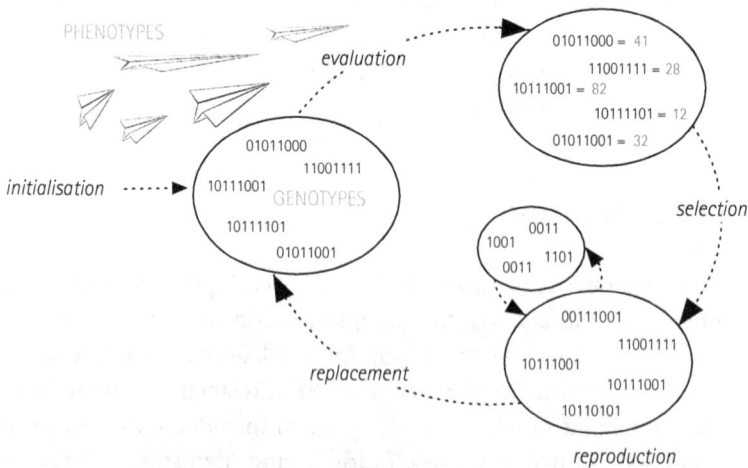

FIGURE 13.3 Outline of a genetic algorithm for evolving paper airplane designs.

A fitness landscape should be smooth, meaning that very similar genotypes result in very similar fitness scores (two similar planes are likely to fly similar distances): the more it looks like Mount Fuji, and less like the Manhattan skyline, the easier it is to make small and gradual steps upwards towards fitter solutions. Too many 'foothills' and the GA may never arrive at the best designs, but get stuck at *local optima*. A vast literature within the field of evolutionary computation discusses how the efficacy of search (navigating this landscape) can be improved by different approaches to designing population structure (Collins and Jefferson 1991; Husbands 1993), selection method (e.g. Eiben and Smith 2003; Mitchell 1998), replacement schemes, and other factors. In practice, the best choice of genetic representation, operators, and evaluation functions are often problem-specific and can be best illustrated by example.

13.2.2 Using Evolutionary Algorithms in Music

Evolutionary algorithms (EAs) have been applied in a wide range of musical contexts and applied at levels of the compositional process from the harmonization of Bach chorales to evolution of bebop improvisers, (see e.g. Burton and Vladimirova 1999; Miranda and Biles 2007). Here we focus on a few specific examples in order to illustrate the key design issues. We first consider the evolution of synthesis algorithms, where evolution offers a practical solution to the problem of searching for new sounds in an often unintuitive space.

Take a software synthesizer such as Native Instrument's FM8. It boasts well over 1,000 parameters, representing a vast, high-dimensional search space. The synth's 960 named presets offer a means for users to intuitively access sounds, but this does not help in the discovery of novel sounds, which must be done by manual parameter tweaking. Although this is the default approach to controlling digital musical instruments, it is not necessarily efficient. Assuming a musically sensible design, evolutionary computation offers a practical alternative.

MutaSynth (Dahlstedt 2001) and Synthbot (Yee-King and Roth 2008) are two of many projects exploring the application of artificial evolution to synthesis. MutaSynth was designed as a general-purpose tool for evolving programmable hardware synthesizers and was later integrated into and distributed with Clavia's popular Nord Modular G2 series of programmable hardware synths as PatchMutator, shown in Figure 13.4. Any of the control knobs available to the user can be encoded to form part of the genotype for the system's evolutionary search. The sonic results of eight variations of the given parameter set are presented in a grid to the user, who auditions them and picks their favourites. The system then creates further variations, which are presented to the user ad infinitum, until a desired sound is achieved.

Synthbot similarly enables an automatic search of a synthesizer parameter space, but rather than requiring feedback from the user, it is designed to search automatically for a match to a given target sound, which the user provides as input.

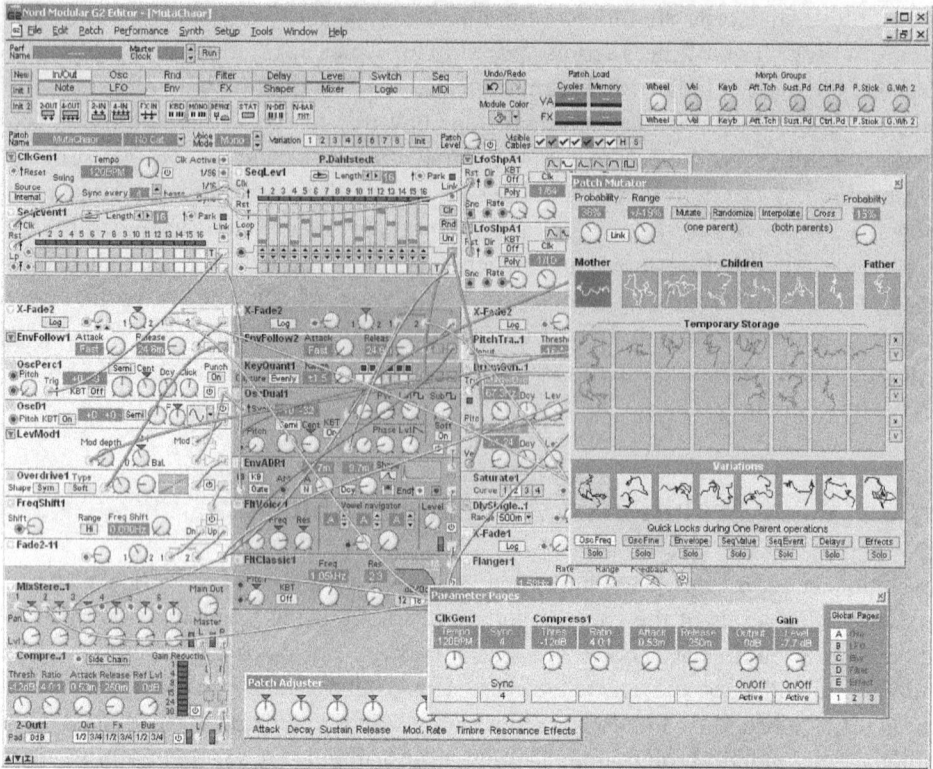

FIGURE 13.4 The Nord Modular editor with Patch Mutator on the right hand side.

13.2.3 Approaches to Genetic Algorithm Design

MutaSynth and Synthbot are used here to illustrate some of the issues around designing genetic representations, operators, and fitness functions, and how these influence
the fitness landscapes. Imagine, for example, designing a GA to search for harmonious
sounds in the space of all possible frequency modulation (FM) synthesis parameters
(Chowning and Bristow 1987).

Encoding Schemes. The genome might include a representation of the synthesis graph,
specifying the configuration of modulator and carrier oscillators, the frequencies and
amplitudes of the component oscillators, and the modulation depths. Just as carefully
designing mappings from parameters to the knobs and sliders on a graphical user interface can increase usability, so applying some domain-specific knowledge in the design
of the encoding scheme can greatly enhance the power of search for a prespecified task.
The search for pleasing harmonic sounds, for example, could be expedited by encoding
the modulation frequency *relative* to the carrier frequency, rather than as an absolute
value (as we know that the ratio between these values dictates the harmonicity of the
sound). A successful representation scheme will shape the fitness landscape in musically meaningful ways. We can think of this in terms of stretching or magnifying areas of

greater potential interest in the fitness landscape. In MutaSynth, for example, Dahlstedt uses nonlinear (exponential, logarithmic, or cubic) mappings from genotype to phenotype in order to make the most musically useful values more probable, while allowing for the possibility of more extreme values.

Representing the synthesis graph, rather than the parameters of fixed graphs, allows for the evolution of different configurations, as in Garcia's use of genetic programming (GP) (Garcia-Almanza and Tsang 2006). In GP (Koza 1992), another type of EA, the genotype takes the form of a mathematical function, which itself comprises a set of nested subfunctions, represented as a tree. Genotypes in this form can be mutated and crossed over in ways that grow or shrink the overall size of the function, and can result in dramatic new designs in a single operation. The rules for the genetic operations can be designed such that the resulting functions will always be mathematically valid. This provides a potentially powerful mechanism to evolve code, but is not without issues. One challenge is that pruning and grafting of GP subtrees can lead to dramatic discontinuities in the fitness landscape, obstructing intuitive evolution.

Genetic Operators. Variation in the population is introduced by the genetic operators (mutation and crossover). In our paper airplane example, we randomly modified the fold lines. Domain-specific knowledge can be beneficial here too. In MutaSynth, these standard operators are mixed with a morphing process by which offspring can be created in a more intuitive way by interpolating genotype parameters between two parent genotypes, providing a continuous 'cross-fade' between two points in the phenotypic space of evolved sounds.

Fixed Fitness Functions. The design of fitness functions suitable for creative application of EAs has been the topic of much research in the field. Synthbot aims to evolve parameters of a synth to match a given target sound: a *fixed* similarity measure is provided by a perceptually motivated timbral measure, Mel-Frequency Cepstral Coefficients (MFCCs). This works well for optimization tasks; however, when a more general 'aesthetic' quality is desired, it can be difficult to formalize the desired target, despite efforts to measure aesthetic quality in music and visual art (e.g. Birkhoff 1933; Romero and Machado 2008).

Interactive Genetic Algorithms. Richard Dawkins tantalized a generation of artists with the creative power of evolution in his simple *Biomorphs* program (Dawkins 1986), which he used to illustrate how easily human selection could lead to a diverse array of lifelike forms. This use of human aesthetic judgement in place of a formalized fitness function (as used in MutaSynth above) has been explored extensively in the application of EAs in visual art (Sims 1991; Todd and Latham 1999) and is known as an 'interactive genetic algorithm' (IGA). As a stochastic search method, EAs typically require large populations to be explored over many generations. A key issue for this method then, is the amount of time required to perform the fitness judgements, described as the *fitness bottleneck* (Biles 1994). This is an issue for music in particular, as its fundamentally temporal nature obviates presentation of multiple individuals in parallel. User fatigue becomes a significant constraining factor, and various approaches to overcoming this have been explored.

Distributed IGAs. One natural response to the fitness bottleneck is to distribute the IGA amongst multiple users via the Internet. Distributing the selection process places a lighter burden on a single user, but presents a challenge in organizing how multiple users might work together to produce evolved outputs: differing preferences may equally result in serendipitously creative outcomes, or a directionless tug-of-war. The potential population size of candidate solutions also becomes overwhelming. This approach, which has been applied more extensively in the visual arts, has been used in Draves's animated screensaver, *Electric Sheep Project* (Draves 2005) and Secretan's *PicBreeder* (Secretan et al. 2008), both of which show some promise for arriving at complex structures that wouldn't have come about through a traditional human design process.

Artificial Critics. Another alternative to either fixed or interactive fitness functions is to train a machine learning system such as a neural network to perform fitness assignment. Early attempts suggested that this is merely deferring the problem: if a fitness function is hard to formalize for something as ineffable as a good jazz solo, then it will also be hard to train a machine learning system to take the place of that fitness function (Biles 1994). However, as machine learning achieves more impressive results this may still prove to be a fruitful approach in the future.

Coevolutionary Approaches. Rather than using evolution as a *convergent* optimization tool or interactive search mechanism, the power of evolution as a *divergent* engine for generating novelty and diversity has also been explored by many musicians. Taking further inspiration from the natural world, *coevolutionary* approaches have been explored in which populations of solutions are evaluated by populations of critics, which are themselves evolving. By virtue of this dynamic coupling, coevolution can also produce diversity within a population. Synchronic diversity can be generated through sexual selection, leading to speciation—splitting the population into subpopulations of individuals with distinct traits and preferences (see e.g. Todd and Latham 1991). Coevolution can also amplify the diversity of novel forms over time, causing rapid evolution of traits as in predator-prey 'arms race' models (e.g. Futayama and Slatkin 1983).

Taking inspiration from a model of the evolution of birdsong (Todd and Werner 1998), this coevolutionary approach has become a popular paradigm within computational creativity research, both for the generation of music and art, and as a modelling tool in the simulation of creative societies (Dahlstedt and Nordahl 2001). These interactions between critic and composer in the coevolutionary approach are seen as a proto-social behaviour (Miranda 2002a) and discussed in the context of multi-agent systems in section 13.3.

Broadening the metaphor from the Darwinian evolution of isolated genotypes to the digital specification of entire ecosystems, the *computational ecosystems* discussed in section 13.5 develop the standard evolutionary algorithm by embedding the population in an environment where they interact with each other and with other environmental elements and spatial constraints. In these cases, 'fitness' is no longer an explicitly pre-specified function as in standard EC, but defined *implicitly* in the interaction between individuals and their shared environment.

13.2.4 Variations on a Theme

The examples above aim to give a flavour of the different approaches to the design of EAs in musical applications. As is clear, many aspects of music making don't fit neatly into an engineering optimization framework and there are many interesting and promising cases where standard EAs have been 'hacked' for idiosyncratic aims. A common theme is the adaptation of classical EA components for the generation of diversity and variation. For example, Magnus and Waschka II both take advantage of the structural changes which take place in the population through time (Magnus 2006; Waschka II 2001). Rather than employing EAs as a search mechanism and listening to the 'winning' individual at the final generation, the evolutionary process itself is sonified, conveying the changes that occur in the population across generations. Kiefer describes a GP-like variant designed for live performance situations that can be used to generate and interactively explore synthesis graphs on the fly (Kiefer 2014). Here again, no fitness function is specified but the representation scheme is adopted as a means for a user to rapidly search through a vast space of possibilities.

13.2.5 Evolution and Usability

This highlights the importance of human-computer interface (HCI) design aspects of evolutionary approaches. That IGAs present unique HCI issues has long been recognized (Todd and Latham 1999); novel interfaces which allow users to mix their own multi-objective fitness functions for building structures have been explored (Bentley and O'Reilly 2001). Dahlstedt's MutaSynth also has a strong user-centred focus, aiming to understand and tackle the common limitations of IGAs in functional end-user software. For example, MutaSynth is designed to be operated with one hand so that a user can play a keyboard and constantly audition sounds whilst controlling the algorithm. Dahlstedt also experimented with the visualization of genotypes, addressing issues of recall and rapid evaluation: 'This visual representation was not a faithful representation of the actual sound. Rather, it was derived from the parameter values of the synthesiser, which were used as length and angle values for a multi-segment line, scaled to fit the window. A small change to a parameter value would cause a small change to the visual representation' (Dahlstedt 2001, 90). The visual representation of sounds also aided users in recalling and organizing the sounds they had discovered. Also with MutaSynth, since the 'genes' are also the synthesis parameters available to the user, the user can get in and tweak any evolved sounds, committing their modifications to the evolving population.

Through the commercial availability of his software, Dahlstedt was also able to gain user feedback. For example, users reported that the software supported working with increasingly complex patches where the relationship between individual synthesis parameters and good sounds became increasingly obscured. It has also been suggested that IGAs are most applicable in areas where the creator lacks either mastery or a strong conceptual model of the type of entity being created (Takagi 2001), which supports the use of EAs in synthesizer programming.

Many of these recent developments—and also much of the frustration with lack of progress, despite powerful tools—in the use of evolutionary computation in creative tasks suggest that this focus on user workflows and basic interaction design, incorporating evolution in sensible ways into existing practice, is where fruitful advances can be achieved. The need for computational creativity to embrace interaction design has been emphasized recently (Bown 2014; McDermott, Sherry, and O'Reilly 2013).

13.3 MULTI-AGENT COMPOSITIONS

A diverse group of researchers in mathematics, physics, and several branches of biology view self-organization alongside natural selection as a complementary mechanism of evolution (Bak, Tang, and Wiesenfeld 1988; Camazine et al. 2001; Kauffman 1993; Nicolis and Prigogine 1977). 'Self-organization' refers to the process whereby an observed complex macro-level structure emerges from a series of local interactions between relatively simple agents: insect swarms, animal markings, and even high-level neural maps (Kaschube et al. 2010) and gross physical movement (Kelso 1995) arise through local interactions in the absence of top-down control. Self-organizing phenomena have been extensively studied across disciplines such as ALife (Langton 1989), artificial chemistry (Dittrich, Ziegler, and Banzhaf 2001), ecology, sociology, and neuroscience using agent-based simulations. The aesthetic potential of such emergent behaviour has been explored across many artistic domains, including sound and musical composition (Beyls, Bernardes, and Caetano 2015; Blackwell 2003; Miranda 1995).

13.3.1 Agent-Based Modelling

Agent-based simulations are used to explain high-level complex structures in terms of the interaction of a collection of simple agents in a shared environment, and as such are a key tool for understanding self-organization and other emergent phenomena. Important early work in this area includes Von Neumann's description of self-replicating machines: devices capable of creating copies of themselves by precisely following a set of detailed instructions (Neumann 1966). These ideas were developed by Ulam, as a collection of cells on a grid, and evolved into what we now know of as cellular automata. A cellular automaton (CA) can be conceived of as a regular n-dimensional grid of cells. In common with most agent-based modelling systems, a set of 'agents' is situated in a shared environment; each agent can exist in a finite number of internal states and act according to update rules which are in turn contingent upon the states of surrounding agents—their 'neighbours'. For certain rule sets, an astonishingly complex range of global dynamics emerges from simple, local interactions. CAs have been explored in a wide variety of contexts to model phenomena across ecology (Hogeweg 1988), biology (Ermentrout and Edelstein-Keshet 1993), and sociology (Epstein 1996).

This dynamic complexity has also been extensively explored in compositional processes. Xenakis used CAs in the mid-1980s to create the 'complex evolution of orchestral clusters' in *Horos* (Hoffmann 2002, 122), and various computer scientists and composers have followed suit (e.g. Beyls 1991; Burraston, Edmonds, Livingstone, and Miranda 2004; Millen 1990; Miranda 1995 for historical and technical reviews).

Beyond the more abstract example of CAs, many models look at population behaviour in animals and humans. Thomas Schelling's study of patterns of segregation in urban geography was one of the first agent-based models to have significant impact in the social sciences (Schelling 1971). Through the 1980s, agent-based models were developed across a wide variety of domains to explore game theory (Axelrod and Hamilton 1981), evolutionary processes (Hinton and Nowlan 1987), and as generative tools in computer graphics (Reynolds 1987). Reynolds's *boids* model demonstrated that the phenomenon of flocking and swarming behaviour in bird, fish, and insect species could be achieved in a simple computer model where each agent adjusted its update vector in relation to near neighbours according to three rules: cohesion, alignment, and separation (Reynolds 1987). Although making no claims to biological plausibility, the model demonstrates that coherent and robust flocking can arise through a process of self-organization.

13.3.2 Emergence in Human and Artificial Music

Besides their more abstract pattern-generating properties, multi-agent models appeal to musicians through their potential to explore new artificial notions of ensemble behaviour. Self-organizing processes had been explored by postwar experimental composers and pioneers of free improvisation. The text-score of Cornelius Cardew's *Paragraph V* of *The Great Learning* (Cardew 1971), for example, instructs singers to start on a random note and proceed with each breath by picking a note sung by their neighbours. From a random beginning, a unified harmony emerges. The concept provokes algorithmic composers to this day: what are the timbral, harmonic, rhythmic, and structural possibilities of self-organization? How can humans and machines interact within such a framework?

Exploring such questions, Blackwell and Young (Blackwell 2003) used swarming as a model of free improvised ensemble performance. They considered a symbolic (MIDI) version, in which particles swarm around a 3D (pitch, density, volume) space, and a granular audio version, in which particles swarm in a space of grain parameters.

The classic swarm algorithm is made interactive: the signal from an improvising musician is analysed and used to define points in the swarm space which act as attractors to the particles, creating a cycle of interaction between the human and the multiple swarm particles. The nature of the swarming algorithm means that the particles are loosely coupled in their movement: the resulting trajectories are likely to be similar, generally moving in parallel, providing a form of coordination that may result in counterpoint, repetition with interesting variation, canons, or novel harmonization. Swarm models have also been used in a more literal manner, using 3D audio to create the effect of a multiplicitous swarming of sounds around the listener (Kim-Boyle 2005).

In general, the self-organizing dynamics of agent-based systems have been applied to good effect where the time-evolution of the system state is mapped in a sonically meaningful way. But how meaningful can such abstract systems be as models of human musicality? Does this kind of self-organizing system capture aspects of human musical intelligence? On the one hand, various creatively productive musicological accounts have focused on the dynamic surface structure of music (Toch 1977), using metaphors from Newtonian physics such as gravity, inertia, and momentum (Larson 1997) in describing how melodies play out, interact and are perceived. On the other hand, such models are seen to contribute only a limited amount to our understanding of human musical behaviour. Nevertheless, such experiments help explore the nature of music in an in-between space where the computer behaviour is neither that of a mere static object, nor of a sophisticated cognitive nature. In that sense, it may be reasonable to think of modelling musical behaviour without modelling human-level cognition, a topic discussed in section 13.4.

13.3.3 Modelling Musical Communities

Another class of models looks at human interaction and self-organization in social behaviour. Miranda's models of musical interaction in virtual agents tackle both creative and empirical questions in an interesting synthesis of methodologies (Miranda 1999, 2000, 2002a, 2002b). Following approaches in ALife and in the evolutionary models of language made famous by Kirby and Hurford (1997), Miranda presents agent-based modelling as a means to study the evolution of musical behaviour in human and artificial societies.

In language evolution, Kirby and Hurford challenged the expectation that we should seek adaptive functional explanations for the features of language, using a proof-of-concept model (e.g. Kirby and Hurford 1997). They showed that certain aspects of language structure could emerge from a process of iterative learning, that is, through self-organization, without the need to implicate evolved functions. Iterative learning is a process whereby one agent produces some form of output and another agent is presented with that output and learns its structure—typically using artificial unsupervised learning. The information content is just passed from one individual to the next—basically like Chinese Whispers—and we can look at the long-term coevolution of content and learned cognitive structure. Over successive iterations, whatever is most 'copyable' is what gets copied; the iterative learning shapes the structure and usage of the artificial language, leading Kirby and Hurford to comment that 'it appears that languages adapt to aid their own survival over time' (Kirby and Hurford 1997, 2). Miranda, Kirby, and Todd (2003) adapted such models to musical contexts and developed musical works using a similar scheme in which a population of agents first establishes a common vocabulary and then performs with it.

Models of human performance interaction have also been derived from agent-based models of *game theory*—the study of interactive decision making amongst agents. In some scenarios there is a clear mutual benefit to joint cooperation, but in countless real-world cases there is some individual benefit to being selfish—assuming you can get away with it—and the greatest reward comes from being selfish whilst somehow ensuring that your co-player chooses to act cooperatively (often referred to as 'freeloading'). This scenario is known as the Prisoner's Dilemma, imagining two criminals who are being separately interrogated, with respective, mutually dependent payoffs for confessing or refusing to confess. The dilemma is that if both you and your co-player choose to be selfish, then you are both worse off than if you had both cooperated. This simple form of analysis has proven to have great power in understanding individual behavioural strategies, and has been successfully applied to questions of the biological evolution of social behaviour (Maynard Smith 1982).

Didovsky developed some of the earliest multi-agent game works during the ascendency of ALife. *Lottery* (Didkovsky 1992) presents a model society competing for access to a limited resource determined by a lottery process. Performers join this virtual society in playing the game. Harrald (2005) similarly uses game theoretic dynamics to establish rules for indeterminacy in the tradition of the 1950s New York School, specifically the work of Christian Wolff, using 'unpredictable chains of performance situations that could arise only through the act of performance' (Harrald 2005, 69).

The latter uses the *iterated* Prisoner's Dilemma. If you have a history of interaction with your co-player it is possible to develop strategies for cooperation. Axelrod and Hamilton showed that a very simple strategy, called 'tit-for-tat', would suffice to establish mutual cooperation, without being prone to exploitation (Axelrod and Hamilton 1981): begin with cooperation, then simply copy what your co-player did in the previous turn. This became the major paradigm for understanding reciprocal altruism in ecology (e.g. Wilkinson 1984). Harrald's composition ENSEMBLE establishes Prisoner's Dilemma competitions between players. The players are programmed with different strategies, such as tit-for-tat, and the results of games at each round are used to dictate musical following behaviour by assigning different behaviours to defectors and cooperators.

Murray-Rust, Smaill, and Edwards (2006) created a system of intelligent musical agents that interact using 'musical acts', a concept based on speech act theory (Searle 1969). This sets out to devise a sensible protocol by which agents can communicate amongst each other, opening up the possibility of allowing richer emergent and creative behaviour in multi-agent music systems, that might reflect emergent behaviour in humans. Murray-Rust used Terry Riley's celebrated minimalist work *In C* (1964) as a case study for how the agents could be put to task. The score consists of a series of short phrases; each player may proceed through each at their own pace. In this implementation the players are required to stay within two or three patterns of each other, must listen to each other and occasionally drop out, and must aim to merge into full unison at least once or twice during the piece. The application of the musical acts to this task demonstrated how a system of communication between multiple agents could be practically devised and structured with a specific outcome in mind.

Eigenfelt has produced a number of works (Eigenfeldt 2007a, 2007b, 2008) based around the design of multi-agent systems where the interacting agents collaboratively develop compositional content by listening and responding to each other. The behaviour of agents evolves over time, with agents 'reflecting' on their behaviour according to a number of pre-programmed personality traits.

13.3.4 Agent-Based Models at Different Scales

As well as their application in virtual systems running on a single machine, agent-based software models also have relevance to how we approach performance with networked multicomputer systems and multiple performers. The League of Automatic Music Composers and later The Hub (Gresham-Lancaster 1998) are widely cited as pioneering the practice of networked computer music performance, and thus inherently initiating forms of experimentation into the modelling of multiple musical agents. The Hub experimented with audience participation over the internet as early as 1989, creating works in which multiple human users and algorithms acted on a shared memory space to produce musical output, and in doing so encountered questions of how to structure massive multi-user musical constructions.

A new project by Eigenfeldt, Bown, and Casey (2015) attempts to create a very general purpose distributed multi-agent architecture, the *Musebot Ensemble*, that allows developers to easily work together and attempt to exploit the emergent properties of musical interaction by running their distinct musical agents together in the same environment.

Multi-agent strategies need not only model the network interactions between individuals. In AI a number of theories and methods are based on a model of cognition in which competing hypotheses battle it out for recognition, as determined collectively or by some central attention system (Minsky 1988). This has proven to be an effective strategy in music information retrieval as well, specifically in the case of beat tracking (e.g Dixon 2000; Large and Palmer 2002), based on the idea that rhythm perception uses multiple resonant oscillators and determines which resulting oscillation gives the strongest signal. This can lead to multiple harmonic rhythmic oscillations that may explain our metrical perception. Wiggins (2012) has proposed a multi-module model of musical creativity which frames an understanding of music information dynamics and expectation in terms of parallel mental computation.

13.4 ADAPTIVE COLLABORATORS

Only the environment can design a brain. (Ashby 2008, no. 151)

Understanding and modelling adaptive behaviour was of primary concern to cyberneticians—indeed for Ashby the 'problem' of life boiled down to the problem of adaptation—and is a critical concept in contemporary cognitive science and AI. In general, a system that is adaptive is able to provide appropriate responses to situations it encounters in its environment. This may be as simple as an autonomous vehicle correcting its path in response to an obstacle or a kitten learning to avoid a fire, or as profound as the emergence of a new species under evolution by natural selection. In contrast to the founding assumption of symbolic AI, numerous early cybernetic accounts and devices (introduced below) powerfully illustrated that the appearance of what looks to an observer like intelligent behaviour does not necessarily require sophisticated internal mechanisms. Such ideas inspired the development of ALife approaches to autonomous robotics and continue to influence contemporary philosophy of life and mind (Di Paolo 2003; Froese and Stewart 2010).

Some of these core principles were illustrated by the Machina Speculatrix, built by neurophysiologist and robotician William Grey Walter (Walter 1950). Walter built two electromechanical 'tortoise' robots, named Elma and Elsie, in which the output of a pair of light sensors controlled the robot's wheel motors. Putting two robots together (or in front of a mirror), each with a light on top, he achieved phototaxis (light-seeking behaviour), but also compelling 'lifelike' dances. From such work we have the seeds of the later formalized principles of *situatedness* and *embodiment*. 'Situatedness' refers to the fact that the relevant behaviour is apparent only once the agent is placed in the relevant, rich environmental context. 'Embodiment' refers to the fact that the behaviour is contingent upon a physicality: a physical body that has myriad interactive affordances beyond those which may have been consciously implemented by the designer. The term 'lifelike' has been used widely in the ALife literature to express the notion that both organic and artificial systems may, specifically through their adaptive behaviour, take on the appearance of a living system, inviting 'attributions of intentionality'.

Biologically inspired adaptive systems provide opportunities to explore novel modes of interaction and decision making in a musical context. Much of the work in intelligent music systems is implicitly geared towards modelling human musical behaviour. But as we have seen in other examples above, computers allow us to explore musical interaction in new ways, creating novel and possibly hybrid interaction scenarios that weave in exotic computational behaviours. Computer music makers have incorporated simple or complex adaptive behaviours into their software systems, drawing on a variety of models, and applied to a variety of compositional goals. These scenarios place the performing musician in a new relationship with the musical dynamics where they are not necessarily in direct control of the outcomes, but act as negotiators with the machine. The process becomes one of mutual adaptation.

13.4.1 Adaptive Behaviour and Musical Interaction

For musicians, the idea that simple, situated, biologically primitive behaviours might underlie complex pattern making in music is compelling. Some musicians working in this area have taken a very Alife-inspired view of music itself as a 'dynamical complex of interacting situated embodied behaviours' (Impett 2001). From this perspective, there is arguably greater potential for human and machine to take on equivalent roles in co-creation. Each contributes to a shared collaborative musical environment which in turn affords musical opportunities for the other.

This conversational model of musical interaction was explored in the 1950s by cybernetician Gordon Pask in a quirky experimental analogue audiovisual improvisation system, Musicolour (for an overview, see e.g. Pask 1971). The system listened to a performing musician and responded with patterns of coloured lights. But by Pask's design, the system would become bored if the input was always the same, and would adapt its behaviour to provoke a new response. The system was not particularly sophisticated in its behaviour, and had limited scope, but it began to chart how we might design interactive music systems that maintained a compelling engagement with a musician; players touring music halls with it in the 1950s reportedly engaged with it much like another musician (Haque 2007).

13.4.2 Behavioural Objects

In earlier work we have explored a number of biologically inspired models which display adaptive behaviour, including various neural models and homeostasis. We proposed the term *behavioural object* to refer to the ways in which software can act as the focus of interaction. This extends the traditional model of interaction inherited from acoustic instrumental performance, which focuses solely on musical interaction at performance time (for details, see Bown, Eldridge, and McCormack 2009) to include other types of interaction: the social interaction between communities of developers; the interaction between developer and software; and the musically significant interactions between software elements themselves, incorporating all of the biologically inspired forms of inter-agent interaction discussed in this chapter. To frame these distinct forms of interaction, we distinguished two senses in which a behavioural object could have agency: performative agency (in performance time) and memetic agency (out of performance time, i.e. spanning multiple performances or acts of musical production). Performative agency refers to the ability of a software system to influence the outcome of a specific musical performance. For those primarily interested in human-computer collaborative improvisation systems, performative agency is directly synonymous with the quality of musical interaction it affords.

Eldridge (2005) explored an Ashbian model of homeostasis as a core organizing mechanism for electroacoustic improvisation. Ashby (1960) addressed a fundamental conundrum: how can a system (biological or mechanical) be at once state-determined and yet adapt to a changing environment and learn? He proposed that the key mechanism underlying adaptive behaviour is homeostasis—the maintenance of key internal variables in the face of external perturbation. As a good cybernetician, he built a physical device to demonstrate his theoretical notion of *ultrastability*: the homeostat. This was an electromechanical device, but the critical elements can be described in the abstract and consist of: a *physical system* which is capable of interaction with its *environment* such that it perturbs the value of *an essential variable* with a specified boundary of viability (e.g. body temperature, blood pressure, etc.). When this boundary is exceeded a *selector* is triggered that specifies a break, or change, in the system (e.g. starting to shiver when cold or sweat when hot) such that it *randomly* changes its organization until a new set of parameters is arrived at under which the essential variable is brought back within limits. This conceptualization of homeostasis and ultrastability has influenced advances in autonomous robotics and contemporary theories of cognition (Di Paolo 2003; Froese and Stewart 2010; Maturana and Varela 1987).

Eldridge simulated the device in a neural-network-style model of interconnected nodes which displayed the key homeostatic and learning behaviour demonstrated by Ashby's machine: from an initially random state, oscillatory dynamics emerge. Small perturbations cause temporary disturbance followed by a return to the initial state; larger perturbations push the system into a new trajectory where it settles into a different cyclic attractor. In Ashby's original conception, the essential variables represented the nervous system in interaction with the environment; in an electroacoustic performance setting, the model provides a conceptual vehicle and algorithmic means for collaborative human-computer interaction. The model was implemented in an audio-visual improvisation system where the homeostat received input from a visual display, the essential variables being used to 'remix' samples taken live from an acoustic improvisor which in turn generated visuals, closing the loop. The homeostat acted to recompose earlier musical elements, influencing the performer's subsequent improvisations to create an electroacoustic dance-like experience (Eldridge 2005).

Bown took inspiration from the minimal cognition work of Beer (e.g. Beer 1997), in which generic network architectures were evolved from first principles to learn simple cognitive tasks in an elementary virtual world. The dynamics of continuous-time recurrent neural networks (CTRNNs) were evolved with fitness functions defined in highly abstract musical terms (Bown and Lexer 2006). With creative search rather than musical competency in mind, these fitness functions were designed to steer the evolved structures towards some degree of complexity and responsiveness of behaviour, rather than specific musical outcomes of rhythm and harmony. Fitness properties consisted of notions such as 'always output cyclical patterns' or 'if the input changes, the output should change'—compositional instructions which may or may not be evolved literally. Bown showed that it was relatively simple to evolve rich, complex, interactive behaviours that could then be applied in a variety of generative or interactive musical contexts. He also introduced a music-specific variation to the CTRNN algorithm, in the form of a sinusoidal transfer function, that enabled individual nodes in the network to oscillate on their own, boosting the overall oscillatory behaviour of the network.

This work was followed up by Zamyatin (Bown 2011), an improvising agent in which a decision tree, which exhibits feedback by influencing an internal state array, in turn influences the decision tree's decisions. The decision tree was shown to have a number of traits that made it more practical in musical use than the CTRNN. For example, it outputted discrete rather than continuous data, which proved to be more convenient for mapping to event-based musical control.

13.4.3 Subsumption Architectures for Musical Agents

A significant strand of work approaches the study of musical interaction through the development of software performance systems, explicitly adopting the design philosophy of roboticist Rodney Brooks.

The 1980s saw a resurgence of interest in cybernetics, which fuelled a new approach to autonomous robotics. Brooks was one of many who eschewed the representationalist approach of AI—or Good Old Fashioned AI (GOFAI) as it was dubbed—arguing that AI was too focused on advanced cognitive behaviour (such as the logic of playing chess) rather than fundamental intelligence, or minimal cognition (such as picking up and moving the chess pieces). Under the dictum 'the world is its own best model' (Brooks 1999, 115), Brooks developed a subsumption architecture approach to robotics in which real-time sensory information is coupled to action selection in an intimate, bottom-up fashion, rather than being guided by symbolic mental representations of the world (Brooks 1991). This approach, Brooks argued, enables agents to respond quickly and appropriately to changes in an unpredictable world, a task which is fundamental in the real world but which challenged the task-specific GOFAI robots of the time.

Brooks's approach was parsimonious and incremental, following observed biological behaviours. Starting with the basic behaviours (move forward, avoid obstacle), he built robots comprising sensors, motors, and the simplest possible 'brain' (designed by hand). Once debugged and tested in the real world, additional behavioural layers were added, mimicking the phylogeny of real creatures. Layers worked in parallel and generated outputs which might be signals to inhibit or suppress other layers or issue commands to actuators. This deviated from the more common top-down designs produced on GOFAI principles, aligning instead with the emerging evolutionary perspective of cognition as compartmentalized into efficient but narrow domain-specific competencies (e.g. Barkow, Cosmides, and Tooby 1992).

Bryson (1995) was the first to apply such techniques to music performance, which she also treated as an empirical investigation into what level of cognitive sophistication music operates at (many of the systems described here implicitly pose this question, but Bryson explicitly noted the epistemological potential of this approach). Her reactive accompanist modelled the capacity to derive chord structure from a melody (Bryson 1995, 2) and followed Brooks's subsumption architecture principle. Bryson showed that the reactive accompanist could perform to a relative degree of sophistication using this principle, demonstrating the efficacy of subsumption and action selection as part of a software design method in creative computer music.

The subsumption approach to handling input and output is highly suggestive of the tight coupling between listening and playing (input and output) of free improvisers, who exhibit a robust, flexible approach to dealing with unpredictable changes in the sonic environment (Clarke 2005; Sudnow 1978). Inspired by this observation, Linson developed an artificial improvising agent based on subsumption architecture, *Odessa*, in order to carry out research into the complex dynamics of musical improvisation as a 'situated psychosocial and embodied cognitive practice' (Linson, Dobbyn, Lewis, and Laney 2015). His design follows Brooksian parsimony at every level. Just three competing behaviours are implemented: the ability to spontaneously produce output ('Play'); to respond to musical input ('Adapt'); and to disregard input, introducing silence, and initiate endings ('Diverge'). Within these, the simplest conceivable algorithms are deployed. Despite this, Linson demonstrated that complex interactive behaviour, subject to evaluation by experts who improvised with it, can emerge from interactions between modules, supporting the idea that in-the-moment inferences, based on behavioural cues, perceived in realtime, can lead to the attribution of intentional agency in musical machines. Linson's work addresses our understanding agency and autonomy in interactive computer music, a core theme which is revisited at the close of this chapter.

13.5 CREATIVE ECOSYSTEMS

The term 'ecosystem' was first used in Arthur George Tansley's paper on vegetational concepts to pronounce his conviction that organisms cannot, fundamentally, be considered in separation from 'the environment of the biome—the habitat factors in the widest sense ... with which they form one physical system' (Tansley 1935, 299). Tansley wished to comprehend not only communities of organisms, but also to bring the complex interactions of biotic and abiotic factors surrounding them into focus. By giving a name to this tightly coupled collection of biotic and abiotic organisms and processes, he sought to establish the ecosystem as 'a recognizable self-contained entity' (228).

The concept has gained popularity in recent years as a metaphorical cornerstone in a range of musical contexts: as a framework for understanding the influences and elements of *performance ecosystems* in theory and practice (Bowers 2002; Waters 2007); as an approach to composition based on acoustic and adaptive systemic feedback in *audible ecosystems* (Di Scipio 2003); and as a vehicle for creative discovery, in *computationally creative ecosystems* (McCormack 2012). Although differing in basic materials, all three approaches take Tansley's original conception as a central organizing principle and theorize or construct situations in which organisms (human musicians in the world, agents in simulation, or some electroacoustic hybrid) and their environments are irrevocably coupled via feedback processes, such that any musical agency of the system can sensibly be understood only as originating from the system as a whole. Systemically, concern shifts from understanding embodied, situated agent behaviours in an environment, to conceiving of the environment and organisms within it as a super-entity.

13.5.1 Audible Ecosystems

Di Scipio's *Audible Eco-Systemic Interfaces* is a series of works which integrate aesthetic, philosophical, and technical aspects of ecosystemic thinking. Technically, Di Scipio creates software components that monitor, adapt to, and transform their ambient acoustic environment. For example, a signal processing module might be set up to automatically alter its internal configuration according to changes in the input sound, in turn altering the acoustic environment (see Figure 13.5). Algorithmically, the idea is fairly simple and could be described as 'adaptive audio feedback'. But in altering the acoustic environment, sound itself determines the conditions and boundaries for its own transformations. This has a practical creative value as the real world has a dynamic richness that is hard to reproduce in software; conceptually it establishes the same conditions in which the human agent is situated in a music performance context.

Philosophically, his approach references autopoiesis (literally, 'self-creation'). The term was coined in the 1960s by the biologists Humberto Maturana and Francisco Varela (for the first English publication, see Varela, Maturana, and Uribe 1974) to convey their conviction that the root of biological autonomy lies in the circular organization of living organisms, a feature they saw to be both necessary and sufficient for life. The use of sonically situated adaptive processes can be seen as a *structural coupling* of software and environment. This is a critical construct in autopoietic theory used to describe a process of engagement which effects a 'history of recurrent interactions leading to the structural congruence between two (or more) systems' (Maturana and Varela 1987, 75).

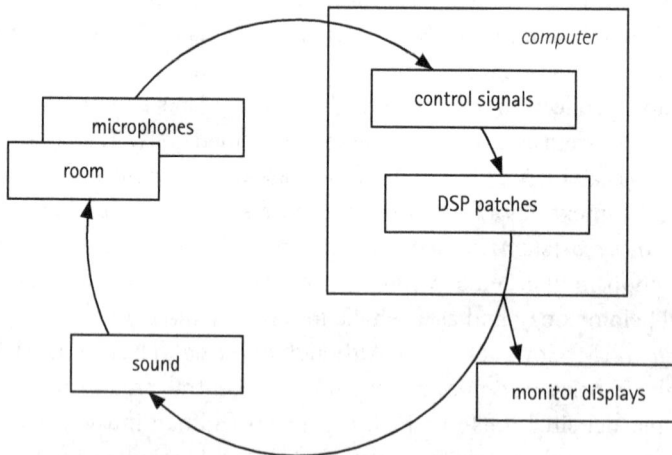

FIGURE 13.5 Basic design of Di Scipio's *Audible Eco-Systemic Interfaces*.

Di Scipio's systems cannot be seen to literally be 'alive', but reference to autpoietic theory promotes a fresh perspective on agency and hence interaction in a systemic sonic practice and leads to a productive framing of (non-biological) autonomy in terms of self-determination. This breed of 'self-determined' acoustic ecosystem has grown into a genre in its own right within electroacoustic composition (Bown 2009; Eldridge 2008, 2013; Green 2011; Sanfilippo 2013).

As with other practitioners described in this section, Di Scipio takes as his starting point a critique of the notion that 'basic' interactive music systems place the autonomy of the performer at their centre: 'agent acts, computer re-acts. ... The only source of dynamical behaviour lies in the performer's ears and mind' (Di Scipio 2003, 270). By adopting an ecosystemic approach, where the software system can directly respond to and alter its environment, the works demand consideration of the different notions of 'interaction' as applied to human performers and software components differentially and in combination.

A number of other cybernetically minded practitioners at around the same time shared similar sentiments in critiquing interactive art and creative systems across sonic, visual, and kinematic arts including Simon Penny, Usman Hacque, Bert Bongers, and Jon McCormack. The term 'interaction' becomes a battleground, upon which a potentially richer and more organic set of interactions with machines is at stake, compared to the standard material of HCI. Pask's conversation model of interaction (outlined above in section 13.4.1), for example, continues to be cited as a goal for what constitutes meaningful interaction.

Other ecosystemic approaches are more focused on the construction of narrative than on emergent complexity. Billed as 'a musical composition that grows in the same way as a forest ecosystem', *Living Symphonies* is a recent work by Jones and Bulley which toured UK forests in 2014. The work comprises a spatial multi-agent system where agents represent organisms of different species of flora and fauna—beetles, birds, mammals, fungi, and plants. The population densities of species are parameterized by data collected from each site, reflecting the local biodiversity and community structure and further influenced by local weather conditions (wind speed and direction, sunlight, rainfall) streamed from an onsite weather station. Different species are sonified by distinct musical motifs, composed from fragments of acoustic instrumental recordings and generated according to the current simulation state. The simulation runs at quasi realtime, such that the movement of agents across the simulated world are mapped across the real world, heard as movement across speakers. The audience experience the real-time dynamics of a simulated ecosystem which portrays the changing activity of the forest.

13.5.2 Computational Ecosystems for Creative Discovery

The prospect that an ecosystemic approach to algorithm design and implementation might afford a richer, creatively productive means of interaction motivates a revised view of human-machine co-creativity, described by McCormack in the language of computational creative discovery: 'where computer processes assist in enhancing human creativity or may autonomously exhibit creative behaviour independently' (McCormack 2007, 1). The intention is to develop ways of working with technology that achieve creative possibilities unattainable from any existing (software) tools or methods. McCormack has looked at extending the EAs and agent-based systems described above, incorporating adaptive behaviour at the ecosystem level. Thus, rather than thinking of evolution as applying to the optimization of a specific task (or satisfaction of a specific human preference goal), as in many of the approaches outlined in section 13.2, it is treated as more of a free-for-all in which self-organization may also play a significant role in the process, evoking Darwin's image of the 'tangled bank, clothed with many plants of many kinds' (Darwin 1861, 524).

In works such as the audiovisual installation *Eden*, McCormack develops upon earlier adaptive agent-based models (Holland 1999) and ALife art, which deployed IGAs in exhibition contexts (e.g. Sommerer and Mignonneau 1997) to create an evolutionary agent-based system in which physical presence of audience members (monitored via IR cameras) becomes a resource for a population of virtual agents, whose behaviour then seeks to best maintain their attention (McCormack 2001). Later, McCormack and colleagues engaged in a more systematic study of ecosystemic approaches to algorithmic art. The standard EA was extended from modelling isolated genotypes to the digital specification of an environment and agent's interactions with it and each other; rather than explicitly or interactively defining a fitness function, evolutionary pressure was exerted via competition for an abstract resource, which could in turn alter the environmental structure. Within this framework, various evolutionary processes and principles were explored, including resource recycling, mutualism, evolution of generalism and specialism, and niche construction.

That heterogeneity of resources in an environment can give rise to complex agent behaviour at the population level was established in early agent-based models (Epstein 1996). Other evolutionary models consider what conditions are required to establish and sustain species diversity and interspecies interaction. In musical applications we have seen one example already in the competitive coevolution of 'singers' and 'listeners' (see section 13.2.3). Organism interactions are not limited to direct competition however: the myriad of symbiotic relationships observed in the natural world provide rich inspiration for achieving synchronic diversity in creative system design. The sonic ecosystem Filterscape (Eldridge and Dorin 2009) was built to explore the conditions under which energy recycling can lead to the emergence of cooperative as well as competitive survival strategies, increasing the *behavioural* diversity of a generative system as well as environmental heterogeneity.

The ecological concept of niche was also explored as a means to investigate the evolution of generalism and specialism. A niche describes the role of an organism: the ways in which it interacts with and depends upon other elements in its environment (Elton 1927). The structure of individual niches and the relationships between them provide a means to understand competition and analyse the species composition of a community, as well as its stability. The *width* of a niche describes the degree of specialization of an individual or species: a specialist has a narrow niche, it occupies a limited and particular habitat or set of activities; a generalist occupies a broad niche and can make use of a wider diversity of actions and habitats. Niche overlap provides a measure of competition between species. Drawing from early work in theoretical ecology, Eldridge and Dorin demonstrated that the trajectory of an evolving population through phenotype space could be controlled by explicitly manipulating the width of the niche, biasing the evolution of generalist or specialist survival strategies (Eldridge, Dorin, and McCormack 2008). This provided a means to support the generation of complexity, whilst providing a simple, intuitive 'handle' with which a user could exert some degree of influence over an otherwise intractable complex system.

McCormack and Bown subsequently implemented a form of niche *construction*—the process whereby organisms, through their activities and choices, modify their own and each other's niches—as a means to increase environmental heterogeneity and so more complex structures in music and art. For example, the line-drawing agents of Annunziato and Pierucci (2000) were adapted to include a niche construction element within an evolutionary framework (McCormack and Bown 2009). Agents draw different types of lines (e.g. varying curvature or density) and have a preference for local environments which contain certain line structures. Line drawing tendencies and preferences are both encoded genetically. Coevolution of tendency and preference in localized areas acts to create local variation and so greater global diversity. Bown and McCormack also developed various sonic ecosystem models that made use of the acoustic environment as a niche, following Di Scipio, that could both be modified and have an influence on the evolution of agents (Bown and McCormack 2010). They then looked at general principles for designing creative ecosystems, including design-focused principles such as maximizing the amount of knowledge you have about the system, and more specific issues such as understanding when and how boundaries might emerge in spatial networks of agents (Bown, McCormack, and Kowaliw 2011).

The experiments begin to show how ecological principle can be applied in the development of software for creative discovery to increase complexity of structure or behaviour, whilst allowing some degree of user influence, in some cases. Like many of the projects described here, such work offers new perspectives on how we conceptualize creative collaboration as we increase the generative autonomy of software.

13.6 LOOKING FORWARD

The principles of cybernetics and ALife established during the twentieth century continue to feed into experimental practice in algorithmic music, which is itself evolving new tools and methods. But a considerable challenge remains—and this is as much a challenge of HCI as of algorithm design—to find effective and usable ways to exploit these principles in compositional practice. Biologically inspired computational creativity has found its way into commercial software in a small number of cases, but has not yet come of age.

Likewise, the vision of virtual worlds autonomously spawning the kinds of rich complexity seen in natural systems remains an elusive goal in ALife research (Bedau et al. 2000). This may simply be because we have not arrived at the order of magnitude required, because there are still missing elements to our algorithmic toolkit, or perhaps because there is a more fundamental difference in kind between digital computers and the biological world. Some argue that if it is possible in nature it should be possible in virtual worlds; others assert that computation has nothing to do with what goes on in brains and other biological systems (Harvey 1997).

But although inspired by the sciences of the artificial, biologically inspired algorithmic composition does not share the same aims and is not hampered by what some see as disappointments. The snapshot of practice we have given in this chapter is symptomatic of the creative fluidity found throughout algorithmic music: the algorithmic musician is seen weaving together equations from complex systems, biology, social behaviour, and so on, to create new artistic works that employ nature in their behaviour, rather than merely representing nature. The changing technological environment also seems to be shifting towards one in which biologically inspired and agent-based approaches may have new grounds for making relevant contributions; massively distributed computing networks, increasingly high-level application program interfaces, the widespread scriptability of advanced software platforms, the connectivity of web programming, and the rich ecosystems of the Internet of Things, all potentially offer new creative environments for the application of the methods described in this chapter.

As this practice progresses, we have argued, there is no need to remain true to biology, social behaviour, or human musicality. Indeed, perhaps the greatest value of a biologically inspired approach is that it offers a radically different starting point for the coevolution of new forms of music making. In this sense we see two key themes for future exploration: a further advancement in understanding the role of software in the co-creation of music making, through the concepts of autonomy and agency, and the value of biological metaphors in supporting the creative development and cultural sharing of computer music.

Autonomy and Agency. Autonomy can be usefully understood as the ability of a system to control its own future state rather than to be controlled by external factors. This has proven challenging to measure, and it is arguably the case that qualitative, narrative approaches to understanding autonomy may be especially relevant in the application area of music. The most common narratives surrounding autonomy relate to two primary forms of adaptation: evolution and learning. Although it is hard to witness the autonomy of the slowest of organisms, including plants and trees, we understand that species are autonomous by virtue of their evolutionary inception; they need no help in surviving. Likewise, learning creates a situation in which the actions of an individual may be the result of an inductive analysis of their environment. We might therefore appreciate that learning is one way for a machine to acquire behaviour which has not been programmed directly, and recent advances in deep learning have driven home this notion to a wider public (Devlin 2015).

Closely related to autonomy, and more central to questions of creativity, is the notion of agency: the ability to have influence on the world, as in the production of an artwork. How and when we attribute creative agency to software systems is a question being asked by many of the researchers mentioned in this field. Autonomy and agency may not take the obvious forms associated with the simulation of human behaviour, and an important challenge is to come to terms with broader notions of creative software agency, for which biological and social science understandings of action and interaction will inevitably act as fertile sources of ideas, as well as philosophical accounts of intentionality (Dennett 1987) and expressivity (Linson 2013).

The Poetics of Biologically Inspired Algorithms. Algorithmic music in general utilizes a plethora of computer models derived from across the spectrum of scientific disciplines, but in many ways biological models are the most exotic and evocative, posing conceptual challenges for how we think about music, and linking the most hypermodern of activities to the most primitive. Biologically inspired models offer an alluring narrative for audiences and computer music makers. Beside epistemological, technical, and creative opportunities then, biologically inspired models hold potential as vehicles for both cognitive and cultural engagement.

In the forward to *Visual Complexity*, Lev Manovich highlights data art as a new medium for critical reflection on the world: 'Figurative artists express their opinions about the world by choosing what they paint. ... Now artists can also talk about our world by choosing which data to visualize' (Lima 2006). In an analogous way, the choice of model or system design in algorithmic composition becomes a vehicle for expression and comment—poetic, philosophical, or political—about our world and our relationship to it (Eldridge 2012). In offering a familiar narrative frame, biological models and metaphors support engagement for both creator and audience. During the design and development of new works, such metaphors support rich 'system stories' (Whitelaw 2005), providing a cognitive scaffold for the coder. Similarly, for the audience, familiar narratives provide a 'way in' to algorithmic music, which can otherwise be less than approachable to wider audiences (Garnett 2001; Stubbs 2009). At a time when technology develops as fast as our natural environment is threatened, biologically inspired models offer a valuable vehicle for reflecting upon the relationships between the biological, cultural, and technological worlds in which we live, and for sharing complex concepts and aesthetics with audiences.

BIBLIOGRAPHY

Annunziato, M., and Pierucci, P. 'Emergent Relationships'. In *Proceedings of the ALife VII*. Portland, OR, 2000.

Ascott, R. 'Behaviourist Art and the Cybernetic Vision'. *Cybernetica: Journal of the International Association for Cybernetics* 10, no. 4 (1967): 25–56.

Ashby, W. R. *An Introduction to Cybernetics*. London: Chapman and Hall, 1956.

Ashby, W. R. *Design for a Brain: The Origin of Adaptive Behaviour*. London: John Wiley, 1960.

Ashby, W. R. 'Ashby's Card Index of Aphorisms'. *Ashby's Notebooks*. http://rossashby.info/aphorisms.html. 2008.

Axelrod, R., and Hamilton, W. D. 'The Evolution of Cooperation'. *Science* 211, no. 4489 (1981): 1390–1396.

Bak, P., Tang, C., and Wiesenfeld, K. 'Self-Organized Criticality'. *Physical Review* A 38, no. 1 (1988): 364.

Barkow, J. H., Cosmides, L., and Tooby, J. *The Adapted Mind: Evolutionary Psychology and the Generation of Culture*. New York: Oxford University Press, 1992.

Bedau, M., McCaskill, J., Packard, N., Rasmussen, S., Adami, C., Green, D., Ikegami, T., Kaneko, K., and Ray, T. 'Open Problems in Artificial Life'. *Artificial Life* 6 (2000): 363–376.

Beer, R. D. 'The Dynamics of Adaptive Behavior: A Research Program'. *Robotics and Autonomous Systems* 20 (1997): 257–289.

Bentley, P. J., and O'Reilly, U. 'Ten Steps to Make a Perfect Creative Evolutionary Design System'. In *GECCO 2001 Workshop on Non-Routine Design with Evolutionary Systems*. 2001.

Beyls, P. 'Chaos and Creativity: The Dynamic Systems Approach to Musical Composition'. *Leonardo Music Journal* 1, no. 1 (1991) 31–36.

Beyls, P., Bernardes, G., and Caetano, M. 'The Emergence of Complex Behavior as an Organizational Paradigm for Concatenative Sound Synthesis'. In *Proceedings of the Third Conference on Computation, Communication, Aesthetics and X*, edited by A. Clifford, M. Carvalhais, and M. Verdicchio, 184–199. Porto: Universidade do Porto, 2015.

Biles, J. A. 'GenJam: A Genetic Algorithm for Generating Jazz Solos'. In *Proceedings of the 1994 International Computer Music Conference*, 131–137. San Francisco, CA: ICMA, 1994.

Birkhoff, G. D. *Aesthetic Measure*. Cambridge, MA: Harvard University Press, 1933.

Blackwell, T. 'Swarm Music: Improvised Music with Multi-Swarms'. In *Proceedings of the AISB '03 Symposium on Artificial Intelligence and Creativity in Arts and Science*, 41–49. Aberystwyth: Society for the Study of Artificial Intelligence and the Simulation of Behaviour, 2003.

Bowers, J. *Improvising Machines: Ethnographically Informed Design for Improvised Electro-Acoustic Music*. Master's thesis, School of Music, University of East Anglia, Norwich, 2002.

Bown, O. 'A Framework for Ecosystem-Based Generative Music'. In *Proceedings of the SMC 2009*, edited by G. Gouyon, A. Barbosa, and X. Serra, 195–200. Porto, 2009.

Bown, O. 'Experiments in Modular Design for the Creative Composition of Live Algorithms'. *Computer Music Journal* 35, no. 3 (2011): 73–85.

Bown, O. 'Empirically Grounding the Evaluation of Creative Systems: Incorporating Interaction Design'. In *Proceedings of the Fifth International Conference on Computational Creativity*, edited by S. Colton, D. Ventura, N. Lavrač, and M. Cook, 112–119. Ljubljana, 2014.

Bown, O., Eldridge, A., and McCormack, J. 'Understanding Interaction in Contemporary Digital Music: From Instruments to Behavioural Objects'. *Organised Sound* 14, no. 2 (2009): 188–196.

Bown, O., and Lexer, S. 'Continuous-Time Recurrent Neural Networks for Generative and Interactive Musical Performance'. In *Applications of Evolutionary Computing: EvoWorkshops 2006*, edited by F. Rothlauf et al., 652–663. Berlin: Springer, 2006.

Bown, O., and McCormack, J. 'Taming Nature: Tapping the Creative Potential of Ecosystem Models in the Arts'. *Digital Creativity* 21, no. 4 (2010): 215–231.

Bown, O., McCormack, J., and Kowaliw, T. 'Ecosystemic Methods for Creative Domains: Niche Construction and Boundary Formation'. In *IEEE ALife 2011: Symposium on Artificial Life*, 132–139. Piscataway, NJ: Institute of Electrical and Electronic Engineers, 2011.

Brooks, R. A. 'How to Build Complete Creatures rather than Isolated Cognitive Simulators'. In *Architectures for Intelligence: The 22nd Carnegie Mellon Symposium on Cognition*, edited by K VanLehn, 225–239. Hillsdale, NJ: Lawrence Erlbaum, 1991.

Brooks, R. A. *Cambrian Intelligence: The Early History of the New AI*. Cambridge, MA: MIT Press, 1999.

Bryson, J. 'The Reactive Accompanist: Adaptation and Behavior Decomposition in a Music System'. In *The Biology and Technology of Intelligent Autonomous Agents*, edited by L. Steels, 365–376. New York: Springer, 1995.

Burnham, J. *Beyond Modern Sculpture: The Effects of Science and Technology on the Sculpture of this Century*. New York: George Braziller, 1968.

Burraston, D., Edmonds, E., Livingstone, D., and Miranda, E. 'Cellular Automata in MIDI-Based Computer Music'. In *Proceedings of the 2004 International Computer Music Conference*. Miami, 2004.

Burton, A. R., and Vladimirova, T. 'Generation of Musical Sequences with Genetic Techniques'. *Computer Music Journal* 23, no. 4 (1999): 59–73.

Camazine, S., Deneubourg, J.-L., Franks, N. Sneyd, J., Theraulaz, G., and Bonabeau, E. *Self-Organization in Biological Systems*. Princeton, NJ: Princeton University Press, 2001.

Cardew, C. *The Great Learning Paragraphs 2 and 7*. Deutsche Grammophon and Universal Classics 471 572 (1971). LP.

Chowning, J., and Bristow, D. *FM Theory and Applications: By Musicians for Musicians*. New York: Hal Leonard, 1987.

Clarke, E. F. *Ways of Listening: An Ecological Approach to the Perception of Musical Meaning*. Oxford: Oxford University Press, 22005.

Collins, R. J., and Jefferson, D. R. 'Selection in Massively Parallel Genetic Algorithms'. In *Proceedings of the Fourth International Conference on Genetic Algorithms*, 249–256. San Francisco, CA: Morgan Kaufmann, 1991.

Cope, D. *Experiments in Musical Intelligence*. Madison, WI: A-R Editions, 1996.

Dahlstedt, P. (2001). 'A Mutasynth in Parameter Space: Interactive Composition through Evolution'. *Organised Sound* 6, no. 2 (2001): 121–124.

Dahlstedt, P., and Nordahl, M. G. 'Living Melodies: Coevolution of Sonic Communication'. *Leonardo* 34, no. 3 (2001): 243–248.

Danforth, C. M. 'Chaos in an Atmosphere Hanging on a Wall'. *Mathematics of Planet Earth*, 17 March 2013. http://mpe.dimacs.rutgers.edu/2013/03/17/chaos-in-an-atmosphere-hanging-on-a-wall/.

Darwin, C. *On the Origin of Species by Means of Natural Selection, or the Preservation of Favoured Races in the Struggle for Life*. London: John Murray, 1861.

Dawkins, R. *The Blind Watchmaker: Why the Evidence of Evolution Reveals a Universe without Design*. London: Penguin, 1986.

DeLanda, M. (2002). 'Deleuze and the Use of the Genetic Algorithm in Architecture'. http://www.cddc.vt.edu/host/delanda/pages/algorithm.htm. Accessed 20 June 2017.

Dennett, D. *The Intentional Stance*. Cambridge, MA: MIT Press, 1987.

Devlin, H. 'Google Develops Computer Program Capable of Learning Tasks Independently'. *The Guardian*, 25 February 2015. https://www.theguardian.com/technology/2015/feb/25/google-develops-computer-program-capable-of-learning-tasks-independently.

Di Paolo, E. A. 'Organismically-Inspired Robotics: Homeostatic Adaptation and Teleology beyond the Closed Sensorimotor Loop'. In *Dynamic Systems Approach to Embodiment and Sociality*, edited by K. Murase and T. Asakura, 19–42. Adelaide: Advanced Knowledge International, 2003.

Di Scipio, A. 'Sound is the Interface: From Interactive to Ecosystemic Signal Processing'. *Organised Sound* 8, no. 3 (2003): 269–277.

Didkovsky, N. '"Lottery": Toward a Unified Rational Strategy for Cooperative Music-Making'. *Leonardo Music Journal* 2, no. 1 (1992): 3–12.

Dittrich, P., Ziegler, J. C., and Banzhaf, W. 'Artificial Chemistries: A Review'. *Artificial Life* 7, no. 3 (2001): 225–275.

Dixon, S. 'A Lightweight Multi-Agent Musical Beat Tracking System'. In *PRIC 2000: Topics in Artificial Intelligence*, edited by R Mizoguchi and J. Slaney, 778–788. Berlin: Springer, 2000.

Draves, S. 'The Electric Sheep Screen-Saver: A Case Study in Aesthetic Evolution'. In *Applications of Evolutionary Computing: EvoWorkshops 2005*, edited by F. Rothlauf et al., 458–467. Berlin: Springer, 2005.

Eiben, A. E. and Smith, J. E. *Introduction to Evolutionary Computing*. Berlin: Springer Science and Business Media, 2003.

Eigenfeldt, A. 'The Creation of Evolutionary Rhythms within a Multi-Agent Networked Drum Ensemble'. In *Proceedings of the 2007 International Computer Music Conference*, 267–270. Copenhagen, 2007a.

Eigenfeldt, A. 'Drum Circle: Intelligent Agents in Max/MSP'. In *Proceedings of the 2007 International Computer Music Conference*, 9–12. Copenhagen, 2007b.

Eigenfeldt, A. 'Emergent Rhythms through Multi-Agency in Max/MSP'. In *Computer Music Modeling and Retrieval Symposium: Sense of Sounds*, edited by R. Kronland-Martinet, S. Ystad, and K. Jensen, 368–379. Berlin: Springer, 2008.

Eigenfeldt, A., Bown, O., and Casey, B. 'Collaborative Composition with Creative Systems: Reflections on the First Musebot Ensemble'. In *Proceedings of the Sixth International Conference on Computational Creativity*, edited by H. Toivonen, S. Colton, M. Cook, and D. Ventura, 134–141. Park City, Utah, 2015.

Eldridge, A. 'Cyborg Dancing: Generative Systems for Man-Machine Musical Improvisation'. In *Proceedings of the Third Iteration*, edited by T. Innocent, 129–142. Clayton: Centre for Electronic Media Art (CEMA), 2005.

Eldridge, A. 'Self-Directed Feedback Loops–OSX Application'. 2008. http://ecila.org/. Accessed 20 June 2017.

Eldridge, A. 'The Poietic Power of Generative Sound Art for an Information Society'. In *Proceedings of the 2012 International Computer Music Conference*, edited by M. Matija, M. Kaltenbruger, and M. Cigar, 67–81. Ljubljana, Slovenia, 2012.

Eldridge, A. 'Improvising with Self-Observing Systems: A Duet for Cellist and Adaptive Delay Network'. In *Proceedings of the first conference on Computation, Communication, Aesthetics and X*, edited by M. Verdicchio and M. Carvalhais, 297–301. San Francisco, CA: ICMA, 2013.

Eldridge, A., and Dorin, A. 'Filterscape: Energy Recycling in a Creative Ecosystem'. In *Applications of Evolutionary Computing: EvoWorkshops 2009*, edited by M. Giacobini et al., 508–517. Berlin: Springer, 2009.

Eldridge, A., Dorin, A., and McCormack, J. 'Manipulating Artificial Ecosystems'. In *Applications of Evolutionary Computing: EvoWorkshops 2008*, edited by M. Giacobini et al., 392–401. Berlin: Springer, 2008.

Elton, C. S. *Animal Ecology*. London: Sidgwick and Jackson, 1927.

Epstein, J. *Growing Artificial Societies: Social Science from the Bottom Up*. Washington, DC: Brookings Institution, 1996.

Ermentrout, G. B., and Edelstein-Keshet, L (1993). 'Cellular Automata Approaches to Biological Modeling'. *Journal of Theoretical Biology* 160, no. 1 (1993): 97–133.

Froese, T., and Stewart, J. 'Life after Ashby: Ultrastability and the Autopoietic Foundations of Biological Autonomy'. *Cybernetics and Human Knowing* 17, no. 4 2010): 7–49.

Futayama, D., and Slatkin, M. *Coevolution.* Sunderland: Sinauer, 1983.

Garcia-Almanza, A. L., and Tsang, E. P. K. 'Simplifying Decision Trees Learned by Genetic Programming'. In *2006 IEEE Congress on Evolutionary Computation,* 2142–2148. IEEE, 2006.

Garnett, G. 'The Aesthetics of Interactive Computer Music'. *Computer Music Journal* 25, no. 1 (2001): 21–33.

Green, O. 'Agility and Playfulness: Technology and Skill in the Performance Ecosystem'. *Organised Sound* 16, no. 2 (2011): 134–144.

Gresham-Lancaster, S. 'The Aesthetics and History of The Hub: The Effects of Changing Technology on Network Computer Music'. *Leonardo Music Journal* 8 (1998): 39–44.

Haque, U. 'The Architectural Relevance of Gordon Pask'. *Architectural Design* 77 (2007): 54–61.

Harrald, L. 'Fight or Flight: Towards the Modelling of Emergent Ensemble Dynamics'. In *Proceedings of the Australasian Computer Music Conference 2005,* edited by T. Opie and A. R. Brown, 68–74. Fitzroy: Australian Computer Music Association, 2005.

Harvey, I. 'Cognition Is Not Computation: Evolution Is Not Optimisation'. In *Artificial Neural Networks—ICANN '97 LNCS,* 685–690. Berlin: Springer, 1997.

Hinton, G. E., and Nowlan, S. J. 'How Learning Can Guide Evolution'. *Complex Systems* 1 (1987): 495–502.

Hoffmann, P. 'Towards an "Automated Art": Algorithmic Processes in Xenakis' Compositions'. *Contemporary Music Review* 21, nos. 2–3 (2002): 121–131.

Hogeweg, P. 'Cellular Automata as a Paradigm for Ecological Modeling'. *Applied Mathematics and Computation* 27, no. 1 (1988): 81–100.

Holland, J. *Adaptation in Natural and Artificial Systems.* Ann Arbor: University of Michigan Press, 1975.

Holland, J. H. 'Echoing Emergence: Objectives, Rough Definitions, and Speculations for ECHO-Class Models'. In *Complexity: Metaphors, Models, and Reality,* edited by G. A. Cowan, D. Pines, and D. Meltzer, 309–342. Boulder, CO: Perseus, 1999.

Husbands, P. 'An Ecosystems Model for Integrated Production Planning'. *International Journal of Computer Integrated Manufacturing* 6, nos. 1–2 (1993): 74–86.

Husbands, P., Copley, P., Eldridge, A., and Mandelis, J. 'An Introduction to Evolutionary Computing for Musicians'. In *Evolutionary Computer Music,* edited by E. R. Miranda and A. Biles, 1–27. London: Springer, 2007.

Impett, J. 'Interaction, Simulation and Invention: A Model for Interactive Music'. In *Proceedings of ALMMA 2001 Workshop on Artificial Models for Musical Applications,* 108–119. Cosenza: Bios, 2001.

Kaschube, M., Schnabel, M., Löwel, S., Coppola, D. M., White, L. E., and Wolf, F. 'Universality in the Evolution of Orientation Columns in the Visual Cortex'. *Science* 330, no. 6007 (2010): 1113–1116.

Kauffman, S. *The Origins of Order: Self-Organization and Selection in Evolution.* New York Oxford University Press, 1993.

Kelso, J. *Dynamic Patterns: The Self-Organization of Brain and Behavior.* Cambridge, MA: MIT Press, 1995.

Kiefer, C. 'Interacting with Text and Music: Exploring Tangible Augmentations to the Live Coding Interface'. In *Proceedings of the International Conference for Life Interfaces.* Lisbon, 2014.

Kim-Boyle, D. 'Sound Spatialization with Particle Systems'. In *Proceedings of the 8th International Conference on Digital Audio Effects (DAFX-05),* 65–68. Madrid, 2005.

Kirby, S., and Hurford, J. 'Learning, Culture and Evolution in the Origin of Linguistic Constraints'. In *Fourth European Conference on Artificial Life*, edited by P. Husbands and I. Harvey, 493–502. Cambridge, MA: MIT Press, 1997.

Koza, J. R. *Genetic Programming: On the Programming of Computers by Means of Natural Selection*. Cambridge, MA: MIT Press, 1992.

Langton, C. 'Artificial Life'. In *Artificial Life: Santa Fe Institute Studies in the Sciences of Complexity*, edited by C. Langton, 2–3. Boulder, CO: Addison-Wesley, 1989.

Large, E. W., and Palmer, C. 'Perceiving Temporal Regularity in Music'. *Cognitive Science* 26 (2002): 1–37.

Larson, S. 'Musical Forces and Melodic Patterns'. *Theory and Practice* 22 (1997): 55–71.

Lima, M. *Visual Complexity: Mapping Patterns of Information*. Princeton, NJ: Princeton Architectural Press, 2006.

Linson, A. 'The Expressive Stance: Intentionality, Expression and Machine Art'. *International Journal of Machine Consciousness* 5, no. 2 (2013): 195–216.

Linson, A., Dobbyn, C., Lewis, G. E., and Laney, R. 'A Subsumption Agent for Collaborative Free Improvisation'. *Computer Music Journal* 39, no. 4 (2015): 96–115.

Lorenz, E. N. 'Deterministic Nonperiodic Flow'. *Journal of the Atmospheric Sciences* 20, no. 2 (1963): 130–141.

Magnus, C. 'Evolutionary Musique Concrete'. In *Applications of Evolutionary Computing: EvoWorkshops 2006*, edited by F. Rothlauf et al., 688–695. Berlin: Springer, 2006.

Maturana, H. R., and Varela, F. J. *The Tree of Knowledge: The Biological Roots of Human Understanding*. Boston, MA: New Science Library/Shambhala, 1987.

Maynard Smith, J. *Evolution and the Theory of Games*. New York: Cambridge University Press, 1982.

McCormack, J. 'Interactive Evolution of L-System Grammars for Computer Graphics Modelling'. In *Complex Systems: From Biology to Computation*, edited by T. Bossomaier and D. Green, 118–130. Amsterdam: IOS, 1993.

McCormack, J. 'Eden: An Evolutionary Sonic Ecosystem'. In *Advances in Artificial Life: 7th European Conference, ECAL 2003*, edited by W. Banzhaf, T. Christaller, P. Dittrich, J. T. Kim, and J. Siegler, 133–142. Berlin, Springer, 2001.

McCormack, J. 'Artificial Ecosystems for Creative Discovery'. In *Proceedings of the 9th Annual Conference on Genetic and Evolutionary Computation*, 301–307. New York: ACM, 2007.

McCormack, J. 'Facing the Future: Evolutionary Possibilities for Human-Machine Creativity'. In *The Art of Artificial Evolution: A Handbook on Evolutionary Art and Music*, edited by J. Romero and P. Machado, 417–451. New York: Springer, 2008.

McCormack, J. 'Creative Ecosystems'. In *Computers and Creativity*, 39–60. Berlin: Springer, 2012.

McCormack, J., and Bown, O. 'Life's What You Make: Niche Construction and Evolutionary Art'. In *Applications of Evolutionary Computing: EvoWorkshops 2009*, edited by M. Giacobini et al., 528–537. Berlin: Springer, 2009.

McCormack, J., and Dorin, A. 'Art, Emergence and the Computational Sublime'. In *Proceedings of Second Iteration: A Conference on Generative Systems in the Electronic Arts.*, 67–81. Melbourne: CEMA, 2001.

McDermott, J., Sherry, D., and O'Reilly, U.-M. 'Evolutionary and Generative Music Informs Music HCI—and Vice Versa'. In *Music and Human-Computer Interaction*, edited by S. Holland, K. Wilkie, P. Mulholland, and A. Seago, 223–240. London: Springer, 2013.

Mendel, G. (1866). *Versuche über Pflanzenhybriden*. Vol. 4. <u>Brünn</u>: Verlage des Vereines, 1866.

Millen, D. 'Cellular Automata Music'. In *Proceedings of the 1990 International Computer Music Conference*, 314–316. Glasgow, 1990.

Minsky, M. *Society of Mind*. New York: Simon and Schuster, 1988.

Miranda, E. R. 'Granular Synthesis of Sounds by Means of a Cellular Automaton'. *Leonardo* 28, no. 4 (1995): 297–300.

Miranda, E. R. 'Towards an Artificial Life Approach to the Origins of Music'. In *Proceedings of the XIX Annual Congress of the Brazilian Computing Society / VI Brazilian Symposium on Computer Music*. Rio de Janeiro: Brazilian Computing Society, 1999.

Miranda, E. R. 'On the Music of Emergent Behaviour: What Can Evolutionary Computation Bring to the Musician?' *Leonardo* 36, no 1 (2003): 55–59.

Miranda, E. R. 'Emergent Sound Repertoires in Virtual Societies'. *Computer Music Journal* 26, no. 2 (2002a): 77–90.

Miranda, E. R. 'Sounds of Artificial Life'. In *Proceedings of the ACM SIGCHI International Conference: Creativity and Cognition*, 173–177. New York: ACM, 2002b.

Miranda, E. R., and Biles, A. *Evolutionary Computer Music*. Berlin: Springer, 2007.

Miranda, E. R., Kirby, S., and Todd, P. 'On Computational Models of the Evolution of Music: From the Origins of Musical Taste to the Emergence of Grammars'. *Contemporary Music Review* 22, no. 3 (2003): 91–111.

Mitchell, M. *An Introduction to Genetic Algorithms*. Cambridge, MA: MIT Press, 1998.

Murray-Rust, D., Smaill, A., and Edwards, M. 'MAMA: An Architecture for Interactive Musical Agents'. *Frontiers in Artificial Intelligence and Applications* 141 (2006): 36–40.

Neumann, J. V. *Theory of Self-Reproducing Automata*. Champaign: University of Illinois Press, 1966.

Nicolis, G., and Prigogine, I. *Self-Organization in Nonequilibrium Systems: From Dissipative Structures to Order through Fluctuations*. New York: Wiley, 1977.

Pask, G. (1971). 'A Comment, a Case History and a Plan'. In *Cybernetics, Art and Ideas*, edited by J. Reichardt, 76–99. New York: New York Graphic Society, 1971.

Reichardt, J., ed. *Cybernetics, Art and Ideas*. New York: New York Graphic Society, 1971.

Reynolds, C. W. 'Flocks, Herds and Schools: A Distributed Behavioural Model'. In *SIGGRAPH '87: Proceedings of the 14th Annual Conference on Computer Graphics and Interactive Techniques*, 25–34. New York: ACM, 1987.

Romero, J. J., and Machado, P., eds. *The Art of Artificial Evolution: A Handbook on Evolutionary Art and Music*. New York: Springer, 2008.

Sanfilippo, D. 'Turning Perturbation into Emergent Sound, and Sound into Perturbation'. *Interference* 3 (2013): 1–16. http://www.interferencejournal.org/turning-perturbation-into-emergent-sound/.

Schelling, T. C. 'Dynamic Models of Segregation'. *Journal of Mathematical Sociology* 1, no. 2 (1971): 143–186.

Searle, J. R. *Speech Acts: An Essay in the Philosophy of Language*. Cambridge: Cambridge University Press, 1969.

Secretan, J., Beato, N., Ambrosio, D. B. D., Rodriguez, A., Campbell, A., and Stanley, K. O. 'Picbreeder: Evolving Pictures Collaboratively Online'. In *CHI '08: Proceeding of the Twenty-Sixth Annual SIGCHI Conference on Human Factors in Computing Systems*, 1759–1768. New York: ACM, 2008.

Shiffman, D., Fry, S., and Marsh, Z. *The Nature of Code*. Privately printed, 2013.

Sims, K. 'Artificial Evolution for Computer Graphics'. *Computer Graphics* 25, no. 4 (1991): 319–328.

Sommerer, C., and Mignonneau, L. ' "A-Volve": An Evolutionary Artificial Life Environment'. In *Artificial Life V: Proceedings of the Fifth International Workshop on the Synthesis and Simulation of Living Systems*, edited by C. G. Langton and K. Shimohara, 167–175. Cambridge, MA: MIT Press, 1997.

Stubbs, D. *Fear of Music: Why People Get Rothko but Don't Get Stockhausen*. Ropley, UK: John Hunt, 2009.

Sudnow, D. *Ways of the Hand: The Organization of Improvised Conduct*. Cambridge, MA: MIT Press, 1978.

Takagi, H. 'Interactive Evolutionary Computation: Fusion of the Capabilities of EC Optimization and Human Evaluation'. *Proceedings of the IEEE* 89, no. 9 (2001): 1275–1296.

Tansley, A. 'The Use and Abuse of Vegetational Concepts and Terms'. *Ecology* 16 (1935): 284–307.

Toch, E. *The Shaping Forces in Music: An Inquiry into the Nature of Harmony, Melody, Counterpoint and Form*. New York: Dover, 1977.

Todd, S., and Latham, W. 'Mutator, a Subjective Human Interface for Evolution of Computer Sculptures'. Technical report, IBM, United Kingdom Scientific Center Report, 1991.

Todd, S., and Latham, W. 'The Mutation and Growth of Art by Computers'. In *Evolutionary Design by Computers*, edited by P. Bentley, 221–250. 2nd ed. San Francisco, CA: Morgan Kaufmann, 1999.

Todd, P., and Werner, G. 'Frankensteinian Methods for Evolutionary Music Composition'. In *Musical Networks: Parallel Distributed Perception and Performance*, edited by N. Griffith and P. Todd, 173–190. Cambridge, MA: MIT Press, 1998.

Varela, F. G., Maturana, H. R., and Uribe, R. 'Autopoiesis: The Organization of Living Systems, Its Characterization and a Model'. *Biosystems* 5, no. 4 (1974): 187–196.

Von Bertalanffy, L. 'An Outline of General System Theory'. *British Journal for the Philosophy of Science* 1, no. 2 (1950): 134–165.

Wallace, A. R. (1858). 'On the Tendency of Varieties to Depart Indefinitely from the Original Type'. *Proceedings of the Linnean Society of London* 3 (1858): 53–62.

Walter, W. G. 'Imitation of Life'. *Scientific American* 182, no. 5 (1950): 42–45.

Waschka, R. II 'Theories of Evolutionary Algorithms and a "New Simplicity" Opera: Making Sappho's Breath'. In *Artificial Life Models for Musical Applications*, 79–86. Cosenza: Editoriale Bios, 2001.

Waters, S. 'Performance Ecosystems: Ecological Approaches to Musical Interaction'. In *Proceedings of Electro Acoustic Studies Network*. 2007.

Whitelaw, M. *Metacreation: Art and Artificial Life*. Cambridge, MA: MIT Press, 2004.

Whitelaw, M. 'System Stories and Model Worlds: A Critical Approach to Generative Art'. In *Readme 100: Temporary Software Art Factory*, 135–154. Norderstedt: BoD, 2005.

Wiggins, G. A. 'The Mind's Chorus: Creativity before Consciousness'. *Cognitive Computation* 4, no. 3 (2012): 306–319.

Wilkinson, G. S. 'Reciprocal Food Sharing in the Vampire Bat'. *Nature* 308, no. 5955 (1984): 181–184.

Xenakis, I. 'Free Stochastic Music'. In *Cybernetics, Art and Ideas*, edited by J. Reichardt, 124–142. New York: New York Graphic Society, 1971.

Yee-King, M., and Roth, M. 'Synthbot: An Unsupervised Software Synthesizer Programmer'. In *Proceedings of the 2008 International Computer Music Conference*. Belfast, 2008.

CHAPTER 14

PERFORMING WITH PATTERNS OF TIME

THOR MAGNUSSON AND ALEX MCLEAN

14.1 INTRODUCTION

> The process of creating music involves ... a working knowledge of all the processes of transformation which can aesthetically be applied to [patterns of sound]. Beyond these there needs to be a practised awareness of how such materials and operations, and the specific characteristics of each, relate to and influence each others' potentials. (Spiegel 1981)

VARÈSE famously got so tired of people's comment that his work was 'interesting', immediately followed by the question 'but is it music?', that he decided to call the outcome of his practice 'organised sound' (Varèse 1966, 18). A related definition explored here is of music as patterns of sound. These patterns are about relationships between sonic events (whether pitched or not) taking place in time, a strongly mathematical domain which has been extensively explored (for example Burack 2005; Du Sautoy 2003; Fauvel, Flood, and Wilson 2006). It is also a cognitive domain: in music psychology, pattern recognition is seen as a principal feature of human cognition, perhaps explaining the fascination many humans have in the repetitive nature of music, where the *re-cognition* of patterns forms units or words that build up a larger musical meaning (Sloboda 2005, 18). It is fitting that in one of the key texts on computational music, Taube's *Notes from the Metalevel*, we find a reference to the philosopher Alfred North Whitehead, who says: 'Art is the imposing of a pattern on experience, and our aesthetic enjoyment is recognition of the pattern' (Taube 2004, 233).[1]

Musical patterns are not simply the sounds heard: they also refer to the embodied actions performed when playing music. Musicians describe how entrained repetitive practice of performing patterns becomes embodied, tacit knowledge of scales, chords, or arpeggios; motor-movement patterns based on (sometimes unarticulated) musical theory and their incorporation into motor memory (Hayles 1999, 199; Merleau-Ponty 2002, 168; Parente 2015; Sudnow 1993). In musical performance, the instrumentalist relies on past practice and this relates equally to written and improvised music. These embodied patterns resulting in musical performance are often mirrored—albeit not isomorphically—in how people respond to music. Diverse dance forms, from ballet to pogo, with other gestural types such as headbanging, *luftguitar*, head nodding, foot tapping, orchestra conducting, and so on, are all embodied interpretations of musical patterns. We can look to neuropsychology for a biological perspective, for instance in the work of Patel and Iversen (2014), who find neuroimaging evidence to support the hypothesis that human beat perception is supported by two-way interaction between motor-planning and auditory areas of the brain. This suggests that patterns we hear are strongly informed by simulated patterns of movement.

When talking about patterns in this chapter, we are primarily thinking of the temporal patterns of musical performance. As mentioned, these patterns can be understood through the embodied mode of dancing as well as represented algorithmically through a formula. We celebrate the fact that a sequence of events can be represented with different patterns in different languages or systems. Indeed, two people might describe, perceive, or understand the same sequence as having very different pattern structures. Issues of translation, transduction, and transmission become interesting in this context.

If pattern links sound with movement in perception, the present chapter is concerned with how a third element may enter this relation: symbolic notation—and how this can be written in live performance. Notation systems allow music to be expressed in a format suitable for preservation and sharing, where musical patterns may manifest as visual patterns through diverse systems of scoring. For example, in staff notation pitch is described vertically on a horizontal timeline, with secondary notation for dynamics, articulation, and accents. In this case, ornamentation, timbre, and many other articulations are often left for interpretation, and abstract relationships structuring the composition are not made explicit. The profound challenge that we hold in hand for algorithmic music is to notate structural and multidimensional aspects of musical pattern by using algorithmic representations that go beyond the usual dimensions of music notation, while still allowing expressive use by a composer. Furthermore, notation allows us to explore the rich interferences which emerge from combining pattern transformations, where notation becomes a process of live exploration, rather than description.

14.2 BACKGROUND TO MUSICAL PATTERNS

We approach the expansive topic of pattern in algorithmic music with a clear question: how can we directly express musical pattern with computer code? And how can this be achieved, for example within the constraints of a live performance? For our purposes, the *patterning* of music is where a composer represents *and* transforms source material using a set of strategies. This is an inclusive definition, but the context of algorithmic music gives us a special focus on where strategies are expressed in a programming language. By notating sequences and their transformation through code, the abstract structures of pattern can be made explicit, analyzable, and shareable. Code also allows us to generalize aspects of pattern making, on one hand allowing a transformation to be made on multiple scales and timbral dimensions, and on the other allowing transformations to be combined in diverse ways, creating an explosion of possibilities to explore.

This chapter can be read as a response to a call to arms made by Laurie Spiegel thirty-five years ago in her paper 'Manipulations of Musical Patterns' (Spiegel 1981). Spiegel listed twelve classes of pattern transformation, such as transposition, reversal, rotation, and repetition. What is striking about Spiegel's short description of each is how open to interpretation and implementation they are. Each stands for a huge range of possibilities and interpretations, as we will see later in considering the apparently straightforward concept of *reversal*.

Many programming languages designed for computer music include libraries for pattern representation and transformation, such as HMSL, SuperCollider, and Common Music. However, these libraries do not often live up to the promise of a comprehensive library of pattern transformations proposed by Spiegel, perhaps relying on an underlying general purpose language for much of the functionality. Indeed, much of the operation of code can be considered in terms of pattern manipulation. For instance, we can look at basic bit-level operations, loop constructions, data flow rules, and functional mapping and define those as elements in pattern building. This challenges our ability to compare or even standardize pattern libraries; the programming paradigm at play (e.g. functional, logical, object-oriented, mixed) impacts on how patterns are represented, and therefore on the musical constraints and affordances that the composer works with.

The reason for the diverse pattern libraries comes down to affordance: by defining a function of pattern generation or transformation—such as Fibonacci or inversion—the system creator designs an addition to the vocabulary of their compositional language, thus expanding the musical search space immediately available to the end-user. The pattern function becomes an abstraction of a computational process that may be trivial or complex, but by naming it and including it in a set of other functions, a coherent vocabulary is built up: one that will influence the music or style made with the system.

14.3 PATTERNS IN MUSIC

Organizing sound, planning events in time, arranging pitched patterns: most definitions of music outline some kind of description of rules that define its temporal nature and the melodic, harmonic, and rhythmic elements therein. In the following we explore a range of musical forms where algorithmic rules of pattern have been defined as compositional heuristics. We will jump over much musical history and geography, but mention a few traditions that serve as historical underpinning to the way algorithmic music is now composed and produced with computers.

14.3.1 Fugue/Counterpoint

In the form of seventeenth-century contrapuntal fugues, as perfected by J. S. Bach, we find a musical subject which is introduced as the first voice. The second voice appears shortly afterwards, responding to the first, and this voice is called 'countersubject'. A third voice appears. Each voice states a subject—a melody that references the first voice, but the art of the counterpoint is to use transformations such as *stretto* (overlapping melodies, often where one starts after the other or they differ in lengths), *inversion* (turning the pattern upside down), *augmentation* (lengthening the note durations) and *diminution* (shortening the note durations), *retrograde* (playing the melody backwards), *retrograde inversion*, and further combinations of the above. Counterpoint composition is a highly mathematical task, and it is no wonder that musical systems in computational creativity have been highly successful in the production of new fugues. A good example is Kemal Ebcioğlu's system *CHORAL* (Ebcioğlu 1988), which is perhaps even *too* Bach-like to be convincing to Bach experts.

14.3.2 Serialism

Serial music is also characterized by a strongly mathematical approach to pattern. This is a music that lacks tonal and, at times, metric centres; it is composed through rather rigid rule sets of fixed permutations of tone rows. The music of Schoenberg and Webern are good examples of the serial technique, where a musical row is created with all twelve notes in the octave (thus the descriptive label of the *twelve-tone technique*) and no note is repeated in a single voice before all twelve notes have been played. A row may then undergo four different permutations: *prime*, with the 'original' ordering of tones and intervals of nonrepeating notes; *retrograde*, which is a version of the intervallic structure reversed; *inversion*, where the intervallic structure is inverted (up instead of down and vice versa); and *retrograde-inversion*, where the intervallic structure is simultaneously reversed and inverted. This results in twelve transpositional levels and thus forty-eight possible forms.

14.3.3 Minimalism

Terry Riley's 1964 piece *In C* is a good example of the minimalist approach to melodic and rhythmic patterning of sounds. Riley's piece consists of fifty-three phrases between half a beat and thirty-two beats long. The piece can be performed by an infinite number of players, but Riley suggests at least thirty-five. The fifty-three phrases are performed in order (although phrases can be skipped). The performers can choose a phrase that they repeat until they decide to move on to the next phrase. This brings aleatoric and improvisational elements to the music (something Cage had of course explored earlier).

Many of the minimalists would reject the label, but prominent composers working with sonic materials in an approach that might be defined as being minimalist would include La Monte Young, Terry Riley, Philip Glass, Phill Niblock, Tony Conrad, Louis Andriessen, Henryk Górecki, Arvo Pärt, John Tavener, and perhaps the archetype of the genre, Steve Reich. Reich, influenced by African drumming and polymetric structures, is known for exploring 'phasing' musical material, for example by playing two tape loops in sync but slowing the speed of one; or repeating patterns in a musical score where one part is subsequently delayed at regular intervals. His pieces *It's Gonna Rain* and *Clapping Music* are examples of the respective approaches.

The minimalist approach has not resulted in defined methods such as those we find in contrapuntal or serial music, but there is a clearly identifiable compositional method that can be characterized by steady pulse, repetition, gradual transformation of sequences, and a harmony that is built up of fast melodic progressions, often gradually evolving rather than drastically changing in chord or key.

14.3.4 Electronic Music

The compositional approaches taken in twentieth-century precomputer electronic music owe much to the musical affordances of the hardware available at the time. Early equipment consisted of synthesizers and tape (often mapped with a sweeping generalization onto the factions of German *elektronische Musik* and French *musique concrète*), where the focus was on sonic materials, typically arranged using the primitive methods of cutting and pasting slices of tape. The physical materiality of this work process resulted in music where patterns were largely absent: they could be created with tape loops, but this could be achieved only with some difficulty.

Repeating sequences were introduced into electronic music with electronic sequencers that would output currents to voltage-controllable oscillators. Jumping quickly over to computer-based hardware, we find mass-manufactured sequencers, whose design has been aimed at the many. Although the RCA Mark II used punch cards to control voltages, cheaper and more popular sequencers, such as the Moog 960, used potentiometers to set the voltage values. Eight- or sixteen-step sequencers were the most common devices and this tradition continued into the design of software sequencers.

In modern musical software we typically find systems that derive their design metaphors—both in terms of interface and interaction design—from past traditions, such as the musical score, piano rolls, and the hardware sequencer. Musical patterning in such software is therefore still often subject to the hardwired mechanisms of historical physical equipment.

In the light of the constraints imposed by hardware, and the software simulating it, there is no wonder that musical and audio programming languages have opened up multiple doors for creative musical exploration and expression (Stowell and McLean 2013). What attracts composers and performers here are the open possibilities of defining their own patterns, synths, and hardware instruments; users can write any pattern-generating or -manipulating algorithm conceivable without being bound to the musical constraints of software or physical hardware.

'Underground' electronic dance music has fostered a range of experimentation along the hardcore continuum and beyond, including some in manipulations of pattern. Contemporary software applications include a range of means for arranging patterns, for example arpeggiators with parameters that are often pushed beyond their normal limits to create 'hyperreal' effects in trance music, and software for algorithmic slicing and rearranging breakbeats, such as the methods available in SuperCollider (Collins 2006) and the commercial iZotope Breaktweaker software created by the trance producer BT.

14.3.5 Modular Synthesis

We should also mention in passing the current resurgence of modular hardware synthesizers, particularly the huge range of 'Eurorack' modules now available from many manufacturers, all designed in standard sizes and voltages to be used together. Aside from the great focus on analogue synthesis, there are a great many pattern-generation modules available. One example is the Stoicheia module from Rebel Technology, which generates rhythmic sequences where a given number of events are distributed over a given number of steps, in a manner that resembles the operation of Euclid's algorithm. As a module, it produces impulses intended as trigger signal, which can be plugged into a separate synthesizer or indeed a second rhythm generator to add further complexity. Such configuring and reconfiguring of pattern-generation modules is a very tangible form of live coding (Hutchins 2015).

14.4 PATTERNS IN COMPUTER MUSIC

Returning to the text mentioned above, Laurie Spiegel (1981) offers a library of techniques of the most elementary transformations of musical patterns as they appeared to her at the time. Her aim was to present computer musicians with patterns that were 'tried and true' from the musical tradition. Spiegel describes the twelve pattern operations, many of which come from the domain of traditional musical composition, and with names that are already evocative: (1) transposition, (2) reversal, (3) rotation, (4) phase offset, (5) rescaling, (6) interpolation, (7) extrapolation, (8) fragmentation, (9) substitution, (10) combination, (11) sequencing, and (12) repetition. Under a thirteenth title, 'The Great Unknown', Spiegel discusses the possibility of discovering further patternings in the future.

Any attempt to review pattern languages should focus both on general classes of pattern representation and transformations, and on the detail of implementation, where even small differences can have fundamental results on the music. In computer music it becomes clearer that any musical data is of a numerical nature, so any algorithmic procedure can be applied to a row of numbers. A case study in the difficulty of representing a general agreement in the meaning of pattern-like words could be described by a patterning function such as reversal.

14.4.1 A Case Study: Reversal

Reversal is an example of a patterning function with an operation that seems straightforward to implement, but looking deeper we find a large scope for variety in both implementation and use. This underlines a central point: each of Spiegel's classes of pattern is not a constraint, but a heuristic for guiding us, whose operation varies wildly depending on such things as our conception of time, of an event, the scale(s) at which we are working, and the other elements at play.

Let us begin with reversing a trivial sequence, such as a-b-c-d-. Perhaps the most obvious reversal would be d-c-b-a-, but this already carries a number of assumptions. Firstly, that we are reversing a whole sequence, rather than subsections of it; if the sequence represented two bars, we might instead decide to reverse each bar, rather than the whole sequence. This would make particular sense if we were trying to reverse a sequence of unknown or infinite length, as can often be the case in algorithmic music. In this case, we might end up with b-a-d-c-.

Reversal gets more complex than this. Looking closer at our sequence, we see dashes, which might represent continuations of the previous symbol, or rests. If we reverse the whole sequence, by assuming that a dash is a rest, we would end up with -d-c-b-a. In our original reversal of d-c-b-a-, we quietly assumed that a dash was a continuation, which in turn hides all manner of detail about the nature of an event and its representation. Operationally, what do we think a reversal is doing? One answer might be subtracting the onset of each event from the total duration of its pattern. However, if we are reversing an event, shouldn't the event onsets and offsets be swapped? This really depends on the nature of the pattern you are working with; in music, event durations often have an expressive quality quite separate from those of onsets, and so if we swap onsets with offsets, the result will be incoherent. There are still further issues at play with reversal; for instance, the user might expect that each sound sample should be reversed in time, or that its expressive envelope should be played backwards.

The above example illustrates the problem of signification and interpretation of patterning function names, such as those listed in Spiegel's article. Furthermore, it points to the hermeneutic problem of translation, transmission, and interpretation when pattern functions from one language are written into another. Each programming language creates its own vocabulary that defines a particular way of thinking, which means we can go only so far in generalizing concepts of pattern making across them. This lack of standard contributes to the beauty and diversity in the ecosystem of expressive languages and it should be celebrated.

14.5 PATTERN LIBRARIES IN COMPUTER MUSIC SYSTEMS

Patterns often begin with sequences, and there are many nuances to the question of how to represent musical sequences in a programming language. How can they be notated such that it becomes intuitive and natural for the composer to write music of any genre? How should the traditional notions of pitch and duration be represented, for example? How can the design solve the problem of an event duration with a note length, or sustain, that exceeds the time of the duration? And moreover, since we are writing for synthesized sound, how can do we notate timbre, envelopes, and other synthesis parameters that, clearly, should be controllable from any pattern system?

The SuperCollider language includes an advanced pattern library, but one could argue that it is difficult to read the musical sequences written in it. Having created the following SuperCollider SynthDef called 'piano'

```
SynthDef(\piano, { arg out=0, freq=440, amp=0.1, gate=1,
                    decay=0.8, sustain=0,mix=0.4, room=0.5,
                    damp=0.5;
        var signal, reverb;
        signal = MdaPiano.ar(freq, gate, release: 0.9,
            stereo: 0.3, decay:decay, sustain: sustain);
        reverb = FreeVerb.ar(signal, mix, room, damp);
    DetectSilence.ar(reverb, 0.01, doneAction:2);
    Out.ar(out, reverb * amp);
}).add;
```

one could write a sequence like this, transcribing Mozart's Piano Sonata No 16 in C major:

```
Pbind(
  \instrument, \piano,
  \midinote,Pseq([72,76,79,71,72,74,72,81,79,84,79,
            77,76,77,76],1),
  \dur,Pseq([4,2,2,3,0.5,0.5,4,4,2,2,2,1,0.5,0.5,4]/4,1)
).play
```

Here the Pbind 'binds' values to keys, where the key \instrument is given a symbol (the name of the instrument or synth definition to be used), and other keys, such as \midinote or \dur are given another pattern—Pseq—which is a specifies a sequence as a list, with the number of times the sequence is repeated following the value array. Pseq could then be swapped out with other pattern types, such as: Prand, Pser, Pshuf,

Ptuple, Place, Pslide, Pwalk, and other list patterns whose names try to express their functionality: a pattern that will be randomized, a pattern series, a pattern to be shuffled, interlaced, and so on.

One problem with the above—particularly when compositions become more complex—is that the music is difficult for a human to read; for example, the list of note lengths is not visually aligned with the list of pitches. This could be lined up with spaces, but this quickly becomes too time-consuming to be practical. One solution could be to group the pitch and duration into a subarray, as follows:

```
Pbind(
  \instrument, \piano,
  [\midinote, \dur], Pseq ([[72,1], [76, 0.5], [79, 0.5], [71,
                           0.75], [72, 0.125], [74, 0.125],
                           [72, 1], [81, 1], [79,0.5], [84,
                           0.5], [79, 0.5], [77, 0.25], [76,
                           0.125], [77, 0.125], [76,1]], 1)
).play
```

Some might find this representation more logical, as musical events are here spatially grouped. Of course any parameter in the synth can be controlled, and we could add features such as note sustain and the damp of the reverb:

```
Pbind(
        \instrument, \piano,
        \midinote,Pseq([72,76,79,71,72,74,72,81,79,84,79,77,
                        76,77,76],1),
        \dur,Pseq([4,2,2,3,0.5,0.5,4,4,2,2,2,1,0.5,0.5,4]/4,1),
        \sustain,Pseq([1,0.2,0.2,0.5,0.25,0.25,0.5,1,0.5,
                       0.5,0.5,0.5,0.25, 0.25,2]/4,1),
        \damp,Pseq([0.5,0.4,0.2,0.5,0.5,0.45,0.5,0.3,0.5,
                    0. 5,0.5,0.5,0.25, 0.45,0.5]/4,1),
).play
```

This could also be represented with a data collection called 'Event'. Here, each event contains all the information, such as note or frequency, duration, amplitude, or any other parameter that the user might want to control in the synth. In the example below, it makes sense to use the event system for the right hand, but since the left hand plays notes of the same duration (crotchet) throughout, it can be simply represented with a Pseq:

```
Ppar([
// right hand - using the Event-style notation
Pseq([
        (\instrument: \piano, \midinote: 72, \dur: 1),
```

```
                (\instrument: \piano, \midinote: 76, \dur: 0.5),
                (\instrument: \piano, \midinote: 79, \dur: 0.5),
                (\instrument: \piano, \midinote: 71, \dur: 0.75),
                (\instrument: \piano, \midinote: 72, \dur: 0.125),
                (\instrument: \piano, \midinote: 74, \dur: 0.125),
                (\instrument: \piano, \midinote: 72, \dur: 1),
                (\instrument: \piano, \midinote: 81, \dur: 1),
                (\instrument: \piano, \midinote: 79, \dur: 0.5),
                (\instrument: \piano, \midinote: 84, \dur: 0.5),
                (\instrument: \piano, \midinote: 79, \dur: 0.5),
                (\instrument: \piano, \midinote: 77, \dur: 0.25),
                (\instrument: \piano, \midinote: 76, \dur: 0.125),
                (\instrument: \piano, \midinote: 77, \dur: 0.125),
                (\instrument: \piano, \midinote: 76, \dur: 1)
], 1),

// left hand - array notation
Pbind(\instrument, \piano,
        \midinote, Pseq([60,67,64,67,60,67,64,67,62,67,65,
                         67,60,67,64,67,60,69,65,69,60,67,64
                         ,67,59,67,62,67,60,67,64,67],1),
        \dur, 0.25
        )], 1).play
)
```

This is all good—we are notating music very much as we do in traditional Western notation, but here in the language of the computer. The pattern system in SuperCollider does well what it is designed for, but in this chapter we are wanting to explore pattern transformation and live manipulation of pattern. Such live manipulation of running patterns is a little harder to do in SuperCollider, unless you redefine patterns whilst they are running (using Pdefs, for example) or you use PatternProxys as below:

```
~note = PatternProxy(Pseq([72,76,79,71,72,74,72,81,79,84,
                           79,77,76,77,76], inf));
(
  Pbind(
        \instrument, \piano,
        \midinote, ~note,
        \dur, Pseq([4,2,2,3,0.5,0.5,4,4,2,2,2,1,0.5,0.5,
                    4]/4, inf)
  ).play
  )


~note.source = Pshuf([72,76,79,71,72,74,72,81,79,84,79,77,76,7
                      7,76], inf);
```

In the example above, a pattern sequence (Pseq) has been placed in a proxy whose source can be redefined in runtime. However, the syntax for this is less suitable for live coding or real-time experimentation or composition.

As a domain-specific language for music, SuperCollider provides a comprehensive set of methods of array manipulation which are often of musical nature. These are not explicitly part of the pattern system, but methods that work on lists. The examples below show how list transformations (a form of what we call *patterning* in this chapter) are reintroduced into the source of a playing pattern:

```
a = [72, 76, 79, 71, 72, 74, 72, 81, 79, 84, 79, 77, 76, 77, 76];
b = a+7; // up a fifth

~note.source = Pseq(b,inf);

b = a.reverse; // reverse
~note.source = Pseq(b,inf);

b = a.scramble; // randomize the pattern
~note.source = Pseq(b,inf);

b = a.pyramid; // create a pyramid structure of the pattern
~note.source = Pseq(b,inf);

// or custom made algorithms
b = a.collect({arg note; if(note.even, {note+7}, {note-5})})
~note.source = Pseq(b,inf);
```

Another approach could be to create Pattern definitions which contain keys (such as Pseq or Prand) standing for subpatterns, which can be hot swapped in realtime:

```
Pdefn(\notes, Pseq([72,76,79,71,72,74,72,81,79,84,79,77,76,77
                   ,76], inf));

Pbindef(\x,
   \midinote, Pdefn(\notes),
   \dur, 0.125
).play;

Pdefn(\notes, Prand([72,76,79,71,72,74,72,81,79,84,79,
                    77,76,77,76], inf));
```

As seen above, the SuperCollider pattern system is highly flexible and productive system to work with. It is ideal for the writing of complex algorithmic pieces and it has inspired other musical languages for almost two decades. People equally write sequences by hand or write algorithmic pattern generators. However, the system is not straightforward to compose with, and certainly not in a live performance context such as the one we find in the practice of live coding.[2]

For this reason, this chapter explores the representation of patterns in two systems designed to be written, understood, and manipulated easily in realtime, in particular during improvised live coding performances. We present ixi lang and TidalCycles, explaining how the design of these systems are aimed at live performance, live coding, and real-time sketching in a compositional process. The systems are both high-level, constraining the user possibilities to a higher degree than SuperCollider, but what is gained is the speed of writing music and the 'pleasure' of the constrained system (Magnusson and Hurtado Mendieta 2007).

14.5.1 ixi lang

ixi lang is a high-level minilanguage written on top of SuperCollider. It establishes a language that translates high-level notation into the SuperCollider pattern system. The system aims at fast composition, readability, and high tolerance for syntactical mistakes. ixi lang allows for a convenient way of exploring musical patterns and reverts to prior states effortlessly, either through undoing, or saving the state of the code and the running patterns. This is achieved through creating system agents that are assigned performance scores. Agent behavior can be changed through calling methods (or verbs), which in turn update the code for the agent, turning the code into a form of visual (yet still textual) score.

The key elements of the language relate to melody and rhythm, and these are controlled through the use of agents that are assigned a score:

```
lucy->|q  q  q c q  |
```

Here the agent 'lucy' has been given a rhythmic score (specified by using the pipe '|' symbol) where the characters 'q' and 'c' stand for sampled sounds, a range of which have been mapped to the letters of the roman alphabet. The spaces are silences, so the use of monospaced font is essential in ixi lang. It would be easy to create a polymetre by adding another agent:

```
yoko->|q  q  q c q  |
john->|z  zxz  |
```

A fundamental feature here is to represent features of time and event as close together as possible, to ease cognitive load and speed up the compositional process. In ixi lang the elements of time and sound or note are the most important features and they gain primary representation in the notational language. Secondary parameters, such as amplitude, panning, note length, and so on can then be written behind the score as postfix sequences, which can add further descriptive transformations of the events:

```
john -> |z  z x z  | <1928> // panning from left (1) to right (9)
john -> |z  z x z  | (1442) // note sustain (a whole note, two
                              quarter notes and
                            // a half note)
john -> |z  z x z  | ^4419^ // amplitude
john -> |z  z x z  |!16 // wait sixteen steps before repeating
the pattern
```

and all can be combined, of course

```
john -> |z  z x z  | ^4419^ (1992) <195528>! 12
```

Note how there are six values in the panning argument. Here it wraps around, such that the second time the pattern plays, the first 'z' will be panned 2 to the left. This creates a polyrhythmic effect.

Another mode of ixi lang is the melodic mode. Here the items in the score represent pitches and are therefore numerical:

```
scale minor
paul -> obo[1 2  4  1  2  ]
```

Above we have basic rhythmic and melodic sequences. There are 'actions' that can be applied to the agents, for example

```
swap paul
paul -> obo[4 1 1  2  2  ]
shake paul
```

Other methods include revert, expand, >shift, transpose, and so on, much in the spirit of what we find in Spiegel (1981). These pattern transformations can be set to take place in time, automatically, for example using the future function:

```
future 4b:20 >> shake yoko
```

Here the score of agent yoko will be shaken (scrambled) every four bars, twenty times. An important feature of ixi lang is that the score is updated in the document when it is transformed through code: the code in the document rewrites itself. The way this happens is that the text is highlighted for half a second in a different color and then the score is replaced with a new and running score.

Agents can also receive effects, such as

```
yoko >> distort >> reverb
```

Here the output of agent yoko is routed through a distortion effect and then through a reverb effect unit. The symbol << removes all effects. Here we find another example of ixi lang's graphical design, where the idea is that the >> operator is a visual reference to jack cables used with electric guitars.

In terms of tempo, it is clear how easy it is to create polymetre in ixi lang:

```
yoko->|q  q  q c q  |
koyo -> |a  a s a  |
```

However, for polyrhythm, the calculations would have to be slightly more advanced:

```
yoko->|q  q  q c q  |
koyo -> |a  a s a  |*1.333333
```

where agent koyo has now been stretched to the length of yoko.

ixi lang is a notational live coding language. On its own it is not a fully-fledged programming language, but it harnesses the power of SuperCollider for more complex coding. The focus here is on speedy input, redesign, reevaluation, manipulation of agents' scores, and the routing of agents' output through effects.

14.6 TIDALCYCLES

TidalCycles, known as Tidal for short, is a minilanguage for pattern, embedded in Haskell, a pure functional programming language. This functional basis allows Tidal to define patterns within generalized type structures, which in practical terms means Tidal has a very strict, formal model of what constitutes a pattern, yet is highly flexible in how those patterns are expressed and combined together.

Tidal is a domain-specific language, in that it provides an alternative model of computation designed for its domain of patterning. In conventional (i.e., general-purpose, imperative) languages, statements within an algorithm describe a list of steps to be evaluated one after another, over time. In Tidal, a first step produces a sequence of events over time, and then each successive step may transform that pattern. In other words, time in Tidal is represented not by control flow, but by a functional relationship between time and events. A pattern may therefore be transformed in terms of time (e.g. reversing, slowing down, or stuttering), in terms of events (e.g. transposing, inverting), or combined with another pattern through the combination or juxtaposition of events over time. This flexibility results in a set of simple operators and functions which offer an explosion of possibilities in how they may be combined together.

Tidal is really two languages in one, a language for sequencing events, and another for combining and transforming those sequences into a pattern. Some of the sequencing aspects of Tidal are illustrated in Figure 14.1, demonstrating how polyphonic sequences with compound and polyrhythmic time structures can be specified using a terse syntax. The remainder of this section is focussed on combining and transforming these sequences as patterns.

```
"light dark black"
```

(a) Sequences are specified in double quotes, with steps separated by spaces.

```
"light [dark white] black"
```

(b) Steps can be broken down into subsequences using square brackets.

```
"white [dark [white light]] black"
```

(c) Steps within subsequences can be further broken down.

```
"[dark light white][black light]"
```

(d) Steps may be broken down into irregular parts.

```
"[dark light white, light black]"
```

(e) A step may be broken down into more than one subsequence, creating polymeter.

```
"[dark black, [light black]/2 white]*4"
```

(f) A step may be slowed down with the symbol '/', and sped up with '*'.

FIGURE 14.1 Some features from Tidal's terse yet expressive minilanguage for specifying sequences. It is also possible to represent different kinds of polyrhythm, spread sequences over multiple cycles, and add rests and random variation; see documentation at tidalcycles.org for full details.

In Tidal, a `Pattern` type is defined as instance of Haskell's applicative functor type, which simply means that the end-user live coder can treat whole patterns of things as if they were single things. This needs an example:

```
(+) <$> "1 2 3" <*> "4 5 6"
```

The above uses the addition function + to add together the two sequences `"1 2 3"` and `"4 5 6"` to create `"5 7 9"`. Tidal defines which pairs of numbers are given to the + operator in order to construct a new pattern, so that the end user need only think about what combination they want, rather than how it should be achieved. The advantage of this declarative approach to combining patterns becomes clearer in the slightly more complex example below, which combines two patterns with different structures:

```
(+) <$> "1 [2 3] 4" <*> "1 2"
```

This results in the pattern `"2 [3 5] 6"`, demonstrating that the structure of the first pattern is maintained, with the first half having `1` added, and the second half having 2 added. This split extends into the two steps of the middle subsequence.[3]

14.6.1 Timbral Dimensions

So far we have been discussing Tidal in the abstract, even using patterns of colour rather than sound to illustrate its output. The abstract nature of Tidal's approach is also its strength, in that many of its functions are polymorphic, operating on patterns containing values of any particular type. However, let us ground our discussion with a more musical example. The following is a Tidal pattern composed of simple parts, with complex results:

```
jux (iter 4) $ (every 3 (density 1.5) $
               sound (pick <$> "bd can*2 [sn cp] can" <*> (slow 8 $
                      scan 8)))
             #  speed (slow 4 $ (+1) <$> sine1)
             #  delay "1"
             #  delayfeedback "0.7"
             #  delaytime "[0.02 0.01 0.03 0.02]/3"
             #  vowel "[e x a, x i x i]/4"
```

The first two lines of the above centre on the sound pattern, specifying some sound samples (bd, can, sn, and cp), which are combined with a pattern of numbers generated by scan using the pick function, which together steps through variants of those samples in a way that gradually increases complexity. In addition, every third cycle of the pattern has its density increased by 50 percent.[4] The remaining lines combine the sound pattern with effect patterns; the speed of sample playback (i.e., the change in its pitch) follows a sine wave over four cycles, a comb-filter-style delay effect cycles through four values over three cycles, and a vowel filter has a polyphonic pattern of e, a, and i formants applied over four cycles. Finally, the use of jux and iter causes the whole pattern to shift in steps of a quarter-cycle every cycle, but only in the right-hand channel, creating a stereo panning effect.

Although the above example is relatively straightforward, it contains patterns of different types being composed together into patterns of synthesizer control messages, their time structure being manipulated, and functions being selectively applied in terms of time (in this case with every) and space (in this case with jux). The majority of functions in Tidal take one or more patterns as input, and produce another pattern as output, and so it is easy to chain simple transformations together and achieve rich results.

One particularly interesting Tidal function is weave, which in the following case combines three sound patterns using a fourth pan pattern:

```
weave 16 (pan sine1)
  [sound "bd sn cp",
   sound "casio casio:1",
   sound "[jvbass*2 jvbass:2]/2",
   sound "hc*4"
  ]
```

By design, weave offsets each of the sound patterns in time, after applying the pan pattern, which is itself stretched over the given number of sixteen cycles. The end result is that the three patterns are spatialized, each moving between the two (or potentially, multichannel) speakers following a sine-wave pattern, but phase-shifted, so that when two of the patterns are hard left and right, the other two are meeting each other in the centre.

The `weave` function was designed with this spatialization technique in mind, but can be applied to any effect; for instance, it works well to have distortion effects rising and falling across different patterns in different phases. There are, however, surprising affordances which fall out of this generalization. We may use 'weave' in a different way:

```
jux rev $ weave 16 (sound (samples "arpy*8" (run 8)))
  [vowel "a e i",
   vowel "i [i o] o u",
   vowel "[e o]/3 [i o u]/2"
  ]
```

Instead of applying different phases of an effect pattern to a set of sound patterns, the above does the opposite, applying different phases of a sound pattern to a set of different effects. In musical terms, the result is a canon; the `run` function in the above gives us a rising scale of `arpy` notes, and the overlaying of different phases results in a canon that seems to continuously rise. Furthermore, swapping the sound and effect patterns in this way causes the rhythmic structure of each part to come not from the sound pattern but from the effect patterns, so that the combined result is a rich polymetry.

The latter usage of `weave` to create canons was discovered by chance. The structural correspondence between patterns arranged over time in pitch and in space are not surprising, but it is enjoyable to see this functionality appear almost by magic, through the process of generalizing patterning functions.

14.6.2 Levels of Patterning

Tidal's functional approach leads to a multilayered view of pattern making, where each layer builds upon those beneath. On the base layer, we find a view of patterns as *sequences*, which are potentially polyphonic, or polymetric, but are described in a linear, imperative fashion, as we have seen in the SuperCollider examples above.

Also familiar to our common conception of pattern is the concept of *symmetry* in pattern, which in musical terms can be understood in terms of time, for example reversal or rhythmic rotation, or in terms of note values, for example inversion.

Pattern can also be understood in terms of *deviation* from a structure: imperfections, glitches, and confounded expectations, which can be explored through the introduction of random-number generators. For example in Tidal, the `sometimes` combinator can be used to apply a given combinator, but only sometimes (`rarely` and `often` are also available).

We should also not forget the key importance of *composition* to pattern making; of joining together different patterns, where the method of joining itself becoming a core part of the pattern. A special form of composition is *interference*, where constituent patterns combine to form a new pattern, with features not present in the originals. Such interference patterns can be well understood from patterns in weaving, where colour sequences in warp and weft threads, combined with the weave structure at play, form surprising images (Harlizius-Klück, 2008). Such interference is explained in our above description of Tidal's aptly named 'weave' function.

As we climb up these layers of sequencing, symmetry, glitch, composition, and interference, we find that code becomes increasing important to the creative process, in generating surprising results that are otherwise beyond the imagination of the programmer. At this point code becomes more like physical material, with results emerging through continual reaction to sensory feedback, rather than transcription of a pure idea. In practice then, Tidal's focus on higher-order patterns and interference affords an improvisatory approach to music, where language becomes an exploratory environment. Perhaps it is not too controversial to wonder whether interference patterns, which can also be found in ixi lang, SuperCollider, and any other programming language, lie at the very heart of algorithmic music.

14.7 CONCLUSION

An algorithm is a description of a pattern. By defining an algorithm, we engage in a process of generalization and normalization, where certain features are notated into patterns that can be abstracted into a standardized system of notation. With the advent of automated machines the ideal sequence-repeating engine appeared, where diverse input mechanisms, such as the punch card, could be used to represent the patterns. However, with computer code this becomes infinitely more powerful, as we have language constructions that make this very flexible, for example with the use of for-loops and recursion. With a meta-machine controlled by the programming language, we can automatically generate, transform, and analyse patterns from new data, and represent them across different media domains.

High-level pattern languages are useful as they are minilanguages or high-level systems that provide bespoke and often idiosyncratic ways of thinking and performing music. In the design of pattern systems, the naming of the functions suggests affordances: they are linguistic abstractions of processes that may or may not be easy to write in a standard language. There are no standards for pattern languages— so the implementation of Spiegel's methods would hardly ever be unified across different systems. The system's method names thus become semantic entities in the compositional thinking of the composer or performer. They outline the scope of the possible.

The creation of patterns in computer music languages is therefore a two-edged sword: on the one hand it provides a language, a vocabulary, a 'technology for thinking' about music, which enables the composer to build structures through the scaffolding of the system. On the other hand, the patterns are compositional structures on their own, thus influencing, perhaps limiting or directing the compositional thoughts in a way that would not have been the case if the composer had had to write their own pattern systems. For us, this is a question about time, affordances, 'ready-at-handness' of a musical system that should be capable of an engaging performance in the practice of coding in front of a live audience.

We will probably never have high-level, universal definitions of patterns, because interpretations of what the linguistic signifiers stand for can differ greatly. Every system presents its own approach for working with patterns and some might reject the idea of writing patterns completely, expecting that such work is of a compositional nature and should be done by the user.

The two systems discussed in this chapter—ixi lang and Tidal—are constrained, purpose-built live coding systems that are used by people all over the world. The authors have given workshops, presentations, and performances with the systems and they have become relatively well known in the world of computer music performance. Although they have proven to be good tools for musical performance, perhaps the greatest contribution with this research has been to rethink the computer language design and purpose, where performance and the conception of the code as something that be sculpted in realtime is given a high priority.

This chapter has focussed on the computer music language as something we use in sketching and performing real-time music. Although the focus has been on performance, we are not excluding the possibility, well covered elsewhere in this book, that we might start to perform with agents of computational creativity, where creative processes are delegated to AI agents. Indeed, such features are already taking shape in ixi lang's 'autocode' function, where the language begins coding on its own, as well as the application of evolutionary algorithms to Tidal (Hickinbotham and Stepney, 2016), both already resulting in some fine music.

ACKNOWLEDGEMENTS

Thanks to Geraint Wiggins and James Harkins for discussing some topics in this text with us during the writing. Also, many thanks to the editor of the text, Roger Dean, for superb comments and pointers.

NOTES

1. Although note that going to the original reference, we find that this is in the context of Whitehead arguing against quoting soundbites.

2. Although some third-party libraries do extend the Pattern functionality of SuperCollider, noteably JITlib and the PLx quark.

3. If you are having trouble understanding the structure of "2 [3 5] 6", see Figure 14.1b, which shows an example with the same structure.

4. For the purposes of this chapter, we only need a general understanding of this code; for a closer understanding, refer to the tidal tutorial on the Tidal website (tidalcycles.org).

BIBLIOGRAPHY

Burack, J. 'Uniting Mind and Music: Shaw's Vision Continues'. *American Music Teacher* 55, no. 1 (2005): 84–87.

Collins, N. *Towards Autonomous Agents for Live Computer Music: Realtime Machine Listening and Interactive Music Systems*. PhD dissertation, Centre for Science and Faculty of Music, University of Cambridge, 2006.

Du Sautoy, M. *The Music of the Primes: Searching to Solve the Greatest Mystery in Mathematics*. London: Fourth Estate, 2003.

Ebcioğlu, K. 'An Expert System for Harmonizing Four-Part Chorales'. *Computer Music Journal* 12, no. 3 (1988): 43–51.

Fauvel, J., Flood, R., and Wilson, R., eds. *Music and Mathematics: From Pythagoras to Fractals*. Oxford: Oxford University Press, 2006.

Harlizius-Klück, E. 'Arithmetics and Weaving: From Penelope's Loom to Computing' [posters]. *8. Münchner Wissenschaftstage*, 18–21 October 2008.

Hayles, K. *How We Became Posthuman: Virtual Bodies in Cybernetics, Literature, and Informatics*. Chicago: University of Chicago Press, 1999.

Hickinbotham, S., and Stepney, S. 'Augmenting Live Coding with Evolved Patterns'. In *Proceedings of the 5th International Conference on Evolutionary and Biologically Inspired Music, Sound, Art and Design*, 31–46. New York: Springer, 2016.

Hutchins, C. C. 'Live Patch / Live Code'. In *Proceedings of the First International Conference on Live Coding*, 147–151. Leeds: ICSRiM, University of Leeds, 2015.

Magnusson, T., and Hurtado Mendieta, E. 'The Acoustic, the Digital and the Body: a Survey on Musical Instruments'. *Proceedings of the Seventh International Conference on New Interfaces for Musical Expression*, 94–99. New York: ACM, 2007.

Merleau-Ponty, M. *Phenomenology of Perception: An Introduction*. 2nd ed. London: Routledge, 2002.

Parente, T. J. *The Positive Pianist: How Flow Can Bring Passion to Practice and Performance*. Oxford: Oxford University Press, 2015.

Patel, A. D., and Iversen, J. R. 'The Evolutionary Neuroscience of Musical Beat Perception: the Action Simulation for Auditory Prediction (ASAP) Hypothesis'. *Frontiers in Systems Neuroscience*, 13 May 2014. http://journal.frontiersin.org/article/10.3389/fnsys.2014.00057/full.

Sloboda, J. A. *Exploring the Musical Mind: Cognition, Emotion, Ability, Function*. Oxford: Oxford University Press, 2005.

Spiegel, L. 'Manipulations of Musical Patterns'. In *Proceedings of the Symposium on Small Computers and the Arts*, 19–22. Philadelphia, PA: IEEE, 1981.

Stowell, D., and McLean, A. 'Live Music-Making: A Rich Open Task Requires a Rich Open Interface'. In *Music and Human-Computer Interaction*, edited by S. Holland, K. Wilkie, P. Mulholland, and A. Seago, 139–152. London: Springer, 2013.

Sudnow, D. *Ways of the Hand: Organization of Improvised Conduct*. Cambridge, MA: MIT Press, 1993.

Taube, H. K. *Notes from the Metalevel: Introduction to Algorithmic Music Composition*. London: Taylor and Francis, 2004.

Varèse, E. 'The Liberation of Sound'. *Perspectives of New Music* 5, no. 4 (1966): 11–19.

CHAPTER 15

COMPUTATIONAL CREATIVITY AND LIVE ALGORITHMS

GERAINT WIGGINS AND JAMIE FORTH

15.1 INTRODUCTION

LIVE algorithms have been present in Western music since as early as the eighteenth century. *Der allerzeit fertige Menuetten- und Polonaisencomponist* (The always ready minuet and polonaise composer; Kirnberger 1757) allows minuets and polonaises to be generated by choosing random numbers. While probably not necessarily intended for live operation, as opposed to prepared performance, the music is certainly performable in this way.

In the twentieth century, algorithmic processes in music became a feature of modernist composition, with composers such as Philip Glass (Potter 2000) and John Cage (Revill 1993) specifying processes in advance of performance and writing their output down. Live algorithmic music was less common, because of practical limitations, but there are examples, such as the chance-driven processes in Lutosławski's *Jeux vénitiens* (Lutosławski 1961). Perhaps the archetype of human-driven live algorithmic music is Terry Riley's *In C* (Potter 2000; Riley 1964), in which the performers, guided by a conductor, choose the transitions in a prespecified algorithmic sequence.

However, despite technologically interesting and musically successful prototypes which, for example, generate instrumental scores live (e.g. Eigenfeldt 2014; Eigenfeldt, Burnett, and Pasquier 2012), live generation for human performance retains the major drawback that it requires human musicians to perform live, sight-reading and without ensemble rehearsal. While this is common in some very specific genres, it is generally not, and pressure to work in this way puts severe stress on even highly accomplished musicians, making it very difficult indeed to achieve a satisfactory outcome. For this reason, we base our argument on the restricted case where a human is programming a computer, live, to play sounds, which are specified by programmatic means: live coding. Live coding is a very specific microcosm of the broader live algorithms field, and its specificity helps make our model clear. However, the model we develop in this chapter is equally applicable to musical coding which happens not to be live.

Given that a computer is involved, a natural question to ask is 'how involved is the computer?' In the majority of cases, we believe that the computer serves as a very power-ful sequencer, where the specification of the sequence is given in *intensional* terms[1] (that is to say, specified as a generative process) rather than extensional terms (that is to say, specified as a set of notes, or by the positions of a sequence of knobs or switches on an analogue synthesizer). As such, while the scope of such expression is clearly substan-tially broader, and the means of expression fundamentally different, the essential nature of the activity is not different from the complex sequencer-based work of bands such as Tangerine Dream in the 1970s.[2] The nature of the intensional specification of sequence is very clearly exemplified in languages such as Tidal (McLean 2011), which are optimized from the perspective of easily, efficiently, and intensionally specifying operations, live, that map between sequences specified extensionally or intensionally. This difference is crucial to the purposes of this chapter, not so much because of the breadth of expression afforded, but because programs and the numbers that drive them are capable of repre-senting information at more than one level simultaneously. In particular, they are able to represent and reason about themselves, as well as about their outputs, affording the capacity for *reflection* (reasoning about one's own behaviour), which is not available to a hardware sequencer, whose knobs are (literally) hard-wired to whichever functions they control, and whose clock is just that, voltage control notwithstanding. Reflection is a key feature of creative autonomy, and our purpose here is to explore future paths for live coding, in which the computer is given more *creative responsibility* (Colton and Wiggins 2012) for the outputs produced than is the case at present.

McLean and Wiggins (2010b) elicited opinions from practising live coders as to the current and future development of automation in live coding, particularly in respect of creative autonomy of the computer. Of those who responded, 40.7 percent believed that it was possible, at the time of the survey, for live code to modify itself in an artistically valued manner, and some of those who disagreed were optimistic that this would be possible in future. Exactly half of the respondents agreed that a computer agent has been developed that has produced a live coding performance indistinguishable from that of a live coder, or that one such will be developed within five years of the survey. Of the same cohort, however, 34.6 percent believed that such an agent will never be developed.

The aim of the current chapter, therefore, is to begin to lay out the path towards such valued creativity in a live coding agent. We begin by defining the Creative Systems Framework (Wiggins 2006a, 2006b), which will provide the context for our discussion, and illustrating its application with a very simple example concerning an imaginary live coder, and we very briefly introduce Tidal, our live coding language of choice. We then proceed to examine the consequences of following through the various possibilities to foresee a live coding system that might work in creative partnership with a human in a true hybrid creative system.

15.2 CREATIVE SYSTEMS

The Creative Systems Framework (CSF: Wiggins 2006a, 2006b) takes as a starting point the following definition of a *creative system*.

> **Creative system** A collection of processes, natural or automatic, which are capable of achieving or simulating behaviour which in humans would be deemed creative. (Wiggins 2006a, 451)

This definition presupposes, not unreasonably, that creativity is best understood in terms of human behaviour, in that we can meaningfully discuss nonhuman creativity only with reference to behaviour exhibited by humans. However, depending on the vantage point one takes when considering a creative process, alternative conceptualizations of creative systems can emerge. For example, an improvisation context, comprising human and/or artificial agents, may be considered a creative system when viewed from a certain level of abstraction, as a 'black box'. Likewise, the abstraction boundary may be increased still further, resulting in creative behaviour at the level of societal dynamics, or lowered into hypothesized mechanisms underlying individual human creativity or more general cognition (Baars 1988; Wiggins and Forth 2015). Applying the CSF at various levels of abstraction, it becomes possible to separate out the contribution of many disparate elements that together give rise to complex and emergent creative behaviour.

In the practice of live coding, computational systems are predominantly viewed as tools or instruments under the control of the human performer, and thus as means of expressing human creativity. Some live coders tend to view systems more as collaborators, particularly when the systems exhibit behaviour that is complex and challenging (Bovermann and Griffiths 2014). In this case it appears that sense of agency on behalf of the system becomes established by the perception of the system's behaviour in the mind(s) of the performer and/or audience. Taking the live coder together with the system as a basic level of abstraction for applying the CSF, we are able to identify where principal boundaries of responsibility lie for sustaining creative behaviour in the partnership of human and algorithmic processes. Clarifying these distinctions will enable the potential shifting of creative responsibilities and for artistic motivations leading to more inspiring interactive live coding partnerships, but also for motivations in the scientific study of creativity.

15.3 THE MANY LEVELS OF CREATIVITY IN LIVE CODING

Creativity abounds, on multiple levels, within live coding. The software employed, or at least the core set of abstractions used to express musical concepts, are typically developed by the live coder as an integral part of the development of their musical aesthetic. Before a live coder takes to the stage, a long series of artistic and technical challenges have to be addressed, requiring a high degree of ingenuity and technical competence. In the cycles of software development or dedicated rehearsal sessions, the live coder will experiment and become more familiar with the system's idiosyncrasies and explore the potential scope of musical output. In a manner akin to the extended-mind theory of consciousness (Clark and Chalmers 1998), the live coder becomes attuned to thinking with and through the medium of code and musical abstractions, such that the software can be understood as becoming part of the live coder's cognition and creativity. Such fundamental engagement with musical structure through the medium of code leads to a musical aesthetic suffused with algorithmic elegance. This phenomenon is not restricted to live code practitioners: Magnusson (2014, 13) identifies, from an extensive survey of live coding practice, that a common motivation amongst performers is to 'communicate algorithmic thinking'. More generally, Collins characterizes the role of the music analyst when considering computer-generated music as being to seek 'to explain a given output (a production) in terms of the originating program (a source)' (2008, 240). Given the prominence of code projection and other forms of algorithmic visualization during live coding performance—enabling audiences to form, at least to some degree, an appreciation of the music with reference to the processes by which it is being generated—it is reasonable to assume that means of generation are integral to the aesthetic values of live coded musical performance.

Beyond the coupling of live coder and computational system, creative behaviour can be observed in group performance. Group live coding performances typically follow a model borrowed from improvised jazz, where performers interact with fellow performers in an ongoing negotiation of musical development (McLean 2014). Creativity here can be viewed as distributed among the participating creative agents. Audience members, simply by engaging with the performance, can be understood as exhibiting creative behaviour by at the very least making meaning out of the experience, but also potentially influencing the direction of the performance by means of what McLean (2014) identifies as the inherent social feedback involved in live code performance.

15.4 FORMALIZING CREATIVE SYSTEMS

To formalize the idea of a creative system, we first introduce Boden's abstract model of creativity, and show how it can be formalized, to provide a tool set for discussing creative systems. Of course, the creative systems we have in mind here are hybrid ones, composed of a human programmer–composer and an algorithm. In particular, in this section, we introduce some specific properties of creative systems that will be useful in our taxonomy.

15.4.1 Boden's Model of Creativity

Boden's (2004) model of creativity revolves around the notion of a *conceptual space* and its exploration by creative agents. The conceptual space is a set of artefacts (in Boden's terms, *concepts*) which are in some quasi-syntactic sense deemed to be acceptable as examples of whatever is being created. Implicitly, the conceptual space may include partially defined artefacts too. *Exploratory creativity* is the process of exploring a given conceptual space; *transformational creativity* is the process of changing the rules which delimit the conceptual space. Boden (1998) also makes an important distinction between mere membership of a conceptual space and the *value* of a member of the space, which is defined extrinsically, but not precisely.

Bundy (1994) and Buchanan (2001) join Boden in citing reflection, and hence meta-level reasoning, as a requirement for 'real' or 'significant' creativity (though the definition of such creativity is so far left imprecise). Again, it is the capacity to reflect that we consider central here.

For completeness, we mention here that there are other views. Ritchie (2007), for example, presents a completely different account of what is going on in 'transformational' creativity, in which the notion of transformation is not so clearly present. Colton and colleagues (2014) and Colton and colleagues (2015) present the IDEA and FACE models, that attempt to characterize creativity from different perspectives. However, since the current chapter is primarily focussed on the application of Boden's theory to live coding, via our CSF, explained next, we defer discussion of alternative approaches.

15.4.2 The Creative Systems Framework

The central idea of the CSF, the formalism presented by Wiggins (2006a), is that an exploratory creative system in Boden's (2004) terms, may be abstractly represented by a septuple, thus:

$$\langle \mathcal{U}, \mathcal{L}, [\![.]\!], \langle\langle .,.,. \rangle\rangle, \mathcal{R}, \mathcal{T}, \mathcal{E} \rangle.$$

The symbols here are defined in Table 15.1. The function of each is briefly explained below.[3]

$\mathcal{U}$ is the (abstract) set of all possible partial and complete artefacts describable in the creative system being modelled. $\mathcal{R}$ is a set of rules, expressed using the language $\mathcal{L}$, which select an 'acceptable' or 'relevant' subset of $\mathcal{U}$, which corresponds to Boden's (2004) conceptual space. In Wiggins's formulation, selection is *permissive*, in the sense

Table 15.1 The symbols used in Wiggins's (2006a) description of Boden's exploratory–creative system

$\mathcal{U}$	a universe of possible concepts (artefacts), both partial and complete
$\mathcal{L}$	a language in which to express concepts (artefacts) and rules
$[\![\cdot]\!]$	a function generator, which maps a subset of $\mathcal{L}$ to a function which associates elements of $\mathcal{U}$ with a real number in $[0,1]$
$\langle\langle.,.,.\rangle\rangle$	a function generator, which maps three subsets of $\mathcal{L}$ to a function that generates a new sequence of elements of $\mathcal{U}$ from an existing one
$\mathcal{R}$	a subset of $\mathcal{L}$
$\mathcal{T}$	a subset of $\mathcal{L}$
$\mathcal{E}$	a subset of $\mathcal{L}$

that it admits partial artefacts, even some of whose completions may eventually turn out not to be admitted. So applying a selector function generated from $\mathcal{R}$ by $[\![\cdot]\!]$ and a suitable real comparator (e.g. 0.5) gives Wiggins's formalization of Boden's conceptual space:

$$\{c\,|\,c\in\mathcal{U}\wedge[\![\mathcal{R}]\!](c)\geq 0.5\}.$$

The ruleset, $\mathcal{R}$, then, defines what it is to be an artefact of the kind we are interested in creating: a piece of music, a joke, and so on. (Alternatively, the output of $[\![\mathcal{R}]\!]$ might be used directly in a fuzzy selector; we postpone discussion of this for now.)

T is a set of rules which, when interpreted, perhaps along with those in $\mathcal{R}$ and $\mathcal{E}$, by $\langle\langle.,.,.\rangle\rangle$, describe the behaviour of a creative agent as it traverses the conceptual space from known artefacts to unknown ones (much as the standard AI search framework; Wiggins 2006b explains the relationship in detail), and possibly back again. The first argument of $\langle\langle.,.,.\rangle\rangle$ takes a concept/artefact definition ruleset, such as $\mathcal{R}$, above, and the second a rule set such as $\mathcal{T}$, which is the specification of the traversal strategy. The third argument is $\mathcal{E}$, the rules by which *value* is attributed to a created artefact, new or otherwise (see below). $\mathcal{R}$ and $\mathcal{T}$ are included so that it is possible for $\mathcal{T}$ to include reasoning about them, but this is not a requirement; thus, $\mathcal{T}$ can in principle generate artefacts which do not conform to the rules of $\mathcal{R}$ and this can be used to trigger subsequent reasoning and reflection about the creative system under simulation (Wiggins 2006b). There is no explicit equivalent of $\mathcal{T}$ in Boden's writing, though it is implicitly present at all times. To distinguish between transformation of $\mathcal{R}$ and transformation of $\mathcal{T}$ we write '$\mathcal{R}$-transformation' and '$\mathcal{T}$-transformation'.

$\mathcal{E}$ is a set of rules which define the evaluation of the creative outputs resulting from the agent's activity, appropriately contextualized. The formalism does not specify what this context is; it might be the subjective judgement of the creating agent, or the subjective combined judgements of other agents, or comparison with some objective measure. $\mathcal{E}$ allows us to express the notion of value proposed by Boden (1998). For completeness, we mention that we would expect $\mathcal{E}$ to be amenable to transformation, also, in particular ways, especially if this theory were applied in the context of a multiagent system. However, for the moment we leave the interesting question of how usefully to formalize $\mathcal{E}$-transformation to future work.

A further useful mechanism is the function $^{\diamond}$, defined such that

$$\mathcal{F}^{\diamond}(X) = \bigcup_{n=0}^{\infty} \mathcal{F}^{n}(X),$$

where $\mathcal{F}$ is a set-valued function of sets; this allows generation of all the concepts derivable under $\mathcal{T}$ from a given starting concept: below, we will substitute $\langle\langle .,.,, \rangle\rangle$ for $\mathcal{F}$ in this formula to capture iteration across the whole search space. A useful constant will be $\top$, the null (or completely undefined) concept, which inhabits all conceptual spaces.

A brief example may help to clarify the usage of this mechanism. Consider the familiar task (for example Ebcioğlu 1988) of harmonizing seventeenth-century German hymn tunes in the style of J. S. Bach. We can model this case as follows (but note that there are other ways, depending on what one wants to achieve). $\llbracket \mathcal{R} \rrbracket$ selects a subset of $\mathcal{U}$, which might be described as the set of all partial and complete harmonizations of the canon in question. $\mathcal{E}$ then selects those which are considered good, according to criteria that may be related to appropriate rules of music theory, psychological models of music perception, and/or socially inspired metrics designed to quantify aspects of value in relation to existent harmonizations. To see why there is a difference between $\mathcal{R}$ and $\mathcal{E}$, consider the comparison between the harmonizations produced by J. S. Bach himself, and those produced by a first-year music student: the latter are not usually valued as highly as those of the former, because even the best student is unlikely to produce music of the same quality as those Bach harmonizations which have been selected by several hundred years of history.

This same pair of subjects can help understand the need for $\mathcal{T}$, also. An extremely competent and experienced composer and improviser such as Bach will normally have the ability to 'see' a harmonization which is correct in quasi-syntactic terms and of high quality in value terms without too much conscious effort. This is rarely true of beginning composers, who need to develop their intuitions over a period of time, usually through a kind of problem-solving approach. $\mathcal{T}$ allows us to model these behaviours individually, and to study their interactions with the externally defined $\mathcal{R}$ and $\mathcal{E}$. Also, crucially for the evaluation of artificial creative systems, the process by which a system produces new artefacts, as defined by $\mathcal{T}$, is integral in determining the extent to which behaviour may be deemed creative. For example, brute-force search, or a very prescriptive approach based on hand-coded rules, is unlikely to be considered creative, especially compared to a process containing a set of learned, higher-level abstractions enabling the generation of highly valued artefacts with a high degree of efficiency.

Wiggins (2006a, 2006b) gives examples to elucidate how the framework may be used, and shows how *transformational* creativity can be cast as exploratory creativity at the meta level, where the conceptual space is the set of possible rule sets, generated by a given language, as informally suggested by Bundy (1994).

A substantive difference between Boden's formulation and that of Wiggins is the addition of the rule set, T, which describes the actual behaviour of a creative agent as it goes about its business: Boden is not concerned with this level of detail. The difference gives Wiggins's formulation more power to describe the behaviour of *implemented* creative systems. Thus, it may be compared in detail with existing similar methods, such as those of AI state space search. Further, the introduction of T, as an explicit component admits a new kind of transformational creativity, in which an agent modifies its own behaviour by reflective reasoning.

15.4.3 Useful Properties of Creative Agents

The apparent supposition in Boden's work is that creative agents will be well behaved, in the sense that they will either stick within their conceptual space, or alter it politely and deliberately by transformation. It can be argued, however, that this is not adequate to describe the behaviour of real creative systems, natural or artificial, either in isolation or in societal context. This section identifies some situations not covered by the assumption of good behaviour, and gives names to them. The important point is that some of these situations may appropriately trigger particular events, such as a step of transformational creativity, so it is useful to be able to identify them in the abstract. This leaves us with several general classes of small-scale conditions which might be observed in AI systems, of which we can then assess the creative potential.

These characterizations are only descriptively useful unless appropriate responses, categorized by condition, can be specified. This section does so. We assume some appropriate learning mechanism(s) which can adapt the rules (expressed in language $\mathcal{L}$ and categorized into $\mathcal{R}$, T, and $\mathcal{E}$), from positive and/or negative training sets.

15.4.3.1 *Application to Live Coding*

For the purposes of this chapter, it is useful first to calibrate the application of the system, by assigning meanings to the various symbols in the formalism. To illustrate the use of the framework in as transparent a way as possible, we omit the more complex questions, such as interaction between human performers and any aspect of performance by our notional programmer which is not mediated via the live coding system.

First, the universe, $\mathcal{U}$ in our case, is all possible music that could potentially be produced (under any definition of 'music' with respect to a given representation),[4] whether or not by our example live coding practitioner. At the most abstract level, the conceptual space, $\mathcal{C}$, specified by the rule set, $\mathcal{R}$, is the range of live coded music that our practitioner can imagine (which is therefore in all probability a subset of $\mathcal{U}$). T, the transition rules, specify a combination of her craft as a live coder and the music that can be produced by the algorithms that she writes. $\mathcal{E}$, the evaluation rules, express her preferences in the outcomes of this process, and may refer to the quality of the code, or to the music, or both.

It becomes immediately clear that one could more precisely conceptualize this hybrid creative system, in which a human creates a program, which then creates for itself, as two distinct layers within the CSF. There would be two universes, one of live algorithms and one of music, with a mapping between them, corresponding to the execution rule of the relevant programming language. Thus, we express our performer's creativity in programming, and in music, distinctly. Doing so would allow us to consider programming techniques and the design of specialist languages for live coding (McLean and Wiggins 2010a, 2010c), and this is our aim later in the chapter. For now, however, to do so would overcomplicate our example. Therefore, T, in our first example, corresponds to the ability of the *code produced* to traverse C, and *not* with the ability of the programmer to write it. Similarly, our evaluation function, $\mathcal{E}$, corresponds to musical value, and not to value judgements concerning the elegance of code or other such matters of programming. What is more, we focus $\mathcal{E}$ specifically on musical value attributed by our practitioner, and not on that endowed by the approval of an audience, for example. This will come later.

Here and elsewhere, the sonic entity being evaluated may be any of a range of musical structures at various scales, depending on the code being used, and the focus of attention by the listener. We do not make these distinctions in our examples, because they do not add to our discussion: the reader may choose any or all of the possible facets of the generated music as his or her preferred area of interest.

15.4.3.2 *Uninspiration*

There are various ways that a creative system can fail to be creative in a valued way. These ways can be characterized through the rule set $\mathcal{E}$ and its relationship with the other components of the CSF.

Hopeless uninspiration is the simplest case, where there are no valued concepts in the universe:

$$[\![\mathcal{E}]\!](\mathcal{U}) = \varnothing.$$

This system is incapable, by definition, of creating valued concepts, and as such might be termed ill-formed (if such creative behaviour is the intention).

In this case, there is no solution within the specified universe; there is no capacity within the system to solve the problem. Therefore, it is up to the system designer to remedy the problem, like a *deus ex machina*.

For the purposes of our example, we suppose that this case does not arise. It corresponds to the situation where no valued music exists. (With a more specific application of the framework, however, hopeless uninspiration is possible: if we were to take as our universe all live algorithms music, we cannot necessarily assume that $\mathcal{E}$ will accept any members of $\mathcal{U}$.)

Conceptual uninspiration arises when there are no valued concepts in the conceptual space:

$$[\![\mathcal{E}]\!]([\![\mathcal{R}]\!](\mathcal{U})) = \varnothing.$$

We label this form of uninspiration 'conceptual' because it entails a mismatch between $\mathcal{R}$ (which defines the conceptual space) and $\mathcal{E}$ (which evaluates concepts within it, and, more broadly, within $\mathcal{U}$). This condition is contradictory to the purpose of the two-rule sets: if $\mathcal{R}$ is supposed to constrain the domain of a creative process, then it is inappropriate for $\mathcal{E}$ not to select some of the elements it admits. As such, like the hopeless case, conceptual uninspiration indicates ill-formation of the intended-creative system.

Conceptual uninspiration can only be addressed, within the system, by the transformation of $\mathcal{R}$.

In our live-coding example, this situation is where our programmer does not value the kind of music which she conceptualizes. It is probably not, therefore, likely to be an interesting case.

Generative uninspiration occurs when the technique of the creative agent does not allow it to find valued concepts within the space constrained by $\mathcal{R}$:

$$[\![\mathcal{E}]\!]\left(\langle\langle\mathcal{R},\mathcal{T},\mathcal{E}\rangle\rangle^{\circ}(\{\bot\})\right) = \varnothing$$

This kind of uninspiration is less serious than the other two, and does not necessarily indicate an ill-formed creative system: it merely indicates that a creative agent is looking in the wrong place. This raises the question of *why* there is such a mismatch. Boden's underlying assumption seems to be that the conceptual space is in some sense definitive, and, certainly, in a multiagent environment, it is the only place in the formalism where the consensus about a creative domain can logically be represented. Generative uninspiration can be remedied within the framework. Transformational creativity is required. To transform the set $\mathcal{T}$ in a useful way, we need to identify one or more valued concept(s), in the conceptual space constrained by $\mathcal{R}$ (otherwise, we may have aberration, discussed below), and to use it (them) to guide the transformation. However, there is a methodological problem here: there is no clear way to pick the concept(s) automatically, except at random or by use of an oracle. The 'oracle' might in fact be systematic search of $\mathcal{R}$ (assuming this is possible in finite time), or, again, the *deus ex machina* of user intervention.

In the live coding context, this situation corresponds to a programmer who has not written an algorithm that generates music that she values. She must transform her algorithm so that it can do so.

15.4.3.3 *Aberration*

Now, consider the following more interesting set of scenarios, which also concern the relationship between $\mathcal{R}$ and $\mathcal{T}$. A creative agent, **A**, is traversing its conceptual space.

From any (partial) concept(s) in the conceptual space, A's technique will enable it to create another. Suppose now that the new concept does not conform to the constraints required for membership of the existing conceptual space (note that there is no guarantee that it should do so—there is only an assumption in Boden's work), and is therefore not selected by $[\![\mathcal{R}]\!](.)$. In this case, the set $\mathcal{A}$ given by

$$\mathcal{A} = \langle\langle \mathcal{R}, \mathcal{T}, \mathcal{E}\rangle\rangle^{\Diamond}(\{\bot\}) \setminus [\![\mathcal{R}]\!](\mathcal{U})$$

is nonempty. The CSF terms this *aberration*, since it is a deviation from the notional norm as expressed by $\mathcal{R}$. The choice of this rather negative terminology is deliberate, reflecting the hostility with which changes to accepted styles are often met in the artistic world.

The evaluation of this set of concepts is actually slightly more complicated than the single-concept motivating case outlined above. The aberrant but valued subset, which called $\mathcal{V}_A$ here, is calculated thus:

$$\mathcal{V}_A = [\![\mathcal{E}]\!](\mathcal{A}).$$

Because we are working in the extensional limit case, with all the created concepts notionally elaborated, we have to consider the possibility that all aberrant concepts, some aberrant concepts or no aberrant concepts may be valued. The CSF terms these *perfect* $(\mathcal{V}_A = \mathcal{A})$, *productive* $(\mathcal{V}_A \subset \mathcal{A})$ and *pointless* $(\mathcal{V}_A = \varnothing)$ aberration, respectively.

In the case of aberration, there is a choice as to whether to value the result or not, and therefore we have the three categories: perfect, productive, and pointless. Acceptability is determined in terms of evaluation by whatever audience the agent, A, is playing to—our live coder in this case. If a new concept is accepted, then a sensible solution might be to revise the notion of what the correct domain (as constrained by $\mathcal{R}$) is, so as to include the new concept. This, of course, might have consequences: other new concepts might be included and/or existing ones might be excluded along the way. If the new concept is not accepted under evaluation, then a reasonable recourse would be to adapt A's technique, $\mathcal{T}$. This may have similar consequences with respect to added and existing concepts available to A: valued concepts may be lost, and new aberrant behaviour may be made possible.

One approach is to use the sets $\mathcal{A}$ and $\mathcal{V}_A$ to generate training examples to modify $\mathcal{R}$ and $\mathcal{T}$, using our learning mechanism(s), as follows. Note that there are open questions here about some of the training sets required, since that choice is a major factor in the behaviour of the system. The main issue here is a standard one for AI: how much of what an AI program does is simply programming a computer directly to do something, and how much is emergent behaviour which was not directly programmed? In particular, if we first simply train $\mathcal{T}$ to match $\mathcal{R}$, we might be 'coaching' our creative agent too directly, instead of allowing it to develop, and, second, in doing so we might be restricting its creative capability.

Perfect aberration yields new concepts, all of which are valued, and so should be added to $\mathcal{R}$. $\mathcal{T}$ has enlightened us as to new possibilities. We therefore attempt to revise $\mathcal{R}$, by whatever learning methods are available, in such a way that all the concepts in $\mathcal{A}$ (and $\mathcal{V}_A$) are included, so $\mathcal{V}_A$ is a positive training set, and the negative training set is either $\emptyset$ or $\mathcal{U} \setminus [\![\mathcal{R}]\!](\mathcal{U}) \setminus \mathcal{A}$ or some subset of the latter, depending on the effect desired.

In our running example, perfect aberration is the case where the programmer's algorithm generates unexpected music, all of which is valued. Obviously, on defining a hit in a way that she hadn't previously conceptualized, she will want to adapt her notion ($\mathcal{R}$) of what is live coded music.

Productive aberration means that we need to transform both $\mathcal{R}$ and $\mathcal{T}$, because we wish valued concepts to become accepted, and unvalued ones not to be generated. $\mathcal{V}_A$ and $\mathcal{A} \setminus \mathcal{V}_A$ constitute positive and negative training sets for R, since R needs to expand just enough to include only the valued concepts in $\mathcal{A}$. $\mathcal{T}$, on the other hand, needs to be transformed to restrict its coverage: $\mathcal{A} \setminus \mathcal{V}_A$ is a negative training set for $\mathcal{T}$, while, again, a positive training set might be $[\![\mathcal{R}]\!](\mathcal{U})$, or simply $\emptyset$.

For our example live coder, productive aberration is more difficult than perfect. It requires deeper introspection to identify which aspects of the aberrant music should be retained and which should be rejected. She will need to open her mind (R) to the new concepts that she had not previously entertained, while adapting her algorithm so that it no longer produces the aberrant music that was not valued.

Pointless aberration suggests the need to transform $\mathcal{T}$ only, so as to prevent the unvalued aberrant concepts from being generated. There is a negative training set: $\mathcal{A}$. Again, the nature of the positive training set is an open question.

For our example programmer, pointless aberration is an indication of failure. She will need to rewrite her algorithm to preclude the unvalued musical concepts.

15.5 TIDAL

The Tidal programming language (McLean 2011; McLean and Wiggins 2010c) is a real-time embedded domain-specific language for live coding.[5] It consists of a conventional command line interface, which its inventor uses within the Emacs programmable editor, to enable easy reference and reuse of past commands. The language itself is implemented as an extension of the strongly typed functional programming language Haskell (Thompson 2011). Functional languages are particularly well suited to this kind of task, partly because they are symbolic, making it very easy (for the live coder) to associate program fragments with easy-to-remember symbols (that the live coder has chosen); these program fragments, which may be simple constant values or complex sound-generation routines, can then be composed into sequential structures, stacked into simultaneities, or both, and then operated on by high-order combinators, expressed directly in Tidal syntax. For example, one can construct a sequence of drum beats by writing down the names of the relevant sounds in sequence, then reverse it by the application of one simple combinator, and then execute performance of both simultaneously by the application of another.

Importantly from the perspective of live performance, Tidal is a live compiling language. Commands are implicitly looped, and whatever is playing currently continues until a new command has been successfully compiled. What is more, there is a notion of completion, which ensures that execution of a new command begins at a time which is musically appropriate, according to McLean's particular aesthetic. This, coupled with Haskell's very powerful type-checking system, helping the live coder to produce correct code, yields a highly expressive and flexible performance interface.

The final crucial ingredient is synchronized parallelism: Tidal is capable of running several commands at once, and implicit rules ensure that their output is synchronized, again in keeping with McLean's musical aesthetic.

Underlying Tidal is a scheduling system based on Open Sound Control (OSC; Wright and Freed 1997), which means that, ultimately, anything that can be done in Tidal can be done in the reader's favourite generative composition system, given an OSC interface—but probably not as easily. This means that Tidal can form a conceptual framework for the rest of the current discussion, while not limiting its scope, because the modes of expression it affords are general.

15.6 LIVE CODING IN THE CREATIVE SYSTEMS FRAMEWORK

What, then, does the philosophy of computational creativity have to offer the hybrid creative system formed by a live coder and her Tidal performance system? We now consider the components of the hybrid system in terms of the CSF, generalizing from our earlier illustrative example. First, we formalize the representations of the conceptual spaces and the relationship between them. Then we formalize the dynamics of the system. This allows us, finally, to identify where some of the creative responsibility in live coding performance might be shared with the computer.

15.6.1 Intentional and Extensional Representation of Knowledge

Our original universe, $\mathcal{U}$, of all possible musics must be expanded to include Tidal programs, as we now consider these explicitly. We introduce a conceptual space of well-formed Tidal programs, $\mathcal{C}_{TP}$. Since the execution rule of Tidal is deterministic,[6] there is a many-to-one mapping from $\mathcal{C}_{TP}$ to the conceptual space of Tidal music, which we call C_{TM}. The mapping, which we call $\mathcal{X}$ (for 'eXecution') is many-to-one because there is more than one way to express the production of some items of music, with no audible difference (e.g., two bars of four beats or four bars of two beats in a performance that does not emphasize metrical structure). In an intuitive sense, C_{TM} gives semantics to C_{TP}, which potentially opens interesting questions about music similarity as a measure of program similarity, and which will enable part of our proposal, below. Note that these two conceptual spaces are objectively defined by the syntax and execution rule of Tidal. This is illustrated in Figure 15.1a. We also introduce an inverse mapping, $\mathcal{X}'$, from points in C_{TM} to sets of points in C_{TP} such that $\mathcal{X}'(m_i) = \{p_i : \mathcal{X}(p_i) = m_i\}$. This partitions the conceptual space of Tidal programs into equivalence classes on the basis of identical musical output.[7]

Now we move on to the subjective part of the system: the Live Coder, whom we will call Elsie. For simplicity, we assume that Elsie will program without making audible errors—while this would be a big assumption in most programming languages, Tidal is specifically designed not to degrade on error, so it is not unreasonable here. Supposing that Elsie is only human, and therefore not perfect, it is reasonable to assume that her personal conceptual space of Tidal programs is a strict subset of C_{TP}. Equally, the likelihood is that her personal conceptual space of Tidal-produced music will be smaller than C_{TM}. It may also have elements in it that are *not* members of C_{TM}, because the coder's prediction of what her code will do may sometimes be incorrect. So we give ourselves

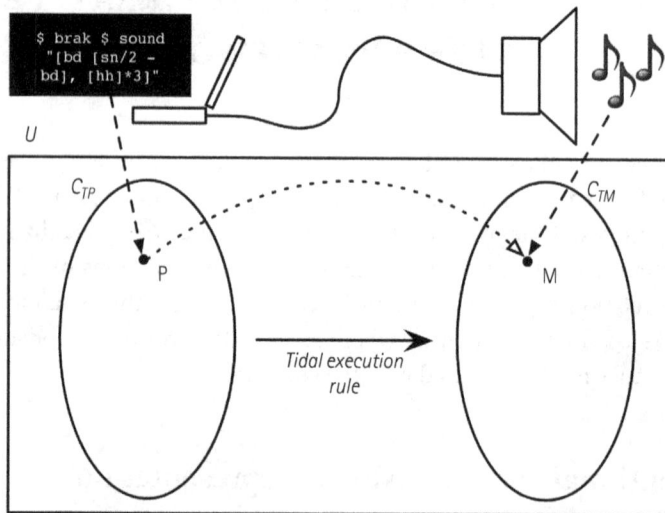

FIGURE 15.1 The defined conceptual space of Tidal programs, and its corresponding conceptual space of music. This structure, represented as a Venn diagram, forms the basis of our argument. The program at point P in the conceptual space of Tidal programs, C_{TP}, corresponds with the music at point M in, the conceptual space of music generated by Tidal programs, C_{TM}. The dashed lines indicate the relationship between the program and the music and their respective points in the conceptual spaces; the dotted line represents the process of execution of Tidal.

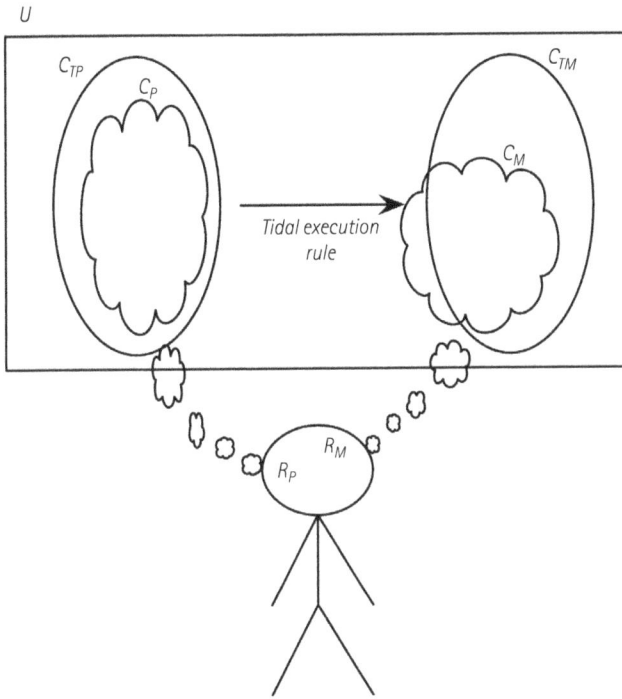

FIGURE 15.2 Elsie's personal conceptual spaces of Tidal programs and Tidal music may not match exactly to the objective spaces. Specifically, C_P is smaller than C_{TP}, and C_M may be smaller than C_{TM} and also include music that is not included in C_{TM}.

the extensional sets C_P and C_M respectively, and the corresponding intensional rule sets $\mathcal{R}_P$ and $\mathcal{R}_M$ respectively, to express these points. The extensional nature of set C_P should not be confused with the intensional nature of its constituent artefacts: Tidal programs are intensional representations of musical sequences, but within the CSF, component sets are considered extensionally. These are illustrated in Figure 15.2.

Because we are focussed on a wider remit than just live coding in this chapter, we omit consideration of Elsie's aesthetic preference regarding coding style, because it complicates our model beyond what is necessary to convey our message. In an equivalent model specifically of live coding, this would be an indispensable component. Tidal is a very concise language, and therefore there is not very much range of expression in this sense. We therefore use the empty set, $\emptyset$, instead of the more predictable $\mathcal{E}_P$.

The formalization starts to become interesting when we add in Elsie's music-aesthetic preference, expressed as a rule set $\mathcal{E}_M$, which selects a subset of $\mathcal{U}$, which may contain some or all of each of C_{TM} and/or C_M. This gives us the arrangement illustrated in Figure 15.3. The different combinations of intersection and nonintersection between C_{TM}, C_M, and the extension of $\mathcal{E}_M$, labelled with lower-case letters in the diagram, indicate areas into which actual or imaginary pieces of music might fall,

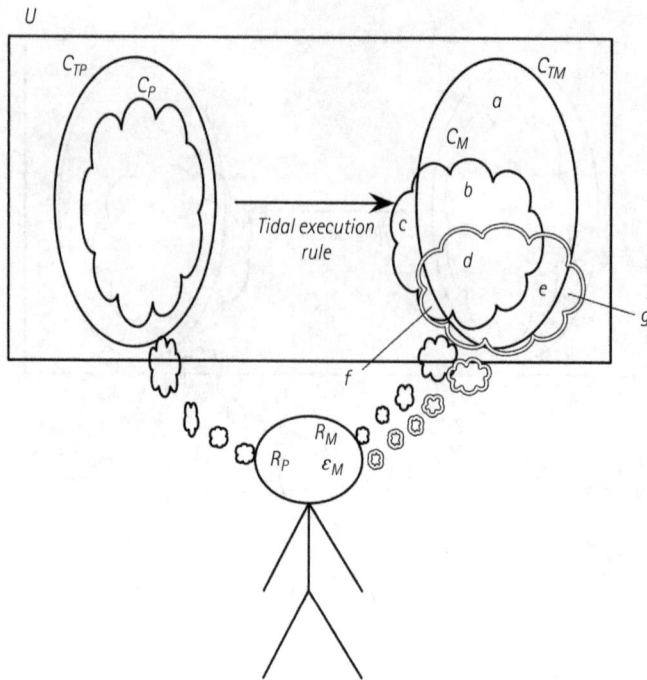

FIGURE 15.3 Elsie's music-aesthetic preferences, $\mathcal{E}_M$, are expressed as a rule set which select a range of the available possibilities. We use a simple yes/no approval rating here for simplicity, but a continuous fuzzy set membership could be used if richer expression were required (Wiggins 2006a; Ritchie 2012).

and each of them corresponds to a different possibility, from the perspective of computational creativity. We now consider each in turn, not in terms of the constructive process necessary to build a program, but in terms of the knowledge and/or imagination required to generate the computational and/or musical concept. The details are summarized in Table 15.2.

15.6.2 Representation of Dynamics of the Hybrid Creative System

Now we have mapped out the landscape of possible outcomes of our human-computer hybrid creative system, we must look at the dynamics. The Tidal techniques envisaged by McLean (2011) involve a somewhat incremental approach to programming, where one often constructs a basic musical structure extensionally (that is, in literal notes or sounds), and then elaborates on it by a mixture of added extensional structures (For

Table 15.2 Analysis of knowledge required to refer to music in one of the labelled areas in Figure 15.1b, in the context of Elsie's aesthetic preference.

a	music achievable through a Tidal programme that is neither imaginable nor liked by Elsie
b	music that Elsie can imagine and that is achievable through a Tidal programme, but that Elsie does not like
c	music that Elsie can imagine but does not like and which is not programmable in Tidal
d	music that Elsie can imagine and likes, and that is programmable in Tidal
e	music that is programmable in Tidal, and that Elsie would like, but that she has not (yet) conceptualized
f	music not achievable in Tidal, but which Elsie can imagine and likes
g	music not achievable in Tidal, which Elsie cannot conceptualize, but which she would like if she could

example, a counterrhythm to be played simultaneously; the approach lends itself well to strict additive process, such as that used in the early work of Philip Glass: Potter 2000), or intensionally, by applying Tidal functions that manipulate the material as part of the performance. This approach lends itself well to description by the CSF, where the function $\langle\langle .,.,.\rangle\rangle$ is envisaged as an enumeration process which traverses a conceptual space, stepping from one concept to the next, in a sequence defined by $\mathcal{T}$ and possibly influenced by $\mathcal{R}$ and $\mathcal{E}$. In our case, there is a more complicated interaction, because the conceptual space being traversed is not that of music, but that of programs. While we would like to argue that Elsie's knowledge and capability are such that she would be able to traverse the space of Tidal music directly, merely selecting the appropriate program to achieve what she wants, any programmer knows that such exactitude may be expected only for trivial cases. Therefore, such a model would be unrealistic.

We model Elsie's traversal of the conceptual space of programs with a rule set, $\mathcal{T}_P$, and the corresponding notional traversal of the space of Tidal music with $\mathcal{T}_\mathcal{M}$. Because, as mentioned above, the execution function, $\mathcal{X}$, of Tidal maps from $\mathcal{C}_P$ to $\mathcal{C}_\mathcal{M}$, many to one, there is an interaction between $\mathcal{T}_P$ and $\mathcal{T}_\mathcal{M}$ which can be partly explained in terms of $\mathcal{X}$. For each program, $p_i \in \mathcal{C}_P$, there is a corresponding musical performance, $m_i = \mathcal{X}(p_i)$. Elsie traverses $\mathcal{C}_P$ by means of application of $\langle\langle\mathcal{R}_P, \mathcal{T}_P, \varnothing\rangle\rangle$ to a vector of programs, $\bar{p}$, which Elsie has already conceptualized. In some cases, this will merely result in selection: Elsie will choose a code fragment that she uses frequently, perhaps to achieve a known effect, or to begin an improvisation sequence with a personal signature. In other cases, she will be generating new programs from old, perhaps by conceptual blending (Turner and Fauconnier 1995) or bisociation (Berthold 2012;

Koestler 1964). At our current level of abstraction, however, the specific function is unimportant: its details are tucked neatly away inside $\mathcal{T}_P$. The application of the function generates a new vector of programs, $\bar{q} = \langle\langle \mathcal{R}_p, \mathcal{T}_p, \varnothing \rangle\rangle(\bar{p})$. At this point, we can identify the nature of the latest product: the Tidal music at point $\mathcal{X}(q_i)$ in $\mathcal{C}_{TM}$ (where $q_i \in \bar{q}$) will fall into one of the areas a, b, d, e, in Table 15.2, which we examine in the next section.

15.6.3 Sharing Creative Responsibility

We are now in a position to describe abstractly, with some precision, the actions that Elsie can take as she performs, and the nature of the resulting outputs, in terms of what she knows and likes, and what is possible in Tidal. The question, then, is: what can be done to change the components of this system, so that some of the creative responsibility in Elsie's performance can be shared with the computer, as is the aim of computational creativity (Colton and Wiggins 2012)?

Clearly, given the hybrid nature of the creative system under discussion, different parts of it are subject to different kinds of modification: Tidal could be enhanced with implementations of one or more of the components of the CSF formalization; or Elsie could be modified in a way that is necessary less straightforward. However, perhaps the computer can help. Essentially, the potential modifications to Tidal are in two categories: generative power and reflection. Generative power, here, refers not to generation of music, but to generation of programs that make music. Reflection, in the current model, refers to evaluation of music and not of programs. Elsie, on the other hand, as a healthy human, can reflect; she already has some notion of what she expects from her programs, and an aesthetic by which she judges them.

In section 15.4.3, we explained some useful tests, under the general headings of *uninspiration* and *aberration*, that can be applied to a CSF formalization. The same ideas will be useful in this extended hybrid formalization. We now consider the cases in turn in Table 15.2. Because we have multiple conceptual spaces represented concurrently, it is important to pay attention to the subscripts in the symbology. We will mostly be thinking in terms of $\mathcal{C}_{TM}$, the space of Tidal music; however, because that objective conceptual space of Tidal music, $\mathcal{C}_{TM}$, can have elements that are not in C_M (that is, they would not generally be considered as music in a wider context), we must necessarily use both spaces.

Area a (in Figure 15.3) is an area of both *hopeless* and *conceptual uninspiration*, in terms of the CSF. This is the case because, even though the objective conceptual space of Tidal music, $\mathcal{C}_{TM}$, has elements here, these elements are not valued, and because C_M does not include it. To remedy the hopelessness would be to change Elsie's aesthetic, so we do not consider this possibility further. To remedy the conceptual uninspiration without addressing the hopelessness would merely produce unwanted music, so we treat it in the same way.

Area b contains music that Elsie does not value, which means that, presumably, she would prefer not to generate it. This entails some kind of filter on the production of code using $\mathcal{T}_P$, and this brings us to the nub of our proposal. We propose two separate ways in which the Tidal system might be enhanced, to allow creative responsibility to be passed to the computer. The first approach is to restrict the syntax that Elsie is able to use in her programs, in such a way as to divert her performance away from the areas of $\mathcal{C}_{TP}$ and $\mathcal{C}_P$ that generate music that she does not value. This enhances the creativity of the system, in so far as it improves Elsie's chances of producing a result that is satisfactory to her; and some of the creative responsibility is definitely passed to the computer. It might be argued that the description of a restricted syntax in terms of a filter within $\mathcal{T}_P$ could instead be modelled as a change in $\mathcal{C}_{TP}$ or $\mathcal{C}_P$, that is, as examples of transformational creativity. However, $\mathcal{C}_{TP}$ is objectively defined by Tidal, so it cannot be transformed. Changes in $\mathcal{R}_P$, resulting in a smaller $\mathcal{C}_P$ (in order to maximize the intersection of $\mathcal{C}_P$ and $\mathcal{E}_M$), could indeed be conceptualized in terms of transformational creativity. However, the core issue here is that $\mathcal{T}_P$ could still take us beyond any restricted subset of $\mathcal{C}_P$, because by definition $\mathcal{T}_P$ traverses $\mathcal{U}$ and is not therefore restricted to $\mathcal{C}_P$, so an explicit modification of $\mathcal{T}_P$ is necessary in either case. Thus, as noted above, the CSF gives us two distinct (but related) notions of transformational creativity. There is a challenging and deep open research question of how such changes may be efficiently and promptly conveyed back to users, by means other than simply listening; because of Tidal's restricted but full formal semantics and expressive syntax, it is conceivable that users might be briefed as to changes in these terms. However, some users might feel that purely aural feedback is best.

The second approach is to replace Elsie's direct manipulation of program code with automated generation of code fragments, which are subsequently selected and/or approved by Elsie. The fragments offered can be filtered in the same way as Elsie's own programs, above, so as to be within Elsie's preferred range. However, both of these approaches entail knowledge about Elsie's preferences that is not currently encoded in Tidal or, indeed, in other systems of which we are aware. To make the hybrid system effectively creative, we need a mechanism for Elsie to feed back approval to Tidal. We return to all these points in the next section, once our analysis is complete.

Areas c, f, and g are imponderable from within the closed system formed by Elsie and her computer. However, given examples of other musics that Elsie values, it would in theory be possible to use the mapping $\mathcal{X}'$ to identify examples within $\mathcal{C}_P$ that would generate music with similar properties. These could then be used to adapt the generation process, away from c, because it is not valued, or towards a member of $\mathcal{C}_M$ that is similar for cases f and g.

Area d is the comfort zone for Elsie: she has conceptualized this music, and she values it. So no action is required in this area, except to gather feedback so that the computer records Elsie's approval.

Finally, area e is interesting because it offers an opportunity to change Elsie's programming behaviour in ways that she will value. In this area, Elsie has not yet conceptualized the music, so it will be surprising to her, and her programming style does not give her access to it; so to have the computer lead her programming towards this area would be of high creative value.

All this reasoning serves no purpose unless a system could be built with the necessary knowledge. In the next section, we identify the capabilities required by a cooperative creative system based on Tidal to enable it to fulfil the potential suggested by our analysis.

15.7   Proposal: A Hybrid Creative System Based on Tidal

In order to fulfil the potential that the above analysis suggests, we need three key ingredients. The first is the ability for our computer to relate the meaning of a program to its syntax. The second is for our computer to have a model of our coder's preference. The third is for our computer to manipulate the syntactic constructs available to our coder so as to take on some of the creative responsibility for the music. We outline the potential to build systems that address each of these in turn, with the intention of raising a challenge to builders of systems for algorithmic music of the future.

15.7.1   Semantics

The key difficulty with any computational art system (or indeed any computational system of any kind) is predicting its output for any given nontrivial program. The theoretical reasons for this relate to Turing's 'halting problem' (Turing 1936) and their detail is beyond the scope of this chapter; however, we may summarize by saying that the only way to understand what a program does is to run it and see—but it, or parts of it, may not terminate, in which case we cannot know what it can do in full.

The upshot of this is that, in general, it is very hard to say what a program means, to give it *semantics*. One way of doing so is to consider the 'answer set', the fixpoint[8] of the set of possible outputs of the program iterated until there are no more available. However, this idea is clearly a hostage to the halting problem, because some outputs may be prevented from appearing by nontermination of code that precedes their output in the executing sequence.

Strongly typed functional programming languages, such as Haskell, on which Tidal is based, are particularly well behaved in terms of understanding their semantics mathematically. That is not to say that they are exempt from the halting problem—they are not—but their strong type checking does make the notion of program well-formation much, much stronger than that in other languages.

Our case, however, is a special one. Tidal is designed to execute programs in loops, and its syntax is designed to work in this way. Specifically, cycles within Tidal are represented by the set of natural numbers, and the principle datatype, `Pattern`, is a map of time to events, which is notionally infinite in length and can be queried given any time interval expressed as a pair of rational numbers (McLean 2014, 64–65). If Elsie restricts her code to operators that are part of Tidal and not part of the underlying language, we can be sure that the programs will halt, and Haskell's type checking confirms that they are well-formed before they are run. This means that it is possible to construct a theoretical space of syntax trees, in which each runnable program is a point. Indeed, it is possible to do this for Haskell programs in general.[9] Such a space is still a representation of syntax, not semantics, but it does allow us to realize an implementation of C_{TP}, as required by our argument above.

The behaviour of Tidal as a means of controlling the generation of sound gives us an exciting way to provide semantics for our programs. There is extensive research in the literature on methods for analysing sound, in terms of *features*—analytical aspects—which may be more or less perceptually motivated: for example, the ISMIR,[10] ICASSP, and WASPAA[11] conferences afford extensive possibilities for the analysis of sound along dimensions that may or may not be salient for a given human listener. These features allow the sonic outputs of Tidal to be represented, more or less approximately, as points in a multidimensional space in which dimensions correspond to perceptually meaningful qualities (Gärdenfors 2000). This feature space constitutes an additional level of representation, or domain of information, within C_{TM}, providing a perceptually motivated space mapping the lower-level acoustic space. Dimensional reduction using standard mathematical techniques such as Principal Component Analysis (PCA: Jolliffe 2002) may be used to throw away features that add little information, for parsimony. Now, we are in a position to enumerate a relation approximating the function, $\mathcal{X}$, introduced above, and thence to compute its inverse mapping, $\mathcal{X}'$, though we must remember that C_{TP} is infinite, and therefore compromise is necessary in doing so: there are various principled ways of limiting search through C_{TP}, based on techniques from genetic programming and static program analysis.

The infinite size of C_{TP} is less of a problem than we might expect, for two reasons. First, C_P is a subset, and, given the finite nature of humans, is probably not infinite. Second, given an initial estimate of C_P, it can be expanded piecemeal as Elsie produces her work, and so exhaustive enumeration becomes unnecessary: instead, the system learns about its user as she uses it. C_M, for noninfinite C_P, can be computed offline, and there is an excellent application here for cloud computing: a shared effort to identify as much as possible of C_{TM} would generate a valuable resource indeed.

15.7.2 Identifying Value

Given the semantic mapping proposed above, it becomes possible to learn Elsie's preference, expressed as rules in $\mathcal{E}_M$, whose evaluation, $\left[\!\left[\,\mathcal{E}_M\,\right]\!\right]$, yields a value which can be viewed as an extra dimension in the extensional set of points in C_M subtended by the range of performances she has made with her system. To do this, feedback from Elsie is required. It may be given in terms of explicit ratings of the music that is currently happening (perhaps by buttons expressing positive, neutral and negative affect, or by a slider over a similar range); alternatively, affective response might be measured indirectly, for example by timing how long Elsie allows a given program to run (assuming that she will replace music she does not value quite quickly), or by measuring physiological responses, such as skin conductance or heartbeat, though the physiological approaches are subject to the drawback that being in a performance situation may cause them to fluctuate. However they are gathered, the responses will allow us to categorize regions of C_M according to how much Elsie values them. Having done so, we can use X' to lead us back to the program(s) that create that music, and it is this possibility that admits computational creativity into our hybrid system. This involves making assumptions about the nature of $\mathcal{E}_M$, and this is an interesting area for further research: would a further PCA of the perceptual features of C_M with the addition of the value dimension, give a different, possibly more useful, reduction applicable to the larger space of C_{TM}?

15.7.3 Transforming Human Creativity

Given an estimate of the value that Elsie places on each piece of music, the computer can analyse each program that Elsie writes, mapping its C_{TM} value via X to its equivalent point in C_M, whose value, given by application of $\llbracket \mathcal{E}_M \rrbracket$, is known. From this, the system could feed back to Elsie, before executing a piece of code, if it will generate music in a region with which she has associated negative value previously. Thus, the computer system detects pointless aberration (see section 15.4.3.3) and is able to apply transformational creativity of its human, by influencing her T_P. This corresponds to rejecting areas a and b in Figure 15.3.

Conversely, and more interestingly, in the event that Elsie is exploring an area of C_{TM} that is new to her, the computer may be able to make predictions from $\llbracket \mathcal{E}_M \rrbracket$ about which nearby points, so far unexplored, are likely to be valued. It is thence possible to map back to corresponding points in C_P, and present Elsie with a range of programs to try. Again, this is a kind of transformational creativity: C_M and C_P are expanded, and $\mathcal{E}_M$ would be modified to reflect Elsie's evaluation of the result. Here, points in areas a or e are being moved into area b or d in Figure 15.3.

15.8 CONCLUSION: TRANSFORMING COMPUTATIONAL CREATIVITY

We now look beyond the analysis possible in the restricted space of the chapter. Any or all of the above operations can in principle lead to changes in C_P and R_P. Given an appropriate metric on C_{TP} (which may also be aesthetically motived), we can consider beginning to traverse C_P automatically, using, for example, genetic programming (GP). At this point, Elsie can begin to relax her artistic control and really work with the system: she can, for example, restrict her 'coding' to telling the computer when to change or to evaluating the outputs, perhaps intervening when things go too far from her artistic preference (and, of course, we do not assume that she must necessarily like every idea she chooses to follow up). If she observes the results in detail, either by listening or by a Tidal-based feedback mechanism, there will be feedback into her own C_P and C_M, which themselves will feed back into her use of the system and thus inform future transformations.

Thus, Elsie's Live Coding achieves not the 'singularity' of science fiction, but a duality in which she is working on an equal creative basis with the computer, with shared notions of artefact and of meaning. It would in principle be possible to estimate Elsie's $\mathcal{E}_M$ as a function, and thus simulate her musical aesthetic. However, we propose that this would be pointless: there is no evidence at all to suggest that a silicon-based computer has qualia, so we suggest that aesthetic response is best left to the entities that seem most likely to be conscious of it.

ACKNOWLEDGEMENTS

The authors are supported by the Lrn2Cre8 project, which acknowledges the financial support of the Future and Emerging Technologies (FET) programme within the Seventh Framework Programme for Research of the European Commission, under FET grant number 610859.

NOTES

1. These terms are borrowed from symbolic logic. An *intensional specification* is one which is couched in terms of properties, such as 'the integers between 0 and 100', while an *extensional specification* lists the set of things referred to. Clearly, intensional specifications can specify infinite things (e.g. 'all numbers greater than 0', where extensional ones cannot. Further, and most important to our purposes, an intensional specification can be a programme that generates things.
2. Two groundbreaking examples of 'live in studio' manipulation of sequencers by Chris Franke can be found in the Tangerine Dream tracks 'Phaedra' (1974) and 'Stratosfear' (1976) from the albums of the same names.
3. Ritchie (2012) presents a slightly different formalization of broadly the same ideas.
4. We will not explore issues of representation at this point, suffice it to say that music, fundamentally a psychological phenomenon, may be represented from multiple perspectives and at various levels of abstraction, such as digital audio signals, score-like discrete representations or in terms of psychological models of musical perception (Babbitt 1965; Wiggins 2012a, 2012b, 2012c).
5. It grew out of the earlier Petrol language, but it is intended to be more sustainable. It may be downloaded from http://yaxu.org/tidal/.
6. Assuming, as we do here, that randomness is not involved.
7. The issue of the representation of $\mathcal{U}$ influences the notion of identity. In this case we may most usefully consider identity in terms of a psychological space, since any such space will typically be smaller than the mathematical space of programme output (Collins 2008, 240).

8. Recall our operator, $\diamond$, which computes this in the CSF, from section 15.4.2. Note that the halting problem does not affect the CSF formulation because there is no actual attempt to enumerate the various sets involved: all the constructs are theoretical.
9. Forth (2012) applies a similar approach to musical-metrical trees, with syntactic considerations drawn from music theory instead of computer science.
10. www.ismir.net.
11. www.waspaa.com.

BIBLIOGRAPHY

Baars, B. J. *A Cognitive Theory of Consciousness*. Cambridge: Cambridge University Press, 1988.
Babbitt, M. 'The Use of Computers in Musicological Research'. *Perspectives of New Music* 3, no. 2 (1965): 74–83.
Berthold, M. R. *Bisociative Knowledge Discovery*. Berlin: Springer, 2012.
Boden, M. A. 'Creativity and Artificial Intelligence'. *Artificial Intelligence Journal* 103 (1998): 347–356.
Boden, M. A. *The Creative Mind: Myths and Mechanisms*. 2nd ed. London: Routledge, 2004.
Bovermann, T., and Griffiths, D. 'Computation as Material in Live Coding'. *Computer Music Journal* 38, no. 1 (2014): 40–53. doi:10.1162/COMJ_a_00228.
Buchanan, B. G. 'Creativity at the Metalevel: AAAI-2000 Presidential Address'. *AI Magazine* 22, no. 3 (2001): 13–28. https://aaai.org/ojs/index.php/aimagazine/article/view/1569/1468.
Bundy, A. 'What Is the Difference between Real Creativity and Mere Novelty?' *Behavioural and Brain Sciences* 17, no. 3 (1994): 533–534.
Clark, A., and D. Chalmers. 'The Extended Mind'. *Analysis* 58, no. 1 (1998): 7–19. doi:10.1093/analys/58.1.7.
Collins, N. 'The Analysis of Generative Music Programs'. *Organised Sound* 13, no. 3 (2008): 237–248. doi:10.1017/S1355771808000332.
Colton, S., Pease, A., Corneli, J., Cook, M., Hepworth, R., and Ventura, D. 'Stakeholder Groups in Computational Creativity Research and Practice'. In *Computational Creativity Research: Towards Creative Machines*, edited by T. R. Besold, M. Schorlemmer, and A. Smaill, 3–36. Amsterdam: Atlantis, 2015.
Colton, S., Pease, A., Corneli, J., Cook, M., and Llano, M. T. 'Assessing Progress in Building Autonomously Creative Systems'. In *Proceedings of the Fifth International Conference on Computational Creativity*, edited by S. Colton, D. Ventura, N. Lavrač, and M. Cook, 137–145. Ljubljana, 2014
Colton, S., and Wiggins, G. A. 'Computational Creativity: The Final Frontier?' In *Frontiers in Artificial Intelligence and Applications*, edited by L. de Raedt, C. Bessiere, D. Dubois, P. Doherty, F. Heintz, and P. Lucas, 21–26. Amsterdam: IOS, 2012. doi:10.3233/978-1-61499-098-7-21.
Ebcioğlu, K. 'An Expert System for Harmonizing Four-Part Chorales'. *Computer Music Journal* 12, no. 3 (1988): 43–51.
Eigenfeldt, A. 'Generative Music for Live Performance: Experiences with Real-Time Notation'. *Organised Sound* 19, no. 3 (2014): 276–285.
Eigenfeldt, A., Burnett, A., and Pasquier, P. 'Evaluating Musical Metacreation in a Live Performance Context'. In *Proceedings of the Third International Conference on Computational Creativity*, edited by M. L. Maher et al., 140–144. Dublin, 2012.

Forth, J. C. *Cognitively-Motivated Geometric Methods of Pattern Discovery and Models of Similarity in Music*. PhD dissertation, Goldsmiths, University of London, 2012.

Gärdenfors, P. *Conceptual Spaces: The Geometry of Thought*. Cambridge, MA: MIT Press, 2000.

Jolliffe, I. T. *Principal Component Analysis*. 2nd ed. New York: Springer, 2002. doi:10.1007/b98835.

Kirnberger, J. P. *Der allezeit fertige Menuetten- und Polonaisencomponist* [The always ready minuet and polonaise composer]. 1757.

Koestler, A. *The Act of Creation*. London: Hutchinson, 1964.

Lutosławski, W. *Jeux vénitiens* [Venetian Games]. Krakow and Celle: Polskie Wydawnictwo Muzyczne, Moeck 1961. Score.

Magnusson, T. 'Herding Cats: Observing Live Coding in the Wild'. *Computer Music Journal* 38, no. 1 (2014): 8–16.

McLean, A. 'Making Programming Languages to Dance to: Live Coding with Tidal.' In *FARM '14: Proceedings of the 2nd ACM SIGPLAN International Workshop on Functional Art, Music, Modeling and Design*, 63–70. New York: ACM, 2014. doi:10.1145/2633638.2633647.

McLean, A. *Artist-Programmers and Programming Languages for the Arts*. PhD dissertation, Goldsmiths, University of London, 2011.

McLean, A., and Wiggins, G. A. 'Bricolage Programming in the Creative Arts'. In *Proceedings of the 22nd Psychology of Programming Interest Group*. Madrid.

McLean, A., and Wiggins, G. A. 'Live Coding towards Computational Creativity'. In *Proceedings of the First International Conference on Computational Creativity*, edited by D. Ventura et al. Lisbon, 2010b.

McLean, A., and Wiggins, G. A. 'Tidal–Pattern Language for the Live Coding of Music'. In *Proceedings of the 7th Sound and Music Computing Conference*. Barcelona, 2010c.

Potter, K. *Four Musical Minimalists: La Monte Young, Terry Riley. Steve Reich, Philip Glass*. Cambridge: Cambridge University Press, 2000.

Revill, D. *The Roaring Silence. John Cage: A Life*. New York: Arcade, 1993.

Riley, T. *In C*. Celestial Harmonies/Temple, 1964. Score.

Ritchie, G. 'Some Empirical Criteria for Attributing Creativity to a Computer Program'. *Minds and Machines* 17, no. 1 (2007): 67–99.

Ritchie, G. 'A Closer Look at Creativity as Search'. In *Proceedings of the Third International Conference on Computational Creativity*, edited by M. L. Maher et al., 41–48. Dublin, 2012.

Thompson, S. *Haskell: The Craft of Functional Programming*. 3rd ed. Harlow: Addison-Wesley, 2011.

Turing, A. 'On Computable Numbers, with an Application to the Entscheidungsproblem'. *Proceedings of the London Mathematical Society* 2, no. 42 (1936): 230–265.

Turner, M., and Fauconnier, G. 'Conceptual Integration and Formal Expression'. *Metaphor and Symbolic Activity* 10, no. 3 (1995): 183–203.

Wiggins, G. A. 'A Preliminary Framework for Description, Analysis and Comparison of Creative Systems'. *Journal of Knowledge Based Systems* 19, no. 7 (2006a): 449–458. http://dx.doi.org/10.1016/j.knosys.2006.04.009.

Wiggins, G. A. 'Searching for Computational Creativity'. *New Generation Computing* 24, no. 3 (2006b): 209–222.

Wiggins, G. A. 'Music, Mind and Mathematics: Theory, Reality and Formality'. *Journal of Mathematics and Music* 6, no. 2 (2012a): 111–123.

Wiggins, G. A. 'On the Correctness of Imprecision and the Existential Fallacy of Absolute Music'. *Journal of Mathematics and Music* 6, no. 2 (2012b): 93–101.

Wiggins, G. A. 'The Future of (Mathematical) Music Theory'. *Journal of Mathematics and Music* 6, no. 2 (2012c): 135–144.

Wiggins, G. A., and J. C. Forth. 'IDyOT: A Computational Theory of Creativity as Everyday Reasoning from Learned Information'. In *Computational Creativity Research: Towards Creative Machines*, edited by T. R. Besold, M. Schorlemmer, and A. Smaill, 127–150. Amsterdam: Atlantis, 2015.

Wright, M., and Freed, A. 'Open Sound Control: A New Protocol for Communicating with Sound Synthesizers'. In *Proceedings of the International Computer Music Conference*, 101–104. Thessaloniki, 1997. https://quod.lib.umich.edu/i/icmc/bbp2372.1997.033

CHAPTER 16

......

TENSIONS AND TECHNIQUES IN LIVE CODING PERFORMANCE

......

CHARLES ROBERTS AND GRAHAM WAKEFIELD

16.1 INTRODUCTION

VARIOUS definitions of the term 'live coding' have been suggested; these variations and their subtleties depict the tensions that riddle the field.[1] Let us begin with a working definition: *in live coding performance, performers create time-based works by programming them while these same works are being executed.* Although many would likely argue for greater specificity, we believe the above definition encompasses most live coding performance practices, and will use the remainder of this chapter to explore the tensions encountered and techniques utilized in live coding that make a one-size-fits-all definition so difficult to achieve. Although the chapter is primarily concerned with musical performance, we would be remiss if we did not mention that live coding is active in visual media as described by Griffiths (2014), Della Casa and John (2014), and Wakefield, Smith, and Roberts (2010), and other domains of performance such as dance (Sicchio 2014).

Modern, canonical live coding performances began in the early 2000s, with performances given using custom live coding systems such as feedback.pl as well as musical programming languages such as SuperCollider (Ward et al. 2004), though the origins of live coding can be traced to earlier dates (Collins 2014). One characteristic of these performances is the writing of algorithms and their subsequent manipulation, and for some this is integral to its attraction: 'the more profound the live coding, the more a performer must confront the running algorithm, and the more significant the intervention in the works, the deeper the coding act' (Collins 2011). While the primacy of the algorithm in live coding performance has been identified by other authors (Magnusson 2011a; Ward et al. 2004), it is not universal, and there is a veritable spectrum of use and modification of algorithmic processes in live coding practice. In the live coding environment LOLC, by Jason Freeman and his research group at Georgia Tech, variation and development of musical pattern emerge from collaborative ensemble practices instead of algorithms. Performers submit musical patterns to a queue that is accessible to all ensemble members, so that each can easily download, modify, and execute musical patterns as they see fit (Freeman and Van Troyer 2011). Although algorithms are not strictly necessary for her work, live coder and choreographer Kate Sicchio employs them both for their creative potential and as aesthetic commentary: 'Some of my work uses generative elements but this is less about typing or transmission and more about the computer providing new possibilities for scores. My work does not need algorithms to be achieved but by using them I am commenting on both coding and choreographing and their similarities in being organisational practices' (Sicchio, personal communication 2015).

To further whittle away at our initial definition, such performances often included (and continue commonly, but not always, to include) projection of the live-edited source code documents. Privileging either the composition or the source code as the primary output of a performance may elide significant complexities; for example, as reported by Collins and McLean: 'Dave Griffiths, of the live coding band slub, considers the music he makes to be a side product, rather than an end-product of his live coding languages, where the visual aesthetics of his interfaces are more important' (2014). Although projecting source code can both reveal activity and potentially intent, there is debate inside the live coding community (and inside individual members of it) about the merit of such projections; it has been suggested that the source code can distract from the artistic focus of a performance (Bruun 2013).

There are dozens of live coding systems available, using a wide spectrum of general-purpose to domain-specific programming interfaces and languages. Where some are widely used in communities far beyond live coding performance, others are so idiosyncratic that they become statements of a personal performance oeuvre. Live coders have been active across many academic disciplines, including education, the psychology of programming, and aesthetics; performances are regularly given around the world in both concert halls and nightclubs.

We have already touched upon some tensions at play in live coding; in the rest of the chapter we use a lens of five recurrent tensions to examine the range of practices found in this young and vibrant community.

(1) How do performers consider risk when performing? An errant algorithm can send an audience running for the doors covering their ears, while a memorized performance risks boredom and inflexibility.

(2) What is the responsibility of the performer to convey algorithmic processes to the audience, and how does this responsibility hinder the immediacy of their interface?

(3) What abstractions are necessary and appropriate in live coding performance? What models, metaphors, and levels of abstraction are performers working with? Does complexity limit or enable freedom?

(4) How do live coders maintain pace and flow in performance? What techniques and notations make this easier—and how do live coders divide their attention over the temporal limits of real-time performance?

(5) What happens at the meeting of the deterministic machine with the indeterminacy of improvisation? And how do live coders deal with the much larger durations of time that code puts within their reach?

16.2 SECURITY, STABILITY, RISK, AND IMPROVISATION

Live electroacoustic performance clearly contains more risk than the unmodified playback of a pre-rendered audio file. While Clowney and Rawlins (2014) argue that such risk is an essential part of all musical performance, the consequences are intensified in live coding performance, where typographical errors can lead to system crashes and performance nightmares. Despite these risks, the live coding community generally seems attracted to uncertainty. Shelly Knotts states: 'I like to perform in circumstances that are risky or have a high cognitive load. ... I'm also interested in how unintentional outputs of algorithmic processes become part of musical structure, so I don't do too much to reduce risk of error in live coding performance' (Knotts, personal communication 2015). In a similar vein, live coder Tanya Goncalves notes that although beginning programmers 'may fear that they will introduce a mistake into their code, and that the audience will recognize it ... this tension is something I personally wait for during a performance and look forward to' (Goncalves, personal communication 2015).

In addition to the attraction that uncertainty and risk hold for performers, Alperson (1984) notes that audiences are more forgiving of performances if they know that significant musical risks are being taken. He also notes the attunement that audiences can feel with an improviser: 'as if the improviser's audience gains privileged access to the composers mind at the moment of musical creation'. This tradeoff is especially relevant to live coding practice, where errors can cause system failure, whereas success can provide unique insight into the improvisational process. Accepting that risk is an integral element of musical performance, the following section examines strategies for managing and encouraging it in live coding performance.

16.2.1 The Allure of Failure

Over the course of many performances, it is almost inevitable that performers will experience crashes. In a notable prototypical live coding performance that took place in

1985, Ron Kuivila accidentally ended a performance with a system failure, described in Curtis Roads's review as follows:

> Ron Kuivila programmed an Apple II computer onstage to create dense, whirling, metric sounds that layered in and folded over each other. Considering the equipment used, the sounds were often surprisingly gigantic in scale. Kuivila had trouble controlling the piece due to system problems. ... The reality is that personalized home-grown hardware and software, for all the freedom of expression they can afford, are especially subject to flakiness. (Roads 1986).

David Ogborn notes that despite the danger that errors in performance pose, risk yields potential rewards for audience members: 'I think too much risk and you could alienate, or even hurt, your audience, and that is something which is hard to come back from. At the same time, my experience as an audience member is that one of the exciting things about live coding are those moments where you can see people really wrestling with decisions—where you see "authentic" moments of decision being taken in front of you' (personal communication 2015). After giving a performance in which one crash occurred and another was narrowly averted, the first author of this chapter noted: 'multiple audience members commented on how the errors lent the performance a sense of danger, and emphasized the improvisatory nature of the performance. Some went as far as to say the errors were their favorite part; we remain unclear what this says about the quality of the error-free segments of the performance' (Roberts, Wright, Kuchera-Morin, and Höllerer 2014).

16.2.2   Practice Makes Perfect

As with traditional instruments, practising is one way of mitigating risk in performances. The amount of preparation accompanying a performance is a tension that each performer must navigate. Developing the fluency of gestures required to improvise in live coding is difficult, and must be potentially balanced with the desire to not 'pre-compose' public performances.

In one experiment, Nick Collins and Fredrik Olofsson committed to an hour of live coding practice each day for a month. Both felt their skills as live coding performers improved 'by introducing various shortcuts, by having certain synthesis and algorithmic composition tricks in the fingers ready for episodes, and just by sheer repetition on a daily basis' (Nilson 2007). Aaron followed this with a more nuanced description of what he believes live coding practice should ideally consist of: 'Preparation for performance should involve activities that are neither original engineering, nor simple repetition' (Aaron, Blackwell, Hoadley, and Regan 2011, 385) Aaron contends that the goal of practice is not to extend the system being used, nor is it to build up to a performance consisting of pre-composed, memorized code. Instead, the goal is to develop 'a fluent repertoire of low-level coding activities [that] will allow the performer to approach performance at a higher level of structural abstraction—based on questions such as where am I going in this performance and what alternative ways are there for getting there' (385–386). Such practice can both reduce the risk of technical errors during a performance and increase the aesthetic freedom of the performer.

16.2.3 Collaborative Performance

One of the simplest methods for managing risk in live coding performance, at least conceptually, is collaboration. If one person is forced to reboot or cease musical output for other reasons, other performers are there to fill in until that person can get up and running again; there is safety in numbers. Although the technical challenges of developing systems for ensemble live coding performances can be daunting, many environments (such as the previously mentioned LOLC) provide affordances for group performances out of the box, including the ability to share musical timing and source code. Jason Freeman, who led the research group creating LOLC, shared the following: 'With LOLC, I think scale is a big mitigating factor to risk. I've done performances with LOLC with 30+ laptops on stage, and so if I as an individual performer crash and burn, or don't generate any sound for a minute or two or three, it's no big deal. In fact, the biggest risk with LOLC tends to be that musicians feel they need to be making sound all the time and never step back to listen and choose wisely when to contribute to the texture' (Freeman, personal communication 2015).

Ensemble live coding performances are not limited to sharing code and timing; users of the Republic extension for SuperCollider can even share laptop speakers, as the extension enables users to execute code remotely on the computers of any ensemble member. The group PowerBooks UnPlugged uses this extension and sits among the audience, in effect creating a spatially distributed instrument that is controlled by all ensemble members simultaneously (Rohrhuber et al. 2007). In addition to executing code on any ensemble member's computer, the live coding environment Gibber also lets performers see and edit everyone's source code and provides a shared clock for rhythmic synchronization (Roberts, Yerkes, et al. 2015). While Republic and Gibber users are limited to specific live coding systems, the extramuros system enables distributed live coding performances using heterogeneous environments (Ogborn et al. 2015). In extramuros, performers program in a central, browser-based editing interface; each can see what their fellow ensemble members are doing. The commands from this interface are then piped to audio synthesis environments, such as Tidal or SuperCollider, pointing towards a promising future where ensembles can easily perform together regardless of the software that individual members use.

We return to another 'safety in numbers' technique to reduce risk in the discussions of scheduled parallelism later in the chapter. The topic of ensemble performance is also discussed in chapter 20, 'Network Music and the Algorithmic Ensemble', by David Ogborn.

16.2.4 Previewing Algorithms

Many of the risks in live coding stem from the relative brittleness of code. Code is rarely fault-tolerant and has the potential for unpredictable and far-reaching consequences. Renick Bell notes that his Haskell-based live coding system Conductive both adds and removes risk from live coding performances. His choice of Haskell, a strongly typed language, removes the risk of type-based errors breaking a performance. At the same time he states: 'The algorithms I use also provide risk, since they use random or stochastic processes to varying degrees and do not give me an audio preview of the output.' (Bell, personal communication 2015)

DJs mitigate risk by *previewing*: experimenting with and evaluating tracks or loops before adding them to a mix, via headphone cue mixes (and in software, via graphical representations of the audio). Yet few live coding environments emphasize previewing material before it is presented to audiences. Perhaps this is because the temporal pressures of coding in a performance, arguably a more intensive task than DJing, preclude patient preview and search? Live coding performances often suffer from a lack of variation, and having a performer previewing generative content instead of immediately presenting it could exacerbate this problem. Moreover, even deterministic algorithms can result in unexpected sounds (Rohrhuber, Campo, and Wieser 2005). Bell (personal communication 2015) notes that he *can* preview the output of algorithms by looking at their numeric output, but 'at this point that output is not as useful as I would like' and 'I just listen to their output in the performance and switch quickly if I do not like the results.'

Previewing can be a useful technique for ensemble performances, where multiple performers help ensure greater variation and provide temporal cover for other members. Some live coding ensembles have given performances where a central computer is projecting output to the audience while each performer submits code to it for execution (Ogborn et al. 2015; Roberts et al. 2014; Wang, Misra, Davidson, and Cook 2005). In such performances it is easy for performers to preview the results of executing code using the audio output of their laptops before sending it to the central computer (although this only works if the code does not heavily rely on the state of the central machine). We will return to these themes in the discussion of 'extended presents' at the end of the chapter (section 16.6.2).

16.3 Legibility and Immediacy for Audience and Performer

Canonical live coding practice encourages performers to project their screens to audience members in order to provide transparency about methods of production. Although, as mentioned in the introduction, there is debate in the live coding community about the necessity and desirability of this, it remains entrenched in live coding culture. Blackwell and Collins (2005) suggest that the audience members are end-users or consumers of the code being produced; but that without knowledge of the language, all code effectively becomes *secondary notation*. Blackwell (2015) later suggests that audience members who concentrate on following the projected code may have a greater cognitive load than performers.

Given this, what is the code of a performance able to convey to an audience? Is it merely reinforcement that content is being created in realtime? Or is there the potential to convey the development of algorithmic processes to the audience? Other possibilities also exist, from political (Bruun 2013) to educational. Shelly Knotts notes: 'One of the more important aspects to me in projecting my screen is revealing mistakes and errors by the performer which I see as having a humanizing effect on electronic music performance' (personal communication 2015). However, she also feels that projecting source code does not typically reveal underlying processes, even when the language is familiar. Regardless of motivation, it is incumbent on performers who project code to consider if and how they make it comprehensible to audiences, and to balance this with the cognitive and temporal costs improving comprehension potentially incurs.

16.3.1 Annotations and Live Visualizations

A first step to improving legibility of source code is the addition of comments and the thoughtful naming of variables. As Brown and Sorensen note: 'we make an effort to use function and variable names that people will recognize and that may assist in their interpretation of the code. Symbol names such as 'outrageous-kick' and 'grunge-it-up' never fail to communicate our intent!' (2009). In the above example there is a tradeoff in verbosity (which increases typing and potentially decreases immediacy) and legibility.

As with many other techniques for increasing the legibility of source code, this requires deliberate action on the part of the performer that incurs both a temporal and cognitive cost. However, the authors clearly feel the extra effort is worth it. In the case of descriptive variable names, there is a mutual benefit to the performer, who does not have to spend time later trying to remember what was stored in a variable named 'x2'. Another technique that improves legibility for both performance and audience is the clear indication of the current cursor location. In other cases, however, increasing legibility for the audience can potentially decrease it for the performer. For example, increasing the font size of a source code document to improve its legibility to the audience means that less code is simultaneously visible to the performer.

A number of environments provide affordances for visualizing the state and timing of algorithms without requiring extra effort by performers. Alex McLean's feedback. pl was the first system for live coding performance to explore this; it used code comments to both present the current state of data structures and also afford manipulation of their contents (Ward et al. 2004). The ability to display the current state of data is also included in Thor Magnusson's ixi lang live coding environment, where algorithmic manipulations of musical patterns automatically update the source code document with their results (Magnusson 2011b). Dave Griffith's Scheme Bricks live coding system flashes visual programming elements to indicate control flow (McLean, Griffiths, Collins, and Wiggins 2010). Gibber inherits elements of all three of these systems, while also revealing the output of functions triggered over time and displaying the state of running programs by dynamically changing font characteristics of source code (Roberts, Wright, and Kuchera-Morin 2015).

While these systems all use the source code itself to indicate state and timing, other designers of live coding environments have experimented with incorporating additional graphical widgets. Impromptu and Extempore both use graphical overlays to reveal information such as timing of musical sequences and progress in audio file playback (Swift, Sorensen, Gardner, and Hosking 2013). These widgets appear primarily intended to inform performers about system state as opposed to the audience (their impact on audiences is not mentioned in papers describing the research), but it is interesting to consider how blurring the line between source code text and user interface elements could improve audience understanding, as it could open up opportunities for comprehension to people who might be intimidated or disinclined to attempt parsing source code text. Lee and Essl (2014) also provide techniques for visualizing state in the urMus system for live coding musical instruments; similar to the work of Swift and Sorensen, audience comprehension is not considered but could be an important side effect.

As mentioned in the introduction, it has been argued that viewing and attempting to comprehend code is at odds with appreciating the musical output of live coding (Bruun 2013). However, we believe visual annotations of source code can improve appreciation of music cross-modally. In the same way that watching the pendulum of a metronome can improve perception of pulse, annotations such as those found in Scheme Bricks and Gibber have the potential to accentuate appreciation of the rhythms running through both algorithmic processes and generated music as they unfold.

16.3.2 Privileging Legibility

A number of live coding syntaxes have been designed to be understandable by audiences (Freeman and Van Troyer 2011; Magnusson 2011b; McKinney 2014). Despite this being a design concern when creating the environment LOLC, Jason Freeman notes: 'Audiences, for the most part, did not understand the code fragments as they were projected on the screen; the natural musical terminology used for durations and dynamics was of little help to them. There is, perhaps, an inherent design paradox here. Performing musicians tend to prefer a concise syntax that requires minimal typing, whereas audiences tend to have difficulty understanding text that is not sufficiently verbose' (Freeman and Van Troyer 2011). Freeman goes on to note that the ensembles using LOLC obtained better audience responses from projecting text messages passed between ensemble members in LOLCs built-in chat room, a practice first performed by The Hub (Brown and Bischoff 2002). In order to help associate performers with their text messages, each LOLC performer wore a baseball cap that matched the color of their messages in the chat room. Such messages can be considered a type of meta-score, where ensemble members are considering and debating the direction of a performance; in this role the chat dialogues can convey a good deal of musical structure to an audience.

Perhaps there is a need for a live coding environment that considers audience legibility its primary motivation, so that the limits of legibility in the genre can be more fully explored. Such an environment would not only consider the language syntax from the perspective of the audience but also require visual annotations for every expression evaluated, as suggested by Georg Essl in a 2014 email to the TOPLAP (live coding) mailing list, where he hypothesized a language in which: 'all run-time states/consequences induced by code must have a visualizable code-side representation'. As a simple yet thought-provoking example, Andrew Sorensen gave an inspiring live coding performance at the 2014 OSCON conference, where he vocally narrated what each block of code was doing as he was writing and evaluating it, successfully conveying the algorithmic processes he developed. Could such a narrative be automatically generated and inserted into the source code as comments immediately above or below the lines of code they are associated with? Alternatively, a running commentary could be generated alongside the source code document via multiple autonomous commentators with different personalities and aesthetic sensibilities.

A recent survey on the visualization of state and timing in source code in Gibber found widespread support for improving visualization of algorithmic processes in live coding environments, with over 96 percent of the 102 participants believing that visualization techniques should be explored in other live coding environments (Roberts, Wright, and Kuchera-Morin 2015).

16.4 ABSTRACTIONS

Many live coders begin performances with a blank screen or empty page; for some this is essential (Swift, personal communication 2015). No matter the degree to which an audience member can understand the code, the relation between the accretion of code and the emergence and development of music makes a dramatic communication of creative endeavour. This may serve a cultural agenda, evoking 'an understanding that anyone else could start from the same place' (Biddle 2014), or may simply be a commitment to improvisation. Regardless, being able to rapidly code from nothing to sound speaks to the remarkable strengths of live coding environments. However, as Bell notes this is hardly 'coding from scratch ... since all coding rests upon layers of *abstractions*' (personal communication 2015); Swift echoes that 'with a few textual characters I can orchestrate *(with the right abstractions)* quite rich output material' (personal communication 2015; emphasis added). With these layers of abstractions—including interface capabilities, language features, and library code—no page is empty, no screen void; more pregnant with potential than an empty canvas, they are volatile and excitable by computation.

In this section we explore the tensions that arise due to the abstractions used in live coding performance, starting from computational abstractions and moving towards musical abstractions. But first, a quick definition: abstraction (literally, 'drawing away') is the process of extracting from a diversity of concrete specific examples general rules, patterns, or concepts no longer dependent upon those examples, by eliding or suppressing some of their specific differences and details.

Computational abstractions suppress lower-level complexities in order to establish higher-level interaction. By doing so, we effectively sacrifice one class of affordances for another. Naturally, high and low are relative terms. For example, zmölnig (personal communication 2015) describes building synths from raw oscillators using the visual programming language PureData as 'low level', but this would be high level compared to a live coder operating on raw audio signals with bitshift operations. Bearing this in mind, we consider the gamut of abstraction levels live coders use, and why. Bell (personal communication 2015) is explicit in his preference for high-level affordances: 'I always seek higher and higher levels of abstraction ... to quickly make sudden changes in my performance with minimal code. ... I do not wish to program at lower levels of abstraction.' Meanwhile, Swift (personal communication 2015) argues for a broader spectrum of affordances: 'I like low-level control of memory since I want to signal-rate stuff (e.g. DSP) as well as note-level stuff. But from a [programming language] perspective, I love higher-order functions/closures, and prefer to map & reduce rather than [to] use a for loop.' Indeed, one of the attractions of higher-level abstractions is their generality: by being so abstract, the very same tools can utilized in a wider diversity of concrete applications. Moreover, to the extent that such high-level abstractions are *composable*—that they can be flexibly combined—they also effectively reduce the number of primitive concepts that need to be borne in mind.

Several live coding environments place quite low-level mechanics of the machine within reach during performance. Brown and Sorensen (2009) agreed that in general 'there is no possible way for us to deal with the complexity of the underlying operating system and hardware without levels of abstraction', yet Sorensen (2014) endeavoured to make Extempore 'efficient enough, and general enough, to make its own audio-stack implementation available for on-the-fly code modification at runtime.' Though coding 'down to the metal' of an audio driver seems unlikely to be needed during a performance, the ability for high-level code to directly manipulate low-level implementations can certainly improve performance through dynamic compilation of high-level algorithms into native machine code,[2] and thus increase the quantity, diversity, and complexity of the algorithms that can effectively be coded live (Sorensen 2014; Wakefield 2012).

16.4.1 Discourse with Models and Metaphors

Abstractions are not merely structural convenience: through their constraints and affordances, abstractions effectively present a *model* of a world with which a live coder maintains discourse (Rohrhuber, Campo, and Wieser 2005). In the case of musical performance, this model serves as 'a scaffold for externalising musical thinking' (Magnusson 2014b).[3] McLean suggests that by matching abstractions to the semantic needs of the problem world 'we are more able to engage the right kind of cognitive load, and become more absorbed in programming as a live experience' (2014a). For Magnusson, the cognitive load should be musical rather than a computational (Magnusson 2014b). Consequently, many live coders have explored embedding mental models of musical structure and process into abstractions at the level of programming languages themselves, resulting in domain-specific mini-languages, and sometimes even mini-languages within mini-languages. McLean's Tidal is a domain-specific language embedded within Haskell (a general-purpose functional language), but itself embeds a further mini-language for polyrhythmic cycles. Magnusson's ixi lang is a DSL for live coding with an embedded mini-language for agent scores, that is written in the audio programming language SuperCollider, itself derived from the general purpose SmallTalk language and implemented in C++. For Magnusson, the underlying power of SuperCollider was a pragmatic shortcut to providing ixi lang with rich sonic material. Although it remains possible to descend to the SuperCollider language live, the primary goal is a level of user-friendliness such that his language 'frees performers from having to think at the level of computer science, allowing them to engage directly with music through a high-level representation of musical patterns' (Magnusson 2010).

To be more easily understood (not *too abstract!*), these abstractions are often grounded in metaphor. Computer science has utilized metaphors of containers, privileges, inheritance, agents, and so on (Travers 1996), while music software has leveraged metaphors relating to instruments and scores as well as studio technology concepts such as mixers and patchable synthesizers. Not surprisingly these reappear in live coding. For example, ixi lang is designed around three simple concepts: *agents* use *instruments* to perform *scores* (Magnusson 2011a). Agents recur in Alive, to which *properties* and *tags* can be associated, and with which *queries* can be made (Wakefield et al. 2014). Tidal's pattern transformations have an explicit relation to weaving and knitting (McLean 2013a, 2014b). Language-level abstractions for live coding also often utilize notational analogies. The score language within ixi lang relates text-space to musical-time, in that 'spaces between the notes represent silence, spatial organization therefore becoming a primary syntax of the language' (Magnusson 2011a). Other examples include the iconic => and =^ operators used in ChucK, or the -> and >> operators in ixi lang, to denote the creation of connections between objects and projections into the future.

By creating model cognitive worlds, rich abstractions substantially influence what kinds of ideas can be expressed, and what discourse can ensue. Collins and McLean

(2014) note how many live coding interfaces incorporate looping and layer-centric patterning typical of dance music. The more closely an interface is fitted to the semantics of a particular model or metaphor, the more likely that it also becomes too rigid to be manipulated algorithmically outside of these assumptions, and thus can be described as *abstraction hating* in the Cognitive Dimensions of Notations framework (Blackwell and Collins 2005; Green 1989). Magnusson (2014b) suggests that, by affording certain practices and preventing others, the highly constrained abstract model of ixi lang should itself be considered 'a compositional form'. For Blackwell, 'every new live coding tool can become the starting point for a miniature genre, surprising at first, but then familiar in its likely uses' (2015, 59). Here becomes evident another tension of abstraction, in terms of *the complexity of freedom*. As Brown and Sorensen (2009) state, such influences have both positive and negative implications. While, as Magnusson notes, 'constraints inherent in the language are seen as providing *freedom from complexity*' (2014b; emphasis added), and that some 'users report that these limitations encourage creativity' (Magnusson 2011a), other live coders prefer interfaces that grant the *freedom to create complexity*. For example, live coder Shelly Knotts prefers the relative complexity of SuperCollider because she wants greater control of detail during a performance: 'Writing simple patterns and sequences can be complicated, but the complexity of what you can achieve is high. Languages such as ixi ... sacrifice the micro level control' (personal communication 2015)

16.4.2 Temporal Abstractions

Let us focus in detail on a domain of abstraction of special relevance to live coding: the representation of time. Oft neglected in computer science (Lee 2009), temporality is vital to the domain of music; not surprisingly many live coding systems present musically oriented abstractions of time. For example, many live coders *measure* time in terms of *meter*. Renick Bell counts beats with floating point numbers in order to represent intervals or events between them; 'relevant functions return a value which is valid at any point within the requested span' (Bell, personal communication 2015). Rather than floating-point numbers, Tidal uses rational numbers in order to capture a musical conception of time that incorporates both cyclic repetitions and linear progressions. For example, 'a time value of 8/3 would be the point that is two-thirds through the third cycle' (McLean 2013b). Meter may also be part of the coding environment itself; for example, code fragments in Gibber can be triggered to automatically execute on the next bar. And in systems where time is primarily represented in terms of seconds or audio sample counts, performers can easily create (or reuse) 'metronome' abstractions in order to trigger things with a metric rhythm (Swift, personal communication 2015).

Parallelism is another important temporal musical concept, found in polyphony, polyrhythm, and ensemble performance, that has been addressed through live coding abstractions. The ChucK live coding environment provides a system for easily launching and running multiple routines concurrently. These routines, or 'shreds', support precisely timed pauses within their execution, expressed in code through the assignment of new values to the variable now—as if the code itself directed the passage of time! In reality the assignment yields the instruction flow (which is thus a *coroutine*[4]) to an underlying scheduler, which will later resume it at the precisely designated time. In LuaAV and Alive precisely scheduled coroutines present a more passive metaphor of 'waiting' or 'sleeping' to achieve the same result (Wakefield et al. 2014; Wakefield, Smith, and Roberts 2010).

The Extempore system and its precursor Impromptu provide a closely related vehicle for concurrency via *temporal recursions*. A temporal recursion is defined as: 'any block of code (function, method, etc.) that schedules itself to be called back at some precise future point in time' (Sorensen 2013). Where the coroutine does this by yielding midfunction, the temporal recursion does this by reinvoking the function's entry point by name, granting an easy opportunity to rewrite the function between recursions. Both approaches support complex procedural flows with precise timing. (With sufficiently low-latency audio drivers, these precisely scheduled intentions in musical time become practically indistinguishable from real clock time.)

The procedural approaches to time described so far distribute the control flow of step-by-step procedures within the flow of musical time. As control passes from one instruction to the next, time passes as interruptions to that control flow; interruptions that are accurately placed in musical time. In Tidal, there is no such control flow, and time is instead treated in such a way that it can be manipulated, for example stuttered or reversed, through successive application of functions. Tidal centres on *patterns*: functions that map a given time span to a set of events that occur within that span (where events themselves occur over time spans). Rather than specify imperative instructions which act upon the current state, behaviour is declared in terms of pure functions that can operate on multiple levels of abstraction. An interesting result of working with time spans rather than instantaneous time points is that the representation works for notionally continuous as well as discrete patterns (McLean 2013b). Declarative statements describe rules, relationships, or ongoing activities that operate over an indefinite, cyclic timeline. Because composing a Tidal pattern is a matter of composing behaviours, the temporal structure and contents of patterns can be manipulated without having to calculate their entirety. Furthermore, these compound behaviours may operate on time distinctly from the specific contents of events, and vice-versa.

As with the 'safety in numbers' of collaboration mentioned earlier, the computational abstractions of parallelism above provide another way of reducing risk in performance. Ideally, the unexpected termination of any individual 'shred' in ChucK, or any temporal recursion in Impromptu or Extempore, will not crash the system as a whole. (The capability is somewhat dependent on the use of code with minimal side effects; temporal recursions that launch and control other temporal recursions are more powerful but potentially more destructive.) Ben Swift, who has used both Impromptu and Extempore in performance, shares: 'The fact that I can *change* things during a performance is important, and the fact that I can *break* things is a corollary of this. Having multiple temporal recursions running in parallel (which I do a lot) mitigates the risk, since if I make an error in one of them the others keep going. This happens not infrequently' (personal communication 2015). Moreover, ChucKs dedicated performance environment provides a user interface for monitoring, starting, stopping and replacing shreds during performances (Wang and Cook 2004), and likewise performers using Extempore can monitor the state of running temporal recursions through graphical overlays (Sorensen and Gardner 2010).

16.5 Flow and Pace

Coding live in front of an audience can clearly be a stressful experience, not least 'at 2am at a dance club after a couple of beers'! (Guzdial 2014) Languages such as ixi lang are explicitly designed to reduce stress by lessening the cognitive load. However, in discussions at the 2013 Dagstuhl Seminar on live coding it was noted that 'cognitive load is not necessarily to be avoided' (McLean 2014a); it can be part of what makes improvisation enticing. For example, in an informal qualitative evaluation, performers described of one of our prior live coding environments as 'stressful', 'frustrating', and 'difficult', *while at the same time* also as 'playful', 'fun', and positively 'live' (Wakefield et al. 2014). Some degree of cognitive load is necessary to enter the deep engagement of *flow* (Csikszentmihalyi 1990), 'a happy medium somewhere between frustratingly difficult on one side, and distractingly boring on the other' (McLean 2014a). This brings us neatly to our next tension topic: the achievement of this happy medium through the meeting of the capabilities of the models of abstraction with the intentions and capabilities of the performer.

In particular, live coders have developed various techniques to rapidly express musical ideas and move from the empty page to the production of sound. In some cases this becomes essential, since 'a long slow build up doesn't work very well if it's a middle-of-the-night set' (Knotts, personal communication 2015). The simplest technique is to make use of materials prepared in advance, such as patterns, instruments, and presets that can be dropped in and reused on the fly. The core tension here is control versus pace. McLean notes that since many interesting-sounding signal-processing algorithms are nontrivial to author, triggering and manipulating pre-made material is 'essential for a performance with any kind of pace.' Nevertheless, overuse of prepared material is dissatisfying: 'with improvisation you can really mold things to fit the sound system and mood, and when I trigger something pre-made it just sounds flat and lifeless, and breaks the flow of the performance' (McLean, personal communication 2015).

16.5.1 Succinct Expression?

We noted above that some live coders design mini-languages oriented to the author's conceptual music spaces, but these languages (and their environments) are also frequently designed to support immediacy in the interface. Both are evident in McLean's summary: 'I have developed Tidal to make it fast to improvise the kind of music I like to make' (personal communication 2015). Language- and environment-level abstractions enable complex articulations to be made far more succinctly. And as language interfaces more closely resemble the abstract (musical) models in play, these succinct statements are (hopefully) the kinds of things performers want to express. In consequence, code edits can more closely correspond to rates of musical decision-making. For example, 'ixi lang is a live-coding system designed with the criteria that it be fast (maximum 5-second wait before some sound is heard)' (Magnusson 2011a), and for Conductive, 'minimal notation is important, and I make efforts to reduce line noise to allow code to be input and edited efficiently' (Bell, personal communication 2015).

Interestingly, succinct expression does not necessarily imply brevity of notation. It can also be achieved in the programming environment by making prepared 'snippets' of code accessible from minimal keypresses (Swift, personal communication 2015). For example, in Extempore and Impromptu, some quick-to-type fragments of code are automatically *macro-expanded* in-place into more detailed pre-prepared implementations. We note that this automation from terseness to verbosity constitutes *a reversal* of the customary use of software abstractions to reduce code length, code repetition, and *conceal* implementation details. (It certainly presents exceptions to the tendency observed in live coding performances that code rarely fills a single-page; Biddle 2014.) From a different perspective, however, it can be seen as simply sharing involvement between coder and machine at a later stage in the translation from abstraction to execution. In fact Brown and Sorensen, who perform together as aa-cell, argue against the encapsulation of abstraction (the affordances of 'minimal code') in favor of *descriptive transparency*, since 'when programmers make a decision to abstract code away into an abstract entity, a black box ... the ramification is that they no longer have the ability to directly manipulate the algorithmic description' (2009). By making more of the mechanics of an algorithm explicit in the code, they expose more affordances to manipulate its ongoing process. But in a counterargument, Bell (2013) points out that higher-level abstractions need not be opaque, referencing the example of higher-order functions taking functions as parameters (which allows internal structures to be redefined later), and furthermore that in many languages abstractions may be accessible for modification during performance without having to have their implementation shown at all times.

Some artists prefer to minimize reliance on prepared materials by instead following an almost developmental approach, in which a complex algorithm is achieved piecewise from simpler but sound-making components that can 'carry the musical progression until the larger algorithm is working', such that 'it starts producing (non)musical events early in its life-cycle' (zmölnig, personal communication 2015). David Ogborn echoes this sentiment, noting that gradual development from low to high levels of abstraction during a performance also 'helps the audience grasp the abstractions ... like changing a harmony by typing it out, then making that a sequence/pattern, then making it a sequence/pattern with some variable elements, then making those elements/sequences/patterns, etc.' (personal communication 2015).

16.5.2 Frantic Expression?

A concrete result of the study by Swift et al. (2014) is that the live coders performed an average of fifteen significant edits per minute. Considering that an average computer user transcribes at around thirty-three and composes text at around nineteen *words* per minute (Karat, Halverson, Horn, and Karat 1999), this is no leisurely activity! Indeed many live coders agree: 'I'm generally coding all the time' (Mclean 2015); 'i usually spend the entire performance coding (unless i have to spend some time debugging)' (zmölnig, personal communication 2015); 'my gut feeling is that I'm almost always on the keyboard, either navigating, inserting or deleting' (Swift, personal communication 2015). For others, however, typing is not so uniformly dense: 'Often I may only type in one line of code every minute, though in other parts of the performance I may be more frantic and type in 8 or 10 lines a minute' (Freeman, personal communication 2015). Of course live *coding* is not just live *typing*: 'I code continuously through a performance [but] I do not consider *thinking time* as not coding' (Bell, personal communication 2015). A performer's subjective attention is necessarily spread across multiple activities. For example, for Freeman, planning and listening form the principal activity (Freeman, personal communication 2015). Similarly for McLean: 'You also have to be fully aware of the passing of time. . . . I might not know what I will change, but I will know when. If I miss a deadline then I have to wait until the next opportunity . . . depending on what feels right' (personal communication 2015).

16.6 TIME

This brings us to our final tension: the open-ended multiplicity of becoming yet ultimate constraint of being within time—slippery concepts, whether regarded from music, philosophy, or computing. For example, we often think of time in terms of the past, the present, and the future; and these concepts might appear to have clearly defined correspondences in languages equipped with temporal abstractions. At the limit, the present is that instantaneous step from one instruction to the next. The future would be the list of scheduled instructions (events in a sequenced pattern, callbacks for temporal recursions, resumptions for suspended coroutines, etc.). In a purely declarative functional language, this encompasses the entire *current* program. The recallable past would be all markings recorded into memory that remain accessible to ongoing or scheduled code, including materials prepared in advance of the performance. Yet already the analogy breaks down, since any *scheduled* future is already a marking made in the past. A less imperative perspective might separate the program *state* as containing the accessible past with the program *code* as a description of the future, since 'a program obviously is a plan of how something is supposed to happen, an anticipation of future events' (Rohrhuber, Campo, and Wieser 2005). This distinction can become complicated when considering systems such as feedback.pl (Ward et al. 2004), ixi lang (Magnusson 2011b), and Gibber (Roberts, Wright, and Kuchera-Morin 2015), in which events that modify code can be thrown into the future, and in which the program code updates itself to continually represent the most recent changes of state (i.e. the past). This brings our attention to the subtlety by which live coding not only operates in multiple *levels* of time, it operates in different *kinds* of time.

16.6.1 Code ≠ Execution

We can usually distinguish between the *program as description*, which is a plan for what happens in a world (i.e. the code), and the *program as process*, which is the actual unfolding history of the consequences of that plan (i.e. the execution). However, when programs are modified during execution, as in live coding, this distinction is subverted.

For example, Rohrhuber showed that this leads to a limit, which he characterizes as *algorithmic complementarity*, in attaining complete access and control over both code and execution, which forces us to choose between privileging continuity of a particular history, or privileging the accuracy of that history's representation in the code. More generally, semantic gaps between description and process can lead to unexpected or unpredictable behaviour.

It is the indeterminacy of interaction, not simply program modification, that creates the formal division between description and process. From a theoretical computer science perspective, a program that modifies itself noninteractively can be reduced to an equivalent non-self-modifying program due to its logical determinism. Only changes coming from *outside the system*, such as live coding edits, constitute irrational cuts that divide islands of rational history.[5] The self-modifying programs mentioned above (feedback.pl, ixi lang) are no exceptions: the effect of rewriting on the execution is deterministic, but the effect on the code *representation* impacts what kinds of actions the performer can and may take. There are theoretical precedents to draw upon here, such as the *interaction machine* model discussed by Wegner, which behaves perfectly like classical Turing machines between each interaction point, but indeterminately across interaction points; and thus becomes *more powerful* than a regular algorithm (and arguably, a regular Turing machine) (Wegner 1997). In fact, Turing proposed an extension of his famous theoretical machine by adding an infinite read-only data stream returning results of a question not computable from within—which he called an 'Oracle machine' (Turing 1939). From the perspective of the computer, the decisions made by live coders (such as, what would be musically interesting to do next?) are exactly this kind of stream.

As Rohrhuber concludes with regard to algorithmic complementarity, tensions between description and process are positive, in that not only do the 'surprises or frictions with intuition or convention inspire creative solutions', but more importantly this 'prevents live coding from becoming a *merely technical* problem' (Rohrhuber 2014a).

16.6.2 Spectra of Extended Presents

Rather than a past and future divided by a singular infinitesimal cut, in many ways our subjective experience of time can be better understood as a superposition of spectra of *presents of varying duration*. This can be illuminated by phrases in the *present progressive* tense such as 'what are you doing?', which could refer to spans as short as a handful of seconds up to presents comprising months. This perspective is particularly relevant to live coding, in which performer's actions can span a longer spectrum of duration than with acoustic instruments.

At the narrower end of the spectrum, most acoustic instruments respond to gesture within milliseconds, but this raw expressive immediacy is largely lost to code. To recoup this immediacy, some live coders have augmented their interfaces with rapid-trigger key-bindings, some have mapped hardware input devices to global symbols in code (Brown and Sorensen 2009), some have collaborated with instrumentalists (such as 'vocalists, thrash guitarists, drummers, and banjo players'; Stowell and McLean 2013), and some have performed with both code and instrument simultaneously (such as Dan Stowell's live coding while beatboxing, David Ogborn's live coding with guitar, or live coding with bio-sensors as performed by Marije Baalman 2013–2014).

At the broader end of the spectrum, whereas the effects of an acoustic gesture might persist for a handful of seconds, live coding can easily spawn processes that are indefinitely extended. In this regard the improvising live coder 'is primarily a composer, writing a score for the computer to perform' (Magnusson 2011a). Improvisation has always involved planning ahead and reflecting back at a range of musico-temporal scales, but rarely through such persistent automation. Swift (personal communication 2015) notes that with a few keystrokes a complex and never-ending sequence of events can ensue without effort, 'whereas on the piano I have to play each note directly. ... I have to stay there or it goes quiet'. This freedom is echoed by Freeman (personal communication 2015): 'I'll sometimes schedule musical content to loop a lot ... so I can get more going without having to type so much.' The connotation is that these extended durations create more space for performers to multitask: 'I gain the ability to affect simultaneous change across of number of processes that would be impossible with any traditional musical instrument' (Bell, personal communication 2015).

What is left unspoken is that, although activity is multiplied, attention is not (Swift, personal communication 2015, teases that he could even 'go and have a beer'). Infinite repetitions, even of repetitive conditionals (Collins and McLean 2014), are not the same as repeated efforts. In this regard, zmölnig mainly avoids the broader spectrum of long durations in order that *performance time* is forced onward through the performer's actions of coding, rather than the code alone (personal communication 2015).

Returning to our first tension, we close with the observation that in any case, throwing algorithmic plans of action into the further future is a risky venture. First, algorithms might not unfold in the way expected due to nondeterministic or probabilistic components or simply the unpredictability of generative complexity. Second, algorithms may be influenced to an unpredictably changing context. We noted earlier the absence of

DJ-style previewing in live coding, which we can identify as largely unrealistic: in comparison to the fixity of a DJ's audio file, the effect of a line of code can be radically different according to the stateful context in which it is executed—thus live coding becomes live debugging (Blackwell and Collins 2005). Third, projected plans might no longer be *musically* appropriate by the time the future comes around, and require immediate modifications to stay relevant. As Sicchio (personal communication 2015) contends, 'it is less about managing the time in terms of authoring and more about sensing what that moment in time needs'. This is compounded within an environment of collaboration: 'I can't predict what other musicians will do between now and the future and their decisions will inevitably make irrelevant whatever far-in-the-future plans I may have had' (Freeman, personal communication 2015). Finally, many live coders acknowledge that projecting code into the further future is just too attentionally demanding: 'I do not plan what kind of algorithms I want far in advance' (Goncalves, personal communication 2015); 'I still struggle to listen, reflect, and create across so much time … it's hard for me to think ahead while still listening to the present' (Freeman, personal communication 2015); 'With the diversity of time scales involved, I will occasionally do a bit of preparation if I have an idea in advance of when I want to put it into practice, but this only happens if I'm feeling very awake!' (McLean, personal communication 2015); and 'I don't tend to do so much future planning in performance … I often feel like I don't have enough time to sit back and listen because of time pressure to keep coding and making changes to the music' (Knotts, personal communication 2015).

16.7 CLOSING REMARKS

Given the tensions we have described and the plurality of practices they generate, perhaps it is useful to close by returning to the why of live coding.

Blackwell and Collins (2005) raised the question why any live performer would choose the challenges of a programming language over the 'comfortable ride' of software such as Ableton Live. A suggested response was to escape the confines of stylistic bias in such mainstream application interfaces, and instead to embrace the vast exploratory potential for experimental music that full-fledged programming languages provide. A decade on however, we see plenty of domain-, style-, and even artist-specific bias in the live coding languages of today. Moreover, it is not clear that the unlimited potential of programming is profoundly utilized during performance. What, after all, differentiates live coding from noncoding in performance? The diversity of interfaces and interface abstractions evident in live coding practice suggests it is better to look to the underlying models of discourse that coding live make possible. Specifically, computational abstractions here include complex conditional control-flows, symbolic reasoning, and the ability to create *new abstractions*. Without these, it could be easily argued that the abstract model of discourse utilized is little different from adjusting patch cables and knobs of a modular synthesizer. But Magnusson notes that, according to such a 'strong' definition of live coding, many so-called live coding performances do not include coding at all (Magnusson 2014a).

Moreover, the contrast with modular synthesizers is unconvincing: not only do many live coders use visual interfaces directly inspired by modular synthesizers, but actual modular synthesis patching can be clearly articulated as a form of live coding, sharing many practical similarities with its digital counterparts (Hutchins 2015). Indeed, many live coders confirm that relatively little performance time is dedicated to the more abstract algorithmic complexities made possible by programming. (It could also be argued that performance of live circuit creation, such as by *The Loud Objects*, is a form of abstraction-free live coding.) For example, Knotts spends 'more time writing immediate updates than on algorithm design in performance, and in general only use very simple algorithmic control' (personal communication 2015). Too much anticipatory algorithmic work is at risk of losing the audience: 'I'd certainly rather watch a performance of parameter tweaking that gets moving quickly than wait minutes for someone to pull off some very complicated algorithm' (zmölnig, personal communication 2015). As Blackwell (2015) observes, complex algorithms are rarely explored in performance because an executing program's structure can only be changed *gradually*.

In contrast to the image of the live coder as modern concerto artist—'the virtuosity of the required cognitive load, the error-proneness, the diffuseness' (Blackwell and Collins 2005)—there is perhaps a greater interest in its simplicity and potential to democratize code in a fun, exploratory fashion. As Biddle (2014) notes, many performers 'want their audience to understand that the process is visible and reproducible: anyone can make music just like this. As the magician says, there's nothing up my sleeve.' Supporting this claim is a growing use of live coding in education. Sicchio (personal communication 2015) relates, 'I also recently have been teaching middle schoolers Sonic Pi to create music. They have performed several times including at their middle school dance,' pointing towards live coding performances that are the antithesis of virtuoso, concert hall performances.

And, why *not* code live? Canonical live coding emerged not long after programming languages such as SuperCollider made it possible to concisely define sounds and hear them expressed close to realtime; which is to say at the point at which experimental approaches to code-based musical composition became possible during live performance. This suggests that programming and live performance have been thought of as separate only *by convention* (Rohrhuber 2014b). As Goncalves simply states: 'I choose to live code because I think writing a musical expression and hearing its result is very powerful' (personal communication 2015). Many composers would agree with this, and there is little reason besides cultural intractability to demote code beneath piano, pen, and paper as a compositional tool.

For those who code, compose, and perform, live coding combines these activities into a single gestalt, and by blurring boundaries it enables them to engage with the coincident juncture of all parts. For example, Freeman (personal communication 2015) states that his goal is to 'abstract the process of creation, transformation, listening, organizing, collaborating, and improvising in a structured way that mimics some techniques we often use to do these activities in other forms of music'. Combining all three of these activities comes with its own costs, but these are potentially appealing in and of themselves.

Since its inception, live coding has progressed from something with technological and cultural promise to what is now a worldwide set of communities, regular performances, workshops, and international conferences. As it continues to evolve, so inevitably will its tensions, providing space for continued experimentation. Live coding performers are free to memorize code employing no algorithms whatsoever, type it verbatim during a performance, and not project their code to the audience. They can also improvise algorithms that in turn improvise music while potentially revealing intricate details of their compositional process to the audience via source code projection. Although some practices might not take full advantage of the medium, this alone does not determine aesthetic results. Which brings us to the members of the audience, who are free to choose how they engage with these performances. They can attempt to understand the algorithms created during a performance or ignore projected code in favor of appreciating the resulting music, leaving live coding performances to be both produced and consumed in an absence of absolutes.

ACKNOWLEDGEMENTS

This chapter was supported by the Social Sciences and Humanities Research Council, and the Canada Research Chairs program, Canada; the Robert W. Deutsch Foundation, and the Media Arts and Technology Program at the University of California Santa Barbara. Special thanks to the live coding performers who completed our original survey: Renick Bell, Jason Freeman, Tanya Goncalves, Shelly Knotts, Alex McLean, David Ogborn, Kate Sicchio, Ben Swift, and IOhannes zmölnig. We'd additionally like to thank Alex McLean for his helpful suggestions in regards to this text.

NOTES

1. See for example the list compiled at http://iclc.livecodenetwork.org/2015/definitions.html.
2. The related principle of 'dogfooding'—building a live coding system's libraries from within its own language—has also shown its value in performance (McLean and zmölnig, personal communication, through online survey: http://www.charlie-roberts.com/live_coding_survey 2015).
3. In a formal study based on the transcription of thirteen live coding performances using Impromptu, Swift, Sorensen, Martin, and Gardner (2014) found that coding actions were relatively consistent between performers, whereas musical activities showed much greater divergence. Although limited in scope, the study may serve to highlight the semantic gap between the coding environment and the musical models of the performers.
4. Coroutines are a procedural representation of collaborative, single-thread multitasking. In computer science terms, they are one-shot continuations. In layman's terms, they create the illusion of a procedural function that can be paused in the middle and resumed later. Note that tasks are not truly concurrent in that only one task can be executing at any given moment.

5. We borrow the term 'irrational cut' from Deleuze's treatment of nonclassical cinema, where it refers to cuts that subvert the otherwise rational narrative flow; an incommensurable link between shots that is not a member of the series that preceded it, nor of that which follows, and thus creates a direct presentation of time (Deleuze 1989). Deleuze appropriated the term from the concept of an irrational Dedekind cut in number theory, which belongs to neither of the sets it produces.

BIBLIOGRAPHY

Aaron, S., Blackwell, A. F., Hoadley, R., and Regan, T. 'A Principled Approach to Developing New Languages for Live Coding'. In *Proceedings of the Conference on New Interfaces for Musical Expression*, 381–386. Oslo, 2011.

Alperson, P. 'On Musical Improvisation'. *Journal of Aesthetics and Art Criticism* 43, no. 1 (1984): 17–29.

Baalman, M. *Wezen-Gewording*. 2013–. https://www.marijebaalman.eu/?p=404.

Bell, R. 'Pragmatically Judging Generators'. In *Proceedings of the Generative Art Conference*, 221–232. Milan, 2013.

Bell, R. Personal communication via online survey. http://www.charlie-roberts.com/live_coding_survey. Survey completed on 5 April 2015.

Biddle, R. 'Clickety-Click: Live Coding and Software Engineering'. In *Collaboration and Learning through Live Coding*. Dagstuhl Seminar 13382, *Dagstuhl Reports* 3, no. 9 (2014): 154–159.

Blackwell, A. F. 'Patterns of User Experience in Performance Programming'. In *Proceedings of the First International Conference on Live Coding*, 53–63. Leeds: ICSRiM, University of Leeds, 2015.

Blackwell, A., and Collins, N. 'The Programming Language as a Musical Instrument'. *Proceedings of Psychology of Programming Interest Group* 3 (2005): 284–289.

Brown, A. R., and Sorensen, A. 'Interacting with Generative Music through Live Coding'. *Contemporary Music Review* 28, no. 1 (2009): 17–29.

Brown, C., and Bischoff, J. *Indigenous to the Net: Early Network Music Bands in the San Francisco Bay Area*. 2002. http://crossfade.walkerart.org/brownbischoff/IndigenoustotheNetPrint.html. Accessed 20 June 2017.

Bruun, K. *Skal vi danse til koden?* Kunsten.nu, 18 November 2013. http://www.kunsten.nu/artikler/artikel.php?slub+livekodning+performance+kunsthal+aarhus+dave+griffiths+alex+mclean+algorave.

Clowney, D., and Rawlins, R. 'Pushing the Limits: Risk and Accomplishment in Musical Performance'. *Contemporary Aesthetics* 12 (2014).

Collins, N. 'Live Coding of Consequence'. *Leonardo* 44, no. 3 (2011): 207–211.

Collins, N. *Origins of Live Coding*. 1 April 2014. http://www.livecodenetwork.org/files/2014/05/originsoflivecoding.pdf.

Collins, N., and A. McLean. 'Algorave: A Survey of the History, Aesthetics and Technology of Live Performance of Algorithmic Electronic Dance Music.' In *Proceedings of the Conference on New Interfaces for Musical Expression*, 355–358. London, 2014.

Csikszentmihalyi, M. *Flow: The Psychology of Optimal Experience*. New York: Harper Perennial, 1990.

Deleuze, G. *Cinema 2: The Time-Image*. Translated by H. Tomlinson and R. Galeta. Minneapolis: University of Minnesota Press, 1989.

Della Casa, D., and John, G. 'LiveCodeLab 2.0 and Its Language LiveCodeLang'. In *Proceedings of the Second ACM SIGPLAN Workshop on Functional Art, Music, Modeling and Design*, 1–8. New York: ACM, 2014.

Freeman, J. Personal communication via online survey. http://www.charlie-roberts.com/live_coding_survey. Survey completed on 18 March 2015.

Freeman, J., and Van Troyer, A. 'Collaborative Textual Improvisation in a Laptop Ensemble'. *Computer Music Journal* 35, no. 2 (2011): 8–21.

Goncalves, T. Personal communication via online survey. http://www.charlie-roberts.com/live_coding_survey. Survey completed on 19 June 2015.

Green, T. R. 'Cognitive Dimensions of Notations'. In *Proceedings of the Fifth Conference of the British Computer Society, Human-Computer Interaction Specialist Group on People and Computers V*, 443–460. New York: Cambridge University Press, 1989.

Griffiths, D. 'Fluxus'. In *Collaboration and Learning through Live Coding*. Dagstuhl Seminar 13382, *Dagstuhl Reports* 3, no. 9 (2014): 149–150.

Guzdial, M. 'Live Coding, Computer Science, and Education'. In *Collaboration and Learning through Live Coding*. Dagstuhl Seminar 13382, *Dagstuhl Reports* 3, no. 9 (2014): 162–165.

Hutchins, C. C. 'Live Patch / Live Code'. In *Proceedings of the First International Conference on Live Coding*, 147–151. Leeds: ICSRiM, University of Leeds, 2015.

Karat, C.-M., Halverson, C., Horn, D., and Karat, J. 'Patterns of Entry and Correction in Large Vocabulary Continuous Speech Recognition Systems'. In *Proceedings of the SIGCHI Conference on Human Factors in Computing Systems*, 568–575. New York: ACM, 1999.

Knotts, S. 2015. Personal communication via online survey. http://www.charlie-roberts.com/live_coding_survey. Survey completed on 31 March 2015.

Lee, E. A. 'Computing Needs Time'. *Communications of the ACM* 52, no. 5 (2009): 70–79.

Lee, S. W., and Essl, G. 'Communication, Control, and State Sharing in Networked Collaborative Live Coding'. In *Proceedings of the Conference on New Interfaces for Musical Expression*, 263–268. Goldsmiths, University of London, 2014.

Magnusson, T. 'Designing Constraints: Composing and Performing with Digital Musical Systems'. *Computer Music Journal* 34, no. 4 (2010): 62–73.

Magnusson, T. 'Algorithms as Scores: Coding Live Music'. *Leonardo Music Journal* 21 (2011a): 19–23.

Magnusson, T. 'ixi lang: A SuperCollider Parasite for Live Coding'. In *Proceedings of the International Computer Music Conference*, 503–506. Huddersfield: International Computer Music Association, 2011b.

Magnusson, T. 'Herding Cats: Observing Live Coding in the Wild'. *Computer Music Journal* 38, no. 1 (2014a): 8–16.

Magnusson, T. 'ixi lang'. In *Collaboration and Learning through Live Coding*. Dagstuhl Seminar 13382, *Dagstuhl Reports* 3, no. 9 (2014b): 150.

McKinney, C. 'Quick Live Coding Collaboration in the Web Browser'. In *Proceedings of the Conference on New Interfaces for Musical Expression*, 379–382. Goldsmiths, University of London, 2014.

McLean, A. 'The Textural x'. *xCoAx 2013: Proceedings of Computation Communication Aesthetics and X*, 81–88. Porto: Universidade do Porto, 2013.

McLean, A. 'Stress and Cognitive Load'. In *Collaboration and Learning through Live Coding*. Dagstuhl Seminar 13382, *Dagstuhl Reports* 3, no. 9 (2014a): 145–146.

McLean, A. 'Textility of Live Code'. In *Torque 1: Mind, Language and Technology*, 141–144. Link Editions.

McLean, A. 'Tidal: Representing Time with Pure Functions'. In *Collaboration and Learning through Live Coding*. Dagstuhl Seminar 13382, *Dagstuhl Reports* 3, no. 9 (2014c): 142–145.

McLean, A. Personal communication via online survey. http://www.charlie-roberts.com/live_coding_survey. Survey completed on 6 April 2015.

McLean, A., Griffiths, D., Collins, N., and Wiggins, G. 'Visualisation of Live Code'. *Proceedings of Electronic Visualisation and the Arts 2010*, 26–30. London, 2010.

Nilson, C. 'Live Coding Practice'. In *Proceedings of the International Conference on New Interfaces for Musical Expression*, 112–117. New York, 2007.

Ogborn, D. Personal communication via online survey. http://www.charlie-roberts.com/live_coding_survey. Survey completed on 7 April 2015.

Ogborn, D., Tsabary, E., Jarvis, I., Cárdenas, A., and McLean, A. 'Extramuros: Making Music in a Browser-Based, Language-Neutral Collaborative Live Coding Environment'. In *Proceedings of the International Conference on Live Coding*, 163–169. Leeds: ICSRiM, University of Leeds, 2015.

Roads, C. 'The Second STEIM Symposium on Interactive Composition in Live Electronic Music'. *Computer Music Journal* 10, no. 2 (1986): 44–50.

Roberts, C., Wright, M., and Kuchera-Morin, J. 'Beyond Editing: Extended Interaction with Textual Code Fragments'. In *Proceedings of the International Conference on New Interfaces for Musical Expression*, 126–131. Baton Rouge, LA, 2015.

Roberts, C., Wright, M., Kuchera-Morin, J., and Höllerer, T. 'Gibber: Abstractions for Creative Multimedia Programming'. In *Proceedings of the ACM International Conference on Multimedia*, 67–76. New York: ACM, 2014.

Roberts, C., Yerkes, K., Bazo, D., Wright, M., and Kuchera-Morin, J. 'Sharing Time and Code in a Browser-Based Live Coding Environment'. In *Proceedings of the International Conference on Live Coding*, 179–185. Leeds: ICSRiM, University of Leeds, 2015.

Rohrhuber, J. 'Algorithmic Complementarity, or the Impossibility of "Live" Coding'. In *Collaboration and Learning through Live Coding*. Dagstuhl Seminar 13382, *Dagstuhl Reports* 3, no. 9 (2014a): 140–142.

Rohrhuber, J. 'SuperCollider and the Just In Time Programming Library'. In *Collaboration and Learning through Live Coding*, Dagstuhl Seminar 13382, *Dagstuhl Reports* 3, no. 9 (2014b): 150–151.

Rohrhuber, J., de Campo, A., and Wieser, R. 'Algorithms Today: Notes on Language Design for Just in Time Programming'. In *Proceedings of the International Computer Music Conference*, 455–458. Barcelona, 2005.

Rohrhuber, J., de Campo, A., Wieser, R., van Kampen, J.-K., Ho, E., and Hölzl, H. 'Purloined Letters and Distributed Persons'. In *Proceedings of the Music in the Global Village Conference*. 2007. http://www.wertlos.org/articles/Purloined_letters.pdf.

Sicchio, K. 'Hacking Choreography: Dance and Live Coding'. *Computer Music Journal* 38, no. 1 (2014): 31–39.

Sicchio, K. Personal communication via online survey. http://www.charlie-roberts.com/live_coding_survey. Survey completed on 22 June 2015.

Sorensen, A. 'The Many Faces of a Temporal Recursion'. 2013. http://extempore.moso.com.au/temporal_recursion.html.

Sorensen, A. 'Extempore'. In *Collaboration and Learning through Live Coding*. Dagstuhl Seminar 13382, *Dagstuhl Reports* 3, no. 9 (2014b): 148–149.

Sorensen, A., and Gardner, H. 'Programming with Time: Cyber-Physical Programming with Impromptu'. *ACM Sigplan Notices* 45, no. 10 (2010): 822–834.

Stowell, D., and McLean, A. 'Live Music-Making: A Rich Open Task Requires a Rich Open Interface.' In *Music and Human-Computer Interaction*, edited by S. Holland, K. Wilkie, P. Mulholland, and A. Seago, 139–152. London: Springer, 2013.

Swift, B. Personal communication via online survey. http://www.charlie-roberts.com/live_coding_survey. Survey completed on 23 March 2015.

Swift, B., Sorensen, A., Gardner, H., and Hosking, J. 2013. 'Visual Code Annotations for Cyberphysical Programming'. In *Proceedings of the 1st International Workshop on Live Programming*, 27–30. Piscataway, NJ: IEEE Press, 2013.

Swift, B., Sorensen, A., Martin, M., and Gardner, H. 'Coding Livecoding'. In *Proceedings of the 32nd Annual ACM Conference on Human Factors in Computing Systems*, 1021–1024. New York: ACM, 2014.

Travers, M. D. *Programming with Agents: New Metaphors for Thinking about Computation*. PhD dissertation, Massachusetts Institute of Technology, 1996.

Turing, A. M. 'Systems of Logic Based on Ordinals'. *Proceedings of the London Mathematical Society* 2, no. 1 (1939): 161–228.

Wakefield, G. *Real-Time Meta-Programming for Interactive Computational Arts*. PhD dissertation, Media Arts and Technology, University of California at Santa Barbara, 2012.

Wakefield, G., Roberts, C., Wright, M., Wood, T., and Yerkes, K. 'Collaborative Live-Coding Virtual Worlds with an Immersive Instrument'. In *Proceedings of the Conference on New Interfaces for Musical Expression*, 505–508. Goldsmiths, University of London, 2014.

Wakefield, G., Smith, W., and Roberts, C. 'LuaAV: Extensibility and Heterogeneity for Audiovisual Computing'. In *Proceedings of the Linux Audio Conference*, 31–38. Utrecht: Hogeschool voor de Kunsten, 2010.

Wang, G., and Cook, P. R. 'On-the-fly Programming: Using Code as an Expressive Musical Instrument'. In *Proceedings of the Conference on New Interfaces for Musical Expression*, 138–143. Hamamatsu, 2004.

Wang, G., Misra, A., Davidson, P., and Cook, P. R. 'CoAudicle: A Collaborative Audio Programming Space'. In *Proceedings of the International Computer Music Conference*, 331–334. Barcelona: ICMA, 2005.

Ward, A., Rohrhuber, J., Olofsson, F., McLean, A., Griffiths, D., Collins, N., and Alexander, A. 'Live Algorithm Programming and a Temporary Organisation for its Promotion'. In *Read Me: Software Art and Cultures*, edited by O. Goriunova and A. Shulgin, 243–261. Aarhus: Aarhaus University Press, 2004.

Wegner, P. 'Why Interaction Is More Powerful than Algorithms'. *Communications of the ACM* 40, no. 5 (1997): 80–91.

zmölnig, I. Personal communication via online survey. http://www.charlie-roberts.com/live_coding_survey. Survey completed on 6 April 2015.

Perspectives on Practice B

....

CHAPTER 17

WHEN ALGORITHMS
MEET MACHINES

....

SARAH ANGLISS

MY primary interest in algorithmic music stems from a desire to play more than one instrument on stage at once, while having only two hands. I'm a composer and roboticist, and over the last few years I've been mixing live instrumental performance with programming and robotics to create a solo act that makes the most of a laptop's capabilities but that has a striking physical presence.

My stage set is unusual as it makes extensive use of roboticized found objects. Typically, I'll be on stage with a roboticized polyphonic carillon (bell player); a roboticized 1930s ventriloquist's dummy who 'speaks' vocal samples; a red handbag which opens and closes automatically, pulsing in time with a heartbeat; and a theremin, which I use both as a classic instrument and a sample scrubber. All, except the theremin, are handmade electromechanical devices which use a combination of servos and solenoids to operate. The carillon has twenty-eight handbells, each of which is percussed by its own servo-driven beater which is spring-mounted so it makes a clear ping. These objects are coordinated and controlled by a Max/MSP patch which runs on a laptop and communicates with the machines via serial or audio signals.

Before I started working on algorithms for these instruments, I spent some time experimenting with entirely laptop-based algorithmic music. In this, the code simply triggers and processes samples and synthesized sounds. This approach to composition proved particularly useful in 2012 when I was working under intense time pressure to write music for theatre. At the time, I had been commissioned to write incidental music for *The Effect*, a new play by Lucy Prebble which was about to have its debut at the National Theatre (Cottesloe). The music would take the form of a recorded soundtrack made of around fifty separate elements, cued throughout the show. Theatre music has to be written and redrafted rapidly as a new play develops and changes in the rehearsal room. Frequently, a scene is altered at lunchtime and new music is needed for it by the following morning. The dramatic context of *The Effect* was also challenging musically as the play was a love affair set on a clinical drugs trial. I decided to use some algorithmic techniques to make this job manageable. Rather than scoring music, I devised algorithms which I could manipulate to create music very swiftly. These algorithms manipulated sounds that I'd sampled from clinical equipment—metal bowls, MRI scanners, and so on. Together, the algorithms and sound samples gave me scope to work with a wide variety of textures, tempos, and moods. They also enabled me to swiftly compose disparate passages of music that had a family resemblance to each other, music that also belonged in the sound world of the clinic. This approach seemed to work but I noticed it led to music with a certain coldness—a quality which suited the themes of the play. It was a fascinating exercise and a necessary one given the time constraints, but it also left me wanting to experiment further with algorithms that are mixed with live instrumental performance.

My live set isn't purely algorithmic, nor is it coded live (unless I'm trying to fix a problem on stage). But many of the instruments in the set are playing music generated by software patches that I've crafted over many performances. Some merely couple sensors to sound. One coupler detects the real-time frequency of a theremin note and uses it to transform the instrument into a highly responsive, general-purpose gestural controller. Using this, I can precisely control the speed of a sample, for example, or the cut-off and resonance of a filter. Other patches are pattern-forming algorithms which generate polyphonic note sequences for the carillon. As the algorithms run, they create shifting and intertwining ostinati, a musical bedrock over which I can improvise with other instruments.

Bringing algorithmic elements into my live set frees up my hands (and mind) from one instrument so I can duet with it on another. It also frees the music from certain physical constraints. The carillon can play polyphonically at lightning speed and can handle any polyrhythms I throw at it (8 against 5 and 13 for example). In this way, it can generate live performances that are outside the capabilities of a human player. This approach also gives the music other inhuman qualities that I find interesting compositionally. Unlike a human performer, an algorithm doesn't favour certain harmonic progressions or melodies that readily lie under a keyboardist's hand (unless it's been programmed to do so). When improvising a duet with the carillon, while it's performing algorithms, I'm often surprised by the harmonic twists and turns of the carillon's music, it's constantly surprising me with bitonal layering of melodies and unexpected harmonic shifts. Of course, all these events have been composed to some extent, hand-coded by me into the underlying algorithm. However, the coupling between the code and the harmonic and melodic structure that unfurls from it is a complex one. And from this come serendipity and musical richness. I now have a few simple patches into which I can add small mutations to create endless subtle variations in the harmonic structure of a piece. This keeps the performance fresh, as it gives me a sense of duetting with an inventive fellow performer.

Although there's a great deal of coding in my act, I wouldn't say the use of code was prominent. Nor would I expect my work to be classed as 'algorithmic music'—it might seem out of place at an Algorave, for example.

However, given the nature of my work, I actually don't find it helpful to make a binary distinction between 'algorithmic' and 'nonalgorithmic' music. Rather, I'd say all music has a degree of preprogramming, of simply following patterns that can be expressed as instructional code—traits I would describe as 'algorithmy'. The music at a live coding event is high in algorithmy but the performer still has considerable scope to select routines on the fly in response to the mood in the room. A chamber orchestra playing a baroque piece such as Handel's 'Zadok the Priest' may seem low on algorithmy but the music is rich in patterns that could be reduced to a few lines of code. Handel was obliged to write his famous arpeggio line longhand but its repeating figures and rising structure could equally be expressed in a looping routine of some kind. Of course, Handel veers from a predictable mathematical pattern. I think that gives the music its kick and the listener an awareness of human authorship. Handel's particular choices elevate his music from dozens of other composers of the eighteenth century (and arguably from machine-composed music in the same style).

Over the last few years, I've focused almost all of my algorithmic composition on the carillon. This enabled me to learn the subtle qualities of the instrument and devise a repertoire of functions that suit it. One such function is an automatic baroque-style 'double'. This takes an algorithmic piece and throws in an extra note on the half-beat, a chosen interval away. Another function plays a copy of the algorithmic piece in retrograde, a fourth or fifth above the original, while the original plays. These routines are computationally simple but highlight and flatter the uneven temperament and metallic harmonics of the bells. I use them extensively. Tellingly, if I voice the carillon's algorithms through another instrument, such as a piano or sawtooth synth, the result sounds chaotic and unpleasantly discordant.

Like any mechanically driven sounding object, the bells of the carillon exhibit a subtle unevenness in timing and volume as they are struck. There are also creaks and movements; slight chaos in spring bounces and sympathetic resonances you'd expect with any instrument with moving parts. These irregularities add an aleatoric charm to the sound, a quality I doubt I could convincingly model in code. Thus, I'd say the carillon's music arises from a fascinating collision between computer code and the physical objects that execute it in the real world. Its music, in turn, steers a larger performance as an extemporising human player attempts to duet with it.

Taking this idea further, I've recently been experimenting with algorithmic music that's played back by physical devices that can move themselves around a room. One example is a set of five handbells on robotic platforms which are free to roam the labyrinthine corridors of Newhaven Fort, Sussex. This work was first exhibited at Fort Process sonic arts weekend in the summer of 2013. Each bell found its own path through the fort, moving in a straight line until it encountered an obstacle. When it found something in its way, the bell was struck by its beater five to eight times, then the bell swerved or reversed so it could continue on its way. From this, a spatialized piece arose. Together, the mobile bells, their algorithms and the audience who were hopping between the bells in motion were 'playing' the building.

Thinking about the richness of the piece at Newhaven Fort, I'm reminded of robotics pioneer Rodney Brooks and his claim that 'the world is its own best model'. When you play an algorithmic piece using tangible objects, you get the real world for free and all the complexity that comes with it. There's something delightful in the mobile algorithmic bells as their operation is both mysterious and obvious. Casual observers soon work out the bells are responding to obstacles. Many had an urge to herd them like musical cats. As an instrumentalist, I found I had to surrender too much of this richness when I worked with purely laptop-based algorithms (a trade I was willing to make when evoking a cold, clinical trial). Yet I do feel I want to experiment further with algorithms embedded in tangible, moving objects. In this domain, there's a lot more playing—and a lot more work—to do.

CHAPTER 18

....

NOTES ON PATTERN
SYNTHESIS

1983 to 2013

....

MARK FELL

WITH the advent of British synth pop in the early 1980s, my interests grew from 'commercial' electronic music such as Human League, Soft Cell, and Depeche Mode, to slightly earlier and marginal activities, including groups such as Throbbing Gristle and Cabaret Voltaire. My first encounters in the production of synthetic sound happened around this time, as did my first classes in computer programming. Having had no orthodox musical training, I was largely unaware of procedural approaches to music composition and electroacoustic practices developed in the twentieth century, including Cage, Stockhausen, Schaeffer, and Tudor. Although I have since overcome the aesthetic prejudices I felt towards experimental and academic musics, I nonetheless consider my practice to be grounded in a *pseudo-normative* vocabulary.

The vast majority of music produced by Throbbing Gristle, while clearly radical in approach, unmistakably references and emerges from popular musical traditions. Similarly, the back catalogue of the Austrian record label Mego is described by its founder Peter Rehberg as *extreme* computer music, and Parl Kristian Bjørn Vester, founder of the Autonomous Music School, describes his work as *radical* computer music. Neither have (to my knowledge) used the term 'experimental' in reference to their own practices. Similarly, I do not use the term 'experimental' to describe my music.

Thus the advent of industrial music in the United Kingdom marks the hypothetical origin of the musical heritage to which I align myself. Through the confluence of histories that constitute electronic music, the path I took drew influences from techno and house musics towards the latter part of the 1980s, and turned to the unconventional musical vocabularies developed by artists like Panasonic, Farmers Manual, Autechre, Russell Haswell, and Florian Hecker from around the mid-1990s. For me, these artists display an awareness of this particular musical lineage and present unique contributions to it. Furthermore, each has adopted systematized approaches to the construction of musical materials, often, but not exclusively, with explicit use of algorithmic processes. The function and classification of algorithmic processes within this independent tradition is greatly overlooked.

For me, the drum machine was a crucial step in the development of my interest in algorithmic processes. I was about fourteen years old when my parents, who were very supportive of my interest in electronic music, bought me a secondhand Boss DR55 drum machine. Unlike its more expensive cousin, the Roland TR808, the DR55 did not include any form of visual pattern display, and data entry was in the form of two buttons: one added an event, and a second added a gap (or rest). Three separate sounds were organized in sequences of sixteen such elements, a structure which became embedded in my musical imagination. I quickly became aware that the DR55's lack of visual feedback and limited interface generated unplanned results that extended my musical vocabulary. Thus I became interested in what it was to understand a given process or system and how this impacted my musical productions.

Another important technical development was the interplay between the drum machine and a borrowed monosynth, where a trigger could be sent from the drum machine to the synthesizer to step through notes in a sequence. I immediately began working with different sequence lengths, creating primitive yet evolving structures. The elementary methods afforded by this simple pairing became central to my work in algorithmic and generative systems (as opposed to the encoding of harmonic or melodic structures, emulations of famous composers, the mapping of biological or mathematical systems to music parameters, the sonification of data, the extension of expressive gestures, or, any other methods typically associated with orthodox approaches to algorithmic composition).

Although my parents supported my interest in electronic music production, there was a very practical limit to the resources they could invest as my father was an unemployed steelworker, and I was able to own only one piece of equipment at a time. A pattern emerged whereby I would save money, sell my current synthesizer, and buy another one. Consequently my productions only ever featured one synthesizer at a time. More fundamentally, this meant that I got to know the characteristics of each in exceptional detail and that my productions foregrounded those characteristics.

My use of these systems really upset people—from the rock and punk rock bands around school to the well-educated music students, each horrified and outraged by the inhumanity of my music, its tools and processes. This seemed inseparable from my love of synthetic sound: their lack of pleasure was part of my pleasure. Thus my musical interests were always oppositional and always necessarily linked to a rejection of other musical and ideological positions: the outdated masculinity of rock, the impoliteness of punk, the smug authority of the music student.

Around 1987 I began to work with computer-based MIDI editing environments, and encountered the now-ubiquitous timeline paradigm. I struggled with this approach until faster hardware and the release of Native Instruments' Generator software allowed me to work with real-time digital signal processing (DSP) and develop unusual synthesis and pattern-generating structures. Consequently, I developed two distinct strands of practice: the first working entirely within a timeline environment and a hardware sound sampler as SND (with Mathew Steel); the second made mostly with Generator as the lesser-known ShirtTrax (with Jeremy Potter). This resulted in two very different records in 1999: *Makesend Cassette*, on the German record label Mille Plateaux, and *Good News about Space*, on Russell Haswell's Or label. While the timeline project (SND) featured extremely repetitive structures with little or no rhythmic development, Shirt Trax, by contrast, was highly erratic, including eclectic and extremely volatile musical structures. Likewise, my solo releases *.h Ep* (Hobby Industries, 2000), *Reproduction* (Bottrop Boy, 2003), and *Ten Types of Elsewhere* (Line, 2004) explored instability through generative techniques. These two technical and aesthetic approaches, of either stable or unstable rhythmic structures, have formed the two fundamentally dichotomous threads of my musical practice.

After the release of the two records (*Makesend Cassette* and *Good News about Space*) I began using Max/MSP to develop systems both as compositional studio tools and as performance systems for music events. My first patches were based around emulations of the Roland TR808 drum machine's pattern entry interface, but modified or extended in a number of ways. I began with the TR808's eight sounds, triggerable over a resolution of sixteen discrete time steps. As an initial modification to this 16×8 grid, I replaced the binary on/off with a value ranging from 0 to 100, and a slider which muted events below a threshold. An alternative version treated the values as probabilities, which the slider globally adjusted. Both methods allowed the density of patterns to be controlled, but unfortunately, this proved to be generally uninteresting.

Reverting to a binary on/off method, I developed a system to scroll individual rhythmic layers—for example the clap could be offset by one or more positions on the grid, and the hi-hats by a different amount, adjustable as they played. This method developed into a system where two distinct rhythmic patterns could be interpolated (or morphed) to create a number of in-between states. I also experimented with polymetric layers of different loop lengths—for example the kick over three steps, a closed hat over two, a clap over sixteen, and so on. These and other procedures formed part of the 2003 record release *Tender Love* by SND, on Mille Plateaux, and were performed extensively around Europe, Japan, and North America from 2003 to 2007.

Around 2007 I decided to integrate the stable and unstable strands of my practice, the outcome of which was *Multistability* (Raster Noton, 2010). The term 'multistability' has two primary uses: first, it describes a system that switches between stable and unstable states, or which has the quality of both stability and instability; second, in descriptions of human perception, it refers to single image that can evoke multiple perceptual forms, such as the orientation of the Necker cube or a vase that may be perceived as faces in profile. My aim was that the structural quality of multistability should be distinct from the stable forms of SND or the unstable forms of my other projects from that time. It began with seven informal guidelines governing its production (although they were deliberately transgressed at points throughout the CD).

Guideline 1: Do not use the 'pencil tool' to enter notes into a grid
Guideline 2: No obvious or fixed tempo or meter
Guideline 3: Limited set of objects and keep patches 'simple'

These first three guidelines all represent a desire to avoid technical complexity and musical structures such as the time line or grid. Although we could argue about what simplicity might mean in this context, I feel that Figure 18.1 clearly illustrates my point.

Here low-level parameters are favoured over high-level procedures, such as those that aim to model preexisting musical vocabularies or abstract musical entities. By contrast, computational structures, user input, and musical output are deliberately held in a highly constrained state. For me, a highly reduced formal structure and the music it outputs are interesting and desirable because they allow me to explore musical constructs that are closely connected to the software's operational logic and to foreground this

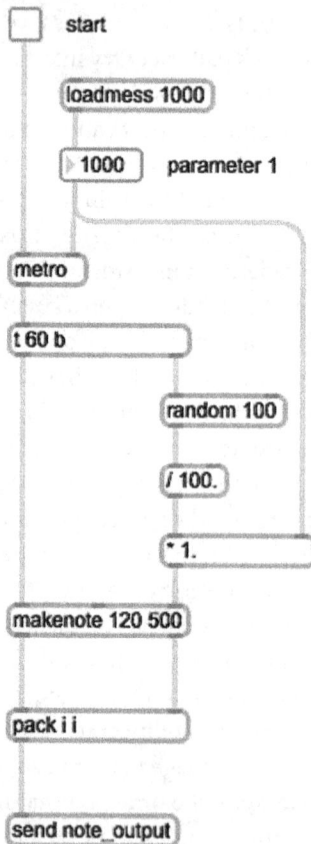

FIGURE 18.1 *Multistability*, track 7, implemented in Max. The system triggers a sound at the rate given by parameter 1, with a duration ranging from 0 to 99 percent of the clock speed. Parameter 1 can be changed, and the whole process can also be started and stopped. These were the only compositional and performance parameters.

connection. The structure of the music mirrors the logic of the software. This position is clearly drawn from my earlier practice, where my productions dealt with the specific character of tools and technologies as opposed to song-type structures.

Guideline 4: Focus on velocity, speed, and length of notes as compositional parameters

The fourth guideline was to explore the use of velocity, speed, and duration of sonic events as primary compositional materials. Pitch and timbre are generally constant throughout the compositions, and sounds are neither introduced or removed. This emphasis responds to how musical events (as encoded within the MIDI specification) are handled in the Max environment. In particular, the 'makenote' object requires the duration of the sonic event to be specified at the time it is triggered, and no reference is made to the sound's tonal envelope: it is merely an event of a predetermined duration and velocity, whose end point must already be known at its onset. This is quite unlike how music is played in realtime. I regard this way of dealing with musical data, along with the speed of automated triggering, as a brute fact of the Max/MIDI paradigm— a conceptual scheme with which the user must implicitly interact, in order to operate within the environment. I therefore aimed to place this scheme in the foreground of the work itself.

A clear example of this emphasis on speed, velocity, and duration is track 1 of *Multistability*. Here a list of ten values, ranging from 0.02994 to 1, is stored (see Figure 18.2). The list is then stepped through one item at a time, at varying speeds. The output is scaled to determine the velocity (i.e. volume) of a kick drum, and the velocity (volume and brightness) and duration of a chord. I was able to speed up, slow down, start

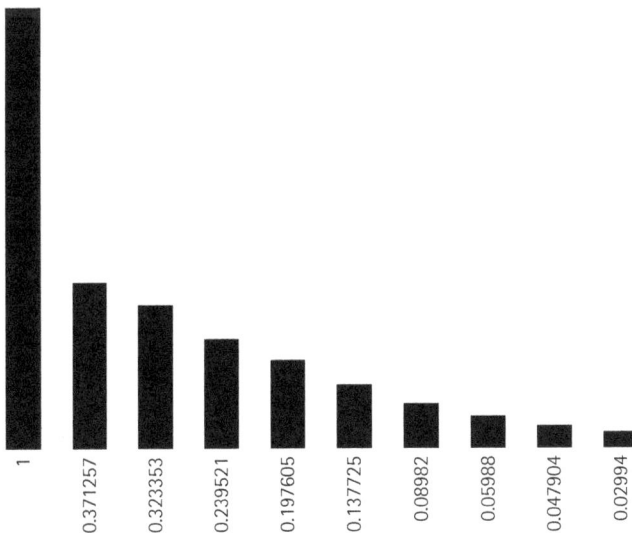

FIGURE 18.2 A list of values used in *Multistability*, track 1.

and stop playback, and also change the values in the list. Typically, this would include increasing an additional value in the list to 1.

It should also be noted that, in this piece, the value controlling the chord is one position behind the value controlling the percussion, therefore the loudest kick is followed one step later by the loudest, brightest, and longest chord. I did not choose to implement this feature; it came about as the result of an unforeseen synchronization issue. Although correcting it would have been relatively straightforward, I found the result appealing and so decided leave the feature in place.

Guideline 5: Synchronic use of 'percussion' and 'chord' layers

I define two categories of sound: (1) percussion sounds, typically a kick drum and clap; and (2) a chord made up of five notes. It is imperative that both sounds happen at the same time, that there is never percussive event without a chord, and vice-versa.

Guideline 6: Percussion sounds

Percussion sounds are generally constrained to a synthetic clap sound and a kick drum. For this project, kick drum sounds are typically derived from the Linn LM-1 drum machine, which has a rather 'sharp' quality. This sound featured heavily on chart hits of the early 1980s, including the Human League's 'Don't You Want Me' (Virgin Records, 1981), and set the sonic agenda for the following years, for example with the distinctive use of the Oberheim DMX kick on New Order's 'Blue Monday' (Factory Records, 1983). To some extent this type of kick drum was adopted beyond overtly 'electronic' musical practices, for example, in Brian Adams's 'Run to You' (A&M Records, 1984). The sharp kick is contrasted with the archetypal 'techno' kick, derived from the Roland TR808 and later TR909. It should be noted that while many early techno productions featured this deeper kick, some, such as Armando's '151' (Warehouse Records, 1988), incorporated the sharper kick paradigm drawn from earlier electronic musics. In these works I wanted make deliberate reference to early and pre-techno production styles, feeling that this switching of dominance (from the sharp to the deep) marked an important shift within electronic musics from one paradigm to another.

Guideline 7: 'Chord' sounds

The second category of sound I refer to as 'chord' sounds. These five-note chords were made using four-operator frequency modulation synthesis, and derived from sounds distributed with the Yamaha DX100 and TX81Z synthesizers. As with the kick drum sound described above, this sonic palette makes specific reference to early house and techno musics with frequent use of the 'LatelyBass' preset and 'JazzOrg', both of which were variously modified throughout the project. Descendants of these sounds remain present in contemporary house music productions.

FURTHER SYSTEMS

Several further techniques were used in the production of the *Multistability* CD, generally concerning the generation of timing data. In one case a table of values determines the speed of a clock that in turn triggers sonic events. My initial test used tables of various lengths (such as that shown in Figure 18.3), typically around five or six items, where tables could be stored and recalled to switch between different groups of rhythmic intervals.

An interesting feature of this method is that it generates rhythmic structures of variable overall duration, producing recognizable structures that can be transformed in an unfamiliar manner. I attempted to work around this dynamic loop length feature and developed a procedure whereby the values represented ratios of a fixed temporal duration: when the duration of step 1 is increased, all the other steps are reduced, so that the overall loop length remains constant. Rather than making it easier for the listener to assimilate temporal changes, I found that this method had the opposite effect and interrupted the flow of rhythmic progression.

A second version, shown in Figure 18.4, reduced the range of values to between 1 and 10. The temporal divisions generated by this version were restricted to multiples of a base value (in the figure, 40 and 50 milliseconds); therefore resulting in more familiar musical structures, but retaining unfamiliar characteristics such as dynamic pattern duration. I liked this combination of familiar and unfamiliar features, and a further development of this system allowed each timing step to be repeated a number of times.

By contrast to loops of indeterminate global duration, I also developed a system to produce patterns of determinate duration. Here a number of events, typically five, are triggered at a given speed, and retriggered at specified intervals (see Figure 18.5).

These overtly 'low-level' or 'simple' approaches are in stark contrast to my early interest in parameters that could drastically transform rhythmic structures from 'relaxed' to 'intense', or 'normal' to 'weird', and so on. I recall imagining a two-dimensional fader,

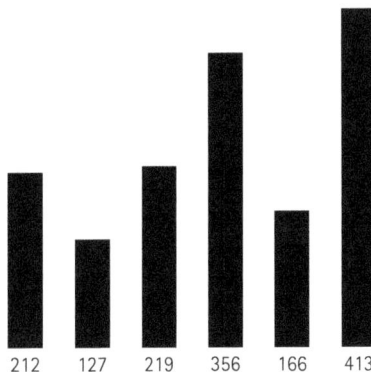

FIGURE 18.3 A list of timing steps with a range from 0 to 500 milliseconds.

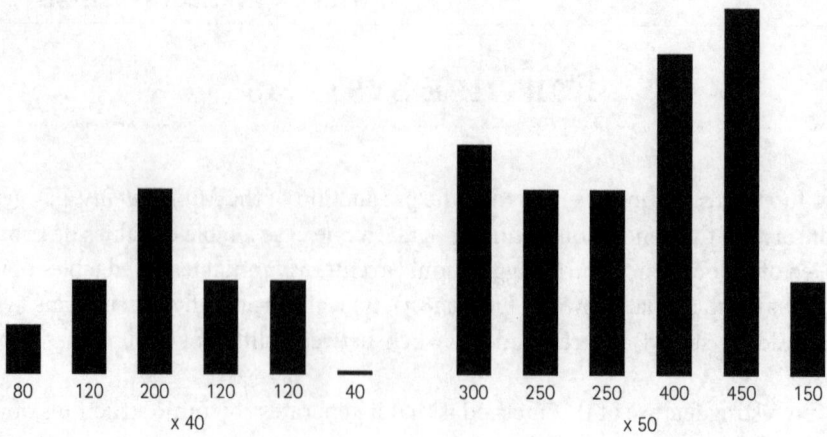

FIGURE 18.4 A range of values from 1 to 10 multiplied to produce longer intervals.

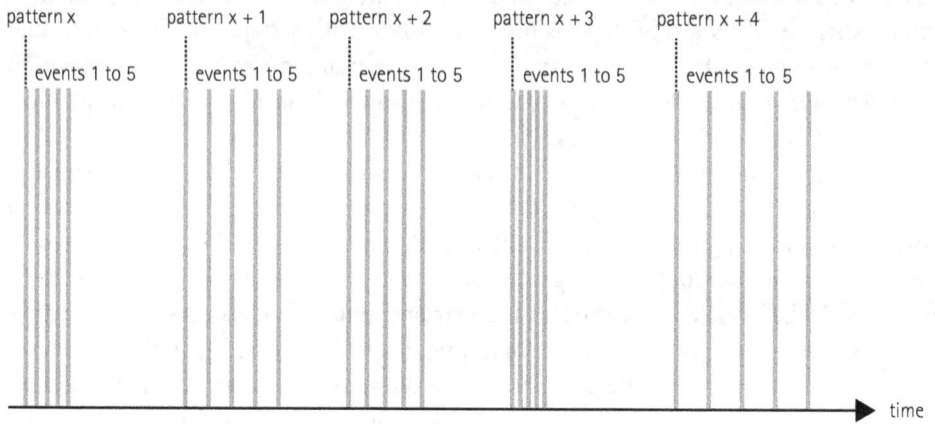

FIGURE 18.5 Event compression and expansion within patterns of determinate length.

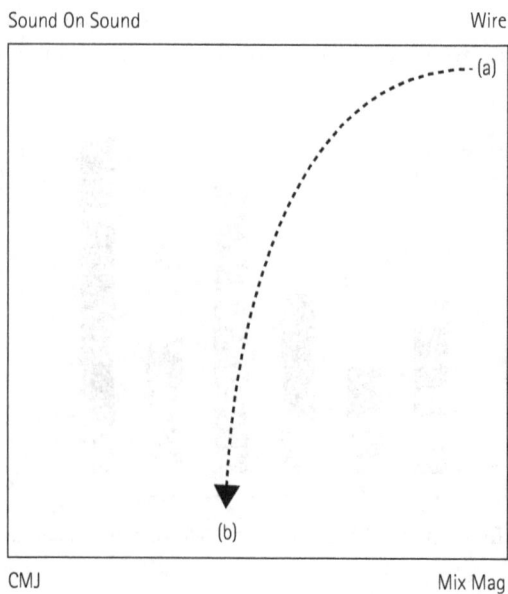

FIGURE 18.6 Magazine Interpolation.

loosely referred to as 'Magazine Interpolation' (see Figure 18.6): if the mouse position was top left this would ensure a good review in *The Wire* magazine, if the mouse was dragged to the right it would result in more favourable reception in *Mixmag*, and so on. Sadly, Magazine Interpolation remains unimplemented to this day, and so, too, its potential for generating favourable reviews.

CHAPTER 19

....

PERFORMING ALGORITHMS

....

KRISTIN GRACE ERICKSON

ALGORITHMS and music are central to my work and understanding of the world.

When I first heard the term 'algorithmic music', I was an undergraduate electronic music student at Mills College. I was in a computer science class learning about algorithms and data structures, which led me to assume that algorithmic music must be a musical form where the time-based procedures contained in each algorithm we studied could be heard as music. I imagined what *for loops*, comparisons, conditionals, and recursion would sound like slowed down into audible frequencies and timbres algorithmically organizing through time. In my mind, algorithmic music was not the sound of a computer program's final output, it was the sound of computation itself, an audible representation of algorithmic processes.

How would a musical logic gate function?

Could a performance calculate the value of π, and if so, to what decimal place?

Now I consider this imagined style of algorithmic music to be a type of sonification. Carla Scaletti defines sonification as 'a mapping of numerically represented relations in some domain under study to relations in an acoustic domain for the purposes of interpreting, understanding, or communicating relations in the domain under study'.[1] My idea of algorithmic music is sonification, where the domain under study is *algorithms*.

I have explored this approach from a variety of perspectives over the years. In the spirit of systems theory, I identify isomorphic relationships between algorithms and collaboration, music, and performance. I extend the boundary of the computer to include systems of people and sound. I extend the definition of music and performance to include the process, the rules, the machines, the execution. Here are some examples of what I mean.

19.1 PERFORMING BUBBLE SORT

The Bubble Sort algorithm is commonly taught in introductory computer science classes. It is one of many algorithms used to sort data stored in an array, and is easy to explain and demonstrate. Bubble Sort has been a useful tool for illustrating a separation between the algorithmic process and the content (data) that the algorithmic process organizes.

To perform Bubble Sort with humans as data, start with a group of people standing in a single line. Consider this an unordered array of people. Now, systematically step down the line from right to left, comparing the heights of adjacent people. When one person is taller than the next, they swap positions in line and the taller person moves to the left. Continue iterating through the entire line until everyone is in ascending order according to height.

By performing Bubble Sort, I had the unexpected realization that this type of performance does not *represent* the algorithmic process, it *performs* the algorithmic process. The people are the data, their positions in line are their addresses in the array, and they compute a real result by physically sorting themselves in space. Each different sorting algorithm reveals a different pattern of formal organization, even though they all compute the same final result.

To make Bubble Sort a musical performance, I replace the height value with musical gestures. For the comparison, adjacent people each generate a *unit* of music, sequentially or simultaneously. The result of each comparison is determined by the preferences of the performers. They swap positions in line only if they want to. The experience of previous iterations informs the swapping decisions of subsequent iterations. Human personality and interaction determine the nature and sound of the sort over time. The probability of a performed comparison generating a specific Boolean result correlates with the aesthetic preferences and negotiations of the performers.

Unlike machines, people can contribute opinions, feelings, and urges to a computational process. With humans as logic gates, algorithms can self-determine their own outcome. People are a strange computer.

19.2 M. T. Brain

In order to communicate complex algorithmic instruction sets to multiple performers in realtime, I developed an audio cueing system called M.T.Brain (Music Theater Brain). In M.T.Brain, discrete channels of audio instructions are distributed from Max/MSP through a multichannel soundcard. The performers wear headphones connected to the outputs of the soundcard by very long cables, so they can move around a space. Large numbers are sewn onto the performers' costumes which represent their soundcard channel number (address) in the initial performance array.

M.T.Brain's main limitation is the long cables used to connect the performers to the soundcard. As the performers move around each other, the stage becomes a treacherous tangle, distracting and dangerous (see Figure 19.1).

In the M.T.Brain software, audio instructions are generated by layering and concatenating chunks of text-to-speech audio, synthesized tones, audio samples, and timing cues.

FIGURE 19.1 M.T.Brain, performed by the Improvised Music Theater class, California Institute of the Arts, 2012.

Using learned rules of sonic syntax and semiotics, the instructions can include spoken stage directions, audio to be imitated, or tightly timed conducting cues. I think of these audio instructions like a programming language, a performer programming language.

The M.T.Brain software operator is the programmer. Lines of program code are executed as the M.T.Brain audio instructions are distributed, heard, interpreted, and responded to by the performers. The performers are a computer, and their performance is a computational medium.

M.T.Brain shifted my thinking about music and sound. Music ceased to be the intended final product of my creative work. Instead, music became a means of communication, a protocol, a transport, a score, a conductor, a script, a tool, a language, a logic.

Sonification became central to my work for a different reason. I became intrigued by the capacity of sound to communicate nonspeech information and by the limitations of human perception and cognition to interpret this embedded information.

19.3 TELEBRAIN

To go wireless, I created a browser-based platform for generating, organizing, and distributing performance instructions called Telebrain, (http://www.telebrain.org).

Telebrain allows for real-time telematic performances to be created over the Internet using personal wireless devices. Audio, image, and text elements are stored and shared on Telebrain and then concatenated and layered into longer instructions and patterns of distribution. The instructions are delivered via multimedia nodal multicasts, which function like chat rooms, except that multiple unique instructions can be sent simultaneously to different people.

19.4 PANDEMIC PERFORMANCE

Pandemic Performances use principles of complex adaptive systems to create emergent behavior in scalable, self-organizing groups of people. Using simple rulesets, interacting, adapting, and cooperating performers 'show coherence in the face of change,'[2] as *units* of expression spread and evolve like a contagious disease. In collaboration with its original performers, under the tutelage of Sara Roberts and David Rosenboom, I developed a Pandemic Performance called *EVOLOVE*, with rules based on a genetic algorithm.

Genetic algorithms simulate complex adaptive systems. Human social groups are also complex adaptive systems. *EVOLOVE* explores what happens when people perform the rules of another complex adaptive system, a genetic algorithm. The people are the agents and their *units* of expression are the genes.

EVOLOVE begins with a group of people moving around in a space. Each person has their own *unit* of expression that they perform when they come face to face with another person. Then, they have three options to choose from for their next interaction. They can keep their current *unit* of expression the same, they can copy the other person's *unit* of expression, or they can mutate their expression by combining the two in some way.

EVOLOVE is tricky to perform. To maintain balance, its rules need to be tweaked based on the individuals, their moods, the space, and the weather. If the system is too closed, if there isn't enough new information coming in, then the performance stagnates; choosing between the options feels like work instead of play. Conversely, if too much new information comes in, the performance spins out of control, resulting in a hyperactive frazzle, and no one can keep up.

Somewhere in the middle, usually after a few failed attempts, one can find a sweet spot. Tweak the rules until the right amount of new information is allowed into the system, and *EVOLOVE* takes on a life of its own. Regurgitated playground patterns reverberate through the tribe like intersecting zippers.

19.5 MANDELBROT MIDI

The first Mandelbrot music I created was at the Workshop in Algorithmic Computer Music (WACM) in 2013 at the University of California, Santa Cruz. I wrote a Lisp program that generated a two-dimensional array of the Mandelbrot set using mono-space ASCII characters to represent the values of the set. The ASCII characters were useful as visual guides to the information, and were easy to explore as inputs for musical expression. To sonify the Mandelbrot set, I gradually changed the input coefficients of the algorithm to zoom in, out, and around the set, generating new ASCII art arrays, like frames in stop-motion animation (see Figure 19.2).

FIGURE 19.2 Fractal from the Mandelbrot set rendered using monospace ASCII characters.

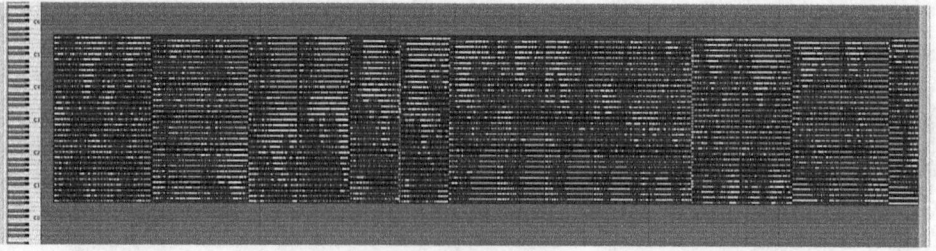

FIGURE 19.3 Screen capture of Mandelbrot MIDI, eighteen frames, all instruments, approximately 10 seconds.

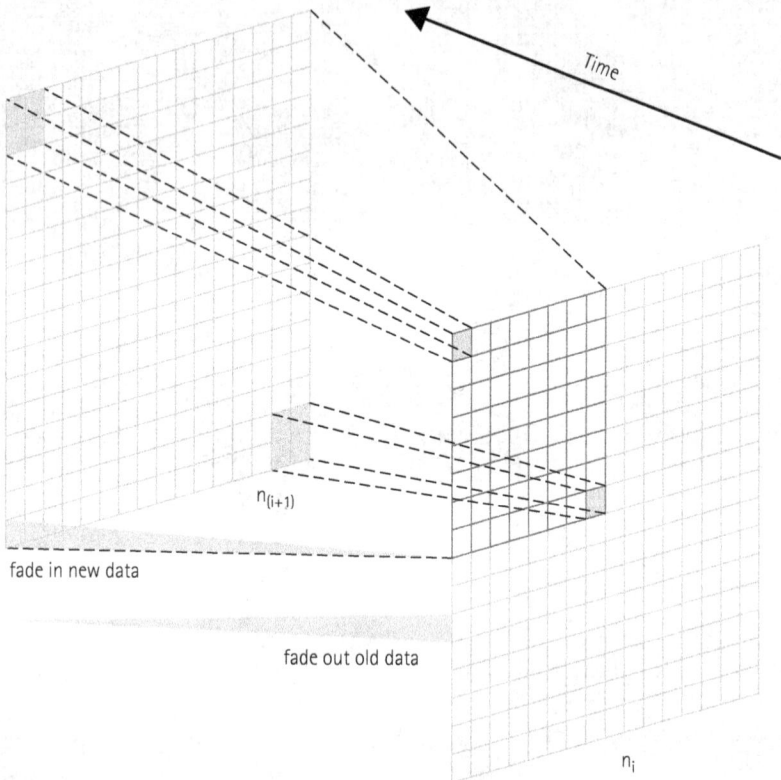

FIGURE 19.4 Dataset zoom proposition, for navigating fractals.

When looking at fractals, our eyes group together colour and brightness information into shapes. Noise in the distance morphs into contour when zooming in. The shapes reference both the larger pattern and the minute detail that may be seen elsewhere in the self-similar patterns.

Sonifying this data is about how information transfers between magnitudes of scale and our boundaries of perception. The challenge for me is in deciding how to map the data to time, while considering the different hierarchical levels of time perception we experience when listening. The pitch-fusion threshold is a perceptual phenomenon where frequencies below 20Hz are perceived as discrete and those above as continuous. So, where in the Mandelbrot set shall the pitch-fusion threshold lie? Where does pitch become timbre and rhythm become phrase?

I wrote a second Lisp program to convert the ASCII values into MIDI data. Each 2D frame was read quickly, like a rhythm, from left to right. I was hoping that shifts in pattern from frame to frame would be heard. Each unique ASCII value was assigned its own timbre. The height of the value in the frame determined its pitch according to the whole-tone scale. Amplitude was determined by the density of similar values: values that were clumped together, with many neighbours of equal value, were played softer, and values with many different neighbours were played louder. In Figure 19.3, you can see how the MIDI looked after being imported into Logic Pro Audio. The colours of the MIDI notes represent amplitude (black is louder than white).

I'd like to further sonify the Mandelbrot set by aligning the zoom axis with the time axis (see Figure 19.4), by fading out the old data while the new and higher-resolution data fades in; a technique similar to the Shepard tone scale could create the illusion of continuity between different scales of time and resolution. Although zooming through the set at a musical rate quickly hits the limit of 64-bit calculation, there must be a decent workaround for that.

19.6 CONCLUSION

Algorithms continue to serve as a foundational and structural aspect of my work. Algorithms inform how I collaborate, how I perform, and what I make. From web-based performance engines, to physically engineered technical environments, to turning my friends into a weird calculator, I have only begun to scratch the surface of what I hope to create from sonifying algorithms.

NOTES

1. C. Scaletti, 'Sound Synthesis Algorithms for Auditory Data Representation', in *Auditory Display: Sonification, Audification, and Auditory Interfaces* (Reading, MA: Addison Wesley, 1994), 224.
2. J. H. Holland, *Hidden Order: How Adaptation Builds Complexity* (Reading, MA: Helix, 1995), 4.

PART III

PURPOSES OF ALGORITHMS FOR THE MUSIC MAKER

CHAPTER 20

....

NETWORK MUSIC AND THE ALGORITHMIC ENSEMBLE

....

DAVID OGBORN

20.1 INTRODUCTION

THE early twenty-first century is marked, among other things, by a proliferation of networking technologies. This proliferation is multidimensional: the sheer quantity of computer networking devices, their distribution over the space of the planet (and the space of everyday life in many but not all places), and the coexistence of different types of networking devices—from the venerable and robust ethernet, to rapidly succeeding generations of wireless signals, to the networks within networks of virtual private networks (VPNs) and other darknets, to tentative gropings with decentralized mesh networks. This proliferation of networking technologies is no less visible, or rather audible, in music than in any other field of human endeavour. The emerging field of network music brings together a wide range of musical experiments that take these readily available (and sometimes not so readily available) contemporary networking technologies and make music with them. Frequently, the design and deployment of algorithms goes hand in hand with this artistic activity.

Network music happens when people make music over, or through, a computer network. On account of the ubiquity of networking technologies in contemporary everyday life, it is perhaps useful to add a further qualification: network music happens when people make music that explicitly depends on the affordances or materiality of computer networking technologies. While it is possible to find precedents for such activity, the field of network music has gained momentum throughout the first decades of the twenty-first century, as evidenced by new hardware and software systems, ensembles and research groups, and festivals, such as the annual Network Music Festival in Birmingham, United Kingdom, or the TransX Transmission Art festival in Toronto, Canada. Network music takes a large and ever-growing number of distinct forms, and network music practices can be broadly characterized along two axes of remote versus colocated and synchronous versus asynchronous collaboration (Barbosa 2003). A musician or ensemble might use network music techniques to project their performance in one location to other, remote locations. A group of musicians may play together although they are distributed geographically, whether to adjacent rooms in the same building or to opposite corners of the world (and everything in between). Network music performances may involve elaborate attempts to construct a sense of co-presence through immersive audio and video streams, while other network music performances will eschew such 'realism' in favour of more 'abstract' forms of musical cooperation. While network music most often involves performing together at the same time, situations where networked collaborators take turns over longer spans of time are also possible.

Networked, algorithmic music is at an exciting turning point in its development. The rise of an energetic live coding movement is happening in parallel with the arrival of more powerful audio and video sharing mechanisms, and the potential of the Web Audio API to make algorithmic music and audio languages widely accessible has only begun to be explored. Alongside regular network music appearances at festivals and conferences, ad hoc network music events are becoming increasingly common (see Figure 20.1). The broad reach of social media and other modern conveniences of connection and communication allows such events to be organized from anywhere in the world and receive an audience. Indeed, one of the potentials of network music most

FIGURE 20.1 Members of the Cybernetic Orchestra, connected via Ethernet, at the 2013 live. code.festival in Karlsruhe, Germany.

generally is to reduce or remove impediments to wide attendance and participation in artistic events (Boyle 2009). See Roberts and Wakefield, chapter 16 in this volume.

This chapter begins with a closer examination of the allure of network music, and then continues with a discussion of the nature and influence of key network music dynamics that are more or less bound to the materiality of the networking technologies: latency (and jitter), bandwidth, and security. Each of these dynamics is modified strongly when it becomes a matter not only of network music, but also of algorithmic music.

20.2 Why Do We Play with Networks?

Contemporary digital computers are themselves networks. A computer consists of a discrete set of elements (memory locations, registers, input and output transducers, etc.), and each of these elements occupies a different point in physical space (for example, a different location on a motherboard or within an integrated circuit). In response to machine-language instructions, electronic signals pass between these points. High-level programming languages tend to conceal this small-scale networking, instead encouraging us to think of the computer as a unified organism (a so-called black box), to think of data as occupying no physical space, and to think of the materials of computation as immediately and absolutely present.

At larger scales, the act of communicating signals from one location to another becomes more clearly recognizable as a matter of networking. When the body of the computing device is pierced by cables of varying lengths, when the transmission from point A to point B becomes much less robust (and also an easier target for surveillance), when the probability of a desired response arriving in return falls—in all such cases, no one would doubt that networking is present. In any case, between what is imagined as internal to a computer and what is clearly networking lies a clear difference in terms of user access—I can choose to connect my ethernet cable to whatever I like, and I can (hopefully) turn off my Wi-Fi connection, but rare indeed is the artist who rewires the connections between their memory chips and their processor (one exception would be circuit-bending performer Jonathan Reus, who intervenes directly in the electronics of older iMac computers to listen to the signals found therein and to distort them).

High-level programming interfaces reinforce this distinction between the individual machine and the network. The assignment '$x = x + 7$' is a basic, trivial, and reliable operation whose comprehension is considered an elementary matter of computing education. As soon as x becomes an entity 'out there' on a network, the operation is no longer basic, trivial, or reliable, and rarely a part of anyone's first steps with writing algorithms.

There is a potential analogy between the way that most computer languages represent time and the way that they represent the network outside of the individual computing machine. For the most part, programming languages have only awkwardly and imprecisely represented times and durations, with the dominant assumption being that instructions execute 'as fast as possible' rather than at some very specific time, which sometimes leads to difficulties for a domain like music, wherein the temporal positioning of events is fundamental. In the same way, programming languages deal clumsily with the fact that our computers are almost always connected/networked to numerous other things that are themselves connected/networked to numerous other things, and so on. Given the recent appearance of music and audio programming languages that represent time in precise ways, such as ChucK (Wang and Cook 2003), Extempore, and Tidal (McLean 2014), perhaps we can also hope for new programming language designs that represent networking no longer as the unreliable outside of the machine, but rather as the everyday reality that it is.

In application, networks are about space and its reconfiguration. In the simplest sense, that is what makes a network a network—that an intensity in one spatial location is somehow carried, more or less systematically, to another spatial location. Our contemporary networks create new forms of space, while previous forms of space are frequently remediated within new network spaces. Popular virtual worlds present a readily comprehensible example of this capacity of networking to create new spaces as folds in between real-world spaces. A prominent network music ensemble, the Avatar Orchestra Metaverse, performs from 'within' the Second Life virtual world, but of course their performances are always somehow projected into the multiple real-world spaces occupied by the performers and their audience.

Almost any deployment of screens and loudspeakers, sensors and input devices, creates a new space, though, and ideas about space are intimately connected to ideas about governance and social dynamics. A new type of public space produces a new type of public sphere, and network music, in producing new variants of spaces for public revelation and action, is in a very fundamental way an imagining, or reimagining, of how people can be connected to each other and live together (Baranski 2010). Networking, then, can be productively elided with musicking, understood by Christopher Small as the establishment of relationships that 'model, or stand as metaphor for, ideal relationships as the participants in the performance imagine them to be: relationships between person and person, between individual and society, between humanity and the natural world and even perhaps the supernatural world' (Small 1998, 13).

If networks are about relationships between people, then they are also about power, control, and governance. The everyday perception that our lives are ever more influenced by hidden algorithms is possible only because networks connect those algorithms to all of us. We might chart a progression from forms of governance based on a centralized authority, to those based on decentralized bureaucracy, to those based on distributed, networked, protocological forms, and any number of contemporary realities might be brought in to testify that this has not been a move from bondage into freedom, but rather a reconfiguration of the way things are decided, a reconfiguration of the way that we perform individually and collectively (Galloway 2004).

Whether consciously or unconsciously, and whether by reinforcing or resisting them, network musicians are playing with the tools and symbols of power in 'our age'. Pioneering network ensemble The Hub, known first as the League of Automatic Composers, explicitly rejected an ethos of control and determinacy in favour of surprise and complexity, connecting their machines to each other in decentralized feedback loops that would never produce the same sounding result twice (Bischoff, Gold, and Horton 1978). From the beginning of the 1980s, their system evolved to feature a central hub computer or 'Blob', which acted as a panoptic shared parameter memory for all of the other computers, connected via custom-built RS232 interfaces. Later, with the adoption of MIDI technologies towards the end of the 1980s, the shared memory was abandoned in favour of a practice of distributing and redistributing events targeted to specific receiving computers (Gresham-Lancaster 1998). One can construct a reasonable allegory of contemporary networked life armed with just these two alternatives: the panoptic database and the viral event.

As the example of The Hub suggests, the history of network music certainly does not begin with Wi-Fi and the Internet. Nor does it begin with microcomputers and MIDI. If it begins anywhere (and the search for beginnings is always suspect, except as a reminder of the many threads that are woven together in our present), it begins with radio and the telephone. While the complex history of radio's proliferation would be out of scope here, we can at least underline that our contemporary networking technologies are descendants of a rich heritage of such technologies, and that, similarly, our network arts (our creative uses of these technologies) are connected to longer traditions of radio art, transmission art, and so on. By roughly the middle of the twentieth century, the radio—that diffuse yet universal networking technology—was largely used in a broadcast fashion, as a way of communicating out from the centres of power to the masses. As a reaction to this, traditions of radio and transmission art have often emphasized dialogic elements. In the 1980s, the Japanese radio artist Tetsuo Kogawa championed mini-FM stations and, more broadly, 'polymorphous media' that do not make 'molar groups' out of their listeners but rather encourage individual connections, based on self-controlled tools (Kogawa 2017). Later, Kogawa became known for encouraging people to build their own FM transmitters.

While radio art has become less common, in parallel with a decline of societal investment in broadcast radio, this dynamic continues to exist. Dreams of the Internet as a distributed utopia for the free exchange of knowledge, culture, and entertainment have by and large been replaced by a centralized flow of information to and from a limited number of centralized sites, offered by large corporations and almost invariably tied in one way or another to advertising, surveillance, or both. Art and music that present an alternate, dialogical configuration for the network thus retain critical potential. Algorithmic network music, moreover, has an additional critical potential: to take a posthumanist approach that does not necessarily accept intimate conversation between two bona fide human entities as the principal measure of success. Algorithmic network music can critique and resist the heavily centralized networks known as 'the Cloud' without substituting for them the romantic salon.

Nothing about the deep involvement of networks with contemporary forms of power and governance contradicts the point that when we make music with networks, we derive pleasure both from that musical building activity and from the element of social togetherness that it produces. Pauline Oliveros, a long-standing exponent of network music in its telematic form (making geographically distributed musicians co-present to each other), points to musical friendships as a primary rationale. In Oliveros's words, 'If you are on the East Coast and the musician you want to perform with is on the West Coast then there is a reason [to make network music]' (2009). In almost the same breath, she points to globalization as another reason.

Indeed, our contemporary, centralized social media platforms use the element of playing together socially as their primary attractor (or rather distractor, as in distracting people from either thinking about the fine print of user agreements or objecting to the torrent of advertising content)! Network musicians play together with networking technologies, simultaneously deriving pleasure, cultivating social relationships, and calling critical attention to the forms that permeate our everyday lives. The nature of the technologies that they use gives rise to a number of 'perennial' network music issues, while the specific context of algorithmic music often introduces additional strategies and challenges that aren't present when network music is aimed simply at making traditional, acoustic musicians telepresent to each other. The following sections review three key material dynamics of network music situations with an eye to the changes introduced by an algorithmic focus: latency and jitter, bandwidth, and security.

20.3   LATENCY AND JITTER

It takes time for any signal to go from point A to point B through any medium. This delay is usually called *latency*. This latency can vary from moment to moment, for example as a consequence of a change in the route a signal travels, and such variation in latency is called *jitter*. Latency and jitter are basic 'problems' encountered by all network musicians, with an effect on numerous musical aspects, including but not limited to synchronization, performer interaction, tempo, and spatial imaging.

The diameter of the planet Earth is 12,742 kilometres, and the speed of light is 299,792,458 metres per second. An electromagnetic signal sent straight through the body of the earth would thus arrive at the other side 42 milliseconds later. In practice, of course, network cabling does not bore straight through the centre of the earth nor take the shortest possible surface route, but rather it runs along its surface in complex diagrams. It is easy to think of possible surface routes that sum to 10,000, 20,000, or more kilometers. When network music events are globally distributed, it's easy for the sheer travel time at the speed of light to add up to a delay that is quite apparent to our perception. In short, at these scales the speed of light feels surprisingly slow.

Additionally, long and short network routes both introduce various stages of buffering—the signal goes from one hop to the next, with a small delay introduced at each hop. Networking hardware and software introduces additional layers of buffering and delay, as does the interpretation or rendering of the transmitted information by application-level software. All of these inevitable delays sum to produce a minimum possible latency for a given network route. In practice, these 'nontransmission' delays can often be orders of magnitude greater than that of the speed of light. For example, the round-trip time to send audio between Hamilton, Canada, and Montréal, Canada, in the tabla and live coding duo very long cat is around 111 milliseconds—about twenty-nine times longer than the approximately 3.8 milliseconds it would take a direct signal at the speed of light to travel the 567 kilometres that separate the two Canadian cities (Ogborn and Mativetsky 2015). This leads to a first commandment for all network musicians: never rely on predictions of what network conditions 'should be'—measure what they really are!

The impact of 'nontransmission' delay (hardware and software buffering, etc.) is particularly evident when network music performances are not globally distributed but localized. For example, recent years have seen the emergence of a number of laptop orchestras, more or less large ensembles of musicians with laptops, often spatially distributed across a single acoustic space (like the large ensembles of any number of historical traditions of music making), with their individual machines connected by some form of local networking, such as Wi-Fi (Trueman et al. 2006; Tsabary 2014). In these situations, the raw electromagnetic transmission time becomes insignificant, and yet with typical networking hardware and software significant delays will still be experienced. To give a common 'challenging' scenario: with a poor-quality Wi-Fi connection and heavy encryption or security measures engaged, one can easily encounter delays around a quarter of second, even with everything in the same room. Simplifying or eliminating the network encryption will reduce the latency but at the cost of inviting security problems (see section 20.5, 'Security', below). More elaborate responses involve algorithms to synchronize the clocks on separate machines, scheduling musical events relative to these synchronization algorithms, and forming local caches of shared musical parameters (Ogborn 2012, 2014; Sorensen 2010).

The inherent latency of network transmission is not a problem when such transmission is in one direction only, from a sender to a distant receiver, as in the case of broadcast radio or of a contemporary Internet consumer streaming a video from a centralized service. Indeed, in such cases the latency is literally imperceptible once the decoding and projection of the streamed transmission has begun. When events become dialogical, however, the latency becomes more perceptible, and when the multiway exchange carries precisely synchronized musical events it becomes more perceptible still. The question thus arises: how much latency is acceptable for a musically satisfying situation?

One common approach to this question begins by comparing network latencies to the delay involved with the mechanical propagation of sound signals. At 20 degrees Celsius and sea level, sound travels at 344 metres per second. On this basis, a small network transmission latency of 5 milliseconds is equivalent to the time it takes sound to travel 1.72 metres (i.e., also 5 milliseconds). A 20 metre distance from a proscenium stage to the middle of the audience might be compared to a 58 millisecond network transmission latency. While such comparisons are helpful, it should not be overlooked that these two types of delay are, in common practice, not mutually exclusive but rather effects that sum to produce a new acoustic situation characterized by (among other things) increased direct wave delay time relative to either network transmission or acoustic propagation taken in isolation.

Another approach to the question of how much latency is acceptable bases itself on selected results of psychoacoustic research. For example, early research on auditory perception showed that a threshold of roughly 20 milliseconds between two distinct sound events was required for a listener to identify which precedes the other (Hirsh 1959). With high-quality local network connections, or Internet connections over shorter distances, it is certainly possible to achieve network latencies below this threshold. This is not the full psychoacoustic story, however. Even when we cannot identify which of those two distinct sound events come first, we can recognize them as distinct from a single, fused sound event down to time differences of around 2 milliseconds. This rather more stringent figure will be no surprise to connoisseurs of digital audio interfaces. The manufacturers of such interfaces aim to produce the lowest possible conversion latencies in order to support real-time monitoring of signals transformed by software, and are cognizant that quite small latencies are perceptible to musicians who are monitoring their own signals (or transformations thereof) during recording sessions.

A more recent strand of research approaches the question of latency not from, or not only from, the standpoint of audience perception, but rather from the standpoint of its objectively measured impact on musical performance actions. In one recent study, with pairs of rhythmic clapping performers separated by calibrated delays between 3 and 78 milliseconds, the effect on ongoing musical tempo was measured. This revealed four distinct phenomena: below 10 ms, the performers tended to accelerate; between 10 and 21 milliseconds, their tempo was stable. Above this and up to 66 milliseconds of latency, they tended to decelerate, due to the readily comprehensible behaviour of waiting for a collaborator's delayed metre. Above 66 milliseconds synchronization deteriorated rapidly (Chafe, Cáceres, and Gurevich 2010).

Latency, in short, has musical effects whose characterization evades a simple threshold of 'good enough' versus 'not good enough'. As the fields of music cognition and music information retrieval advance, their analytical instruments may be used to build an even richer picture of the effect of latencies (acoustic and network) on musical performance. In the meantime, none of these results should be taken as an indication that a particular latency is simply 'good enough' for all time. From a position of very low network latency, longer latencies (alongside other virtual acoustic features) can always be simulated. But the inverse does not hold true: once the signal has been delayed 300 milliseconds from arriving at a given point in a system, it will forever be that 300 milliseconds later. High latencies don't only have direct effects on musical perceptions and performance—they are also constraints on what can be simulated or modelled in musical systems.

The algorithmic, network music context introduces an additional reason to pay close attention to the phenomenon of network latency. The performance of algorithms in a network, distributed space, can give rise to situations where small discrepancies in when a given piece of code executes result in large discrepancies in what the result of such code is. One example of this would be the use of oscillators. If, for example, a live coding artist creates identical sine wave oscillators on distributed machines and then later executes a second piece of code on all those machines that accesses the output of that oscillator, they could get wildly different results depending on the time that has elapsed between the two pieces of code running on each machine, as a consequence of jitter. If the frequency of that oscillator is 125 Hz, then a discrepancy of only 2 milliseconds (a quarter wavelength of a total wavelength of 8 milliseconds) is sufficient to make a difference between the maximum absolute values of the oscillator and the minimum. Jitter can easily reach these magnitudes on dedicated wireless networks, and on the general-purpose Internet jitter typically far exceeds these magnitudes.

At the same time, the algorithmic, network music context also provides additional strategies for reducing the perceptual and musical impact of network latency:

(1) Given relatively synchronized clocks (i.e., a frame of reference), algorithms can be written in such a way as to make things happen at aligned times in the near future. In effect, this is the direct response to the above-mentioned example of the low-frequency oscillator. If the multiple, distributed instances of a low-frequency oscillator start at the same time (given a known frame of reference) then they will have a deterministic result at some later known time (given a known frame of reference). This is also the approach taken by the Ninjam network music software, which takes compressed audio performances from a given node in a distributed ensemble and delays the monitoring of that performance at all other nodes to line up with a subsequent period in the music (for example: my performance in a given four bars is heard by my collaborators as precisely lined up with the next four bars of the music) (Ninjam 2017).

(2) Alternately, since an algorithm already implies a temporal gap between its specification and its realization, this can be exploited to create the illusion of simultaneity. One can deliberately delay the local monitoring of the algorithmic result from a given node in the network to line up with the delayed reception of results from other nodes in the network. There is no limit to how many live coding or algorithmic nodes can be aligned in this way, and it is also possible to include one live audio (i.e., nonalgorithmic) performance in a network ensemble using this technique (Ogborn and Mativetsky 2015).

(3) Events can be structured in such a way that they do not need to be aligned between nodes at different relative latencies (and are 'immune' to jitter). This has been a common strategy in the general evolution of network music—to accept the inevitability of latency and jitter, incorporating them into the musical fabric, or otherwise adapt to it (Cáceres and Renaud 2008; Tanaka 2006). In its simplest form, this can entail avoiding firm metric structures, or including layered drones, textures, or other relatively rhythmically independent elements so that the timing discrepancy from one node to the next does not become obvious. The algorithmic network music context provides an additional variation of this latency strategy: code can be structured in such a way that it can be rendered or realized completely independently on different machines. Provided that the code to be

rendered does not refer to other time-varying functions (or random functions), the result can be identical at different locations even at drastic relative latencies.

An extreme example of using the inherent latency of network music to advantage is provided by the SoundWIRE technique, which uses the delay of network transmission as the delay component of a physical modelling synthesizer, typically a Karplus–Strong model of a plucked string (Karplus and Strong 1983). A system is configured such that audio is sent to a node and then returned to where it came from, and then monitored, filtered, and recirculated on the network (after a low-pass filter). The monitored sound becomes a sonic representation of the underlying network conditions, whereby low latency will produce higher pitch, and low jitter (variation in latency) will produce minimal vibrato (Chafe and Leistikow 2001).

20.4 BANDWIDTH

Driven by an orientation towards telepresence, network music has frequently resorted to the transmission of high-quality digital audio and video signals, creating links between distinct spaces through the medium of video projections and the 'sound screens' that exist in and between arrays of loudspeakers. In addition to latency and buffering issues, these techniques consume very large amounts of bandwidth.

For example, a single, raw mono bidirectional 44,100 Hz digital audio signal at 24 bits per sample will require an absolute minimum of 1 megabits per second (Mbps) in each direction (44,100 times 24 is 1,058,400). In practice, the requirement is greater as some amount of redundancy is required in order to make such a network audio signal robust to long drop-outs. This is especially the case when it is a matter of audio transmission over the Internet.

The JackTrip software is in wide use by network musicians, as a free, open-source, and readily available means to stream audio over networks. JackTrip transmits uncompressed audio (i.e., buffers of linear pulse code modulation [PCM] samples) in order to avoid the additional latency that would be introduced by encoding/decoding operations (Cáceres and Chafe 2009). To compensate for network problems, JackTrip sends redundant copies of the audio data. This tends to demand more, and more reliable, bandwidth than is commonly available in home situations. Results tend to be strongest with either local ethernet networks, or very robust Internet connections such as are sometimes available in universities and other research institutions. A more recent JackTrip server can receive many client connections, mix them, and redistribute mixed signals back to the clients (Cáceres and Chafe 2010).

At the time of writing, extremely robust standards have been put in place for the distribution of lossless Audio over Ethernet (AoE), and devices implementing these standards are increasingly common in high-end institutional settings, such as theatres, concert halls, and recording studios. While there is a large market for these devices (which considerably simplify the physical running of cables in spaces), there is also a confusing competition between both open and proprietary standards. Like JackTrip, the AoE formats generally transmit uncompressed, lossless audio and require reliable network connectivity, such as is rarely found on 'consumer grade' Internet connections. Indeed, the Audio Video Bridging (AVB) standard requires specialized network switches.

A number of formats exist for lossy streaming of audio signals, including the Constrained Energy Lapped Transform (CELT) and its successor, Opus. These formats have the advantage of using significantly reduced bandwidth, and also of being open standards. Opus is supported by the Internet Engineering Task Force (IETF) and has been developed to introduce a relatively small 'algorithmic delay' (i.e., the component of the overall latency due to encoding and decoding of the audio signal). While these haven't yet been widely used in the network music and algorithmic music communities, one can expect that experimental energy will travel down these paths as software integrating these formats with practical, working algorithmic and electronic music systems appears.

However, the algorithmic music context provides an alternative or addition to all three of the above relatively bandwidth-hungry methods of supporting telepresence; code (algorithms in text form) can be transmitted as low-bandwidth text data, and then executed either immediately or (better) on some definite schedule against a synchronized clock. In the extramuros software, for example, any number of live coding performers collaborate on shared code that appears in a web browser interface. Like with the popular Google Docs word-processing platform, each performer's changes to the code are visible to every other performer in realtime. When a performer triggers the evaluation of some piece of this shared code, it is transmitted to any number of 'client' computers for rendering into sound by whichever programming language the ensemble chooses. The combination of a shared editing interface with any number of rendering computers allows the software to be used in diverse network music settings, from workshops, where participants bring nothing but a web browser and connect over a local area network, sharing a single local projector and sound system, to globally distributed ensembles, where each participant renders the audio independently at their own location (Ogborn et al. 2015).

Network music thus involves the transmission of two broad types of musical data, and a given network music performance might involve either type of data in isolation, or it might combine them into a hybrid network music topology. On the one hand, there is data representing continuous audio and video signals, and on the other hand, data representing discrete objects or events, including isolated musical parameters and perceptible notes and events, as well as more highly articulated objects like code structures. There is a long history of musical networking systems based on the transmission of discrete objects and events, from the earliest sequencers, through the standardization and widespread adoption of the MIDI protocol, to the comparatively recent spread of Open Sound Control (OSC; Wright, Freed, and Momeni 2003). It is possible to see the recent trend towards the transmission and distribution of code as an extension and abstraction of these more long-standing networking practices.

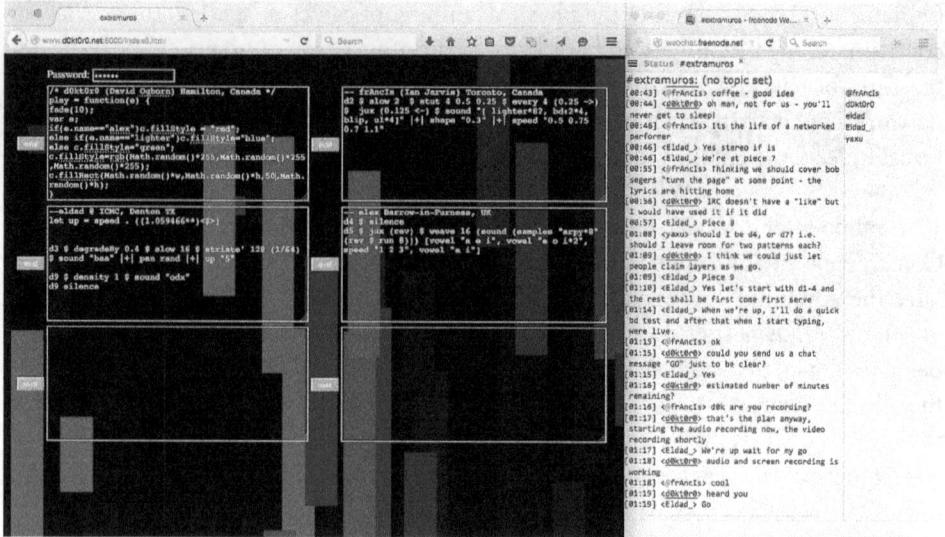

FIGURE 20.2 Members of the "shared buffer group" performing at ICMC 2015 via extramuros, from four different global locations (clockwise from top left: David Ogborn, Ian Jarvis, Alex McLean, and Eldad Tsabary).

Pieces built on these two broad types of musical data are found in the repertoire of the Birmingham Ensemble for Electroacoustic Research (BEER). In BEER's *Pea Stew* continuous audio signals (sent with JackTrip) are recirculated around a network of performers, with phase shift based on Fast Fourier Transforms (FFT) as well as other live-coded transformations applied at each node. In another BEER piece, *Telepathic*, the network is used to share parameters—establishing a shifting, centralized tempo as well as quantizing the redefinition of three different musical layers by individual live-coding ensemble members, so as to produce unified, drastic changes in the resulting texture (Wilson et al. 2014).

Beyond their bandwidth parsimony, systems based on discrete objects and events have the further advantage of there being natural ways to adapt and modify the realization of the transmitted data to reflect 'local' conditions, which could include the presentation of code intentions on Braille devices, screen-readers, or fantastic kinetic sculptures (see Figure 20.2). Live coding, by projecting both the code and its result has always been a kind of projectional editing, where intentions are communicated or 'projected' in multiple ways (Walkingshaw and Ostermann 2014). There is a massive and exciting space for play around the development and exaggeration of these projectional possibilities. The transmission of algorithmic art as discrete events (rather than as audio and video signals) increases the exposure of the art to differences in the execution context. While it is sometimes imagined that the code is equivalent to its realization, this translation is no mere algebraic operation and this can become quickly exposed when code is being simultaneously realized by machines in different places and conditions. Network art can thus provide practical explorations of the mystery and unpredictability of software that is pointed to by software studies (Chun 2011).

A simple demonstration of this is provided by the laptop ensemble context: take a group of computers each connected to its own loudspeaker and then trigger some synthesis events on each of them from a 'central' computer over a network. The result will, at a minimum, demonstrate differences in timing due to the way each computer's network stack works, as well as differences in timbre due to each computer's audio subsystem. While an oscillator at a given frequency is mathematically well defined, it can quickly take on surprising details and differences when rendered by a specific computers connected to specific speakers and ultimately, specific ears. PowerBooks UnPlugged (Rohrhuber et al. 2007) were early explorers of this terrain, using Wi-Fi and just the built-in speakers of their laptops and with a SuperCollider-based system, allowing small pieces of code to be rendered as sound on any of the machines, or any combination of machines, in a spatially distributed ensemble with all of the sound coming out of the small built-in speakers of each laptop.

20.5 SECURITY

If networks are, fundamentally, about power, governance, and public space, it should be little surprise that they can be seen through the lens of security. Indeed, network music events routinely run into practical obstacles that can be directly traced to one or more contemporary information security issues. Despite this, comparatively little attention is paid to such issues in the development of network music software and environments. This is a research and creative space that is ready for considerable expansion.

When network algorithmic music is a matter of a closed group performing on a wired ethernet that they control, the possibility of security problems being deliberately caused by an outside agent is relatively minimal. Even in such an environment, security issues frequently come up as the machines themselves will typically be used in other less secure settings, and will 'import' issues from elsewhere into the closed situation of the group. In laptop orchestras, there are always firewalls—necessary for the student who frequently connects to the Internet in public cafés no doubt—that need to be turned off (or have exceptions added to them). There is occasionally the computer whose web browser has been damaged by malware. Closed, secured Wi-Fi networks introduce additional latencies, while open, unsecured Wi-Fi networks expose ensembles to mischievous bystanders who might interfere with a performance by sending 'unauthorized' OSC packets to shared systems. It is even possible to crash some of the common audio programming applications with the right OSC message, if that message is not sanitized before subsequent processing (Hewitt and Harker 2012).

When the Internet is used as part of a network music event, practitioners frequently run into the firewall structures of home and institutional networks. In the most common default configuration, these firewalls are configured to reject incoming communications but accept (and continue) outgoing communications. This scenario is closely bound up with the recentralization of the Web in 'Web 2.0'. If the primary usage scenario for the network is home users 'surfing' a limited and stable selection of large central content 'providers' (i.e., distributors), then that default firewall model makes a lot of sense. However, it tends to work against individuals temporarily placing their own web servers on the Internet, and thus against ad hoc multinodal network music applications. While not impossible, the grain of the network tends to work against people doing these things without another level of either willpower or experience with the configuration of networking devices.

When algorithmic music environments are deployed permanently on the open Internet as platforms, the security issues become even more significant. Moreover, the specific domain of networked algorithmic music may produce requirements that aren't fully met by the dominant approaches to securing networked appliances (requiring logins or even more robust credentials). A public artwork involving a collaborative text editor on the Internet no doubt requires some protection against real-time spam, but passwords and credentials require coordination, and frequently become stumbling blocks to fluid collaboration, or at least stumbling blocks to beginning that collaboration. In the 'real world' musicians don't require credentials in order to recognize that they share a space and should not destroy the space that they are sharing. There is much potential for the design of spaces for networked collaboration around algorithmic music that take account of particular security requirements: we often want or need people to openly and freely participate, and standard Internet security apparatus could impede that.

With the caveat that much more work does need to be done to make networked algorithmic music production environments robust to security issues (to allow the directors of laptop orchestras to get better sleep the night before a show, among other reasons), it would be a mistake to see information security issues simply from the standpoint of protecting oneself from threats. Indeed, there is significant critical and creative potential in realizing works of art and music that experiment and explore and thus make representable and comprehensible security-related phenomena. We might listen to a distributed denial of service attack, spoof a SuperCollider server, or force a 'shy' performer to share their screen via van Eyk phreaking—giving new meaning to the venerable 'show us your screens' (Ward et al. 2004)! We might imagine live coding battles where one part of an ensemble writes (and shares with the audience) algorithms to identify and counteract 'malicious' code, while another part of the ensemble redoubles its efforts to write code that gets through the filters of the other part of the group.

20.6 Future Directions

Two interesting roads little travelled in network music are: (1) the use of analyses derived from images and signals to defeat latency barriers, such as using visual information to predict when a percussionist will strike a drum ahead of the actual event, with a simulacrum of the sounding result synthesized at a destination site ahead of what latency would otherwise allow (Oda, Finkelstein, and Fiebrink 2013); and (2) the centralized rendering of algorithms connected to facilities that stream back the results, with the possibility of using enormous cloud-pooled processing resources to render things that could not be rendered by any domestic machine (Hindle 2015). The centralized rendering of sonic algorithms could be a promising road to providing accessibility to algorithmic music environments, as presently many of them have significant installation challenges on the diverse machines and operating systems in contemporary circulation. A central rendering and streaming approach could remove the need for substantial installations in these situations—instead, users would connect to a web-based editing environment, a server somewhere would render the result, which they would receive back via streaming.

The recently developed Web Audio API represents another way in which algorithmic music environments can be rendered immediately accessible to anyone with a web browser. With the Web Audio API, all the rendering of the audio takes place within web browsers, which have now become miniature, media-rich, standardized, quite secure operating system/virtual machines. Lich.js and Gibber are two recent projects that clearly display the potential of the Web Audio API to create zero-configuration entry routes into algorithmic music making, with both incorporating networked, collaborative editing (McKinney 2014; Roberts et al. 2015).

As part of the expansion of research around web audio, we can hope for the emergence of algorithmic network music platforms that facilitate the sharing of musical algorithms together with their results, as well as enable the formation of social communities around those tuples. Platforms have played an occasional role in the story of network music thus far, such as the asynchronous, primarily MIDI-based composition platform of NetJam (Latta 1991), as well as Roger Mill's use of the furtherfield.org platform to host a network music ensemble (Mills 2010). Nonetheless, it is a striking lacuna of the contemporary moment: we have algorithmic music languages, and we have maturing network music technologies, and we have platforms for sharing social connections and platforms for sharing media (i.e., Freesound, SoundCloud) but little in the way of platforms for creating and sharing collective algorithm-sounds (things that are simultaneously algorithms and sounds or music). Building and experimenting with such platforms are exciting avenues for future research, as such platforms would have the potential to reach a wide international audience, while also blowing up worn but persistent ideas that the identity of the work consists only in the sound, or only in the sound–video composite, or only in the artist's head, intentions, and so on, that is, only 'in one place or sense' of some sort. Algorithmic art has always represented a healthy challenge to aesthetics that privilege sense intuitions—and new networked platforms for the presentation of algorithmic art would be exciting, multiple, distributed places to be.

BIBLIOGRAPHY

Baranski, S. 'Topographie de la musique en réseau sur Internet'. *Filigrane: Musique, Esthétique, Sciences, Société* 12 (June 2011). http://revues.mshparisnord.org/filigrane/index.php?id=295.

Barbosa, A. 'Displaced Soundscapes: A Survey of Network Systems for Music and Sonic Art Creation'. *Leonardo Music Journal* 13 (2003): 53–59.

Bischoff, J., Gold, R., and Horton, J. 'Music for an Interactive Network of Microcomputers'. *Computer Music Journal* 2, no. 3 (1978): 24–29.

Cáceres, J.-P., and Chafe, C. 'JackTrip: Under the Hood of an Engine for Network Audio'. In *Proceedings of the International Computer Music Conference*, 509–512. Montreal, 2009.

Cáceres, J. P., and Chafe, C. 'JackTrip/SoundWIRE Meets Server Farm'. *Computer Music Journal* 34, no. 3 (2010): 29–34.

Cáceres, J.-P., and Renaud, A. B. 'Playing the Network: The Use of Time Delays as Musical Devices'. In *Proceedings of International Computer Music Conference*, 244–250. Belfast, 2008.

Chafe, C., and Leistikow, R. 'Levels of Temporal Resolution in Sonification of Network Performance'. In *Proceedings of the 2001 International Conference on Auditory Display*, 50–55. Espoo, Finland, 2001.

Chafe, C., Cáceres, J.-P., and Gurevich, M. 'Effect of Temporal Separation on Synchronization in Rhythmic Performance'. *Perception* 39 (2010): 982–992.

Chun, W. H. K. *Programmed Visions: Software and Memory*. Cambridge, MA: MIT Press, 2011.

Galloway, A. R. *Protocol: How Control Exists after Decentralization*. Cambridge, MA: MIT Press, 2004.

Gresham-Lancaster, S. 'The Aesthetics and History of The Hub: The Effects of Changing Technology on Network Computer Music'. *Leonardo Music Journal* 8 (1998): 39–44.

Hewitt, S., and Harker, A. 'Security in Network Connected Performance Environments'. In *Proceedings of the International Computer Music Conference*, 320–324. Ljubljana, 2012.

Hindle, A. 2015. 'Orchestrating Your Cloud-Orchestra'. In *Proceedings of the Conference on New Interfaces for Musical Expression*, 121–125. Baton Rouge, LA, 2015. https://nime2015.lsu.edu/proceedings/244/0244-paper.pdf.

Hirsh, I. 'Auditory Perception of Temporal Order'. *Journal of the Acoustical Society of America* 31, no. 6 (1959): 759–767.

Karplus, K., and Strong, A. 'Digital Synthesis of Plucked-String and Drum Timbres'. *Computer Music Journal* 7, no. 2 (1983): 43–55.

Kim-Boyle, D. 'Network Musics: Play, Engagement and the Democratization of Performance'. *Contemporary Music Review* 28, no. 4 (2099): 363–375.

Kogawa, T. 'From Mini-FM to Polymorphous Radio'. *Subsol*. http://subsol.c3.hu/subsol_2/contributorso/kogawatext.html, accessed 30 June 2017.

Latta, C. 'Notes from the NetJam Project'. *Leonardo Music Journal* 1 (1991): 103–105.

McKinney, C. 'Quick Live Coding Collaboration in the Web Browser'. In *Proceedings of the Conference on New Interfaces for Musical Expression*, 379–382. Goldsmiths University of London, 2014. http://www.nime.org/proceedings/2014/nime2014_519.pdf.

McLean, A. 'Making Programming Languages to Dance to: Live Coding with Tidal'. In *Proceedings of the 2nd ACM SIGPLAN International Workshop on Functional Art, Music, Modelling and Design*, 63–70. New York: ACM, 2014.

Mills, R. 'Dislocated Sound: A Survey of Improvisation in Networked Audio'. In *Proceedings of the New Interfaces for Musical Expression Conference*, 186–191. Sydney, 2010.

Ninjam. http://www.ninjam.com. Accessed 30 June 2017.

Oda, R., Finkelstein, A., and Fiebrink, R. 'Towards Note-Level Prediction for Networked Music Performance'. In *Proceedings of the International Conference on New Interfaces for Musical Expression*, 94–97. Daejeon, 2013. http://www.cs.princeton.edu/~fiebrink/publications/OdaFinkelsteinFiebrink_NIME2013.pdf.

Ogborn, D. 'EspGrid: A Protocol for Participatory Electronic Ensemble Performance'. In *Audio Engineering Society Convention 133*. San Franscisco, CA: Audio Engineering Society, 2012. http://www.aes.org/e-lib/browse.cfm?elib=16625.

Ogborn, D. 'Live Coding in a Scalable, Participatory Laptop Orchestra'. *Computer Music Journal* 38, no. 1 (2014): 17–30.

Ogborn, D., and Mativetsky, S. 'Very Long Cat: Zero-Latency Network Music with Live Coding'. In *Proceedings of the First International Conference on Live Coding*, 159–162. Leeds: ICSRiM, University of Leeds, 2015. doi:10.5281/zenodo.19348.

Ogborn, D., Tsabary, E., Jarvis, I., Cárdenas, A., and McLean, A. 'Extramuros: Making Music in a Browser-Based, Language-Neutral Collaborative Live Coding Environment'. In *Proceedings of the First International Conference on Live Coding*, 163–169. Leeds: ICSRiM, University of Leeds, 2015. doi:10.5281/zenodo.19349.

Oliveros, P. 'From Telephone to High Speed Internet: A Brief History of My Tele-Musical Performances'. *Leonardo Music Journal (Online Supplement)* 19 (2009): 95.

Roberts, C., Yerkes, K., Bazo, D., Wright, M., and Kuchera-Morin, J. 'Sharing Time and Code in a Browser-Based Live Coding Environment'. In *Proceedings of the First International Conference on Live Coding*, 179–185. Leeds: ICSRiM, University of Leeds, 2015. doi:10.5281/zenodo.19351.

Rohrhuber, J., Campo, A. de, Wieser, R., van Kampen, J.-K., Ho, E., and Hannes, H. 'Purloined Letters and Distributed Persons'. In *Proceedings of the Music in a Global Village Conference*. Mücsarnok, Budapest, 2007. http://www.wertlos.org/articles/Purloined_letters.pdf.

Small, C. *Musicking: The Meanings of Performing and Listening*. Middletown, CT: Wesleyan University Press, 1998.

Sorensen, A. 'A Distributed Memory for Networked LiveCoding Performance'. In *Proceedings of the International Computer Music Conference*, 530–533. New York: 2010. http://eprints.qut.edu.au/55715/1/icmc2010.pdf.

Tanaka, A. 'Interaction, Experience, and the Future of Music'. *Computer Supported Cooperative Work* 35 (2006): 267–288.

Trueman, D., Cook, P., Smallwood, S., and Wang, G. 'PLOrk: The Princeton Laptop Orchestra, Year 1'. In *Proceedings of the International Computer Music Conference*, 443–450. New Orleans, 2006. http://www.scott-smallwood.com/pdf/plork_icmc2006.pdf.

Tsabary, E. 'Music Education through Innovation: The Concordia Laptop Orchestra as a Model for Transformational Education'. In *8th International Technology, Education and Development Conference*, 657–664. Valencia, 2014.

Walkingshaw, E., and Ostermann, K. 'Projectional Editing of Variational Software'. In *Proceedings of the 2014 International Conference on Generative Programming: Concepts and Experiences*, 29–38. New York: ACM, 2014. http://web.engr.oregonstate.edu/~walkiner/papers/gpce14-projectional-editing.pdf.

Wang, G., and Cook, P. R. 'ChucK: A Concurrent, on-the-Fly, Audio Programming Language'. In *Proceedings of the International Computer Music Conference*, 219–226. Singapore, 2003.

Ward, A., Rohrhuber, J., Olofsson, F., McLean, A., Griffiths, D., Collins, N., and Alexander, A. 'Live Algorithm Programming and a Temporary Organisation for Its Promotion'. In *Read_me, Software Art and Cultures*, edited by O. Goriunova and A. Shulgin, 243–261. Aarhus: Aarhus University Press, 2004.

Wilson, S., Lorway, N., Coull, R., Vasilakos, K., and Moyers, T. 'Free as in BEER: Some Explorations into Structured Improvisation Using Networked Live-Coding Systems'. *Computer Music Journal* 38, no. 1 (2014): 54–64. doi:10.1162/COMJ_a_00229.

Wright, M., Freed, A., and Momeni, A. 'OpenSound Control: State of the Art 2003'. In *Proceedings of the Conference on New Interfaces for Musical Expression*, 153–159. Montreal, 2003. http://www.music.mcgill.ca/musictech/nime/onlineproceedings/Papers/NIME03_Wright.pdf.

Alan from BEAM. May 2022

said

Touch is A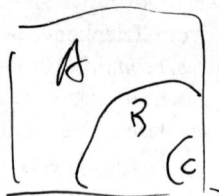

↓

seeing is B

↓

listen   is C

The encounter of two
organisms create sense

See Sight and hearing
came after

↓

text then. speach
then text

CHAPTER 21

SONIFICATION ≠ MUSIC

CARLA SCALETTI

sematic

21.1 CAN SOUND CONVEY MEANING?

THERE is a widely held misconception that words and only words are capable of conveying meaning. When Steven Pinker writes, 'Even a plot as simple as "boy meets girl, boy loses girl" cannot be narrated by a sequence of tones' (Pinker 2009), he seems to imply that the only meaningful information is that which can be conveyed by symbolic language and that if nonspeech audio can't even convey something as 'simple' as 'boy meets girl, boy loses girl', then it must be because sound is an inferior or impoverished language.

Philosopher Mark Johnson addresses this misconception in *The Meaning of the Body: Aesthetics of Human Understanding* (Johnson 2007), where he argues that symbolic language is not the sole channel for human communication and that to restrict ourselves to symbolic language alone is to deny ourselves the full range of human expression, thought, and communication. If your definition of creating meaning is limited to making logical assertions using propositional calculus, then yes, you would be forced to conclude that nonspeech audio is an ineffective way to convey meaning, but in *The Meaning of the Body*, Johnson reminds us of a myriad of other ways that humans create meaning—ways that include spoken and written language but which extend beyond symbolic representation.

Beyond symbolic representation

21.1.1 Conceptual Metaphor, Index, and Morphism

All transmission of meaning relies to some extent on shared experience; the sender and receiver may share a common history and culture or a common physical environment. Are there some aspects of experience that are, if not universal, at least widespread? We and other terrestrial life forms, for example, share the experience of gravity; we can strike things and things can strike us (with varying degrees of force); we can move through three-dimensional space towards food or a mate and away from predators and

3D Space
Move

parallel

pattern is too summative to precise

noxious substances. In their theory of the embodied mind (Lakoff and Johnson 1999), Mark Johnson and cognitive linguist George Lakoff posit that, at a very young, even pre-linguistic, age, we begin to generalize these common experiences, that we start to notice certain repeating dynamic patterns of our interactions with the environment. Later in life, long after these patterns have become well established, we think, understand, and communicate even the most abstract concepts by way of analogy to these basic kinaes-thetic patterns or schemata. Lakoff and Johnson's schemata are not static visual patterns; they usually involve time, space, and muscle sensation. They can be difficult to describe in words alone but easy to communicate with a physical gesture; for example, it can be complicated to define the meaning of the word 'up', but it's easily described with a ges-ture. Johnson and Lakoff call an analogy to one of these basic patterns a *conceptual meta-phor*, which they define as a cross-domain, inference-preserving mapping.

Imagine using a pantograph: as you trace a drawing with the stylus, the panto-graph produces a smaller copy of your outline using a pen coupled to your stylus. Since the pantograph is constructed as two congruent triangles—one large and a smaller one inside of it—a pantograph preserves the shape of an outline while it scales the size and offsets its location. (For a schematic animation, see the Wikipedia entry on 'Pantograph': Wikipedia 2015.) A pantograph is a physical example of what the nineteenth- and early twentieth-century American logician Charles Sanders Peirce would call an *index*; in his semiotic theory of representation, an index links an aspect of one object to an aspect of another object (the signifying element).

Taking his cue from mathematics and category theory, computer scientist Joseph Goguen defines a semiotic *morphism* (a mapping from one sign system to another) as the process of design, and its inverse (inferring properties of the source from properties of the target) as the process of understanding (Goguen 2004).

In Douglas Hofstadter's (2001) view, there is never a case where we are not using metaphors to understand, reason about, and communicate abstract ideas. He maintains that analogy is the core of cognition and that meaning is a morphism.

A data sonification, like a conceptual metaphor, index, or morphism, is a cross-domain, inference-preserving mapping. Data sonification provides a means for under-standing, reasoning about, and communicating meaning that extends beyond that which can be conveyed by symbolic language alone (whether spoken or written).

21.2 DATA SONIFICATION

When geologist Chris Hayward took a series of ground-displacement measurements made during an earthquake and played them back at ten times the original speed in order to listen to them as audio signal, he made a remarkable observation: sounds trans-mitted through the air (acoustic waves) have similar physics to seismic vibrations trans-mitted through the earth (elastic waves). When you listen to these sped-up geophone recordings, your auditory system immediately identifies them as acoustic environments.

And just as we can identify the characteristics of a space, like its size, reflective surfaces, shapes, and materials, based on the acoustic response to an impulse, it is also possible to infer something about the material properties and structure of the rock surrounding a geophone by listening to the way it responds to the impulse of an earthquake or an explosion (Hayward 1994).

In listening to the seismic displacement data, Hayward was creating a cross-domain, inference-preserving mapping: cross-domain because it took movements of the earth and mapped them to changes in air pressure, and inference-preserving because it is possible to draw some conclusions about the earth surrounding the geophone by listening to that sound.

The premise of data sonification (and data visualization) is that inferences we make in the target domain also hold true in the source domain—in other words, that it's possible to map not just points but also the relationships among the points, from a source domain to a target domain.

21.2.1 Definitions

Data sonification has three necessary and sufficient components: a process, a goal, and a loop-back path for interactive iteration and refinement:

- **Process**: Sonification is taking data generated by a model, captured in an experiment, or otherwise gathered through observation, and mapping those data to one or more parameters of an audio signal or sound synthesis model.
- **Goal**: The goal of sonification is to better understand, communicate, or reason about the original model, experiment, or system.
- **Loop-back path**: Typically, any new understanding obtained through sonification generates new questions, which can loop back to the beginning of the process again, this time with a change or refinement to the mapping or perhaps even with a new set of data, a new set of experiments, or modifications to the mathematical model that generated the data.

21.2.1.1 *Exploration to Presentation*

Data sonification can lie anywhere along a continuum from exploration and discovery of previously unknown patterns to presentation of known information; it can range from highly interactive with a tight feedback loop to passive listening, and it can be carried out privately by a single individual, by a small research group, or in the public sphere. A sonification can exist at any point in this continuum of passive–interactive, public–private, known–unknown space, and often, a sonification moves within this space, for example by starting out life in a tight loop of interaction, exploration, and discovery in a private setting and evolving into a public presentation of hard-won discoveries to a larger audience (see Figure 21.1).

the storage device and use a digital-to-analogue converter (DAC) to convert it back to a continuously varying voltage that moves a speaker cone back and forth, producing air pressure variations: sound.

When you perform arithmetic on the stream of numbers, it's called digital signal processing. Things really start to get interesting when you dispense with the microphone and the ADC and generate a stream of numbers using an algorithm, a process known as digital sound synthesis. While the synthesized numbers are streaming to the output, you can modify the parameters of the digital synthesis model in realtime using yet another data stream, this one generated by a controller interface such as a digital keyboard or a game controller, to create a data-driven instrument.

One approach to data sonification is to substitute a stream of data from an experiment or mathematical model for one or more of the digital audio or controller data streams. Instead of using a microphone, for example, you could use a geophone (a device that measures seismic vibrations). In place of a digital recording saved on a storage device, you could substitute a file of time series data. Instead of the game controller or keyboard, you could substitute a stream of numbers collected from an experiment or generated by a mathematical model, using those values to control the parameters of a digital synthesis algorithm.

21.2.1.3 *Mapping*

Listening to a stream of data directly as an audio signal is known as a oth-order mapping. Using a stream of data to modulate a parameter or parameters of a synthesis model is called a first-order mapping. In higher-order mappings, one can use the data to determine the structure of the synthesis model or to modulate subaudio control signals that control the parameters of the audible signal (Scaletti 1994).

oth order: Listening to data as an audio signal For example, if you wanted to study the tides, you might measure the water level at a specific location once per hour over the course of eleven years; you would end up with a time series that has some characteristics in common with a digital audio recording. A digital audio recording is a time series of air pressure measurements made at a single point by a microphone, sampled at evenly spaced time intervals by an analogue-to-digital converter; similarly, your tide data would be a time series of water levels, measured at evenly spaced time intervals at a single location using a meter stick. If you were to play back those eleven years of data as a sound file at a 48 kHz sampling rate, the entire data set would go by in approximately two seconds. Seasonal or yearly cycles would be perceivable as 2–5 hz amplitude modulation, and the twenty-four- and twelve-hour cycles would be perceived as 2000 and 4000 hz tones.

The most common problems encountered when trying to listen to data directly as an audio signal are that there are not enough data points, that the measurements were not taken at regular time intervals, or that the resolution of the measurement instrument wasn't fine enough. When 48,000 data points generate one second of sound, either you have to gather data over a long period of time or you have to sample the data at shorter time intervals in order to end up with more data points per unit of time. If the original phenomenon was not sampled at regular time intervals, the data have to be resampled before you can play them back as an audio signal (and resampling can introduce arte-facts that were not present in the original experiment). We've become accustomed to listening to audio recordings made with measurements accurate to at least 24 bits, so unless your measurements are accurate to within about one part per eight million, you may start to notice degradations in the audio signal due to quantization noise.

Despite all these obstacles, when it works, listening to time series data as an audio signal can be extraordinarily intuitive and revealing. Cycles, reflections, impulse responses, and minute deviations in expected patterns are all things the auditory system has evolved to detect with exquisite accuracy and refinement, and we can perform these signal detection feats with little or no special training.

First order: Using the data to modulate or set the parameter of an audio synthesis model What if, instead of playing back the eleven years of tide data at 48 kHz, you were instead to play them back at forty-eight data points per second? At that rate, each of the cycles would be a thousand times longer: the twenty-four- and twelve-hour cycles would occur at two and four times per second, and seasonal changes would take three to eight minutes to unfold. You might conclude that these cycles would be below the range of human hearing and you would be correct; however, although you cannot hear a four-cycle oscillation as a frequency, you can easily track its variations when it is used to modulate the parameter of a synthesis model.

For example, if you were to take the data stream and use it to control the cutoff fre-quency of a low-pass filter with white noise as its input, you could follow the changes in tide levels as changes in the bandwidth of the noise; it's another form of cross-domain, inference-preserving mapping.

21.2.1.3.1 *Points in a Parameter Space: Each Vector Creates a New Event*

A slightly different approach is to map individual points in the data space to events at time points in sound space. For example, the data sets in Lily Asquith's LHCSound proj-ect are collections of vectors, where each vector represents measured and computed characteristics of a single collision in the Large Hadron Collider at CERN (Asquith and Scaletti 2015). One approach to sonifying these data is to create a single sound event for each of the collision events in the data set. If each collision has been characterized by five values, for example, you could design a sound generation algorithm with five param-eters and map each collision variable to a particular sound synthesis parameter. In other words, instead of imagining a single sound-generating object whose parameters are changing over time under the control of data streams, you can instead imagine that each line of data, each multidimensional data point, corresponds to a single sonic event with its own start time, duration, and other synthesis parameters.

21.2.1.3.2 *Emergent Mapping*

In some situations it can be enlightening to explore more complex mappings; instead of mapping one data parameter to one synthesis parameter, it might be more reveal-ing to map a combination of data parameters to a single sound parameter (or linear combination of sound parameters). Exploring all possible combinations would quickly become overwhelming, but Ludovic Laffineur, Damien Grobet, and Rudi Giot at the Sonification Lab at the Research Laboratory in the field of Arts and Sciences (LARAS) are experimenting with using a matrix of weights, initially seeded with random numbers, as a quick way to interactively test random combinations of input and output parameter mappings, iteratively refining them to arrive at perceptually salient combinations of source variables and target parameters (Grobet Laffineur, and Giot 2014). In this evolutionary approach, the most effective mapping can emerge rather than having to be explicitly specified. In effect, the LARAS technique is taking the feedback-with-refinement path intrinsic to all data sonification and merging it into the mapping stage of the process.

21.2.1.3.3 *Mapping to a Process or a Structure*

Whether a model is external to the synthesis algorithm or becomes part of the synthesis algorithm, the fundamental process of sonification remains that of mapping values from some domain into the domain of sound. As Stephen Barrass and Paul Vickers put it, 'Any time something is represented in a form external to itself, a mapping takes place; an object from a source domain is mapped to a corresponding object in the co-domain (or target domain). Sometimes the mappings are very obvious and transparent, as in parameter-mapped sonifications, but even model-based sonification involves mappings in this general sense as there are still transformation rules that determine how the data set and the interactions combine to produce sound which represents some state of the system. The mappings may not be simple, but mapping is still taking place' (Barrass and Vickers 2011, 153).

For example, imagine modelling a chemical process using differential equations and building an analogue circuit based on those same differential equations (or emulating such a circuit in software). You could connect the output of the circuit to some speakers and listen to sound (oth-order mapping) or you could use the output of the circuit to control the parameters of another sound-generating circuit (first- or higher-order mapping). Emulating a process in sound-generating software or analogue circuitry has all the same characteristics as mapping data from an unknown or external source (whether that source is a mathematical model or a collection of measurements or observations).

As another example, imagine using a data set to construct a gong, using the data to determine its shape, size, and materials. You could then explore the characteristics of the data by tapping on the gong, throwing handfuls of pellets at the gong, bowing the gong at different positions, or setting off a small explosive device next to the gong and listening to how the gong (or a software model of the gong) responds to the explosion (or a software model of an impulsive event) (Hermann 2011). In this example, the data are used to specify the construction of an object (or a computer model of that object) and characteristics of the data may be inferred by listening to the way the finished object responds to impulsive events. In order to listen to the object's response, you could either map some dynamic aspect(s) of the object directly to an audio signal (oth-order mapping) or indirectly to a control signal on the parameter(s) of an independently synthesized audio signal (first- or higher-order mapping). Again, the data are mapped from one domain to a target domain that produces sound influenced in a systematic way by those data.

21.2.1.3.4 *Preprocessing Data*

There is nothing that says you must use the data set in its original form or that you cannot also visualize or otherwise analyse the data in preparation for selecting an informative sonification mapping. If you are sonifying a large data set with the goal of categorizing what the data points represent, for example, it might make sense to first try some cluster analysis techniques on a known set of test data to determine whether you can glean some hints as to which variables or groups of variables are the most relevant to the categorization task, so you can choose to map those to the most psychoacoustically salient parameters when you sonify the actual, unknown data set. All's fair in love, war, and the pursuit of knowledge, and sonification is just one tool among the many that can be used in pursuit of the ultimate prize—new and deeper understanding of the original phenomenon.

21.2.2   A Matter of Time and Space

Episodic memory, the kind of memory that integrates 'what', 'where', and 'when' in the proper temporal sequence, is conserved over diverse species and is thought to convey evolutionary advantages in the form of memory-based prediction, long-term planning, and the maintenance of social relationships and networks (Allen and Fortin 2013). Hence, both the temporal ordering and spatial positioning of sound events are likely to be interpreted by human (or indeed by other mammalian or avian) listeners as meaningful parameters.

Up to this point, we've been tacitly assuming that the data are a time series—a sequence of successive measurements made at equal time intervals. That's a natural assumption given that time series have so much in common with audio signals and control signals; they are all one-dimensional functions of time, so there's a clear and direct mapping from streams of data points to streams of digital audio samples or control values. But not all data sets are time series. A more general description of a data set would be: a matrix where each column represents a variable and each row represents a member of the data set (or a point in the multidimensional space defined by the variables).

One approach to sonifying datasets that have no time dependency is to simply present each member of the set, one at a time, as a single sound event whose parameters are related to the values of the variables in that row. However, presenting the data points in an arbitrary time order can, at best, waste a potential cue for conveying information and, at worst, lead to misjudgements based on an ordering that has no correspondence with the underlying model (Kusev et al. 2011).

21.2.2.1 *Random Access, Spatial Interaction*

An alternative approach to sonifying non-time-series data is to create a means for the 'data explorer' to have random access to the sound points, triggering events interactively by moving through a real or virtual space. The listener explores the data space by exploring the set of sound points based on those data points. For example, imagine a data set with variables x and y, with x mapped to frequency and y mapped to attack and release time; further, imagine that each sound event is associated with a position on a touch screen such that, if you touch that x–y position, you trigger the associated sound event. Instead of associating a sound event with its ordinal position in a temporal sequence, we are associating it with a spatial position. If the spatial position of a sound point trigger in a virtual space is based on some or all of the variable values, listeners can combine their spatial reasoning and memory abilities with their sonic reasoning and memory to create an even stronger sense of embodied engagement with the data space as a physical space. This kind of spatial data mapping is employed by Stuart Smith and Haim Lefkowitz in the Exvis project (Erbacher et al. 1995) and, for higher-dimensional spaces, in Thomas Hermann's (2011) model-based sonifications.

One can imagine taking this concept several steps further, using a headset to create a visual, auditory, and tactile feedback environment of three (or more) dimensions that one could explore by physically walking around in an augmented reality space and interacting with the data points as physical objects.

21.2.2.2 *Mapping a Variable to Time*

The sequence, the arrangement of events in time, is a perceptually powerful cue; by mapping different variables to start time, you can get vastly different perspectives on the same data points. There is a tendency for humans to hear, recall, and describe a sonification as a kind of narrative. For example, if after listening someone observes, 'It started out very sparse with high pitches and short durations and about halfway through, it got very dense and the pitches suddenly dropped,' then you can go back to the data (or back to the experiment) to determine what the midpoint of the time variable was, why there are more events that have values around that midpoint, and why the related variable that was mapped to frequency has smaller values associated with that particular time variable. By mapping one of the variables to start time, you make it easier to hear correlations between the start time variable and each of the other variables.

21.2.3 Sonification and Data-Driven Visualization

Pairing data sonification with data visualization is a synergistic tool for augmenting both visualization and sonification, and can highlight connections or correlations between variables. The best multimodal data mappings use the same source data set to drive both the images and the sounds (not to mention tactile stimulation plus any other data 'sensification' being employed).

The use of data visualization is so commonplace that one often hears a graph or other visualization referred to simply as 'the data' rather than as a mapping of the data to a 2D image or animation. In most cases, it's a conscious linguistic shortcut, but sometimes it indicates an underlying misconception. Sonification is not a mapping from a visualization to sound; it's a mapping of the original, source data to sound. In other words, a sonification is not a map of a map; it's a map of the territory. It's possible to start from the same data source and map aspects of the data that are best conveyed by sound (like timing or sequence) to the sonification while mapping the kinds of variables that are best conveyed through images (like spatial location) to the visualization. For complex data sets, sound can serve as an additional channel of information, a way of widening the data pipe so you can observe more variables in parallel; while the eyes follow a complex visualization, the ears are still open and available for detecting changes or for monitoring parallel data streams.

21.2.4 Reusable Tools

Each question, each experiment, each data set, suggests the design of a particular mapping or mappings. However, over time and over multiple data sets, certain patterns begin to emerge—patterns that are generalizable over a variety of data sets, no matter the original source of the data. Testing for a specific condition, comparing two variables, looking for correlations among variables, are just a few examples of objectives that come up repeatedly in data sonification, irrespective of the actual source of the data. The more recurring patterns we can start to recognize, the more reusable sonification tools we can add to the arsenal, for example:

Comparator: To compare two or more data streams, map them in the same way, but route them to independent audio output channels; all the data streams can be heard simultaneously, with the same parameter mapping, but separated in space. A level control on each channel can bring different combinations of data streams to the foreground.

Marker: To mark the exact point in time when a specific condition is met, try triggering a broadband, short-duration click; a marker can pinpoint conditions like changes in direction or sign, threshold crossings, specific value matches, equality or inequality of variables, and so on.

Sonic histogram: When faced with a data set whose members can be separated into several categories, a sonic histogram can enable you to monitor the changing size of the categories with respect to the variable you have mapped to time. For example, each category can be assigned a sound generator with a unique frequency or identifying timbre and spatial location, and the changing numerosity or magnitude of that category could be mapped to the amplitude of that sound generator. Sonic histograms give an overview or gestalt sense of how the categories are changing, synchronously or independently, over the range of the variable you have mapped to time.

Sonic underlining: This is related to the marker, but instead of triggering a marker event, a sonic underline emphasizes a subset of events by changing one or more psychoacoustically relevant parameters in response to certain conditions in the data. For example, you could scale the duration of any events for which a particular variable is above a threshold value; by making those events last longer than surrounding events, they will tend to stand out against the background, and all of their parameters—frequency, amplitude, spatial position, timbre—can be more easily identified because the listener has a longer interval of time in which to hear and process the event.

Sonic scatter plot: If you're using a sound synthesis environment such as Kyma (http://symbolicsound.com) that can resynthesize an audio signal from spectral data, you can map triples of data variables to time, frequency, and amplitude and save the result as a spectrum file; this can be more computationally efficient than treating the triples as the parameters of individual sound events, especially when you are dealing with large data sets. Spectral resynthesis of mapped data is an effective way to sonically plot three variables against each other for the purpose of hearing correlations among the variables.

21.2.5 Maps are also Mappings

Thus far, we've been using the words map and mapping in the mathematical sense, to mean morphism or function. Is there any relation to what we would call a 'map' in the sense of cartography? Traditionally, we've always thought of maps as representational—as immutable and totally faithful reference sources, like dictionaries. In fact, if you look in a dictionary for the definition of 'map', you find: 'a representation, usually on a flat surface, as of the features of an area of the earth or a portion of the heavens, showing them in their respective forms, sizes, and relationships according to some convention of representation' (Dictionary 2015).

The proliferation of personal GPS navigation devices and on-demand creation of maps on the web is changing our concept of a map from that of a fixed image on a piece of paper into a tool for interactive, exploratory data display. Bruno Latour and his co-authors from the École Polytechnique Fédérale (EPFL) in Lausanne state it explicitly: 'Maps are interfaces to datasets' (November, Camacho-Hübner, and Latour 2010). A map is an interface that allows us to navigate through heterogeneous datasets that are continually refreshed, localized, and refined by our queries, just as data sonification is an interface to data sets that are similarly refreshed and refined by our questions and by the further sets of questions spawned by each answer.

In this new way of thinking about maps, a map is not like a painting; it doesn't resemble the appearance of the world, and it isn't a two-dimensional representation of the world. Neither is data sonification a piece of music, nor does it resemble a natural acoustic soundscape, nor does it seek to mimic natural sounds that may be associated with the model world or system under observation.

What do you want from a map? When you consult a map, you're not looking for a faithful resemblance to the way the world looks. What you want from a map is a way to get from here to there; you're seeking the relevant information on any and all conditions or structural relationships that might impede or facilitate your progress.

Similarly, what do you want from a data sonification? You aren't looking for musical entertainment or a simulacrum of the acoustic ecology of the model world. You are looking for relevant relationships, connections, and patterns in the data (which, in a well-designed, inference-preserving mapping, are also present in the original source of the data).

Geographers John Krygier and Denis Wood write that a map is a proposition (Krygier and Wood 2015). Each map (and I would argue, each sonification) conveys a particular message. Each sonification influences you to think about the underlying model world in a certain way. Sonification design, like map making, is both a way of thinking about the underlying model world and a set of assertions about that world. Good design is thus essential, not just for aesthetics, but also because poor design can obscure or distort the structure and meaning inherent in the original data set.

21.2.6 Sound as an Interface to Data Sets

Like a map, a data sonification is an interface—a means for exploring points, locations, relationships, and connections in the abstract space defined by a data set.

Guidelines for good interface design (and good sonification designs)   Joseph Goguen (1999) defines an algebra of user-interface design in which interfaces, representations, and metaphors are all 'morphisms from one sign system to another'. The key word here is *systems* of signs. A sign system is more than just set of objects; it also includes axioms that define how those objects relate to one another. For Goguen, an interface is a morphism, and the quality of that interface is the degree to which it is a structure-preserving morphism. Preserving the structure, preserving the relationships among the points, is what makes it possible to draw inferences in one sign system that also hold true in the other. In other words, Goguen's structure-preserving morphism is an inference-preserving morphism.

The same source sign system can be mapped in multiple ways, and it's always a partial mapping; not every sign in the source can be represented in the target, which leads to three of Goguen's guidelines for good interface design (and, by extension, good data sonification design):

- The most important signs in the model should map to correspondingly important signs in the interface

Sound parameters are not all equally weighted in terms of the way they are perceived. For example, the human auditory system is more sensitive to minute differences in frequency than it is to small differences in pan position. Therefore, when designing a data mapping, the choice of mapping the most important data variable to pitch, rather than to pan position, serves to preserve more of the original structure in that it makes any changes in that variable more perceptually noticeable.

- If something must be sacrificed, it is better to preserve form than content (the F/C law)

It's more important for a data sonification to give you opportunities to hear connections and establish relationships among elements than it is for the sonification to sound like an imitation of objects or sound sources in the model world.

- The most important axioms should be satisfied in the target

A well-designed data sonification is a map or a morphism that preserves the most important elements, connections, and relations such that sequences of actions and chains of reasoning in the target domain also make sense in the source (and vice versa). For example, if the value of data point y depends on the value of data point x, there should also be a spatial or temporal connection between sound event y and sound event x.

In essence, Goguen is saying that each interface (and I would add, each sonification) is a theory about the underlying model or source of the data and that, since you can't preserve everything, a good design is one that preserves the structure and allows you to reason about the underlying source. There are multiple ways to map the same data to sound; what you choose to map and how you choose to map it have a huge impact on what a sonification reveals and what it obscures in the data.

To illustrate how the choice of conceptual metaphor morphism can facilitate some forms of reasoning while discouraging others, consider the abstract concept of number (Lakoff and Núñez 2000).

- If you map the idea of number to the quantity of objects in a container (as you might have been taught to do as a young child), it's easy to reason about addition and subtraction as putting objects into the container and taking objects out of the container. The number as object metaphor helps you reason about the natural numbers (whole numbers), but it doesn't suggest the existence of or a way to reason about negative numbers.
- Using the number-as-length metaphor, you can reason about addition as laying the 'number strings' end to end and reason about subtraction as cutting the string. The number-as-length metaphor allows for irrational numbers (if you arrange strings as sides of a right triangle or around in a circle). But it, too, fails to suggest or facilitate reasoning about negative numbers.
- The number as movement in a direction along a 'number line' lets you reason about addition and subtraction as moving right or left along a number line. It allows for reasoning about zero as the absence of movement and negative numbers as movement to the left. But it doesn't immediately suggest the possibility of complex numbers, and so on.

21.3 Sonification Is Not Music
(but Music is Sonification)

Throughout history, composers have been known to embed extramusical structures and relationships into their music. But did they do so with the express purpose of understanding and reasoning about the original source of the data? In 1436, Guillaume Dufay composed a motet for the dedication of the cathedral of Santa Maria del Fiore (Il Duomo) in Florence. Several musicologists have documented correspondences between temporal proportions in the motet and architectural proportions found in Il Duomo or in Solomon's Temple (Warren, 1973; Wright, 1994; Trachtenberg, 2001).

Did Dufay echo these architectural proportions in his music in order to improve his understanding of static forces? To assist architects in predicting the structural integrity of their buildings? We have no way of knowing exactly what was in Dufay's mind, but my guess is that he did it to celebrate the beauty and balance of the architecture or possibly to create a metaphorical link between 'Il Duomo' and the Temple of Solomon. That moves beyond the realm of sonification and enters into the domain of data-driven music.

Much as composers in the twentieth century looked to stochastic processes as new sources of complexity and serendipitous discovery in their music, composers today look to data—the outputs of sensors, network monitoring tools, financial analysis, scientific experiments, mathematical models, statistics, online communication, and other data— as new sources of complexity, structure, and dynamic pattern formation; composers seek, in data, new sources of inspiration for the continual evolution and invention of new forms of sound art. Given that we are inundated by notions of data, Big Data, data streams, data mining, data analytics, it seems only natural that these data should find their way into new forms of musical expression.

21.3.1 Important Distinctions between Data-Driven Music and Data Sonification

Despite the fact that sonification and music share much of the same technology and techniques, there are some important distinctions between data sonification and data-driven music, and it is in the interests of both music and sonification to maintain a clear delineation between the two activities.

21.3.1.1 *Teams and the Auteur*

Data sonification tends to work best when it's a team effort. At this point in history, it is rare for one individual to have the level of expertise required to do original scientific research and to be an expert in mapping data to sound parameters and in interpreting the sonic result. There are exceptions of course—people making contributions in both science and in sound synthesis—but even for those individuals, it can be more effective to work as part of an interdisciplinary team.

Electronic musicians are experienced in mapping numbers to sound parameters and being able to hear and recall the way sound parameters change over time in great detail. But listening to a sonification usually raises as many questions as it answers. The musician may hear a sonic relationship, but is the relationship in the original source domain profoundly meaningful or is it trivially obvious to the researchers familiar with the experiment? The dream scenario is when a musician can apply years of experience and ear training to hear something that the other researchers hadn't noticed before, and thus open a new area for further investigation.

Perhaps it is an antiquated or romantic notion, but we tend to think of music composition as a solitary pursuit rather than a team effort. Even on the rare occasions when composers collaborate, there is usually one person who takes the final responsibility for the work—like the director of a film or the author of a book. Whereas data sonification seems to work best as a multidisciplinary collaboration.

21.3.1.2 *Cascades and Context*

As sociologist of science Bruno Latour observes with respect to scientific visualization, a visualization is meaningless in isolation; it is always presented as part of a cascade of inscriptions (November, Camacho-Hübner, and Latour 2010). The same is true of a scientific data sonification: a sonification is meaningless when pulled out of its context, when it is not presented as part of a cascade of text, equations, tables, graphs, captions, legends, and citation of previous work. Each inscription in the cascade is one step of an argument, and it is only the sequence of steps in its entirety that constitutes a proof; removing any one element from the cascade is breaking the chain of reasoning that leads to the scientific conclusion. As Joseph Goguen (1999) wrote, 'A signal that is meaningful in one sign system may not be in another, even though they share a medium'. Just because data sonification and data-driven music share the medium of audio signals, it doesn't mean that a data sonification retains its meaning once it has been pulled out of its context and separated from its supporting scientific inscriptions.

By way of contrast, an artistic work, rightly or wrongly, is typically presented as self-contained; its meaning is in its form. Often the implication is that the work is radically novel and has no precedent; even though artists, too, stand on the shoulders of giants, it is rare for a composer to acknowledge prior work and influences. Some artists even argue against the use of written program notes, insisting that the music must stand entirely on its own, rather than in the context of a cascade of written text and references to the work of others.

21.3.1.3 *Distinct Goals*

Perhaps the most important distinction between sonification and music is the difference in intent. The goal and purpose of data sonification is to aid in understanding, exploring, interpreting, communicating, and reasoning about a phenomenon, an experiment, or a model, whereas in sound art, the goal is to make an audience think by creating a flow of experience for them—sometimes fostering an experience of ecstasy, in the literal sense of being outside one's self.

21.3.2 Is Music a Sonification?

If you accept that it is possible to map changes in a meaningful data variable to changes in a sound parameter and that music is a structured sequence of changes to sound parameters, then is it possible that music is also a sonification? And that the changes in musical parameter values are indexed to changes in something else? And if so, ... what?

Neuroscientist Antonio Damasio defines thought as 'a continuous flow of images many of which turn out to be logically interrelated. The flow moves forward in time, speedily or slowly, orderly or jumpily, and on occasion it moves along not just one sequence but several. Sometimes the sequences are concurrent, sometimes convergent and divergent, sometimes they are superposed' (1999). Damasio's definition of image is not just visual, nor is it static; it's a structured dynamic pattern that includes all the senses—sight, sound, touch, taste, smell—along with a sense of internal state. Damasio's image is not a representation of experience; it is the experience itself.

If you substitute the word 'sound' for Damasio's word 'image', his description of thought becomes a description of music. It seems that the way we experience music is very closely related to the way we experience thought: we experience it directly, without translation into and out of symbols. Music is a sound index, a morphism, a cross-domain, inference-preserving mapping from thought to sound, without the mediation of symbolic language. As such, music is one of the most profoundly meaningful of all human expressions in that it is directly indexed to the flow of experience, to thought, to what it feels like to be a living mind-body interacting with the physical, chemical, and social environment. When we create music, we're creating a sonification of what it's like to be inside our heads, to feel time passing, to move through space, to be alive. And when we listen to music, it's like mind-melding with the person who created that music.

21.3.3 Maintaining the Distinction between Data Sonification and Data-Driven Music

A computer musician is one of the few individuals who has had the benefit of training and experience in designing sounds, in choosing and transforming parameter mappings, and, perhaps most importantly, in listening analytically to abstract sound structures and detecting subtle patterns (patterns that may have gone undetected by others who have not benefitted from the same training and experience). Thus a computer musician has both the skills and the opportunity to pursue both music composition and data sonification. By keeping those two activities separate, by defining them differently, a musician has an opportunity to pursue each of them in different contexts, with knowledge and experience gained from each pursuit informing and enhancing the other.

21.3.3.1 *How It Benefits Sonification*

There is a fairly widespread, though mistaken, expectation on the part of the general public that a sonification should sound like that individual's favorite genre of music-as-entertainment. By defining and maintaining a separation between music and sonification, nonspecialists can begin to listen analytically to abstract sound structures. If nonmusicians expect a sonification to convey meaningful information about the phenomenon under study, then they are more likely to listen carefully, with alertness and without entering the 'trance' state that many people automatically enter when listening to their favourite music as a drug.

Saying that sonification is not music is by no means an argument for perfunctory sound design or low audio quality. Good design includes finding the morphisms that reveal important structure without confusing, obscuring, or annoying the listeners to the extent that they stop paying attention.

Once people hear something meaningful in a data sonification, once they recognize sonification as a tool, as an adjunct to the myriad of other tools they use, sonification will no longer be thought of as an amusing curiosity; it will become a serious tool, widely expected as part of the cascade of scientific inscriptions and taught to kindergarteners, along with visual graphs.

21.3.3.2 *How It Benefits Composers*

By naming it data-driven music, rather than sonification, a composer is freed from the expectation that data-driven music must be somehow instructive or illustrative; it avoids the implication that music is the handmaiden of the great and almighty god of Science. The purpose of sonification is first and foremost to assist the listener in interpreting, understanding, and reasoning about the source of the data. Whereas the purpose of music is to create a flow of experience for an audience and to express a profoundly meaningful sense of what it is like to exist in this temporal, chemical, physical, cultural world in ways that cannot be adequately captured by language alone.

In an ideal scenario, the sound of a sonification fades into the background leaving only the structure of the underlying model; whereas in music the sound and the structure merge and it is the underlying stream of felt experience (thought) that becomes sole focus of attention.

Perhaps the strongest argument for maintaining a clear separation between sonification and music is that, ultimately, sonification must become more normalized and conventionalized in order to become an accepted part of the cascade of scientific inscriptions. Whereas the role of music is to run counter to expectation, to be ineffable, transcendent, disruptive, surprising, and to stand as a radical challenge to the status quo.

21.3.3.3 *Suggestions on How to Create and Maintain a Separation*

Given that it benefits both sonification and music to maintain a clear separation between the two, what are some concrete steps that could serve to emphasize that separation?

Terminology: One of the most potent ways to distinguish two activities is to give them different names. If we use 'sonification' to describe the three-part activity that is part of a cascade of scientific inscriptions, then we can use 'data-driven music', 'data-based sound art', or simply 'music' to describe the use of data as a component of artistic expression.

Sonic source material: To avoid misconstrual as natural ambiences or music, data sonification practitioners might choose to consciously avoid the use of overtly imitative sounds or musical references.

Dumbing down doesn't help anyone: Although most scientific researchers are musically literate in the sense of having performed historical or popular music on an acoustic instrument, far fewer of them have had training in experimental electronic music composition and sound design. It's unusual for anyone to have been exposed to the idea that abstract sound structures like music can be meaningful in addition to being entertaining (not unexpected when the educational curriculum stresses mathematics and language, and views musical training as optional and extracurricular). Nevertheless, it is counterproductive to attempt to placate scientific collaborators by trying to make a data sonification sound more like music. For example, while sacrificing the exquisite resolution of the pitch dimension by quantizing it to a pentatonic scale may put a smile on your collaborators' faces, it may also serve as a signal to them that it's time to go into a music trance rather than to listen carefully and analytically. Sonifications should be designed so as to encourage listening to the data themselves, rather than to the sounds, which are after all, only an interface to the data.

Concerts on conferences: The inclusion of what are labelled 'concerts' on data sonification conferences only serves to confound the expectations of researchers, further muddying the distinction between data-driven music and data sonification. Instead, why not transform these into 'listening sessions' where practitioners can showcase particularly successful examples of data sonification that the audience can evaluate and discuss together? Listening sessions could be used as an opportunity to experiment with analytic sonification techniques, polling the audience to evaluate the results, or to engage the audience in interactive data sonifications followed by discussion.

Clarifying these distinctions does no disservice to composers; instead it stands to benefit them. Computer musicians are uniquely positioned to become part of a data sonification team: their ability to listen analytically, to pick out subtle dynamic patterns, to imagine how a change to the mapping might clarify an obscure structure turns out to be rather rare among the general population. So computer musicians have an opportunity to contribute to interdisciplinary research groups by assisting them with the technical aspects of data sonification, while still pursuing an independent career of creating and performing their own music (including music that is inspired and structured by external data sets!)

21.4 DATA SONIFICATION AND THE FUTURE

It seems that sonification has been the hot new trend for last fifty years, probably even longer. Steven Frysinger (2005) cites published references to sonification research dating from as early as 1954, and one could argue that some of Pythagoras's work with a string stretched around a triangle were sonification experiments. Clearly it's an idea that fascinates people, so what are some of the obstacles that have prevented sonification from having been universally adopted as the natural adjunct to visualization and other tools for analyzing and interpreting data? Are there some steps that could increase the acceptance of data sonification?

21.4.1 Obstacles and Steps

The map is not the territory, any more than a two-dimensional graph is the data. Yet we often say (and feel) that we are directly 'looking at the data' when interpreting a graph—just as it can feel that we are watching the progress of a storm as it moves across the continent even though we are, in fact, watching a visualization of storm data animated and superimposed on a map of the continent that was generated from geographic data, distributed over the Internet, and locally rendered as a graphic image on the screen of a mobile device. This is the kind of shorthand that data sonification has yet to achieve. Apart from some notable exceptions (primarily alert sounds that notify us about events such as 'a text message has arrived', 'dinner is ready', or 'the patient in room 202 is in need of oxygen'), most data sonification is still self-conscious; a quick search of Google news over the past month would confirm that it is still considered a newsworthy novelty. How can we shift the focus from the map to the underlying territory and encourage the use sound as part of a multimodal interface for exploring heterogeneous data sets?

21.4.1.1 *Beyond the Usual Suspects*

While technical exchanges among practitioners are and always will be beneficial and instructive, if sonification is ever to mature beyond its status as an intriguing curiosity, one of our objectives should be to present more examples of sonifications that are integrated into the 'cascade' of scientific papers presented at topic-specific conferences; when researchers hear convincing examples of data sonification presented in the context of supporting graphical and analytical presentations of the data, the focus will shift from the technique (as a novelty) to the actual scientific topic at hand. The more examples of compelling data sonifications that are presented in support of new scientific observations and theories, the more data sonification can start to become normalized and accepted as a valid information channel and the less it will be misused as an attention-seeking gimmick on the part of researchers hoping for publicity in the popular press.

21.4.1.2 *Interchanges at Data Conferences*

In addition to being shared with fellow sonification practitioners, it's important for data sonification results to be presented to broader audiences of data analysts on conferences devoted to data visualization, data mining, and the currently trending 'big data'. Not only might it serve to expose data specialists to sound as an analytical tool, but there are new general data analysis tools currently under development that could benefit sonification practitioners as well.

21.4.1.3 *Introducing Data Sonification to (Very) Young Researchers*

Data visualization conventions like histograms and two-dimensional graphs are so pervasive and they're introduced at such an early age (the standard curriculum in Illinois introduces children to measurement and graphs as early as pre-kindergarten) that when we look at a graph as adults, we are often no longer conscious of the mapping process or the morphism; we feel we are simply 'looking at the data'. As more data sonification conventions emerge and as data sonification is introduced into the curriculum as a valid way to present, understand, and explore measurements and data, future researchers will slip more easily into 'listening to the data' and will use it as an adjunct to 'looking at the data'.

21.4.1.4 *Identifying and Reusing Effective Patterns and Solutions*

The concept of the traditional map as a two-dimensional drawing first began to emerge in the 1500s, when European sailors adopted common technologies and conventions like astrolabes, clocks, and latitude and longitude. At that time, all ship captains were expected to log, survey, and bring back spatial data (an expectation that survived into the twentieth century in the form of the fictional *Star Trek* franchise, where a great deal of time and attention is focused on stellar cartography and gathering data on new phenomena, new worlds, and unknown civilizations). Conventionalization of map data acquisition made it possible to combine data from several different sources in a single map. Settling on conventions like latitude and longitude for maps made transoceanic navigation more manageable by providing information that was portable across both space and time.

More recently, the emergence of graphical user interface conventions has made it easier for people to switch quickly between different operating systems, applications, and web navigation devices. Similarly, the emergence, identification, and teaching of patterns in software design have helped programmers recognize situations and problems that have been encountered before, thus saving on development time by reusing some hard-won solutions in lieu of constantly reinventing the toolbox *de novo*.

These examples argue for continuing the group project of recognizing and documenting additional patterns and situations for which particular kinds of sonification have proven to be useful. Certain patterns are identifiable across many fields of study; the more we can recognize and characterize these patterns, the better the results we can obtain and the easier it will be to put sonification tools directly into the hands of researchers, thus shifting the researchers' expectations away from hearing music and towards the desired expectation of hearing structure and relationships in the data source.

21.4.1.5 *Adaptive Technology for All*

The best argument and most highly motivated advocates for data sonification may in fact be those researchers who cannot use data visualization to analyze and interpret their data (Diaz-Merced 2014). Adaptive technology for presenting graphs and other nonverbal morphisms to visually impaired computer users might pave the way towards incorporating graph-to-sound as an adjunct to text-to-speech services in all data analysis and presentation software or better yet, as an integral part of every computer operating system.

Like a bionic ear, sonification enables us to hear things that would otherwise be inaudible—changes that are too small, too fast or too slow, dynamic patterns that have nothing to do with vibrations of air molecules. Sonification enables us to map from one domain (which may or may not be directly perceivable by humans) to a domain of energies that we can perceive. In this sense, sonification is an adaptive technology enabling us to perceive abstract structures in the same way that we perceive physical and chemical signals in the outside world. Sonification forces us to recognize that we are all perceptually (and conceptually) impaired in varying degrees and that all of us could benefit from adaptive technology that enhances our abilities to perceive, interpret, and understand the dynamic patterns unfolding on various time scales throughout the universe.

21.4.2 The Inevitable Conclusion

The auditory system itself is a cross-domain, inference-preserving mapping: it maps air pressure variations to physical movements of bones in the ear, to vibration of the cochlear membrane, to electrical patterns of neuron firing, and so on up the eighth nerve to temporal electrochemical patterns in the auditory cortex. The auditory system fits our definition of a morphism, a mapping that preserves important aspects of the structure and relationships of the original signal sources.

If that's true, then could we not just short-circuit that interface and map directly from data to neuronal firing patterns? In other words, could we skip over the mechanical portions of the ear to create an interface that maps data from a numerical model directly to electrochemical stimulation patterns of the nerves leading to the brain or to stimulation of the brain itself? Fans of the *Matrix* franchise may ask: how would we know whether this is not already the case?

FURTHER READING

Hermann, T., Hunt, A., and Neuhoff, J. G., eds. *The Sonification Handbook*. Berlin: Logos, 2011.
International Community for Auditory Display (ICAD). Last modified June 2017. http://www.icad.org.
Scaletti, C. Words. http://www.carlascaletti.com/words, accessed 30 June 2017.
Worrall, D. 'An Introduction to Data Sonification'. In *The Oxford Handbook of Computer Music*, edited by R. T. Dean, 312–330. Oxford: Oxford University Press, 2011.

BIBLIOGRAPHY

Allen, T. A., and Fortin, N. J. 'The Evolution of Episodic Memory'. *Proceedings of the National Academy of Sciences of the United States of America* 110 (suppl. 2): 10379–10386.

Asquith, L., and Scaletti, C. *Lhcsound*. 2015. https://lhcsound.wordpress.com, accessed 30 June 2017.

Barrass, S., and Vickers, P. 'Sonification Design and Aesthetics'. In *The Sonification Handbook*, edited by T. Hermann, A. Hunt, and J. G. Neuhoff, 145–171. Berlin: Logos , 2011.

Damasio, A. R. *The Feeling of What Happens: Body and Emotion in the Making of Consciousness*. New York: Harcourt, 1999.

Diaz-Merced, W. 'Making Astronomy Accessible for the Visually Impaired'. *Scientific American*, 22 September 2014. http://blogs.scientificamerican.com/voices/2014/09/22/making-astronomy-accessible-for-the-visually-impaired/.

Dictionary.com. 'Map'. http://dictionary.reference.com/browse/map?s=t., accessed 30 June 2017.

Erbacher, R. F., Grinstein, G. G., Lee, J. P., Levkowitz, H., Masterman, L., Pickett, R., and Smith, S. 'Exploratory Visualization Research at the University of Massachusetts at Lowell'. *Computers and Graphics* 19, no. 1 (1995): 131–139.

Frysinger, S. P. 'A Brief History of Auditory Data Representation to the 1980s'. In *First Symposium on Auditory Graphs*, 410–413. Limerick, 2005.

Goguen, J. A. *An Introduction to Algebraic Semiotics, with Applications to User Interface Design*. Berlin: Springer, 1999.

Goguen, J. A. 'Semiotic Morphisms'. Last modified April 2004. https://cseweb.ucsd.edu/~goguen/papers/sm/smm.html.

Grobet, D., Laffineur, L., and Giot, R. 'KISS2014—Internet Rumbles and Plantification'. *Vimeo*, September 2014. https://vimeo.com/113729841.

Hayward, C. 'Listening to the Earth Sing'. In *Auditory Display: Sonification, Audification, and Auditory Interfaces*, edited by G. Kramer, 369–404. Boston, MA: Addison-Wesley, 1994.

Hermann, T. 'Model-Based Sonification'. In *The Sonification Handbook*, edited by T. Hermann, A. Hunt, and J. G. Neuhoff, 399–427. Berlin: Logos, 2011.

Hofstadter, D. R. 'Analogy as the Core of Cognition'. In *The Analogical Mind: Perspectives from Cognitive Science*, 499–538. Cambridge, MA: MIT Press, 2001. https://prelectur.stanford.edu/lecturers/hofstadter/analogy.html.

Johnson, M. *The Meaning of the Body: Aesthetics of Human Understanding*. Chicago: University of Chicago Press, 2007.

Krygier, J., and D. Wood, D. 'Ce n'est pas le monde (This Is Not the World)'. In *Rethinking Maps: New Frontiers in Cartographic Theory*, edited by M. Dodge, R. Kitchen, and C. Perkins, 189–219. Abingdon: Routledge, 2011. http://makingmaps.owu.edu/this_is_not_krygier_wood.pdf.

Kusev, P., Ayton, P., van Schaik, P., Tsaneva-Atanasova, K., Stewart, N., and Chater, N. 'Judgments Relative to Patterns: How Temporal Sequence Patterns Affect Judgments and Memory'. *Journal of Experimental Psychology: Human Perception and Performance* 37, no. 6 (2011): 1874–1886.

Lakoff, G., and Johnson, M. *Philosophy in the Flesh: The Embodied Mind and its Challenge to Western Thought*. New York: Basic Books, 1999.

Lakoff, G., and R. E. Núñez, R. E. *Where Mathematics Comes From: How the Embodied Mind Brings Mathematics into Being*. New York: Basic Books, 2000.

MacEachren, A. M. *How Maps Work: Representation, Visualization, and Design.* New York: Guilford, 2004.

November, V., Camacho-Hübner, E., and Latour, B. 'Entering a Risky Territory: Space in the Age of Digital Navigation'. *Environment and Planning D: Society and Space* 28 (2010): 581–599.

Pinker, S. *How the Mind Works.* 2nd ed. New York: Norton, 2009.

Scaletti, C. 'Sound Synthesis Algorithms for Auditory Data Representations'. In *Auditory Display: Sonification, Audification, and Auditory Interfaces*, edited by G. Kramer, 223–252. Boston, MA: Addison-Wesley, 1994.

Trachtenberg, M. 'Architecture and Music Reunited: A New Reading of Dufay's "Nuper Rosarum Flores" and the Cathedral of Florence'. *Renaissance Quarterly* 54, no. 3 (2001): 740–775.

Warren, C. W. 'Brunelleschi's Dome and Dufay's Motet'. *The Musical Quarterly* 59, no. 1 (1973): 92–105.

Wikipedia. 'Pantograph'. Last modified 4 May 2017. http://en.wikipedia.org/wiki/Pantograph.

Wright, C. 'Dufay's *Nuper rosarum flores*, King Solomon's Temple, and the Veneration of the Virgin'. *Journal of the American Musicological Society* 47 (1994): 395–441.

CHAPTER 22

COLOUR IS THE KEYBOARD

Case Studies in Transcoding Visual to Sonic

MARGARET SCHEDEL

THE title of this chapter derives from a longer quotation from the artist Wassily Kandinsky, who constantly used the language of music to explain the language of form and colour in art (Selz 1957, 134; see *Three Sounds* by Kandinsky). The entire quote reads 'Colour is the keyboard, the eyes are the hammers, the soul is the piano with many strings. The artist is the hand that plays, touching one key or another purposely, to cause vibrations in the soul' (Kandinsky 1966, 45). Wassily Kandinsky had synaesthesia, a condition where one attribute of an (inducing) stimulus automatically engages the experience of additional (concurrent) features in a different sensory modality (Bor et al. 2014). Composers from Erik Satie and Olivier Messiaen to Amy Beach experienced some form of colour-sound synaesthesia, and the 'colour organs' of the eighteenth century were an early example of a way to turn music into visuals; this chapter will focus on how composers and inventors have used visuals to control sound in the twentieth and twenty-first centuries.

Interestingly, synaesthesia is not often transitive; individuals who see colours when they hear musical tones [colour → pitch] usually do not hear pitches when they see colour [pitch ← colour]. We have no empirical proof that the painter Wassily Kandinsky is one of the few transitive synaesthetes [colour ↔ pitch] (Ione and Tyler 2003), but his synaesthesia has had an undeniable influence on the artists and musicians who followed him. 'Kandinsky's curious gift of colour-hearing, which he successfully translated onto canvas as "visual music", to use the term coined by the art critic Roger Fry in 1912, gave the world another way of appreciating art that would be inherited by many more poets, abstract artists and psychedelic rockers throughout the rest of the disharmonic 20th century' (Ward 2006). This chapter encompasses analogue and digital algorithms that translate information from the visual domain into the audio domain. Most of the algorithmic procedures in this chapter are not reversible; in other words, the visuals cannot be generated from the sound. The instruments and programmes which translate visual material into sound synthesis should not be confused with 'algorithmic synaesthesia', a phrase coined by this volume's coeditor Roger Dean and others to describe multimedia works in which synchronous computer-mediated manipulation of sound and image results in shared features created in two domains (Dean 2009, 294). While both kinds of algorithm deal with light and sound, algorithmic synaesthesia can imply a cross-modal convergence, while the algorithms in this chapter are a specific subset of synaesthetic algorithms, translating image into sound.

The chapter's subtitle contains the word 'transcode'; media artist Kyle McDonald defines transcoding as 'a label for the intuition that information can be translated from one form to another' (McDonald 2007). Because transcoding is a creative, individual act, generalizations of systems are not helpful. Synaesthesia itself is idiosyncratic, the sound and light artist Robin Fox, who creates synaesthetic experiences, understands that 'each synaesthetic person experiences it [synesthesia] in a unique way. It's an intimate condition, born of the interior and unknowable to others, private. By claiming to manufacture it homogeneously in a group of people, am I some kind of cross-modal fascist?' (Fox 2014). One person with synaesthesia might think that the note B♭ is blue, while someone else will see it as orange, and similar differences occur when transcoding between vision and sound. There are some correlations that seem to be natural, such as size and loudness, pitch and height, but some artists deliberately subvert those expectations in their work. In this chapter, I present an analysis of the algorithms in diverse case studies, by which I hope the reader may gain an understanding of the spectrum of techniques and the methods used to convert ocular data into auditory signals. Some of these algorithms are simple analogue electro-mechanical devices, while others are complex programs that perform calculations, process data, and make logical (or even illogical!) decisions. At its most basic definition, an algorithm is a set of instructions; some of the simpler techniques are included for the historical context of translating image to sound.

Writing about transcoding necessitates the use of analogous language; the writer, historian, and philosopher François-Marie Arouet, more famously known by his nom de plume Voltaire, fully accepted the similarity between tones and colours, writing 'this secret analogy between light and sound leads one to suspect that all things in nature have their hidden rapports, which perhaps some day will be discovered' (Hankins and Silverman 1995, 76). While preparing his book on the Newtonian worldview, *Élémens de la philosophie de Neuton*, Voltaire corresponded with the inventor of the ocular organ, Louis Bertrand Castel. Although it was never built, the ocular organ can be seen as a prototypical synaesthetic algorithmic instrument, meant to generate visual and sonic material simultaneously. Castel proposed to change the mechanism of a harpsichord so that 'the pressing of the keys would bring out the colours with their combinations and their chords; in one word, with all their harmony, which would correspond exactly to that of any kind of music' (Castel 1725). Essentially, Castel 'set out to prove that there was an analogy between the phenomena of sound and light, and between tones and colours, such that what had up till then been performed only with sound, that is, arranging different tones in such a way that we appreciate the effect as a form of art, should be equally possible by arranging different colours, so that a whole new form of art would emerge, a music of colours' (Hankins and Silverman 1995, 19). Voltaire then linked Castel's idea to Newton's analogy between the widths of the various colours in the spectrum and the differences in string length for the pitches (Franssen 1991). Although perhaps the most famous, Castel was not the first to propose a synaesthetic instrument; in the late 1500s, Arcimboldo developed an instrument which produced a greyscale relationship between notes and visuals based on applying Pythagorean auditory formulas to paint shades (Caswell 1980).

The difference between analogy and transcoding is subtle; 'normally analogy is about establishing partial equivalence between two different entities. Transcoding is a sort of extreme analogy, where we establish complete correspondence based on trans-formations between entities' (McDonald 2007). Often authors speak of 'mapping' fea-tures from one domain to another. In these case studies, when possible, I will indicate how features in the visual arena control aspects of the resultant sound. For some of the more complex systems or older electro-mechanical machines, there are not enough data available to fully describe the algorithmic process of transcoding from the visual to the sonic.

22.1 LIGHT, THEN AND NOW: THE RHYTHMICON, THERMAL IMAGE, AND THE CANDELA VIBROPHASE

One of the first performable instruments to translate visible light into sound, the Rhythmicon, was invented by Leon Theremin for the composer Henry Cowell in 1934. Theremin had already developed his namesake instrument when he was asked by Cowell to create a new device that would be able to accurately create complex polyrhythms. Cowell was obsessed with the Pythagorean harmonic series, and wanted to integrate the frequency relationships of intervals with rhythmic relationships. So an octave, with the ratio of 1:2, would produce crotchets against quavers, while a fifth, with the ratio of 2:3, would produce quavers against quaver triplets (Cowell 1930). Theremin devised a keyboard which, instead of playing a single note when pressed, turned on a single light inside the machine, which illuminated two black metal disks 20 inches in diameter and ⅛ inch thick, perforated with ½ inch holes—the pitch wheel and the tempo wheel (York 1992). Theremin most likely knew of Opelt's siren, which used a spinning wheel with holes punched in it: a jet of air could produce fifteen simple notes, five different interval scales, and four chords (see Figure 22.1).

Each wheel of the Rhythmicon was controlled by a separate motor, with a rheostat on the outside of the instrument to adjust its speed. The pattern of interrupted light created by the interaction of the two wheels was converted with a photodetector to an equiva-lent electrical pattern that then controlled heterodyning vacuum tube oscillators (York 1992). When a single note was held down, a steady rhythm emerged; holding down mul-tiple notes resulted in the polyrhythms Cowell requested. This system operated in real

FIGURE 22.1 Opelt siren. Collection of Historical Scientific Instruments, Harvard University.

time, and the visual component was invisible to both the audience and the performer because the lightbulbs were housed in the body of the instrument. One of the original machines Theremin built is in an asbestos-contaminated warehouse in the Smithsonian Institution, Washington, DC, and is no longer operational, while one he built in Russia in the 1960s out of spare parts still functions and is now owned by the Theremin Center for Electroacoustic Music in Moscow (in this later model the wheels and lights are exposed: a demo can be seen at https://www.youtube.com/watch?v=-_ngPJoypQ8; see Figure 22.2). Theremin most likely did not think of the Rhythmicon as translating from the visual to the sonic domain; he used the materials at hand to create an instrument to the specifications of a patron. The Rhythmicon is now hailed as the precursor of the drum machine.

Among the earliest ways to control sound without human intervention were the musical automatons of the late sixteenth to early twentieth centuries. These music-making machines could be cranked by hand or wound up and left to run on their own. Most relied on a system of rotating gears, so it is not surprising that similar systems of disk-like control were used in the first electrical systems to control sound, including the Telharmonium, Rhythmicon, and Hammond organ. Composers and inventors are still finding new ways to use cyclic motion to control audio. In 2013 the composer Barry Moon and sculptor Hilary Harp collaborated on *Thermal Image*, a networked electro-mechanical sculpture that used data from Twitter to create a visual thermal display and ambient soundscape (see Figure 22.3). As with the Rhythmicon, light bulbs are used as a control mechanism, in this case not simply via the binary operand of turning the bulb on and off, but through continuous brightness control. The intensity of the light changes based on positive and negative emoticons on Twitter in twenty cities; the frequency of the tweets containing the selected emoticons causes lightbulbs to brighten or dim. The

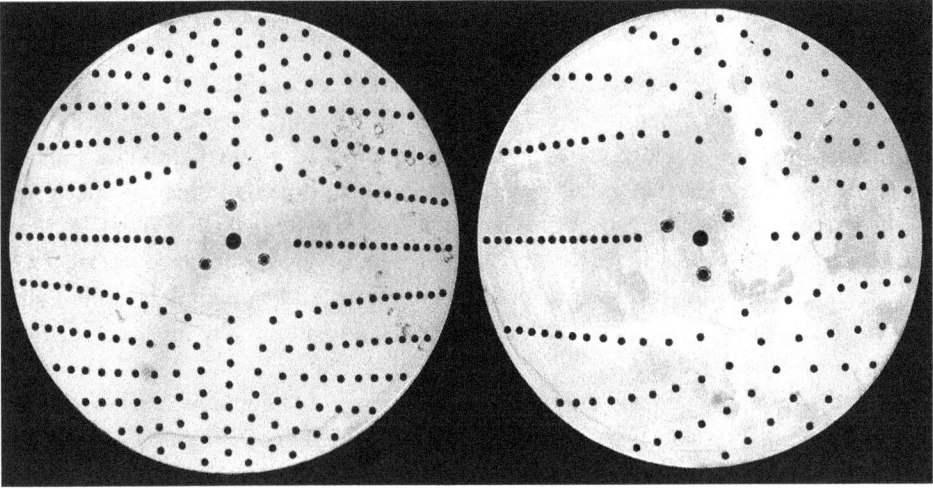

FIGURE 22.2 Rhythmicon disks. Left: pitch disk; right: rhythm disk. Courtesy of Andre Smirnov.

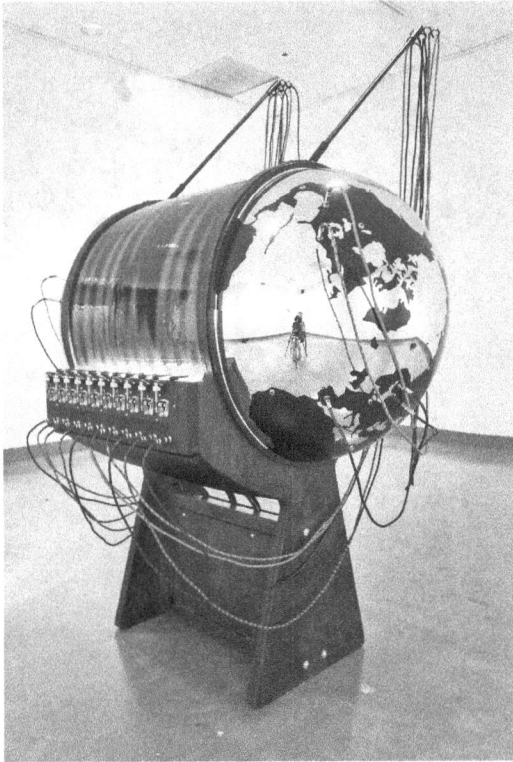

FIGURE 22.3 Hilary Harp and Barry Moon, *Thermal Image* (2013). Image: Suzie Silver.

more tweets with emoticons, the brighter the light. These lightbulbs are arrayed under a slowly turning drum covered in thermochromic film, which changes colours according to the brightness of the bulbs, creating residual trails of colourful shifting mood data. This drum sits between hemispheres of a globe, with female plugs at each of the cities. Viewers can use provided cables to link up to ten cities to the lightbulbs. Finally, each bulb also has a corresponding photo-sensor, which directly influences the speed at which ten small music boxes play (Harp 2017). As the frequency of tweets diminishes, the music boxes play more slowly. As can be seen at https://vimeo.com/69295896, *Thermal Image* is a multistage algorithmic sculpture, translating digital data into light. The byproduct of light makes heat which creates colour, and finally those colours are used to control the speed of the playback of analogue sound.

While the Rhythmicon used only the photo properties of lightbulbs, *Thermal Image* used both photo and thermal properties of light bulbs. Zachary Vex's guitar pedal takes it one step further, with the Candela Vibrophase (see Figure 22.4), which uses photo, thermal, and electric properties of light to modify the sound of its input signal. The pedal is powered by a single tealight candle which produces: (1) power for the electronic effect via two high-efficiency solar cells; (2) power for the spinning modulation disc via a miniature Stirling heat engine; and (3) signal light to activate the photocells in the audio

FIGURE 22.4 Candela Vibrophase. Courtesy of Zachary Vex.

circuit when it shines through the spinning modulation disc (Vex 2016). The audio circuit with the modulation disk, which he encourages users to customize, and photocells are a clear descendant from Cowell's Rhythmicon.

22.2 SOUND ON FILM: FISCHINGER, McLAREN, WHITNEY BROTHERS, SPINELLO, SHOLPO

The Rhythmicon is a very simple photosensitive machine, but more complex photo-optical methods for sound synthesis flourished in the twentieth century, particularly with the advent of optical soundtrack technology. Starting in 1919, the American inventor Lee De Forest was awarded several patents that led to sound on film; in his system, the soundtrack was photographically recorded on to the side of the strip of motion picture film to create a composite, or 'married', print. This 'Phonofilm' used a variable density optical sound recording and reproduction system, the operation of which was designed to be synchronous with a moving image (Enticknap 2006, 275).

> The technology underpinning Phonofilm marked a radical departure from all the audio technology previously used in conjunction with film in two respects. First, it used a microphone to capture the signal, which was then amplified for recording and reproduction electronically, unlike the other [mechanical] systems of the 1900s and 1910s, in which a horn was used to capture a signal that was then engraved on a wax disc or cylinder. Second, the Phonofilm signal was recorded as a photographic (optical) analogue waveform, exposed onto raw 35mm film stock, not as grooves in a wax surface. Unpublished biographical notes in the San José collection suggest that there were three principal technical problems he had to overcome: optimizing the sensitivity of the light source used to expose the sound record to electrical modulations in the input signal, developing a photosensitive cell that was sensitive enough to reproduce the modulations as the film passed between it and a light source ... and damping the intermittent movement of the film in the projector in order to reduce wow and flutter to acceptable levels. By August 1920, he claimed to have recorded and reproduced a clear enough signal that 'with what grim satisfaction I first definitely determined whether or not the film was being run backwards.' (Enticknap 2006, 277)

During the twentieth century there were many other systems for creating audio tracks, most notably Fox's Movietone. These audio tracks were usually used to record audio directly, but experimental artists soon realized they could manipulate shapes directly onto the audio track and synthesize sound from image. Oskar Fischinger was one of the first people to realize the possibilities of synthetic soundtracks. Already recognized as a masterful visual animator, Fischinger had a conceptual breakthrough in the spring of 1932, realizing that the kind of 'ornaments,' abstract designs that he used in his films, were not substantially different from the sort of patterns that were generated by sounds on the optical soundtrack ... [he prepared] and shot hundreds of trial soundtrack images ... by studying pre-recorded soundtracks, he quickly mastered the calligraphy of conventional European music, drawing out ... a nursery rhyme ... and other simple melodies ... it quickly became a massive undertaking—and not an easy one: when he picked up his first reels of *Ornament Ton* [Ornament Sound] from the laboratory, and had them play the film on their test projector, the technicians were horrified by the weird sounds, and feared that any further such reels of noise might damage their equipment! (Moritz 2004, 42–43)

Fischinger created his Ornament Sounds by filing open the soundtrack aperture on his camera and shooting abstract images directly onto the film (see Figure 22.5). Pictures from the newspaper coverage often showed him with large 'sound scrolls' that were painted just for the 'purpose of having a flashy object for press photos—and possibly to deceive anyone who would try to mimic Oskar's work for his own profit' (Moritz 2004, 44). Although he never reached the level of control he hoped for, Fischinger foresaw the possibilities of synthesis writing:

If you look at a strip of film from my experiments with synthetic sound, you will see along one edge a thin strip of jagged ornamental patterns. These ornaments are drawn music—they are sound: when run through a projector, these graphic sounds broadcast tones of a hitherto unheard of purity, and thus, quite obviously, fantastic possibilities open up for the composition of music in the future. Undoubtedly, the composer of tomorrow will no longer write mere notes, which the composer himself can never realize definitively, but which rather must languish, abandoned to various capricious reproducers. (Fischinger 1932)

Fischinger was not the first to synthesize sound directly on optical film strips; that honor of first 'artificial soundtrack' is most likely awarded to the Russian Arseny Avraamov in 1930, who wrote predesigned phonograms directly onto a soundtrack (Kaganovsky and Salazkina 2014, 22). Fischinger is best remembered today as the progenitor of animated 'visual music'. The definition has expanded since Fry saw Kandinsky's paintings and coined the term, and it has now come to refer to 'visuals composed as if they were music, by using musical structures, or a visualization of music, using the structures of the underlying composition. Examples of visual music include works using manual, mechanical, or algorithmic means of transcoding sound to image, pieces which translate image into sound, abstract silent films, and even live performance painting and other types of live cinema' (Schedel, Fox-Gieg, and Keefer 2012, 97).

Norman McLaren, a Canadian animator, realized that using preconstructed pitch and amplitude templates would be a faster method of synthesizing sound than Fischinger's process of drawing every shape manually. 'McLaren created and catalogued dozens of index cards, each painted with a pattern of stripes whose spacings produced notes in the chromatic scale. He would then mask these stripes with cutout amplitude-envelope

FIGURE 22.5 Oskar Fischinger, display card from Ornament Sound experiments. © Center for Visual Music, Los Angeles.

cards, in order to produce sounds with differing attacks and decays' (Levin 2000, 31). McLaren still positioned the templates by hand, but his templates were the beginning of a generative method of synthesizing audio from visuals. In generative art, the artwork is generated at least in part by some process that is not under the artist's direct control (Boden and Edmonds 2009, 21). Both McLaren and Fischinger created algorithms that used sequenced photographs of shapes to generate synthesized sound.

The filmmakers John and James Whitney, commonly known as the Whitney Brothers, were inspired by Fischinger, but they rejected his practice of visual composition synchronized to already existing music and were determined to create nonobjective films, using an autonomous visual generative grammar as sophisticated and rational as that of serial music itself (James 2005, 262). Five years after its completion, the Whitney Brothers won the 1949 prize for best sound at the Brussels Film Festival for *Five Abstract Film Exercises*;

these short films are now in the collection of the Guggenheim in New York (Karlstrom 1996, 230). The soundtrack that the Whitneys produced for these 'audiovisual musics', as described in their artistic statement, 'was an experiment in synthetic sound … sound is inscribed directly on the film. … Rather than recording and re-presenting an external source, the pendulum created patterns that could generate sound through the projector, making motion audible' (Patterson 2009, 39). By 'slowly advancing the film past the shutter while the pendulums swung back and forth … periodic bands of darkness and lightness onto the film's optical soundtrack [were exposed] … these bands would then produce audible sine tones when played back at a higher speed by the film projector' (Levin 2000, 32). Instead of taking still pictures and setting them into motion, the Whitney Brothers actually captured physical motion and translated it into photographs, which were then translated into sound. In the 1940s, 'before the perfection of recording tape, these sounds, with exotic "pure" tone qualities, mathematically even chromatic glissandos and reverberating pulsations—were truly revolutionary and shocking' (Haller 1998, 65). Accounts as to exactly how this machine worked are varied, but in a 1959 article John Whitney described how the machine operated:

> Our subsonic sound instrument consisted of a series of pendulums linked mechanically to an optical wedge. The function of the optical wedge was the same as that of the typical light valve of standard optical motion picture sound recorders. No audible sound was generated by the instrument. Instead, an optical sound track of standard dimensions was synthetically exposed onto film which after processing could be played back with a standard motion picture projector. … Due to the design of the mechanical linkage any number of pendulums could be played simultaneously. The linkage in effect 'mixes' sinusoidal oscillations without undue distortion. Composing for an instrument with the thinness of tone spectra as ours had determined a need to exploit our resources with ingenuity and to their fullest. … As a formal point then, we chose to tune the instrument to a serial row that would be different with each composition … a vertical note mixture (not a chord) would be produced, the timbre of components of which could be continuously varied by bringing in and out different groupings of frequencies … the attack and decay of the tones … could be controlled by literally starting and stopping the pendulums either abruptly or slowly … it was possible to start and stop a sequence of perhaps 20 pendulums within one frame … [and] establish a continuum from rhythm to pitch. … Second, third, and fourth records were exposed on the sound track at different recording speeds … in this way if became possible to conceive still another facet of the interrelationship of time and pitch. The act of performing on this instrument—essentially starting and stopping the pendulums and controlling their amplitude—could be governed by the instrument time (i.e., frame speed) or by the constant clock time … thus pitch ratios and time ratios were drawn still closer together and became more accessible as compositional elements. (Whitney 1960, 63–64)

In a 1973 reprint of the article, Whitney added a parenthetical note, in which he explained that the 'continuum of pitch, timbre, and rhythm relationships of this machine was unprecedented in Western musical resources and anticipates the application of computer technology to musical composition' (Whitney 1980, 154). In the brothers' notes to the 'Art in Cinema' screening in San Francisco they describe the coordination between the audio and the visual:

> A 15-second visual sequence is begun every five seconds, after the fashion of canon form in music. This constitutes the leading idea, a development of which is extended into three different repetitions. The establishment of complex tonal masses, which oppose complex image masses, builds upon the section. The durations of each are progressively shortened. (Stauffacher 1968, 61)

This quotation is notable because of the mixture of visual and musical imagery; the analogies between the two mediums are extensive. John Whitney was a serious musician and had studied twelve-tone composition with René Leibowitz, and each of the five films uses a different serial row. 'Because they had been forced to work at such slow speeds, the Whitneys were able to precisely synchronize the temporal relationship between sounds and their visual graphic imagery, with a parallel to the transpositions, inversions and retrogressions of the twelve-tone row technique' (Milicevic 2005, 3). The brothers were able to create a 'score', where each note was related to the speed of the pendulums and the speed of the film over the light source. 'The result was a sophisticated additive synthesizer able to produce a wide variety of timbres from reasonably pure components. ... Effectively, the soundtrack was animated i.e. the temporal framework of sound production was very similar to the process of film animation. ... In this way, the Whitney brothers were able to deal with sound and image synthesis, exploiting the relative novelty of both the forms whilst controlling the output with the same level of detail simultaneously' (Grierson 2005, 86). Like Fischinger, John Whitney saw the possibilities of new technology; some fifty years after his experiments with pendulums he wrote, 'the computer [is] the only instrumentality for creating music inter-related with active color and graphic design, and though the language of complementarity is still under-examined and experimental, it foretells enormous consequences and offers great promise' (Whitney 1994, 46). It should be noted that the projected images were not created by the pendulum system, and the optical soundtrack was not meant to be seen. The system is notable because the mechanical algorithm created the visual soundtrack in realtime, albeit at a much slower pace than would be used for the eventual playback instead of the frame-by-frame creation of other experimental artists.

Experimental artists used analogue optical soundtracks well into the latter half of the twentieth century; in the 1970s Barry Spinello was able to specify thousands of sound parameters with a single substance (Pres-Tone adhesive strips with various densities and gradations of half-tones printed on it), achieving sounds which would be nearly impossible to produce by hand-drawn means (Levin 2000). Patterns of lines created square waves, dots sine waves, and diamonds sawtooth waves. The closer the spaced patterns the higher the tones; more space resulted in lower pitches. Spinello's work is an example of synaesthetic algorithm, because he worked 'with sound and picture at the same time, in the same way. [His] dream was to squeeze sound and picture out of the same tube—to weave a cloth with warp as sound, woof as picture, and meaning the fabric itself' (Rothmans 2013). In this case, sound is still created from image, but the image is created using the same method. Spinello was inspired by the writings of the polymath artist László Moholy-Nagy, whom he quotes extensively in his *Canyon Cinema News*:

> Only the inter-related use of both sight and sound as mutually interdependent components of a purposeful entity can result in a qualitative enrichment or lead to an entirely new vehicle of expression ... To develop creative possibilities of the sound film, the acoustic alphabet of sound writing will have to be masters; in other words, we must learn to write acoustic sequences on the sound track without having to record real sound. The sound film composer must be able to compose music from a counterpoint of unheard or even non-existent sound values, merely by means of opto-acoustic notation. (Moholy-Nagy 1947, 277)

Spinello believed that most abstract film and sound combinations are choreography instead of transcoding:

> Film and music have gone their separate ways so that the only conciliation of the two seems to be—in one form or another—in 'synchronization,' namely: an existing musical passage to which the film is composed/ or an existing film sequence to which music is composed. This is really choreography of one art form—technology— thought sequence—to another ... it's not what true audio-visuality can be. The synchronization process (even when both music and visual are made by the same person—but separately) is like two people closely collaborating to write one story— with one person providing the verbs, the other nouns. Why not an audio-visual mix that is conceptually a unit. (Spinello 2008, 123)

Until the advent of digital sound for film in the 1990s, the optical sound track was the prevailing format for sound on film; there was not a single standard because there was a large variety of film formats (16mm, 70mm, Super 8, etc.) and sizes, and even the number of tracks wasn't standardized. Companies had their own methods of encoding and decoding sound, and some, such as RCA/Disney's Fantasound, even allowed for four-channel sound (Garity and Hawkins 1941). Essentially there were multiple algorithms to translate from the visual to the auditory domain, including the variable-area and variable-density soundtracks; in each technique the lightness values of the source frame are averaged, and then amplified to produce the audio waveform.

> Optical soundtracks are printed onto film rolls as fluctuating patterns of light and dark, occupying a narrow strip next to the images. ...The projector sonifies this soundtrack by means of an optical sound head. ... An exciter lamp shines through the film onto a photocell, filtered by narrow horizontal slits on either side of the film. As the film passes across this thin band of light, it produces a fluctuating voltage which is processed and output as the audio signal. Due to the need for continuous film speed when producing sound, as opposed to the stopping and starting required when projecting images, the optical sound pickup in a 16mm projector is placed 26

frames ahead of the lens. Thus, assuming a playback rate of 24 frames per second, the audio on any point of an optical soundtrack will be heard a little over a second before its adjacent image is seen. (Dupuis and Dominguez 2014)

This is why Spinello had to realign the image and audio track in Figure 22.6, otherwise the audio would not match the image in the still representation. It is important to note that all of the circuitry in the optical-film tracks is analogue. Recently, Alexander Dupois created a digital 'virtual optical soundhead' in the programming language Max/MSP for his piece *No-Input Pixels*, which can be seen at https://vimeo.com/77643568. His code, and some examples, can be downloaded from http://www.alexanderdupuis.com/code/opticalsound.

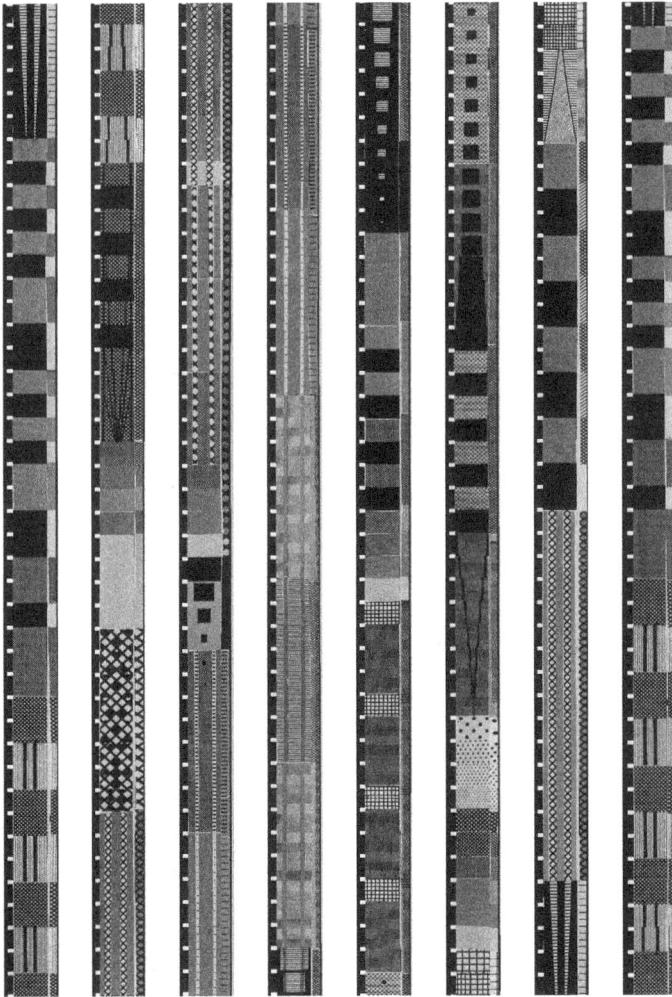

FIGURE 22.6 Segment of Barry Spinello, *Six Loop Paintings* (1970), with audio track visually aligned to projected track. Courtesy of the artist.

Until recently, not much was known about early synthetic sound in non-Western countries. Thanks to the efforts of Andrey Smirnov at the Theremin Center for Electroacoustic Music in Moscow (the same organization that owns the only working Rhythmicon), early Russian contributions to visual synthesis have been publicized. Smirnov breaks these efforts into four main trends: (1) hand-drawn ornamental sound (Avraamov, early Yankovsky); (2) hand-made paper sound (Voinov, Ter-Gevondian, and Konstantinov); (3) automated paper sound: Variophone as a sort of proto-wavetable synthesis (Sholpo, Rimsky-Korsakoff); and (4) spectral analysis, decomposition, and resynthesis technique (Yankovsky). His articles are worth reading, in this chapter I will cover only the Variophone, which was developed by Evgeny Sholpo in 1930 with help from Georgy Rimsky-Korsakoff, grandson of the famous composer Nikolai Andreyevich Rimsky-Korsakov, and later the ANS Synthesizer (Smirnov 2009).

The Variophone, invented by the Russian Evgeny Sholpo, was an optical synthesizer that utilized sound waves cut onto cardboard disks rotating synchronously with a moving 35mm movie film (see Figure 22.7). The advantages of the Variophone were in its flexible and continuous pitch control and vibrato. Sholpo continuously refined his instrument, and by 1936 the arsenal of musical and acoustical means of the second version was enriched highly with possibilities of free glissando with a speed of up to four octaves per second, flexible and exact control over dynamics, and options

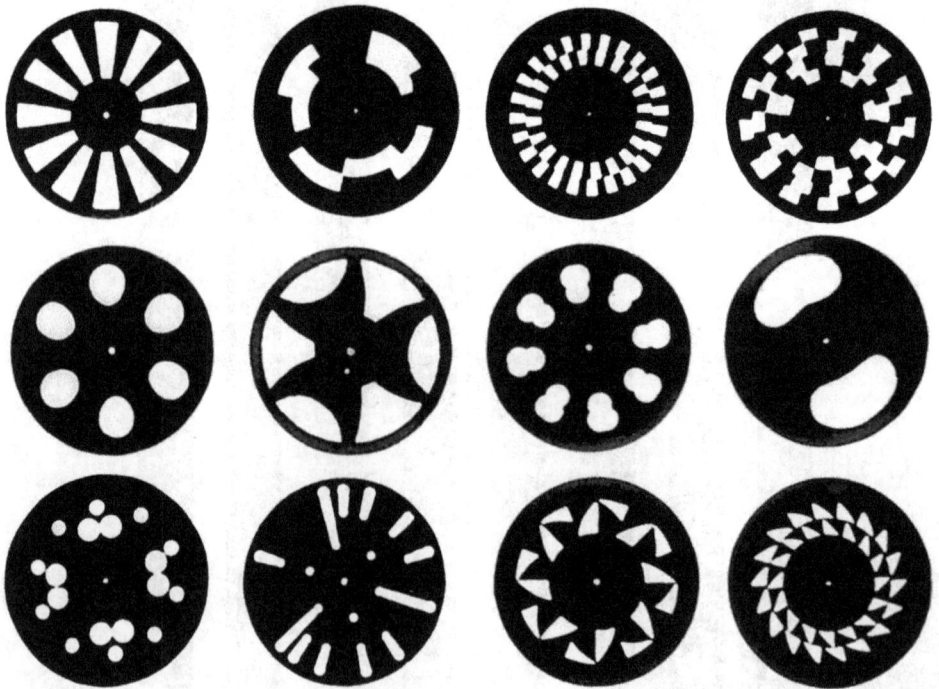

FIGURE 22.7 Variophone disks with cut wave shapes; Version 1, 1932. Courtesy of Andrey Smirnov.

for deep vibrato for pitch, volume, and timbre. The Variophone could produce polyphonic soundtracks with up to twelve parallel voices, and was able to simulate more subtle variations in tempo, such as rubato, rallentando, and accelerando. In besieged Leningrad in 1941, Sholpo and the composer Igor Boldyrev created the soundtrack for the animated film *Sterviatniki* (Vultures). The Variophone was unfortunately destroyed by one of the last enemy shells to hit the city, at the very end of the blockade (Smirnov and Pchelkina 2011).

22.3 Drawing Sound: Oramics, UPIC, and Metasynth

A number of other early electronic instruments used similar optically controlled tone generators, including the Welte Light-Tone (1936), the Singing Keyboard (1936), the Optigan (1971), and the Photosonic Instrument (1972). But of all optical synthesis methods, the Variophone is 'perhaps the closest in terms of its functional characteristics and underlying design philosophy' to that of the Oramics Machine created by British composer Daphne Oram (Manning 2012). Oram holds two US patents, one from 1964 for variable electric resistances and one from 1969 for digitally controlled waveform generators (Oram 1994, 1969), but the earliest technical drawing in the Oram archive is dated December 1951. This diagram sketches an optical playback system consisting of two loops of threaded film, a light source, and a photocell, with the necessary tension for each loop of film maintained by an associated pulley (Manning 2012). The digital signal control was added later, as transistors became more affordable. Oram wanted to draw audio control directly onto film, inverting the experience she had when she sang into a microphone and saw the resultant waveform on the screen of an oscilloscope. As excerpted in Manning's article, her logbook from 1961 details her needs for her instrument:

(1) To have complete control of timbre, pitch, dynamics, vibrato, reverberation, attack, decay, timbre changes within the note;
(2) To control these characteristics in a visual form so that all alterations within the aural comprehension of the human ear and mind have an easily recognizable counterpart in the visual medium;
(3) To achieve this controlled complexity of waveform whilst keeping all parameters within the scope of written waveforms;
(4) To obtain sounds which are more 'musical' than those achieved by electronic devices and which have a greater range of timbre.

Luckily, unlike the Variophone, the Oramics machine still exists. Dr. Mick Grierson, director of the Daphne Oram Collection, traced the machine and Oram's archives to a

FIGURE 22.8 Daphne Oram and the Oramics machine, with a 'neume cutting block'. © Daphne Oram Trust and Fred Wood.

French barn in 2009, and brought it back to London, where it served as the centrepiece to the London Science Museum's exhibit 'From Oramics to Electronica', which ran from July 2011 to June 2015. In addition to the exhibit, Goldsmiths College and the Science Museum funded a PhD student, Tom Richards, to research the machine. In Oram's notes Richards found a succinct description of Oramics:

Oramics is a three-step process utilizing analogue and hybrid digital electronics:

1. Define a range of four timbres or wave shapes by drawing them on glass slides. (Direct Analogue Process)
2. Define the outline melody by drawing groups of black dots on clear 35mm film. (Digital Symbolic Process)
3. Define other analogue parameters (envelope shapes of the four different timbres, pitch vibrato, and reverb mix) by drawing graphs on clear 35mm film. (Direct Analogue Process)

Similar to the Variophone, the Oramics machine used a set of waveshapes to interrupt light, but instead of cut-out cardboard disks, the wave shapes were painted onto glass slides (see Figure 22.8). The pitch was defined by one set of five film strips; four strips to control the pitch with groups of black dots (or pieces of square tape) and the fifth reserved for freely drawing vibrato curves. Timbre was defined by a second set of five film strips, all of which were drawn on freely; four to control the amplitude of the four waveshapes and a fifth that she called the 'reverberation room'. Oram constantly refined how the pitch mapping worked, and could even reprogramme the system to use microtones. Richards was able to study the original machine and found that essentially:

In the pitch control system, digital information in the form of groups of drawn ink spots (referred to by Oram as Neumes) were used to control a bank of bistable flip-flops (discrete transistor based logic circuits) which in turn switched in and out various relays, which then controlled a resistor/capacitor network determining the pitch of an analogue sawtooth wave oscillator. There are 16 light dependent resistors (LDRs) in the pitch control sensor system … eventually only 12 of the 16 sensors were utilized to control pitch changes. (Richards, personal communication 2015)

Once the pitch was determined, the wave scanners turned the sawtooth wave into four complex waveforms or timbres to be utilized in the overall composition. The sawtooth waveform defined the repetitive scanning rate on the x-axis of four cathode ray tubes (CRT, similar to an oscilloscope) inside the 'commode' part of the Oramics machine. Each CRT was combined with a photomultiplier (thermionic valve-based photoelectric component) and a feedback circuit, which forced the y-axis (amplitude) to follow the contours of the drawn waveforms that were inserted between the CRT and the photomultiplier (Richards, personal communication 2015). Richards has not been able to figure out precisely how the lower set of five film strips worked because, like the Variophone, the Oramics machine was a constant work in progress. Earlier images of the device showed that Oram needed only a simple thin line on the bottom tracks to control the audio. Later photographs indicate opaque filled-in envelope shapes that reach the bottom edge of the film strip. 'The LDRs which are still installed in the machine are wired to a set of co-axial cables which just tail off to jack plugs, bare wire and ring terminals which might indicate that the audio mixing stage happened externally, in equipment which has not made it into the parts of the machine held in the collection of the Science Museum in London' (Richards, personal communication 2015). In 2001 the Daphne Oram collection released an iPhone app by Drs. Mick Grierson and Parag Mital under the moniker Strangeloop Limited, where users are able control pitch, vibrato, envelope and reverb, and timbre by drawing shapes on the screen of the portable device. A promotional video showing how the app works can be viewed at https://vimeo.com/25301328. Tom Richards, as part of his practice-based PhD, has built a mini-Oramics machine based on Oram's original sketches (see Figure 22.9).

Xenakis's UPIC system is another example of a compositional system developed by a singular composer over a number of decades. 'The UPIC (Unité Polyagogique Informatique CEMAMu) is, to put it simply, a complex system of computers and peripherals designed to facilitate direct access to sound and musical material by the user. Unlike the Oramics machine, UPIC is fully digital. The focal point of the system is a graphics tablet on which you draw-design-all necessary parameters and sound information. This is then immediately calculated and transformed into sound by the computer' (Lohner 1986, 42). It is believed that Xenakis had the idea to create UPIC after finishing his composition *Metastasis*, which he composed on graph paper and had to translate to staff paper; he wanted a more immediate and universal way to express musical thought. Xenakis collaborated with the engineer Patrick Saint-Jean on the first release of UPIC, which was introduced to the public in 1977 but which had been completed some

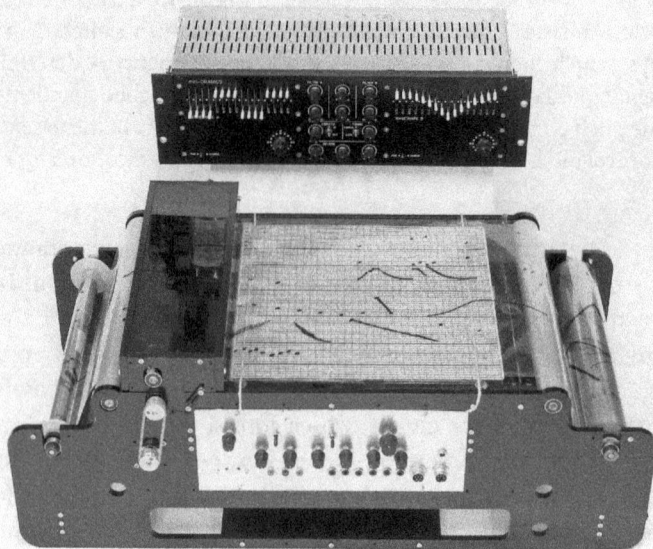

FIGURE 22.9 Mini-Oramics machine, reimagined and built by Tom Richards, 2016. An interpretation of a design for musical hardware by Daphne Oram and John Emmett, circa 1976.

time in 1975. The second version, which ran on two Intel 8086s, was released in 1983, and real-time capabilities were added in 1988. The first version to run on commercial software was released in 1991, and was presented at the 1990 International Computer Music Conference (Marino, Serra, and Raczinski 1993, 259). All of the early versions were housed in France in Xenakis's CEMAMu (Centre d'Études de Mathématique et Automatique Musicales), except one that Gerard Pape bought in 1989 for $50,000 (Makan 2003, 21).

The CEMAMu expanded in 1985 with the creation of CCMIX (Center for the Composition of Music Iannis Xenakis) and was renamed CIX in 2000 to 'redefine the goals of the association, focusing on the preservation, promotion and dissemination of the intellectual legacy of Iannis Xenakis's work' (Delhaye, Bourotte, and Kanach 2014). In 2010 the CIX moved to the University of Rouen, and continued Xenakis's legacy. The French Ministry of Culture allocated a budget for IanniX, a version of UPIC which was initially written as externals for Max/MSP created by La Kitchen and Thierry Coduys and licensed under Creative Commons (Bourotte 2012). The newest version is a standalone program which runs on Mac Os, Linux, and Windows and can be downloaded from http://www.iannix.org/. Another official implementation was approved in 2013, UPIX2014 +, a joint project with the CIX and Computer Science Department at the University of Rouen, creating yet another version of the UPIC software (Delhaye, Bourotte, and Kanach 2014).

Xenakis intended UPIC to be an intuitive interactive system used to transcode drawing into sound. The original hardware interface uses an electromagnetic stylus and a

60cm × 75cm tablet. A conductive pad underneath the tablet calibrates the coordinates of the pen to within 0.25mm. Next to the drawing area are arrays of command functions and memory access. When an operator draws on the tablet, the result is displayed on two screens—one graphic and one alphanumeric. This hand-drawn information is transcoded into sound using a 16-bit computer and sixty-four oscillators (Lohner 1986). As in most computer music systems, time was represented on the x-axis while pitch was represented on the y-axis. Interestingly, the mappings were not fixed; a page could take a few seconds or many minutes. Xenakis used this flexibility in *Mycenae-Alpha*, the first work composed on the UPIC system. Xenakis published the score, a printout of the graphic representation, of the 9′ 38″ composition—some of the pages (or as Xenakis called them, 'arcs') last less than one minute, while others are spread out over two minutes.

Xenakis thought that the UPIC system had tremendous pedagogical implications, writing 'with the UPIC, music becomes a game for the child. He writes. He listens. . . . He corrects immediately. . . . He can imagine the timbres. And, above all, he can devote himself right away to composition' (Xenakis, Brown, and Rahn 1987, 22). This desire for immediacy is fairly common for users and creators of transcoding algorithms, but Xenakis also wanted to bring a human element into his work. Surprisingly for a composer who pioneered the use of advanced mathematical models, he felt that calculation has limits.

> It lacks inner life, unless very complicated techniques are used. Mathematics gives structures that are too regular and that are inferior to the demands of the ear and the intelligence. . . . The hand, itself, stands between randomness and calculation. It is both an instrument of the mind—so close to the head—and an imperfect tool. . . . [Y]ou can always recognize what has been made industrially and what has been made by hand. Industrial means are clean, functional, poor. The hand adds inner richness and charm. (Xenakis, Brown, and Rahn 1987, 23)

After UPIC there were a tremendous number of programmes which used sound-image relationships to permit the generation or control of sound by mark making, but using the mouse instead of a specialized pen (Franco, Griffith, and Fernström 2004). Most keep the convention of time on the x-axis, with left being the past and right being the future, and pitch on the y-axis, with higher frequencies on the top of the screen and low pitches on the bottom. These programmes include (in alphabetical order) Audiosculpt, Aurora, Floo, FMOL, Hyperscore, Loom, Monalisa, Music Sketcher, Photosounder, Phonogramme, SPEAR, Videodelic, Warbo, and Yellowtail, along with a host of other independent research projects. The most famous program to combine image and sound is perhaps MetaSynth, the 'electronic music application and sound design environment, featuring the ImageSynth application that transforms user paintings to sound based on color and brightness' (Chambel, Neves, Sousa, and Francisco 2010). Metasynth started as a private project by Eric Wenger in the early 1990s; he then shared version 1.0 with IRCAM members in the mid-1990s. The first publicly available release was 2.0, published by Arboretum in 1998. Like UPIC, Metasynth is still available; in 2015 version 5.4 was published by U&I software (Metasynth 2015).

Users of Metasynth start with a sound source and an image source.

> At the level of sound design, Metasynth performs a Fast Fourier Transform (FFT) on a source sound (waveform, noise, sample, etc.) and produces a frequency-domain representation that can be altered and manipulated by applying any PICT file on it. … In this sense, Metasynth functions as a subtractive synthesis tool, i.e. it starts from a spectrum rich in frequencies (e.g. noise) and uses pictures as filters to produce the desired sound result. (Giannakis and Smith 2000)

As with UPIC, time is represented on the horizontal axis, while the vertical axis represents pitch. Users can modify the results of the filtering using a pen to draw on the screen.

> MetaSynth's unique feature is its adaptation of 'brushes' of various types and sizes to the sonic context. For instance, the 'pen brush' creates hard-edged rectangles that will yield abrupt attacks. The 'air brush' has rounded edges for smooth attacks and decays. A 'spray brush' creates grainy textures. A 'note brush' specifies quantized pitches. Visual effects such as smearing, cloning, cutting, pasting, and so on all have correlating effects on the sound. (Greenlee 2008)

Interestingly, in Metasynth the vertical axis is scalable (unlike UPIC), supporting the representation of linear, logarithmic, or harmonic scales (Thiebaut, Healey, and Kinns 2008). Metasynth also uses colour information in its algorithms, mapping pixel hue to spatialization. Barry Moon (of *Thermal Image* fame) even created an homage to Metasynth in Max/MSP called appropriately Metasynthy; the tutorial can be viewed at https://www.youtube.com/watch?v=rnERzPwRa4g&lr=1.

22.4 GLITCH: MUSICA SIMULACRA, AND PIXEL PLAYER

All the projects described so far used the computer or electronics to create precise compositions, Xenakis wanted to introduce 'richness and charm' by hand drawing, but he did not want unanticipated results. The Japanese composer Yasunao Tone is more interested in 'glitches, cracks and unstable systems for sound production, all of which use a measure of indeterminacy and chance' (Stuart 2003). Tone is perhaps most famous for his Fluxus performances with prepared or wounded CDs, but he also created several pieces using algorithms to translate image into sound. In this work 'textual source materials manipulated by computers, and images and sounds based on them, form an intense and coherent relationship' (Tone, personal communication 2015). To generate the sounds in *Musica Simulacra* (1997) Tone sonified the 4,516 poems found

In the original Chinese characters:

新 年乃始乃 波都波流能 家布敷流由伎能 伊夜之家餘其騰

In syllabic Japanese:

Atarashiki toshino-hajimeno hatsuharuno kyofuru-yukino iyashike- yogoto

And an English translation:

The first snow, only doubling its felicity, on today the first new day of the New Year.

FIGURE 22.10 Translation of Yasunao Tone, *Man'yōshū*, from *Musica Simulacra* CD-ROM.

in the *Man'yōshū*. These poems were written before the Japanese had their own written language; instead, they used Chinese characters to transcribe Japanese syllables. In modern Japanese the 107 possible Japanese pronunciations can be inscribed with forty-eight Japanese kana, but the *Man'yōshū* required almost 2,400 Chinese characters. Tone researched the meaning of the logographics of the Chinese images and once he was satisfied with his readings of the origin of Chinese characters, he sought appropriate images from a variety of magazines, books, and encyclopedias (Tone, personal communication 2015). For example, one of the poems in the collection is shown in Figure 22.10.

The first syllable of the third word is 'ha', and the character for ha in Chinese is 波 (wave), which can be divided into the constituent elements of 氵 (water) and 皮 (skin). According to Tone's research, 皮 (skin), as stated in Shuowen Jiezi, is an image of the hide torn off an animal. To represent this syllable, Tone and his wife found an image of a farm animal (a sheep) being skinned (Tone, personal communication 2015). Tone used a similar technique in *Musica Iconologos*. 'What we are hearing is not the picture of a Chinese character but the picture of Tone's interpretation of that character in photo-images from our visual experience. … Tone has "translated" the Chinese character for us, not into words … but into its signifier, both in its form … and in its literal trace as a word or combination of signs in the Chinese language' (Kendall 1993). For *Musica Simulacrum*, Tone added an additional step. The programmer Ichiro Fujinaga, hired by Harvestworks, translated the found pictures into sound using the scripting language Lingo within Macromedia Director (Fujinaga, personal communication 2015), and these sounds were then randomized for a CD-ROM experience. The result is the sonic equivalent of infinite monkey theorem—some random sequence of the sonification of characters will match an original poem, but 'Tone deliberately uses a process that obliterates the information contained in the images: given a sound wave as input, there is no algorithm that will return anything close to the original picture as output' (Ashley, Dekleva, and Marulanda 2007, 85). *Musica Simulacra* was released as a boxed set by Atak in 2011 with four contents:

1. *Musica Simulacra* CD-ROM (1996–2010)
   Sound art by converting whole texts from 4,516 poems found in the *Man'yōshū* (Collection of Ten Thousand Leaves, 759, the oldest existing collection of Japanese

poetry. Compilation attributed to Otomo no Yakamochi). More than 2000 hours of recording time. Playable on Mac and Windows.

2. *Musica Simulacra* CD

Audio CD version including 12 tracks taken from 2000 hours of MUSICA SIMULACRA CD-ROM, produced and edited by Keiichiro Shibuya. Playable on regular CD player and computer.

3. *Musica Simulacra* Text

10,000-character (Japanese), 4,000-word (English), and 48-page commentary booklet written by Yasunao Tone. Provided in Japanese and English.

4. 500 Copies of Limited Edition with Autograph

The autograph of Yasunao Tone enclosed within 500 copies of limited edition exclusively. Tone signed with a magic marker on a piece of an old book which is owned by him. (Tone, personal communication 2011)

Tone then used the original CDs of *Musica Simulacra* in his work *Man'yo Wounded* (2001), which won the 2002 Prix Ars Electronica. By 'wounding' the compact disks with tape and scratches, he performed with his sonifications, creating another layer of indeterminacy on top of the original algorithm. The entire algorithmic process for *Man'yo Wounded* spans centuries. The oldest poems in the collection date from 456 AD, and the poems were compiled into a written collection between 759 and 794 AD, using Chinese symbols to represent Japanese syllables. Images representing the meaning of the Chinese symbols were selected by the artist; these images were turned into very short sounds using MacroMedia Director, and the sounds were randomized, resulting in 2,000 hours of audio. Seventy-six minutes of audio were selected and burnt onto a CD, which was then 'wounded' with tape and scratches, and played in performance. At each stage of the process, error was introduced into the transcoding procedure.

Another artist interested in the expressive possibilities of glitch is Antonio Roberts. His programme Pixel Player allows the simultaneous sonification of up to four images, based on the RGB values of individual pixels. Users upload pictures into a PureData patch and can control the overall volume, speed, and the pitch scale for each picture (see Figure 22.11). The software can be downloaded from http://www.hellocatfood.com/pixel-player/. Roberts was inspired by Neil Harbisson's Eyeborg, a system encompassing a webcam, a computer, a pair of headphones, and software that translates colour into sound. Adam Montandon created the system for Harbisson, who has achromatopsia— he cannot see colours at all. The Eyeborg allows Harbisson to experience colour. He has worn his system continuously since 2004, even while asleep, and is now recognized as a cyborg by the British government (Warwick et al. 2014). Harbisson lives continuously with a transcoding algorithm implanted into his biology, one that can be experienced by using Roberts's software.

22.5 SLIT SCANNING: PHONOPAPER, ANS

In Tone's, Roberts's, and Harbisson's systems the computer scans an image pixel by pixel, while in Alexander Zolotov's app Phonopaper, a user moves a handheld device (iOS

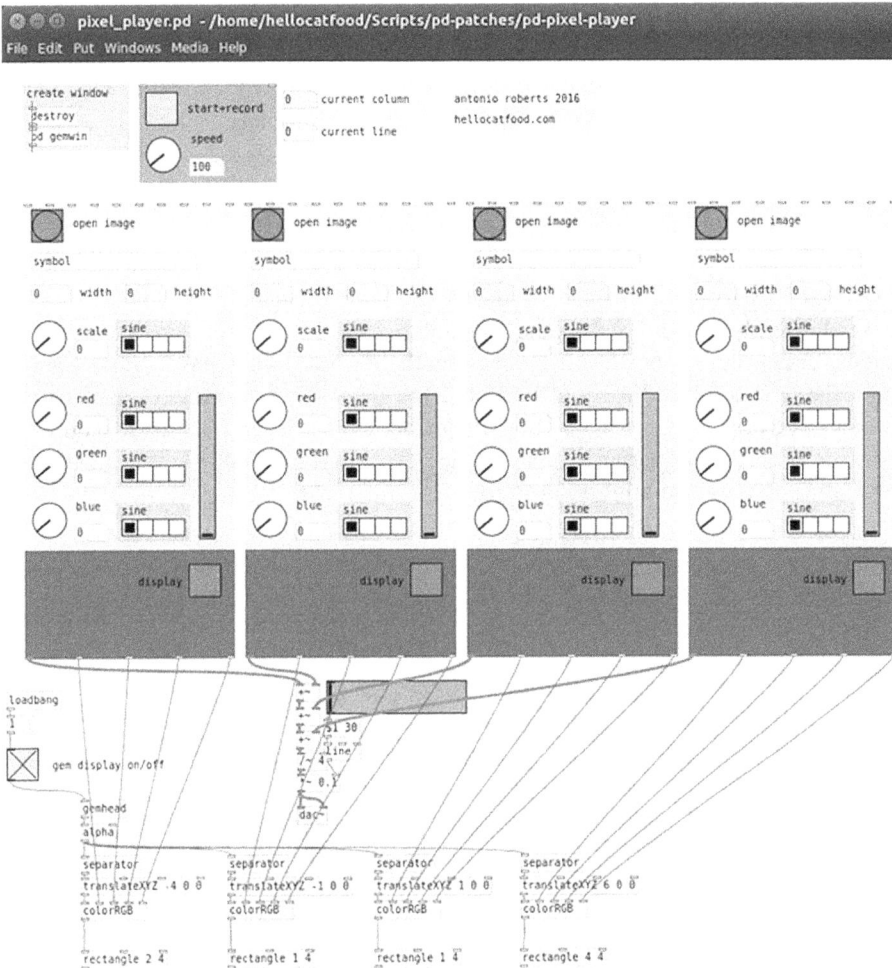

FIGURE 22.11 Antonio Roberts, Pixel Player. © Antonio Roberts.

or Android) over a graphical representation of the sound; the computer then analyzes shapes drawn on the paper and creates sound. Users can scan the paper at any speed and in any direction. Even though Phonopaper can turn any photo into sound resulting in a glitched output, it was really created as a way for people to draw accurate representations of sound, and then play them back—thereby creating a graphic way to send an audio message. The software, released in 2014, and available from http://www.warm-place.ru/soft/phonopaper/, is based on the Virtual ANS engine—Zolotov's emulator of the unique ANS microtonal Russian synthesizer.

The Russian audio engineer Evgeny Murzin had the idea for a microtonal optical synthesizer in 1938 and the finished instrument first generated sound in 1958. It is named after a composer who was fascinated with synaesthesia—Alexander Nikolayevich Scriabin. For a while, the (still functional!) synthesizer was housed in the Theremin Center for Electroacoustic Music, but it has since been moved to the Glinka State Central Museum of Musical Culture (Kirn 2013). The ANS uses a photo-optic generator with 144 phonograms of pure tones etched in glass. Murzin used five sets of these disks rotating at different speeds to produce 720 pure tones, down to the interval of one-sixth of a semitone. The input to the tone wheels is a glass plate covered with an opaque non-drying gummy black paint or mastic. This 'coding field' moves past a narrow aperture with photoelectric cells, at a variable speed. The speed controls only duration, not pitch. To compose, users simply scrape off the mastic, making it possible for the light to pass through and activate the photoelectrics (see Figure 22.12).

> Scraping off a part of the tar-like non-drying mastic at a specific point on the plate makes it possible for light from the corresponding optic phonogram to penetrate the reading device and be transformed into a sound. The non-drying mastic allows for immediate correction of the resulting sound. The glass is then cranked (by hand or by motor) across the light beams. The performance tempo depends upon the score-reading rate and can be varied without changing the pitch and timbre of the sounds. (Smirnov, personal communication 2016)

It is easy to edit the score; unwanted sounds can be smeared over and new sounds etched in the paint. 'All this makes it possible for the composer to work directly and materially

FIGURE 22.12 The score and coder of the second version of the ANS. Courtesy of Andrei Smirnov.

with the production of sound' (Kreichi 1995, 59). The most famous use of the ANS synthesizer is in Edward Artemiev's score of Andrei Tarkovsky's film *Solaris*, while the most prolific user of the synthesizer is by the composer Stanislav Kreichi.

22.6 Image as Control: Graphic Converter, Augur, and Light Pattern

There are now hundreds of apps and programs which translate image into sound, all using slightly different algorithms, and in most of these systems, as with the ANS synthesizer, the visual aspect is a symbolic representation of the sound. Fernando von Reichenbach's Graphic Converter 'represents a significant shift in that the drawn forms can be flexibly applied throughout the system ... [as] a direct analog to fluctuations in volume' (Greenlee 2008). The system, built in 1967, turned pencil on paper drawings into electronic control signals through a closed-circuit television. The drawings could be mapped to various inputs to analogue synthesizers such as Moog and Buchla to control envelope generators, filters, frequencies, and amplitudes (Kröpfl 1997, 27). Reichenbach was awarded a patent for his system, in which 'a horizontal straight line produces a determined frequency ... an inclined line produces a glissando. If the line is drawn free hand, aleatoric functions are produced, subject to a certain degree of control' (Von Reichenbach 1973). The algorithmic system included a paper band in which the control signals were drawn, a television camera that read the paper band, a convertor, and a voltage generator which simultaneously controlled two synthesizer functions, such as oscillators, modulators, or filters (Dal Farra 2006, 343). This interest in hand-drawn functions predates Xenakis's intentions with the UPIC system. The first piece created using the graphic converter was *Analogias Paraboloides* by Pedro Caryevschi, composed in 1970 (Dal Farra 2004).

Shawn Greenlee studied many of these historic transcoding algorithmic systems before creating his own, Augur, to control music with visuals. He breaks down machine conversion of images to sound into three categories: transduction, translation, and interpretation.

> The approaches of transduction and translation involve establishing a system of rules for converting the energy (transduction) or meaning (translation) of the image into an analogous energy or meaning in sound. By learning the correlations between visual specification and resultant sound, a language is learned that can be utilized to evoke predictable results. ... In this case [interpretation], the visual image is not so much a composed, origin point for sound synthesis, but is instead a guiding influence on a process already underway. Here, the performer may access the variables of conversion, and is therefore responsible for determining the sound of the visual as an instrumental practice. (Greenlee 2013, 288)

In his own practice, Greenlee places the most emphasis on interpretation: 'Augur is a system for solo audio performance that combines the action of drawing with methods for generating digital sound from graphic patterns stored within the computer. There are two primary activities: 1) the live action of drawing sensed and mapped to signal processing, and 2) the conversion of previously composed graphics as instructions for sound synthesis' (Greenlee 2008). Like Yasunao Tone, Greenlee is interested in chance: 'the emphasis here is on the discovery of new sounds and the circumstances that bring them into play, rather than on a concrete, ascertainably correct result.' Greenlee began developing Augur in 2005; his current system uses four Wacom tablets, two spinners, MIDI dials, a keypad, and a live video camera (see Figure 22.13). The system is programmed in Max/MSP/Jitter, and utilizes what Greenlee calls 'graphic waveshaping', which can be understood as 'non-linear distortion synthesis with time-varying transfer functions stemming from visual scan lines. ... In graphic waveshaping, the transfer function is time-varying. It fluctuates according to the rate of navigation through the image' (Greenlee 2013). Greenlee uses pre-drawn images as input to the computer during the real-time performance, exploring the effect these images have on sound synthesis and audio effects. His performance is very physical, he moves large sheets of drawings, repositions cameras and lights, spins oversized controller disks, and utilizes multiple tablets. Like the Rhythmicon, Augur is truly a performance instrument, in this

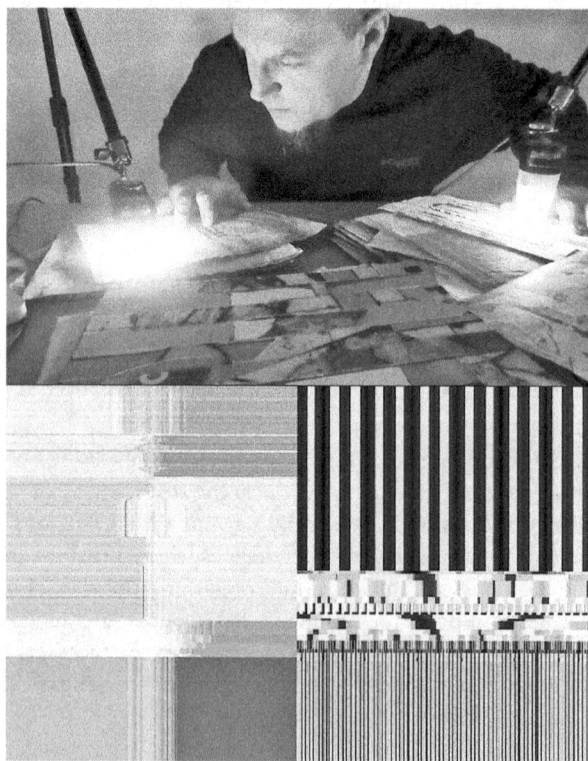

FIGURE 22.13 Shawn Greenlee, *Impellent*: microscopic image capture and screenshot. © Shawn Greenlee.

case made possible by the real-time capabilities of computation and subtle algorithms, which translate visual and controller data into an audiovisual production.

Daniel Temkin takes the concept of visual control of audio in a completely different direction. In his work, photos are used to programme the computer directly; the programme then generates the audio synthesis. His programming language, Light Pattern, like other programming languages, is a list of rules, a grammar to communicate with a compiler. It can be downloaded from http://lightpattern.info. Light Pattern is an 'esolang' (esoteric programming language), a class of languages made for reasons other than practical use. Esolangs have been compared to art practices such as Fluxus, Oulipo, and so on, but many esolangers did not necessarily view their work from an art perspective, but more as a hobby. However, their work does overlap with an arts practice, albeit an outsider arts practice, due to the conceptual complexity of their work (Temkin 2014). In his bio, Temkin identifies as an artist, influenced by Fluxus, and says he is interested in inherently broken patterns of thought.

In Light Pattern, these broken patterns are accessed through source code made up of photographs. Instead of using words, keystrokes, or physical controllers to communicate with the machine, Temkin uses photos to programme the computer; specifically he tracks changes in colour and exposure from one image to the next. Instead of merely translating pixel data directly, commands are determined by sets of three-digit histograms. Each number represents the delta from one photo to the next, and represents changes in one of three attributes: colour, aperture, and shutter speed (see Figure 22.14). In 2015 Temkin used photos of himself mouthing the words 'Hello World' to programme the computer to speak the words using a speech synthesis engine. The pictures could have been of anything; they just needed to have the correct changes. As one can imagine, it is very difficult to programme in the Light Patterns language. Temkin takes delight in the errors created by feeding his algorithm incorrect data, considering them integral to his aesthetic.

22.7 Live Video and Design: Hearing Red, Ocusonics, and the Giant Theremin

What Temkin does in code, the artist collective LoVid, Tali Hinkis, and Kyle Lapidus do in hardware. Their hybrid video–sound sculptures generate live signals instead of relying on stored data. LoVid's work 'revolves around a rethinking of the role of technology in the art-making process ... deconstruct[ing] our relationship with wires and screens.' They were the final artists in residence at the Experimental TV Center in upstate New York before the programme ended in 2011. They prefer to make their own tools because 'a tool influences what you can do creatively. Someone has built it with certain ideas and constraints; they made certain sacrifices and decisions along the way. When it comes to tools, everything that enables you constrains you, and everything that limits you empowers

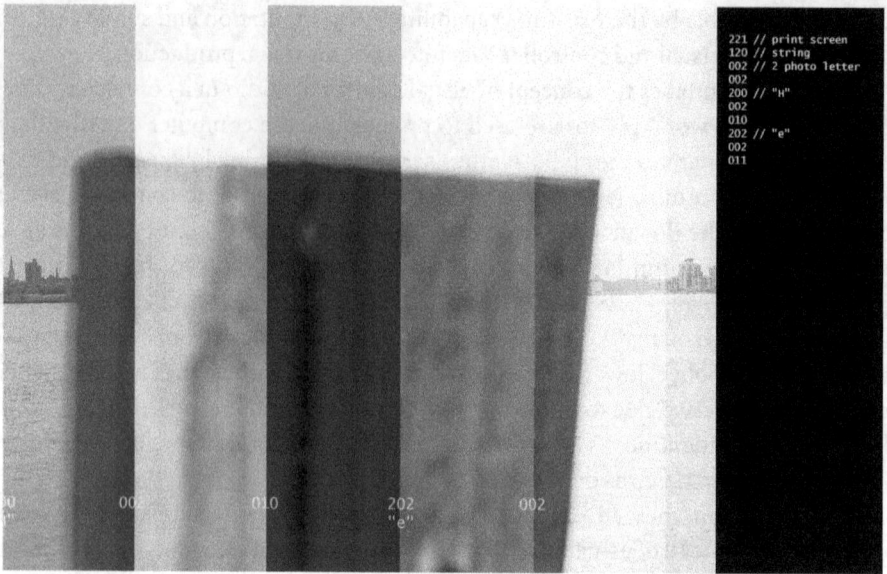

FIGURE 22.14 Daniel Temkin, Light Patterns. Courtesy of Daniel Temkin, licensed under the Creative Commons Attribution 4.0 International License.

you' (High, Hocking, and Jimenez 2008, 184). In their work *Hearing Red*, a pure red video signal is generated, and then played through a speaker, taking its form from the frequencies of the video signal (see Figure 22.15). The resultant drone that fluctuates slightly as the analog circuitry reacts to the environment, is mainly a static presence in the gallery. The artistry in the work is in the algorithm: the hardware needed to generate the video signal, translate the signal to the audio domain, and transduce the electrical signal into sound. This piece is unique in these case studies because the sonic result is not the primary aesthetic consideration; rather the concept behind the transcoding and the final physical form of the sculpture take priority over the sound of the work.

There are a huge number of pieces that translate real-time video into audio. LoVid's *Hearing Red* is an example of synthetic control, but many other algorithms take live camera data as input. An early piece using video tracking uses the eye itself as a controller. Andrea Polli began working with eye tracking in 1996 and she found she could track the pupil of her eye through some simple computer vision. In *Gape*, her first piece with eye tracking, she was able to trigger one of nine words through moving her eye. She didn't have complete control over her own movement, so the result was a 'collage of spoken text with multiple meanings' (Polli 1999, 407). The tracking measured the x–y coordinates of the pupil and mapped those numbers to prerecorded sound files. By using STEIM's BigEye software connected to Max, Polli developed a system she dubbed Ocusonics (see Figure 22.16). She used Ocusonics in a number of improvisational performances. Interestingly, she noticed that when playing the instrument she 'felt completely unaware of seeing anything at all, but was purely focused on the sound. The visual image became nothing more than a blur of colour and form, and the sense of hearing took precedence

FIGURE 22.15 LoVid, *Hearing Red*. Photograph by David La Spina, courtesy of LoVid.

over the sense of sight' (407). In her work, the organ that senses visible light becomes a controller of audio, involuntary motions create cascades of sound, and visual distractions are heard in the music.

While Polli tracked only a single eye, Australian artist Robin Fox tracked up to eight bodies for his twenty-first-century interpretation of the original iconic electronic instrument, the Theremin. Fox is comfortable being described as a sound artist, although his work also encompasses 'live digital media in improvised, composer and installation settings' (Eltham 2009). He most often cross-wires sound and visuals; what he loves 'about working in this way is you connect the sound electricity and the light electricity, and so you experience the sound as a particular vibration, and then that same vibration is being visualized. So you're seeing and hearing the same signal at the same time, so it creates a synaesthetic response' (Arnott-Hoare 2011). Fox is famous for his large-scale installations, and in 2011 the city of Melbourne asked him to create a Giant Theremin. Instead of using the electromagnetic system used in Theremin's original instrument, which he thought might microwave people, Fox used a fish-eye camera mounted at the top of a pyramid to track motion on the sidewalk. The system, programmed in Max/MSP, could individually track up to eight people, with volume controlled by distance from the sculpture and pitch controlled by lateral motion (NPR 2011). Some would describe the piece as a soundscape,

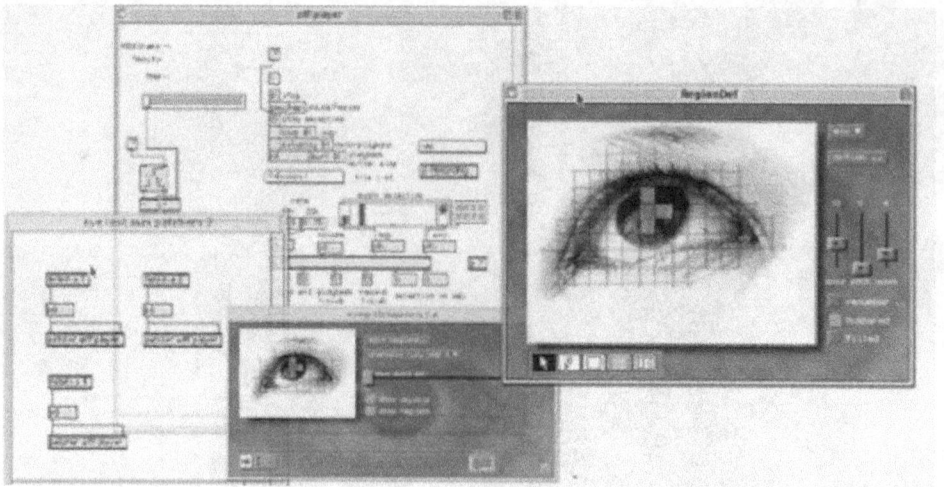

FIGURE 22.16 Andrea Polli, Ocusonics.

but researcher Jordan Lacey thinks Giant Theremin is 'firmly located within the domain of sound-art rather than soundscape design, as [it] translate[s] an idea onto a space that is not dependent on the existing sonic conditions of space; there is no reason that [this work] could not be located within multiple spaces across a city' (Lacey 2014, 6). In addition to functioning as a piece of public sound art, at the opening Stephanie Lake choreographed a dance to activate the sculpture to elicit a specific sonic response. This work is unique in the case studies because the public, rather than a specialized composer or performer, interacts with it. Even though UPIC was designed to democratize composition, it was still not a public work. In Fox's artist statement he explains,

> the ubiquitous 'i' devices that allow us to exchange photos by bumping phones together and the theremin-like controls of soap dispensers and taps across all major airport bathrooms has surely changed the way we interact with machines and our understanding of our position in relation to technology. But despite all this, there is still something simple and quite magical about an instrument that we can play from a distance, with no physical contact whatsoever. What I have attempted to do with this interactive instrument is extend the idea of the Theremin in both scale and function. It is designed to make people move and to make people listen, not only to their own sound but to the sound of others engaging with the instrument as well. Unlike the original Theremin, which was monophonic, the Giant Theremin is polyphonic. So people can play this instrument together, shifting their position in space in order to shift the pitch and loudness of their sound. When many are playing, it may be difficult to discern who has which voice: far from a problem this simply changes the nature of the game. (Fox 2011)

The algorithm behind Giant Theremin is intriguing because when only one person is in the range of the camera the relationship between visual and sound is fairly obvious, but as the number of people interacting with the sculpture increases, the transcoding becomes more complex.

The parameters that relate to x-axis, y-axis, and area are assignable. The creator of the sound design (or sound map) can control volume, pan, various filters and effects, frequency, loop speed, and more, depending on the mode used.

The instrument has three primary modes or voice settings:

(1) Waveforms

This follows the original Theremin instrument, employing only simple waveforms. The first is the classic sine wave or pure tone, the second is the square wave (only odd harmonics ... think the clarinet ...), the third is the triangle wave, and the fourth the sawtooth wave. These basic wave shapes and their combinations form the basis of most simple synthesis techniques.

(2) Samples

This mode uses one loadable file for each voice. This file can be any length and any sound and will be manipulated by the movement of the blobs on both the x and y axes and in terms of blob area. How the files are treated is modular. Each interactive parameter can be mapped to each interactive point (loop size, loop speed, loop direction, filter type, frequency and Q, and so on). This particular methodology is perhaps best suited to opening the Theremin up to invited designs from interested sound artists.

(3) MIDI

Finally, to facilitate a more traditional musical design, the Theremin is also capable of addressing MIDI instruments. This allows designers to bring their own sound fonts to the process, as well as sound files. Kontakt has been used to bridge the MIDI sources and sound files to the Giant Theremin output.

It is possible to assign any of the three environments mentioned above to eight discrete voices. In this way, voice one can produce a classic Theremin tone, voice two can manipulate a sample, voice three can control a midi instrument and so on.

Fox had to subtract the background from the camera image, detect moving objects in each frame, and associate the detections corresponding to the same object over time using computer-vision algorithms, and finally choose how to map that data to the audio. Fox also allowed other composers to define their own mappings, changing the sonic result of the tracking algorithm based on their own aesthetic choices.

22.8 INVERTING THE TROPE: PAPER SPEAKERS

In all of the projects so far, sound has been the result of a process assumed to end before the amplifier. Unlike the Ondes Martenot, with its three different types of diffusers (traditional, reverb spring, and harmonic gong), or the rotating speakers for the Hammond organ's Leslie speaker, most composer–inventors are content to not specify the details

FIGURE 22.17 Jess Roland, *Paper Speakers*. Courtesy of the artist.

of how exactly the sound will translate into the physical domain. The Whitney Brothers cared what their tools looked like (Whitney 1960), and installation artists LoVid and Robin Fox select their own speakers for their installations, but their decisions are pragmatic. Artist and scientist Jess Rowland wondered if there are 'alternative models of sound production that can lead to a different understanding of how audition and vision, physical material and the phenomenological realm, can be construed to create our experience' (Rowland 2013, 33). Rowland creates flat, flexible speaker arrays to put sound itself at the forefront of the artistic experience. Interestingly, the speakers do not have a flat frequency response, the visual aspect of the speaker is the primary consideration, the speaker is both the sounding object and a graphic score. These speakers function both as the algorithm which transcodes from vision to sound, changing the original audio signal during playback, but also as the sound production device itself (see Figure 22.17).

22.9 Conclusion

There are a number of different terms which can be applied to the translation from one domain to another: synaesthetic algorithm, sensory substitution, transcoding, cross-modal, intermedia, and so on. Mitchell Whitelaw teases out these differences in his article 'Synesthesia and Cross-Modality in Contemporary Audiovisuals'. His argument is subtle, but can be summarized thus:

> sensory substitution operates by mapping an otherwise absent modality into an existing one; absent vision into existing hearing. … However for most, audiovisual transcoding links two modalities, 'channels' already in perceptual use. Secondly, sensory substitution involves long-term integration and interaction with the environment; … there are some striking parallels, and transcoded AV certainly hints at artificial synesthesia and a rewired sensorium, but as bounded aesthetic objects these works cannot realise that perceptual transformation. (Whitelaw 2008, 268)

While synaesthesia is an extreme form of stimuli becoming interconnected, human thought is quite generally founded on the concept of connectivity and comparison. Language overflows with metaphors and analogies precisely because humans learn best by comparing new concepts with established ones; integrating new thoughts as reformulations of older ones. Algorithmic transcoding is thus a potent method for illuminating both inputs and outputs. As Nietzsche described it: 'Everything which distinguishes man from the animals depends upon this ability to volatilize perceptual metaphors in a schema, and thus to dissolve an image into a concept' (Nietzsche 1979, 84). The musicians, artists, and inventors in these case studies conceived of metaphors of expression, created algorithms to transcode data, and thus dissolved images into sound. As data become easier to accumulate, sort, and rearrange, we can expect continued exploration of the power, and limitations, of transcoding from the visual arena into the sonic domain.

BIBLIOGRAPHY

Arnott-Hoare, B. 'Robin Fox and the Giant Theremin'. *ArtsHub*, 2 December 2011.

Ashley, R., Dekleva, D., and Marulanda, F. *Yasunao Tone: Noise Media Language. Critical Ear* vol. 4. Dijon: Presses de Réel, 2007.

Boden, M. A., and Edmonds, E. A. 'What Is Generative Art?' *Digital Creativity* 20, nos. 1–2 (2009): 21–46. pg. 21

Bor, D., Rothen, N., Schwartzman, D. J., Clayton, S., and Seth, A. K. 'Adults Can Be Trained to Acquire Synesthetic Experiences'. *Scientific Reports* 4 (2014). https://www.nature.com/articles/srep07089.

Bourotte, R. 'The Upic and Its Descendants: Drawing Sound 2012'. Proceedings of the international symposium 'Xenakis: La musique électroacoustique / Xenakis: The Electroacoustic Music', Université Paris 8, May 2012.

Castel, L. B. *Mercure de France*, November 1725, 2562–2565.

Caswell, A. B. 'The Pythagoreanism of Arcimboldo'. *Journal of Aesthetics and Art Criticism* 39, no. 2 (1980): 155–161.

Chambel, T., Neves, S., Sousa, C., and Francisco, R. 'Synesthetic Video: Hearing Colors, Seeing Sounds'. In *Proceedings of the 14th International Academic MindTrek Conference: Envisioning Future Media Environments*, 130–133. New York: ACM, 2010.

Cowell, H. *New Musical Resources*. New York: A. A. Knopf, 1930.

Dal Farra, R. *A Journey of Sound through the Electroacoustic Wires: Art and New Technologies in Latin America*. PhD dissertation, University of Montreal, 2006.

Dal Farra, R. *Latin American Electroacoustic Music Collection*. Fondation Daniel Langlois, 2004. http://www.fondation-langlois.org/html/e/page.php?NumPage=556.

Dean, R. T., ed. *The Oxford Handbook of Computer Music*. Oxford: Oxford University Press, 2009.

Delhaye, C., Bourotte, R., and Kanach, S. 'The Centre Iannis Xenakis's Establishment at the University of Rouen'. In *Proceedings of the ICMC/SMC. Music Technology Meets Philosophy: From Digital Echos to Virtual Ethos*, edited by A. Georgaki and G. Kouroupetroglou, 1822–1827. National and Kapodistrian University of Athens, 2014.

Dupuis, A., and Dominguez, C. 'Digitally Extending the Optical Soundtrack'. In *Proceedings from the ICMC/SMC. Music Technology Meets Philosophy: From Digital Echos to Virtual Ethos*, edited by A. Georgaki and G. Kouroupetroglou, 133–139. National and Kapodistrian University of Athens, 2014.

Eltham. B. 'A "Game of Nomenclature"? Performance-Based Sonic Practice in Australia'. *Art Monthly Australia* 224 (2009): 441–446.

Enticknap, L. 'De Forest Phonofilms: A Reappraisal*'. *Early Popular Visual Culture* 4, no. 3 (2006): 273–284.

Fischinger, Oskar. 'Sounding Ornaments'. *Deutsche Allgemeine Zeitung* 8 July 1932.

Fox, R. 'Giant Theremin'. 2011. http://www.melbourne.vic.gov.au/AboutMelbourne/ArtsandEvents/GiantTheremin/Pages/GiantTheremin.aspx.

Fox, R. 'The Exploded Infant'. *Mona Blog*, 14 July 2014. http://monablog.net/tag/robin-fox/.

Franco, E., Griffith, N. J. L., and Fernström, M. 'Issues for Designing a Flexible Expressive Audiovisual System for Real-Time Performance and Composition'. In *Proceedings of the 2004 Conference on New Interfaces for Musical Expression*, 165–168. National University of Singapore, 2004.

Franssen, M. 'The Ocular Harpsichord of Louis-Bertrand Castel'. *Tractrix* 3, no. 1991 (1991): 15–77.

Garity, W. E., and Hawkins, J. N. A. 'Fantasound'. *Journal of the Society of Motion Picture Engineers* 37, no. 8 (1941): 127–146.

Giannakis, K., and Smith, M. 'Auditory-Visual Associations for Music Compositional Processes: A Survey'. In *Proceedings of International Computer Music Conference ICMC2000*. Berlin, 2000.

Grierson, M. S. *Audiovisual Composition*. PhD dissertation, University of Kent at Canterbury, 2005.

Greenlee, S. E. *Erratic Interpretation: Drawn Sound in Augur*. PhD dissertation, Brown University, Providence, RI, 2008.

Greenlee, S. E. 'Graphic Waveshaping'. In *Proceedings of the International Conference on New Interfaces for Musical Expression*, 287–290. Daejeon, Korea, 2013.

Hankins, T. L., and Silverman, R. J. *Instruments and the Imagination*. Princeton, NJ: Princeton University Press, 1995.

Harp, H. 'Thermal Image'. 2017. http://hilaryharp.com/project/thermal-image/.

Haller, R. A., ed. *First Light*. New York: Anthology Film Archives, 1998. pg. 63–69. pg. 65.

High, K., Hocking, S. M., and Jimenez, M., eds. *The Emergence of Video Processing Tools: Television Becomes Unglued*. Bristol and Portland, OR: Intellect, 2014.

Ione, A., and Tyler, C. 'Was Kandinsky a Synesthete?' *Journal of the History of the Neurosciences* 12, no. 2 (2003): 223–226.

James, D. E. *The Most Typical Avant-Garde: History and Geography of Minor Cinemas in Los Angeles*. Berkeley: University of California Press, 2005.

Kaganovsky, L., and Salazkina, M., eds. *Sound, Speech, Music in Soviet and Post-Soviet Cinema*. Bloomington: Indiana University Press, 2014.

Kandinsky, W. *Concerning the Spiritual in Art*. New York: George Wittenborn, 1966.

Karlstrom, P. J. *On the Edge of America: California Modernist Art, 1900–1950*. Berkeley: University of California Press, 1996.

Kendall, C. 'Technical Notes'. CD notes to Yasunao Tone, *Musica Iconologos*. Lovely Music, 1993.

Kröpfl, F. 'Electronic Music: From Analog Control to Computers'. *Computer Music Journal* 21, no. 1 (1997): 26–28.

Kirn, P. 'ANS—Amazing, Eerie Russian Optical Synth—Now on Every OS [Megaguide to ANS Old and New]'. *CDM*, 10 October 2013. http://createdigitalmusic.com/2013/10/ans-amazing-eerie-russian-optical-synth-now-ios-links-links-vids/.

Kreichi, S. 'The ANS Synthesizer: Composing on a Photoelectronic Instrument'. *Leonardo* 28, no. 1 (1995): 59–62.

Lacey, J. *Rupturing Urban Sound(scape)s: Spatial Sound Design for the Diversification of Affective Sonic Ecologies*. PhD dissertation, RMIT University, Melbourne, 2014.

Levin, G. 2000. *Painterly Interfaces for Audiovisual Performance*. PhD dissertation, Massachusetts Institute of Technology, 2000.

Lohner, H. 'The UPIC System: A User's Report'. *Computer Music Journal* 10, no. 4 (1986): 42–49.

Manning, Peter. 'The Oramics Machine: From Vision to Reality'. *Organised Sound* 17, no. 2 (2012): 137–147.

Makan, K. 'An Interview with Gerard Pape'. *Computer Music Journal* 27, no. 3 (2003): 21–32.

Marino, G., Serra, M.-H., and Raczinski, J.-M. 'The UPIC System: Origins and Innovations'. *Perspectives of New Music* 31, no. 1 (1993): 258–269.

McDonald, K. 'A Transcoding Framework'. *Computer and Information Science* (2007): 11–27.

Milicevic, M. 'Experimental/Abstract Film and Synaesthetic Phenomena 1725–1970'. 2005. http://myweb.lmu.edu/mmilicevic/pers/_PAPERS/exp-film.pdf.

Moholy-Nagy, L. *Vision in Motion*. Chicago: Theobald, 1947.

Moritz, W. *Optical Poetry: The Life and Work of Oskar Fischinger*. Bloomington: Indiana University Press, 2004.

Nietzsche, F. 'On Truth and Lies in a Nonmoral Sense'. In *Philosophy and Truth: Selections from Nietzsche's Notebooks of the Early 1870s*, edited and translated by D. Breazeale, 79–97. Atlantic Highlands, NJ: Humanities Press, 1979.

National Public Radio (NPR). 'A Giant Theremin Is Watching You Down Under'. *All Things Considered*, 8 December 2011. http://www.npr.org/2011/12/08/143363953/a-giant-theremin-is-watching-you-down-under.

Oram, D. 'Variable Electric Resistances'. US Patent 3156890 A. 1964.

Oram, D. 'Digitally Controlled Waveform Generators'. US Patent 3478792 A. 1969.

Patterson, Z. 'From the Gun Controller to the Mandala: The Cybernetic Cinema of John and James Whitney'. *Grey Room* 36 (2009): 36–57.

Polli, A. 'Active Vision: Controlling Sound with Eye Movements'. *Leonardo* 32, no. 5 (1999): 405–411.

Rothmans, P. 'Event: Center for Visual Music Presents an Evening with Barry Spinello'. *HPlus Magazine*, 9 May 2013. http://hplusmagazine.com/2013/05/09/event-center-for-visual-muisc-presents-an-evening-with-barry-spinello-los-angeles-area/.

Rowland, J. 'Flexible Audio Speakers for Composition and Art Practice'. *Leonardo Music Journal* 23, no. 1 (2013): 31–36.

Schedel, M., Fox-Gieg, N., and Keefer, C. 'Editorial'. *Organised Sound* 17, no. 2 (2012): 97–102.

Selz, Peter. 'The Aesthetic Theories of Wassily Kandinsky: And Their Relationship to the Origin of Non-Objective Painting'. *Art Bulletin* 39, no. 2 (1957): 127–136. pg. 134.

Smirnov, A. 'Sound out of Paper'. Last modified, 29 January 2009. http://asmir.info/gsound1.htm.

Smirnov, A., and Pchelkina, L. 'Russian Pioneers of Sound Art in the 1920s'. In *Red Cavalry: Creation and Power in Soviet Russia between 1917 and 1945* [exhibition catalogue], edited by Rosa Ferré, 210–232. Madrid: La Casa Encendida, 2011.

Spinello, B. 'Letter from Oakland, California'. In *Canyon Cinema: The Life and Times of an Independent Film Distributor*, edited by S. MacDonald, 123–125. Berkeley: University of California Press, 2008.

Stauffacher, F., ed. *Art in Cinema: A Symposium on the Avantgarde Film*. No. 21. New York: Arno, 1968.

Stuart, C. 'Damaged sound: Glitching and Skipping Compact Discs in the Audio of Yasunao Tone, Nicolas Collins and Oval'. *Leonardo Music Journal* 13 (2003): 47–52.

Temkin, D. 'Esolangs as an Experiential Practice'. *Esoteric.Codes: Programming Languages as Experiments, Jokes, and Experiential Art*, 16 December 2014. http://esoteric.codes/post/105355658088/esolangs-as-an-experiential-practice.

Thiebaut, J.-B., Healey, P. G. T., and Kinns, N. B. 'Drawing Electroacoustic Music'. In ICMC. 2008. http://www.eecs.qmul.ac.uk/~nickbk/ papers/ThiebautHealeyKinns-ICMC2008.pdf.

Vex, Z. 2016. 'The Candela Vibrophase'. *Zvex*, 18 February 2016. http://www.zvex.com/about-the-candela-vibrophase/.

Von Reichenbach, F. 'Process and Apparatus for Converting Image Elements to Electric Impulses'. US Patent 3719777, 6 March 1973.

Ward, O. 'The Man that Heard His Paintbox Hiss'. *The Telegraph* 10 June 2006.

Warwick, K., Shah, H., Vedder, A., Stradella, E., and Salvini, P. 'How Good Robots Will Enhance Human Life'. In *A Treatise on Good Robots*, edited by K. Tchoń and W. W. Gasparski, 3–18 1 New Brunswick, NJ: Transaction, 2014.

Whitelaw, M. 'Synesthesia and Cross-Modality in Contemporary Audiovisuals'. *The Senses and Society* 3, no. 3 (2008): 259–276.

Whitney, J. 'Moving Pictures and Electronic Music'. *die Reihe* 7 (1960): 61–71.

Whitney, J. H. *Digital Harmony: On the Complementarity of Music and Visual Art*. Peterborough, NH: Byte Books, 1980.

Whitney, J. 'To Paint on Water: The Audiovisual Duet of Complementarity'. *Computer Music Journal* 18, no. 3 (1994): 45–52.

Xenakis, I., Brown, R., and Rahn, J. 'Xenakis on Xenakis'. *Perspectives of New Music* 25, nos. 1–2 (1987): 16–63.

York, R. 'The Rhythmicon: An Electronic Rhythm Machine, Division of Musical Instruments'. Catalog No. 66.502. Washington, DC: Smithsonian Institution, 1992.

CHAPTER 23

....

DESIGNING INTERFACES FOR MUSICAL ALGORITHMS

....

JAMIE BULLOCK

23.1 INTRODUCTION

....

THE standard literature on the creation of software for musical applications has little to say regarding the topic of user interface or interaction design. Key texts such as
Designing Audio Effect Plug-Ins in C++ (Pirkle 2014) and *The Audio Programming Book*
(Boulanger and Lazzarini 2011) make almost no mention of requirements or techniques
for designing satisfying user interfaces, and instead focus primarily on audio processing
techniques. There are many books about digital signal processing (DSP) and programming in computer music, with significant duplication between texts, yet little has been
written about the general principles of designing interfaces for musical algorithms. One
might conclude from this that the value placed on user interface design within the music
software community is low, and that it is widely accepted that a synthesizer or an effect
can be regarded as a set of parameters to be directly controlled with sliders and knobs, or
connected together with patch cords. Much of this chapter will therefore be concerned
with making a case for the importance of interface design and its centrality in constructing and interacting with musical algorithms.

Interfaces for creating, manipulating, and controlling musical algorithms occupy an
esoteric position within the already niche field of musical interface design. It is tempting
to consider design criteria for such systems as somehow special and therefore subject to
design considerations that are outside the scope of normal human-computer interface
(HCI) practices (McDermott, Gifford, Bouwer, and Wagy 2013; Stowell and McLean
2013). Whilst this is true in part, such an approach risks overemphasizing 'difference'
at the expense of applying basic design practices. Put another way: novel requirements
do not necessarily imply novel HCI methods. I argue that all interfaces have the potential to be improved (often radically) through the application of standard approaches to
interface design, and in particular through collaboration and engagement with design

specialists. It is this balancing of often complex, specialist human requirements on the one hand with the need for carefully designed, usable, even beautiful interfaces on the other, which is at the centre of any successful user interface. The rich, open task of music making requires a rich, open interface (Stowell and McLean 2013); that interface also needs to be discoverable, learnable, instantly gratifying, and aesthetically pleasing. Domain-specific approaches to interfaces should build on and not replace well-established design principles. In the first part of this chapter I therefore present an overview and critical reflection on the broad topic of design within the development process, calling out salient principles for musical interface design. In the second part of the chapter I discuss a range of common design idioms found in software for musical algorithms and illustrate these with a selection of a number of software applications.

23.2 Designing Musical Interfaces

23.2.1 What Is Design?

In a 1972 interview, architect and designer Charles Eames defined design as 'A plan for arranging elements in such a way as to best accomplish a particular purpose' (Neuhart and Neuhart 1989). Whilst this may be useful as an all-encompassing definition, it is rather too general for our purposes. Schön (1992) theorized that designers are in a 'transaction' or 'reflective conversation' with the design situation. Drawing on Goodman's notion of 'worldmaking' (Goodman 1978), Schön situates the design 'transaction' as sensory and material (including both real and virtual worlds), whereby the designer determines what is relevant for the purposes of design, thereby creating a 'design world' in which to function. This 'design world' may be unique to a designer or shared amongst a design community, but in either case he regards design as a primarily social, communicative activity in which individuals are called upon to decipher each other's design worlds (Schön 1992). Schön views designers as being in a recursive process within a visual medium (e.g. drawing) in which the designer draws, sees in context, and draws again in relation to what is seen, thereby informing further designing (Schön and Wiggins 1992).

Schön's 'reflective conversation' bears strong similarity to the concept of sketching, which Buxton calls the 'archetypal activity of design' (Buxton 2007). Again, the focus is on *design as process* for the creation of new knowledge through what Buxton also calls a 'conversation': a cyclic action of sketching (creating), reading, and sketching (Figure 23.1). For Buxton, the purpose of the sketch is to suggest rather than confirm, with its value being as a catalyst to desired behaviours, conversations, and interactions. Qualities of sketches include: incompleteness, ambiguity, timeliness, disposability, and clear design vocabulary (Buxton 2007). The wide availability of cheap programmable micro-controllers and high-level domain-specific programming languages has enabled designers to sketch not only with pen and paper but also with fully functional code. The

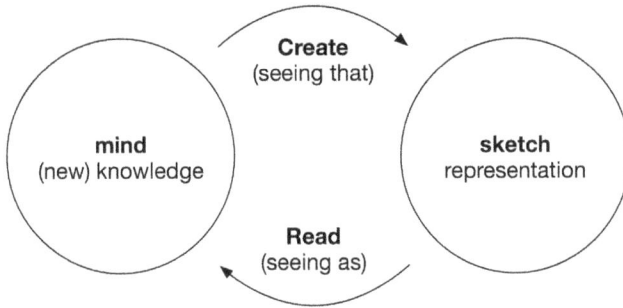

FIGURE 23.1 A 'conversation' between a sketch and the mind of the designer (adapted from Buxton 2007, 114). A sketch is created from current knowledge (top arrow). Reading or interpreting the current representation (bottom arrow) in turn creates new knowledge.

Arduino and Processing projects were initiated to explicitly facilitate this, even calling their programs 'sketches'(Reas and Fry 2014). In this regard, design is becoming increasingly inseparable from programming (coding), hacking, and technology mashups (Hartmann, Doorley, and Klemmer 2006). Or rather: design is the process; code, paper, and hardware are the media through which the process operates.

According to Norman (2013, 218), the role of designers is not to solve the problems directly presented to them but to uncover the 'real issues' implicit in the problem space. Once the true problem is identified, a range of possible solutions are explored before the designer converges on an optimal one. This process, referred to by Norman as 'design thinking' was formalized by the British Design Council as the double-diamond design process model (Design Council 2007), as shown in Figure 23.2.

It is also helpful to consider interface design as a form of communication between designer and user (Crilly, Good, Matravers, and Clarkson 2008). This communication is de-situated in the sense that the context of interface *use* may differ greatly from that of interface *design*, or may simply be different from the intended context. As designers, we therefore need to consider the user experience holistically, rather than looking at the interface in isolation:

> doing design requires more than making meaningful objects; it requires crafting whatever it is about objects that lets them participate in the creation of meaningful experiences. According to this view of meaning, the sense of an object cannot be separated from the experience that the object simultaneously sits in and helps to create. (Rheinfrank and Evenson 1996, 69)

Thus, in algorithmic music, the designed interface could be regarded as a communications medium, mediating the experience of musical creation and/or performance (Manovich 2013). The nature and design of the software are therefore critical to an understanding of the ways in which it mediates the experience of musicians. For example, well-designed software can lead to a pleasant and productive experience; poorly designed software can lead to real human frustration (Picard 1999). Furthermore,

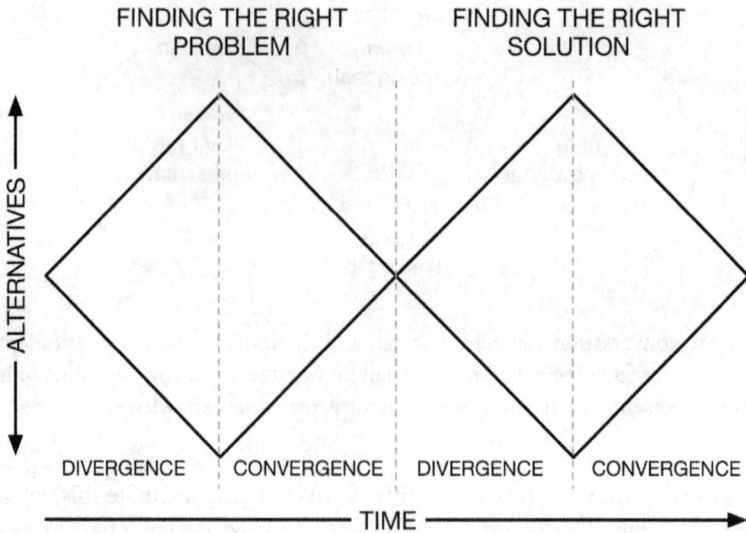

FIGURE 23.2 British Design Council double-diamond process model, as adapted in Norman 2013, 220.

it should be noted that no user interface is 'neutral' with respect to mediation; even a blank screen has social and cultural associations.

23.2.2 Usability

The field of usability provides a set of concepts, frameworks, and methods for both measuring how 'usable' an interface is and ensuring that user needs are embedded in the design and development process. Within the context of extant literature on interaction design for musical algorithms, many authors refer to the 'usability' of interfaces instinctively, assuming tacit understanding from the reader without giving any formal definition of the term. Others, for example Wanderley and Orio (2002), have devised domain-specific usability criteria without explicit reference to prior usability frameworks or established HCI methods. I will therefore focus here on key aspects of usability that may serve as helpful tools in the design of interfaces for musical algorithms, offering pointers for further reading.

The term 'usability' gained currency with the rise of personal computing in the 1980s, the broad goal being to ensure users were satisfied by designed computer systems and software. The widely cited ISO 9241 standard defines usability as 'the extent to which a product can be used by specified users to achieve specified goals with effectiveness, efficiency and satisfaction in a specified context of use' (ISO 1998). Shackel developed the concept of usability within a wider framework for system acceptability, his original four criteria being: effectiveness, learnability, adaptability, and attitude (Shackel 1986).

Nielsen extended Shackel's framework for system acceptability and formalized usability as having the following five attributes:

- Learnability: The system should be easy to learn so that the user can rapidly start getting some work done with the system.
- Efficiency: The system should be efficient to use, so that once the user has learned the system, a high level of productivity is possible.
- Memorability: The system should be easy to remember, so that the casual user is able to return to the system after some period of not having used it, without having to learn everything all over again.
- Errors: The system should have a low error rate, so that users make few errors during the use of the system, and so that if they do make errors they can easily recover from them. Further, catastrophic errors must not occur.
- Satisfaction: The system should be pleasant to use, so that users are subjectively satisfied when using it; they like it. (Nielsen 1993, 26)

Nielsen subsequently developed a widely used set of '10 Usability Heuristics for User Interface Design', which the reader is encouraged to explore (Nielsen 1994). However, as will be discussed later in this chapter, the dimensions of usability are not restricted to the criteria of Shackel or Nielsen, and a range of other methods may be applied in usability evaluation, particularly those that are non-task-oriented or that take account of context of use (cf. Greenberg and Buxton 2008).

HCI concepts such as usability and approaches to interface design for musical algorithms have evolved somewhat independently since the origins of algorithmic computer music in the 1950s (Hiller and Isaacson 1959). Only recently have authors started to formally enquire into the co-evolution and cross-fertilization between these two fields (Holland, Wilkie, Mulholland, and Seago 2013). Much of the literature that relates HCI practices to musical algorithms focuses on the development of digital musical instruments (DMIs), physical controllers, or control-synthesis mappings (Fiebrink et al. 2010; Kiefer, Collins, and Fitzpatrick 2008; McDermott et al. 2013; Wanderley and Orio 2002). However, it could be argued that many of the design considerations (in achieving musically meaningful interaction) can be adapted to musical interfaces in general. For example, Wanderley and Orio (2002) propose a set of usability criteria specifically for the evaluation of musical controllers, namely: learnability, explorability, feature controllability, and timing controllability, which may also be applied to tool-like or primarily linguistic interfaces such as live coding environments.

23.2.3 User Experience

Usability has more recently been subsumed into the wider field of 'user experience' (UX). UX takes a holistic approach to products, systems, and services, encompassing aspects of branding, visual design, identity, usability, interaction design, information architecture, discoverability, and socio-cultural context. However, at its most fundamental level, UX is concerned with affective notions of how systems make people *feel* and not only how efficient they are in facilitating productivity (Garrett 2003, 2–8). UX as we understand it today has its origins in the 1980s and 1990s HCI literature, an early source being Laurel's influential essay 'Interface as Mimesis', in which she writes:

> an interactive computer program may be intended to enable its user to do a variety of different things—find information, compose and format a document, play a game, or explore a virtual world. The user's goals for a given application may be recreational, utilitarian, or some combination of both, but it is only through engagement at the level of the interface that those goals can be met. An interface, like a play, must represent a comprehensible world comprehensibly. That representation must have qualities which enable a person to become engaged, rationally and emotionally, in its unique context. (Laurel 1986, 69)

Relating usability to UX, Cooperman and Lam (2014) suggest that making software usable is analogous to making food edible: edibility is the bare minimum that is required in preparing food for a customer—so usability should be regarded as a fundamental baseline in software design, not the ultimate goal. Thus, rather than focusing only on 'task', 'productivity', and 'ease-of-use', UX tends to be concerned with 'engagement', 'enjoyment', 'emotion', and even 'beauty' (Norman 2007).

23.2.4 User-Centred Design

'User-centred design' (UCD) could be regarded as a set of design principles and approaches aimed at achieving high standards of usability and user experience (Bevan 1999; Garrett 2003; Ritter, Baxter, and Churchill 2014, 44). The fundamental essence of UCD is an iterative process of understanding user needs, design/sketching, implementation, and evaluation. This process is shown in Figure 23.3. Many variations on this diagram can be found in the UCD literature, for example Norman (2013) provides a simplified version based on 'observation', 'idea generation', 'prototyping', and 'testing'. The version here incorporates Buxton's notion of sketching (Buxton 2007) and Ritter and colleagues' central theme of understanding user needs (Ritter, Baxter, and Churchill 2014). In the case where the sketching medium is code (Evans 2011, 17–32; Reas and Fry 2014), then 'sketch' and 'implement' essentially collapse into a single process element.

UCD has evolved to encompass a range of techniques from usability, Agile development, user experience, ethnography, and participatory design. The International Standards Organization defines a set of six core principles intended to be used for managing design processes in order to ensure interactive (hardware and software) systems are designed to enhance human–system interaction:

1. The design is based upon an explicit understanding of users, tasks, and environments.
2. Users are involved throughout design and development.
3. The design is driven and refined by user-centered evaluation

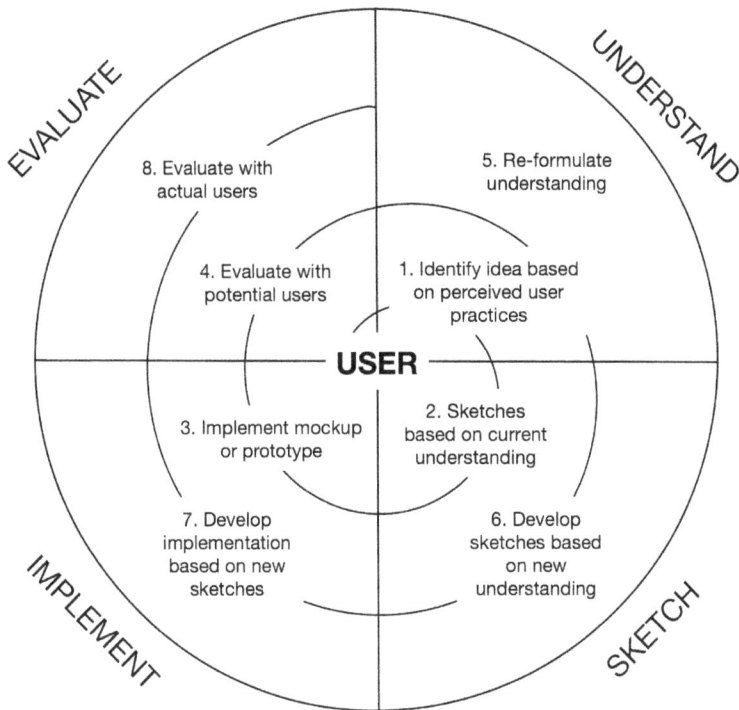

FIGURE 23.3 Iterative user-centred design process.

4. The process is iterative.
5. The design addresses the whole user experience.
6. The design team includes multidisciplinary skills and perspectives.(ISO 2010, 5–8)

Thus, UCD foregrounds the requirements, needs, and psychology of 'the user' as distinct from being centred on the needs of the developer, or an exploration of the capabilities of technology. UCD therefore requires a model, which represents the qualities of a typical user, or more often a set of archetypes that represent exemplars for user types (often called 'actors' or 'personas'). Lowdermilk (2013) highlights that whilst designs can be constructed against archetypal personas, testing and evaluation should be conducted using real users. It is widely accepted that up to thirty representative users should be used to surface significant design issues, but that often as few as five users is sufficient (Lowdermilk 2013).

Evaluation and testing methods employed in UCD need not follow the Schackel/Nielsen acceptability model described in section 23.2.2. Other qualitative, ethnographic, contextual, informal, reflective methods may also be used with equal value (Gould and Lewis 1985). The key in UCD is that a valid and appropriate evaluation method is employed iteratively and objectively and that methods are not applied arbitrarily just for their own sake or to give the impression of being scientific (Greenberg and Buxton 2008): The best evaluation method is the one that serves the needs of the interface in question.

23.2.5 Activity-Centred Design

'Activity-centred design' (ACD) is a philosophy and set of design processes that steers away from the user/system/task focus of UCD, and instead focuses on the *activity* being undertaken (Gay and Hembrooke 2004). ACD draws upon Activity Theory (Gay and Hembrooke 2004; Kaptelinin and Nardi 2012) in that the focus is on the combination of tasks and flows that construct an activity rather than individual tasks in and of themselves. Examples of activities include playing a musical instrument (Saffer 2009, 36), listening to music, and music production—hence ACD could be regarded as highly applicable to the design of interfaces for musical algorithms. Indeed, Small argues that 'music' should be considered not in terms of 'objects' (musical works, recordings, performances, scores), but as an activity:

> The fundamental nature and meaning of music lie not in objects, not in musical works at all, but in action, in what people do. It is only by understanding what people do as they take part in a musical act that we can hope to understand its nature and the function it fulfills in human life. (Small 1998, 8)

He therefore proposes that 'music' should not be viewed as a noun, but as a verb, 'to music', which he defines as follows:

> To music is to take part, in any capacity, in a musical performance, whether by performing, by listening, by rehearsing or practicing, by providing material for performance (what is called composing), or by dancing. (Small 1998, 9)

Thus, ACD in the context of music can be understood (at least in part) in terms of Small's concepts. ACD shares many traits with UCD, including the iterative design cycle shown in Figure 23.3 (Gay and Hembrooke 2004, 12); however, in general, ACD takes a more holistic approach encompassing activity, social/cultural context, and the mediating effects of technology (see also section 23.2.1). Thus, activity-centred design is to user-centred design as user experience is to usability, that is, a 'third paradigm' HCI process commensurate with the modern 'ubiquitous' design space. Most practical design methods draw on and combine elements from UCD, ACD, usability, and UX as appropriate to context. The common themes are an early and continual focus on users, tasks, and/or activities, empirical measurement, and iterative design. This may seem like common sense, but as found in a large-scale study by Gould and Lewis (1985), these seemingly common-sense principles are often not valued or followed in practice.

23.2.6 Music-Centred Design

Due to the largely experimental and nonmainstream nature of algorithmic art (cf. Taylor 2014, 249–267), many systems for musical algorithms arise out of a process of formal or informal 'action research' (McNiff 2013) and/or developer-centred design. However, if the algorithmic music software community is to find wide social and cultural impacts for its work, the scene is ripe for approaches that arm developers with appropriate concepts, tools, and methods to improve software design, such that results are more appealing to end users. I therefore introduce the concept of 'music-centred design'.

Music-centred design is a domain-specific design philosophy and method-ology, which draws upon user-centred design and activity-centred design, usability, and user experience, whilst emphasizing a number of music-specific approaches. These include the need to provision for artistic 'flow' in the creative process (Nash 2012; Nash and Blackwell 2015), through techniques such as progressive disclosure and layered abstraction (Nielsen 2006; Victor 2011) and a greater emphasis on the experience of non-user personas (or 'actors') such as members of live music audi-ences as highlighted in Small's theory of 'musicking' (Small 1998; see section 23.2.5). Borrowing Small's notion of music as a verb, music-centredness is the degree to which the requirements and needs of those engaged in music *activity* are fulfilled through experience design. Music 'actors' from all backgrounds deserve well-designed sys-tems that are tailored to these specific requirements; in particular, they need software that has embedded in it their needs, assumptions, and contexts of practice. According to Stowell and McLean:

> Much development of new musical interfaces happens without an explicit connec-tion to HCI research, and without systematic evaluation. Of course this can be a good thing, but it can often lead to systems being built which have a rhetoric of generality yet are used for only one performer or one situation. With a systematic approach to HCI-type issues one can learn from previous experience and move towards designs that incorporate digital technologies with broader application— e.g. enabling people who are not themselves digital tool designers. (Stowell and McLean 2013, 148)

This notion of 'enabling' is important, but I would suggest that we need to go further and not only enable the use of digital tools but inspire and empower music creation through the design of engaging and fulfilling experiences based around specific music activi-ties. Music-centred design should seek to accelerate serendipity in the process of music creation and performance. To achieve this, the way in which software mediates the algorithmic music experience should be centred on the knowledge, cultural practices, activity domains, and emotional needs of composers, performers, producers, and audi-ences. User interaction flows should be structured around achieving common func-tional tasks quickly and easily, but yet still allow for deeper exploration and expressivity. The need for richness and openness should be balanced with immediacy and transpar-ency. In a sense, algorithmic music is about music as process, so in the case of designing interfaces for musical algorithms, a key consideration is how musical processes are rep-resented and how this representation interrelates with music as activity. This is consid-ered in detail in sections 23.3.1–6.

23.2.6.1 *Designing for Flow*

'Flow' is a psychological theory concerning the notion of 'optimal experience' developed by Mihaly Csikszentmihalyi between the 1960s and 1990s, culminating in the seminal work *Flow: The Psychology of Optimal Experience* (Csikszentmihalyi 2002). In general terms, 'flow' is a mental state where a person becomes completely immersed in an activity for its own sake. Consciousness of time and self disappear and levels of engagement and concentration are high. The focus of Csikszentmihalyi's work has been on so-called autotelic experiences—activities that are an end in themselves as distinct from serving a specific goal. In order to investigate this, he conducted a series of studies in which he asked practitioners in autotelic activities—athletes, musicians, surgeons, chess players, dancers, rock climbers, etc.—to explain in detail their 'reasons for enjoying' the activities in which they engaged (Csikszentmihalyi 1977, 13). His findings led to the enumeration of the following characteristics of flow experience:

1. A challenging activity that requires skills
2. A merging of action and awareness
3. Clear goals and feedback
4. Concentration on the task at hand
5. Feeling of personal control
6. The loss of self-consciousness
7. A condensed sense of time (Csikszentmihalyi 2002, 49–66)

A number of approaches have been proposed for incorporating the notion of flow into a practical HCI framework and/or user interface design guidelines (Novak, Hoffman, and Yung 2000; Ghani and Deshpande 1994). However, Finneran and Zhang (2002) suggest a number of problems with such approaches, arguing that due to the conceptual vagueness of 'flow' it is difficult to operationalize in empirical HCI evaluation. Citing Ellis, Voelkl, and Morris (1994), they note that the 'skill' and 'challenge' constructs are complex and multidimensional, encompassing emotional, mental, or physical challenges and skills. Therefore, unidimensional scales may not serve as valid measures for these (Finneran and Zhang 2002).

The applications of flow within the context of algorithmic music have been explored by Nash and Blackwell, whereby a large-scale study was conducted into virtuosity and flow within a tracker environment (Nash 2012; Nash and Blackwell 2011, 2012, 2015). Nash's work foregrounds motor skill acquisition as a key to achieving virtuosity, and challenges the notion that musician-friendly software should be created through extended use of visual metaphor (as is the case with conventional digital audio workstation software and audio plugins). Instead, he advocates a concise textual representation, which he argues supports an engaging user experience, supporting flow, virtuosity, and the see–hear–understand learning cycle as described in Leman (2008). Nash and Blackwell (2015) present the concept of 'virtuosity-enabled' systems as those which enable a novice user to become expert. To support the designer in achieving this, they propose five design heuristics:

(1) Support learning, memorization, and prediction (or 'recall rather than recognition')
(2) Support rapid feedback cycles and responsiveness
(3) Minimize musical (domain) abstractions and metaphors
(4) Support consistent output and focused, modeless input
(5) Support informal interaction and secondary notation (Nash and Blackwell 2015, 392–396)

Of these, heuristic 3 is particularly worthy of note. Nash and Blackwell are suggesting that designers should not attempt to match a target user's conceptual model by 'abstracting' underlying functionality, for example, through a commonly understood metaphor, but instead should provide access to underlying 'low-level' functions that can be combined such that users may create their own abstractions (the implications of this will be explored further in sections 23.3.1 and 23.3.2). One problem in relating Csikszentmihalyi's characteristics of flow experience to music-centred design is that of musical time. Csikszentmihalyi observed that those within a 'flow state' experience a 'transformed' sense of time, in which time no longer seems to pass the way it normally does (Csikszentmihalyi 2002). But what does this mean in the context of music performance, where a performer may need to be acutely aware of immediate rhythmic placement whilst also envisaging or conceiving wider temporal structures?

The level of abstraction at which an interface is presented is also a crucial question in music-centred design. If the abstraction level is too high then the potential for individual expression becomes limited, but if the interface abstraction level is too low, we approach the infamous Turing Tar Pit, where 'everything is possible but nothing of interest is easy' (Perlis 1982, 10). One solution to this is to build interfaces where the level of abstraction is not fixed, but where multiple levels of abstraction are designed into the system, and users can move freely between them (cf. Victor 2011). McDermott and colleague (2013) introduce the idea of 'layered affordance' in relation to DMI design, whereby interface restrictions added initially to facilitate learnability could be lifted as the user became more proficient, thereby introducing structured challenge or reward, and facilitating flow. This concept could also be applied to noninstrument interfaces for musical algorithms, yet it need not be used to introduce additional difficulty, but rather to introduce functionality to the user only as and when required. This technique is known as 'progressive disclosure' (Jones 1989), and is commonly used to keep the comprehensiveness of information presented to users within manageable bounds. In many musical traditions, for example some forms of punk and folk music, social inclusiveness may be given precedence over feats of personal virtuosity (in the classical sense). In these instances, tiered abstraction and layered affordance become even more important as design patterns. Here, interfaces may need to facilitate 'flow' in the more general sense: by enabling openness and collaboration, and lowering barriers to participation.

23.2.6.2 *Easy and Expressive*

Like Nash, McDermott and his colleagues (2013) question the idea of 'ease of use' (as described in Nielsen's usability attributes) in relation to music interaction, arguing that this criterion is not formulated in terms that are appropriate to musical practice. Issues they identify include:

- Musicians don't 'use' instruments to be 'productive', they play them for enjoyment, entertainment, art, and so on (i.e. music making is an autotelic experience and not task-oriented);
- Many existing acoustic instruments would fail if tested against usability criteria;
- In order for instruments to allow for virtuosity and flow (e.g. Nash and Blackwell 2011), they need to have a built-in difficulty curve that allows progression over time, and therefore should not be 'easy' to learn.

Drawing upon McDermott and colleagues (2013), Nash and Blackwell (2011), Stowell and McLean (2013), and Small (1998), it could be concluded that in designing interfaces for musical algorithms we need to aim for interfaces that allow for virtuosity, flow, richness, and openness, but that are framed in terms of musical *activities* such as performing, composing, and dancing. Furthermore, in order to be music-centred, interfaces also need to be underscored by high standards of usability and experience design drawing on established HCI approaches. After all, the same criticisms that some authors (McDermott et al. 2013; Nash and Blackwell 2012) make in relation to Nielsen's usability heuristics (e.g. that they are not tailored to autotelic applications) could be levelled at Nash's heuristics, in that they are derived from and tailored towards the domain of tracker software and don't necessarily generalize to other algorithmic music approaches. The key is to employ heuristics if and when appropriate and not to treat them as rigid dogmas. Music-centred interfaces can be easy to use at one level and allow for structured challenge, exploration, or expressivity at another. This doesn't mean resorting to studio metaphors as suggested in Nash (2015), but instead conceiving new, as yet unforeseen approaches to interfaces and interaction based on a comprehensive understanding of the requirements of music activities and actors. The field of music interaction is in its infancy, and as we develop a theory of music-centred design it may be pertinent to look to the established field of computer game interface design for inspiration. Games share many traits with music-based interfaces, including the need for a base level of usability coupled with the need for extended 'play' within a consistent 'design world' (Bernhaupt 2010; Isbister and Schaffer 2008). Note that not all musical interfaces are instrument-like, and some may require capabilities for virtuosity, others may not. Other models for interfaces for musical algorithms include linguistic approaches (Aaron and Blackwell 2013; McLean 2014; Sorensen and Gardner 2010; Wang and Cook 2003), tool-like interfaces (Berg 2003; Intermorphic Ltd 2014), and combined instrumental, linguistic, and/ or tool-like approaches (Bullock, Beattie, and Turner 2011; Klügel, Friess, Groh, and Echtler 2011; Magnusson 2014). These will be covered in depth in section 23.3.

23.3 INTERFACE DESIGN CASE STUDIES

In following sections I present a range of common interface idioms used in software for musical algorithms. I focus particularly on interfaces and interface elements pertaining to stochastic, evolutionary, and conditional branching procedures, as these could be considered more significant with regard to the creation of new musical aesthetics. The range of interfaces associated with such musical algorithms is diverse. One common approach is that of 'meta-software': the output of the program is in itself a piece of software. This may take the form of a set of rules, or 'patch', that is then used to either generate musical output autonomously, or provide a user interface for further interaction. In software and hardware interfaces for musical algorithms, interfaces are often multifaceted. They may need to:

(1) Show the musical algorithm,

(2) Allow manipulation of the algorithm,

(3) Allow live control of the algorithm's parameters,

(4) Provide a visualization of the algorithm's output

(5) Facilitate communication between the algorithm and other software,

(6) Enable the algorithm and its data to be stored in a portable sustainable format.

Given that any conceivable user interface can be mapped to any conceivable sound-producing algorithm, it is possible (even likely), that makers and experimenters in new interfaces for musical expression will create incoherent interfaces that have a mismatch between the affordances of the user interface and the algorithm being controlled (Paradiso and O'Modhrain 2003). This lack of coherence may be particularly acute if the user has 'a priori' ideas relating to the musical algorithm that manifest in starkly different terms from those expressed in the interface. An example would be an onscreen 'piano' keyboard being used to trigger drum loops by clicking on the keys with a mouse.

In the following sections I examine a number of software applications in relation to their primary interface types: patchers, text-based, graphical, node-based, physics-based, and 'No UI'. Case studies have been chosen as exemplars of particular design approaches, or because they have elements of illustrative benefit.

23.3.1 Patchers

By far the most popular and widely used software interfaces for musical algorithms are so-called patcher environments, namely Cycling 74's commercial Max application (http://cycling74.com 2015) and its open-source cousin Pure Data (http://puredata.info 2015). Max (Figure 23.4) is available for Mac OS X and Windows and is written using the JUCE framework in conjunction with native components, thus combining the familiarity of

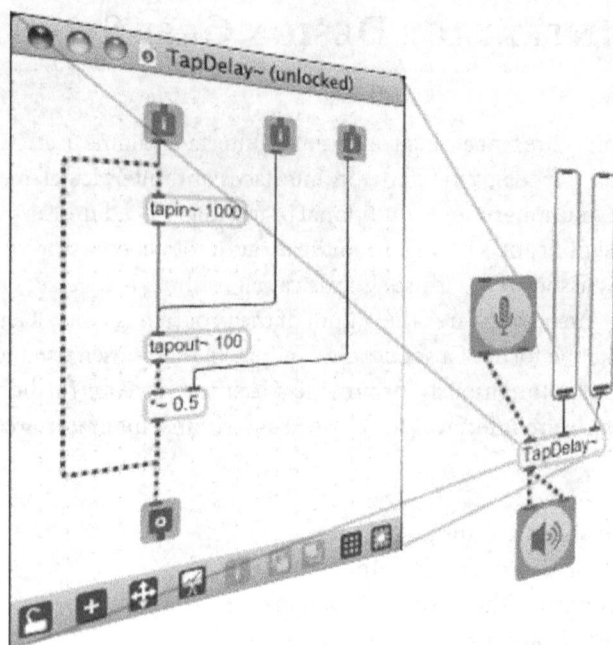

FIGURE 23.4 The anatomy of a Max abstraction. The rightmost object, 'TapDelay~', is a user-created abstraction with two sliders, and audio input/output connect to its inlets and outlets. The left of the diagram shows the 'inside' of the abstraction, where the action audio-processing network for the TapDelay has been created.

platform-native user interface (UI) 'chrome' with a primary interface that is consistent between platforms. Pure Data comprises an audio server written in C and a TCL/Tk UI. Pd is available for many platforms including Windows, Mac OS, Linux, and other Unix variants. It is also available on mobile platforms such as iOS and Android via the libPd library.

The interface design for both Max and Pure Data can be traced back to the Patcher software developed by Miller Puckette in the mid-1980s as a means of controlling IRCAM's 4X computer via a MIDI connection (Puckette 1988). The early development of the Patcher provides an example of a music-centred design approach (see section 23.2.6), and could be regarded as a response to the difficulties encountered by musicians in engaging with technology at computer music research centres in the early 1980s. By providing an interface that was easily comprehensible, the Patcher sought to bridge the gap between scientific research and musical practice at IRCAM (Born 1995, 215–219). Writing in 2002, seventeen years after the initial development of Patcher, Puckette writes:

> In my experience, computer music software most often arises as a result of inter-actions between artists and software writers (occasionally embodied in the same person, but not in my own case). This interaction is at best one of mutual enabling and mutual respect. The design of the software cannot help but affect what computer music will sound like, but we software writers must try not to project our own

musical ideas through the software. In the best of circumstances, the artists remind us of their needs—which often turn out quite different from what either of us first imagined. To succeed as computer music software writers, then, we need close exposure to high-caliber artists representing a wide variety of concerns. Only then can we can identify features that can solve a variety of different problems when in the hands of very different artists. (Puckette 2002, 31)

He subsequently highlights the importance of testing 'in context' within artistic practice based on actual musical works and performances:

it was Philippe Manoury's *Pluton*, whose production started in Fall 1987 and which premiered in July 1988, that spurred Max's development into a usable musical tool. The Pluton patch, now existing in various forms, is in essence the first Max patch. (Puckette 2002, 34)

The primary interface to both Max and Pure Data is a blank graphical 'canvas' onto which the user places 'text boxes', which represent either 'objects' that encapsulate specific pieces of functionality, or 'messages' that can be used to pass data to objects. Each object has zero or more inlets and/or outlets that enable it to communicate messages to other objects via 'connections' drawn on the canvas as lines between inlets and outlets. These connections are analogous to 'patch cords' traditionally found in analog synthesizers, hence the term 'patching environment'. Max and Pure Data additionally provide a range of user interface widgets and controls (such as sliders) that can be connected to object inlets and outlets and used for generating or visualizing message data. Finally, Max and Pd allow users to write their own objects in the Max and Pd languages, thus providing a facility for abstraction. Figure 23.4 shows the anatomy of a Max 5 abstraction.

Max is widely regarded as the lingua franca of interactive 'live' electronic music. By introducing a programming language, which through its use of a quasi-graphical elements doesn't 'feel' like a programming language, Max has facilitated an era of self-sufficient musician-programmers and a culture of open-ended experimentation. It could be argued that the simple, reusable design of Max has enabled a generation of musicians to work independently without the need for technical support or imposing a priori artistic limitations on their creative process. Like Nash (see section 23.2.6.1), Puckette argues that a 'reified' interface, free from developer-defined abstractions, is best for music creation and that 'expressiveness is enhanced by concreteness (the opposite of abstraction), directness, and straightforwardness' (Puckette 2002, 37). However, despite Puckette's aim of neutrality, Max encodes and communicates a set of assumptions that were relevant in the time, place, and context of its development. Whilst Max's reified design provides an open and experimental environment for artistic expression, it also restricts access to many users through a steep learning curve and requirement for low-level DSP knowledge. The 'DIY' culture encouraged by Max and Pure Data also leads to widespread duplication of effort, whereby many users reinvent common high-level components, distracting needlessly from musical goals (Bullock, Beattie, and Turner 2011).

23.3.2 Text-Based

Text has been one of the primary interfaces to musical algorithms since the origins of computer music in the 1950s, and makes use of the natural availability of keyboards as input devices to computation. Text-based environments that support algorithmic operations include the MUSIC-N derivatives (e.g. Csound), as well as SuperCollider, ChucK, and Lisp-based software such as Common Music. Typically in text-based environments, users enter code in a domain-specific language, using a standard text editor. That code is then either compiled using a provided compiler or sent to an interpreter on the fly. Therefore text-based environments require users to learn a number of separate interfaces: the music or audio programming language, the text editor (which they may already know), and the configuration interface to the compiler, interpreter, or server. Users may also have an additional interface to documentation (e.g. accessed through a web browser or Unix 'man' pages). Culturally this approach follows the Unix philosophy, whereby functionality is divided up between separate executable components, thus providing opportunities for reuse of tools (Gancarz 1995). An alternative approach is to provide users with an integrated environment that includes 'everything' in a single application context. The recent Sonic Pi software is a good example of the latter approach (http://sonic-pi.net 2015). Sonic Pi is a free audio-based live coding environment designed to support computing and music lessons within schools (Aaron and Blackwell 2013). It is available (with easy installers) for Raspberry Pi, Windows, and Mac OS X, and comes with a range of musically interesting examples that enable users to not only get started quickly, but also learn the language to the point that complex musical results can be created. Sonic Pi is thus an example of a 'virtuosity-enabled' environment with good potential for 'flow' (Nash and Blackwell 2015). Sonic Pi is written in the Qt framework and whilst it has some native controls, it is overall nonnative in its look, feel, and behaviour. Sonic Pi could be regarded as an example of *inclusive* design (Norman 2013): in creating a simplified, friendly, accessible, and usable environment for schoolchildren Aaron has succeeded in designing an environment suitable for adult users also.

Another interesting approach to text-based UI is the Texture live coding software by McLean (2011). Texture is a hybrid visual-textual interface to the Tidal library for live coding of musical pattern (McLean and Wiggins 2010, 2011). Texture provides a novel approach by combining textual language with a graphical geometric notation that carries syntactic meaning. This is distinct from 'patcher' software like Pure Data (described in section 23.3.1), in which the visual configuration of the graphical elements (boxes, lines) bears no effect on the interpretation of the program. Texture explores this fuzzy distinction between 'text-based' and 'graphical' programming. According to McLean:

We find that Texture also has high closeness of mapping, as the visual representation of trees within trees corresponds well with the hierarchical structure of the pattern that is being composed. This echoes the tree structures common in music analysis, and indeed we would expect significant correspondence between the Texture structure and the listener's perception of it. (McLean and Wiggins 2011, 627)

McLean (2014) subsequently surfaces the design challenges associated with the development of live coding environments for algorithmic music in which he conducted a survey of fifteen Tidal users in order to elicit findings related to the software's user experience. The results showed highly positive results, with users indicating that the software helped them be 'more creative'. However, as noted in the paper, these results can't necessarily be regarded as a reflection of the wider Tidal community. It would also be interesting to see the survey repeated with a user group more diverse with respect to their a priori experience—for example including users with no previous experience of Tidal, or indeed of programming environments in general.

23.3.3 Live Visualization

There are often multiple actors engaged in algorithmic music making, each experiencing interfaces for musical algorithms from a different perspective. In the case of algorithmic software design intended for live performance, audience members could be considered as such actors. Audience members perceive musical performances visually as well as aurally, and may have many preconceptions such as an expectation of a causal link between performer movement and sound output (as is the case with acoustic instruments). A key consideration in designing interfaces for musical algorithms is therefore what (if anything) should be displayed to the audience. According to Paradiso,

> one aspect of musical performance that is often overlooked in the design of electronic musical instruments is that of the audience's understanding of how the instrument is played. An artist playing an acoustic instrument usually exploits a mental model that the audience has of the instrument's action-to-response characteristics, allowing virtuosity to be readily appreciated. In contrast, electronic controllers, especially those with overly complex high-level mappings or relatively hidden interfaces (e.g., a laptop keyboard or bioelectric sensors) can often confuse an audience, who often can't follow and relate to what the performer is doing. (Paradiso and O'Modhrain 2003, 4)

A computer-based system could be said to present an 'anxious object' (Rosenberg 1973) from the perspective of the audience, whereby they are unsure as to the purpose, function, or intention of the computer in the musical work. In the case of live algorithmic music, the same anxieties may also apply to the *performer* using the system, particularly in a case where the performer is not the system designer. Furthermore, we may have assumptions about computers, software, and technology in general; thus computer-based algorithmic systems may present the following uncertainties for performer and/or audience:

- What is the system doing?
- Is the system working?
- Were outputs from the system intended?
- Who is doing what?

According to Tanaka (2000) and Gurevich (2015), 'cause and effect' between physical input and sound output is essential to audience understanding and enjoyment of a performed work. By contrast, the presence of an 'anxious object' (which defies comprehension) has the capacity to leave the audience feeling baffled, bewildered, angered, or bored (Gablik 2004). Extending these anxieties to potential nondeveloper users of systems for musical algorithms, creators of such systems are faced with weighty design challenges requiring significant skill to resolve. According to O'Modhrain:

> The challenge, therefore, is to determine how audiences disentangle judgments about performance error from judgments concerning instrument failure—both result in a breakdown between a performer's intent and the outcome of an action, but the source of the former is a mistake by the player, whereas the source of the latter is a failure of the technology. (O'Modhrain 2011, 33)

Of course, not all computer-based interfaces for musical performance are instrument-like. In the tradition of 'laptop performance', it is usual to make no pretence that the performer is using the laptop 'as an instrument'. Instead, the performer typically uses the laptop, and possibly a range of input devices and controllers, to initiate and modify musical processes. However as foregrounded by Collins (2003) this approach can sometimes leave the audience anxious as to whether the performer is indeed 'any good' or simply 'watching TV' or 'checking email'. As a reaction to this approach, the TOPLAP movement (a group of programmer-musicians and visual artists interested in live coding) developed a manifesto explicitly setting out to addresses this by insisting 'Obscurantism is dangerous. Show us your screens' (TOPLAP 2015). In practice this has mainly been implemented literally: live coders have projected the contents of their screen such that it is visible to the audience. Taking a music-centred design approach (and taking the perspective of audience members within the music activity), we should first seek a detailed understanding of audience perceptions of extant live coding practices. Anecdotal evidence from McLean (McLean and Wiggins 2011) suggests such an understanding has not yet been reached by the live coding research community, but that audience reactions to the projection of screens have been varied.

One option is to project something other than the contents of the live coder's screen, for example an alternative representation of the running state of the program and interactions of the coder. Max (section 23.3.1) has a Presentation Mode that shows to the performer only those user interface elements they have determined to be relevant for live performance.

Similarly, an audience representation could be devised that expresses the system in more familiar terms and aesthetics for the general audience. Laptop performers within dance music traditions have been creating intricate music-centred visualization systems for decades, leading to the evolution of a video jockey (VJ) tradition, working alongside DJs who are responsible for the music element. Prominent examples of live visual presentations and manipulations in dance music include Coldcut—who helped pioneer the VJ movement, Optique Vid Tek, and Étienne de Crécy—one of the early artists to incorporate projection mapping into a dance music set (cf. Barrett and Brown 2009; Faulkner, D-Fuse 2006). A design vocabulary for audience-facing laptop screen projections, drawing on the rich VJ culture and tradition offers fertile ground for further exploration.

23.3.4 Node-Based

Node-based interfaces represent musical algorithms using a visual notation comprising a graph of connected nodes (usually circles), each of which has associated note or instrument assignments and/or rules. The graphs in node-based interfaces can typically be created, executed, and manipulated in realtime, where the user adds new nodes, connections, and rules interactively by clicking or tapping on the elements in the graph. Node-based interfaces therefore represent a good example of a user interface based on direct manipulation of the running algorithm (Norman and Draper 1986, 87–124). Whilst node-based interfaces don't imply any particular musical style, they are well suited to pattern-based approaches and the generation of varying but consistent textures. I have found they are less well suited to building architectonic musical structures with strong sectional contrasts.

One of the most successful node-based interfaces is Nodal by Jon McCormack (McIlwain, McCormack, Dorin, and Lane 2006). In Nodal, each node in the graph (network) represents a musical note, and the length of each connection (edge) represents the amount of musical time between each note. Edges can be one-way or bi-directional, and nodes can 'signal' their edges sequentially, in parallel, or randomly. These concepts are all indicated visually on the network, for example a node that fires in parallel has two parallel lines drawn on it. Nodal is written using the Qt framework using primarily native widgets and UI chrome, with a custom-drawn central canvas. It is very easy to install and comes with comprehensive documentation and examples. However, one aspect of the design that could prove problematic for users favouring all-inclusive systems is that Nodal (like early versions of Max) works as a generator of control information and therefore provides only a rudimentary General MIDI sound set. It is left to the user to explore more musically satisfying control-to-synthesis mappings using a synthesizer or sampler of choice.

Mcilwain and McCormack's comments on the design of Nodal are salient and hint at a music-centred design approach:

> Software of this kind can potentially be too complex for a user to configure in a meaningful way. For this reason the overriding design constraint was that the software be

as simple and as intuitive to use as possible. Therefore a significant proportion of the software development process was given over to addressing user interface issues. This design constraint also determined that the software would present of a limited set of possibilities while still providing a flexible and multifaceted network environment. This raises the question of how to select design features that fall within the design constraint. Part of the process that was adopted was to look at how an existing piece of music might be created and represented as a real-time nodal network. This should then provide clues as to what features and designs may be the most useful. The nursery rhyme Three Blind Mice was selected for its apparent simplicity. This selection was made with the rationale that if a nodal network was not able to generate a simple melody then it would be hardly likely to generated anything more sophisticated. (McIlwain and McCormack 2005, 97)

As with Sonic Pi (discussed in section 23.3.2), Nodal therefore serves as an example of *inclusive design* (Norman 2013) and through its simple but open approach makes simple things easy and complex things possible. In general Node-based interfaces offer an elegant solution to several of the design problems highlighted in sections 23.3.1–3 because they allow for open-ended programmability whilst having a discoverable and highly learnable visual syntax.

23.3.5 Physics-Based

Physics-based interfaces are interfaces that are influenced by or incorporate elements of real-world physics, including mechanics (dynamics, kinematics, kinetics) and 'digital physics', incorporating input from sensors such as accelerometers and touch screens (Bourg and Bywalec 2013). Interest in physics-based interfaces for algorithmic music has grown rapidly with the widespread availability of mobile touchscreen devices equipped with high-resolution displays and a variety of sensor inputs. Introducing physics and quasi-physicality can enhance the interface's sense of tangibility in touch-screen interaction (Löwgren 2009). Such an approach can greatly improve the immediacy of an interface by creating a close coupling between intention and effect. Löwgren refers to this coupling as interface 'pliability', which he defines as follows:

> The use of a digital artifact is characterized as pliable if it feels like a tightly connected loop between eye and hand, between action and response. A pliable interaction is one where the user is drawn into a sense of shaping the digital information with her fingertips, even though the actual artifact might employ standard, nontactile interaction techniques such as mouse, keyboard and a display monitor. (Löwgren 2006, 3)

The degree of interface pliability is therefore likely to be highest when the tactile, visual, and auditory cues are all aligned to a common model, thus minimizing potential 'dissonance' between the modalities. However, achieving a high level of interface pliability shouldn't be confused with using physics to model realism. The aim is often instead to create a believable sense of digital reality that goes beyond the real, using physical modelling as a kind of illusion, to 'draw in' the user. Wigdor (2011) refers to this as 'super realism', an intuitive extension of the real that simultaneously feels 'natural' and pushes beyond what is physically natural so that experiences do more than is possible in the real world.

The application of physical behaviours varies widely in interfaces for musical algorithms. A common idiom is that of the 'bouncing ball', whereby a circle is animated around a 2D canvas and made to 'bounce' off other objects (lines, edges, other circles) when the objects' edges touch. Circles may be given artificial inertia such that they drift around indefinitely with direction determined by the angle at which they 'hit' other objects, or may 'fall' under artificial gravity. A good example of this style of design is the Soundrop iOS app by Max Weisel. Soundrop expresses its musical algorithm through an arbitrary number of balls (represented as small white circles) which are generated sequentially and fall under artificial gravity from a 'hole' on the screen. The user can interact with the software by drawing coloured straight lines of any length. When a ball hits a line, a sound is produced and the ball bounces off with inertia, with line length corresponding to musical pitch and line colours corresponding to different timbres. Thus a range of musical rhythms and textures can be created by drawing different combinations of lines and changing the physics settings and/or timbre assignments. Many algorithmic music 'apps' exist for mobile platforms, employing a range of physics-based metaphors and a full survey of these is beyond the scope of this chapter. The reader is encouraged to refer to chapter 34 in this volume which discusses these in more detail.

23.3.6 No UI

Software that implements machine listening in order to respond 'live' to audio input could in one sense be regarded as an example of a 'No UI' approach to interface design. 'No UI' is a user interface meme that represents a way of thinking about design situated as the 'anti-' of prevalent screen-based interfaces. The idea that 'the best interface is no interface' is an attempt to introduce a counter-narrative that focuses on activity-centred design (see section 23.2.5) and emphasizes processes, data, sensing, connectedness, and minimalist aesthetics to create highly positive user experiences. Whilst the idea of 'No UI' has been popularized in Krishna (2015), the core design values associated with it can be traced back to influential thinkers in the UI design space such as Don Norman who, in an 1990 interview said the following:

> The real problem with the interface is that it is an interface. Interfaces get in the way. I don't want to focus my energies on an interface. I want to focus on the job . . . I don't want to think of myself as using a computer, I want to think of myself as doing my job. (Laurel and Mountford 1990, 210)

'No UI' is also strongly related to Basden's notion of 'proximal' user interfaces, derived from Polanyi's concept of 'proximal tacit knowledge' (Fuchs 2001):

> A proximal user interface (PUI) is one that is so 'natural' that it does not 'get in the way'. It embodies the norm that the software should be able to become, as it were, part of the user or, as Polanyi [...] stated, 'proximal'. By contrast, conventional UIs are 'distal', relating to the user via a 'dialog' of commands, messages, clicks, menus, etc. To achieve proximity, the UI should not consume the user's thinking and attention nor interrupt the flow of thinking within the task. (Basden 2003, 29)

Machine listening approaches to interacting with musical algorithms (aka 'live algorithms') could be regarded as exemplars of the No UI or PUI approaches. The 'user', in this case an instrumental performer, simply 'plays' their instrument and listens to the computer-generated response through a loudspeaker system. The interface in the case of such machine listening systems is therefore sound itself (Bullock 2009). The field of machine listening is very much a research area, with systems often developed on an ad hoc basis for specific research projects or musical works in environments such as Max (see section 23.3.1) and SuperCollider. Some of the better-developed machine listening systems include OMax, by the IRCAM Music Representations Team, and Jnana, developed at Stanford University's CCRMA group. Jnana is written in JavaScript and implemented as plugin for Ableton Live, using the Max for Live extension. Machine listening systems offer a promising music-centred design approach for instrumental performers with little technical knowledge since traditional modes of screen-based interaction are mostly avoided. In the ideal case the performer 'just starts playing', and the system 'responds' in a musically meaningful manner, offering potential for immersion and flow (see section 23.2.6.1). A complimentary 'No UI' approach involving the application of machine learning systems in facilitating human musical expression is discussed in chapter 12 by Fiebrink and Caramiaux.

It is worth noting that the concepts of No UI or 'seamless' computing, where the interface is 'invisible' to the user, are by no means unchallenged within the interface design community. There exists an 'anti-No UI' sentiment as characterized by the writings of designers and academics such as Timo Arnall, who problematizes No UI in his widely cited blog post 'No to No UI' (Arnall 2013). Furthermore, Ratto (2007) argues that what he terms 'seamless infrastructures' (the No UI philosophy applied to connected systems) may even be 'ethically problematic' due to their invitation to a passive relationship between people and aspects of their social and material environment. It may be the case that the concept of No UI has served as a placeholder for something that we, as designers, do not yet fully understand—the realm of immersive sensor-based interaction, as distinct from traditional screen-based computing. In the case of sensor-based environments, the UI may often be 'invisible', but it doesn't 'not exist'. We therefore need to begin developing new design concepts and frameworks that enable us to reason effectively about these new technologies.

23.4 CONCLUSIONS

I have explored in this chapter a range of approaches to designing interfaces for musical algorithms. We have seen that there is great diversity in both interface design patterns and quality of execution. The experimental nature of generative and algorithmic music has inspired similarly experimental approaches to the look, feel, and workings of musical software. However, the value of such novelty is sometimes lost on end users due to a poor user experience. If the overall experience of software is negative, due to factors which may be considered of low priority (such as website, branding, visual design, installation, documentation, and so on), then many users will simply give up and discard the software. There is, after all, no shortage of alternative programs and 'apps'. The lessons from this chapter are therefore twofold: to impress upon the reader the importance and value of design in and of itself, and to suggest that iterative design practices should be consciously and pragmatically incorporated into research and development in order to meet the needs of algorithmic music as an activity.

New music technologies have the capacity to open up previously unforeseen musical possibilities, to create new relationships between body and sound, and to empower and inspire new forms of cultural expression with potential to transform musical and aesthetic ideals (Greene and Porcello 2005, 3–9). However, in order to achieve tangible, widespread, and lasting benefits to creative practice, novel and innovative approaches need to be underscored by a meticulous attention to the needs of users throughout the design process. By combining an awareness of music-specific requirements with user-centred and activity-centred design principles, music-centred design is a means to achieving this.

BIBLIOGRAPHY

Aaron, S., and Blackwell, A. F. 'From Sonic Pi to Overtone: Creative Musical Experiences with Domain-Specific and Functional Languages'. In *Proceedings of the First ACM SIGPLAN Workshop on Functional Art, Music, Modeling Design*, 35–46. New York: ACM, 2013. doi:10.1145/2505341.2505346.

Arnall, T. 'No to NoUI'. *Elastic Space*, 13 March 2013. http://www.elasticspace.com/2013/03/no-to-no-ui.

Barrett, L., and Brown, A. R. 'Towards a Definition of the Performing AudioVisualist'. In *Proceedings of the Australasian Computer Music Conference*, 46–55. Brisbane: Australasian Computer Music Association, 2009.

Basden, A. 'Guidelines and Freedom in Proximal User Interfaces'. *Human-Computer Interaction: Theory and Practice*, edited by J. A. Jack and C. Stephanidis, 1:28–32. Mahwah, NJ: Lawrence Erlbaum, 2003.

Berg, P. 'Using the AC Toolbox: A Tutorial'. The Hague: 2003. http://kc.koncon.nl/downloads/ACToolbox/files/AC_Toolbox_Tutorial.pdf.

Bernhaupt, R. *Evaluating User Experience in Games*. Edited by R. Bernhaupt. London: Springer, 2010. doi:10.1007/978-1-84882-963-3.

Bevan, N. 'Quality in Use: Meeting User Needs for Quality'. *Journal of Systems and Software* 49, no. 1 (1999): 89–96. doi:10.1016/S0164-1212(99)00070-9.

Born, G. *Rationalizing Culture: IRCAM, Boulez, and the Institutionalizaion of the Musical Avant-Garde*. Berkeley: University of California Press, 1995. doi:10.1525/california/9780520202160.001.0001.

Boulanger, R. C., and Lazzarini, V. *The Audio Programming Book*. Cambridge, MA: MIT Press, 2011.

Bourg, D. M, and Bywalec, B. *Physics for Game Developers: Science, Math, and Code for Realistic Effects*. Sebastopol, CA: O'Reilly Media, 2013.

Bullock, J. *Implementing Audio Feature Extraction in Live Electronic Music*. PhD dissertation, Birmingham City University, 2008.

Bullock, J., Beattie, D., and Turner, J. 'Integra Live: A New Graphical User Interface for Live Electronic Music'. In *Proceedings of the International Conference on New Interfaces for Musical Expression*, 387–392. New York: NIME, 2011.

Buxton, B. *Sketching User Experiences: Getting the Design Right and the Right Design*. Burlington, MA: Morgan Kaufmann, 2007.

Collins, N. 'Generative Music and Laptop Performance'. *Contemporary Music Review* 22, no. 4 (2003): 67–79. doi:10.1080/0749446032000156919.

Cooperman, H., and Lam, J. *Making Things Special: Tech Leadership from the Trenches*. Book 1. Edited by S. Berkun. Seattle, WA: Jackson Fish Market, 2014. http://makingthingsspecial.com/wp-content/uploads/2016/08/Making-Things-Special-Book-1.pdf.

Crilly, N., Good, D., Matravers, D., and Clarkson, P. J. 'Design as Communication: Exploring the Validity and Utility of Relating Intention to Interpretation'. *Design Studies* 29, no. 5 (2008): 425–457. doi:10.1016/j.destud.2008.05.002.

Csikszentmihalyi, M. *Beyond Boredom and Anxiety*. San Francisco, CA: Jossey-Bass, 1977.

Csikszentmihalyi, M. *Flow: The Psychology of Optimal Experience*. New York: Random House, 2002.

Design Council. *Eleven Lessons: Managing Design in Eleven Global Companies Desk Research Report*. London: Design Council, 2007.

Ellis, G. D., Voelkl, J. E., and Morris, C. 'Measurement and Analysis Issues with Explanation of Variance in Daily Experience Using the Flow Model'. *Journal of Leisure Research* 26, no. 4 (1994): 337–356.

Evans, B. *Beginning Arduino Programming*. Berkeley, CA: Apress, 2011.

Faulkner, M., D-Fuse, ed. *VJ: Audio-Visual Art + VJ Culture*. London: Laurence King, 2006.

Fiebrink, R., Trueman, D., Britt, C., Nagai, M., and Kaczmarek, K. 'Toward Understanding Human-Computer Interaction in Composing the Instrument'. In *Proceedings of the International Computer Music Conference*. New York: International Computer Music Association, 2010.

Finneran, C. M., and Zhang, P. 'The Challenges of Studying Flow within a Computer-Mediated Environment'. In *Proceedings of the Eighth Americas Conference on Information Systems*, 1047–1054. Dallas, 2002.

Fuchs, T. 'The Tacit Dimension'. *Philosophy, Psychiatry, and Psychology* 8, no. 4 (2001): 323–326. doi:10.1353/ppp.2002.0018.

Gablik, S. *Has Modernism Failed?* New York: Thames and Hudson, 2004.

Gancarz, M. *The UNIX Philosophy*. Woburn, MA: Elsevier, 1995.

Garrett, J. J. *The Elements of User Experience*. San Francisco, CA: Peachpit, 2003.

Gay, G., and Hembrooke, H. *Activity-Centered Design: An Ecological Approach to Designing Smart Tools and Usable Systems*. Cambridge, MA: MIT Press, 2004.

Ghani, J. A., and Deshpande, S. P. 'Task Characteristics and the Experience of Optimal Flow in Human–Computer Interaction'. *Journal of Psychology* 128, no. 4 (1994): 381–391.

Goodman, N. *Ways of Worldmaking*. Indianapolis, IN: Hackett, 1978.

Gould, J. D., and Lewis, C. 'Designing for Usability: Key Principles and What Designers Think'. *Communications of the ACM* 28, no. 3 (1985): 300–311. doi:10.1145/3166.3170.

Greenberg, S., and Buxton, B. 'Usability Evaluation Considered Harmful (Some of the Time)'. In *Proceedings of the Twenty-Sixth Annual CHI Conference*, 111–120. New York: ACM, 2008. doi:10.1145/1357054.1357074.

Greene, P. D., and Porcello, T. *Wired for Sound*. Middletown, CT: Wesleyan University Press, 2005.

Gurevich, M. 'Skill in Interactive Digital Music Systems'. In *The Oxford Handbook of Interactive Audio*, edited by K. Collins, B. Kapralos, and H. Tessler, 315–332. Oxford: Oxford University Press, 2015.

Hartmann, B., Doorley, S., and Klemmer, S. R. 'Hacking, Mashing, Gluing: a Study of Opportunistic Design and Development'. *Pervasive Computing* 7, no. 3 (2006): 46–54.

Hiller, L. A., and Isaacson, L. M. *Experimental Music: Composition with an Electronic Computer*. New York: McGraw-Hill, 1959.

Holland, S., Wilkie, K., Mulholland, P., and Seago, A., eds. *Music and Human-Computer Interaction*. London: Springer, 2013. doi:10.1007/978-1-4471-2990-5.

Intermorphic Ltd. 'Noatikl User Guide'. 10 December 2014. http://intermorphic.com/noatikl/guide/pdf/intermorphic_com_noatikl_guide.pdf.

International Standards Organization (ISO). *Ergonomic Requirements for Office Work with Visual Display Terminals (VDTs)—Part 11: Guidance on Usability*. ISO 9241-11:1998.

International Standards Organization (ISO). *Ergonomics of Human-System Interaction—Part 210: Human-Centred Design for Interactive Systems*. ISO 9241-210:2010.

Isbister, K., and Schaffer, N. *Game Usability*. Burlington, MA: CRC Press, 2008.

Jones, M. K. *Human-Computer Interaction*. Englewood Cliffs, NJ: Educational Technology, 1989.

Kaptelinin, V., and Nardi, B. 'Activity Theory in HCI: Fundamentals and Reflections'. *Synthesis Lectures on Human-Centered Informatics* 5, no. 1 (2012): 1–105.

Kiefer, C., Collins, N., and Fitzpatrick, G. 'HCI Methodology for Evaluating Musical Controllers: a Case Study'. In *Proceedings of the International Conference on New Interfaces for Musical Expression*, 87–90. Genoa: NIME, 2008.

Klügel, N., Friess, M. R., Groh, G., and Echtler, F. 2011. 'An Approach to Collaborative Music Composition'. In *Proceedings of the International Conference on New Interfaces for Musical Expression*, 32–35. Oslo: NIME, 2011.

Krishna, G. *The Best Interface Is No Interface*. San Francisco: New Riders, 2015.

Laurel, B. K. 'Interface as Mimesis'. *User Centered System Design: New Perspectives on Human-Computer Interaction*, edited by D. A. Norman and S. W. Draper, 67–85. Hillsdale, NJ: Lawrence Erlbaum, 1986.

Laurel, B. K., and Mountford, S. J. *The Art of Human-Computer Interface Design*. Boston, MA: Addison-Wesley Professional, 1990.

Leman, M. *Embodied Music Cognition and Mediation Technology*. Cambridge, MA: MIT Press, 2008.

Lowdermilk, T. *User-Centered Design*. Sebastopol, CA: O'Reilly Media, 2013.

Löwgren, J. 'Pliability as an Experiential Quality: Exploring the Aesthetics of Interaction Design'. *Artifact* 1, no. 2 (2006): 85–95.

Löwgren, J. 'Toward an Articulation of Interaction Esthetics'. *New Review of Hypermedia and Multimedia* 15, no. 2 (2009): 129–146. doi:10.1080/13614560903117822.

Magnusson, T. 'Improvising with the Threnoscope: Integrating Code, Hardware, GUI, Network, and Graphic Scores'. In *Proceedings of the International Conference on New Interfaces for Musical Expression*, 19–22. London: NIME, 2014.

Manovich, L. *Software Takes Command*. London: Bloomsbury, 2013.

McDermott, J., Gifford, T., Bouwer, A., and Wagy, M. 'Should Music Interaction Be Easy?' In *Music and Human-Computer Interaction*, edited by S. Holland, K. Wilkie, P. Mulholland, and A. Seago, 29–47. London: Springer, 2013.

McIlwain, P., and McCormack, J. 'Design Issues in Musical Composition Networks'. In *Proceedings of the Australasian Computer Music Conference*, 96–101. Brisbane: Australasian Computer Music Association, 2005.

McIlwain, P., McCormack, J., Dorin, A., and Lane, A. 'Composing with Nodal Networks'. In *Proceedings of the Australasian Computer Music Conference*. Adelaide: Australasian Computer Music Association, 2006. http://acma.asn.au/conferences/acmc2006.

McLean, A. 'Making Programming Languages to Dance to: Live Coding with Tidal'. In *The 2nd ACM SIGPLAN International Workshop*. New York: ACM, 2014. doi:10.1145/2633638.2633647.

McLean, A., and Wiggins, G. 'Tidal-Pattern Language for the Live Coding of Music'. In *Proceedings of the Seventh Sound and Music Computing Conference*. Barcelona, 2010. http://smcnetwork.org/resources/smc2010.

McLean, A, and Wiggins, G. 'Texture: Visual Notation for Live Coding of Pattern'. In *Proceedings of the International Computer Music Conference*, 621–628. Huddersfield: International Computer Music Association, 2011.

McNiff, J. *Action Research*. London: Routledge, 2013.

Nash, C. *Supporting Virtuosity and Flow in Computer Music*. PhD dissertation, University of Cambridge, 2012.

Nash, C., and Blackwell, A. F. 'Tracking Virtuosity and Flow in Computer Music'. In *Proceedings of the International Computer Music Conference*, 575–582. Huddersfield: International Computer Music Association, 2011.

Nash, C., and Blackwell, A. F. 'Liveness and Flow in Notation Use'. In *Proceedings of the International Conference on New Interfaces for Musical Expression*. Ann Arbor, MI: NIME, 2012. http://nime.org/archives.

Nash, C., and Blackwell, A. F. 'Flow of Creative Interaction with Digital Music Notations'. In *The Oxford Handbook of Interactive Audio*, edited by K. Collins, B. Kapralos, and H. Tessler, 387–404. Oxford: Oxford University Press, 2015.

Neuhart, J., and Neuhart, M. 'What Is Design? An Interview with Charles Eames'. In *Eames Design: The Work of the Office of Charles and Ray Eames*, edited by J. Neuhart and M. Neuhart, 14–15. New York: Harry N. Abrams, 1989.

Nielsen, J. *Usability Engineering*. San Francisco, CA: Morgan Kaufmann, 1993.

Nielsen, J. 'Heuristic Evaluation'. In *Usability Inspection Methods*, edited by J. Nielsen and R. L. Mack, 25–62. New York: John Wiley & Sons, 1994.

Nielsen, J. 'Progressive Disclosure'. *Nielsen Norman Group*. 4 December 2006. http://www.nngroup.com/articles/progressive-disclosure.

Norman, D. A. *Emotional Design*. New York: Basic Books, 2007.

Norman, D. A. *The Design of Everyday Things*. New York: Basic Books, 2013.

Norman, D. A., and Draper, S. W. *User Centered System Design: New Perspectives on Human-Computer Interaction*. Hillsdale, NJ: Lawrence Erlbaum Associates, 1986.

Novak, T. P., Hoffman, D. L., and Yung, Y.-F. 'Measuring the Customer Experience in Online Environments: a Structural Modeling Approach'. *Marketing Science* 19, no. 1 (2000): 22–42.

O'Modhrain, S. 'A Framework for the Evaluation of Digital Musical Instruments'. *Computer Music Journal* 35, no. 1 (2011): 28–42.

Paradiso, J. A., and O'Modhrain, S. 'Current Trends in Electronic Music Interfaces'. *Journal of New Music Research* 32, no. 4 (2003): 345–349.

Perlis, A. J. 'Epigrams on Programming'. *SIgPLAN Notices* 17, no. 9 (1982): 7–13.

Picard, R. W. 'Affective Computing for HCI'. In *HCI International 1999: Proceedings of the 8th International Conference on Human-Computer Interaction*, edited by H.-J. Bullinger, 829–833. Hillsdale, NJ: Lawrence Erlbaum, 1999.

Pirkle, W. *Designing Software Synthesizer Plug-Ins in C*. Boca Raton, FL: CRC, 2014.

Puckette, M. 'The Patcher'. In *Proceedings of the International Computer Music Conference*, 420–429. San Francisco, CA: International Computer Music Association, 1988.

Puckette, M. 'Max at Seventeen'. *Computer Music Journal* 26, no. 4 (2002): 31–43. doi:10.1162/014892602320991356.

Ratto, M. 'Ethics of Seamless Infrastructures: Resources and Future Directions'. *International Review of Information Ethics* 8 (2007): 20–27.

Reas, C., and Fry, B. *Processing*. Cambridge, MA: MIT Press, 2014.

Rheinfrank, J., and Evenson, S. 'Design Languages'. In *Bringing Design to Software*, edited by T. Winograd, 63–85. New York: ACM, 1996.

Ritter, F. E., Baxter, G. D., and Churchill, E. F. *Foundations for Designing User-Centered Systems*. London: Springer Science and Business Media, 2014.

Rosenberg, H. *The Anxious Object: Art Today and Its Audience*. New York: Collier, 1973.

Saffer, D. *Designing for Interaction*. San Francisco, CA: New Riders, 2009.

Schön, D. A. 'Designing as Reflective Conversation with the Materials of a Design Situation'. *Research in Engineering Design* 3, no. 3 (1992): 131–147. doi:10.1007/BF01580516.

Schön, D. A., and Wiggins, G. 'Kinds of Seeing and Their Functions in Designing'. *Design Studies* 13, no. 2 (1992): 135–156. doi:10.1016/0142-694X(92)90268-F.

Shackel, B. 'Ergonomics in Design for Usability'. In *Proceedings of the Second Conference of the British Computer Society, Human Computer Interaction Specialist Group on People and Computers: Designing for Usability*, edited by M. D. Harrison and A. F. Monk, 44–64. New York: Cambridge University Press, 1986.

Small, C. *Musicking: The Meanings of Performing and Listening*. Middletown, CT: Wesleyan University Press, 1998.

Sorensen, A., and Gardner, H. 'Programming with Time: Cyber-Physical Programming with Impromptu'. *ACM Sigplan Notices* 45, no. 10 (2010): 822–834. doi:10.1145/1869459.1869526.

Stowell, D., and McLean, A. 'Live Music-Making: A Rich Open Task Requires a Rich Open Interface'. In *Music and Human-Computer Interaction*, edited by S. Holland, K. Wilkie, P. Mulholland, and A. Seago, 139–152. London: Springer, 2013.

Tanaka, A. 'Musical Performance Practice on Sensor-Based Instruments'. *Trends in Gestural Control of Music* 13, nos. 389–405 (2000): 284.

Taylor, G. D. *When the Machine Made Art*. London: Bloomsbury, 2014.

TOPLAP. 'ManifestoDraft'. *Toplap*. Last modified, 14 November 2010. http://toplap.org/wiki/ManifestoDraft.

Victor, B. 'Up and Down the Ladder of Abstraction: A Systematic Approach to Interactive Visualization'. *Worry Dream*, October 2011. http://worrydream.com/LadderOfAbstraction.

Wanderley, M. M., and Orio, N. 'Evaluation of Input Devices for Musical Expression: Borrowing Tools from HCI'. *Computer Music Journal* 26, no. 3 (2002): 62–76.

Wang, G., and Cook, P. R. 2003. 'ChucK: A Concurrent, on-the-Fly Audio Programming Language'. In *Proceedings of the International Computer Music Conference*. Singapore: International Computer Music Association, 2003.

Wigdor, D., and Wixon, D. *Brave NUI World*. Burlington, MA: Elsevier, 2011.

CHAPTER 24

ECOOPERATIC MUSIC GAME THEORY

DAVID KANAGA

24.1 INTRODUCTION

THE word 'game' is used in a very broad sense throughout this chapter, having as much to do with the intuitive free play of improvising musicians as with the strict rule-abiding and goal-bound rational play associated with the optimal strategy-functions, winners and losers of mathematical game theory. The word is used to mean 'formalized play' of any sort. The chapter's focus is on computer games as algorithmic musical forms, and the broad meaning of 'game' is adopted because computer games are not games in the game theorist's formal sense; they are interactive objects of a much less specific sort, but nonetheless strictly formalized playspaces. When musical games are created without first recognizing that the economism of the goal-pursuing rational agent is not a necessary component of a game's structure, the forms which emerge tend to limit the potential for player improvisation in overly controlling ways. An example is *Rock Band*, one of the most commercially successful musical computer games ever released. Despite its robust modular sampling system, which allows multitrack stems of studio recordings to be recombined piecemeal at runtime, the flexible improvisatory potential of this form is completely ignored. Every 'off-note' that otherwise could function as a 'creative misreading' of the original song and a goad to the expansion of the improvisatory imagination via strange material reconfigurations, is instead reduced to a Boolean 'mistake'. The game imposes an economistic attitude on the player; instead of responding to off-note or off-time input by triggering sample playback at the wrong time or the wrong pitch, which would afford the player a wide variety of spatio-temporal freedoms, it instead plays a scratchy 'mistake' sound effect, thus treating the player as a creature who needs to be told what to do—an insubordinate labourer, or one of Pavlov's dogs—this, as opposed to a musician with an individual voice and unquantifiable creative potential.

The breadth of computer games must not be thought to be limited in any way by game theoretical formalisms and definitions of games. The medium has the capacity to encompass and integrate *all playable forms*—all interactive algorithms—which computers are able to embody. This is a totalizing effect which requires different metaphors. The best may be the image of games as opera, following George Lewis's theorization of interactive computer music, writing that 'interactivity suggests a new model for the Gesamtkunstwerk, one which is wary of hubris and disinclined to over-weening centralization strategies' (Lewis 2009, 460). A musical approach to computer games, which deals not only with the sound of game, but with the totality of its form, including the organic-mechanical conversation between human and machine, proceeds from an understanding of this sort. An opera achieves an integral synthesis of parts and wholes. The word is the plural of *opus*; it is translated from the Latin as 'works'. Operatic form can be considered essentially pluralistic, the 'multi' in multi-media. This book you are holding contains descriptions of many algorithmic forms, and it is possible for one game to implement a synthetic ecosystem composed of just as many such diverse algorithms. 'The many become one and are increased by one' (Whitehead 1978, 21) Algorithms are combined freely in a game, like notes or themes in a piece of music. The many components which are combined to create a musical game can be explicitly musical or ostensibly 'nonmusical'. They may be overtly playful (Sicart 2011) or apparently boring and utilitarian (Bogost 2007). One computer game might be synthesized from a Frankenstein-like patchwork of many instrumental 'tool' mechanics borrowed from the likes of Ableton, Photoshop, Excel, Facebook, and so on; mixed in with 'toy' mechanics modeled on bouncing balls, silly putty, and finger paints; situated in architectural spaces informed by cities, parks, wilderness; propelled forward by 'narrative' mechanics modeled on chatroom bots, AI agents, artificial organisms, and so on. And such 'nonmusical' texture in a game may be interwoven with musical algorithms which create playable form resembling the formal patterning and affordances of instruments like keyboards and flutes; or which resemble musical abstractions, such as serial twelve-tone rows and the circle of fifths; or which imitate structural invariances of compositional forms like sonata-allegro, fugue, and so on. These diverse forms may be combined into strange new hybrids, such as a keyboard that is cyclically transposed through the circle of fifths every time a note is played, and which automatically plays an orchestral accompaniment in a nightmarish sonata-allegro style, allowing for any note to be played from the keyboard, instrumentation changing as time goes by, according to the proportions of that form, tempo determined in a constant flux by the relative density or sparseness of the keys pressed by the player. All of these forms and many more can be freely combined. With computer games, there are no strict lines between instruments, compositions, embodied theories, artificial musicians. Nor is there is a strict line between musical and nonmusical games. Musical games simply magnify latent rhythmical–structural–harmonic—and crucially, improvisatory—tendencies existing in the temporal flows of 'nonmusical' interactive forms at large. Much of the musical potential of games is to be found in a marriage of explicit musical form with supposed nonmusical activity.

The remainder of this chapter is organized into two block sections which attempt to draw various connections between algorithmic patterning of musical and nonmusical form in games. The first deals with particular practical and speculative strategies for composing interactive game music, dealing with a variety of specific situations in turn. The second is attempts to encapsulate games and music into a formal generalization which includes both as instances of a broader class of shifting possibility spaces, such that music and game forms may be studied as formal isomorphisms of one another, as played forms; it considers the ways in which this chapter's formalized musical approach to games contrasts with that of game theory, which has inspired composers such as Xenakis to regard games as essentially rule-bound, goal-pursuing structures; game theory's abstract economic approach is contrasted with a broader musical analysis's concrete ecological approach.

24.2 MUSIC AS A TANGIBLE PROCESS

In his book *Audio-Vision*, Michel Chion distinguishes between three modes of listening, one of which is intimately related to the experience of agency afforded by music games. He calls this mode causal listening, and defines it as 'listening for the purpose of gaining information about the sound's cause' (1994, 25) A bouncing ball drops, bounces, rises, and repeats, and we hear the gradual accelerando of successive bounces speeding up to a buzz, illustrating the loss of potential energy and corresponding diminution of vertical height caused by each bounce. The Earth's gravity plays this piece with the material of the ball. Play is causal influence. Chion's account of causal listening allows listeners to identify players, as it were, but music games emphasize the first-person experience of a player's direct participation with a cause, implicating themselves as a listener + player in a cascading chain of causality—music games allow players to kick the bouncing ball.

The process of composing music games is one of mapping musical parameters either directly or indirectly to free variables controlled by inputs, with which the player may tangibly affect the outcome of the music, in a fully real instance of causal influence. There are obviously many possible approaches to designing music games. The possibilities suggested by the choice of inputs alone (microphone, qwerty keyboard, mouse, MIDI keyboard, MIDI control change knobs, etc.) is enough to fill many books. I would like to narrow the focus, then, and highlight two approaches to compositional form, where the first presupposes a 'nonmusical' game-space which becomes musical by way of its dynamic soundtrack, and where the second speculates as to the possibility of 'adapting' existing musical objects into games, in a process conceptually analogous to that of adapting books to screen, though mechanically very different.

24.2.1 Designing Soundtracks and Composing Games

For the first approach, we begin by assuming the existence of a game which is nearly complete but as yet has no sound added to it. This is our blank slate. Any sounds whatsoever can be added, and it is the job of composer and sound designer to decide how this is to be done. Composing a soundtrack, and designing its interactions to a degree, is the kind of work that a musician is most likely to be hired for in a collaborative game development setting. This topic is the subject of books such as Karen Collins's *Playing with Sound* (Collins 2013), and Winifred Phillips's *A Composer's Guide to Game Music* (Phillips 2014), which cover orthodox scoring and sound design practices in some detail, from those shared with film scoring to those simple dynamic processes which are native to the games medium.

The process of soundtracking can be as simple as putting a piece or sequence of background music in the game, but this will do little to make the music and activity of play feel causally related to one another. It will do little to take advantage of Chion's *synchresis*. The music can become more integrated with the nature of the game-play, however, if the composer first tunes into causal relations and rhythmic events existing in the algorithmic movements of the game as a visual and physical–tangible thing, and then treats these events as triggers in a musical space resembling a vast musical instrument which is performed in part by the player. In this latter approach, many of the grounding algorithms for composition, rhythmic and structural, can be understood to be provided by the game as ready-made. The game functions as a metric scaffolding upon which an open-form musical composition is built.

For instance, in a game where the player is given the power to jump, a variable can be defined that measures the player's distance from the ground. This variable can, in theory, be attached to any musical parameter whatsoever, and its range scaled to map neatly onto the desired range of a musical effects. It could control the pitch of a simple oscillator or speed of an audio file, mapped such as to create a loopy bend up and down when the player jumps and lands. Or, it could be attached to two (or more) virtual volume sliders, *a* and *b*, controlling several different looping audio files, such that when the player is 'on the ground', file *a*, thick with bass tone, plays at 100 percent, and as the player approaches the top of the jump, *a* fades out to 0 percent as *b*, with a light floating texture, fades to 100 percent.

Alternately, in developing a score for this same jump, the distance from the ground could be ignored. The press which triggers the jump could at the same time trigger a single sounding event, with the ground collision of the landing yet triggering another, the way sound effects work. This event-triggering method could be made richer, and less repetitive, by triggering one event from an array of possible events, where the selected event is determined by the position of the player, or by the previous event played, or any number of other parameters. The event triggered need not be a single sound file. It might be a change in state of some more global aspect of the soundtrack. Consider, a map of harmonic interrelations like those explored in *A Geometry of Music* (Tymoczko 2011) could be used as a graph which the game moves through in a stepwise fashion, such that every time the player jumps, the accompanying music modulates around a rich harmonic space.

This approach can be thought of as a kind of 'musical mimesis', in that it mimics how objects in the physical world behave, making some amount of sound when interfered with. The traditionally differentiated tasks of 'sound design', the job of which is to mimic, and 'composition', with the job of making music, are dissolved into a whole. In film scoring, this approach is called 'mickey-mousing', and it is often derided as a ludicrous overscoring of the obvious. In a chapter on the bad habits of film composition, Eisler and Adorno deem musical mimicry an 'unfortunate duplication' of what's already obvious (Eisler and Adorno 1994). But this critique of mickey-mousing in the movies does not apply as sensibly to games, because games are tangle forms, unlike movies. They are not wholly illusory in a causal sense, as with musical sound design, but indeed partially exemplify fully real causal relations between human and machine and have thus a tendency to become musical instruments, to a degree, the affective power of which musical mimesis greatly amplifies. Only when game events, both direct and indirect, are 'mickey-moused' do they actualize this innate tendency; otherwise they remain as 'silent instruments', akin to MIDI controllers which are not yet hooked up to control anything.

As a subset of the imitative musical space, we notice a fundamental distinction between game events which are directly caused by the player, such as the pressing of a button to jump, and those which the player only indirectly influences, such as vertical positioning after a jump. The latter indirect event is enabled by the player but is not directly caused by the player alone; it is just as much caused by the game's code, which defines how high the jump reaches at its apex and how long it takes to get there (how strong is gravity in relation to the player-object's 'weight'). The press resembles the directness of a musical instrument, while the partial autonomy of the rising and then falling y-position begins to resemble the mechanical determination of a linear composition. And then, there is yet another class of event that can drive musical change, that which is wholly uncaused and uninfluenced by the player. For instance, imagine our jumping player is set next to a jumping nonplayer character (NPC)—controlled by a simple random-walk-style algorithm—who pays no need whatsoever to the player's activity. This NPC's jumps can be scored in any of the ways we have already discussed, triggering individual hits or driving continuous changes. This kind of event most resembles the mickey-mousing of movies, in that the player does not affect its outcome.

Now, each of the methods discussed may be used on their own, or they may be combined. Combination may be accomplished simultaneously or sequentially. When used simultaneously, a jump would both trigger a one-off instantaneous musical event and also continuously trigger changes to the musical parameter affected by the player's y-position. This has the effect of thickening the musical texture, or vertical aspect, of this compositional moment. When used sequentially, a jump might trigger instant sound effects in one room with no continuous recognition of jump height, and in a different room, silence this causal response, instead tracing a melodic line, with its frequency determined by the player's vertical y-position. This has the effect of adding to the variation of the game's horizontal musical structure. In this way, a composer can begin to think about games in terms of their musical texture and form.

Given the immensity of moving parts in many existing games, a visit from the spirit of Laplace's Demon may incline composers to try and attach musical parameters to all moving variables in a game, in order to fashion an ambitious mimicry of the causal richness of the real world. It is a sublime thought in its own way, but serious pursuit of this thankless task seems to me somewhat misguided. The goal of the composer may more profitably be directed towards emphasizing the musical effects of certain meaningful, relevant, or interesting activities at the expense of others, and creating a texture which approaches simplicity amidst the complexity of the total situation.

Rez and *Electroplankton* are two games which have been widely celebrated for their embrace of a musical approach to game design. Each emphasizes certain activities and musical forms at the expense of others.

Rez takes a familiar genre, the 'rail shooter', and quantizes its potential for rhythmic input and environmental movement to a semiquaver grid. The method of quantizing events to a grid has become a very common tactic for making something musical, having since been employed in the *Bit.Trip* games, the 'rhythm violence game' *Thumper*, and others. These grids are often static in the scale of quantization they use, settling for the semiquaver, rarely venturing into triplet time, let alone polyrhythms, mixed metres, or changing tempos. There is much exciting room for the development of grid-based games which explore more varied and nuanced rhythmic palettes, in addition to further means of moving between palettes. An example of a more varied game form in germ can be experienced by playing with a free variable attached to a knob controlling an arpeggiator, which moves between quantization values of $\frac{1}{4}$, $\frac{1}{6}$, $\frac{1}{8}$, $\frac{1}{12}$, $\frac{1}{16}$, $\frac{1}{24}$, and so on, affording easy movement through a rich line of duple- and triple-time relationships. Such a one-dimensional form could be made into a space with two dimensions by affording the player one further degree of freedom that would move by doubling or halving within duple time or triple time exclusively, such that at position $[\frac{1}{8}]$, the player could move to $[\frac{1}{4}]$ or $[\frac{1}{16}]$ along this new axis, and at position $[\frac{1}{12}]$, the player could move to $[\frac{1}{6}]$ or $[\frac{1}{24}]$. Further dimensions of control could be added which allow a player to change tempos by relations such as $[\frac{1}{12} = \frac{1}{4}]$, where the time in milliseconds of an quaver triplet in tempo A would be used as the time in crotchets of tempo B. Creating a variety of relationships in this way, and mapping them to further dimensions of control, could create a highly dynamic and intuitive rhythm modulator. Such a space could be controlled using n knobs, where n is the number of dimensions of control available; or instead, these 'knobs' could exist in the background, not afforded direct control by the player, but instead functioning as the ambience of a compositional terrain which the player's movements could affect by way of influence rather than direct cause.

Electroplankton takes a different approach by modelling itself as a kind of 'album' of ten mini-games, each a unique composition exploring a different algorithmic space. One of the more hypnotic games allows for the reconfiguration of the leaves of a plant which 'plankton' are being shot at and bounced off, the spatial orientation of leaves changing the angle of the bounce and thus the speed of the plankton and the rhythm of the resulting music. Other mini-games are playful explorations of signal flow, digital signal processing (DSP), and other topics in computer music. This album format allows for a diversity of algorithmic processes, each compartmentalized so as to avoid interfering with others. But it is also possible to weave wildly disparate algorithms together into a new whole. Such is one interpretation of an 'operatic' approach to composing with algorithms, in the sense of opera's combinatorial pluralism. Computer games which are not designed with a specific musical objective in mind are often structured in this way. An action-adventure game like those in the *Zelda* series combine a wide variety of processes: open-world spatial exploration with its varied psycho-geographical moods; the rhythmic ballet of combat with its varied articulations according to which weapons are being used against which enemies; the mini-games which are playable in the towns; the structured side-quests that a player can take as an interruption of her main journey; the boss fights structured like ABACAD song forms; and so on. Each of these components is not accessible from the main menu, as in an album of games, but is rather nested within a very complex topology which describes the connectedness of game forms, the ways one is able to move between mini-games. Such a topology serves as a map of the high-dimensional musical space, much like the metre of conventionally notated music, but existing in many potential dimensions with elastic temporalities instead of merely the one always-forward-moving time dimension of the classical score. Composers of musical games may eschew the narrative form of games like *Zelda* while still employing the kinds of labyrinthine topologies which a narrative world demands, and which affords interesting spatial relations and dependencies between musical forms. A global game form might be structured such that a certain musical mini-game is accessible only by playing a different mini-game up to a certain point, and then transitioning from this point into the new one. There might be games which are neighbours of one another, such that the player can move from one to the other rapidly and at will. The opening 'Shrovetide Fair' scene of Stravinsky's *Petrushka* is a simple example of how such a form might sound: hopping from one modular set of blocks to another, returning to the first for a shorter stint, and back to the other, which has changed in the interim, and repeating this process with dozens of games in the neighbourhood forming with one another a rhythmic mesh of patterned spatial relations.

The means of achieving interactive musical effects in computer games are readily available using many tools. Common ones include development engines such as Unity or Unreal, and interactive music 'middleware' engines such as Wwise and FMOD. These platforms allow for additional scripting, and some can be hacked so as to allow for integration of existing music programming languages, which some computer musicians are already familiar with. In Unity, for instance, it is possible to integrate Pure Data patches into a game environment using open sound control, such that any algorithms designed in Pd can imported into the game's codebase, 'attached' to game parameters, and played by the player. This effect was achieved by Henk Boom and Richard Flanagan in the game *FRACT OSC*. Possible approaches to the ambitious mode of 'Laplacean' sound design using Pd are richly illustrated by the myriad examples and theory of Andy Farnell's book *Designing Sound* (Farnell 2010), which focuses on procedural synthesis of natural processes, many ideas from which could be reapplied for more explicitly musical purposes.

24.2.2 Adapting Musics

Not only can game forms be treated as musical forms, but so too the roles can be reversed, and existing pieces of music can be studied as games. There are at least two ways in which any piece of music can be formally deconstructed in search of its play aspect. In the first, any composition (or otherwise fixed form) is studied as an imaginary play-through of a game which could have turned out differently. In the second, a performer is considered as the player of game, the rules of which are the instructions of the composition. The first privileges nontemporal or 'eternal' relations in the music, its informational content. A possibility space or game is inferred by imagining the formal-material conditions which gave rise, or theoretically could have given rise, to this particular state of affairs. The second privileges the various real-time contingencies which compositions afford a player by way of the incompleteness of their instructions, whether intentional or accidental. These two approaches identify, in turn, two different operant levels of freedom within a music space, which we can call *composed freedom* and *performative freedom*.

The first sort, composed freedom, is associated with the free play of fixed materials or 'constants' in a linear piece of music. Melodies dance, harmonies drift, rhythms shift, textures expand and contract, and so on. These are all qualities which can be represented quantitatively. We find play in the variability of numbers. The pattern '1298887342346662727' *plays* more than the pattern '22222222'. What is moving or varying in a piece of music is 'playing'≠ these moments of variation are perhaps related to what James Tenney calls 'structural entropies' in his *META Meta ≠ Hodos* (Tenney 2000). These do not represent the freedoms of actual time in its present-flow, but rather freedoms which have been expressed in the past and fixed in place. For instance, when I typed the two strings of numbers above, with the second string, I was determined to repeat the digit '2', the only freedom I allowed myself being how many times to repeat it; whereas, with the first string, I was not sure what I would type, and was free to bang out something quite randomly. But an account of my particular subjective experience is not required to qualify the first as more free; the freedom is embodied in the pattern. Composed freedoms are 'memories' of the past which have been frozen or fossilized into place. They are the material repercussions of events which have solidified into *objective* forms. Works of art are objective manifestations of past freedoms in this way. It is not possible to recover the exact 'game' which produced these artworks, but projects such as David Cope's 'Experiments in Musical Intelligence' (EMI) attempt a kind of reverse-engineering of this sort, recreating from an ensemble of fixed objects a more generalized possibility space which, when *played*, either by a human operator or a random-number generator, is capable of triggering events such that not only the original object might be created, but also any number of other 'sibling' objects, seeded by the same genes, but having played and grown up differently (Cope 2001). Cope's EMI does not allow for real-time play with the games which have been 'inferred', but we can see that they could be. For instance, imagine a computer game hooked up to a MIDI keyboard that affords a simple freedom to that player. Any key can be pressed, and this note will immediately be harmonized in the style of the composer module being used and in the context of what has already been played, and this event will trigger a cascade of automatic material composed in the appropriate style and in appropriate response to the player's disturbance. This interaction could be afforded at the downbeat of every measure, with the automatic play of the game holding a fermata on the final notes of each measure until the player triggers the next harmonic space. Or, more interestingly, the game could perform continuously and automatically while listening for interruptions from the player, who is free to provide monophonic input at any point. This game would resemble some chimeric hybrid of the original composer's style with the patterns offered by the player. It would surely be ugly by some standards, but this ugliness would be worth trying to understand and love. It is likely only by pushing through such barriers of ugliness and apparently profane reinterpretations of fixed masterworks that a new kind of beauty might be discovered in this form of adaptation.

The second sort, performative freedom, rather more resembles the kind we have been discussing, concerning the free play of 'free variables', those values which the composition does not fix and which are left indeterminate up to the moment of performance, being decided by a player of some sort. It is impossible that everything in a composition be made constant. Even a strictly notated piece by Bach might not specify tempo or dynamics, or instrumentation, and the performer can play freely with these variables. The guidelines of a collective improvisation offer a looser example. Even if the 'rules' have not been written down as a score that we can study, the invariant form of such a game can be intuited by listening to two different takes of a recording of loose compositional form, like 'Enter, Evening' by Cecil Taylor or 'Ascension' by John Coltrane, and analysing what remains constant between the two performances amidst the flux of the improvisation. This *constancy* is the game, whereas the performance variability is the *play*. All music is played to one degree or another. Even a recording, supposedly as fixed as an Platonic Form, can *in performance* be played on cheap computer speakers at a low volume or on a massive car stereo, and the listener can choose which and where. Further, the listener can treat the recording itself as playspace, clicking around on an MP3 player's track timeline in order to remix it freely on the spot, or, if it is being played on a vinyl record, slowing and speeding up and reversing playback to turn it into a raw material for 'scratching'. This kind of relation in which the variability of music is dominant is described in Bruce Benson's musical phenomenology (2003), which considers all engagement with music to be essentially improvisatory, in Christopher Small's (1998) concept of 'musicking', which likewise describes music as always an active process, and others. It is an approach which takes on utopian musical hues in Adam Harper's (2011) imagining the next millennium of musicking, and in Jacques Attali's (1985) 'age of composition'.

Composers such as Iannis Xenakis and John Zorn are notable for having worked with free forms which they consciously regarded as games, as with Xenakis's *Duel* and *Strategie*, and Zorn's *Cobra* and others from his series of game pieces from the late 1970s. These works afford performative freedoms at the same time that they embody composed freedoms which performances are constrained by. And, though they may not call their works 'games', a much broader spectrum of composers, too, can be understood to already work with the properties of game forms we are concerned with, and indeed computer games in particular, without calling them such. George Lewis describes a game-like composition of his as follows: 'In *Voyager*, improvisors engage in dialogue with a computer-driven, interactive "virtual improvising orchestra." A computer program analyzes aspects of a human improvisor's performance in realtime, using that analysis to guide an automatic composition (or, if you will, improvisation) program that generates both complex responses to the musician's playing and independent behavior that arises from its own internal processes' (2000).

Improvised musics performed with computers, such as *Voyager*, are, taking a broad view of things, *already* computer games which are simply not mass-distributed, which are only playable by one or several musicians who have access to the software. An alternate history of computer games is awaiting articulation by way of the twentieth century's musical history, specifically in the interplay of its improvised musics and algorithmic techniques. In light of this, it would seem that the attention paid to widespread *distribution* and *accessibility* is as important as any in determining the popular conception of a piece of software as being a musical game. It may be that the process of 'composing computer games' is simply to compose music in ways similar to how it's already being composed, but to distribute it in such a way as to make clear that it is not a *recording* or a *performance*, which is considered the final relation between composer and listener, but rather it is the *game* which is the final relation between composer and listener, or, what is more accurate, composer and player. A major aid in establishing this relation is providing the listener with software that 'just works', that does not require expertise of any sort to set up, as is the case with, say, a pure data patch; playing a game should be as accessible as reading a book or putting on a record. In seeking such accessibility, there is an implied aesthetic turn away from the demand for professionalism from a performer in interpreting a composition, in favour of celebrating musical amateurism—this, in both the negative sense signifying a somewhat lazy dilettantism and the positive sense of its etymology, meaning lover. Both laziness and love affirmed.

In this embrace of the listener-cum-amateur-player, computer games propose a solution to what is not quite a *problem* in computer music, but which can nevertheless be a persistent source of tension and occasional angst—a problem which we might call *process opacity*, which is characterized by the *causal listener* (in Chion's sense) becoming alienated by way of not being able to identify a sound's cause. Nonelectronic folk musics, as a counterexample, have an appealing transparency of process. Most listeners are at least loosely familiar with the means of producing vocal song, and many are familiar with the means of producing percussive music and guitar strums, such that listening causally to these forms naturally evokes an imaginary environment in which the listener is virtually *playing them*. But computer music's tapestries of pinched sounds, impulse pops, stochastic clouds, granular storms, and FFT (using the fast fourier transform) freakouts are often perceived merely as special effects to nonacclimatized ears, sometimes enjoyed, and ever more so when there is visible body movement of some sort connected to the sound-making process, but not yet fully appreciated as the embodied, down-to-earth folksy, haptic constructions that they can be from the first-person point of view of the musicians involved in creating them, in tweaking the knobs or otherwise engaging the interface that translates bodily movement into these strange sounds. For an experience of transparency with computer music, there must be some intimacy with the material cause of the sounds. Process-centric computer music is all too often felt to be impenetrable from the perspective of mainstream audiences, who have not spent time patching together worlds in Max/MSP or even cutting up sounds in a simple audio editor. Many listeners become dismayed to find themselves at a laptop concert if there is not visual ornamentation of some sort happening, or ideally a body moving in such a way as to demonstrate causal influence over the sounds. Understanding the means by which a given sound is created is a key to feeling meaning in that sound, and most people today are not familiar with computer musics' varied and intricate means of production.

Working directly with computer games as a musical medium offers the composer the possibility of designing forms in such a way that a direct causal experience play or *touch* is established as the default relation between the 'audience' and the piece. Critically, this approach aims to distribute such compositions to listeners (players) on a mass scale, serving a potentially (though by no means necessarily) democratizing purpose for interactive algorithmic music, which is analogous to the purpose that recordings or take-home piano scores serve for a piece of linear music.

24.3 FORMALIZED GAMES

24.3.1 Shifting Possibility Spaces

The analogue of the timeline form of a linear composition (embodied equally in recordings and classical scores) is the general structure of an n-dimensional possibility space, the sort of form in which indeterminate activity happens. The rules of a game and the rules of correct voice-leading over a cantus firmus and the material constraints of a saxophone are all equally exemplary of the sorts of mechanical-algorithmic atoms that give rise to this generalized concept.

The notion of a possibility space is one which is by no means native to music or game thinking. An inkling of the form is intuitively entertained in the most mundane circumstances of everyday life whenever we are confronted with a decision point, a branching pathway, physical or mental. It is felt in a more hazy sense when we look at a distant landscape, for instance, and imagine ourselves there, or imagine the lives of whoever is presently there. The sense of possibility is poetic and vague before it becomes formal and narrow. Its formal conceptualization can be described using mathematics.

The formal idea of a possibility space is already present in the simplest instance of a logical-mathematical variable. An algebraic expression is an example of a highly formalized and very simple possibility space. If we write $3 < x < 6$, then we know that x lies somewhere between 3 and 6, but we do not know where. This simple expression describes a one-dimensional possibility space, having only one free variable. An algebraic equation like $x + y + z = 10$ relates three variables to one another, but does not determine their value, it only determines the relational space of possible values. The number of variables are called the 'degrees of freedom' of the space, and the number of degrees of freedom in a given space establishes its dimensionality. Considered as a totality, the possibility space is an n-dimensional geometrical form, or manifold, but a space with its dimensions 'extending' into abstract dimensions of logico-mathematical possibility, as opposed to the three spatial dimensions of our physical space-time. The applications which bridge the continuum between logical-metaphysical form and the materialism of everyday life are filled in by the natural sciences, and to this end, Manuel DeLanda catalogues and describes a series of natural possibility spaces in his book *Philosophy and Simulation* (DeLanda 2011).

Game designers often speak in this way about the totalizing 'possibility space' of a game, in the same way a music theorist might speak of a piece's form (e.g. sonata-allegro, fugue), but what is lost in this global analysis, especially in the case of musical games, is an acknowledgement of the temporal flux of shifting possibilities, based on the contingent value of what is possible for a player at a given moment. *Playing* is a process of moving through possibility spaces. Considered locally, the experience of a possibility space is not that of a solid object but rather of a morphing form, with shifting presences and absences of free variables corresponding to shifts of local dimensionality.

The complexity theorist Stuart Kauffman (2000) uses the concept of 'adjacent possibility' to describe the movement organisms in their environments. This is a useful approach to thinking about game play in general, which opposes the universalizing tendency to think of a possibility space zoomed all the way out, as an object. The adjacent possible is simply the set of whatever is within the immediate sphere of possible moves afforded to the player. To return to our algebraic example, $3 < x < 6$; a player might be afforded the capacity to determine the precise value of x, first by selecting a value between 3 and 6, and second by adding or subtracting 0.1 to this value. Thus, if the player starts at $x = 3.5$, there are two adjacent possible values which could be moved to next, 3.4 and 3.6. If, on the other hand, the player were allowed to move by intervals of 0.1, 0.5, or 1, then the set of adjacent possibles would triple accordingly. Notice that in these examples, the infinite holding capacity of the real number line is now off limits, because the player is not afforded the means of determining a value with infinite precision.

The dimensionality of a local adjacent possible is characterized by its degrees of freedom, the value of which shifts with time. These can be controlled by players operating at a variety of hierarchical levels. One player can control many degrees of freedom, like one body controlling ten fingers dancing across a piano's keyboard, or one player might control only a single degree of freedom, such as a determining the value of a single x variable by way of a MIDI control change slider. We can think of our body as one player, or we can think of it as many players (joints, muscles, nutrition, hydration, etc.). A piano can be played by one player or by a rotating cast of many players. A player with a piano can become one with the instrument by way of her intimacy with it. The process of individuating a 'player' is a matter of chunking several or many parts together at different scales and counting them as units with freedom. In perhaps the most abstract sense, a logical free variable itself can be thought to represent an atomic 'player' of a metaphysical sort.

The 'dimensionality' of freedoms in our everyday lives is incomprehensible, approaching and perhaps actualizing some kind of infinity, or at least indefinable largeness. The human skeleton alone has several hundred joints and these are only scratching the surface of the freedoms of the human experience. Besides, it is not the singular body alone that accounts for our freedoms. A human body tied to a tree or otherwise disabled does not benefit from those several hundred freedoms. Our freedoms are always afforded by our body's relation to other bodies, other humans, nonhuman animals, plants, inorganic materials—houses, neighbourhoods, social groups, musical instruments, games, and so on.

24.3.2 Formalized Computer Games

When software is run on a computer, the activity of the whole functions as a body, which is strictly determinate in some sense, always following the rules according to its algorithmic form. However, when a computer listens for input from a player, even though it has been instructed to do so, it thus invites indeterminism into its body. Computer games are distinct from noninteractive algorithmic forms in that they are composed both of modules which are deterministic (as is the exclusive case of nongame algorithms) and of those which are nondeterministic, affording varying degrees of freedom to a player.

The precise dual form of determined versus free algorithms was defined by Alan Turing at the advent of modern computing as follows:

> If at each stage the motion of a machine ... is completely determined by the configuration, we shall call the machine an 'automatic machine' (or a-machine). For some purposes we might use machines (choice machines or c-machines) whose motion is only partially determined by the configuration.... When such a machine reaches one of these ambiguous configurations, it cannot go on until some arbitrary choice has been made by an external operator. (Turing 1936, 232)

A computer game is formally built of both a-machines and c-machines, with the a-machines forming the deterministic boundaries which enclose the playing field, and the c-machines opening up the space of possibilities which allow for play enclosed by these boundaries.

A player's freedoms ripple throughout the formal space of the game's logic by way of the presence of at least one free variable x attached to a formal c-machine on one end and a physical input device such as a MIDI controller or keyboard or mouse or microphone on the other. A c-machine's x value may be controlled or 'played' by another algorithm, such as a random number generator, thus producing a generative artwork, as in the stochastic process music of Cage, Xenakis, and others, but an x variable means something very different in the hands of a human operator than it does in the hands of a random number generator. Meaning emerges from the process of touching the x-variable, and this process is fundamentally a bodily one which is not reducible to an algorithmic form in the way a machine's processes are. The human is an organic component of the indefinably complex biosphere, an animal in her environment, before she is a 'computing mind' or, as game theorists and neoclassical economists would insist, a 'rational agent'. The relationship formed between human and computer allows for filtered echoes of the biosphere to enter into the x values of the machine's indeterminate configurations. As Marc Leman describes it: 'If the human body and mediation technology are hooked into each other, then it is possible to conceive the digital domain as a natural extension of the physical domain. The human mind will then extend its activity range to this digital environment in a natural way' (2008, 235).

A computer game is not just a hunk of dead formal code. When it is running, it is a half-living thing, a material-energetic creature with sense organs and conceptual movements and expressions analogous to those of an organism. Its sensory inputs—buttons, knobs, joysticks—correspond to an animal's eyes, ears, mouth. Its expressive outputs—flashing lights, vibrations, pulsing sounds—correspond to organic song, dance, speech. And its internal algorithmic architecture in general corresponds to a creature's guts, skeleton, musculature, nervous system, and so on. When we play a computer game, we become the 'environment' which this machine-organism lives 'within'. The output of our play provides the inputs or sense-data for the game, what it 'knows' of its external world, a bizarre inversion of the classical human-centric empiricism in which sensory experience inscribes ideas onto our mental *tabula rasa*. We become the machine's environment. At the same time that we 'immerse' ourselves in the software, this allows it to become our own environment. The relation of organism to environment is parallel to that of the relation between player and playspace.

24.3.3 Ecological and Economic Games

Such an image of games evokes a theoretical approach to the medium which is radically different from that of game theory, with its reductive psychological economism assuming that players improvise in efficiency-obsessed, rational, ways, and whose founders described it as 'the proper instrument with which to develop a theory of economic behavior' (Von Neumann and Morgenstern 1953). Thus, whereas musical works like Xenakis's game-theory-inspired compositions *Duel* and *Strategie* are characterized by payoff functions resembling those of goal-oriented competitions amenable to game theoretical analysis like chess or basketball, computer games have a more general relation to algorithms, one which does not ask of them whether they are more or less optimal or efficient, but which rather accepts all algorithms for what they are, as raw materials, musical players rhythmically churning their patterned textures forward through time. In Xenakis's language, a game of this sort is called a 'false' or degenerate game, 'one in which the parties play arbitrarily following a more or less improvised route, without any conditioning for conflict, and therefore without any new compositional argument' (Xenakis 1971, 113). For

Xenakis, a 'true' game is one in which the players, too, become algorithmic, submitting themselves to a 'compositional argument' and performing optimally in its defense, like a cook following the dictates of a recipe, hoping to make it exactly as advertised. But this chapter has mostly concerned the 'degenerate' game form which has no expectations as to what the player should be doing. In a computer game, the deterministic content is provided by the machine, there is no need to employ the player with a given job.

The relation of the player to the machine and the machine's algorithms in relation to one another as parts and in relation to their collective environmental totality is the subject matter of a non-game-theoretical 'theory of games' which has a musical quality and which is rather more ecological than economical. In the sense that the human player is never truly beholden to any particular task by the machine, an ecological approach to games is better suited than the economism of game theory to deal with the particular materiality of computer game form.

This contrast of economic and ecological form can perhaps begin to differentiate the properties of an apparent dualism at the heart of all game form, computer and otherwise, and including musical works in general. This is that there are two distinctly different sorts of constraints on player movement which establish the boundaries of a game: (1) rules, which are abstract, immaterial, nonactual instructions for operating on material things; and (2) forces, which are concrete, material-energetic, actual, the things themselves. We can call the first sort of constraint 'economic' and the second sort of constraint 'ecological'. Computer games are manifestations of economical rules being transmuted into ecological force.

The root of both 'eco-' words, οἶκος, is Greek for 'household', where 'economy' can be translated to 'rule of the household', and where 'ecology' can be translated to 'ground of the household'. If, for the sake of example, all games are considered as a kind of 'playing house', then economics deals with the abstract legal guidelines managing the expected and allowable behaviours in this house, such as chores and regulations and optimizations of dishwashing, and ecology deals with the concrete energetic actualities, which are impossible to change without forcing a radical transformation of material conditions, actualities such as water temperature, gravity holding furniture to the ground, musculature which allows house members to stand and walk, lightness or darkness of rooms, and so on.

Musically speaking, instruments function as 'ecological games', or energetic forces, in that they do not insist on a particular mode of interfacing with them, even if professionalization does demand such rule-based interfacing, or economizing. Performative compositions, on the other hand, function as 'economic games', being as they are a set of notated or otherwise prestated rules that the labouring duo of musician + instrument must subject themselves to in order to work in accordance with the composer's intentions and in harmony with the trajectory of the musical ensemble as a whole. While game theory provides a robust analytical tool kit for interrogating economic forms of the relation between a rational player and games like chess and warfare, the ecological aspect of games, the raw energetic relations of influence and resonance between organism and playspace/environment, is much less studied in the context of games.

Throughout this chapter, the word 'afford' has been often used, in the sense of 'makes possible'. The theory of affordances, borrowed from the aptly named discipline of ecological psychology and popularized in many design communities, can serve to bridge the patterns of physical energies to their fluxes and invariances as experienced from the first-person perspective of the organism, serving as a grounding for an ecological theory of games. James Gibson describes the concept thus: 'The *affordances* of the environment are what it *offers* the animal, what it *provides* or *furnishes*, either for good or for ill. The verb *to afford* is found in the dictionary, but the noun *affordance* is not. I have made it up. I mean by it something that refers to both the environment and the animal in a way that no existing term does. It implies the complimentarily of the animal and the environment' (2014, 119). When an animal touches anything in its environment, for instance a squirrel holding an acorn, or a dog swimming in a lake, the relations between toucher and touched, such as 'holding' or 'swimming', are illustrative of what it means for an environment to afford some activity, to open up a space of possibilities. In the case of computer games, all mechanical interactivity within the game space is afforded, as described in the previous section, by Turing c-machine modules connected to input devices. The concept of affordance can help us treat these inputs and choice-machines in a way that does justice to the objectivity of the space of possibilities and the player, and crucially, to the relationship between the two.

24.3.4 Operaism

One of the most compelling aspects of Gibson's ecological psychology is its dissolution of subjectivity and objectivity: 'an affordance is neither an objective property or a subjective property; or it is both if you like. An affordance cuts across the dichotomy of subjective-objective and helps us to understand its inadequacy.... It is both physical and psychical, yet neither' (Gibson 2014, 121). Following from this subject-object dissolution, ecological psychology points towards, if it does not explicitly adopt, a speculative panexperiential or panpsychist cosmology (Whitehead 1978), the hypothesis that everything has an experience or 'mentality'. Regardless of the legitimacy of this perspective, it seems to me to offer a pragmatic stance for creative work, affording more potently strange and enchanted mindsets from which to engage with computational materials than that of the game theorist's ceaseless striving for efficiency and its associated positivist metaphysic which views computer algorithms as just rules. For the panpsychist, the computational material is allowed to live in its own inorganic way, as 'vibrant matter' (Bennett 2010), becoming a half-living collaborator in our work and play. The algorithm is not reduced to its abstract rules, but is regarded as a concrete creature participating in the world amongst other creatures—playing, working.

Returning to the concept of opera—works—it takes on the hues of a natural philosophy. In an apparent inversion of Vedanta Hinduism's concept of *lila*, the divine-play aspect of the world, we begin to regard everything as aesthetic work. The world is an opera, or many operas; opera is what happens. The labourer works, the musician looks for a job. 'Because energy can move we may harness and channel it to do work. ... Work is a change of energy, also measured in Joules. So, another definition of energy is the ability to do work. It can cause things to get hotter, or move things, or emit light and radio waves. One way it can move is as sound, so sound can be thought of as changing energy' (Farnell 2010, 10). Sokal and Bricmont would scoff predictably at any fuzziness of scalar reference which treated labour and thermodynamic work and opera within one breath as all of a kind, but other sources such as Darwin, Marx, and Wagner (Barzun 1958) are correct to identify a field of relations shared between ecological, economic, and musical thought. Indeed, as described in Georgescu-Roegen's (1971) work on ecological economics, the physicist's conception of work itself is a product of its economic times—the steam engine and its objectification of what had previously been the province of labour power, horse power. And so, too, computer game theories exert an economic influence in relation to work. This is seen in the spheres of both 'gamification' (Eyal and Hoover 2014; McGonigal 2011), which attempts to convince labourers to happily perform otherwise boring tasks by couching them in addictive game mechanics, and automation, which attempts to dispense with the labourer altogether, by converting an already mechanical task which once required organic labour power into pure mechanism, in a process analogous to the way in which a chess computer game automates the upholding of the rules such that what was contingent on an implicit contract between players agreeing to play by a shared value-system has become enforced by way of the ecological affordances which resist any change to this contract.

It should be emphasized that 'ecological' form cannot by any means be equated with 'good', and 'economic' with 'bad'. Ecological form resists freedoms at the same time as it affords them. Crucially, economies of music, the directed jobs or goals of players, ought not be ignored to the degree which I've largely been guilty of throughout this chapter, supposing as I have that the player has been free to do as she pleases. Though I disagree with Xenakis's description of undirected play forms as 'degenerate games', there is admittedly a sense in which the ostensible apolitical stance of free play—no goals—meshes with the 'anything goes' philosophy of the anarcho-capitalist or libertarian corporatism which has risen to ideological prominence during the same years as those which compose the history of computer games. Free play, free markets: 'There is no alternative'. There is something degenerate indeed about the freedom implied by this perspective! It ought to be asked in what ways a musical 'compositional argument' in Xenakis's language could help aid conceptualizing and implementing in musical microcosm a good economy. Algorithmic automation used to allow economists to dream of a future with no work. Keynes speculated in 1930 as to the character of life once 'the economic problem' had been solved, 'for the first time since his creation man will be faced with his real, his permanent problem—how to use his freedom from pressing economic cares, how to occupy the leisure, which science and compound interest will have won for him, to live wisely and agreeably and well' (Keynes 1963, 367). Though a contemporary dream may not look just like Keynes's, it seems that this quest for leisure, for the positive freedom which arises from being able to work as a choice and not forced by necessity, is one with continued relevance, and one intimately related to games and the question of whether we are playing the game or working it as a 'playbourer' (Bigge 2010). In the face of all of this, a musical approach to games considers an alternative way to conceptualize what it means to be a game. It is a small gesture, but one which may have something to contribute by way of dealing explicitly with many of the materials and concepts which must be engaged with in even the larger questions—economies, ecologies, freedoms, possibilities, necessities, work, play, and so on.

BIBLIOGRAPHY

Attali, J. *Noise: The Political Economy of Music.* Translated by B. Massumi. Minneapolis: University of Minnesota Press, 1985.

Barzun, J. *Darwin, Marx, Wagner: Critique of a Heritage.* Rev. 2nd ed. Garden City, NY: Doubleday, 1958.

Bennett, J. *Vibrant Matter: A Political Ecology of Things.* Durham, NC: Duke University Press, 2010.

Benson, B. E. *The Improvisation of Musical Dialogue: A Phenomenology of Music.* Cambridge: Cambridge University Press, 2003.

Bigge, R. 'How We Fell Out of Love with Slacking'. *The Star,* 6 February 2010. http://www.the-star.com/news/insight/2010/02/06/how_we_fell_out_of_love_with_slacking.html.

Bogost, I. 'Persuasive Games: Why We Need More Boring Games'. *Gamasutra,* 21 May 2007. http://www.gamasutra.com/view/feature/129850/persuasive_games_why_we_need_more_.php.

Chion, M. *Audio-Vision: Sound on Screen.* New York: Columbia University Press, 1994.

Cohen, P. J. *Set Theory and the Continuum Hypothesis.* New York: W. A. Benjamin, 1966.

Collins, K. *Playing with Sound.* Cambridge, MA: MIT Press, 2013

Conway, J. H. *On Numbers and Games.* London: Academic, 1976.

Cope, D. *Virtual Music Computer Synthesis of Musical Style.* Cambridge, MA: MIT Press, 2001.

DeLanda, M. *Philosophy and Simulation: The Emergence of Synthetic Reason.* London: Continuum, 2011.

Deleuze, G., and Guattari, F. *A Thousand Plateaus: Capitalism and Schizophrenia.* Translated by B. Massumi. Minneapolis: University of Minnesota Press, 1987.

Eisler, H., and Adorno, T. W. *Composing for the Films.* London: Athlone, 1994.

Eyal, N., and Hoover, R. *Hooked: How to Build Habit-Forming Products.* London: Portfolio Penguin, 2014.

Farnell, A. *Designing Sound.* Cambridge, MA: MIT Press, 2010.

Georgescu-Roegen, N. *The Entropy Law and the Economic Process.* Cambridge, MA: Harvard University Press. 1971.

Gibson, J. J. *The Ecological Approach to Visual Perception.* London: Psychology Press, 2014.

Haraway, D. *The Haraway Reader.* New York: Routledge, 2004.

Harper, A. *Infinite Music: Imaging the Next Millennium of Human Music-Making.* Winchester: Zero Books, 2011.

Heap, S. H., and Varoufakis, Y. *Game Theory: A Critical Introduction.* London: Routledge, 1995.

Huizinga, J. *Homo Ludens: A Study of the Play-Element in Culture.* Boston: Beacon, 1955.

Kauffman, S. *Investigations.* Oxford: Oxford University Press, 2000.

Keynes, J. M. 'Economic Possibilities for Our Grandchildren'. In *Essays in Persuasion*, 358–373. New York: W. W. Norton, 1963.

Klee, P. *Pedagogical Sketchbook*. Translated by S. Moholy-Nagy. New York: F. A. Praeger, 1953.

Leman, M. *Embodied Music Cognition and Mediation Technology*. Cambridge, MA: MIT Press, 2008.

Lewis, G. E. 'Interactivity and Improvisation'. In *The Oxford Handbook of Computer Music*, edited by R. T. Dean, 457–466. Oxford: Oxford University Press, 2009.

Lewis, G. . 'Too Many Notes: Computers, Complexity, and Culture in *Voyager*'. *Leonardo Music Journal* 10 (2000): 33–39.

McGonigal, J. *Reality Is Broken: Why Games Make Us Better and How They Can Change the World*. New York: Penguin, 2011.

Phillips, W. *A Composer's Guide to Game Music*. Cambridge, MA: MIT Press, 2014.

Schüll, N. D. *Addiction by Design: Machine Gambling in Las Vegas*. Princeton, NJ: Princeton University Press, 2014.

Sicart, M. 'Against Procedurality'. *Game Studies* 11, no. 3 (2011). http://gamestudies.org/1103/articles/sicart_ap.

Small, C. *Musicking: The Meanings of Performing and Listening*. Middletown, CT: Wesleyan University Press, 1998.

Swink, S. *Game Feel: A Game Designer's Guide to Virtual Sensation*. Amsterdam: Morgan Kaufmann / Elsevier, 2009.

Tenney, J. *META ≠ HODOS* and META *Meta ≠ Hodos: A Phenomenology of 20th-Century Musical Materials and an Approach to the Study of Form*. 2nd ed. Lebanon, NH: Frog Peak Music, 2000.

Turing, A. 'On Computable Numbers with an Application to the Entscheidungsproblem'. *Proceedings of the London Mathematical Society* 2, no. 42 (1936): 230–265.

Tymoczko, D. *A Geometry of Music Harmony and Counterpoint in the Extended Common Practice*. New York: Oxford University Press, 2011.

Von Neumann, J., and Morgenstern, O. *Theory of Games and Economic Behavior*. 3rd ed. Princeton, NJ: Princeton University Press, 1953.

Xenakis, I. *Formalized Music: Thought and Mathematics in Composition*. Bloomington: Indiana University Press, 1971.

Whitehead, A. N. *Process and Reality: An Essay in Cosmology*. Corrected ed. New York: Free Press, 1978.

CHAPTER 25

....

ALGORITHMIC
SPATIALIZATION

....

JAN C. SCHACHER

> We must stress an important distinction concerning [the] spatial dimension in acousmatic art. The 'internal space' is formed within the work itself, made of reflections of the sonic contours, of the movement of entities, presenting itself to the hearing as a sensation of composed volume. To this we contrast 'external space', with completely different effects, no longer concerned with the work but with the configuration of the space wherein it is heard, with its particular peculiarities (often undesirable or from time to time exploited). (Bayle 2007, 243)

THIS chapter approaches sound spatialization in musical practices that use algorithms as process and structure generators. The topic presents some complexity because we are dealing not just with a single domain but with a number of intertwined layers that are situated between acoustics and perception, between architectural spaces (Blesser and Salter 2007) and the sound events situated therein. In addition, the practice of spatial audio is a wide-ranging one: it begins for example with a recording engineer's concern with reproducing the sound stage, continues with 'acousmatic' and electroacoustic multichannel compositions, and finally includes artistic applications in games and installations that construct artificial sound spaces. Algorithms as a source of structure or process may be used within only a limited number of these activities, but the implications in these contexts of using spatial audio processes remain critical.

This chapter attempts to give a very brief historical summary as well as an overview over perceptual and technical issues of spatial audio and music; it then discusses the use of algorithms as compositional and performance tools for spatialized sound, in order to finally look at the difficulties and pitfalls of spatialization.

With this sequence we hope to provide the anchor points necessary to explore the question of how to fruitfully use algorithms for audio spatialization and spatial music.

One of the central, yet sometimes ignored aspects of electroacoustic and electronic music is that it needs to be heard in an actual space through loudspeakers or to be delivered to our ears through headphones. Although the dominant mode of playback of music in everyday situations remains the stereo field (and 5.1 is the new standard for films), to use two speakers in order to mirror our two ears is by no means compulsory. Since the beginnings of electrically amplified music the number of speakers used for spatial (re-)constitution has been one of the aspects with which people have experimented. With the advent of electronics the number of channels used has increased, going from one to an arbitrary number. All of these arrangements attempt to mitigate the fact that the inherent spatial and enveloping quality of sound in the lived world collapses into 'flat' representations through loudspeakers, which need to be read or heard as if they were a two-dimensional image. Recording, encoding, and diffusion techniques have evolved sufficiently in tandem with the acquisition of listening skills in particular for recognizing acoustic spaces, in order for the illusion of spatial sound to become credible in many musical and acoustic situations. Nevertheless, the suspension of disbelief remains a necessary pre-condition for this effect to work.

In many musical practices the spatial disposition and, in particular, room acoustics have always played a role, but in a circumstantial rather than deliberate fashion, often dictated by the acoustical spaces where the music has been performed. In recorded music for instance, with the aforementioned limitations, the notion of soundstage has been used extensively to emphasize instrumental relationships, for example between the instruments of a band and the singer; the different instruments of an orchestra; in cinema between the dialogue, the music, and the soundscape within which the narrative is located; or in electronic music between sound and artificial reverberation. These practices can be called spatialized audio, but not necessarily spatialized music, since they deal with the space in an auxiliary manner, not as a core element of musical composition work. Nonetheless, the technical developments made by sound engineers for music recording and film sound (nowadays blending over into videogame sound) are a big contributing factor to the increased focus on space as musical dimension in electroacoustic and electronic music.

The availability of multichannel diffusion systems beyond stereo has led to a musical appropriation not just with the goal of perfectly simulating the way the natural world sounds but in order to use space and its attributes as an additional musical dimension to compose with. The convergence of techniques for dealing with the two complementary aspects of sound can be observed in fields of both spatial audio and music, in surround audio's use in videogames as well as in the newly invigorated virtual reality field. Spatialization means to work on the one hand with acoustic spaces or rooms, which are perceived via both direct and reverberant cues (Bregman 1994), through interaural time and level differences, as well as spectral filtering due to the interaction of the sound waves with our head's and ears' morphology. And on the other hand, it means to work with sound scenes and object-based scene (re)construction methods, which enable, through symbolic operations and a modelling approach, the generation of synthetic spatial audio.

25.1 SPATIAL SOUND CONCEPTS

Sound spaces and spatial sound diffusion are central topics of 'acousmatic' music, elec-troacoustic music, and composed twentieth-century contemporary music, as seen for example in early works by Charles Ives (Cowell and Cowell 1969), Edgard Varèse (Varèse 1966), or in Xenakis's *Polytopes* (Serken 2001), and in the spatial distribution of orchestral groups in Stockhausen's *Kontakte* (Stockhausen 1995), Boulez's *Répons* (Boulez 1998), and Luigi Nono's *Prometeo* (Oehlschlägel 1985). 'In these traditions, the localisation of sounding physical and perceptual space, as well as the creation of senses of virtual space and sonic spatial movement and evolution both between and within sound-objects (Chowning 1977), are harnessed to aesthetic ends either as part of the desired musical effect or as a primary element in compositional imagination. Like "pitch space" formalism, this ... discourse of space prominent in electro-acoustic and com-puter music invokes notions of spatial and musical autonomy' (Born 2013, 11–12).

In addition to compositional musical work using sounds in space, the development of the soundscape perspective of acoustic ecology in the 1970s (Schafer 1993) had a profound impact not just on sound art (Neuhaus 2000) but also electroacoustic music (Truax 1999; Westerkamp 2002). Converging with this development are the composi-tional processes of the stochastic synthesis methods defined by Xenakis (1992) and the expanded sonic possibilities in different time domains that constitute what is now known as granular synthesis (Roads 2001). Despite the rise of a spatial audio diffusion practice since the 1970s, proper formalization has been achieved only in recent years. The categorizations of sound types, as proposed by Lachenmann (1966) for contempo-rary music, can be seen as a complement to Schaeffer's *objets sonores* (Schaeffer 1966). Ihde's phenomenology of listening (Ihde 1976) in turn provides the foundations for Smalley's understanding of sound shapes (spectromorphology; Smalley 1997), which finally leads to the concepts of sound spaces (spatiomorphology), which are essential for a spatialized music practice (Smalley 2007; see Born 2013 for a more comprehensive overview).

The spatial concepts offered by Smalley range from the gestural, the ensemble, and the arena spaces to the proximate and distal spaces that generate the listening perspective by defining the foreground, the midground, and the background, to the social perspec-tives of the intimate, the personal, the social, and public spaces (Hall 1966). Particularly interesting is Smalley's statement that 'sounds in general, and source-bonded sounds in particular, ... carry their space with them—they are space-bearers. ... Source-bonded spaces are significant in the context of any acousmatic musical work ... in musical con-texts where I imagine or even invent possible source bonds based on my interpretation of behavioural space' (2007, 38). The multimodal entwinement of these spaces leads to a perception of the aesthetic configurations of the music through the 'enactive' capabilities provided by our sensori-motor skills (Gallagher 2005) and through 'underlying spatial attributes: texture has space, gesture operates in spaces integrated into the gestural task, cultural and natural scenes are spatial, the highs and lows and motions of sound spectra evoke space. But sense experiences are also rooted in the physical and spatial entity of the human body, which is always at the focal centre of perception—as utterer, initiator and gestural agent, peripatetic participant, observer and auditor' (Smalley 2007, 39).

25.2 MILESTONES OF SPATIALIZATION

A look into the past permits us to examine concepts, technical achievements, and milestone applications of spatial audio and music in order to better understand current practices.

Spatialization or spatial sound diffusion with any practicality became feasible in the late 1950s and 1960s, and was tied to the development of more sophisticated electronics, mainly through the advent of magnetic tape machines and ultimately the development of semiconductors. Earlier applications were tied to sound-scene transmissions via telephone lines (Rumsey 2001, 10).

One of the first examples of a large channel-count sound diffusion system was Edgard Varèse's *Poème électronique* in the 1958 Brussels World's Fair. Here, within the parabolic architecture of the Philips Pavilion designed by Xenakis, the Philips-built multichannel sound diffusion system complemented the architectural space and visual projections (Zouhar et al. 2005). In 1959 electronic music pioneer Karlheinz Stockhausen developed the rotation table, a mechanical device used to generate rotating sounds (Braasch, Peters, and Valente 2008). A decade later, he presented a spherical auditorium in the German pavilion of the Osaka World's Fair in 1970 with fifty speakers surrounding the space in vertically arranged layers. The audience was seated on a lattice floor in the median plane of the sphere, the conductor was placed at the centre, and ensemble positions were dispersed around the space (Stockhausen et al. 1978).[1] In 1971, the Experimentalstudio of the German Südwestrundfunk, which was and still is in charge of performing live electronics for Luigi Nono's pieces, developed the Halaphon, a controllable signal matrix used for spatial sound diffusion (Parra Cancino 2014, 39). In 1974 François Bayle designed the Acousmonium, an eighty-speaker 'orchestra' located at Radio France's research laboratory Groupe de Recherche Musicale (GRM), which is still in use today (Bayle 2007).

In San Francisco, a historical multichannel sound diffusion theatre called the Audium exists in a dedicated space and has been in operation since 1967 (Shaff 2014). More recent multichannel musical spaces are located in Karlsruhe, with the Zentrum für Kunst und Medientechnologie's Klangdom (Brümmer et al. 2014); at University of California Santa Barbara, with the Allosphere (Amatriain et al. 2007); in Queen's University Belfast, with the Sonic Arts Research Centre's Sonic Laboratory space;[2] among others (for a survey of these spaces, see Normandeau 2009). In Paris at IRCAM, the Espace de Projection concert hall provides varying spatial modes through movable panels that can modulate the room acoustics and is equipped with a hemispherical speaker dome that is combined with a large wave-field synthesis array (Noisternig,

Carpentier, and Warusfel 2012); in Graz, the Institut für Elektronische Musik's IEM-Cube (Zmoelnig, Sontacchi, and Ritsch 2003) and the Mumuth concert hall provide regular and irregular multichannel speaker arrays and dedicated spaces for spatial audio (Eckel 2011). There are wave-field synthesis arrays at the Technical University in Delft in the Netherlands (Boone and Verheijen 1993), where this technique originated, at the Technical University in Berlin (Baalman 2010), as well as at an increasing number of venues worldwide.

25.3 PRINCIPLES OF SPATIALIZATION

Spatialization could be defined as the act of placing sounds in a both virtual *and* real acoustic space or room, or the act of creating, extending, and/or manipulating a sound space. The process therefore needs to deal with the spatial attributes of sound sources, but also with the acoustical properties of the space itself. Some practices focus exclusively on the former, building on the notion of an abstract sound scene that is populated by sound objects, while others focus mainly on the latter, modelling the perceived acoustic properties that carry the spatiality of the sounds. In any practical musical situation, both domains need to be taken into account.

In the 'acousmatic' practice of speaker orchestras, the sounds are routed directly to actual speakers distributed in space, either as single-source channels or grouped in 'stems [that] constitute the submixes or—more generally speaking—discretely controllable elements which mastering engineers use to create their final mixes' (Wilson and Harrison 2010, 245). The speakers can have different sonic qualities, thereby influencing the colouring of the diffusion; they are given the role of different instruments in an orchestra. This channel-based placement is also the technical method of cinema surround-sound, where the content, in particular the dialogue, is routed to a dedicated speaker. Only with the recent advent of object-based audio in systems such as Dolby Atmos (Dolby Laboratories 2014) has this mode of operation been extended.

Object-based or abstract sound placement methods can be considered as simulating a sound scene. These simulation methods are built on the premise that all the elements of an acoustic scene can be constructed one by one, and that by assembling the abstract elements a convincing acoustical space can be generated. We will see that this is not always the case, since by working with sound objects in an abstract space, a geometric mode of thinking is emphasized whose visual paradigm doesn't always translate into perceivable auditory results (Couprie 2004).

Sound objects in a sound scene are conceptually independent from the specifics of the audio reproduction system. They are modelled first in the abstract space before being rendered in the concrete venue. As soon as a sound source needs to be placed at a location that falls *between* the diffusing speakers, the term 'phantom imaging' (Lennox 2009, 261) or 'virtual source' is used. In the simplest case this involves panning a source between a stereo pair (pairwise panning), but this can be extended to an arbitrary number of speakers and even pass through a simulation of an entire wave front of a sound, as is the case in Ambisonics or wave-field synthesis.

Sources in sound scenes have geometric properties such as position, orientation, and size. They are often considered as mere points in space, sometimes with added spatial extension. Sources in a scene also have acoustic qualities and spatial attributes, such as directivity or diffusion pattern, that is, the way sound is projected into space, for example the narrow sound beam exiting the bell of a trumpet versus the diffuse sound-waves originating from the drumheads of the timpani. Sound objects in a scene are also subjected to the acoustical properties of space. These affect spatial perception and are modelled using acoustic cues such as distance attenuation (falloff of sound intensity with increasing distance), spectral air absorption with distance (high-frequency components of sounds are filtered by air moisture), doppler shifts of moving sound sources (pitch changes due to compression or dilation of soundwaves when moving towards or away from the listener), and reflections from elements in the sound scene such as walls.

In addition to these source-bound properties, certain spatialization processes introduce additional cues that reconstruct either psychoacoustic effects, such as interaural time difference, pressure difference, and filtering effects by the anatomy of the head, or other processes that add global spatial effects, such as reverberation, components of which might be localized or which might reconstitute the acoustics of an actual space by convolving an impulse response obtained in a real space, or might reconstruct the field of the soundwave as it existed in real acoustics.

An entirely different mode of musical thinking with spatiality of sound is the deconstruction or combination of sounds in an artificial manner, which doesn't intend to simulate an existing sounding space. The aim of these techniques is to generate different senses of envelopment and engulfment of the listener (Lynch and Sazdov 2011; Paine, Sazdov, and Stevens 2007). Through blending or fragmentation (decorrelation) of sound elements, spatial effects are generated that have no correspondence in the natural world. This can occur in the temporal, spectral, or spatial domains. In the temporal domain the construction of auditory cues is manipulated by placing events close together on the temporal threshold of the auditory system. In the spectral domain the spatial coherence of a sound gets extended or suppressed by splitting and displacing frequency components of the sound (Parry 2014). In the spatial domain sounds can be spread across groups of speakers, usually combined with some manipulation of the signal such as filtering. The listener's auditory processes provide the basis for this creative play with the boundaries of perception. More subtle processes that fall into this category are also applied when manipulating spatial properties of sounds via traditional sound-engineering techniques such signal matrixing.

Many electroacoustic composition and diffusion practices involve the use of techniques that deal with the distribution of preproduced sound elements on speaker arrays (Wilson and Harrison 2010) through a variety of compositional principles (Lyon 2008). Since these sound groups carry their own spatial imagery (Kendall 2010), even through metaphorical connections (Bayle 2007), overlaying these subspaces and combining their gestural presence generate a different sense of spatiality and tangibility (Barrett 2015).

25.4 SPATIALIZATION ALGORITHMS

There are two meanings of the term 'algorithm' that need to be distinguished in a discussion about algorithmic spatialization.

The first is applied to mathematical formulas that process and synthesize those audio signals that carry spatial information to the listener's ears. They are called spatialization algorithms or, in analogy with computer-graphics, spatial audio *rendering* algorithms.

The second meaning is used to denote rule-based operations that generate structure from (sometimes) symbolic elements. These algorithms are used in compositional operations with elements that are part of an abstract sound scene or a symbolic space.

This separation is not always strictly enforceable, in some rendering processes there are parameters that can also serve for symbolic operations (see Figure 25.1, processing layers 3 and 4).

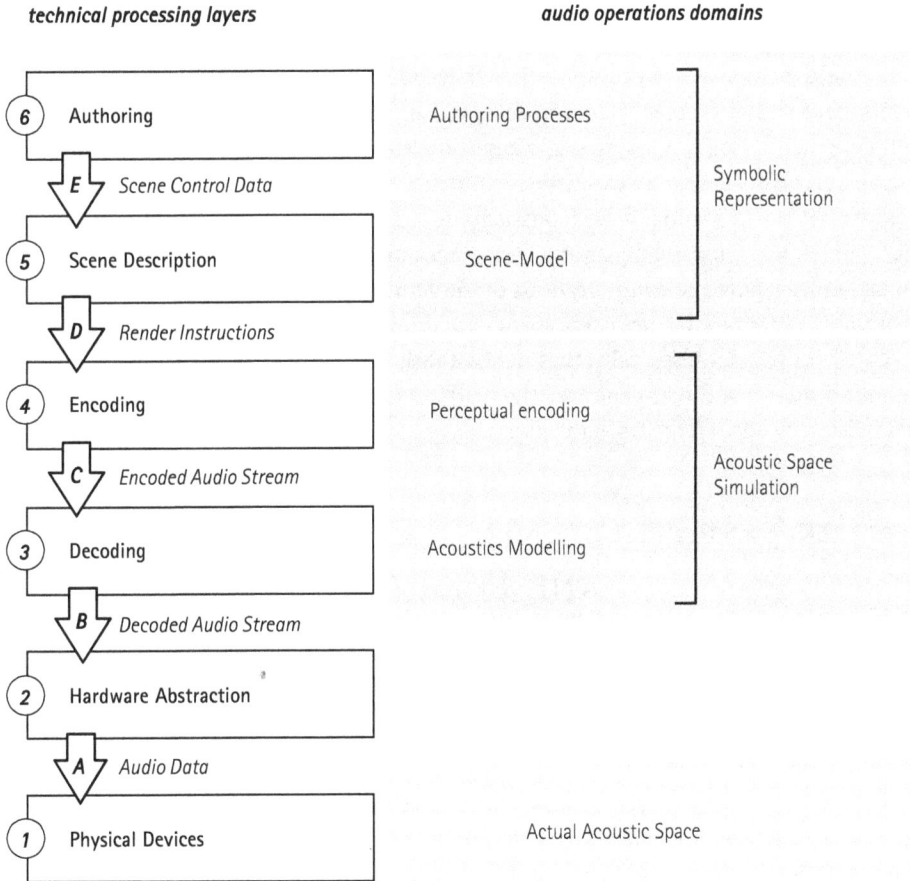

technical processing layers　　　　　　　　　**audio operations domains**

technical processing layers	audio operations domains	
6 Authoring	Authoring Processes	Symbolic Representation
E Scene Control Data		
5 Scene Description	Scene-Model	
D Render Instructions		
4 Encoding	Perceptual encoding	Acoustic Space Simulation
C Encoded Audio Stream		
3 Decoding	Acoustics Modelling	
B Decoded Audio Stream		
2 Hardware Abstraction		
A Audio Data		
1 Physical Devices	Actual Acoustic Space	

FIGURE 25.1 Spatial audio processing layers and compositional operation domains.

Within the first category, the 'rendering' algorithms sometimes represent mere multichannel panning processes, but at other times they involve many layers of sound processing in order to generate the acoustic and psychoacoustic cues necessary for convincingly simulating spatialized audio. Commonly used rendering algorithms are Vector Base Amplitude Panning (VBAP; Pulkki 1997) and, derived from that, Distance Based Amplitude Panning (DBAP; Lossius, Baltazar, and Hogue 2009); the more complex and powerful Ambisonics (Gerzon 1985) and Higher Order Ambisonics (Daniel 2000); wavefield synthesis (WFS; Berkhout, De Vries, and Vogel 1993); the virtual microphone techniques (ViMiC; Braasch 2005); and binaural rendering (for headphones; Bedini 1985; Noisternig et al. 2003). Each one of these audio-processing algorithms offers specific controls over the spatiality of sound. Some of the controls of these signal-processing methods may even become part of a composition system's parameter space, for example the spread factor offered by VBAP that changes apparent source width or the order factor used in Ambisonics that describes the angular resolution of the sound image.

In a blending of the two paradigms, spatialization needn't be concerned only with objects in a sound scene, it could equally be dealing with creating sound spaces in general with a mix of acoustic elements coming, for example, from field recordings or artificial spaces. To some extent all (electroacoustic) music inherently takes the spatial effect of its sound elements into account, since there is no dissociation possible between the sound space and the sound image (Bayle 1993; Kendall 2010).

In general, the topic of using sonic environments is a less explored area of electroacoustic composition and by extension of musical forms developed with algorithms. In most spatial audio practices the acoustical properties of a chosen space are configured once and left static for the duration of the piece and the performance. We will present a few examples where the configurations of the acoustic spaces themselves become compositional operations and are carried out with the aid of algorithms.

When working with spatialization the first task is to decide which dimension of spatial sound, audio, or music generates the material for the compositional operations and/ or provides the core elements of the musician's activity.

25.5 SPATIALIZATION PROCESS LAYERS AND DOMAINS

When looking at a workflow for spatialization (Peters et al. 2009), making the following subdivisions can help to distinguish the domains we operate in and the types of representations and dimensions that are in play (see Figure 25.1). The technical processing layers represent necessary steps of a workflow; a different category of a data flows from one layer to the next, and each layer contains conceptually similar and unique classes of functionalities. These provide 'services to the layer above it and receive services from the layer below it' (Peters et al. 2009, 220).

- The Authoring Layer: This layer contains all software tools for the end-user to create spatial audio content without the need to directly control underlying audio processes.
- The Scene Description Layer: This layer mediates between the Authoring Layer above and the Decoding Layer below through an abstract and independent description of the spatial scene.
- The Encoding Layer: Here the source signals are encoded acoustically for the first time. Some spatialization algorithms process the encoding and decoding in one step, whereas others implement it in two or more steps. All the perceptually relevant sound cues are encoded here.
- The Decoding Layer: In this layer the sounds are assembled into a coherent virtual acoustical sound space or scene. In this step additional acoustics modelling and simulation are applied to the source sounds.
- The Hardware Abstraction Layer is located with the operating system's audio drivers.
- The Physical Devices are the speakers needed to make an audio signal audible in a physical space.

Juxtaposing this technical model with the operations done in the compositional domain can help to clarify how these operations are related to each other. This is particularly relevant when reflecting on the distinction between operations that modify a sound scene and those that modify the acoustic space. The two main sections differentiate between symbolic operations in an abstract (parameter) space applied to discrete properties of abstract sound objects and signal operations directly affecting the acoustic qualities and properties of the sounds that will be projected and heard. As with all categorizations, there are exceptions that straddle the divide, as we discuss below with regard to spectral operations (section 25.7.3).

Authoring processes deal with placements, movements, and groupings, as well as with time organization of the scene. The processes themselves are embedded in the algorithms that are used to shape the evolution of sound objects over time within the sound scene. The scene model is an abstract representation of a space evolving over time. This space can maintain its state as a container for spatial audio operation, but can also become the object of operations itself (Wozniewski, Settel, and Cooperstock 2007).

Acoustic space simulation deals with all the processing necessary to produce the audio signals that we will hear as containing spatial audio. This includes positioning a source around the listening position, giving it distance cues, movement cues, and directivity cues, in short, constructing all the necessary auditory cues for the perceptual encoding of a source in space. In addition, the processes may include the acoustical modelling of a space, for example by simulating the reflections a sound source would produce in an architectural space.

Finally, working with the physical devices themselves, that is, working with the speakers, is a necessary part of controlling the effect of the actual physical space on the simulated acoustical space that is being projected. In some practices this is leveraged for interesting creative effects, for example in 'acousmatic' interpretations on a speaker orchestra, whereas in other settings the influence of the actual space is eliminated as much as possible in order to obtain as 'pure' a simulation of a virtual space as possible (this is of course only really possible in anechoic conditions).

25.6 STORAGE AND TRANSMISSION

One of the challenges of working with spatialized audio is the storage and transmission of pieces, and in particular of in-progress and nonfixed sound compositions. Traditionally, an 'acousmatic' composition is either stored as a rendered version for a dedicated speaker setting (an eight-channel circle, a 5.1 mix for DVD, etc.) or, the same way as work in progress, the components of the composition are stored individually. The spatial placements, transformations, and manipulations that constitute the piece are stored in the session formats of the digital audio workstation (DAW) software that was used, and as sound files containing single tracks or stems (grouped tracks). However, storing a sound scene and all its constituting elements so that all the relevant aspects remain editable is only beginning to be possible in commercial environments (e.g., Dolby Atmos, MPEG-H) and still represents an important hurdle in a composer's workflow. Several initiatives have tackled this issue in the past, including standards bodies such as the MPEG group (Scheirer, Vaananen, and Huopaniemi 1999), the production format Audio Definition Model endorsed by the EBU (2014), and software projects intended to generate a unified framework for audio spatialization (Geier, Ahrens, and Spors 2010).

The SpatDIF project group, of which the author forms part, approaches this task in a pragmatic manner by defining and implementing the Spatial Sound Description Interchange Format. 'SpatDIF provides a semantic and syntactic specification for storing and transmitting spatial audio scene descriptions ... a simple, minimal, and extensible format as well as best-practice implementations' (Peters, Lossius, and Schacher 2013, 11). In this syntax, the sound scene and its embedded entities have descriptors that represent as many relevant properties as necessary in order to describe and at a later stage reconstruct the scene. The descriptors with their values are stored in human-readable form in text files or transmitted in network packets for real-time applications and joined with the sound files or streams that make up the content of the work. In the SpatDIF concept, the authoring and the rendering of spatial scenes may occur at separate times and places using tools whose capabilities are unknown. It is a syntax rather than a programming interface or file format and can therefore be represented in any of the structured markup languages or message systems that are in use today or in the future.

In addition to specifying the syntax and format, the SpatDIF group is developing reference implementations that show best-use applications, and it also provides a software library for easy integration in various audio software (Miyama, Schacher, and Peters 2013). This library has been embedded in code plugins (externals) for the MaxMSP and

Pure Data environments, and is currently being integrated into a new version of the Zirkonium software (Wagner et al. 2014), providing it with SpatDIF import and export capabilities and opening up possibilities for interchanging compositions between different software environments and venues.

25.7 SPATIALIZING WITH ALGORITHMS

Spatialization as defined earlier deals with placing sounds in an acoustic space and/or creating and modifying such a space. Evidently algorithmic spatialization does this by using rule-based processes. Selecting which of the elements are generated, controlled, or transformed between the abstract sound scene and the simulated room determines which algorithmic operations are possible. Since algorithms in this context are defined as being rule-based processes organizing elements and structures of a musical work, in the case of composition, or as processes that directly affect the timbral, temporal, and spatial qualities of the music, those two domains need first to be considered separately before we can find overarching processes that affect both simultaneously.

25.7.1 Point Sources

As discussed earlier, the objects in a sound scene as well as the scene-defining acoustic elements possess various parameters useful for creating musical work. The most immediate and spatially most intuitive aspects of the objects are their locations and displacements in space. Algorithms for generating, controlling, and transforming the movement trajectories are quite common and are closely related to traditional panning automations. Beginning with the earliest multichannel works based on computational processes, working with point sources and transforming their geometrical as well as acoustical properties has become the most common way of composing and transforming a sound scene.

The 1972 composition *Turenas* by Chowning (1977) created at Stanford's nascent Center for Computer Research in Music and Acoustics (CCRMA) is a four-channel piece that for the first time simulated several aspects of spatial sound diffusion beyond source panning, such as doppler, reverb, and air absorption. *Turenas* represents an important step in the context of algorithmic thinking, since the source movements are derived from mathematical functions rather than subjective drawings or placements, and the model for connecting the perceptual and the compositional aspects are highly formalized (Chowning 2011). In this piece Lissajous formulas serve as algorithms that describe source movements, resulting in expressive trajectories (see top left of Figure 25.2).

Composing by choreographing sounds with geometric shapes and trajectories within the frame of space is further explored conceptually by Wishart (1996). He proposes an

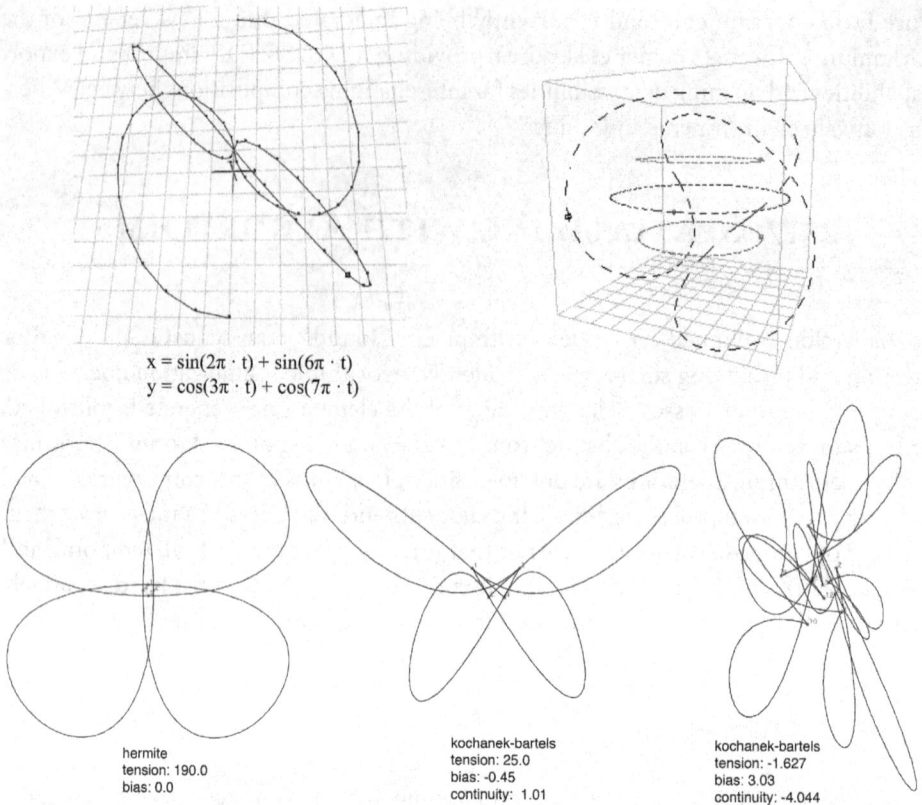

$$x = \sin(2\pi \cdot t) + \sin(6\pi \cdot t)$$
$$y = \cos(3\pi \cdot t) + \cos(7\pi \cdot t)$$

hermite
tension: 190.0
bias: 0.0

kochanek-bartels
tension: 25.0
bias: -0.45
continuity: 1.01

kochanek-bartels
tension: -1.627
bias: 3.03
continuity: -4.044

FIGURE 25.2 Top left: Visualization and formulas of the *Turenas* 'insect' Lissajous trajectory in a three-dimensional view. Top right: A three-dimensional view of three circular paths, and a closed cyclical path made of Bézier-curve segments. Bottom row: Wishart's cyclical cloverleaf, butterfly and irregular oscillating motions.

entire typology of movements oriented in the space around the listener. The spatial movements constitute (musical) gestures, and he investigates how the spatial motion of sound objects relate to each other in what he calls 'spatial counterpoint', and how these 'gestures can be used independently of other musical parameters or in a way which reinforced, contradicted or complemented other gestural features of the sound-object' (195). The frame of reference formed by the listener enables the distinction between purely geometric and symmetrical spatial forms, orientations, and directions, which are biased by psychological and aesthetic aspects of spatial perception. Sounds are heard, for instance, most clearly when we turn our face towards them, which emphasizes frontal positions, whereas a unidentifiable sound originating from a rear direction may have, for evolutionary reasons, a threatening or frightening effect. In his typology Wishart considers continuous motion paths in only two dimensions. His catalogue enumerates many direct paths: centre-crossing straight lines, edge-hugging straight lines, centre-crossing arc movements, forward- or backward-moving diagonal paths, centre-hugging diagonal paths, movements towards and away from the centre. For circular motions he distinguishes between cyclical (repeated), central or eccentric circular motion, spiral paths, figure-of-eights and S-curves. By combining and overlaying these shapes, various zig-zags and looping movements arise that exhibit patterns of movement progressing through space that range from oscillatory rotating loops to cloverleaf and butterfly pathways. Further elements are localized and unlocalized irregular motions generated by random or brownian processes, as implemented for example in the ICST's 'ambicontrol' methods (Schacher and Kocher 2006), that can be centre-bound or corner-bound and offer the possibility to be overlaid and combined into compound paths. For defining the behaviour of a motion in time as well as space, Wishart adds time contours that define speed, acceleration, and deceleration, and that generate perceptual forms that transport 'intent' or physical behaviour, such as elastic, bouncing, or throwing movements. These behaviours give rise to the perception of a sound object's material properties or the type of handling by an (unseen) agent. He emphasizes how changing the time contours of a given spatial gesture can influence the aesthetic impact of a spatial motion. Of course all of the principles described by Wishart can be generated, controlled, and transformed through algorithmic processes (see the bottom row of Figure 25.2 for three examples of looping movements generated by applying different spline formulas).[3]

There are research projects developing terminologies, methods, and tools for the *notation* of spatial sound aspects. Thoresen's analysis of Schaeffer's sound objects (Thoresen and Hedman 2007), as well as the sound patterns and form-building patterns both in the temporal and spatial dimensions that he categorizes (Thoresen 2010), have led to an extension to GRM's 'acousmatic' music notation software, the Acousmographe (Geslin and Lefevre 2004). The 'Spatialisation Symbolic Music Notation' project at ICST in Zurich also works towards defining a standard taxonomy of spatial motions (Ellberger and Perez 2013) and a set of trajectory 'gestalts' that are applicable to both sound sources and room aspects, with the goal of representing them as symbols in standard music notation (Ellberger et al. 2014). In these systems the taxonomies of shapes, patterns and relations, and semantic organization of discrete sound elements serve to identify those elements as compositional materials that are equivalent to other musical parameters (in a mode of postserialist compositional thinking).

25.7.2 Source Clusters

The method of dealing with discrete 'sound pixels' in an abstract sound scene is extended when working with clusters of sound elements, or 'ensembles' (Rumsey 2002). These sometimes large groups of objects follow general rules and might appear as more or less diffuse sound objects in the sonic space. Granular synthesis techniques are particularly apt for spatial distribution of large numbers of objects, where each grain potentially occupies a different location in space and together they form a sound mass that can occupy a sector or the entire sound space (Wilson 2008).

A combination of these techniques with generative principles, for example by giving each cluster element emergent spatial behaviours by using flocking concepts such as the perennial 'Boids' algorithm (Reynolds 1987), provides a higher level of handling the entities forming the cluster (Kim-Boyle 2006). These agent-based systems, thanks to their self-organizational properties, permit the generation of complex group or cluster behaviours with a reduced number of semantically relevant parameters. In the case of a Boids flock, for example, moving the attractor point will manoeuvre the entire cluster in a loose cloud whose spatial extension is controlled by the shared cohesion parameter. These agent-based algorithms represent a special case of control algorithms, by offering dynamic, self-organized domain translations that are useful for spatialization as *direct parameter mapping*, since the agents can be modelled as objects in Euclidian space and their location therefore directly translated to spatialization source positions (Schacher, Kocher, and Bisig 2014, 52).

More generic algorithmic models can generate complex behaviours as well, even in interactive settings, for example through the use of hierarchical, nested swarms controlling both visual and sonic surround renderings (see Figure 25.3) (Schacher, Bisig, and Neukom 2011) or the implementation of rules that operate not in the spatial domain but rather on the object's physical attributes, for example on spring forces, mass or damping parameters in physical models (Bisig, Schacher, and Neukom 2011).

Particle systems provide a similar type of high-level cluster control for the dynamic distribution of large numbers of point sources with a few control parameters, in this case exerted as force-fields on particles. 'One of the attractive qualities of particle systems is their ability to model or visually mimic natural phenomena' (Kim-Boyle 2008, 3). Simulating natural phenomena within such a system generates emergent properties for clusters of sound-objects, for example by implementing spatial evasion through sensing of the proximity of another particle or by exerting forces along directional lines, thus orienting the movements of the objects.

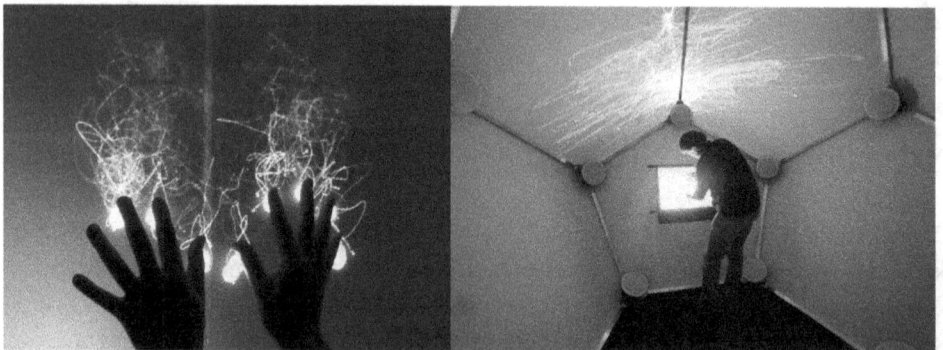

FIGURE 25.3 Touch-based interactions with three hierarchically linked flocks in 'Impacts' in the interactive generative installation *Flowspace* (2009–2010). In this piece, visual and sonic outputs originate from the flocking simulation, which generates musical structure by analysing agent behaviour and by triggering and spatially positioning sound events in a dodecahedral twenty-channel speaker array. Photographs by Jan Schacher © 2010.

An example of the combination of a traditional synthesis technique with a dynamic spatialization is shown by Schumacher and Bresson (2010) in their 'spatial additive synthesis: (a) a harmonic spectrum is generated ... and additional partials (micro-clusters) added around each harmonic; (b) a set of envelopes is used to control both sound synthesis and spatialization parameters; (c) two manually defined ... trajectories are interpolated over the number of partials. Each partial is assigned an individual trajectory.' A similar method by Topper, Burtner, and Serafin (2002) describes a separation process for spatialization purposes as 'taking an existing synthesis algorithm and breaking it apart into logical components' and then '[assembling] the components by applying spatialization algorithms'. In this application the method consists of 'separating the modes or filter the output of a physical model and applying individual spatial processing on each component'.

25.7.3 Spatial Spectral (De-)composition

These decomposition techniques are also applicable to the spectral or timbral domain of (re-)synthesized sound. Different ways of cutting up the spectrum of a sound and spreading these components in the sound-space exist. By fragmenting the sound spectra amongst a network of speakers 'the entire spectrum of a sound is recombined only virtually in the space of the concert hall. ... It is not a conception of space that is added at the end of the composition process ... but a truly composed spatialisation' (Normandeau 2009, 278).

Changing the temporal as well as the spatial location of fragments of a sound's spectrum further de-correlates it and leads to a different type of diffusion within the acoustic space. 'Delaying the resynthesis of individual FFT bins of a short-time Fourier transform can create musical effects not obtainable with traditional types of delays. When those delays are applied to sounds reproduced through the individual channels of a multi-channel playback system, unique spatialization effects across spectral bands can be realized' (Kim-Boyle 2008, 1).

This 'spectral splitting' as a decorrelation technique can also occur involuntarily when using nonhomogeneous speakers that emphasize certain frequencies and thus distribute the spectrum unevenly across a speaker array's sound space. Combining this effect with granulation approaches that determine routing in relation to the input amplitude or spectral characteristics has the potential to create an expanded perceived spatial size of the cluster, for example by spreading 'from the front to the back of the space as the amplitude increases' (Wilson and Harrison 2010, 248).

25.7.4 Manipulating Sound Spaces

A different and subtle way of changing the timbre of sounds throughout the acoustic space is using different room simulations for individual stems that are then overlaid and assigned to different sectors of the space. These artificial acoustic situations can suggest a volume of space through implied spatial occupation (Barrett 2002). Further creative use of spatial zones, as implemented for example with the virtual microphone techniques ViMiC, might not even cover an entire venue homogeneously, but use overlapping virtual acoustic spaces in different parts of the physical space, thus leveraging the effect of the real acoustics to generate a hybrid spatiality (Peters, Braasch, and McAdams 2011, 180)

Similar concepts can be explored by employing rendering processes that do not necessarily generate a unified sound-field. In these processes, stems can be assigned to subspaces or speaker groups in what is effectively a hybrid between 'acousmatic' interpretation in the style of the Acousmonium and signal-processing-based multichannel diffusion methods. Using DBAP (Lossius, Baltazar, and Hogue 2009), for example, in particular by using speaker subsets and partial groups, nonrealistic representations of distributed sounds can be created. In this pragmatic approach the perception of sound *placements* and the local activities of sonic elements, rather than of *trajectories* provides the central characteristic (Baltazar and Habbestad 2010).

The Ambisonics spatialization processes offer yet another way of algorithmically manipulating virtual acoustic space. In this concept all sound events are first encoded into an intermediate abstract sound space—the B-format stream—which consists of the spherical harmonics of a sonic wave-field that covers the full 'periphonic' space, that is, the entire three-dimensional sphere around the listener. This technique originates from a microphone technology that is used to record a full 3D sound-field, but the mathematics of this process have subsequently been implemented for virtual sound encoding and decoding as well. Ambisonics enables the placement of sound objects in the periphonic space (on the unit sphere), but more interestingly permits the manipulation of the sound-field itself (Lossius and Anderson 2014). By changing aspects of the algorithm and introducing transformation of the signals within the intermediate B-format domain, manipulations such as zooming in, pushing out, emphasizing and rotating the entire sound-field become possible (see Figure 25.4).

25.8 Spatialization in Live Situations

Manipulating spatial audio distribution in realtime during live performance poses a few unique problems. To begin with, the musician's listening position is not always centred, and therefore does not always provide the ideal sound image. In 'acousmatic' concerts with surround-sound, the mixing-desk position will be centred in the hall to avoid this problem. In a frontal performance situation, however, surround monitoring is necessary to provide the performer with the same spatial perception as the audience. Replacing this by prelistening over headphones is difficult, unless an additional binaural simulation is implemented in the monitoring paths.

Controlling spatial distribution of a large number of sound sources in realtime (with or without the aid of algorithms) demands a representation of parametric controls that can be understood and handled directly. The challenge and limitation of parametrically controlling a large number of sound sources in realtime are reasons for using higher-level algorithms for control. A mapping strategy that implements one-to-many connections (Arfib et al. 2002) represents the first type of algorithmic control structure. For live situations, higher-level abstracted controls need to be implemented that can be manipulated with lower-dimensional controls, be it directly on single-dimension controllers such a faders, or on compound controllers that encapsulate spatial information such as joysticks or camera-based gesture-recognition systems. Algorithms that contain autonomous, independent components and provide high-level control such as agent or particle systems are particularly suited for real-time control. But any algorithm that is capable of being manipulated through a few variables works. By overlaying several dimensions of control, for example by combining spatial- and temporal-control variables, for example in granular or spectral processes, the overall gestalt of the sounds can be performed with relatively few interactive controls.

This applies to studio and offline processes as well. When composing with algorithms that shape any aspect of a sound scene, be it through placements and trajectories, clustering, and spectral and temporal processes, the composer needs simple methods to interact with rule-based processes in order to judge the results. The principal difference is that these processes can be repeated, layered, and edited in ways which are not possible during performance.

Depending on the context, be it an electroacoustic concert, a live-coding session, a theatre production with real-time sound processing, or a gestural performance (Schacher 2007) in a club or festival, a strategy needs to be devised to maintain expressive control over the spatialization without getting overwhelmed by the complexity of the spatial and algorithmic processes.

25.9 'IMPACTS': AN INTERACTIVE ALGORITHMIC COMPOSITION

In order to show how some of the aspects described above can be applied in practice, an interactive and algorithmic composition provides us with an example. The musical and visual composition 'Impacts' forms part of the *Flowspace* installation (Schacher, Bisig, and Neukom 2011). Within a dodecahedral frame the sound is spatialized on twenty speakers that sit in its corners. The upper faces of the 4m platonic solid serve as rear-projection screens for the real-time graphics, and a touch-sensitive surface provides the interaction modality to the visitor (see Figure 25.3).

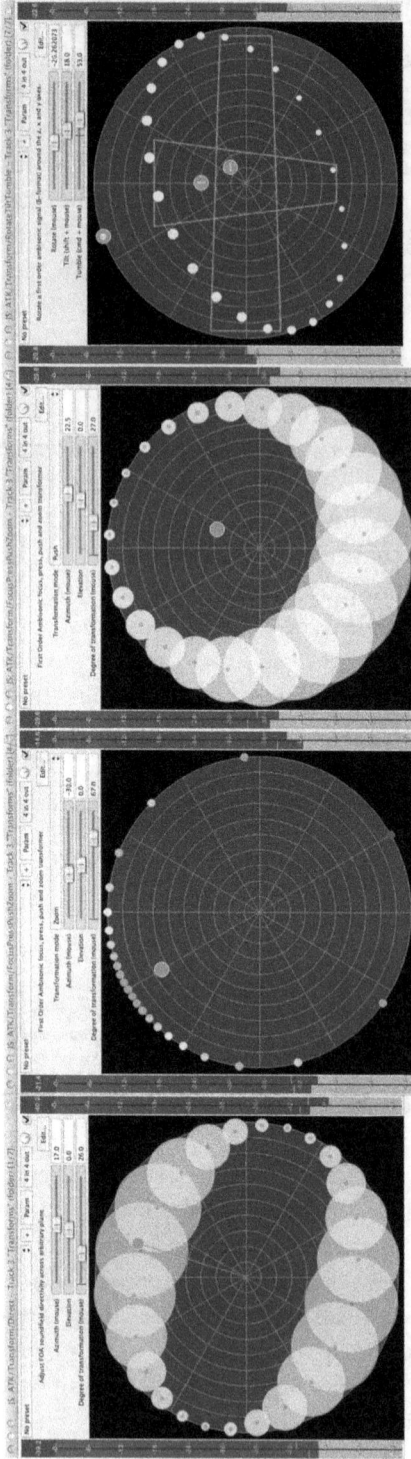

FIGURE 25.4 Four views of Ambisonic sound-field transformation implemented by the ATK in the Reaper plugins by Lossius and Anderson (2014). The processes change the directivity, and zoom, push, or rotate the sound-field. Screenshots used by permission.

The algorithms at the heart of this piece explore hierarchical relationships between three flocks, and represent their interdependence within the ecosystem of the piece. Three types of entities are present in an abstract algorithmic domain: the first are attractor 'touch' points that are controlled by the visitor's actions; the second are agents in a flock that react to the attraction forces of the 'touch' agents as well as those of their own kind; the third flock is subjected to the forces exerted by the second swarm and those of its own peers. The behaviours of the agents within the second and third swarms are based on the classic attraction-evasion-alignment paradigm (Reynolds 1987), and are parameterized to create dynamic motion patterns.

In a next step, perceptually significant events are extracted from the continuous motions of flocking agents in order to provide key impulses for the music. The impacts or (near) collisions between agents are treated as expressive events in the scene that trigger the musical events. In contrast, reaching the farthest points on the escape trajectory from the point of impact triggers a second type of event. The collision events trigger piano samples on impact and granular echoes of the same pitches at the escape points, and thus constitute the musical gestalt of the composition.

A simple state machine tracks the level of engagement of the visitor and controls the choice of pitches accordingly: the higher the level of interaction, the fuller and more dissonant the pitch sets will be. These sets are divided into eight groups, one for each agent in the primary 'touch' flock. The secondary swarm activates the lower-register notes on impact, whereas the third swarm initiates the higher pitches at the escape points. Being repeatedly triggered during the escape trajectory, the expanding granular 'shadows' engender a noticeable perceptual widening of the pitch and surround space. Since the note events are spatialized according to the geometrical positions of agents, the swarm clusters are perceivable as note clusters in different sectors of the surround field. Vertically, the spatial positions of the swarm agents are stretched onto the surround sphere in order to make height perception more evident.

The mixture of all of these elements, arising from the dynamics of events that the agents encounter, generates the sonic texture which is characteristic of this piece. The ebb and flow of density found in the musical domain reflects the state of the underlying model, and even if no global control is applied to the sound producing algorithms directly, the way visitor interactions propagate through the layers of algorithms influences the overall musical result.

A third principal element of the piece, the real-time graphic visualization, reinterprets the idea of impacts and escape points by connecting points into dynamically changing and triangulated 'Delaunay' meshes, and by triggering concentric, rippling circles for each of these events. The graphical language works with rules of its own that affect colours, scaling, and visibility of elements. These algorithms are also controlled by the visitor's engagement level, and provide through graphical means an interpretation of the processes occurring in the underlying hierarchical ecosystem of the piece.

25.10 CHALLENGES, MISCONCEPTIONS, AND PITFALLS OF SPATIALIZATION

It is important to be aware of the subtle and not so subtle ways sound spatialization can fail to fulfil expectations. Since acoustic space represents a complex environment with many factors at play, getting everything right in (re)creating a *believable* spatial sound scene is quite challenging. The degree to which this needs to be achieved depends on the desired outcome. If the perfect simulation of a sonic environment is the goal, criteria come into play that are harder to fulfil than if the goal is compositional work in a creative manner. In the former case great care has to be taken to reconstitute the acoustic space with all the correct localization cues, whereas in the latter case completely artificial spatial combinations are possible. In both cases the sound processes are subjected to the laws and principles of our spatial auditory perception.

Kendall and Ardila (2007) investigate and explain in detail why things don't always work as expected. They give 'three reasons why the spatial potential of electro-acoustic music is not always realised: 1) misconceptions about the technical capacities of spatialisation systems, 2) misconceptions about the nature of spatial perception, especially in the context of such systems, and 3) a lack of creative engagement, possibly due to the first two issues' (2007, 126). According to them, some of the elements responsible for these problems are: the precedence effect (Brown, Stecker, and Tollin 2015; Wallach, Newman, and Rosenzweig 1949), sweet-spot misalignment (Peters 2010), plausibility and comprehensibility issues, time-delay differences from the speakers between small and large venues, image dispersion dependent on transient and spectral characteristics of the source, cross-talk when playing back binaural signals over speakers, and the failure of spectral decomposition to be recognized as separate objects, which can be achieved only by desynchronizing the partials or adding contradictory vibrato patterns on the individual components (Kendall and Cabrera 2011).

A conceptual problem which is often ignored is the fact that sounds and sound objects are not pixels or abstract points in space. The dominant thinking in spatialization is based on a purely geometrical conception in Euclidian space, and most software tools provide a visualization in that paradigm, be it through points or trajectory paths on a visual display. This is misleading for several reasons: our spatial perception and the way sounds are embedded within an acoustic space do not provide by default the sharp point sources imagined; the grouping and stream-segregation principles applied both spatially and temporally by our auditory system (Bregman 1994) do not provide separation of sources in the same way a visual display does; the spatial resolution of our auditory system is not homogeneous in all directions: on the horizontal plane, the frontal localization blur covers +/−1 degree at certain frequencies, with a more typical blur of +/−5 degrees; to the sides the blur increases to +/−10 degrees; and above or below the listener and slightly to the back this blur reaches up to +/−22 degrees (Blauert 1983); a further problem is the front–back confusion, in particular with binaural headphone-rendering without head-tracking, as well as the cone of confusion on which it is impossible to determine where a sound is located (Röttger et al. 2007); phantom images on the side have a tendency to collapse, which leads to confused spatial perception, and finally, without the correct environmental cues, we have limited capabilities for judging the distance of sound objects (Oechslin, Neukom, and Bennett 2008).

It is fair to say that geometrically constructed sound scenes that operate with abstract point sources rarely produce a coherent or convincing spatial scene; for this to occur, additional acoustic and psychoacoustic cues need to be introduced. Therefore, those algorithmic processes that merely manipulate symbolic sound objects without respecting the psychoacoustic reality might not produce the desired effect. The auditory system's 'fault-correction' is capable of presenting the most plausible element as relevant, even if it is not mathematically correct or compositionally intended. Nevertheless, for creative applications that do not expect to produce a 'natural' sounding scene and space, algorithmic spatialization processes can generate interesting and sometimes surprising results.

Notes

1. K. Stockhausen, 'Kugelauditorium', *Medien Kunst Netz* http://www.medienkunstnetz.de/werke/stockhausen-im-kugelauditorium/bilder/4/, accessed 30 June 2017.
2. 'The Sonic Lab', *Sonic Art Research Centre*, http://www.sarc.qub.ac.uk/sites/sarc/AboutUs/TheSARCBuildingandFacilities/TheSonicLab/, accessed 30 June 2017.
3. ICST Ambisonics is now available as a MAX package, accessible from the Cycling74 website https://cycling74.com/ for use with MAXMSP, accessed 28 August 2017.

Bibliography

Amatriain, X., Kuchera-Morin, J., Hollerer, T., and Pope, S. T. 'The AlloSphere: Immersive Multimedia for Scientific Discovery and Artistic Exploration'. *IEEE Computer Society* 16, no. 2 (2007): 64–75.

Arfib, D., Couturier, J. M., Kessous, L., and Verfaille, V. 'Strategies of Mapping between Gesture Data and Synthesis Model Parameters Using Perceptual Spaces'. *Organised Sound* 7, no. 2 (2002): 127–144.

Baalman, M. A. J. 'Spatial Composition Techniques and Sound Spatialisation Technologies'. *Organised Sound* 15, no. 3 (2010): 209–218.

Baltazar, P., and Habbestad, B. 'Unruhige Räume—Spatial Electro-Acoustic Composition through a Collaborative Artistic Research Process'. In *Proceedings of the International Computer Music Conference*, 538–541. New York, 2010.

Barrett, N. 'Spatio-musical Composition Strategies'. *Organised Sound* 7, no. 3 (2002): 313–323.

Barrett, N. 'Creating Tangible Spatial-Musical Images from Physical Performance Gestures'. In *Proceedings of the International Conference on New Interfaces for Musical Expression*, 191–194. Baton Rouge, LA, 2015.

Bayle, F. *Musique acousmatique: Propositions ... positions.* Paris: Institut National de l'Audiovisuel INA and Editions Buchet/Chastel, 1993.

Bayle, F. 'Space, and More'. *Organised Sound* 12, no. 3 (2007): 241–249.

Bedini, J. C. *Monaural to Binaural Audio Processor*. US Patent 4555795. 1985.

Berkhout, A. J., De Vries, D., and Vogel, P. 'Acoustic Control by Wave Field Synthesis'. *Journal of the Acoustical Society of America* 93, no. 5 (1993): 2764–2778.

Bisig, D., Schacher, J. C., and Neukom, M. 'Flowspace: A Hybrid Ecosystem'. In *Proceedings of the Conference on New Interfaces for Musical Expression*. Oslo, 2011.

Blauert, J. *Spatial Hearing: The Psychophysics of Human Sound Localisation*. Cambridge, MA: MIT Press, 1983.

Blesser, B., and Salter, L.-R. *Spaces Speak, Are You Listening? Experiencing Aural Architecture*. Cambridge, MA: MIT Press, 2007.

Boone, M. M., and Verheijen, E. N. G. 'Multichannel Sound Reproduction Based on Wavefield Synthesis'. In *Audio Engineering Society Convention 95*. October 1993. http://www.aes.org/e-lib/browse.cfm?elib=6513.

Born, G. *Music, Sound and Space: Transformations of Public and Private Experience*. Cambridge: Cambridge University Press, 2013.

Boulez, P. *Répons and Dialogue de l'ombre double*. A. Damiens and Ensemble InterContemporain, conducted by P. Boulez. Berlin: Deutsche Grammophon, 1998. CD.

Braasch, J. 'A Loudspeaker-Based 3D Sound Projection Using Virtual Microphone Control (ViMiC)'. In *Audio Engineering Society Convention 118*. New York: Audio Engineering Society, 2005.

Braasch, J., Peters, N., and Valente, D. L. 'A Loudspeaker-Based Projection Technique for Spatial Music Applications Using Virtual Microphone Control'. *Computer Music Journal* 32, no. 3 (2008): 55–71.

Bregman, A. S. *Auditory Scene Analysis: The Perceptual Organization of Sound*. Cambridge, MA: MIT Press, 1994.

Brown, A. D., Stecker, G. C., and Tollin, D. J. 'The Precedence Effect in Sound Localization'. *Journal of the Association for Research in Otolaryngology* 16, no. 1 (2015): 1–28.

Brümmer, L., Dipper, G., Wagner, D., Stenschke, H., and Otto, J. A. 'New Developments for Spatial Music in the Context of the ZKM Klangdom: A Review of Technologies and Recent Productions'. *Divergence* 1, no. 3 (2014).

Chowning, J. 'The Simulation of Moving Sound Sources'. *Computer Music Journal* 1, no. 3 (1977): 48–52.

Chowning, J. '*Turenas*: The Realization of a Dream'. In *Proceedings of the 17es Journées d'Informatique Musicale (JIM '11)*. Saint-Étienne, 2011.

Couprie, P. 'Graphical Representation: An Analytical and Publication Tool for Electroacoustic Music'. *Organised Sound* 9, no. 1 (2004): 109–113.

Cowell, H., and Cowell, S. R. *Charles Ives and His Music*. Oxford: Oxford University Press, 1969.

Daniel, J. *Représentation de champs acoustiques, application à la transmission et à la reproduction de scènes sonores complexes dans un contexte multimédia*. PhD dissertation, University of Paris VI, 2000.

Dolby Laboratories. *Dolby® Atmos® Next-Generation Audio for Cinema*. White paper, 2014. http://www.dolby.com/us/en/technologies/dolby-atmos/dolby-atmos-next-generation-audio-for-cinema-white-paper.pdf.

European Broadcast Union (EBU). *Tech 3364, Audio Definition Model*. Geneva, January 2014. https://tech.ebu.ch/docs/tech/tech3364.pdf.

Eckel, G. *Random Access Lattice*. 2011. http://iem.at/~eckel/download/RAL-description.pdf.

Ellberger, E., and Perez, G. T. *SSMN Taxonomy*. 2013 http://blog.zhdk.ch/ssmn/files/2012/10/SSMN_Taxonomy.pdf.

Ellberger, E., Perez, G. T., Schütt, J., Zoia, G., and Cavaliero, L. 'Spatialization Symbolic Music Notation at ICST'. In *Proceedings of the Joint International Computer Music and Sound and Music Computing Conference*. Athens, 2014.

Gallagher, S. *How the Body Shapes the Mind*. Oxford: Clarendon Press, 2005.

Geier, M., J. Ahrens, J., and Spors, S. 'Object-Based Audio Reproduction and the Audio Scene Description Format' *Organised Sound* 15, no. 3 (2010): 219–227.

Gerzon, M. A. 'Ambisonics in Multichannel Broadcasting and Video'. *Journal of the Audio Engineering Society* 33, no. 11 (1985): 859–871.

Geslin, Y., and Lefevre, A. 'Sound and Musical Representation: The Acousmographe Software'. In *Proceedings of the International Computer Music Conference*, 285–289. Miami, 2004.

Hall, E. T. *The Hidden Dimension*. New York: Anchor, Doubleday, 1966.

Ihde, D. *Listening and Voice: Phenomenologies of Sound*. 2nd 2007. Albany: SUNY Press, 1976.

Kendall, G. S. 'Spatial Perception and Cognition in Multichannel Audio for Electroacoustic Music'. *Organised Sound* 15, no. 3 (2010): 228–238.

Kendall, G. S., and Ardila, M. 'The Artistic Play of Spatial Organization: Spatial Attributes, Scene Analysis and Auditory Spatial Schemata. In *International Symposium on Computer Music Modeling and Retrieval CMMR*, 125–138. Copenhagen, 2007. LNCS Series, Volume 4969, Cham, Switzerland: Springer International.

Kim-Boyle, D. 'Spectral and Granular Spatialization with Boids'. In *Proceedings of the 2006 International Computer Music Conference*, 139–142. New Orleans, 2006.

Kim-Boyle, D. 'Spectral Spatialisation: An Overview'. In *Proceedings of the International Computer Music Conference*. Belfast, 2008.

Lachenmann, H. 'Klangtypen der neuen Musik'. In *Musik als existentielle Erfahrung*, edited by J. Häusler, 1–20. Wiesbaden: Breitkopf and Härtel, 1966.

Lennox, P. 'Spatialisation of Computer Music'. In *The Oxford Handbook of Computer Music*, edited by R. T. Dean, 258–273. Oxford: Oxford University Press, 2009.

Lossius, T., and Anderson, J. 'ATK Reaper: The Ambisonic Toolkit as JSFX Plugins'. In *Proceedings of the Joint International Computer Music and Sound and Music Computing Conference*. Athens, 2014.

Lossius, T., Baltazar, P., and Hogue, T. de la. 'DBAP: Distance-Based Amplitude Panning'. In *Proceedings of the International Computer Music Conference (ICMC)*, 489–492. Montreal, 2009.

Lynch, H., and Sazdov, R. 'An Ecologically Valid Experiment for the Comparison of Established Spatial Techniques'. In *Proceedings of the International Computer Music Conference*, 130–134. Huddersfield, 2011.

Lyon, E. 'Spatial Orchestration'. In *Proceedings of the Sound and Music Computing Conference*. Berlin, 2008.

Miyama, C., Schacher, J. C., and Peters, N. 'SpatDIF Library—Implementing the Spatial Sound Descriptor Interchange Format'. *Journal of the Japanese Society for Sonic Arts* 5, no. 3 (2013): 1–5.

Neuhaus, M. 'Sound Art?' In liner notes for *Volume: Bed of Sound*. New York: P.S. 1 Contemporary Art Center, July 2000. http://www.max-neuhaus.info/soundworks/sound-art/SoundArt.htm.

Noisternig, M., Carpentier, T., and Warusfel, O. 'ESPRO 2.0: Implementation of a Surrounding 350-Loudspeaker Array for Sound Field Reproduction'. In *4th International Symposium on Ambisonics and Spherical Acoustics / Audio Engineering Society 25th UK Conference*. York, 2012.

Noisternig, M., Sontacchi, A., Musil, T., and Hóldrich, R. 'A 3D Ambisonic Based Binaural Sound Reproduction System'. In *Audio Engineering Society Conference: 24th International Conference: Multichannel Audio, The New Reality*. Banff: Audio Engineering Society, 2003.

Normandeau, R. 'Timbre Spatialisation: The Medium Is the Space'. *Organised Sound* 14, no. 3 (2009): 277–285.

Oechslin, M., Neukom, M., and Bennett, G. 'The Doppler Effect: An Evolutionary Critical Cue for the Perception of the Direction of Moving Sound Sources'. In *International Conference on Audio, Language and Image Processing: ICALIP 2008*, 676–679. Shanghai: IEEE, 2008.

Oehlschlägel, R. *Klanginstallation und Wahrnehmungskomposition: Zur 'Nuova Versione' von Luigi Nono's 'Prometeo'*. Cologne: MusikTexte, 1985.

Paine, G., Sazdov, R., and Stevens, K. 'Perceptual Investigation into Envelopment, Spatial Clarity, and Engulfment in Reproduced Multi-Channel Audio'. In *Proceedings of the Audio Engineering Society Conference*. London, 2007.

Parra Cancino, J. A. *Multiple Paths: Towards a Performance Practice in Computer Music*. PhD dissertation, Academy of Creative and Performing Arts, Faculty of Humanities, Leiden University, 2014.

Parry, N. 'Exploded Sounds: Spatialised Partials in Two Recent Multi-Channel Installations'. *Divergence* 1, no. 3 (2014).

Peters, N. *Sweet [Re]production: Developing Sound Spatialization Tools for Musical Applications with Emphasis on Sweet Spot and Off-Center Perception*. PhD dissertation, McGill University, Montreal, 2010.

Peters, N., Braasch, J., and McAdams, S. 'Sound Spatialization Across Disciplines Using Virtual Microphone Control (ViMiC)'. *Journal of Interdisciplinary Music Studies* 5, no. 2 (2011): 167–190.

Peters, N., Lossius, T., and Schacher, J. C. 'The Spatial Sound Description Interchange Format: Principles, Specification, and Examples'. *Computer Music Journal* 37, no. 1 (2013): 11–22.

Peters, N., Lossius, T., Schacher, J. C., Baltazar, P., Bascou, C., and Place, T. 'A Stratified Approach for Sound Spatialisation'. In *Proceedings of the Sound and Music Computing Conference*, 219–224. Porto, 2009.

Pulkki, V. 'Virtual Sound Source Positioning Using Vector Base Amplitude Panning'. *Journal of the Audio Engineering Society* 45, no. 6 (1997): 456–466.

Reynolds, C. W. 'Flocks, Herds and Schools: A Distributed Behavioral Model'. In *ACM SIGGRAPH Computer Graphics* 21, no. 4 (1987): 25–34.

Roads, C. *Microsound*. Cambridge, MA: MIT Press, 2001.

Röttger, S., Schröger, E., Grube, M., Grimm, S., and Rübsamen, R. 'Mismatch Negativity on the Cone of Confusion'. *Neuroscience Letters* 414, no. 2 (2007): 178–182.

Rumsey, F. *Spatial Audio*. Oxford: Focal Press, 2001.

Rumsey, F. 'Spatial Quality Evaluation for Reproduced Sound: Terminology, Meaning, and a Scene-Based Paradigm'. *Journal of the Audio Engineering Society* 50, no. 9 (2002): 651–666.

Schacher, J. C. 'Gesture Control of Sounds in 3D Space'. In *Proceedings of the 7th International Conference on New Interfaces for Musical Expression*, 358–362. New York: ACM, 2007.

Schacher, J. C., Bisig, D., and Neukom, M. 'Composing With Swarm Algorithms: Creating Interactive Audio-Visual Pieces Using Flocking Behaviour'. In *Proceedings of the International Computer Music Conference*. Huddersfield, 2011.

Schacher, J. C., and Kocher, P. 'Ambisonics Spatialization Tools for Max/MSP'. In *Proceedings of the International Computer Music Conference*. New Orleans, 2006.

Schacher, J. C., Kocher, P., and Bisig, D. 'The Map and the Flock: Emergence in Mapping with Swarm Algorithms'. *Computer Music Journal* 38, no. 3 (2014): 49–63.

Schaeffer, P. *Traité des Objets Musicaux*. Paris: Seuil, 1966.

Schafer, R. M. *The Soundscape: Our Sonic Environment and the Tuning of the World*. Rochester, VT: Inner Traditions/Bear, 1993.

Scheirer, E. D., R. Vaananen, R., and Huopaniemi, J. 'AudioBIFS: Describing Audio Scenes with the MPEG-4 Multimedia Standard'. *IEEE Transactions on Multimedia* 1, no. 3 (1999): 237–250.

Schumacher, M., and Bresson, J. 'Spatial Sound Synthesis in Computer-Aided Composition'. *Organised Sound* 15, no. 3 (2010): 271–289.

Serken, S. 'Towards A Space-Time Art: Iannis Xenakis's Polytopes'. *Perspectives of New Music* 39, no. 2 (2001): 262–273.

Shaff, S. 'Audium—Sound-Sculptured Space'. *Divergence* 1, no. 3 (2014).

Smalley, D. 'Spectromorphology: Explaining Sound-Shapes'. *Organised Sound* 2, no. 2 (1997): 107–126.

Smalley, D. 'Space-Form and the Acousmatic Image'. *Organised Sound* 12, no. 1 (2007): 35–58.

Stockhausen, K. *Kontakte: Für elektronische Klänge, Klavier und Schlagzeug*. Kürten: Stockhausen-Verlag, 1995. Score.

Stockhausen, K., Schnebel, D., von Blumröder, C., and Misch, I. *Texte zur Musik, 1970–1977*. Vol. 4. Cologne: DuMont, 1978.

Thoresen, L. 'Form-Building Patterns and Metaphorical Meaning'. *Organised Sound* 15, no. 2 (2010): 82–95.

Thoresen, L., and Hedman, A. 'Spectromorphological Analysis of Sound Objects: An Adaptation of Pierre Schaeffer's Typomorphology'. *Organised Sound* 12, no. 2 (2007): 129–141.

Topper, D., Burtner, M., and Serafin, S. 'Spatio-Operational Spectral (SOS) Synthesis'. In *Proceedings of the International Conference on Digital Audio Effects DAFx '02*. Hamburg, 2002.

Truax, B. 'Composition and Diffusion: Space in Sound in Space'. *Organised Sound* 3, no. 2 (1999): 141–146.

Varèse, E. 'The Liberation of Sound'. *Perspectives of New Music* 5, no. 4 (1966): 11–19.

Wagner, D., Brümmer, L., Dipper, G., and Otto, J. A. 'Introducing the Zirkonium MK2 System for Spatial Composition'. In *Proceedings of the Joint International Computer Music and Sound and Music Computing Conference*, 823–829. Athens, 2014.

Wallach, H., Newman, E. B., and Rosenzweig, M. R. 'A Precedence Effect in Sound Localization'. *Journal of the Acoustical Society of America* 21, no. 4 (1949): 468–468.

Westerkamp, H. 'Linking Soundscape Composition and Acoustic Ecology'. *Organised Sound* 7, no. 1 (2002): 51–56.

Wilson, S. 'Spatial Swarm Granulation'. In *Proceedings of the International Computer Music Conference*. Belfast, 2008.

Wilson, S., and Harrison, J. 'Rethinking the BEAST: Recent Developments in Multichannel Composition at Birmingham ElectroAcoustic Sound Theatre'. *Organised Sound* 15, no. 3 (2010): 239–250.

Wishart, T. *On Sonic Art*. Revised ed. Newark, NJ: Harwood Academic, 1996.

Wozniewski, M., Settel, A., and Cooperstock, J. R. 'Audioscape: A Pure Data Library for Management of Virtual Environments and Spatial Audio'. In *Proceedings of the Pure Data Convention*. Montreal, Canada, 2007.

Xenakis, I. *Formalized Music: Thought and Mathematics in Composition*. 2nd ed. Hillsdale, NY: Pendragon, 1992.

Zmoelnig, I. M., Sontacchi, A., and Ritsch, W. 'The IEM-Cube, a Periphonic Re-/production System'. In *Audio Engineering Society Conference: 24th International Conference: Multichannel Audio, The New Reality*. Banff, AB: Audio Engineering Society, 2003

Zouhar, V., Lorenz, R., Musil, T., Zmölnig, J., and Höldrich, R. 'Hearing Varèse's *Poéme électronique* inside a Virtual Philips Pavilion'. In *Proceedings of the International Conference on Auditory Display*, 247–252. Limerick, 2005.

Perspectives on Practice C

Perspectives on Structure

CHAPTER 26

FORM, CHAOS, AND THE NUANCE OF BEAUTY

MILEECE I'ANSON

On being asked to write for this volume, I began to investigate the phenom-enon of music as a sort of 'living language', how it can be broken down into component parts and applied to algorithms. It brought me to theoretical considerations on music itself, then reflection on how these relate to my algo-rithmic and generative musical works to date. I review these thoughts in that order and hope that they elicit more questions than propose to answer!

IN general, musical forms iterate from axioms of pitches and intervals organized by a set of principles. By these defining characteristics, we may think it possible to lay oper-ations upon variables and produce successful musical oeuvres. Indeed, from the ancient Greeks through to modern times, formulaic and even algorithmic compositional sys-tems have been successful in creating 'desirable' music.

It is reasonable then to infer that the same can be achieved via algorithms pro-grammed in a computer. Yet when describing music as 'desirable' or 'effective', we point not to its form, but to our experience of it; in successful music, composers move their lis-tener through a series of emotive states. It is given that these states are predetermined by the composer, who employs formula to aid in their manifestation. However, if we then deduce that music should require little other than the arrangement of numbers within procedural structures, we equally suggest that we are merely beings of order and pre-dictability, automata of sorts. This is arguably contrary to perhaps the most significant reward music provides: a key to engaging the elusive nucleus of our psyche.

Music, like almost nothing else, overrides our most stagnant emotions, reorients our mental states, and even propels us into spontaneous, physical movement. While this effect also tempts appropriation into sonic propaganda, such as in didactic marches or corporate retail environments, music primarily exists as the gateway to beyond the barriers of our patterned systems of perception, the usher of the nebulous manifests of poetry to the depths of our being.

I propose that this juncture between inciting a routine response versus intangible exaltation is a matter of dipping pattern and form into the infinite pool of the nonre-peating, uncontrollable, unmeasurable source, chaos. That it is through flirting with the untouchable that a functional yet forgettable song can transcend to the status of master-ful oeuvre. More specifically to this query regarding algorithms, it is a matter of weaving chaos into the patterns themselves, and thereby creating true beauty through math.

It is perhaps reasonable to surmise and restrict this proposition as being a question of nuance. This is to say that in all instances (composition, performance, timbre, sound design, etc.) fine variances are where a dance with form and chaos manifests itself. Nowhere is this quite so starkly in evidence than in music generated by machines; it can be manifestly cold if left to iterate solely as strict formula. Yet quite paradoxically, by examining algorithmic computer music architecture, we arrive at a unique ability to peer into the workings of where formula entwines with fuzzy particularities.

As with the excitement of the use of algorithms in the 'Turing Machine', stochastic pro-cesses offer an otherwise impossible exploration into the iteration of music purely through numbers and logic. Although compositionally this is perhaps not such a wide departure from nondeterministic or aleatoric processes already employed at least from the common-practice period, computing power applied to sound design can offer a probe from math-ematics into the heart of what tips structured rota into a more organic beauty.

It is no surprise that forms of algorithmic processes where self-referential decisions interplay with dynamic variables, especially those based on mimicking natural patterns, can be categorized as low-level artificial intelligence. As a defining element of nuance in nature, we point to how 'outside' forces become distorting influences on the replication or iteration of otherwise stagnant patterns and constricted formulae; to where the sheer density of variables can equate to a dose of chaos, resulting in recognisable yet unique forms. Within algorithmic processes, we can declare such functions and operations as partially nondeterministic navigational or even divinatory tools (depending on how you wish to perceive them), allowing for structures to generate autonomously much in the same way as a plant or any other living entity would. It can be a question of finely parameterising chaos[1] to introduce varying digressions and extraneous elements so as to start with a set of principles and 'organically grow'. These are the logical foundations that generate form with such nuance, whereby a certain glimmer of life, even in the inorganic, emerges.

Notwithstanding a vast array of other relevant subjects in general, there are several angles of consideration to investigate where and how nuance permits a departure from formula in algorithmic music and sound design, but it serves to limit them to a few particular manifestations I have implemented in my own work within SuperCollider, providing a properly assessed commentary to remain within the boundaries of this contribution.

The first series of algorithmic pieces I wrote were essentially deterministic processes containing stochastic elements. They were based on the premise of programming com-puter music that did not seek to mimic other acoustically or electronically created sounds or methods. Notably it was an exercise in restricting the number of harmonics on tones which otherwise unavoidably manifest in the acoustic domain through the resonation of physical objects as well as the timbral colorations from room or environment dynamics. In this respect, it was an experiment in *removing* nuance, which in itself brings up an interesting point regarding the so far discussed element of chaos in the formation of beauty; the purity of the sine wave, and our natural affinity towards it, proposes that there are exceptions where the opposite may also hold true.

Consider the voice of Emma Kirkby versus that of Amy Winehouse. Kirkby's restrains itself from distorting harmonic variance, emanating a sort of clarity of tone we often ascribe to angels, whereas Winehouse's 'overtoned' raspiness (noise) elicits intrigue and something unpredictable, even naughty. Both can be equally as pleasurable to listen to whilst constituting opposite ends of the question of chaos's role in nuance. The former 'angelic' voice offers solace and comfort by 'reining in' the indeterminate universe, whilst the latter becons it.

Similarly to Kirkby's voice, a sine wave with limited harmonics creates an honest and undemanding sound. In my earlier works, this allowed for the more basic forms of nuance to be the source of musical depth. The envelope, amplitude, and decay of the fundamental and harmonics were partly modulated through algorithmically parameterized randomisers. This created simple but engaging flows where interplay with functions based on natural patterns (divisions of frequencies) modulated by an additional set of chance operators, formed the cadence of gently nuanced sine tones.

In one piece called *Formations*, such patterns iterated as sections cycled upon themselves with precise division, which with the above characteristics created a self-generating stochastic music within a linear composition. It was originally designed for a diffusion system as four stereo pairs with each cycled section delegated to a combination of the pairs with the individual channel distribution relying on nondeterministic operators.

For me, the exciting possibility of such a music is the referential sonification of some of the core mechanisms behind formulaic growth in natural systems and the subsequent creation of a recognisable 'piece' of music, where random or chaotically driven nuance in timbre and cadence coalesce to form a sort of fluttering livingness. Moreover, these factors dictate that each iteration exist only once; whereby enforced by its own nature it remains eternally unique like an individual tree or snowflake.

Some curious side notes of the project were the unintended manifestations of working with such constrained timbral properties, such as acoustically generated sum and difference tones, binaural beat frequencies, and, more relevant to this angle of consideration, the wide amplitude dynamics which led much of the piece to be relatively quiet and thus subject to digital encoding distortion. This sort of uniform, even harmonic distortion carries an unappealing quality to it, whereas with analogue signal generators and processors, the stochastic interplay of 'rogue' electrons draws us in though their 'warm' feeling, odd harmonics. In this instance, the question arises as to whether the structure of this form of chaos sways towards odd numerical ratios to determine if we feel it a 'beneficial' nuance.

Subsequent to working with music generated by stochastic process embedded within a deterministic format, I inverted the procedure by using biological and sensor-derived data as controllers in formulating what can be defined as a continuous stochastic process. Here a live stream of variable data, such as from bioelectrical signal generated by a living plant, directly animates a series of set parameters. Only a few deterministic processes employ algorithmic functions; mostly it's a free-form program, allowing direct modulation of elements within maximum and minimum thresholds, set by the limitations of what is reasonable in terms of synthesis and the computational capacity of the computer. This creates a music where the timbres and cadence are collectively generated in reciprocation with an organic input.

Compositionally, both mechanisms and angles of approach form compositional templates which have a discernible motif, or recognizable formula, based on overarching musical principles, but whose more captivating qualities are derived through the gentle engagement with the nuance of chaos. Composing using algorithmic operators with carefully choreographed 'chaos' is essentially the practice of weaving structure with the stochastic to transcend formula and manifest what feels almost a living sort of beauty that nuance of this sort can infuse.

NOTE

1. Of course pertaining to algorithmic computer music we are not usually referencing pure chaos, but rather functions that generate relative randomness; true randomness by definition cannot iterate from a function and therefore within a computer program. They are, however, more than adequate to offer strikingly similar renditions of natural modulations.

CHAPTER 27

....

BEYOND ME

....

KAFFE MATTHEWS

1988. My radical girls band split and by chance I land a job in a progressive studio in the early days of acid house music. I'm cutting beats for Graeme Park of the Hacienda and struggling with understanding MIDI, and loving to make humanly impossible rhythmic patterns banged out with Notator on Atari 1040STFMs and Akai S900s and the best bits are when things crash or lock up spewing super sounds I would never think of. A door opens. Not only does digital technology enable the making of other forms of music, but it can provide trajectories beyond me. The hook becomes not only how to find more of these situations to make new music, but how to use what they offer.

Around seven years later I'm sipping coffee in a Manhattan loft with the host of my previous night's first MIDI violin solo in New York. A compilation CD that had just come in the post is playing on the huge sound system and this stunning sound emerges that I am swamped by. It seems synthesized yet creates one of those epiphanic moments when you realize you don't need to do it any more as the job is done. This sound is it. Then of course it becomes a mission and so I go, pre-Internet, to seek out the maker.

It's Alan Lamb, simply amplifying and recording 30m long steel wires stretched across the Australian outback that expand and contract with the changing temperature of the day, so altering the frequencies at which they vibrate as they're struck by insects land-ing, winds blowing, twigs touching, humming, until without warning their resonant frequencies are found and they lift off. An invisible, tumultuous choir, only heard when amplified, and entirely acoustic. In essence, Alan and I are both working with strings but doing the opposite thing: me using a steel wound string of known pitch that I play, break down, and transform in purposeful ways through daily hours of experiment and digital processing; him using untuned steel wires pinned outside, their music created by the unpredictable and ever-changing patterns of the weather.[1]

Having been using microphones in nearby spaces to provide chance sounds for solo shows for years, the reality of my using not just sounds but systems of unpredictable patterns as source to play or from which to make sound, became a possibility a year later on the uninhabited Scottish island of Sanda. With a bunch of digital artists, writers, and world champion kite flyers, we embark for a week of experiments to catch the wind speed, direction, and shifting intensities of light through a huge kite carrying sensors hovering 30 metres above the top of the highest hill. Beneath it we crouch in a transparent tent with laptops, receiving the kite's gathering weather data via radio, with me turning the streams into MIDI through Max/MSP and feeding it into my live sampling gadget LiSa, to start and stop sampling of the fat, taut kite string, its playback or looping, determining its filters and pitch shifting. The weather in effect is doing what I would normally do on stage and the result is sublime. Noisy and patterned and otherworldly. Combined with weather-processed text from Shakespeare's *The Tempest*, I make *Weather Made* (1999).[2]

Ten years later I am standing with Jon Shelley (Environment Agency) and artist Laura Harrington inside a glorious noise fest that is the sound of the river Tyne at Riding Mill, Northumberland. Jon is telling us how they'd been counting and measuring the fish and the river since 1996. 'All that data!' I yell. ' Imagine! We could make wild tunes with it and take them to children in riverside schools and make songs together. It's got to be the migrating wild salmon journey we follow. We could even make an incredible opera together!' And so we did. Working with local coder Adam Parkinson and Max/MSP, we make melodies using 2010 data charting the number of wild salmon coming up and down the Tyne to breed then depart to Greenland, determining the rhythm from the river's flow, while the whole piece is shaped by the variety of the seasons and the river's route through the earth's shifting topography.[3]

The results are astonishing and precise in their seasonal variety. Springtime fish-count data over rushing springtime flow really does make high-pitched energetic tunes; summertime slowing and becoming more chilled; sensuous, autumn lush and groovy; and winter, still and rich in monochromatic greyness. Using this direct sonification of river data to pitch and rhythm, the tunes created are also surprisingly tonal and singable for groups of eleven-year-olds or just inspirational for their own songwriting. The interesting point is that we had not only found a way to make music we never would have otherwise made, but that it provided an irresistible way into listening, awareness, and thinking for children who have this river rushing past their back gardens that until that point they had failed to see beyond its surface, least of all to write their own songs inspired by it.[4]

For me, having been a maker who had not written or wanted to write a tune since 1986, this technique also produces a meaningful and quite unexpected way of reconsidering melody and rhythm as worthwhile components of new music making again.

Earlier that year I spend a week underwater with hammerhead sharks (*Sphyrna lewini*) in the Galapagos archipelago, being mesmerized by their ancient beauty and power as they swoop and cruise in circles around me.[5] On land I meet shark scientists who, over the ensuing two years, give me the traces of six of them that they tag and track over twenty-four-hour durations. The sharks are out night-time hunting. Swimming alone in straight lines at depths as great as 300 metres, maybe going as far as 4 km, then turning round and coming back in straight lines, finding their way in the dark, the researchers think by reading the shift in the Earth's magnetic crust through their underhead phenomenal electroreceptor system (see Figure 27.1).

The traces I'm given are as three-dimensional data; latitude, longitude, and depth, as well as ocean temperature and shark speed. If I build a multichannel system in 3D,

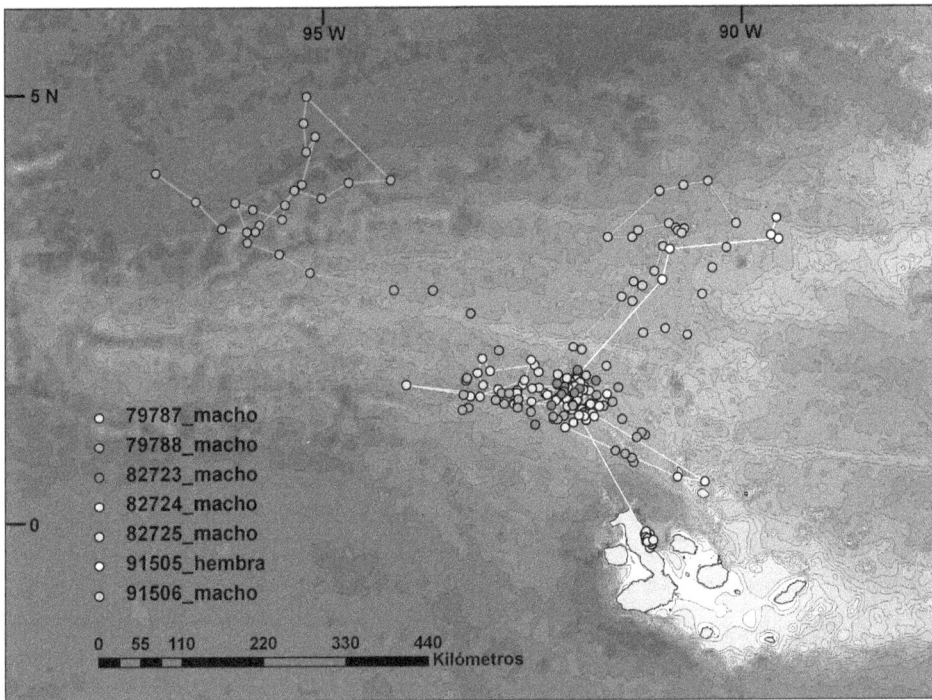

FIGURE 27.1 Six hammerhead shark routes taken in the sea north of Wolf Island. Tracking and map courtesy of James Ketchum, Dr Alex Hearn, and Dr Pete Klimley, Biotelemetry Laboratory, University of California, Davis.

I could finally make a truly spatial and architectural composition from it. Invite visitors to come lie in it on a centrally placed wooden platform in the dark that will sympathetically vibrate through bass transducers strapped underneath. So that they can feel and hear the sharks pass through and swoop over them in blasts of sudden fat noise action or hover still, gently entwining or swirling in tonal patterns you or I or a machine could never articulate. A sense of another world?

So I call Adam, and we work with Max/MSP to develop a system of eighteen digital oscillators, three per shark, through which we feed these traces. Through a variety of settings mixed with underwater recordings, played through an eight-channel system plus stereo subs and two channels for six bass transducers positioned in an exact cube around the platform, the results are extraordinary. And it reaches and communicates to such wide-ranging audiences from babies to the elderly. 'This is maybe how it feels to see with electrical impulses in deep ocean'.[6]

From this moment, data streams from action in nature or grabbing patterns that exist simply provide parameters, like a gesture with a violin bow, to add to my toolbox of music-making techniques. But unlike my violin playing, this information has a relevance and a link to systems outside of me and my ego, and is therefore much more interesting and relevant. So there continue to be other data-driven projects, for example *The Lock Shift Songs* (2014). In this, from the rise and fall of a cross-England canal path and my seven-day walk along it I make a seven-verse song. Its surprisingly lyrical melodies are created from the direct sonification of the land rising and falling over distance and are sung by a choir gathered from the final city's community.[7] (Note: A canal, unlike a river, is a human-made structure, whose wobbly or straight-lined direction is also a reflection of the finance available to the maker.)

Today my focus is making music that moves and changes through spaces, taking the performer and audience with it. I am doing this worldwide and collectively, making varieties of operas and audioscapes completely determined by the context for which we make them. This is the world of *sonic biking*, in which the audience can cruise through landscapes on bicycles mounted with speakers. The audience moves through the sound, music, noise, narratives that play, changing them dependent on where they go. To date, I've made ten of these pieces, understanding more each time about how to use the shifting context, audio landscape, passing street and its changing physical and social architectures, acoustics—oh for miserable empty suburban streets or downtown Houston at night reflections!—weather, time of day, people/animal, absence/presence, social demagogue to determine, to enable the audio material used, made, positioned, and played.

Apart from receiving GPS data to enable the sonic bike to interact with its location, so playing a specific sound file, there is no numerical data being read to create this music. However, the site, the moving street I realize is the algorhythm. Providing an invisible source that cannot be quantified or written down, but that utterly feeds and determines what and how things work and what therefore is made.[8]

It's a year ago and I'm sonic biking through central Brussels, hearing a guy from Guinea softly talking above the bullish traffic about how it was only when he left Africa that he realized that the world is so big. That the world is too big for him. That he had had no idea it is like this here in Europe. And I'm riding amongst cars and buses, past varieties of shops and busy humans marching up and down, sirens kicking and tears come and I am understanding something of his experience. Later I'm recording Jimmy from the Congo not wanting to talk but instead to play djembé. He plays and sings with a passion and freedom I've not heard since Senegal 1986. It's amazing. Later in the studio I'm listening to it thinking, 'but how do I use this material?' This is 80s world music, this is not my thing. But a few days later it's on the bike and it hits as I ride into a soulless city street and it jumps life now and it's the best thing. And totally contemporary.

NOTES

1. http://diy.spc.org/annetteworks/html/outback/.
2. http://londonfieldworks.com/Project-1-Syzygy.
3. C. Roberts, 'Background', http://www.kaffematthews.net/salmon/background/.
4. Kaffe Matthews, 'Music', March 2010, http://www.kaffematthews.net/salmon/m-u-s-i-c/, 'Where Are the Wild Ones?', http://www.kaffematthews.net/salmon/.
5. Kaffe Matthews, 'Background', 16 April 2009, http://www.kaffematthews.net/sharks/background/.
6. 'You Might Come out of the Water Every Time Singing', http://www.kaffematthews.net/sharks/.
7. 'The Lock Shift Songs', http://www.kaffematthews.net/works/the-lock-shift-songs/.
8. http://sonicbikes.net/.

CHAPTER 28

··

PERSPECTIVE ON PRACTICE

··

JAN BERAN

28.1 INTRODUCTION

MATHEMATICAL studies of music go far back in history. Some of the earliest documented sources are due to the Pythagorean school in ancient Greece (see e.g. van der Waerden 1979). Gottfreid Wilhelm Leibniz (1646–1716) called music the 'arithmetic of the soul'. Indeed, for centuries composers have been searching for ideal logically coherent forms. Beethoven's famous sketchbooks exemplify this tormenting search for a 'perfect algorithm'. Musical forms such as the sonata, symphony, or fugue may be regarded as examples of 'perfect'—though not fully specified—solutions. On the other hand, while the conceptual similarity between mathematics and 'musical logic' has been well recognized, until the beginning of the twentieth century mathematics was hardly ever at the origin of a composition. In 1899 the German mathematician David Hilbert published his book *Grundlagen der Géométrie* (Hilbert 1999). In retrospect, this is regarded as the dawn of modern mathematics. In the decades that followed, axiomatic foundations of mathematics and general theories that emphasized the role of mathematics as a science of abstract structures were developed. This widened the range of mathematical applications far beyond physics and other natural sciences. In the theory of music, modern algebra and geometry turned out to play a key role (see e.g. Andreatta 1997; Assayag and Fiechtinger 2002; Beran 2003; Jedrzejewski 2006; Mazzola 1990, 2002). In addition, in the second half of the twentieth century, technological progress made it possible to implement mathematical ideas directly. Early examples are for instance compositions by Stockhausen, Xenakis, Eimert, Boulez, and others. In this short chapter, the application (and applicability) of mathematical music theory to musical composition is illustrated by selected examples. Due to limited space, we focus on the notion of musical transformations in an algebraic context.

28.2 SYMMETRIES

Essential building blocks of musical composition are musical scales, motifs, chords, modulation, and variation. It is therefore not surprising that one of the best-developed branches of mathematical music theory deals with exactly these notions. Historically, transformations in composed and improvised music include, for instance, transposition, arpeggio, retrograde, and (vertical) inversion in the pitch-onset plane. In the twentieth century, more exotic transformations were added. A famous and frequently discussed transformation is for example Herbert Eimert's rotation by 45° combined with a dilatation by $\sqrt{2}$ (see Eimert 1964). In serialism, onset time, pitch, duration, and other musical parameters are treated as equal so that transformations that exchange parameters play a central role. The introduction of an increasing number of possible transformations together with the abandonment of the tonal system led to the question of how to define suitable structural principles to ensure the intrinsic coherence of compositions. A mathematical framework that proved to be very useful in this context is outlined in Mazzola (1990, 2002). In this approach, musical transformations are viewed as affine transformations in algebraic modules. To understand the theoretical background of this approach, the following algebraic definitions are needed (see e.g. Gilbert 2002):

Definition 1 *A nonempty set G with a binary operation "+" is called a group, if $a+b \in G$ for all $a,b \in G$, "+" is associative, there is a zero element (denoted by o) and for each $a \in G$ there is an inverse element (denoted by -a). The group is called abelian, if $a + b = b + a$ for all all $a,b \in G$. If the operation is written as multiplication "·" then the zero and the inverse element are denoted by e and a^{-1} respectively.*

Definition 2 *A nonempty set R with two binary operations "+" and "·" is called a ring, if (R, +) is an abelian group, and the following holds. For all $a,b \in R$, $a \cdot b \in R$, $(a \cdot b) \cdot c = a \cdot (b \cdot c)$ (associativity) and $(b+c) \cdot a = b \cdot a + c \cdot a$ (distributive law).*

Example 1 *Simple examples of rings are $\mathbb{R}$ (real numbers), $\mathbb{Q}$ (rational numbers) and $\mathbb{Z}$ (integers), with "+" and "·" denoting usual addition and multiplication of numbers. Another example is $\mathbb{Z}_{12} = \{0,1,\ldots,11\}$ where all integers that differ by a multiple of 12 are considered to be the same (i.e. $\mathbb{Z}_{12}$ are integers modulo Yl). In music, this ring is of particular interest when considering the standard well-tempered tuning.*

Definition 3 *Suppose that $(R,+,\cdot)$ is a ring and M is a nonempty set with binary operation "+". Then M is called an R–module, if the following holds: (a) (M, +) is an abelian group, (b) for all $r \in R, a \in M : r \cdot a \in M$, (c) for all $r,s \in R, a,b \in M$ we have $r \cdot (a+b) = r \cdot a + r \cdot b, r \cdot (s \cdot a) = (r \cdot s) \cdot a$ and $(r+s) \cdot a = r \cdot a + s \cdot a$.*

Example 2 *A well-known example of a module is the three-dimensional space* $\mathbb{R}^3$. *It consists of three-dimensional vectors*

$$x = (x_1, x_2, x_3)^T = \begin{pmatrix} x_1 \\ x_2 \\ x_3 \end{pmatrix}$$

with real valued components x_1, x_2, x_3. *Identifying* $(R, +, \cdot)$ *with* $(\mathbb{R}, +, \cdot)$, $M = \mathbb{R}^3$ *is an* $\mathbb{R}$ *–module. In linear algebra, this is also called a (three-dimensional) vector space.*

Definition 4 *A mapping g between two R–modules* M_1 *and* M_2 *is called a (module-) homomorphism, if for every* $a, b \in M_1$ *and* $r \in R$ *the following holds:* (a) $g(a + b) = g(a) + g(b)$ *and* (b) $g(r \cdot b) = r \cdot g(b)$. *If g is a one-to-one mapping, then it is called an iso-morphism, and the two modules are called isomorphic. Moreover, if* $M_1 = M_2$, *then it is called an automorphism.*

Definition 5 *Given a homomorphism g between two R–modules* M_1 *and* M_2, *and an element* $c \in M_2$, *the mapping* $h : M_1 \rightarrow M_2$ *defined by* $h(a) = c + g(a)$ *is called an affine transformation. If* $M_1 = M_2 = M$, *then h is called a symmetry.*

It is important to note that the notion of 'symmetry' used here (symmetry = affine transformation) is much more general than the colloquial use of the word. The algebraic theory of groups, rings, and modules turned out to provide a natural starting point for many questions in music theory (see e.g. Andreatta 1997; Assayag and Fiechtinger 2002; Beran, Goswitz, Mazzola, and Mazzola 2014; Beran and Mazzola 1999; Jedrzejewski 2006; Lewin 1987, 1993; Mazzola 1990, 2002; Noll 1997). One of the simplest applications is the classification of musical scales. For instance, in the twelve-note equal-tempered system, all frequencies ω can be represented as $\omega = \omega_0 \cdot 2^{x/12}$, where $x \in \mathbb{Z}$. In logarithmic representation we have $\log \omega = \log \omega_0 + \frac{x}{12} \log 2$. Here, ω_0 is a reference frequency (such as for instance 440Hz). If transposition by one or several octaves is ignored (i.e. trans-position by an octave leads to the 'same note'), then the space of all possible pitches can be identified with the group $\mathbb{Z}_{12} = \{0, 1, 2, \ldots, 11\}$ with '+' defined as addition modulo 12. Musical scales (in the twelve-note equal-tempered system) are subsets C of $\mathbb{Z}_{12}$ char-acterized by group-isomorphisms (also called inner symmetries of C) that leave C (or certain subsets of C) invariant. Two scales with the same inner symmetries are said to belong to the same class. Classification of scales using inner symmetries is a very useful general method that is not restricted to the well-tempered system. Given any musically meaningful group of pitches, all 'logically possible' scales can be found, scale-preserving transformations can be characterized, relationships between chords can be investigated, and a general theory of modulation can be derived (see e.g. Mazzola 1990, 2002).

More generally, one may consider higher-dimensional musical spaces that include other coordinates such as onset time, duration, or loudness. For instance, let t denote onset time (restricted to some range $0, 1, \ldots, n$) and p pitch in $\mathbb{Z}_{12}$. A musical motif (in the onset time and pitch space) is a subset $S = \{m_1, m_2, \ldots, m_k\}$ of the $\mathbb{Z}$–module $M = \mathbb{Z}_n \times \mathbb{Z}_{12}$, where $m_i = (t_i, p_i)^T$ with t_i =onset time and p_i = pitch. As before, the sets S can be classified using symmetries (see e.g. Straub 1989). Two motifs S_1, S_2 are 'equivalent', if there is an isomorphism mapping S_1 to S_2. Higher-dimensional motifs involving onset time, pitch, duration, and loudness can be defined as subsets $S = \{m_1, m_2, \ldots, m_k\}$ of the $\mathbb{Z}$–module $M = \mathbb{Z}_{n_1} \times \mathbb{Z}_{n_2} \times \mathbb{Z}_{n_3} \times \mathbb{Z}_{n_4}$. Here

$$m_i = (t_i, p_i, d_i, l_i)^T = \begin{pmatrix} t_i \\ p_i \\ d_i \\ l_i \end{pmatrix}$$

with d = duration and l = loudness. For software implementations of the module $M = \mathbb{Z}_{71}^4$ (i.e. $n_1 = n_2 = n_3 = n_4 = 71$); see, for example, Mazzola and Hofmann (1989) and Milmeister (2009). Here are some typical standard transformations in $\mathbb{Z}_n^4$:

Example 3 *(Pitch) transposition in $M = \mathbb{Z}_n^4$ by the amount $a \in \mathbb{Z}_n$ is given by the affine transformation $g(m) = m + (0,a,0,0)^T$ (where $m = (t,p,d,l)^T$). An onset-pitch arpeggio (with fixed hyperplane $p = p_0$) is defined by $g(m) = (t + p - p_0, p, d, l)^T$. For computational as well as theoretical reasons, it is useful to write this in matrix notation:*

$$g(m) = \begin{pmatrix} t + p - p_0 \\ p \\ d \\ l \end{pmatrix} = \begin{pmatrix} -p_0 \\ 0 \\ 0 \\ 0 \end{pmatrix} + \begin{pmatrix} 1 & 1 & 0 & 0 \\ 0 & 1 & 0 & 0 \\ 0 & 0 & 1 & 0 \\ 0 & 0 & 0 & 1 \end{pmatrix} \begin{pmatrix} t \\ p \\ d \\ l \end{pmatrix}.$$

Exchanging the pitch and duration coordinates corresponds to

$$g(m) = \begin{pmatrix} t \\ p \\ d \\ l \end{pmatrix} = \begin{pmatrix} 1 & 0 & 0 & 0 \\ 0 & 0 & 1 & 0 \\ 0 & 1 & 0 & 0 \\ 0 & 0 & 0 & 1 \end{pmatrix} \begin{pmatrix} t \\ p \\ d \\ l \end{pmatrix}.$$

It should be emphasized that all multiplications and additions are carried out in $\mathbb{Z}_n$, i.e. modulo n.

28.3 APPLICATIONS

The mathematical theory of algebraic symmetries is very elegant. In the context of musicology, many fundamental results can be derived within this framework, including harmonic analysis, counterpoint, modulation, and performance analysis (see e.g. Beran et al. 2014; Beran and Mazzola 1999; Jedrzejewski 2006; Fiore, Noll, and Satyendra 2013; Mazzola 1990, 2002; Noll 1997). How far does the theory open up new possibilities for composers? Composers are well aware of the fact that a wider choice of possibilities does not necessarily lead to better or more interesting music. Igor Stravinsky formulates the principle of parsimony as follows: 'The more controlled the art, the more free ... And the composer must find unity in multiplicity, choose the reality of a limitation over the infinity of a division' (Straus 2004, 44). The following examples and comments may serve as an illustration of this principle in the context of module-symmetries.

To fix ideas, consider the following schematic subdivision of a compositional process:

(1) *local definition*: creation of a set $A = \{a_1, a_2, \dots\}$ of basic musical objects a_i (e.g. motifs, harmonies, rhythms);

(2) *variation*: transformation of $a \in A$, i.e. mapping each a to $g(a)$; such that the set B of all transformed motifs is not identical with A;

(3) *global definition*: definition of global structures based on elements from $A \cup B$ and some fundamental structural principles (e.g. principles of modulation, counterpoint, or specific general forms such as sonata, canon, fugue, etc.).

Steps (1)–(3) may be viewed as a formalization of Stravinsky's view of musical composition: 'The elements which the imagination receives must be passed through a sieve ... and, like the sounds of nature, become music only after they have been organized, or controlled' (Straus 2004, 44). Here, A may be interpreted as the set of 'elements which the imagination receives' whereas steps (2) and (3) represent the '*sieve*'. In practice, steps (1), (2), and (3) are often carried out repeatedly and in an arbitrary sequence, and sometimes it may not even be clear how to classify some of the 'actions'. In the literature, (1) and (2) are also called 'local composition' whereas step (3) is referred to as 'global composition' (see e.g. Mazzola 1990, 2002). Suppose now that the basic set A is given. Creating a composition out of A leads to the question of which transformations g we would like to use in (2), and which general principles are to be applied in (3). The choice of transformations is not independent of the choice of fundamental principles. For instance, using atonal inversion or parameter change may be rather problematic within the framework of tonal music. On the other hand, for completely free atonal music, no affine transformation can be excluded a priori. The number of possible affine transformations is huge, even if one restricts them to small spaces such as $\mathbb{Z}_{12}^4$. However, in practice most transformations turn out to be unsuitable, even if no restrictions such as tonality or other structural demands are imposed. Unfortunately, up to date, there is no general theory that would provide a reliable assessment of the usefulness (or 'musicality') of a transformation. 'Guidelines' for composers therefore rely mostly on case studies.

We conclude the discussion by some examples. One of the first compositions where the theory of higher-dimensional module-symmetries (in $\mathbb{Z}_{71}^4$) was applied directly is the author's first (electronic) piano concerto recorded on the CD *Immaculate Concept* (Beran and Mazzola 1992). Starting with an initial motif from Beethoven's 'Hammerklavier' Sonata op. 106 (i.e. the initial set A in step [1] consists of one element a only), a limited number of transformations g_i ($i = 1, 2, \ldots, k;$) in the four-dimensional space $\mathbb{Z}_{71}^4$ was applied. Since most of the transformations are nonstandard, the tonal setting of the initial motif is extended to what one may call 'transformed tonality'. A natural flow and intrinsic coherence of the piece is ensured by restricting nonstandard transformations to a small fixed set.

A completely different composition created in the $\mathbb{Z}_{71}^4$—module is *Šānti* (Beran 2000). The impression of 'vast delicately crafted tonally balanced soundscapes over deep sustained pedal points, gradually changing their physiognomy' ('Erleuchtete Tasten' 2001) is created by imposing strong tonality restrictions on the onset-pitch projection of each transformation. An even more parsimonious set of transformations was applied in *Rêverie* (Beran 2014; also see Mazzola 2014). The initial set A consists of three elements a_1, a_2, a_3. Two elements (say a_1, a_2) are newly created motifs, while *as* is the main motif from Franz Liszt's *Liebestraum* No. 3 (Figure 28.1a). The core transformation is a vertical inversion together with repeated shifts of individual notes in the onset direction. The transformed motifs $g_i(a_3)$ are then used (together with a_1, a_2) for the global composition (Figure 28.1b). Thus, the remaining creative process can be subsumed under point (3), the main principles being modulation, cadenzas, and local standard transformations such as rhythmic shift, transposition, and vertical and horizontal dilatation, all applied within a tonal framework. Due to the extreme restriction of transformations, the composition is calm and meditative, bordering on minimalism.

While relatively simple basic principles are used in many compositions, the creative work is often more complex and consists of many little steps. Transformations are hardly ever applied globally. Instead, local structures of varying size are transformed in various ways. To illustrate this, consider for instance the motif in the upper system of the first half of bar 16 in *Rêverie*. This 'local composition' was obtained by gradually transforming subsets of the second half of bar 5 in Liszt's *Liebestraum*. Figure 28.2 shows the sequence of transformed motifs. First, the minim duration of C is stretched to a dotted minim, thus removing the following B♭. Throughout all subsequent transformations, this dotted minim is kept fixed. We may thus ignore it in the following. The initial material for the next transformation steps is defined by the sequence C, D♭, E♭, G, E♭, D♭, with quaver durations. In $\mathbb{Z}_{12}$ this corresponds to 0, 1, 3, 7, 3, 1. In the language of algebraic modules, the initial motif to be transformed may be written as a subset A_0 of $\mathbb{Z} \times \mathbb{Z}_{12}$ with the coordinates 'onset time' and 'pitch' (or $\mathbb{Z}_6 \times \mathbb{Z}_{12}$ if one wants to treat onset times in a cyclic manner). Note that here, onset-time steps of size 1 correspond to quaver steps in the original score. More specifically,

$$A_0 = \big\{(0,0),(1,1),(2,3),(3,7),(4,3),(5,1)\big\}.$$

The next transformation is a vertical inversion of A with fixpoint 3 (i.e. E♭). This yields G♭, F, E♭, B, E♭, F; i.e.

$$A_1 = \big\{(0,6),(1,5),(2,3),(3,-1),(4,3),(5,5)\big\}$$
$$= \big\{(0,6),(1,5),(2,3),(3,11),(4,3),(5,5)\big\}$$

FIGURE 28.1 (a) Franz Liszt, *Liebestraum* No. 3, bars 1–11 (Leipzig: Fr. Kistner, [1850]). Public domain.
(b) Jan Beran, *Rêverie*, bars 1–23 (Bern: Müller & Schade, 2014).

Rêverie

I. Lento con espressione

Jan Beran

FIGURE 28.1 Continued.

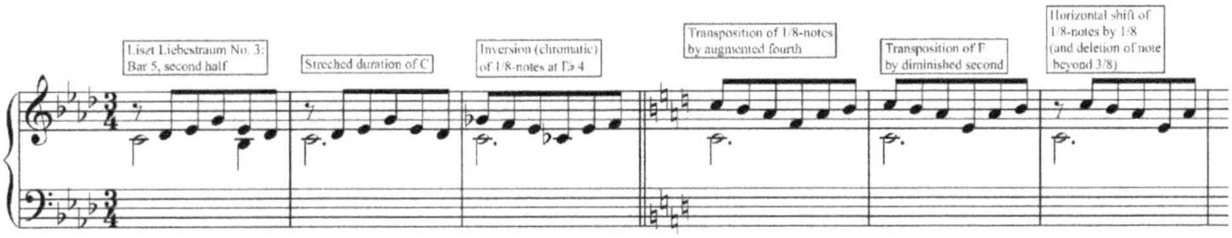

FIGURE 28.2 Creation of Beran, *Rêverie*, bar 16 (first half) from Liszt, *Liebestraum* No. 3, bar 5 (second half), by applying a sequence of transformations.

This motif is transposed by an augmented fourth, i.e. we obtain C, B, A, F, A, B, or

$$A_2 = \{(0,12),(1,11),(2,9),(3,5),(4,9),(5,11)\}$$
$$= \{(0,0),(1,11),(2,9),(3,5),(4,9),(5,11)\}.$$

Transposing F downwards by a diminished second (which corresponds to a vertical shift by −1 of the element $(3, 5)$), we have

$$A_3 = \{(0,12),(1,11),(2,9),(3,4),(4,9),(5,11)\}$$
$$= \{(0,0),(1,11),(2,9),(3,4),(4,9),(5,11)\}.$$

The final motif in Figure 28.2 is obtained by removing the last note (i.e. $(5,11)$) and applying a horizontal shift by one unit:

$$A_4 = g(A_0) = h_4(A_3) = h_4 \circ h_3 \circ h_2 \circ h_1(A_0)$$
$$= \{(1,12),(2,11),(3,9),(4,4),(5,9)\}$$
$$= \{(1,0),(2,11),(3,9),(4,4),(5,9)\}.$$

Together with the previously fixed dotted minim C, this is the first half of bar 16 in *Rêverie* (in the upper system).

Finally, it should be noted that transformations are often involved even if they are not chosen explicitly. In particular, the choice of specific scales or sets of chords is equivalent to the choice of symmetries. A typical example is Messiaen's notion of 'modes of limited transposition'. Translated into the mathematical framework, these are subsets of $\mathbb{Z}_{12}$ defined by invariance with respect to certain transpositions (see e.g. Mazzola 2002). For instance, $M_5 = \{0, 1, 2, 6, 7, 8\}$ (e.g. C, C♯, D, F♯, G, G♯) is invariant under the transformation $g(a) = a + 6$, since

$$M + 6 = \{0+6, 1+6, 2+6, 6+6, 7+6, 8+6\} = \{6,7,8,0,1,2\} = M.$$

An interesting property of these scales is that the complement M^c shares the same property. In the example, this is $M^c = \{3,4,5,9,10,11\}$. Because of this duality, it is sufficient to consider modes with at most six elements. It can be shown that, apart from the trivial scale $\mathbb{Z}_{12}$, there are exactly ten types of modes of limited transposition.

BIBLIOGRAPHY

Andreatta, M. *Group-Theoretical Methods Applied to Music*. PhD dissertation, University of Sussex, 1997.

Assayag, G., and Fiechtinger, H. G., eds. *Mathematics and Music: A Diderot Mathematical Forum*. New York: Springer, 2002.

Beran, J. *Šânti*. col legno, WWE 1CD 20062, 2000. CD.

Beran, J. *Statistics in Musicology*. New York: Chapman and Hall, 2003.

Beran, J. *Rêverie*. Berne: Müller und Schade, 2014.

Beran, J., and Mazzola, G. *Immaculate Concept*. Zürich: SToA Music, 1CD 1002.92, 1992. CD.

Beran, J., and Mazzola, G. 'Analyzing Musical Structure and Performance: A Statistical Approach'. *Statistical Science* 14, no. 1 (1999): 47–79.

Beran, J., Goswitz, R., Mazzola, G., and Mazzola, P. 'On the Relationship between Tempo and Quantitative Metric, Melodic, and Harmonic Information in Chopin's Prélude Op. 28 No. 4: A Statistical Analysis of 30 Performances'. *Journal of Mathematics and Music* 8, no. 3 (2014): 225–248.

Eimert, H. *Grundlagen der musikalischen Reihentechnik*. Vienna: Universal Edition, 1964.

'Erleuchtete Tasten' [CD review of *Šânti*]. *Frankfurter Allgemeine Zeitung*, 24 August 2001.

Fiore, T., Noll, T., and Satyendra, R. 'Morphisms of Generalized Interval Systems and PR-Groups'. *Journal of Mathematics and Music* 7, no. 1 (2013): 3–27.

Gilbert, W. J. *Modern Algebra with Applications*. New York: Wiley, 2002.

Hilbert, D. *Grundlagen der Géométrie*. Stuttgart: Teubner, 1999. Originally published in 1899.

Jedrzejewski, F. *Mathematical Theory of Music*. Paris: Delatour France and IRCAM, 2006.

Lewin, D. *Generalized Musical Intervals and Transformations*. New Haven, CT: Yale University Press, 1987.

Lewin, D. *Musical Form and Transformation: Four Analytic Essays*. New Haven, CT: Yale University Press, 1993.

Mazzola, G. *Geometrie der Töne*. Basel: Birkhäuser, 1990.

Mazzola, G. *The Topos of Music*. Basel: Birkhäuser, 2002.

Mazzola, G., and Hofmann, G. R. 'Der Music Designer MD-Z71: Hardware und Software für die Mathematische Musiktheorie'. In *Musik—Gehirn—Spiel*, edited by H. Petsche, H. Beiträge zum 4. Herbert-von-Karajan-Symposium. Basel: Birkhäuser, 1989.

Mazzola, P. *Idyll und Refugium*. Gallo CD-1422, 2014. CD.

Milmeister, G. *The Rubato Composer Music Software: Component-Based Implementation of a Functorial Concept Architecture*. New York: Springer, 2009.

Noll, T. *Morphologische Grundlagen der abendländischen Harmonik*. Bochum: Brockmeyer, 1997.

Straub, H. *Beiträge zur modultheoretischen Klassifikation musikalischer Motive*. Diploma thesis, ETH Zürich, 1989.

Straus, J. N. *Stravinsky's Late Music*. Cambridge: Cambridge University Press, 2004.

van der Waerden, B. L. *Die Pythagoreer*. Zürich: Artemis, 1979.

....

CHAPTER 29

THOUGHTS ON AN
ALGORITHMIC PRACTICE

....

WARREN BURT

In 1969–1970, I started working with the Moog CEMS system at Albany State University (Chadabe 2015) and one of the first things I did was to add the outputs of two or three analogue sequencers running asynchronously to explore the results of what that additive process would be. It was the spirit of the times, and also the result of reading a lot of John Cage's writings and the example of my teacher, Joel Chadabe. I was hooked. And although I've done all sorts of composing in the past forty-seven years, my main mode of thinking compositionally ever since has been the creation and application of all sorts of somewhat nondeterministic processes to music, video, and verbal composition. In fact, probably 90 percent of the several hundred pieces I've written have involved algorithmic processes of one kind or another. (Some examples: *Studies for Synthesizer* [Burt 1982]; *39 Dissonant Etudes* [Burt 1996a]; *The Animation of Lists and the Archytan Transpositions* [Burt 2006a].) I've written about some of these processes in the past (Burt 1996b) but what I want to write about here is a change of attitude that has occurred in me in the past five to ten years, a change in how I approach the use of processes.

This change might be characterized as a freer and more intuitive use of processes—and a very un-self-critical, nonintrospective use of them. That is, I no longer think about which process I want to use, I just pick one, as spontaneously as I can, and use it, more with an eye to exploring what will be produced than trying to direct the output to a desired outcome. This might be the result of having composed so long with these things that I instinctively feel what the rough result of using a process will be, or it might be that as I've gotten older, I no longer feel under an obligation to 'craft' works, based on some kind of imagined idea of what musical 'quality' or 'profundity' might be. I want to be just involved with doing it, with as few glances in the rearview mirror (of history) as possible.

It could also be that over the years, I've become more and more involved in improvisation. At first, I used algorithmic processes as instruments in a free improvisation context. And in fact, I still do this on occasion. But more and more, an improvisatory aesthetic seemed to be taking over, and I find that by now, my composition activities resemble at first an out-of-time kind of free improvising followed by real-time improvising.

That is, my algorithmic working has become progressively more improvisatory—really made in the moment of composition, without structural reflection or much pre-planning. I find I'm freely combining elements, observing their outcomes, and if I don't like the outcome, simply making another process. I don't want to give the impression that my making work with these processes is unthinking. Quite the opposite: in the real-time listening and recording phase of the work, I sometimes make quite subtle adjustments to parameters.

I'll discuss three recent algorithmic pieces, and then make some observations on the context in which those works exist.

29.1 THREE MOVEMENTS IN MEMORIAM PAUL PANHUYSEN

From January to February 2015 I wrote a piece as a memorial to Paul Panhuysen, Dutch composer, sound and visual artist, art organizer, and friend. I had heard about Paul's passing in an email, and I wrote the piece on the commuter trains to and from work, listening over headphones. In the piece, I used a kind of exhaustive permutation process that I thought Paul would have found interesting. I posted the piece on my website (Burt 2015). I heard from several friends, again via email, that they liked the work. I was especially gratified that Helene, Paul's life-partner, and Rene van Peer, a good friend of his, both found the work appropriate and beautiful.

In this piece, I wanted to work with a series of twelve-note just-intonation microtonal scales I had developed back in 2004–2006, scales which derived from Ervin Wilson's 'The Scales of Mt. Meru' series of papers (Wilson 1993). As part of this work, I came up with a family of approximately 268 scales, which were generated algorithmically by taking the first twelve unique (odd or nonoctave reducible) members of an additive sequence, treating them as members of the harmonic series, and reducing them to within an octave. Different additive sequence rules and seeds were used for each scale (Burt 2006b). Additionally, I made a series of eleven additive sequence generators for ArtWonk and MusicWonk (Dunn 2015), each of which uses a different set of rules to make an additive sequence, which can then be divided using a modulo of any desired number. These additive sequence rules are the same ones I used to generate the scales. (And in 2014, I upgraded this set of sequence generators by adding another six rules, bringing the total of different additive sequence rules available to seventeen.)

Frankly, I've found the resources available in this set of scales overwhelming. Since making the catalogue, I've been using the scales regularly in my music, but not in any systematic way. In fact, in line with the improvisatory methods discussed earlier, I've often chosen the scale for a particular piece at random, without even reading its list of pitches beforehand. I use the piece to find out what it will sound like, rather than choosing it for a particular sound.

Because these are just-intonation scales, using whole-number ratios for their tuning, the concept of interval equivalence under inversion does not apply. That is, in twelve-note equal temperament, it is assumed that a kind of harmonic equivalence obtains between a minor third and a major sixth, and that all intervals of, say, three scale degrees will be harmonically equivalent to intervals of nine scale degrees. In a just-intonation scale, this is not the case. For example, in one of my scales, the interval between scale degrees 0 and 3 is 303.199 cents, while the interval between scale degrees 1 and 4 is 281.359 cents, and the interval between scale degrees 2 and 5 is 266.871 cents. Therefore, in each twelve-note scale, there are 131 unique dyads. The idea of a totally exhaustive sequence of artistic elements is one which Paul frequently used in his pieces. Therefore, I decided to write a piece for him which would consist of a totally exhaustive nonrepeating sequence of the 131 dyads available in my scales. That is, the sequence would consist of one and only one instance of each of the dyads available in the given scale.

I listed all 131 dyads of a twelve-note (any twelve-note) scale in two arrays (one for the bottom notes, one for the top notes). Using ArtWonk, a control was set up to manually scatter the contents of an array which controlled the order of both dyad-containing arrays before each run of the piece. This ensured a unique, but totally exhaustive, structure of the dyads for each run through the piece. A second 131 element array was set up, which each time through simply sets up a unique order of the numbers 1, 2, and 3. These are used to select between three different octave values for each dyad. So in each run of the piece, there will be a different order of the 131 dyads and they will be in a unique set of octave transpositions.

Velocity for each dyad is determined by a probability distribution generator in which MIDI velocities are selected. The MIDI velocities 72, 84, 96, and 108 were chosen with percentages, respectively, of 17, 24, 38, and 21. Durations are selected by a shuffle generator, which generates a continually changing unique ordering of 0, 1, and 2. This selects from three durations, 48, 72, and 96 ticks, in ever-changing orders. So while order and register of the dyads are a single pass through an ordering of 131 elements, both velocities and durations are 'random'—velocities from a probability generator, and durations from an unending set of permutations.

The first piece uses just one run-through of the dyads. The second uses two simultaneous run-throughs of the process, each with a different ordering of the dyads and registers. Since the durations are made by a permutation process, permuting a very small set of durations (which are all multiples of each other), the two different sets of dyads will have different durations. The third piece uses three of the processes simultaneously. This means that potentially, up to six different notes of the chosen twelve-note scale will be present in any one chord. (Although, because of the different randomly chosen 131-dyad-sequences being combined in the piece, the choice of which six-note chords are heard is not controlled, and is definitely not exhaustive. Again, durations may or may not differ from moment to moment in the three dyad rows, meaning that varying rhythmic complexes will be set up.

Each of the three pieces uses a different twelve-note scale from my collection. All three of these scales were generated with the Lucas Triangle, which uses the seed (2,1). The three scales were Lucas D Left, Lucas H Right, and Lucas J Left (Table 29.1). Again, these were chosen spontaneously, with no knowledge beforehand of the exact harmonic content of the scales.

To realize the pieces, I used the Steinway Piano sample set from the Garritan Personal Orchestra. All three movements were played at the same tempo. I tried having the three movements at different tempi, but that was somehow inconsistent with the monolithic structures Paul had used for some of his works. So the end piece uses algorithmically generated scales, and an algorithmic process to scatter the order of the 131 possible dyads in the scale, and probabilistic processes to control dynamics and durations in the piece. And all—scales, dyad sequences, choice of dynamic levels, and durations (which are subjected to probabilistic choosing processes) are chosen arbitrarily and freely.

29.2 WORD SWIRL TWO

Another recent piece, *Word Swirl Two*, a sound poetry piece, had a similar origin. I received a text message from Sjaak de Jong, a sound-poet friend, saying that he was putting together a CD of a collection of sound poetry works, and would I be interested in contributing? I texted back, 'Of course,' and then proceeded to make the work using a number of interesting (to me) processes. When finished with the work, I (at his request) burned it to CD and mailed it to him.

In this piece, I generated the text algorithmically, then recorded the text in the environment, so the sonic background was beyond my control, then proceeded to electronically modify the text using an interactive algorithmic sound-modification program.

ArtWonk has a number of modules for manipulating text. One of these is StrRand, which generates a string of alternating randomly chosen consonants and vowels. This generates nonsense words which, because of their consonant–vowel alternations, look like they could be words, in some language or another. There are also a number of modules where a given text can be fragmented in many different ways. For example, a set of lines can be reordered in any number of ways.

I had an earlier patch in ArtWonk from a video-sound piece, *Word Swirl*, in which these two processes, StrRand and reordering the order of a set of words, ran simultaneously. *Word Swirl* had rearranged the word order randomly, and the word-list source (146 lines long, with each line consisting of one, or sometimes two, words) included the names of some famous literary hoaxers. For purposes of *Word Swirl*, the appearance of names and repeating words were a nice feature of the piece. For *Word Swirl Two*, however, the repetitions and the presence of the names seemed to grate a bit. So I removed the names by hand from the text, and also any word repetitions were replaced by words from the source list not already used. I realized what I was doing was using the logic of the Shuffle module with my word order, so eventually (after I completed *Word Swirl Two*, using the hybrid generated-then-altered-by-hand text), I made a new patch in

Table 29.1 Three microtonal scales used in *Three Movements in Memoriam Paul Panhuysen*

Lucas (2–1) scale D Left: twelve tones
Rule: $D_n = D_{n-4} + D_{n-1}$
Seed string from triangle: 1, 2, 2, 2
Resulting sequence: $D_n = D_{n-4} + D_{n-1}$: 1 2 2 2 3 5 7 9 12 17 24 33 45 62 86 119 164 …

Scale

0	1/1	0.000	unison, perfect prime
1	33/32	53.273	undecimal comma, al-Farabi's ¼-tone
2	17/16	104.955	seventeenth harmonic
3	9/8	203.910	major whole tone
4	5/4	386.314	major third
5	41/32	429.062	
6	43/32	511.518	
7	45/32	590.224	diatonic tritone
8	3/2	701.955	perfect fifth
9	7/4	968.826	harmonic seventh
10	119/64	1073.781	
11	31/16	1145.036	thirty-first harmonic
12	2/1	1200.000	octave

Lucas (2–1) scale H Right: twelve tones
Rule: $H_n = H_{n-5} + H_{n-3}$
Seed string from triangle: 2, 0, 0, 1, 0
Resulting sequence: $H_n = H_{n-5} + H_{n-3}$: 2 0 0 1 0 2 1 0 3 1 2 4 1 5 5 3 9 6 8 14 9 17 20 17 31 29 34 51 46 65 80 80 116 126 …

Scale

0	1/1	0.000	unison, perfect prime
1	65/64	26.841	thirteenth partial chroma
2	17/16	104.955	seventeenth harmonic
3	9/8	203.910	major whole tone
4	5/4	386.314	major third
5	23/16	628.274	twenty-third harmonic
6	3/2	701.955	perfect fifth

(*continued*)

Table 29.1 Continued

7	51/32	806.910	
8	7/4	968.826	harmonic seventh
9	29/16	1029.577	twenty-ninth harmonic
10	31/16	1145.036	thirty-first harmonic
11	63/32	1172.736	octave–septimal comma
12	2/1	1200.000	octave

Lucas (2–1) scale J Left: twelve tones
Rule: $J_n = J_{n-6} + J_{n-1}$
Seed string from triangle: 1 2 2 2 2 2 3 5 7 9 11 13 16 21 28 37 48 61 77 98 …
Resulting sequence: $J_n = J_{n-6} + J_{n-1}$: 1 2 2 2 2 2 …

Scale

0	1/1	0.000	unison, perfect prime
1	9/8	203.910	major whole tone
2	37/32	251.344	thirty-seventh harmonic
3	77/64	320.144	
4	5/4	386.314	major third
5	21/16	470.781	narrow fourth
6	11/8	551.318	undecimal semi-augmented fourth
7	3/2	701.955	perfect fifth
8	49/32	737.652	
9	13/8	840.528	tridecimal neutral sixth
10	7/4	968.826	harmonic seventh
11	61/32	1116.885	
12	2/1	1200.000	octave

which the word order was determined by the Shuffle module. The Shuffle module has several modes—one is to have a single shuffled order repeating, another is to have a newly shuffled order generated at the end of each sequence. For this patch, I decided to use the repeating shuffled order. Since in this new patch, I'm alternating the output of the consonant–vowel generator and the text picking, and that alternation is controlled by a random generator, this means that I'm now getting 'holes' in the repeating random sequence, which are substitutions by the consonant–vowel generator. So, for example, two statements of the repeating sequence on a short (nineteen-word) array looks as below (Table 29.2). The nonsense words are all upper case, the repeating randomized-order sequence words are upper and lower case. The nineteen-word long repeating

Table 29.2 Sample sound poetry output from *ArtWonk* word permutation program

BIRALA

Trout

Effervescence

Riddled

LOHORETI

TI

FO

Magnified

Manufactures

LOHIWI

Of

DOSEF

Sunrises

Speckled

ME

Nun

SAPI

Greatcoat

That

The

Trout

ME

Riddled

Efflorescences

Within

Holds

Magnified

Manufactures

Miniaturized

Of

Malfeasance

Sunrises

TA

Molluscan

Nun

Fringe

Greatcoat

That

sequence is broken up by randomly substituting some of the StrRand nonsense words for the words in the repeating sequence.

Wearing a pair of binaural microphones, I sat on the front porch of my house, which is on a rather busy highway, and recorded a reading of the text, accompanied by the sounds of the afternoon and its traffic. I then took this recording into my studio, where I modified the sound in the programs AudioMulch and Glitch2. Glitch2 allows you to make sequences of nine different effects units. Each effects unit can have its parameters randomized, and it will set up random sequences of the effects units. Naturally, I used the randomization features, both on each individual effect and to make sequences of effects. I made eight different sequences like this. Many of the sounds generated were sounds I wouldn't normally choose to make myself. This was good, as it forced me to stretch my sense of what would be appropriate. In making the final recording, I alternated improvisationally between the eight sequences of randomized sound modifications. The result was a noisy and quite fun piece, in which the sounds of a nonsense text, traffic, and various effects-based sounds alternated and commingled with glee.

29.3   A New Piece for Electronic Tones and Tuning Forks

I would like to discuss a new work in progress, as an example of how I'm structuring my work in a freely combinational way these days. Recently, using the iPad app *Wilsonic* (Satellite 2017), I made a CoPrime Grid scale of seventy-four just pitches to the octave. The CoPrime Grid is Erv Wilson's generalization of the Harry Partch and Julian Carillo 'tonality diamond' concept (Wilson 2003), After playing with the scale in the iPad app, I thought that an algorithmic process would be a good way to articulate the very close pitches of the scale—a way to get beatings and pure intervals in a fairly balanced way. Obviously, with seventy-four pitches to the octave, there will be a lot of very closely tuned intervals to work with. But there are also a lot of pure, more traditional larger intervals in the scale as well. I tuned up an oscillator in Vaz Modular (Fay 2012) to play the scale with timbres close to sine waves. In ArtWonk, I had previously incorporated the Latoocarfian fractal into its family of fractal functions. The Latoocarfian fractal was created by Clifford Pickover in his book *Chaos in Wonderland* (Pickover 1995) and it's a fractal that produces beautifully elaborate patterns, and if scaled properly, musical gestures similar to the structure of traditional melodies. I simply set up a Latoocarfian fractal to control pitch with the x-axis output, and velocity with the y-axis output. I scaled the outputs appropriately, of course. The seventy-four pitches of the scale produced just a little over an octave and a half with MIDI notes 0–127, so for pitch, scaling the x-axis output up to 0–127 was quite adequate. For velocity, I wanted notes always to be present, just with some amount of change in volume, so scaling the y-output up to 0–63 and adding 64 to the output produced velocities between 64 and 127. For durations, I simply set up a random walk module, generating outputs between 2 and 25, with a relatively small range of value choice allowed. When multiplied by a tempo factor (about 25), this produced a series of durations which 'clustered' around particular values—that is a series of shorter durations which change to a series of longer durations, and so on. But the thing is, I didn't spend a lot of time evaluating fractal or random functions before I chose the Latoocarfian fractal, I simply spontaneously chose it (based on some experience with it in the past, and as stated above, I *did* write the function for ArtWonk as well), and then followed through with listening to its results. When I heard the results, I thought, 'Well, three of these, operating asynchronously might be interesting.' Why did I think that? Instinct, I guess. Lately, I've been doing a lot of pieces in which I combine multiple passes of a particular process (as in the last movement of *Three Movements in Memoriam Paul Panhuysen*). So I set up three versions of the same process, which would work at the same tempo, but asynchronously, and with slight differences in the initial starting parameters of the Latoocarfian fractal. The result was quite nice. And I later added some slight timbre changes and panning to the individual voices and got quite a nice spread of beating intervals, and larger pure intervals, now spread over stereo space and with a small amount of timbral variation (somewhat like a low-pass filter, but more like a 'brightness' control than a filter would be. Now the question arises: yes, it's nice, but what do you *do* with it? In this case, the answer was easy—use it with my tuning forks. These are aluminium forks, made in 1985–1986 at the CSIRO in Melbourne, which are tuned into a nineteen-note just-intonation scale, and which ring for 30 seconds with a very sine-like timbre (Burt 1987). And I would have an opportunity to perform the piece. It was performed with Simon Edwards and myself on tuning forks accompanying the recording described here, on 29 April 29 2015 as part of the Performance Seminar series at Box Hill Institute. So at least *this* piece had a chance to be played before live human beings and was not just consigned to ear-buddery and Inter-nettery. And this points out a problem of context with much of my recent work.

29.4 A PROBLEM OF CONTEXT

I realize that the algorithmic processes by which I structured these pieces will probably be of primary interest to readers of this anthology, but what is also of interest and urgency to me, just now, is the social context within which these works exist. That is, I've worked with algorithmic ideas for so long that by now they are second nature to me. What concerns me now is the fact that during the making of these and other recent works, no face-to-face human contact occurred at any time, and none is likely to occur at any time in the future. The exception to this is the last piece discussed above, which I'm intending to be used as an accompaniment to an improvisation for two people with my tuning forks, in a live presentation.

But, for example, the *Three Movements in Memoriam Paul Panhuysen* came about through receiving an email. I put the pieces on my website, and received feedback about them via email. And that was it. No live performance, no distribution on CD, no reviews, no dialogue with colleagues, just a making of the work in a private medium and a distribution on a forum which, while seemingly public, is actually about as removed from the public as one can get, as far as I can see. As someone who has spent his whole life participating in the creation of alternative arts centres and building musical communities, this is quite an adjustment to make.

Word Swirl Two began with receiving a text message. I made the piece, and mailed it off to Sjaak, and received a text message from him saying he liked to work, and was looking forward to putting out the CD. Again, no face-to-face contact was involved in the making or distributing of the piece. When the CD is launched, I guess there will be a launch party of some kind. This will be the first time in the life of the work, where there will be face-to-face communication with fellow artists. I hope I'll be able to attend.

Recently, I was asked by the National Film and Sound Archive of Australia to submit work to them for their webcast series Lateral Listening. I did so happily, knowing that this was a very nice series with interesting sound artists contributing (NFSA 2017). Probably quite a few people will hear my music there—more than at almost any live venue at which I could present my work. Without being ungrateful for this honour in any way, I still ask, why does this not fulfil my need for artistic communication? Why does presenting my work under ideal listening conditions to an audience of twenty or thirty friends feel so much better than having my work 'out' on a recording or on an internet site? This is not just a phenomenon of the Internet age—it's been around since the birth of sound recording. Why do I feel so frustrated that most of my music making is now occurring in this non-personal-contact manner, despite the fact that, potentially, many more people are able to hear it than would ever be able to hear it live?

I remember reading that people like Edgard Varèse and Milton Babbitt, in talking about electronic music in the 1930s to 1950s, made the analogy that with electronics, a composer could now be like a painter, and music making could now be a studio-based art. But every painter, while in their studio, thinks about the possibility of being at the showing of their work. The art opening is the visual art world's equivalent of the live performance in the music world. In my current work, I seem to have the studio practice Varèse and Babbitt longed for, but without the potential emotional payoff, if you will, of either the opening or the live performance event.

Perhaps the situation is analogous to writing. For example, this chapter was mostly written on trains in response to an email request and was emailed to an editor and a publisher. No human face-to-face contact will be involved in it. There will undoubtedly be email exchanges with the editors, but it's all going to occur in what my wife Catherine Schieve calls 'the fragmented, online, disembodied world of the internet'. And although it's true that we have increased electronic access to others, at least in my life the amount of personal contact with colleagues has shrunken alarmingly. How did we get to this state?

At this point, I should say that I do have a live practice—improvising live with algorithmic processes, usually involving an iPad. I manage to do a performance every four or five months in this manner, usually with friends, in small venues. And I do seven live (stand-up) performances a week, to captive audiences—those are called lectures, and the reason I do them is to fund my composing and pay my rent—but that which ostensibly matters the most to me, composing works with serious structural content, has become mostly a hermetic, non-face-to-face-contact-based activity. How did this happen?

A possible answer, without theorizing; just stating facts. In Australia, for new music, the activity is mostly self-organized and occurs in large urban centres. There are very few organizations which take a curatorial, overarching, non-stylistically biased approach to presenting new music concerts. So usually, if you want to have public contact with your music, you have to organize it yourself. Real estate prices in Australia have shot up amazingly since the turn of the millennium. It becomes harder and harder for one person on a middle- or lower-middle-class income (the income an academic in Australia would have) to actually afford to rent a place in an inner-city art-active area. In my case, this has meant living in Daylesford, 107 km from Melbourne, and supporting myself and my family by working a very time-intensive, demanding job at Box Hill Institute, a two-hour commute away from home. There is only one train–bus combination back from Melbourne to Daylesford in the evening, and that leaves at 5:51 pm. So unless one can make arrangements to stay in Melbourne for the night, weekday evening art activities become, if not impossible, extremely difficult. And then, family obligations also make travelling to the city on the weekends problematic. So we've got a trap here—my job does not pay enough to live in the city, and living in the country, where I can afford it, mandates such an enormous commute that I can only with great difficulty arrange to be at art events, which, in the Australian context, happen almost exclusively in urban contexts. The free and active circulation within a large group of colleagues who all see each other at a constant succession of art (and other) events, which in the past was how careers were established and a series of gigs were able to be self-arranged, becomes almost impossible in the current economic and employment contexts. So a web-based presence for art making, which might, in other circumstances provide a complementary sphere of activity to a lively personally immediate career, becomes, instead, the only game in (or out of) town. Given that, thank goodness for the Internet and netbook computers. At least, even in these circumstances, I *can* continue composing, even if it's not in a context I would prefer.

So, at the moment, unless I can find a way to organize my own live presentations of my work, and find the time and venues, and the economic means, to do these presentations, this situation of relative artistic isolation will continue. But we keep going on, nonetheless, hoping that this phase of life will, too, pass.

BIBLIOGRAPHY

Burt, W. *Studies for Synthesizer*. Scarlet Aardvark Records, SA-022. North Carlton, Australia: Scarlet Aardvark, 1982. CD.

Burt, W. 'A Set of Justly-Tuned Aluminium Tuning Forks'. *Experimental Musical Instruments* 2, no. 5 (1987): 12–13.

Burt, W. *39 Dissonant Etudes*. Tall Poppies, TP-093. Glebe, Australia: Tall Poppies, 1996a. CD.

Burt, W. 'Some Parentheses around Algorithmic Composition'. *Organised Sound* 1, no. 3 (1996b): 167–172

Burt, W. 'A Catalog of 12 Note Scales Made from Additive Sequences Obtained from the Diagonals of Number Triangles'. 2004–2006. http://www.warrenburt.com/microtonal_resources/algorithms_microtonality_performance_materials/Proliferating_Infinities_Scale_Catalog.pdf.

Burt, W. *The Animation of Lists and the Archytan Transpositions*. XI Records, XI-130. New York: XI, 2006. CD.

Burt, W. 'Paul Panhuysen, 1934–2015'. *Warren Burt*, 10 February 2015. http://www.warrenburt.com/journal/2015/2/10/paul-panhuysen-1934-2015.html.

Chadabe, J. 'The Role of Apps in Electroacoustic Music'. *Organised Sound* 20, no. 1 (2015): 99–104.

Dunn, J. 'ArtWonk' and 'MusicWonk'. *Algorithmic Arts*. Last modified 2016. http://algoart.com/.

Fay, M. 'Vaz Modular'. *VAZ Synths*. 2015. www.vaz-synths.com/vmIndex.html.

National Film and Sound Archive (NFSA), Australia. 'Lateral Listening'. *Soundcloud*. https://soundcloud.com/nfsaaustralia, accessed 30 June 2017.

Pickover, C. *Chaos in Wonderland: Visual Adventures in a Fractal World*. New York: St. Martin's Press, 1995.

Satellite, M. 'Wilsonic'. *iTunes Preview*. Last modified 29 May 2017. https://itunes.apple.com/app/wilsonic/id848852071?mt=8.

Wilson, E. 'The Scales of Mt. Meru'. 1993. http://anaphoria.com/meruone.PDF.

Wilson, E. 'The Co-Prime Grid'. *The Wilson Archive*, 1999–2003. http://www.anaphoria.com/wilsondiamondcoprime.html.

PART IV

ALGORITHMIC CULTURE

CHAPTER 30

....

THE AUDIENCE RECEPTION OF ALGORITHMIC MUSIC

....

MARY SIMONI

30.1 AUDIENCE RECEPTION THEORY AND RESEARCH

....

IT seems implausible that that we know so little about why an audience responds favorably to a musical composition. After all, the fashion industry has honed the practice of audience research by marketing to particular niches. The myPersonality Facebook application collected six million psychometric test results from four million individual Facebook profiles, and made the database available to researchers. Why was there critical acclaim for the premiere of Leonard Bernstein's *Mass*, laden with its blasphemous symbolism, yet the premiere of Stravinsky's *Rite of Spring* evoked a riot? Wasn't Stravinsky's boisterous musical language just as socially revolting? Or was it that society was not yet ready to hear the bared soul of humanity communicated with such pungent ferocity? It's these kinds of questions that spur audience reception theory: a culturally situated approach to the interpretations and preferences of choices by a group of people.

Following Marshall McLuhan's influential aphorism 'the medium is the message', Hans-Robert Jauss proposed reception theory as a way to understand a reader's response to a literary text (Jauss 1982). Stuart Hall, a founder of the Birmingham School of Cultural Studies, extended Jauss's theory to include media and communication studies (Hall 1993). According to reception theory, a communicator conveys an encoded message. The recipient observes the encoded message and decodes it based on their cultural context and personal experience. Thus, in reception theory, the intent of encoding is to convey a message and the process of deriving meaning from a message is decoding. In media studies, the audience is an active interpreter and producer of meaning. Reception theory recognizes that a media object can be decoded several different ways, depending on the backgrounds of the audience. Each member of the audience negotiates their personal perspective with the message in order to derive meaning (Goldstein and Machor 2008). When the sender and receiver have interacted before, the encoding-decoding process generally results in greater meaning. Therefore, processes of encoding-decoding are practised and, with practice, the derivation of meaning usually improves and the meaning of the message is understood in the way it was intended.

Since the decline of the troubadours in the fourteenth century, the roles of composer, performer, and audience have become increasingly distinct. With the emergence of romanticism in the nineteenth century, this tripartite ontology is characterized as the composer as creative genius—a person who constructs the medium and mode for the message. The composer communicates through a highly stylized artefact such as a musical score. This musical score is presented to skilled performers, whose role in the ontology is to interpret the composer's message through musical performance. The audience, oftentimes enveloped in the cultural formalism of the concert hall, is a passive recipient of the interpreted message. Although feedback loops are possible between and among the roles of composer, performer, and audience, the cultural evolution in Western classical music has been to relegate feedback to accepted non-real-time norms: Critics write about the composer and performers after the concert; performers rarely share their impressions of the music with the composer, who is usually deceased; and the behaviour that characterizes the ideal audience response is limited to applause.

But since the mid-twentieth century, composers have been putting computers to work for all sorts of compositional tasks. Laptops and PDAs were quickly deputed to simultaneously enhance and blur the romanticized ontology of composer, performer, and audience. Ensembles of laptops and mobile devices sprung up on college campuses and nightclubs, using live coding and improvisation during performance, further eroding the distinction of composer and performer. Laptops and controllers were introduced into a number of real-time processes aimed at expanding the capabilities of the performer.

One could argue that up to this point, much algorithmic music has been composed for a coterie—a community of like-minded individuals, many of whom are practising composers themselves—who understand the motivations, technologies, and processes employed to create the work. In audience reception theory, the coterie is considered the ideal audience to decode an encoded message: an audience of professionally competent critics. The paradox of composing and performing music for the coterie is that the feedback loop between composer/performer and audience is limited to the experiences of the elite, assuring with near certainty that the music will not be understood beyond the coterie.

In order to develop an audience for new algorithmic concert music, one must wrestle with why this music does not currently have mass appeal. Lillehaug asked the question, 'Is there a gap between many contemporary composers and the general audience today?' After interviewing over two hundred people in schools and colleges across the

United States, he concluded that there was little doubt about a significant gap between composers and their audiences. There was a divergence of opinion on how best to close the gap, including recommendations for increased education across all age groups, composer–performer symposia, and improvements in the quality of the performance of new music (Lillehaug 1969).

Beaumont's research was aimed at bridging the gaps among the composer, performer, and audience through repeated, impeccable performances. He conducted an empirical study of a newly commissioned work by Jonathan Harvey: *The Riot* for flute, bass clarinet, and piano. This work, commissioned by Het Trio of Amsterdam, received two performances at 'The Intention, Reception and Understanding of Musical Composition' symposium held at Bristol University. Beaumont moderated a discussion that included the composer, performers, and members of the audience following both performances. He concluded that if the compositional style is a good match for the performers, the result is a good performance that in turn improves audience reception (Beaumont 1998).

A number of researchers have explored audience engagement using real-time interactive or generative systems (Dillon, Adkins, Brown, and Hirche 2008). The roles of composer and performer become one as an individual simultaneously creates and performs the music. Methodologies that assess the efficacy of human-computer interface (HCI) are typically designed to objectively measure task completion. Since interfaces that create and perform music have concurrent and sometimes convergent creative, affective, and cognitive tasks, it is difficult not only to differentiate the type of task but also to design the concomitant methods that objectively measure task completion. Talk-aloud protocols, although useful in HCI research, create competition in the sonic ecology of music by introducing human verbalization about the process into the same environment that produces the music. Discourse analysis has been successfully employed, but using a constrained vocabulary in an attempt to reduce the labor involved in transcription and analysis (Camurri et al. 2003).

Landy identifies the 'something to hold onto' factor (SHF) as necessary in order for meaning to be decoded in electroacoustic music. The SHF includes musical attributes such as pitch, rhythm, homogeneity of sound and the search for new sounds; textures not exceeding four sound types at once; and programs, which could be natural sounds, processed familiar sounds, or acousmatic tales (Landy 1994, 2006). Weale (2006) contributed to this research by developing a methodology that measures a composer's intention (e.g. encoding) against a listener's response (e.g. decoding) across levels of experience.

Audience reception of dance presents some challenges beyond those found in music, since dance is multidimensional and multimodal. Real-time data that measure audience engagement with a semi-improvised solo work have been collected using PDAs. The audience members record their engagement with the dance; engagement meaning that they were 'compelled, drawn in, connected to what is happening, interested in what will happen next' (Schubert, Vincs, and Stevens 2013; Stevens, Glass, Schubert, Chen, and

Winskel 2007). There is compelling evidence that when the choreography fulfills audience expectations, the audience is more likely to be engaged. This association between aligning audience expectation and engagement is conceptually similar to Landy's SHF, indicating an audience preference for decoding new works that draw from prior experience, creating a framework of expectation facilitating the decoding meaning.

Musical Turing Tests, modeled after the theory of machine intelligence proposed by Alan Turing, are considered by many to be the Sangraal of algorithmic composition (Turing 1950). These tests are designed to create such convincing music that the listener, analogous to an audience member, cannot tell the difference between a human and a computer. Musical Turing Tests have ranged from computer-generated repertoire modeled after composers such as J. S. Bach and Mozart to the highly publicized yet bewilderingly irreproducible *Experiments in Music Intelligence* (Cope 1991). More recent work employs computer-based beat tracking in comparison to a live performer (Stowell, Robertson, Bryan-Kinns, and Plumbley 2009). It is critical that researchers who employ the Turing Test as validation of artificial intelligence thoroughly document the research so that it is extensible.

Bartel sought to identify relationships between personality and education and the cognitive-affective response of individuals to three different styles of music—traditional classical, jazz, and country music—with 146 college undergraduates. In this study, a cognitive response to a musical stimulus is dominated by objective observation and classification. An affective response is dominated by subjective observations of emotions or aesthetic interpretations (Bartel 1988). These responses become part of musical memory and converge or diverge into the meaning and understanding of a composer's intent. Bartel concluded that the greater the musical training, the greater the likelihood of a cognitive response. Although Bartel's work is not a longitudinal study, his approach using cognitive-affective response to music influenced the methodological design for this chapter.

Since the time Brian Eno coined the term 'generative music' (Eno 1996), computers have been programmed to generate music, enhance musical performance, and even to listen to a performance and render decisions about what it hears. The goal of these program is not akin to the Turing Test but instead to foster an ontological reductionism that oftentimes means that the composer, performer, and audience member is the same person.

This fantastic union of humanity and technology not only hastens the feedback loop and rouses creativity, it also helps us better understand ourselves. One could argue that this social reductionism will unwittingly turn a generation of musicians into cyborgs, preferring process over product. Yet many reject this claim, espousing the import of evolutionary practice in advancing art. If we take the long view with a glance over our shoulder to the past, generative practice is analogous to music making in the Middle Ages, when the composer and performer were usually the same person; music was seldom written down but instead passed on through the oral tradition. In both secular and sacred music, audience reception was critical to a musician's livelihood. Without question, we are in a period of social, technological, and cultural flux, propelled forwards by

a goulash of Moore's Law mixed with artistic inquiry. But fear not: things will sort themselves out ... just as the centuries of the Dark Ages ceded to the Baroque.

30.2 HYPOTHESIS AND RESEARCH QUESTIONS

The premise of this study is that reception theory may be applied to the audience reception of algorithmic music using a cognitive-affective model to understand the process of
the decoding of meaning over time. Specific research questions are:

1. Do repeated interactions with an algorithmic composition increase a listener's
   ability to accurately 'decode' the work?
2. Does encoding followed by decoding assist in reception of algorithmic
   composition?
3. Does cognitive understanding influence aesthetic appreciation?
4. Does the 'something to hold on to' factor evoke a cognitive or affective response
   or both?

30.3 METHODOLOGY

This research employs a comparative analysis in a longitudinal study to understand
the audience reception of algorithmic music. The four compositions that are studied
are Barry Truax's *Riverrun*, Elliott Carter's *Canon for Three Equal Instruments*, Mara
Helmuth's *Abandoned Lake in Maine*, and Krzysztof Penderecki's *Threnody for the
Victims of Hiroshima*. These pieces were either composed for fixed media (Truax and
Helmuth) or are recordings of performances (Carter and Penderecki). The author
uses fixed media in order to reduce the variability that may occur during repeated live
performances.

The primary mode of data collection was through the language of the subject, including individual and group verbal and written responses. The think-aloud protocol was
modified to an observe-write protocol to capture thought processes during active listening (Yang 2003). A discourse analysis was applied to written and verbal responses made
while listening to or recalling algorithmic music. Both qualitative and quantitative analyses occurred during four sessions that took place over a period of a month. A group
of eight students were recruited to be the subjects of an audience. The responses of the
audience were used to discover, classify, and label cognitive-affective responses. Analysis
of the labeled cognitive-affective responses of three of the eight members of the audience,
called the case study subjects, was used to examine the change in the reception of algorithmic music over time.[1] An overview of the methodology is found in Figure 30.1.

	Instrument	Elapsed Days	No. of Subjects	Method
Demographics	Survey-1	0	8	On-Line Survey -Individual Response
During First Listening	Survey-2	0	8	Observe-Write Protocol- Individual Response
After the Second Listening of each composition	Survey-3	4	8	Observe-Write Protocol- Individual Response
After the Second Listening of all compositions	Survey-4	26–29	3	Dialogue- Individual Response

FIGURE 30.1 Overview of methodology.

30.4 SURVEY-1

Survey-1 captured basic demographic data about the eight subjects. Of the five male and three females, one was eighteen, four were nineteen, one was twenty, one was twenty-one, and one was twenty-three years of age. Four subjects (50 percent) identified as Caucasian and four (50 percent) identified as Asian/Pacific Islander. All of the subjects had prior musical training: four (50 percent) had studied music for more than ten years, two (25 percent) studied for seven to nine years, and two (25 percent) studied for four to six years. All of the subjects had studied a traditional acoustic instrument or voice at least four years and all of them had participated in a traditional musical ensemble. Most of them (75 percent) had never played in a popular music band. A majority of the subjects (63 percent) had previously used computers to compose or produce music. Half of the subjects had programmed computers to implement algorithms.

When the subjects were asked about how they learned about new music, about half reported using online services such as Pandora and Spotify or by talking with their friends. Six of the subjects (75 percent) reported enjoying music more if they knew something about the composer, performer, or composition. All of the subjects (100 percent) reported that they had found that whether or not they like a piece can change over time.

Three of the eight subjects will be highlighted as case studies. One of the case study subjects is **Anthony**; a Caucasian nineteen-year-old male who is a sophomore majoring in computer science. He has played cello for eight years, studying privately for seven years and playing in an orchestra for eight years. His favourite kinds of music are classical and art music. Anthony has been programming computers for about four years and has used computers to explore algorithmic composition to create music in the style of the Bach cello suites. He has analysed pitch sequences, created a transition table, and used a Markov chain to output notes. Although he described the experience as fun, he was disappointed with his result, saying the output lacked 'musical coherence'. Anthony does not use online music listening services to discover new music nor does he learn about new music from his friends. Instead, he attends concerts or downloads libraries and listens when his schedule allows. He enjoys learning about music and offered that he did not initially care for Jean Sibelius (1865–1957), but found that after about five listenings and access to program notes, he now appreciates Sibelius. When he was asked if he enjoys learning about new music, the term 'new music' was interpreted as a genre, and he quickly offered that he puts new music as a 4 on a scale of 1–10, stating that it's 'not as interesting as the classics'.

Justin is a twenty-three-year-old Asian/Pacific Islander who is a senior majoring in architecture. He has taken private lessons on organ for twelve years. Currently, Justin serves as an organist and music director, a post he has held for two years. Next year, Justin will pursue graduate study in organ performance at Yale University. Justin has never programmed a computer nor does he compose or produce music. His musical preferences are the broadest of any of the case study subjects and include ambient, classical and art music, religious music, and stage and screen music. Justin does not use online listening services to discover new music. Instead, he learns about new music through his faculty. His favourite composer is Oliver Messiaen (1908–1992), particularly the composer's organ works. Justin enjoys learning about music and says that his enjoyment of music can change over time. He describes himself as a visual thinker, so if he can see the score or a live performance he tends to learn about the music more quickly. He stated that repetition does not contribute to whether or not he likes a composition. In fact, repetition may cause him to dislike music.

The last case study subject is Theresa. She is a nineteen-year-old Caucasian sophomore studying chemical engineering. Theresa has performed in vocal ensembles for over ten years and has studied piano privately for six years. She is enrolled in the university's choir and also participates in an a cappella group. In addition to classical and art music, Theresa enjoys jazz and screen and stage music. She plays video games recreationally and enjoys analysing the form and recording processes used to create the soundtracks. She offered that studying music theory helps her understand music, and that she applies her knowledge to help her a cappella group during rehearsal. She has studied computer programming for one year and has used computers to record her voice and 'mess around with it'. She voluntarily uses music notation software to complete her music theory assignments 'because it is fun'. Although Theresa does not use online listening services to discover new music, she learns about new music from her friends. As with all of the subjects, she enjoys learning about music and has discovered that her perception of music varies with repeated listenings.

The audience as well as the three case-study subjects may be characterized as musically literate and open to learning about new music, and predisposed to changing their perspective of music over time. This audience is not the coterie of algorithmic music; however, they do exhibit an aggregated demographic profile that likely reduces variability in the research.

30.5 SURVEY-2

The purpose of Survey-2 was to promote active listening of algorithmic music by requesting that each subject write unstructured observations while listening. Prior to listening, subjects were provided only with written definitions and a brief lecture on the terms *algorithmic composition, soundscape composition, acoustic music*, and *acousmatic music*. The subjects reported that prior to this study they were not familiar with any of these terms except acoustic music.

Algorithmic composition is defined as the application of a set of rules or a sequence of operations for the purpose of creating a musical composition or some aspect of a musical composition. Algorithmic composition may or may not make use of a computer. Following a brief discussion of the definition, subjects were given an introduction to twelve-tone composition, to which none of them had previously been exposed, as an example of algorithmic composition.

Soundscape composition is music that creates an immersive sonic environment to invoke a listener's associations, memories, and imagination related to one or more places. Subjects were asked to consider paintings of landscapes as a visual analogue to soundscape composition.

Acoustic music is music that is performed on traditional instruments such as strings, woodwinds, brass, percussion, or the human voice. Since all of the students have had prior musical training, they quickly understood the definition of acoustic music.

Acousmatic music is a form of electronic music designed specifically for presentation using loudspeakers. Acousmatic music does not include human performers and often exists only as audio recordings. Sometimes acousmatic music will use prerecorded acoustic music or sounds. Subjects understood that acousmatic music does not correlate with a particular musical style in much the same way that landscape paintings are not exclusive to a particular period in art history.

Following the presentation of these terms, subjects listened to four selected compositions in the order listed below. Although the program notes are included to orient the reader, subjects were not provided with any information about the pieces prior to the first listening. They did not know the title or duration, had not read program notes, or even knew if the piece was an example of algorithmic composition.

(1) *Riverrun* by Barry Truax was composed in 1986 with a duration of 19′44″ (Truax 1986). This work is algorithmic on both a micro and macro formal scale. According to Barry Truax, '*Riverrun* creates a sound environment in which stasis and flux, solidity and movement co-exist in a dynamic balance. The corresponding metaphor is that of a river, always moving yet seemingly permanent. From the smallest rivulet to the fullest force of its mass, a river is formed from a collection of countless droplets and sources. So too with the sound in this composition which bases itself on the smallest possible "unit" of sound in order to create larger

textures and masses. ... *Riverrun* is entirely realized with the method of sound production known as granular synthesis' (Truax 1986).

(2) *Canon for Three Equal Instruments: In Memoriam Igor Stravinsky* by Elliott Carter was composed in 1971 with duration of 1′15″ (Carter 1992, score and CD). This algorithmic work employs serialism. The composition 'was composed for one of the Stravinsky memorial issues of the English music magazine *Tempo*, at the request of David Drew, its editor. In proposing the canonic form to a group of European and American composers, he was following the lead of Stravinsky himself, who wrote a number of commemorative pieces in this form' (Carter 1992). This recording is performed with three trumpets using different mutes.

(3) *Abandoned Lake in Maine* by Mara Helmuth was composed in 1997 with a duration of 12′12″ (Helmuth 2007). This through-composed work uses algorithms in the production and processing of sound. The composer writes, '*Abandoned Lake in Maine* explores relationships between humanity, technology and nature. The compositional process uses technology to return the listener closer to nature, and into new relationships with nature. The composer's software instruments and algorithmic programs were used to create much of the material. The sources for this piece are recorded sounds of the loon in its environment, and a naturalist's voice. ... Recorded source sounds of Maine loons and voice are from *Voices of the Loon*, produced by William Barklow' (Helmuth 2015).

(4) *Threnody for the Victims of Hiroshima* by Krzysztof Penderecki was composed in 1960 with a duration of 9′15″ (Penderecki 2012). This 'brief work is written for string orchestra, using unconventional notation such as a black triangle to indicate the highest possible note for an instrument, though of no determinate pitch. There is much use of *glissando* scales, sliding through a succession of quartertones. Instead of time signatures and bar lines, the composer organized his music into units of sound lasting a specified length of time from four to 30 seconds. *Threnody* has been described as atonal program music, and as such it powerfully suggests the hair-raising terror of the moments just prior to the dropping of the atomic bomb. Without a doubt the concentrated, piercing wall of string sound at the climax graphically conveys unspeakable horror as the terrible instrument of war plunges from the sky and explodes' (Philadelphia Orchestra 1999).

These algorithmic compositions were selected because they exhibit several characteristics. They were composed during a relatively short period in music history (1960–1997)—a time in the history of music when computers were first being used to produce algorithmic music (although not all of them make use of a computer in the compositional process). The compositional durations range from 1′15″ to nearly twenty minutes to see if duration has an effect on the capacity of a subject to decode meaning. The composers realized the composition using either acoustic sounds (Carter and Penderecki) or electronic sounds (Truax and Helmuth), or both. The method of notating the score spans traditional notation (Carter) and experimental notation (Penderecki), or the composition does not have a score (Truax and Helmuth) and it has been analysed using spectrograms (Truax) (Helmuth 2006). The researcher purposefully selected compositions without a discernible rhythmic structure to avoid the 'something to hold on to' factor. It was important that at least one of the composers be available for a teleconference so the subjects could interact with the composer. All of the compositions were played using stereo playback in a classroom setting with each subject sitting at a desk.

A discourse analysis was applied to Survey-2 to observe the frequency of word choices by the subjects to discover a cognitive-affective classification system for their responses. A response is classified as 'cognitive' if it exhibits an expression of knowing objective information. All cognitive responses are valid from the point of view of the subject, but were labeled as correct or incorrect by the researcher. Therefore, cognitive responses are categorized as 'correct' or 'incorrect' (labels = Cognitive-Correct or Cognitive-Incorrect). For some responses, a subject expresses uncertainty about a cognitive observation, indicated by a question mark or an expression of unknowing (label = Cognitive-Inquiry). The discourse analysis was not automated, allowing the greatest possible freedom in the type of responses each subject made and beginning to develop a vocabulary and schema that could lead to automated discourse analysis in future research.

A response is labeled as 'affective' if the subject communicates a subjective expression of feeling. Affective responses are descriptive of a subject's emotion or aesthetic judgement. All affective responses are valid from the point of view of the subject and the researcher. Affective Responses are categorized as affective if a subject states an aesthetic judgement (label = Affective-Aesthetic) or reports that they experienced an emotion while listening to the composition or speculated that the composer was trying to evoke an emotion in the listener (label = Affective-Emotion). Although it is conceivable that a subject may be uncertain during an affective response, for Survey-2 no subject expressed affective uncertainty (label = Affective Inquiry).

After all four surveys were completed, the subjects were asked to classify the compositional genre based on their recently acquired knowledge of algorithmic composition, soundscape composition, acoustic music, and acousmatic music. Figure 30.2 displays the majority responses of the subjects.

Recall that the subjects were not informed that all of the compositions made use of algorithms. Subjects were also informed that a composition may fit in more than one category. Regardless of this instruction, all subjects selected only one category for each composition. Eighty-eight percent correctly identified Carter as acoustic music that

		Column 1	Column 2
		Algorithmic Composition	Soundscape Composition
Row 1	Acoustic	Carter (2) – 88%	Penderecki (4) – 88%
Row 2	Acousmatic	Truax (1) – 75%	Helmuth (3) – 75%

FIGURE 30.2 Majority responses to genre classification of the four compositions.

employs an algorithm. That same percentage also identified Penderecki as acoustic and realized the composer's intent to create a soundscape. Seventy-five percent identified Truax as algorithmic acousmatic composition but did not realize Truax's intent was to create a soundscape. It's possible that the duration of *Riverrun* confounded the listener's ability to perceive it as a soundscape composition. A clear majority of subjects recognized Helmuth's composition was an acousmatic soundscape. Clearly, most subjects were able to apply the terminology they had learned to the first listening and many were able to correctly discern the composer's compositional process or intent.

30.6 SURVEY-3

After three days, the subjects engaged in a second listening of the same four compositions, but the order was altered: Helmuth, Truax, Penderecki, and Carter. This listening was preceded by reviewing the work's title, composer, year of composition, duration, and program notes followed by a brief lecture on one or more aspects of the composition. After the second listening, each subject completed Survey-3, which was designed to recall the musical memory of the first listening and the subject's response to the composition now that they understood more of the composer's intent, and to ascertain if they were curious about aspects of the composition that had not been discussed. As with Survey-2, a discourse analysis was applied to the responses and those responses were labeled as Cognitive-Correct, Cognitive-Incorrect, Cognitive-Inquiry, Affective-Aesthetic, Affective-Emotion, and Affective-Inquiry.

30.7 SURVEY-4

Survey-4 took place four weeks after the start of the study and was administered to the three case study subjects. This survey was not preceded by an additional listening. The survey was completed during a one-on-one dialogue with the researcher. The purpose of Survey-4 was to further elucidate prior educational and musical training, formation of musical attitude, musical memory, and musical attitude towards the four compositions in particular and algorithmic music in general. As with the previous surveys, a discourse analysis was applied to the transcripts.

30.8 *RIVERRUN* BY BARRY TRUAX

For the discourse analysis of Survey-2 of *Riverrun*, several subjects reported hearing dripping water or rain, noting that the sound gradually increased in intensity while sounding random and chaotic. One subject reported that the initial sounds of the composition were discontinuous, yet converged into a continuous sound. Several other subjects noted convergence, stating that the sounds converged into pitches. One subject described the composition as 'clouds of sound' and another remarked that there was 'no clear rhythm/beat'. One subject described the form of the composition using adjectives such as 'calm → chaotic + loud → one note (calm)'. The subjects' affective responses included mention of 'hysteria', 'a bit harsh', and 'dark and evil'. One subject visualized 'an image of [a] water wave, moving faster and faster'. It is remarkable that on the first listening, a few subjects observed the water analogy as a metaphor, although no one specifically mentioned a river. At this point in the study, the subjects had no knowledge of granular synthesis, yet remarks such as 'clouds of sound' point to the perception of terminology often associated with this synthesis technique.

An observe-write protocol was segmented and categorized for each of the three case study subjects. Any punctuation written by the subject was replaced with a semicolon. The researcher further segmented the responses so that there would be only one label for each observation. No punctuation is applied to expressions that have been segmented by the researcher. For the purpose of this chapter, a complete observe-write protocol will be presented only for Theresa (Table 30.1). The other case-study subject transcripts are available from the author.

While listening to the work by Truax, Anthony's responses were equally divided into two categories: Cognitive-Correct (50 percent) and Affective-Aesthetic (50 percent). Justin's responses were more varied, arrayed across Cognitive-Incorrect (17 percent), Cognitive-Correct (33 percent), Affective-Emotion (25 percent), and Affective-Aesthetic (25 percent). He did not record any uncertainty in his observations (Cognitive-Inquiry), but incorrectly identified the composition as atonal. Theresa's responses are the most varied of all of the subjects, noting Cognitive-Incorrect (11 percent), Cognitive-Correct

Table 30.1 Observe-write protocol for Barry Truax, *Riverrun*

Theresa	Response	Categorization
Truax-1	sounds kind of like someone clicking their tongue or dripping water	Cognitive-Correct
Truax-2	definitely no tonal centre	Cognitive-Correct
Truax-3	now a new sound-synthesized flute?	Cognitive-Inquiry
Truax-4	now percussion	Cognitive-Incorrect
Truax-5	gradually increasing in intensity	Affective-Aesthetic
Truax-6	randomized	Cognitive-Correct
Truax-7	chaotic	Affective-Aesthetic
Truax-8	kind of unsettling	Affective-Emotion
Truax-9	algorithmic? Maybe	Cognitive-Inquiry

(33 percent), Cognitive-Inquiry (22 percent), Affective-Emotion (11 percent), and Affective-Aesthetic (22 percent). Her response that there was percussion was initially interpreted by the researcher as traditional percussion instruments that was subsequently validated by the subject and was thus labeled incorrect.

Prior to the second listening of *Riverrun*, subjects were provided with the title, composer, year of composition, duration, and the program notes. The program notes included a colour spectrogram of a time segment of *Riverrun*. Subjects were given a brief overview of what a spectrogram is and how to interpret one. The majority of subjects had previous experience interpreting spectrograms. Following the discussion of spectrograms, subjects were given a brief lecture on the technique of granular synthesis, with interactive examples of real-time granular synthesis implemented in Max/MSP.

During the second listening, subjects viewed a video projection of a succession of discrete time-stamped spectrograms presented in thirty-second intervals that coincided with the audio playback. After the second listening, the subjects completed Survey-3. Six of the eight subjects (75 percent) used the word 'water' or 'bubbles' while recalling their first listening, indicating a strong musical memory. Justin initially thought the water sounds were recorded acoustic samples but realized that they were generated digitally as the piece progressed. Several of the subjects described the form of the piece, recognizing the river analogy. Their remarks included 'the sound of water drops becoming faster and faster', 'digital water droplets and rain-like noise; gets louder and more busy', and 'popping bubbles; rushing water; rising and falling from and into chaos; some parts are overwhelming'. One subject said he enjoyed the 'water drop sound of the beginning; the sound is getting faster and faster, it imitates the flow very well', but did not extend this observation to the flow of a river. Theresa said the composition started out 'sounding like clicking tongues or water and continually grew more intense, with low bass tones and percussion', an observation that is similar to her initial impressions recorded in Survey-2. Justin noticed that there are 'different distinct scenes', but wondered, 'How are they related? Are they supposed to be related?' Justin also remarked that he had difficulty interpreting the spectrogram: 'What is creating the distinction of percussive sounds vs. pitch-based sounds? I couldn't tell the difference from the spectrogram.' The additional information about the composition and a second listening clearly sparked Justin's curiosity.

Only one of the subjects applied what was learned about granular synthesis to their aesthetic interpretation of the piece, remarking, 'It is somewhat pleasing when the grains reach a point of synthesis in which no unique grains can be perceived.' Anthony remarked that he now understands that 'the constant tones are made up of smaller parts which I now know is granular synthesis.' Two of the subjects reported finding the sustained tones aesthetically pleasing, referring to them as 'low bass sounds' or 'string-like sounds'. Anthony was again unsettled by what he refers to as 'white noise' and said that his favorite parts were 'when I can audibly discern the different noises and pitches—the beginning and the end'. Theresa also likes the beginning of the composition, saying, 'I think the beginning of the piece with the "droplets" is my favorite. It sounds similar to dripping water but not identical—almost too artificial to be just the sound of water; it's neat. It's also cool when a distinct pitch can be heard.' Justin's aesthetic orientation had a more abstract orientation, discussing 'the concept of the powerless droplet theory. [It] is compelling and watching this on the spectrogram is fascinating.' Justin also mentioned a specific time reference in the composition when he says, 'At 610, when the percussive sounds start to slow down to a point, you can hear pitches.'

But three of the subjects were confused by the introduction of pitched material. One stated that the 'definable pitches don't seem to fit into the piece.' Anthony observed that 'some parts did not fit into the theme, [notably the implied] harmonic progressions.' Theresa wondered, 'Why suddenly drop to a bass tone w/ percussion at 5–6 minutes? It doesn't really sound like running water anymore at that point.' Another subject was confused when his perception of pitch did not align with his interpretation of spectrogram, stating, 'I can see the frequencies align in the spectrogram but I can't perceive them.' One subject was confused by what she referred to as the 'clock sound [a perfect fifth] at 14:14'. And in a written stream of consciousness, Justin wonders, '[Are there] any overarching motives, musically? Should I even be listening for that?' Then he answers his own inquiry, bridging Affective-Inquiry and Affective-Emotion, by going on to say, 'Yes, increasing amplitude and density increases excitement—I get it. Is there anything more to this type of music?'

Many of the subjects described some sounds as jarring. The subject who mentioned the 'clock sound at 14:14' pencilled a sketch of the spectrogram and pointed to the place that she said it sounding 'jarring'. Two subjects described the high-pitched sounds either 'jarring' or 'static'. Anthony was unsettled by the harmonic progression, describing it as 'the least aesthetically pleasing part of the composition'. Justin remarked that 'the piece is interesting in a scientific way' and that he enjoyed 'understanding how sound works', but when it comes to aesthetics, he says, 'it did not do anything musically or emotionally for me' and that he did not understand 'the artistic idea'.

Only five of the eight subjects had questions they'd like to ask Barry Truax. Two subjects aimed their inquiry at compositional intent. One of these subjects had a very specific question: 'At about 17:20 to the end, why did you combine a constant low voice with a high voice together? What's the relationship of this part to the river?' And another subject wondered about symbolic representation in relation to compositional intent: 'Do all of the "grains" represent the droplets in the river?' Two subjects turned their attention to questions about granular synthesis technique, specifically, 'What sound clips (sources for granular synthesis) did you use to create this?' and another wondered 'Why not use natural sounds & sampling to determine "rain patterns"?' Justin had an interesting question for the composer, again bridging the cognitive and affective: 'When composing a piece of this abstract nature, is it even necessary to reconcile aesthetics with process?'

During Survey-4, Anthony recalled that there was a 'constant tone made up of smaller parts which [he now knows] is granular synthesis'. He repeated his observation that the 'sounds are like white noise' and that he found the piece had some 'interesting parts' but that he 'couldn't see how it fits into structure of piece'. Justin recalled the 'bubbles imagery' and his initial thought was that it was 'not digitally produced' and that he 'initially thought it was acoustic'. But as the piece progressed, he 'realized it was digital'. He felt the composer 'exhausted the same technique' which became 'repetitive' and that he did not grasp the 'artistic idea'. Theresa offered that she was not sure what she heard at the beginning but said she thought she recalled writing down it sounded like a 'tongue clicking'. She offered that 'when harmonic sounds same in, more pleasing to listen to'. She recalled that about five to six minutes into the composition, there was a 'wave of noise' and that the introduction of the 'bass pitches were better'.

30.9 *CANON FOR THREE EQUAL INSTRUMENTS* BY ELLIOTT CARTER

The second piece the subjects listened to was Elliott Carter's *Canon for Three Equal Instruments*. After a discourse analysis, a few subjects correctly identified the instrumentation as a brass ensemble, although others were confused by the use of mutes and thought perhaps woodwind instruments had been used. Surprisingly, three subjects (33 percent) correctly identified the piece as a canon. The majority of subjects (67 percent) identified the compositional process as algorithmic, two subjects (25 percent) used the term 'atonal', and two subjects (25 percent) observed that the algorithmic process was likely twelve-tone composition. The subjects had minimal affective responses to this composition, with one subject (13 percent) describing it as 'lots of dissonance'. The observe-write transcript for Theresa is given in Table 30.2.

Similar to his responses to Truax, Anthony's responses Carter are grouped as Cognitive-Correct (60 percent) with only one Affective-Aesthetic observation (20 percent). Anthony's inclusion of wind instruments while describing the instrumentation is interpreted as woodwinds and is Cognitive-Incorrect (20 percent). Justin's responses are equally divided between cognitive and affective: Cognitive-Correct (43 percent), Affective-Emotion (14 percent), and Affective-Aesthetic (29 percent). Justin's Cognitive-Inquiry (14 percent) is focused on whether or not this composition uses

Table 30.2 Observe–write transcript for Elliott Carter, *Canon for Three Equal Instruments*

Theresa	Response	Categorization
Carter-1	It's a brass group, I think	Cognitive-Inquiry
Carter-2	I don't recognize the piece, though	Cognitive-Correct
Carter-3	probably acoustic music?	Cognitive-Inquiry
Carter-4	sounded coherent enough	Affective-Aesthetic
Carter-5	not very random[2]	Cognitive-Correct

the twelve-tone technique. His aesthetic judgements are that the music does not move him but he enjoys listening to a nuanced performance using real instruments. Theresa makes two responses that are labelled as Cognitive-Correct (40 percent). Her responses classified as Cognitive-Inquiry (40 percent) are concerned with the instrumentation. Her only affective response (20 percent) was that the composition sounded 'coherent enough'.

Prior to the second listening, subjects were provided with the title, composer, year of composition, duration, the program notes, instrumentation, and the score. In addition, to these materials, subjects received a brief lecture with a handout, 'Introduction to Pitch Class Analysis'. In this handout, subjects were informed of the process of assigning integers to pitch-classes in an equal-tempered scale (e.g. C = 0, C#/Db = 1, etc.). Subjects were also introduced to the concept of octave equivalence (C3 = 0 and C4 = 0) and modular arithmetic (mod 12) used to complete a pitch-class analysis. The introduction of the concepts of pitch class, octave equivalence, and modular arithmetic was to facilitate further analysis of the *Canon*. Subjects were given an historical and musicological context for the dawn of serialism with references to the socio-political environment that characterized the end of the First World War. Additionally, subjects were given a brief tutorial on extreme chromaticism and how it contributed to the rise of serialism. The subjects assigned pitch classes for voice 1 of the *Canon*, notating the pitch classes on each of their scores. All of the subjects knew what a canon was, and some students quickly observed that voice 1 was a tone row and that voice 3 was the same row, in canon. Without prompting, the subjects notated the tone row for voice 3 on their score. When the subjects were queried about voice two, one subject quickly identified it as the tone row in inversion and transposed up a tritone. The subjects notated the pitch classes for voice 2 in their score. For the second listening, subjects followed along using their own score.

For Survey-3, two of the subjects recalled that they thought the piece was a canon and four remarked on the use of traditional acoustic instruments. Justin was happy to hear acoustic instruments, saying, 'Acoustic instruments, yeah!', but after that, he described this piece as the 'least memorable' in relation to the others. On the first listening, two of the subjects identified the piece as serial composition with one of these subjects stating, 'Canon, serial composition'. Oddly enough, on the second listening, none of the subjects used the word 'atonal' or 'dissonant' as was the case on the first listening. Two subjects recorded new observations about the rhythmic organization of the piece, one stating, 'The rhythmic overlaps are interesting, syncopated moments', and another saying, 'It seems to fit together a lot'. One of the subjects noted that what he found most aesthetically pleasing was that 'a lot of the intervals are harmonically consonant'. Two of the subjects enjoyed listening to acoustic instruments and appreciated the complexity of 'trying to make music out of something different'. Two subjects found the canon aesthetically pleasing to listen to, with one subject noting specifically the 'second voice in inversion'. Theresa seemed to engage aesthetically on an intellectual level, saying, 'The mathematical patterns are definitely the coolest part'. Most of subjects were confused by the composition, saying that the 'arrangement of the notes [in the row] seems random'. Others remarked about the absence of a melodic flow, saying that there are 'many leaps' or that 'some of the jumps don't sound good'. The absence of a tonal centre was unsettling to Anthony, who stated confusion by 'the way it never reaches a home in the scale' and called it 'a bit disorienting'. Justin asked, 'What's the point of serial composition?' When asked about what they found least aesthetically pleasing, there were three remarks about harmonic dissonance, stating that the composition was 'not really harmonically compatible'; another claimed that each line was not really a melody but simply followed the row. One subject took particular offence to voices 1 and 2 of the last bar of the composition (the score in notated in C), saying that the 'ending has a D and a D♭ at the same time which is quite dissonant'. Theresa simply put, 'The song doesn't sound that great without knowing the patterns used to create it'. The composition evoked the least number of questions for the composer (25 percent). This absence of cognitive or affective inquiry may be attributed to the subjects' cognitive engagement with the compositional process coupled with minimal affective response.

During Survey-4, all of the subjects reiterated that they were delighted to hear acoustic instruments. Anthony said, 'I liked that this piece used traditional instruments' and it 'sounds like twentieth-century music'. Justin said that the piece was the 'least memorable' of all of the compositions and that he was 'disappointed that by the time is had ended, it didn't go where he had expected'. Theresa exclaimed, 'Actual instruments that I recognize!', and she recalled that it sounded like 'random notes' but that she was 'familiar with the sounds'. She went on to say that she 'did not notice it was serial [on the first listening], it seems obvious now that it was'.

30.10 *ABANDONED LAKE IN MAINE* BY MARA HELMUTH

While completing Survey-2 during the first listening to *Abandoned Lake in Maine*, six subjects (67 percent) used the word 'birds' while recording their observations and one of these six subjects correctly identified the bird as a loon. Half of the subjects recognized the compositional genre as soundscape and two (25 percent) offered that the piece was acousmatic. One subject noted that the 'composer layered synthetic-sounding pitches on top of nature sounds', and another remarked that the composer was using 'digitally altered natural sounds'. One subject described the form as 'manually written' with 'parts filled in algorithmically'. Three subjects (33 percent) were unsettled by the introduction of the human voice, with two of these three subjects describing the sound as 'creepy'. Theresa's observe-write protocol is given in Table 30.3.

Helmuth's composition presents the greatest amount of cognitive certainty of any of the four compositions: the majority of responses were Cognitive-Correct for Anthony and Justin (68 percent) as well as Theresa (58 percent). It's also interesting to note that Theresa was the only subject among the three case study subjects to offer aesthetic judgements (25 percent).

Table 30.3 Observe-write protocol for Mara Helmuth, *Abandoned Lake in Maine*

Theresa	Response	Categorization
Helmuth-1	well, that screech at the beginning scared me	Affective-Emotion
Helmuth-2	I hear wind and birds	Cognitive-Correct
Helmuth-3	atmospheric	Affective-Aesthetic
Helmuth-4	creepy almost but in a good way	Affective-Aesthetic
Helmuth-5	crickets	Cognitive-Correct
Helmuth-6	more birds	Cognitive-Correct
Helmuth-7	definitely a loon call in there somewhere	Cognitive-Correct
Helmuth-8	so many birds	Affective-Aesthetic
Helmuth-9	now they're in reverb	Cognitive-Correct
Helmuth-10	now they're pitched down	Cognitive-Correct
Helmuth-11	now they're pitched up	Cognitive-Correct
Helmuth-12	oh, a human voice now?	Cognitive-Inquiry

Prior to the second listening of *Abandoned Lake in Maine*, subjects read the title of the composition, composer, year of composition, duration, and the program notes. The program notes include the quote spoken by the naturalist:

it is with a long cry in the still of the night,
that the loon authenticates the northern lake.
The cry is made with the neck stretched forward,
and it is a sound that seems to have come up a tube from an unimaginably deep source.

Following the second listening, the subjects responded to Survey-3 with remarkable recollection. The number of subjects who used the words 'bird' or 'birds' increased from six on the first listening to seven (88 percent) in the second listening. Theresa, who initially identified the bird as a loon, repeated this observation after the second listening. One subject who did not initially identify the bird as a loon later stated that he knew it was a loon. Two of the subjects, one of whom was Theresa, correctly identified the sound of the human voice in the first listening and repeated that observation in the second listening. The majority of subjects reinforced their understanding of soundscape composition by adding more detailed explanations in comparison to Survey-2. Four of the seven subjects who reported that they were confused by the naturalist's voice, also stated that they found the human voice the least aesthetically pleasing. One subject offered, 'The human voice was a bit frightening. After listening to it a second time, it sounded better than when I first heard it.' Although Anthony was also confused by the introduction of the human voice, he said that the part that was the least aesthetically pleasing was the 'loud collections of many of the different animal sounds'. Justin stated that the 'throbbing bass seems too machine-like for a soundscape'.

By this time, the subjects knew that a videoconference with Mara Helmuth had been scheduled for the next session, so that the last prompt, questions they'd ask the composer, should have special meaning. Oddly enough, two of the subjects (25 percent) could not think of a question to ask the composer, even though all subjects reported being confused by some aspect of the composition. Both Justin's and Theresa's questions could be classified as Affective-Inquiry: Justin wondered, 'Do you find this piece horrific? Did you have sinister visions while composing?', while Theresa inquired, 'Did the loon calls have a particular sound you were looking for or do you just really like loons?' One subject wanted to learn more about the type of algorithms used. Not surprisingly, two subjects were curious about the naturalist: 'What's the meaning of the naturalist's voice?', while another asked, 'What's the significance of what the naturalist is trying to say?'

During a thirty-minute interview with Mara Helmuth, subjects learned that the title *Abandoned Lake in Maine* was actually a recording of loons in Wisconsin. They all agreed with Helmuth that a title of *Abandoned Lake in Wisconsin* does not roll off the tongue like *Abandoned Lake in Maine*. During the interview, Theresa followed up on her question about how the composer felt about the sound of the loon and why that motivated her to create the piece. Since many subjects were confused by the introduction of the naturalist's voice, several questions were posed to the composer: 'What is the significance of what [the naturalist] is saying?' Helmuth cited the specific language about how the loon makes its sound and that it 'come[s] up a tube from an unimaginably deep source', and that her composition aimed to create this imagined sound. When one subject inquired about the unintelligibility of what the naturalist is saying, Helmuth described that she was trying to juxtapose the unnaturalness of humanity in a naturally pristine environment.

Prior to the videoconference, subjects were made aware that Helmuth had completed a thorough analysis of *Riverrun*, so they proceeded to ask several questions about Truax's approach to composition using granular synthesis in comparison to her own. Subjects learned that Helmuth's process in working with granular synthesis is iterative and at a smaller time scale than Truax. She explained that her compositional process involves generating hundreds of synthesized sounds that she then explores, freely discarding sounds that do not meet her aesthetic criteria. Subjects were not surprised by her approach, understanding that a multitude of aesthetic judgements would be necessary to create this piece. The subjects thoroughly engaged with the composer interview and appreciated Helmuth's comparative discussion of her compositional process in relation to that of *Riverrun*.

During Survey-4, Anthony thought that the piece accurately conveyed being in a wooded area, thus 'abandoned lake'. He recalled that he thought that the animals were very active and created an unrealistic amount of sound. Justin's reaction was one of extended memory and association. He remembered another soundscape piece that was created in one of his architecture courses four years before and was pleased to have expanded his understanding of soundscape composition. He described Helmuth's composition as meditative but said that the sections interrupted by the loon created a haunting and somewhat horrific effect. Theresa remembered that she realized the bird sounds were loons from the first listening. Since speaking with the composer, she understood that many of the sounds come from the same source material. She repeated that she thought the naturalist's voice coming in was weird.

30.11 *THRENODY FOR THE VICTIMS OF HIROSHIMA* BY KRZYSZTOF PENDERECKI

For Survey-2, the majority of subjects (56 percent) recognized that the performing medium for *Threnody for the Victims of Hiroshima* was orchestra, with several remarks on the nontraditional use of stringed instruments. One subject remarked, 'I appreciate the effectiveness [of the composer] because of the constraints of the musical resources.' Two of the subjects (25 percent) used the words 'horror' to describe the composition and two of the subjects used the word 'soundscape' during their observations. Table 30.4 shows Theresa's observe-write protocol.

Both Anthony and Justin correctly observed a number of objective attributes of *Threnody for the Victims of Hiroshima* (60 percent and 67 percent respectively). These observations are in sharp contrast to Theresa, who was not able to correctly identify any objective attributes (Cognitive-Correct = 0 percent), although she did offer an incorrect observation (Cognitive-Incorrect = 20 percent). This piece is the only one of the four where the three subjects did not express uncertainty or a question about the composition. *Threnody for the Victims of Hiroshima* had the highest percentage of affective responses (Affective-Emotion = 16 percent and Affective-Aesthetic = 32 percent) for a percentage of nearly half (48 percent) of the responses.

Table 30.4 Observe-write protocol for Krzysztof Penderecki, *Threnody for the Victims of Hiroshima*

Theresa	Response	Categorization
Penderecki-1	Ah yes, my favourite use of violins (not)	Affective-Emotion
Penderecki-2	that percussion noise sounded either like an actual drum group	Cognitive-Incorrect
Penderecki-3	or something falling down stairs	Affective-Aesthetic
Penderecki-4	this is like a horror movie soundtrack	Affective-Aesthetic
Penderecki-5	jump scares and all	Affective-Emotion

Prior to the second listening, subjects were provided with the title, composer, year of composition, duration, the program notes, and the instrumentation. Subjects engaged in a brief discussion on the history of the end of the Second World War, citing the bombings of the Japanese cities of Hiroshima and Nagasaki. Subjects were given a brief tutorial on graphic notation and the musical motivations that gave rise to the practice, as well as on extended performance techniques. Following this brief lecture, subjects were shown examples of graphic notation and discussed how it was similar to and different from traditional music notation. For the second listening of the Penderecki, subjects followed an animated score that is available on YouTube (Gerubach 2015).

On Survey-3, two of the subjects related the high-pitched strings at the beginning as a way to immediately invoke horror, describing the convergence and divergence of glissandi as 'interesting'. One subject described the piece as 'eerie suspense everywhere, chaos [and] shuffling'. This subject also noted the 'noise of an airplane'. Anthony remarked that this is 'an interesting atonal piece' in contrast to his initial impressions of 'dissonant', 'slow', and 'chaotic'. He went on to say that he did not remember much about the piece other than 'confusion'. Theresa recalled that she does not enjoy this type of string writing and recounted her observation of 'jump scares'. About midway through the piece, she realized that she'd heard the beginning of this piece before and wondered why she did not recognize it on the first listening. Justin offered new insights during his second listening, stating that 'time is very elastic' and there are 'repeated motives'. He further remarked that it would be 'incredibly difficult to coordinate this with an orchestra: Who takes what notes?'

One subject made connections between the performance techniques and the composer's intent: 'The high pitches show people's horror. Knocking on the cello and bass imitate people running. The violin imitates alarm sounds'. Another subject reported, 'You can hear the terror', and a related remark, 'I don't think any part was supposed to be pleasing in order to show the chaos and panic of the bombing'. Justin discussed the aesthetics of the score saying, 'The notation is beautiful'. He went on to say, 'The visual element was more compelling to me than the sound component; seeing the huge black tutti end bar was extremely powerful'. He further remarked that he thought the compositional process was 'arbitrary' and described his overall response to the piece as 'very critical'.

One subject reported that he was confused because he 'did not know that all of those sounds could come from a string orchestra'. Another subject was looking for a correlation between the events of Hiroshima and the macro formal organization of the composition: 'I'm not sure how this is supposed to relate to Hiroshima, section by section'. Theresa said, 'Nothing about it is confusing to me. I may not like the way it sounds, but I understand why it was written to sound that way.'

Oddly, when prompted about the least aesthetically pleasing attributes of the composition, the subjects used may of the same adjectives that they used when describing what was most aesthetically pleasing: 'horror' and 'terror' achieved by 'the opening shrieking of the high-pitched strings'. One subject speculated that 'The sound would be jarring and unpleasant for people who don't know the background'. Theresa reported that the piece made her physically 'cringe'.

If the subjects could ask Penderecki any question, most of the questions would be focused on rehearsal and performance techniques. One subject asks, 'How did you get the performers to understand how you wanted this piece to be played?' Justin wonders, 'How does each performance of this piece align with the vision you had while composing it? Did you imagine a sound, or is it left up to the performance?' Theresa inquired, 'Why did you decide to use such an atypical notation style? Couldn't [the piece] sound pretty different each time it's performed?' The subject who previously wondered about the macro formal structure of the composition asked, 'Can you explain how each part relates specifically to Hiroshima?' Anthony observed that 'anger is the predominant emotion' and went on to ask, 'Why so much sadness and sorrow?'

For Survey-4, Anthony's response was that he didn't 'remember much except confusion'. Justin's recollection was 'complete chaos' and went on to say that he 'really did not like the piece'. He defended his aesthetic judgement by wondering 'if any thought was put into the making of that piece, it seems arbitrary'. Theresa immediately recalled her physical reaction of 'cringing' and that she didn't 'really like string noises like that'. She said that the piece made her 'jump a few times' and it 'sounds like a horror movie. Not pleasant.' Oddly enough, none of the case study subjects correlated their strong affective responses during Survey-4 to properly decoding the composer's intent.

30.12 COMPARATIVE ANALYSIS

In a comparative analysis of the four composition using Survey-2, *Abandoned Lake in Maine* had the greatest cumulative average of Cognitive-Correct responses (63 percent) and *Riverrun* has the greatest cumulative average for Cognitive-Incorrect (11 percent). Overall, the subjects expressed the greatest Cognitive-Inquiry for the *Canon for Three Equal Instruments* (18 percent); specifically, they asked whether the composition was an example of twelve-tone acoustic music. These observations about Carter are in contrast with Penderecki, where there were no expressions of either cognitive or affective inquiry. *Threnody for the Victims of Hiroshima* had the greatest number of emotional and aesthetic affective responses (48 percent). The only composition that approached this affective response was *Riverrun* (45 percent).

Figure 30.3 presents the cumulative categorical responses by the three case study subjects for all four compositions using Survey-2. Clearly, Anthony has the highest percentage of Cognitive-Correct observations (58 percent) but he also has 0 occurrences of an Affective-Emotion response. Justin, who described himself as a 'visual thinker', recorded the most Affective-Emotion responses (24 percent) and the most Cognitive-Incorrect observations (9 percent) of any of the case study subjects. Yet all of the subjects made cognitive errors, with the maximum variance among the subjects being only 4 percent. Theresa exhibited the greatest Cognitive-Inquiry (16 percent)—approximately four times greater than her peers.

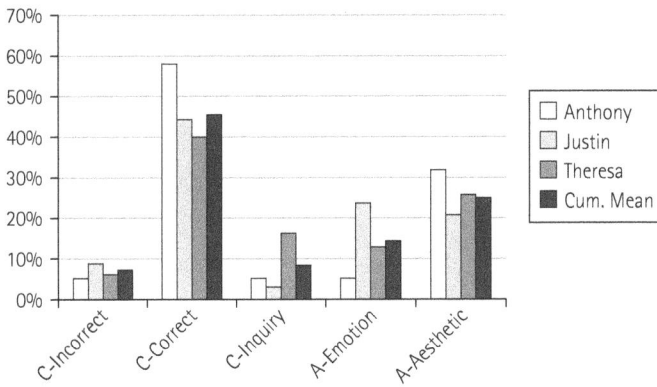

FIGURE 30.3 Survey-2 cumulative responses by subject.

The longitudinal analysis of cognitive-affective responses of each case study subject was aggregated across Survey-2, Survey-3, and Survey-4 for all four compositions. Anthony's cognitive responses were the greatest of all of the case study subjects on Survey-2 (68 percent) and the least for Survey-3 (12 percent). In contrast, his affective response was the greatest for Survey-3 (88 percent) and the least for Survey-2 (32 percent). In other words, the proportion of cognitive and affective responses flipped between Survey-2 and Survey-3. Anthony migrated towards a greater balance between cognitive and affective responses on Survey-4 (27 percent Cognitive; 73 percent Affective). As with Anthony, Justin recorded a large percentage of cognitive responses (63 percent) on Survey-2 and his cognitive responses continued to sharply decline as the study progressed: Survey-3 (19 percent) and Survey-4 (6 percent). Meanwhile, Justin's affective responses significantly increased during the study: Survey-2 (38 percent), Survey-3 (81 percent), and Survey-4 (94 percent). As Justin learned more about each piece over time, the more likely he was to offer an affective response. Like the other subjects, Theresa's cognitive responses were the greatest on Survey-2 (61 percent), declined during Survey-3 (31 percent), and restored cognitive-affective balance in Survey-4 (45 percent). Her cognitive trajectory during the study is similar to that of Anthony. Theresa's affective response was the greatest of any of the case study subjects for Survey- 2 (39 percent) and increased for Survey-3 (69 percent), but not as high as Anthony. Theresa's Survey-4 is the most balanced of all the case study subjects (Cognitive-45 percent and Affective-55 percent). Figure 30.4 shows the change in cognitive and affective responses of the three case subjects during the study. The line graph demonstrates the relative change in cognitive and affective response for Anthony and Theresa but masks the increasing affective and decreasing cognitive responses of Justin.

Prior to the completion of Survey-2, subjects were given very little to 'hold on to'. It's possible that the absence of assistance in encoding the composer's message inclined the subjects towards cognitive observation. As the subjects learned more about the compositions, it became easier to decode meaning that may have contributed to an increase in

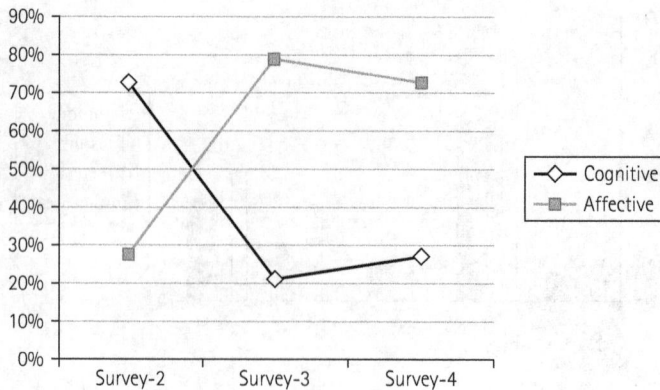

FIGURE 30.4 Cumulative average of cognitive and affective responses.

affective response for Survey-3. By the time Survey-4 was administered, musical memory of cognitive and affective responses became intermingled and migrated towards cognitive-affective balance.

All case study subjects reported that their favourite composition was *Abandoned Lake in Maine*. They recognized the sounds of the loon and the human voice, which gave them 'something to hold onto' as early as the first listening. One subject reported that the composer interview was critical in helping her understand the piece. Two of the case study subjects reported that their least favourite composition was *Threnody for the Victims of Hiroshima*. This composition had the strongest affective response of any of the compositions for both Survey-2 and Survey-4. The subjects simply did not enjoy hearing the chaos, anger, confusion, and profound human suffering communicated through this piece.

30.13 SUMMARY AND FUTURE RESEARCH

Using reception theory as a framework coupled with discourse analysis of an observe-write protocol provides a robust methodology for the longitudinal analysis of the cognitive-affective response to music, particularly algorithmic music. This study demonstrates that repeated interactions with algorithmic music increases a listener's ability to decode meaning, but that decoding the composer's intent does not necessarily lead to the aesthetic appreciation or acceptance of a work. The researcher observed a tendency towards affective response when subjects were provided with information to facilitate decoding; and in the absence of such information, subjects were inclined towards cognitive responses. Cognitive understanding increased in all subjects over the duration of the study but the increase did not necessarily lead to aesthetic appreciation. An unexpected finding is that the 'something to hold onto factor' may be applied to acoustic music since subjects identified musical attributes not only in electroacoustic music but also acoustic music. Even if the subjects have 'something to hold onto' cognitively or affectively, it does not necessarily translate into acceptance or aesthetic appreciation of a work. For future research, it may be possible to automate the discourse analysis without unnecessarily constraining audience vocabulary, and that should increase both the quantity and quality of research in audience reception.

NOTES

1. The Institutional Review Board has categorized this research as exempt.
2. 'Random' could have cognitive or affective meaning and is interpreted as cognitive in this context.

BIBLIOGRAPHY

Bartel, L. R. *A Study of the Cognitive-Affective Response to Music*. PhD dissertation, University of Illinois, Urbana-Champaign, 1988.

Beaumont, A. "Expectation and Interpretation in the Reception of New Music: A Case Study." In *Composition—Performance—Reception*, edited by W. Thomas, 93–104. Brookfield, VT: Ashgate, 1998.

Camurri, A., Mazzarino, B., Ricchetti, M., Timmers, R., and Volpe, G. 'Multimodal Analysis of Expressive Gestures in Music and Dance Performances.' In *Gesture-Based Communication in Human-Computer Interaction*, edited by A. Camurri and G. Volpe, 20–39. Berlin: Springer, 2003.

Carter, E. *Canon for 3 in Memoriam Igor Stravinsky*. New York: Associated Music Publishers, 1992. [Musical score.]

Carter, E. *Canon for Three Equal Instruments, In Memoriam Igor Stravinsky*. Capriccio Records. 1992. CD.

Collins, N. 'Musical Form and Algorithmic Composition'. *Contemporary Music Review* 28, no. 1 (2009): 103–14.

Cope, D. *Computers and Musical Style*. Madison, WI: A-R Editions, 1991.

Dillon, S., Adkins, B. Brown, A., and Hirche, K. 'Communities of Sound: Examining Meaningful Engagement with Generative Music Making and Virtual Ensembles'. *International Journal of Community Music* 1, no. 3 (2008): 357–74.

Eno, B. 'Generative Music'. *In Motion Magazine*, 8 June 1996. http://www.inmotionmagazine. com/eno1.html.

Gerubach. 'Threnody for the Victims of Hiroshima'. Animated score. Accessed 28 February 2015. https://www.youtube.com/watch?v=HilGthRhwP8.

Goldstein, P., and Machor, J. L., eds. *New Directions in American Reception Study*. New York: Oxford University Press, 2008.

Hall, S. 'Encoding, Decoding'. In *The Cultural Studies Reader*, edited by S. During, 90–103. London: Routledge, 1993.

Helmuth, M. *Abandoned Lake in Maine*. Centaur Records, 2007. CD.

Helmuth, M. *Abandoned Lake in Maine*. Program notes. Private communication, 15 February 2015.

Helmuth, M. 'Barry Truax's *Riverrun*'. In *Analytical Methods of Electroacoustic Music*, edited by M. Simoni, 187–238. New York: Routledge, 2006.

Jauss, H. R. *Toward an Aesthetic of Reception*. Translated by T. Bahti. Minneapolis: University of Minnesota Press, 1982.

Landy, L. 'The "Something to Hold on to Factor" in Timbral Composition'. *Contemporary Music Review* 10, no. 2 (1994): 49–60.

Landy, L. 'The Intention/Reception Project'. In *Analytical Methods of Electroacoustic Music*, edited by M. Simoni, 29–54. New York: Routledge, 2006.

Lillehaug, L. A. *Examining the Composer-Audience Gap in Contemporary Music*. PhD dissertation, Augustana College, Sioux Falls, SD, 1969.

Miranda, E. R. *Composing Music with Computers*. Oxford: Focal Press, 2001.

Penderecki, K. *Threnody for the Victims of Hiroshima*. Nonesuch Records, 2012. CD.

Philadelphia Orchestra. 'The Centennial Collection Program Notes. Historic Broadcasts and Recordings, 1917–1998'. Philadelphia, PA: Philadelphia Orchestra Association, 1999.

Schubert, E., Vincs, K., and Stevens, C. J. 'Identifying Regions of Good Agreement Among Responders in Engagement with a Piece of Live Dance'. *Empirical Studies of the Arts* 31, no. 1 (2013): 1–20.

Stevens, C., Glass, R., Schubert, E., Chen, J., and Winskel, H. 'Methods for Measuring Audience Reactions'. In *Proceedings of the Inaugural Instructional Conference on Music Communication Science*, edited by E. Schubert, K. Buckley, R. Eliott, B. Kobroff, J. Chen, and C. Stevens, 155–158. North Ryde, NSW: ARC Research Network in Human Communication Science, 2007.

Stowell, D., Robertson, A., Bryan-Kinns, N., and Plumbley, M. D. 'Evaluation of Live Human-Computer Music-Making: Quantitative and Qualitative Approaches'. *International Journal of Human-Computer Studies* 67 (2009): 960–975.

Truax, B. *Riverrun*. Cambridge Street Records, 1986. CD.

Truax, B. *Riverrun*. 1986. http://www.sfu.ca/~truax/river.html.

Turing, A. M. 'Computing Machinery and Intelligence'. *Mind* 59 (1950): 433–60.

Weale, R. 'Discovering How Accessible Electroacoustic Music Can Be: The Intention/ Reception Project'. *Organised Sound* 11, no. 2 (2006): 189–200.

Yang, S. C. 'Reconceptualizing Think-Aloud Methodology: Refining the Encoding and Categorizing Techniques via Contextualized Perspectives'. *Computers in Human Behavior* 19 (2003): 95–115.

CHAPTER 31

......

TECHNOLOGY, CREATIVITY, AND THE SOCIAL IN ALGORITHMIC MUSIC

......

CHRISTOPHER HAWORTH

31.1 INTRODUCTION

......

IT is probably an understatement to say that algorithmic music does not normally con-
jure an image of music as a social practice. Although endowed with a vast body of literature relative to its scale as a genre, the social ecologies that sustain it—the audiences or
publics that listen to and discuss algorithmic music; the industries that provide for its
production and dissemination; the social practices that are central to how algorithmic
music is learned, practised, and circulated; the cultures and politics of the algorithms
and computers as technologies—are rarely discussed. Iannis Xenakis's *Formalized
Music* (1992), possibly the canonical book on the subject, is characteristic in its disciplinary sweep: music theory, mathematics, computer science, and philosophy form the
core framework, while the social and historical determinants of algorithmic music are
evaded. This constellation of uneven forces reflects a certain self-understanding of algorithmic music that goes beyond discourse, participating in the aesthetics of the genre
itself. When, in the late 1990s, algorithmic composition made its way into popular electronic music, it was a machinist aestheticism that prevailed: in sound, accompanying
artwork, interviews, and promotional literature, nonrepresentational imagery was
favoured over pictures of the musicians, with artists favouring cryptic cyborgian moni-
kers over their real names. Ben Watson's admonishment of 'laptop cool' for sublating
'the contribution of human labour' into a Romantic aesthetics of the sublime therefore
captured some of the underlying rationality of laptop and algorithmic music;[1] however,
the critique is not new. Algorithmic music inherits an ethics and aesthetics that finds
its fullest expression in 'absolute music' and the idea that art and aesthetics might exist

'autonomously' from the political, social, and economic conditions of their production and experience.

Faced with these processes of social 'erasure', sociological analyses of art have traditionally advanced strongly constructivist approaches, showing how taste, experience, meaning, and genre are not transcendent universals but are accomplished through socio-historical processes of differentiation. In what, from the perspective of traditional art theory, might amount to a project of disenchantment, exposure, or revelation (Inglis 2005), 'art worlds' were shown by the sociologist Howard Becker (1984) to comprise the same mundane and ordinary assortments of people, institutions, and things as any area of the social world, and to be governed by the same hierarchical structures of power. Against the notion that art works are produced in spontaneous acts of unmediated creation, springing forth from singularly talented individuals, Becker pragmatically defined them as 'joint products of all the people who cooperate via an art world's characteristic conventions to bring works like that into existence' (1984, 34). Characteristic in this 'classic' sociology of art was the deferral of any engagement with the art object in its specificity. As Antoine Hennion (2003) notes, where art theorists and philosophers took creation, genius, and the works 'in themselves' as their object, sociologists in this first incarnation were more interested in how these categories came to appear as such. Questions concerning the ontology of art or analyses that incorporated the aesthetic judgements of the sociologist were left more or less intact as sociologists went about the business of 'filling in' the social-shaped gaps. Furthermore, the conception of creativity that inhered was a predominantly human-centred one based on social organization and the division of labour. In recovering the social fabric 'behind' the artifice, it illustrated how a coordinated network of people comes together to co-create the conditions for art to appear. As Becker writes, 'art worlds consist of all the *people* whose activities are necessary to the production of the characteristic worlds which that world, and perhaps others as well, define as art' (1984, 34, my italics). The special role technical artefacts, texts, material infrastructures, markets, and other nonhuman entities play in enabling human creativity thus went largely untheorized.

A recent 'postcritical' turn has, however, questioned the implicit presumptions upon which the earlier in sociology of art was based (Prior 2011, 121). Responding to challenges laid down by sociologists like Vera Zolberg and Janet Wolff, some theorists have turned their attention away from ambiguous forces like 'social context' and onto the 'missing' aesthetic object itself (Born 2010, 13). In Georgina Born's recent work, the art object is conceived as a condenser and mediator of the social relations entailed in its production. As she writes, 'all cultural production constructs and engages relations not only between persons, but also between persons and things, and it does so across both space and time' (13). In this view, the path the art-thing takes as it is released into the world, picked up, comprehended, canonized or rejected, going on to influence other works or not, constitutes both a necessary vector of meaning in art and an immanently social realm—one that neither aesthetics, in its disciplinary favouring of immanent, transhistorical meanings, nor the social sciences, in their deferral of any engagement with the art object, can overlook (Born 2005, 16).

And this shift can be seen as a subset of wider changes in the way the social has been conceived and studied. To put it in Bruno Latour's terms, it consists in a move from a 'sociology of the social' to a 'sociology of associations'; from various shades of social determinism to a theory of social mediation—one that takes seriously the contribution of 'nonhuman actors'. Many writers besides Born have taken this direction in cultural musicology, among them Tia DeNora (2000), who has analysed music as a 'technology of the self' in everyday life and the management of subjectivity; Antoine Hennion (2010), who has explored popular music and the mediation of taste; Benjamin Piekut (2011, 2014), whose work on 'actually existing' experimentalism has traced the path of intermediaries, cultural operators and even books in 1960s London and New York; and Nick Prior (2008), who (not far from the concerns of this chapter) has studied electronic music genres such as 'Glitch' and the way in which breakdown, error, and misuse ends up affording new and unforeseen uses for musicians. These writers' positions have, to varying extents, developed alongside and in critical dialogue with the body of literature associated with Actor-Network Theory (ANT), and so it is from here that I will depart in this chapter. An important theory of mediation,[2] ANT famously and controversially grants agency to 'things'—a maxim that effectively translates into a refusal to privilege human intention when analysing how social assemblages cohere and endure. Being presaged on the 'alien' agencies of algorithms, algorithmic music therefore represents a fertile ground for an ANT-style analysis. Indeed, it affords more than just analysis of creative practices, offering insights into the dynamic relationship between industries, technologies, and musical action, as well as into the ways in which the sometimes fragile assemblages of genre hold together.

Yet while providing a useful framework by which to analyse certain facets of algorithmic music, ANT also suffers from some important shortcomings. It fails to offer a means by which to understand how some actors can become more powerful than others and is equally blind to the role played by time and history in these processes (cf. Piekut 2014). For these reasons, ANT provides only a partial account of the social world it purports to analyse. It is for this reason that I turn to other theories of mediation later in the chapter.

What follows is an introduction to ANT followed by two case studies: first, the algorithmic music pioneers The Hub and, second, the live-coding community. The case studies serve to illuminate the affordances and constraints of ANT: both extending the analysis of these algorithmic music scenes by recourse to ANT and, through the analyses, illuminating some of ANT's limits.

31.2 ACTOR-NETWORK THEORY: A SOCIOLOGY OF MEDIATION

The first thing to say about ANT is that it does not represent a coherent body of theory or philosophical system, but is instead a *negative* methodology for studying the social world. Although it does come furnished with a bare set of conceptual tools for understanding social mediation (what Latour 1988 has called its 'infralanguage'), Latour deems it crucial, for 'scientific, political, and even moral reasons', that 'enquirers do not in advance, and in place of the actors, define what sorts of building blocks the social world is made of' (2005, 41). As such, ANT advances a potentially maddening scepticism towards any and all pre-made abstractions. 'Knowledge', 'science', 'languages', 'theories', 'the social', and 'capitalism' all come in for the charge of 'not existing' in the vast and dispersed literature on the subject (Law 2000), a move that is intended to puncture the false stability and staticism that they confer on the social world. In the ANT view, the world that actors inhabit does not lie there passively and continue into the future unchanged. Rather, it is actively 'performed', 'negotiated', and 'made' by those actors that support its continuing functioning.

Now, there is a sense in which this shift from the passive to the active and performative follows the pragmatic approach to art worlds advanced by the earlier-cited Howard Becker. Becker described 'systems of collective action' in which the individuals that are necessary to the world's functioning work collectively to produce things they call art (1976, 704). In a passage that could very nearly have come from the pen of an ANT theorist, Becker gives a list of actors necessary for a symphony orchestra concert to take place:

> For a symphony orchestra to give a concert, for instance, instruments must have been invented, manufactured, and maintained, a notation must have been devised and music composed using that notation, people must have learned to play the notated notes on the instruments, times and places for rehearsal must have been provided, ads for the concert must have been placed, publicity must have been arranged and tickets sold, and an audience capable of listening to and in some way understanding and responding to the performance must have been recruited. (Becker 1984, 3)

In Becker's account, 'music' is the accomplishment of a collective endeavour in which the people and things that are necessary to the functioning of that world come together to produce it. The art thing is therefore not circumscribed in advance, with some worlds worthy of consideration and others not; rather, it is that entity that results from this dynamic system of interactions. Indeed, Becker even includes nonhuman actors in his account: the invention, manufacture, and upkeep of instruments, and the necessity of a system of notation all enter into the mix, interacting with human actors to assemble and perform the social.

Differentiating this from an ANT type of analysis may at first seem like nitpicking, but it is important. Simply put, ANT recognizes the *two-way* exchange that is incurred when humans and nonhumans interact. Humans may delegate tasks and roles to objects, a feature that is recognized by Becker, but this is never the uncomplicated transfer of action it seems to be. An action is always mediated—*translated* in the ANT terminology—by the elements it comes into contact with, which means that the resulting effect cannot be wholly reduced to the person, or group of persons, that set it in motion. ANT therefore raises the status of objects from their common role—evident in Becker—as uncomplicated 'carriers' or 'transporters' of a determinate action. Instead they become *collaborators* and *mediators*; entities endowed with the capacity to change the course of events.

31.2.1 Skill, Agency, Creativity, Technology

By way of demonstration, let us consider Thor Magnusson's *Confessions of a Live Coder* (2011) in which he gives an autoethnographic account of his own learning processes when taking on a new programming language.[3] Magnusson is a computer musician whose favoured programming language is the object-oriented environment SuperCollider. In 2007 he began the process of learning a new environment, Andrew Sorenson's Impromptu, in order to answer two questions, practical and methodological: first, how would the shift from an 'object-oriented' to a 'functional' programming language influence his own computer music practice; and second, how is it possible to reflect upon and analyse the process of 'technological conditioning' that is incurred in this process in such a way as to provide shared insights for the community. Now, the view implicitly propagated by most computer music texts is that this process of skill acquisition would make no difference at all. The software implementation of the terse mathematical equations and signal flow diagrams that we find in periodicals like *Computer Music Journal* is usually left up to the individual user, suggesting strongly that, whether implemented in Supercollider, Csound, or something else, an FM synthesizer is an FM synthesizer: it is the signal-processing mathematics, computer processor, audio card, loudspeaker equipment, and room acoustics that count, because the software itself exerts no audible agency.[4] Yet as Magnusson notes, even 'secondary' aesthetic differences from environment to environment make a difference on the reader or writer of the code. Text inlining, syntax colourization, capitalization, font size, and special symbols all 'begin to condition how the artist thinks' (2011, 3). Getting even further from function, Magnusson finds that the very discursive and social culture one enters into when learning a new programming language does its own work, enhancing or suppressing one's engagement with it. The extent to which a community readily shares projects and code or is more secretive, helps new users or not, and the clarity and availability of documentation and help files, participates in the success or failure of the individual user learning to operate it.

Magnusson's account challenges the 'neutrality' thesis of technology by showing how both the functional and the aesthetic aspects of a programming language, as well as the discursive community in which it is enmeshed, influenced his music making. He finds that thinking in terms of the 'flow' and 'process' of the Impromptu language, as opposed to the 'objects', 'prototypes', 'properties', and 'methods' of the Supercollider one, changed the way he worked with melody in his music. 'I would write dynamic functions to populate lists with note values and recursively through other functions, empty those lists during playing, until they needed populating again. There was never a static entity one could denote as the piece's "melody" ' (Magnusson 2011, 5). Later, Magnusson discovers that the process affords him larger-scale insights into distinct communities of practice that centre on music software. 'SuperCollider users focus largely on synthesis, signal processing, and generative audio, Impromptu users operate more on the more traditional compositional level (*sic*)' (5).

The effects described here are clearly not reducible to the operator's own purposeful intention. Indeed, Magnusson actually quotes Latour in the article, imagining his acquisition of new software-related skills as giving birth to a 'new kind of hybrid, making fresh creative decisions' (Magnusson 2011, 3). In other words, SuperCollider and Impromptu became 'actors' in the ANT sense, The universe of functions and sound generators that they offered, the visual representations that they provided, the modes of interaction they afforded, the support communities that built around them, and the musician-programmer who navigated them all colluded and interacted to produce something unique—it would not be the same were it Max or Csound.

31.2.2. Nonhuman Agency

Now, to return to ANT, this 'something'—a distributed action set in motion across a network of people and things and irreducible to any single one—is what ANT understands as 'agency'. In what perhaps amounts to ANT's strongest ontological claim, agencies that do not correspond to actual effects are rejected entirely. If an actor is not producing socially available traces and information then, according to Latour, 'it is invisible and nothing can be said about it' (2005, 31). If, on the other hand, 'it is visible, then it is being performed and will then generate new and interesting data' (31). It should be clear that this bare methodological axiom—an insistence on *performance* as the minimal condition for agency—immediately reduces in priority and import the 'human-ness' or 'non-human-ness' of the respective agent. So long as it makes a difference, and the course of events would be significantly different were it removed, then it does not matter who or what an agent is presumed to be. To this end, ANT deliberately fosters uncertainty about the 'full' nature of an actor.[5] It is the means through which given assumptions about who or what can 'count' as an agent—a computer musician rather than the software she uses—can be left behind, and 'true' empirical inquiry can proceed.

But does this minimal notion of agency not leave ANT theorists in danger of representing a world in which people and things become interchangeable, with no distinguishing characteristics assigned to either? This remains the most common criticism of ANT—its seemingly amoral stance. But whilst the rhetoric can be overweening, this charge tends to be born of a misunderstanding. As Latour writes, granting agency to objects does not mean that these participants determine action, only that:

> there might exist many metaphysical shades between full causality and sheer inexistence. In addition to 'determining' and serving as a 'backdrop for human action', things might authorize, allow, afford, encourage, permit, suggest, influence, block, render possible, forbid, and so on. (Latour 2005, 71)

So whilst a nonhuman actor can actually be more powerful or significant than a human one, its agency is not to be understood as *isomorphic* with human agency. Indeed, as Sayes points out, ANT offers *no* general theory of agency at all; to reiterate, it aims to provide a negative methodology rather than a substantive theory (2014, 142).

Ultimately, the goal of an ANT analysis is to produce richer empirical investigation into the precise nature of the human-technical ensembles that manifest actions, without foreclosing their nature in advance.[6] Brought to bear on algorithmic music, this allows us to engage in a serious and thorough way with the mutable instruments, changing technological infrastructures, self-sustaining music systems, and laptop crashes that participate in and shape its social environment.

31.3 A HISTORICAL CASE STUDY: THE HUB

In the following, I bring an ANT analysis to bear on the practices of The Hub, pioneers of algorithmic music and the first computer-networking group.

Emerging out of the avant-garde music scene of the San Francisco Bay Area, The Hub is in many ways an archetypal product of the region's distinctive mix of high-tech research and bohemianism. The group embodies principles of antihierarchical organization and collectivism whilst at the same occupying a precarious space right at the vanguard of new technology adoption. Their name was conceived as a generic placeholder for a dynamic constellation of people, things and processes. It names at least three components: (1) the composer-performers associated with the project, including Scot Gresham Lancaster, Mark Trayle, John Bischoff, Chris Brown, and Phil Stone; (2) the hardware and software that they used; and (3) the practice of generating shared information which underlay their work. Clearly this managed uncertainty between people, things, and processes, all drawn together by a concept of 'network', bears more than a passing resemblance to ANT, yet, commensurate with the ideas of the time, it was the conceptual armature of cybernetics and information theory that informed The Hub's practice.[7]

Gresham-Lancaster associates the very origins of the group with a technological and economic development: the advent of MIDI. MIDI had 'a major impact, enabling often-impoverished performers/composers to utilize these new, affordable instruments' (1998, 41). In the early Hub performances, the group utilized a blackboard system for sharing data between the distributed computers. A central memory space housed the active components of the piece, which each computer was able to access remotely. This determined the style of communication between computers, and hence, the form of their interactions. One-to-one communication was not possible; instead, all contributed to, and drew from, a shared data resource.

A paradigmatic example of this period in The Hub's history is the piece *Perry Mason in East Germany*. In it, each of the six members of The Hub runs a program that constitutes a self-sustaining musical process, but which is able to send out and receive variables from the memory source in order to control one another's programs. As Graham-Lancaster notes (1998, 42), these were completely asynchronous interactions. The lack of a shared clock led the group in the direction of a more procedural approach, sharing the tradition of Cage and Tudor.

When OpCode Systems released their Studio 5 MIDI interface, the group opted to redesign The Hub around this new system. Each participant in the network could now directly 'play' the set-up of any other participant, which had not been possible previously. The new Hub was a decentralized peer-to-peer network, which granted more autonomy to each player and also more direct interaction among them. *Waxlips*, a piece composed by Tim Perkis, is considered the canonical work of this period. Here, the pre-written algorithms of *Perry Mason in East Germany* are gone as the network interaction is reduced to its most fundamental and basic form so as to allow the emergent structure to be revealed more clearly.

Gresham-Lancaster notes the precariousness of this dynamic media ecology. Utilizing 'the new possibilities the changing technological context brought to the work' whilst also maintaining a repertoire of works is depicted as a fragile balancing act, with 'the shifting context of hardware and software constantly (updating) the sound of the ensemble' (Gresham-Lancaster 1998, 43). In this sense, The Hub dramatizes the essential 'problem' that ANT tries to solve: that is, how to understand innovation and organization without resorting to accounts that portray either technological development or society as the primary drivers of change, cancelling out the respective other. Each new innovation is typically presaged with a change in hardware or software that, in most instances, radically transforms the way the members conceptualize their compositions and organize their interactions. A hierarchical client/server architecture, where all interaction is mediated by a central data resource, is replaced by a 'flat' peer-to-peer network that allows direct intercommunication, the latter having direct and irreversible effects on the sound. However, not all of the system updates The Hub implement take hold. Both Matthew Wright (2005) and Gresham-Lancaster have independently written of a failed attempt to create a Hub based on Open Sound Control (OSC) to perform over the Internet. Here, the problem was twofold: both that the new OSC-based system was so complex that the group was 'unable to reach a satisfactory point of expressivity', and that the wider network of the Internet required different strategies and aesthetics than The Hub's creative methods afforded. Rejecting the update led to a reinforced sense of who The Hub 'is': a computer network music group with the 'form and function of a conventional musical ensemble' (Gresham-Lancaster 1998, 44).

31.3.1 The Problem of Agency in Technologically Mediated Music Making

We start to see now why Latour and the ANT theorists object to reified categories. What Gresham-Lancaster's account displays very concretely is the sheer dynamism, hybridity, and, at times, instability, of the ensemble of players, software and hardware systems, telecommunication protocols, and other entities that the moniker 'The Hub' forecloses. When the group moved from the blackboard system to the MIDI hub, the previous repertoire was left more or less obsolete and an entirely new set of material had to be produced based on a different model of interaction. At danger of overemphasizing the notion of nonhuman agency, we might compare The Hub's technological revisions to the cycles of change and renewal in line-up that rock bands can undergo whilst maintaining the same moniker—to paraphrase Latour, 'change an element in the network and you change the actor' (Latour et al. 2012, 593). However, without an analysis of the specific agencies that assemble and supervise the new network we end up with the rather banal observation that every element in the chain produces effects: a kinect sensor is different from a computer keyboard, which is in turn different from a MIDI keyboard, and so on. These immediate mediations are important in providing a materialist account of creativity, but the risk inherent in ANT and related approaches is that they are taken to comprise the entire nexus of possible mediations. Absent from the analysis are the larger-scale commercial and political dynamics that sustain the ecology of electronic music making, but whose logics of change and development are dictated by markets, technical standards, and other nonmusical agencies.

Gresham-Lancsaster's account portrays a constant negotiation between two poles of mediation—as Agostino Di Scipio (1995, 37) puts it, formulating the problem in question form: '[H]ow can I use the available existing task-environment to realize my own idea of composition?' or 'How can I design the tools that are necessary to realize my own idea of composition?' No doubt all musical practice falls somewhere between these two poles, rather than at one or the other, but it is clear that, in the pre-Hub days of the League of Automatic Music Composers, the group slide towards a largely self-maintained paradigm, whereby bespoke self-authored tools are produced to realize their own idea of composition. '[E]ach new piece conform[ed] to a uniquely designed software/hardware configuration'. However, with the adoption of Opcode systems' Studio 5 interface they moved towards the use of an 'existing task environment', enjoying the 'simplicity and clarity' that the changing technological context brought to the work (Gresham-Lancaster 1998, 40–43), but sacrificing agency if this interface was changed or discontinued. The negotiation was therefore between an infinitely reconfigurable set of techniques devised and maintained by the artists themselves, and the standardization of techniques in technical systems whose preservation and development is 'autonomous' (as in, not commensurate with the immediate creative goals of The Hub).

To further probe this 'problem of agency', and the challenges it poses to analysis, I want to turn to a controversy that briefly surrounded the music notation software Sibelius. In 2012, the software's community of users rose up against the Avid technology company in the wake of the closure of the company's UK offices. Fearing the discontinuation of the software they knew and loved, they petitioned the company to sell it back to the two developers that originally wrote the program. 'Sibelius is far more than just code, it lives and breathes in the hearts and minds of its inventors and developers. Remove them, and Sibelius eventually becomes roadkill', they wrote (Williams 2015). From the standpoint of the original authors of Sibelius, or the community of users that speak in their name, the software's transformation beyond its original intention and eventual decline was an obvious failure. However, looked at without interest, from the largely managerial perspective advanced by Latour and ANT, what we have is simply a case of the network 'growing' in directions that exceed the will of the developers and user base.[8] As new, stronger actors—the Avid Company—who pursue independent interests, are enrolled within it, the network drifts. What the signees of the 'Sell Sibelius' cause were petitioning for, then, was a form of technological democracy, where the communities that the changes will affect have a say in the systems that they rely upon.

This mediation of creative agency by autonomous technical systems raises the question of who governs, and whose interests govern, technological change. Those musicians and artists with the time and knowledge to resist profit-motivated disruptions in the technological ecology of digital music making, as in the Sibelius case, may wish to maintain older software and operating systems or build their own systems using Open Source software, but it is more often the case that electronic musicians absorb these disruptions into their practices—or do a mixture of both (as with The Hub). An ANT analysis can disassemble these larger economic and political dynamics to uncover the complex chains of agencies that collaborate to make a concrete difference in music making, but one has to ask whether it is really desirable to perform this operation on every grouping or abstraction we encounter—must we account for the countless mediators that contribute to, for example, class, race, Korg, the ECM label, and so on? Georgina Born's theory of social mediation takes off from the opposite starting point. Rather than see the social world as flat, as in ANT, she posits different 'orders' or 'scales' of mediation—scales that are nonexclusive, and that interpenetrate, but that nevertheless have a positivity denied by ANT. She writes:

> The first order equates to the practice turn: here music produces its own socialities in performance, in musical ensembles, in the musical division of labour, in listening. Second, music animates imagined communities, aggregating its listeners into virtual collectivities or publics based on musical and other identifications. Third, music mediates wider social relations, from the most abstract to the most intimate: music's embodiment of stratified and hierarchical social relations, of the structures of class, race, nation, gender and sexuality, and of the competitive accumulation of legitimacy, authority and social prestige. Fourth, music is bound up in the large-scale social, cultural, economic and political forces that provide for its production, reproduction or transformation, whether elite or religious patronage, mercantile or industrial capitalism, public and subsidized cultural institutions, or late capitalism's multipolar cultural economy forces the analysis of which demands the resources of social theory, from Marx and Weber, through Foucault and Bourdieu, to contemporary analysts of the political economy, institutional structures and globalized circulation of music. (Born 2010, 232)

To start from the assumption that there are scales of mediation necessarily means sacrificing some of the rich analytical detail that ANT can afford; yet, at the same time, it also acts as a panacea to the kind of indiscriminate empiricism that can result from keeping track of every human and nonhuman mediator. Either way, it is clearly Born's fourth order of mediation that lends explanatory power to the case under discussion— the large-scale social, cultural, economic, and political forces that provide for music's production, reproduction and transformation. In the next study, I develop the analysis of algorithmic music by reference to Born's theory.

31.4 A CONTEMPORARY ACTOR NETWORK: LIVE CODING

This second section of this chapter makes a substantive and methodological leap forward in time, considering algorithmic music in the context of the present day. Responding to the earlier-cited criticisms of the sociology of art and the constructivist project of demystification and exposure relative to the social, I consider in this section how algorithmic music's own methods and aesthetics might be employed to analyse it. Using algorithmic digital methods designed for online ethnography, the analysis aims to occupy the same meshwork of human and nonhuman actors as the subject itself.

Live coding is an interesting and complex social form. Usually defined by reference to practice, sociality, and technique rather than any coherent musical style, writers generally agree that it constitutes the activity of writing, listening to, and modifying a computer program in realtime before an audience. Lifting the curtain on the 'hidden' instrumentality of advanced computer music—the embodied activity of writing algorithms, auditioning materials, and moving around code—live coding purports to disclose computer music practice in its elemental state. Indeed, rawness, primitivism, and the associated qualities of 'danger' and 'closeness to the source' are often cultivated as an aesthetic, via the projection of the screen to the audience and the deliberate imposition of performative constraints. Echoing the 'truth to materials' principle of modernist architecture—form follows function, and ornament is crime—Collins, McLean, Rohrhuber, and Ward write:

With commercial tools for laptop performance like Ableton Live and Radial now readily available, those suspicious of the fixed interfaces and design decisions of such software turn to the customisable computer language. ... [We] do not wish to be restricted by existing instrumental practice, but to make a true computer music that exalts the position of the programming language, that exults in the act of programming as an expressive force for music closer to the potential of the machine—live coding experiments with written communication and the programming mind-set to find new musical transformations in the sweep of code. (Collins et al. 2003, 322)

This 'true' computer music is far from being a technological determination, however. If it were, then, as Collins and colleagues wryly acknowledge, live coding concerts would entail the performer building a driver or DSP engine from scratch in the back of a venue over a number of nights, 'before finally emerging with a perfect heartfelt bleep on Sunday evening' (Collins et al. 2003, 321). Instead, authentic live computer music is a mutable concept, one that enrolls technical devices (the use of text-based programming languages over readymade graphical interfaces), social expectations (the insistence on openness and transparency over secrecy and opacity), politics (the use of Open Source tools over black-boxed commercial software), and ontology (the insistence on 'liveness', sometimes enforced by starting from a blank screen) in an open-ended negotiation.

31.4.1 Charting the Development of the TOPLAP Manifesto

We see the 'authentic computer music' rhetoric most clearly in the infamous 'ManifestoDraft' that the TOPLAP organization has featured on its website since its initiation.[9] TOPLAP was founded in the Changing Grammars symposium in Hamburg 2004, by a group of practitioners and students which included Alex McLean, Nick Collins, Julian Rohrhuber, Renate Weiser, Alberto de Campo, Adrian Ward, and Fredrik Olofsson, in order to explore and promote the new artform (Ward et al. 2005).

Alongside the manifesto, it hosts concerts, events, pedagogical resources, videos, academic papers, and other related items. When the site debuted, 'ManifestoDraft' was the first item a visitor to the site would encounter. It outlined the conceptual, performative, technological, and philosophical conditions live coders should meet or engage with, performing the dual function of materializing and speculatively positing an idea of authentic live computer music in the form of ten short commandments (see Figure 31.1).

Within the space of a year, the manifesto draft had stabilized to the form it assumes presently (see Figure 31.2). Looking at the development of the manifesto, what we see is a shift from the explicit designation of materials ('no predefined sequences'), programming languages ('languages approved by TOPLAP'), and software ethics ('sole use of Open Source software tools'), to a more strongly worded and ironic, yet less prescriptive, specification of what live coding is. A product of this latter development is the shift in emphasis from 'code' to 'algorithms': a pluralizing move, perhaps, in the sense that it does not explicitly prohibit graphical programming environments and 'live patching' as performance methods, but also an important conceptual shift from the materiality of code to the quasi-immateriality of the algorithms. 'Algorithms are thoughts,' they write, not 'tools': a Cartesian assertion that posits the writing of algorithms as being 'closer' to the abstract musical idea than the use of tools. Transparency is the enduring demand in the manifesto, though, appearing three times ('code should be seen as well as heard', 'obscurantism is dangerous', 'give us access to the performer's mind'). Alongside the taboo on the use of 'backup' material, the emphasis on programming algorithms, and the mention of manual dexterity and the glorification of the typing interface, these elements coalesce to create an image of an idiomatic computer music, one that is 'live' in the performative sense, and 'realtime' in the computing one. In this, it conveys an

ManifestoDraft

HomePage | RecentChanges | Preferences

/very much a work in progress - please fix!/

- Code is music
- Code is meta music (?)
- But Code is not only music. Live coding is not only about live music.
- Code should be seen as well as heard

  ```
  - But it is not necessary for an audience to understand the code to appreciate it,
  much as it is not necessary to know how to play guitar in order to appreciate
  watching a guitar performance.

  - On the other hand, live coding may or may not involve an impressive display of
  manual dexterity. Or even mental dexterity.
  ```

- Programs are instruments that can change themselves
- The program is to be transcended - Language is the way.
- Live coding is not about tools. Algorithms are thoughts. Chainsaws are tools. That's why algorithms are sometimes harder to notice than chainsaws.
- Typing is an ideal human-computer interface for coding
- Live coding is a future of computer music
- TOPLAP live coding requires:

  ```
  1. No predefined sequences {comment: this is not possible due to the referencial
  character of languages}

  2. Sole use of languages approved by TOPLAP {this is absolute must.} (Comment: there
  could be thousands.)

  3. Live interaction through the act of programming

  4. Sole use of Open Source software tools (Comment: Hmm, this one might seem a bit
  prescriptive and might exclude some projects that can't be done another way. And
  does "tools" refer to language a project is written in or the project itself?)
  ```

HomePage | RecentChanges | Preferences
Edit text of this page | View other revisions
Last edited February 25, 2004 7:55 GMT (diff)
Search:

FIGURE 31.1   TOPLAP ManifestoDraft (April 2004).

ontological politics of live computer music (Born 2013; Mol 1999), one that is positioned against two dominant tendencies in electroacoustic and computer music: one, electroacoustic art music, where fixed-media music is played back in concert halls over loudspeakers; and two, the club-based laptop performance of the early 2000s, where audiences watched performers from behind their laptop screens, and the performativity of the spectacle was largely taken on faith.

Looking at the TOPLAP site today, in 2015 (Figure 31.3), it is clear that the identity of live coding no longer hinges on the political manifesto. Slipping from the homepage to a subpage, its relegation indicates that the field has stabilized, its clauses having been either

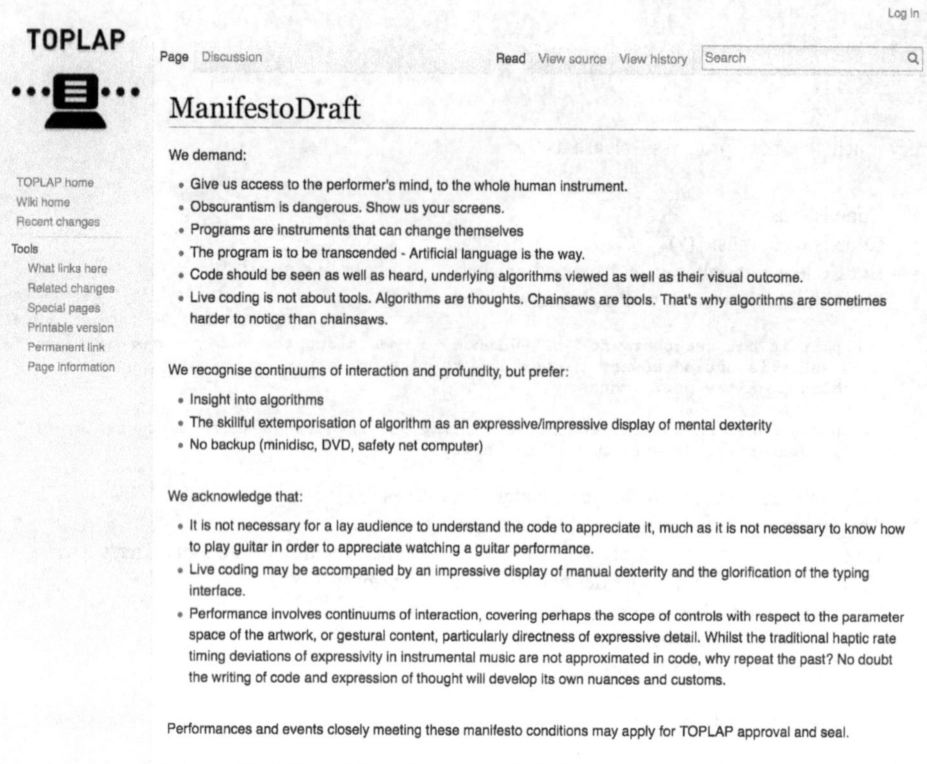

TOPLAP

Page   Discussion   Read   View source   View history   Search   Q

ManifestoDraft

TOPLAP home
Wiki home
Recent changes

Tools
  What links here
  Related changes
  Special pages
  Printable version
  Permanent link
  Page information

We demand:

- Give us access to the performer's mind, to the whole human instrument.
- Obscurantism is dangerous. Show us your screens.
- Programs are instruments that can change themselves
- The program is to be transcended - Artificial language is the way.
- Code should be seen as well as heard, underlying algorithms viewed as well as their visual outcome.
- Live coding is not about tools. Algorithms are thoughts. Chainsaws are tools. That's why algorithms are sometimes harder to notice than chainsaws.

We recognise continuums of interaction and profundity, but prefer:

- Insight into algorithms
- The skillful extemporisation of algorithm as an expressive/impressive display of mental dexterity
- No backup (minidisc, DVD, safety net computer)

We acknowledge that:

- It is not necessary for a lay audience to understand the code to appreciate it, much as it is not necessary to know how to play guitar in order to appreciate watching a guitar performance.
- Live coding may be accompanied by an impressive display of manual dexterity and the glorification of the typing interface.
- Performance involves continuums of interaction, covering perhaps the scope of controls with respect to the parameter space of the artwork, or gestural content, particularly directness of expressive detail. Whilst the traditional haptic rate timing deviations of expressivity in instrumental music are not approximated in code, why repeat the past? No doubt the writing of code and expression of thought will develop its own nuances and customs.

Performances and events closely meeting these manifesto conditions may apply for TOPLAP approval and seal.

FIGURE 31.2   TOPLAP ManifestoDraft (as of 14 November 2010).

absorbed ('show us your screens'), developed upon (the emergence of audiovisual coding languages built especially for live coding), or in some cases overturned, as in the now-prevalent use of external controllers and live instrumentation within the scene. These developments become even clearer later.

31.4.2 Art-Pop Uncertainties

Viewed from the perspective of ethnographic practice, one of the most interesting things about the live coding community is its propensity for self-documentation. Alongside the manifesto, the scene is fastidiously documented, with films about live coding, screen captures of performances, and pedagogical resources all very easy to access. Most of all, live coding is enshrined in dozens of exegetical texts elaborating upon its own practice and theory. These developments were consolidated in 2015, when the first-ever International Conference on Live Coding was held at University of Leeds—an initiative that is set to continue on an annual basis.[10] Often written by the practitioners themselves, this literature is strikingly interdisciplinary, offering perspectives from computer science, software studies, performance studies, philosophy, pedagogical research, and

FIGURE 31.3 TOPLAP.org homepage as seen in April 2015.

computational creativity. More marginally, writers have looked at live coding from the perspective of embodiment and autoethnography.[11] How does one study a community when the community studies itself! As Born and others have noted, theoreticism can provide an important index of a scene's experimentalism and avant-gardism. It comes to play an increasingly significant role in modernism, with books and articles taking 'on the ambiguous role of exegesis and criticism, of proselytizing and publicity, of both expounding and legitimating practice' (Born 1995, 42). In a recent survey article, the live coder Thor Magnusson seems to follow this thread when he roots the art form's beginnings in postwar avant-gardism. It is 'inevitable', he writes, that live coding draws from modernist practices, because formal experiments—linked here to modernism and avant-gardism—are a 'necessary aspect of the exploration of a new medium' (2014, 9). Magnusson quotes approvingly the art critic Clement Greenberg, whose version of modernism had content 'dissolved so completely into form that the work of art ... cannot be reduced in whole or in part to anything not itself' (9). But he conveys a narrative of hybridization and diversification beyond the self-referentialism of formalist modernism as the form develops. Once naturalized, the new medium evolves into a much more diverse set of practices, and the historical circumstances of its birth (such as the manifesto) are internalized or forgotten. Indeed, this diversity is alluded to in the article's title, 'Herding Cats': 'Live coding does not have a particular unified aesthetic in terms of musical or visual style', Magnusson asserts (8).

Magnusson's account can be considered an instance of what the musicologist Eric Drott has dubbed the 'decline of genre' thesis; the narrative that, during modernism, the categories that had once shaped the production, circulation, and reception of Western art declined in relevance, as the vanguard heroically rejected tradition and convention in a wave of aesthetic renewal (Drott 2013). By emphasizing these qualities of theoreticism, formalism, and the lack of any kind of aesthetical coherence over other ones, Magnusson aligns live coding with art music and the avant-garde, despite the fact that, in most of the artists he surveys, a clearly audible dialogue with popular forms of electronic music is being conducted: namely, electronic dance music, glitch, and noise. Now, at first blush this can be seen as a simple outcome of the precedence afforded to technological and theoretical issues over musical ones. Musicality is not really discussed at all in the article, a tendency not unknown to highly technologized musics (cf. Waters 2007). But it is also an outcome of the modernist propensity, if not to directly *oppose* popular musics, then to suppress their influence and instead to root the genre's origins in the aesthetic and technological developments of the neo-avant-garde. Nick Prior came to a similar conclusion in respect to glitch music, noting that:

> In most cases, glitch's support writers are themselves directly involved in the unfolding of the style, and their interventions are either internalist in content—fulfilling aesthetic, formalist or stylistic criteria—or posit glitch as somehow outside the field through the maintenance of a cool distance from pop. (Prior 2008, 307)

But an important subset of live coding, documented in Nick Collins's and Alex McLean's work (Collins and McLean 2014), is the format of the 'algorave'. Referencing the famous 'Anti EP' by Autechre, where the duo engaged with the then-pending Criminal Justice Bill designed to criminalize raves,[12] the article defines Algorave as being 'made from "sounds wholly or predominantly characterized by the emission of a succession of repetitive conditionals"' (McLean 2015). Audible in the live coding of Norah Lorway, Sick Lincoln, Canute, Alex McLean, and Benoît and the Mandlebrots is the undeniable influence of electro, ambient, trance, techno, IDM, electronica, and other electronic dance music subgenres. Indeed, live coders often practise an ironic refusal of the hegemony and prestige of art music, as in Nick Collins's work (under the pseudonym Click Nilson) *Acousmatic Anonymous*, a text-score piece that features an 'acousmatic' who must 'only omit high art for the remainder of the performance' (Nilson 2013). All the same, the ironic attacks on modernist art that take place within the confines of a club or performance venue tend to dissolve into due deference in the more sober context of the peer-reviewed academic article. Writing on electronic dance music, the same author wrote that 'musicians on the fringes of dance music soon enough looked backward to discover the great history of experimental electronic music and automatically merged to become part of that progression (even had they not looked, they could not have helped the latter)' (Collins 2009, 339). Here and elsewhere (Emmerson 2001), popular electronic music's indebtedness to the European avant-garde is emphasized over other equally salient influences, such as its relationship to African American music and the gay subcultures of 1980s (Taylor 2014, 67).

31.4 ACTOR-NETWORK THEORY 2.0

As already noted, live coding is a furiously active scene online. Its web practices extend beyond the usual techniques of publicity, network building, documentation, and promotion to the social, technical, and performative aspects of the scene itself. For example, Charlie Roberts's audiovisual live coding environment, Gibber, runs in a regular Internet browser, facilitating advanced creative coding online; whilst the network music axis of live coding—inheritor of the practices of the earlier discussed The Hub—involves whole performances being carried out online. Code, sounds, images are passed back and forth over the network, as listeners tune in via their own home connections. This is far from an instrumental use of the web; rather, the web enters into live coding's distributed instrumentarium, becoming a medium in its own right.

   A useful social sciences tool for studying aspects of these myriad online socialities is the Issuecrawler (Rogers 2002). Developed in the Department of Science and Technology Dynamics at the University of Amsterdam, the Issuecrawler is a webcrawler tool for visualizing networks using a technique called 'co-link analysis'. The project has links to Latour and ANT,[13] and was specifically designed to help with the problem of 'controversy mapping' online. Given a science and technology controversy, for instance 'government mandate on childhood vaccinations', the issue network would display who (or whose website) in government, business, and civil society is linking to whom, therefore affording insights into how the debate is being framed by key actors. Obviously there are important flaws with such a method. Not all the powerful actors in a given issue are represented by a website, and the Internet in general is an unstable and incomplete resource when viewed as an archive of social associations. Moreover, historical actors become dead links, meaning that the method is heavily biased towards present-day issue networks. Nevertheless, applied to music, the Issuecrawler represents a useful tool for conceiving of genres as social assemblages, thereby releasing the inquirer from the somewhat maddening task—identified by Magnusson—of trying to distinguish and classify genres by reference to a stable set of stylistic features.

Similarly to ANT, the Issuecrawler method is fluid and incomplete by definition. A given actor ('Alberto de Campo', say, or the SuperCollider programming language) may appear in any number of other 'networks' (electroacoustic music, 'extreme' computer music, live coding), an ontological premise that, when applied to genre, means that the longer the list of actors, the more the genre emerges in its distinctiveness. (In other words, more complexity produces greater differentiation.) Furthermore, the method is reversible. True to ANT, any actor is also conceived as a network, so just as de Campo participates in the live coding network, live coding participates in *his* network. It would appear alongside the university he works at, the school he went to, the friends and associates he works with, and so on.[14]

Turning to the results (Figure 31.4 and Table 31.1), the clearest finding is live coding's heterogeneous array of human and nonhuman actors. In a sense, this is a simple outcome of the method. A website can represent a person, an event, an animal, a building, and so on—co-link analysis simply follows the ANT method in making no a priori distinctions between them. However, it is the mix of artists and programming languages—Supercollider, Max, Chuck, and so on—that is distinctive. It illustrates the fact that, in the live coding scene, the instrumentarium represents an extension of the human, the two inseparable. Within the many technological actors that appear we find an interesting mix of free and open source and proprietary software (F/OSS). Alongside Supercollider and Chuck, and the alternative copyright licensing organization Creative Commons, Cycling 74 and Arduino both feature; the latter two suggesting that a diversified politics of software has emerged since the early emphasis on 'code'. And although software like Ableton Live and Reason do not feature, the prominence of Arduino is evidence of hybridization beyond the 'glorification of the typing interface' identified in the manifesto.

Importantly, there are very few record labels, distributors, or record stores represented on the Issuecrawler map, an intriguing omission given the centrality of the independent label within popular electronic music genres. There are at least two reasons for this. First, and most obviously, live coding is centred on performance. As already discussed, it practises a virulent ontological politics of live computer music. So although many of the artists produce physical commodities—sometimes in unconventional formats, where the code is shared with the listener and made available for further hacking and recombination—it is clear from the prominence of festivals and events that the scene is oriented towards, and based around, the live event.

The second reason for the lack of commercial outlets on the map relates to how live coders subsidize their activities. As is clear from the dominance of academic institutions, research groups, conferences, and funding bodies, live coding largely takes place within and in relation to institutionalized sites of music production, with most if not all of the practitioners that feature holding some kind of university affiliation. Many are early career researchers on fixed-term practice-based, practice-led, and 'research-creation' projects; some are doctoral students; whilst others hold positions in music

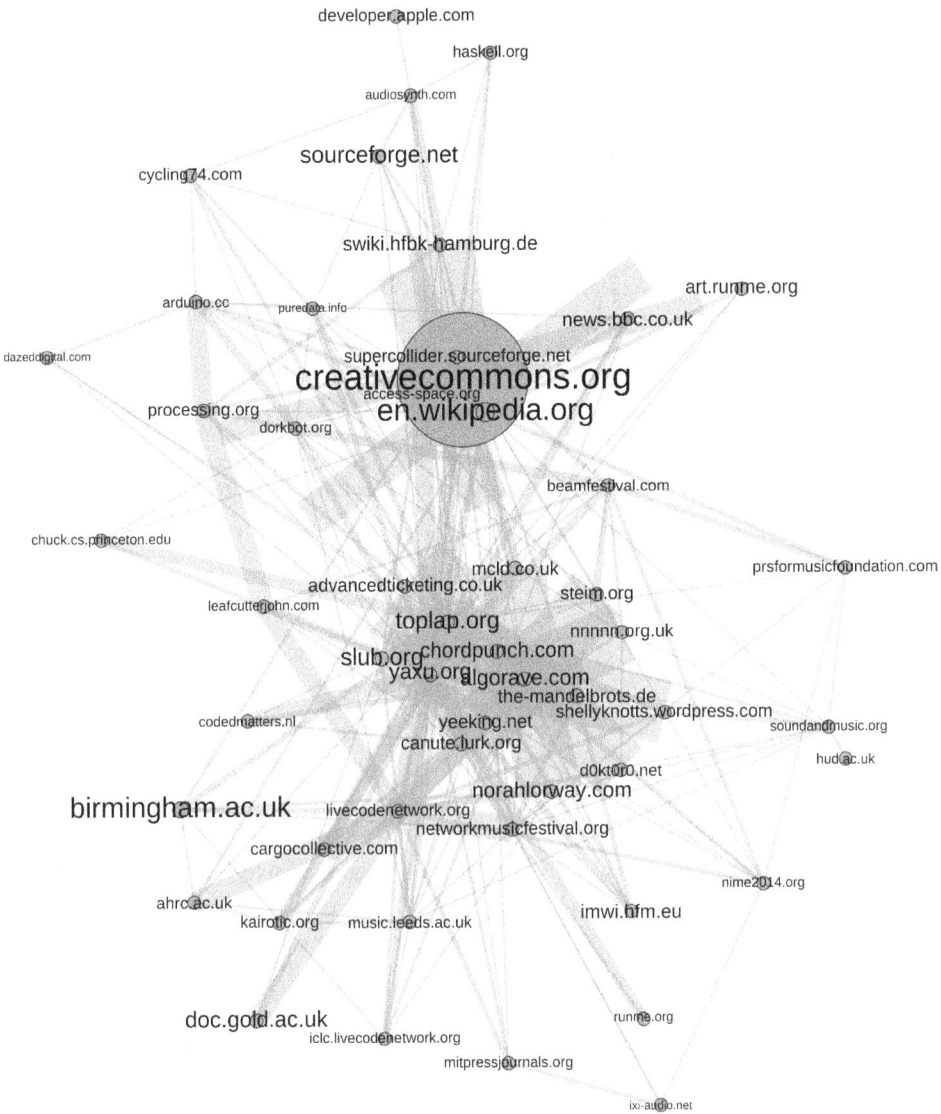

FIGURE 31.4   IssueCrawler map of live coding.

and computing departments. Although live coding draws heavily on popular styles, the strong presence of institutions of higher education makes for a distinct and complicated geography. At the centre of the map we see dynamic patterns of enthusiastic interlinkage amongst the artists, nonprofit organizations, commercial festivals, and F/OSS software communities, each actor participating, via the medium of the hyperlink, in the mutual exchange and accumulation of validation and recognition. But at the edges of the map are the institutions. Affording performance spaces, technological resources, and financial support, they are essential actors in this ecology, yet they do not reciprocally link. This outlier status is illustrative of the somewhat sober public faces institutions of higher

Table 31.1 List of live coding actors by category

Artists	Alex McLean (Yaxu), Leafcutter John, Matthew Yee-King, Shelley Knotts, Norah Lorway, Benoît and the Mandlebrots, David Ogborn (d0kt0r0), Canute, Alexandra Cárdenas, Slub, Ryan Jordan, Dan Stowell (MCLD)
Software and hardware	Cycling 74, Supercollider, Chuck, Haskell, Processing, Arduino, Xcode, Sonic PI
Festivals and events	Noise = noise, dorkbot, Beam Festival, Network Music Festival, live.code.festival, Algorave
Funding bodies	AHRC, PRS for Music Foundation, Sound and Music
Academic conferences	NIME 2014, ICMC 2011, International Conference on Live Coding
Academic institutions	Goldsmiths (EAVI research group), University of Leeds, University of Huddersfield, University of Birmingham, University of Sussex
Nonacademic institutions and organizations	STEIM, Access Space, TOPLAP, Coded Matters, Creative Commons
Labels and publishing	Chordpunch, runme.org (repository for software art)
Literature	Live coding article, BBC News feature, Wikipedia article on live coding, *Computer Music Journal* special issue on live coding
Miscellaneous	Advanced ticketing, Weaving project

education present to the world, yet it may also be indicative of the burgeoning nature of live coding's institutionalization. A similar map produced five years later may tell a different story.

This backdrop of academic and nonacademic institutions gives an institutional context to live coding's delicate negotiation of art and popular electronic music histories. Being subsidized by arts and engineering grants that support such initiatives as interdisciplinarity, science in the arts, code literacy and pedagogy, and innovation with digital technologies, the earlier-cited emphasis on novelty and formal experimentation (to the detriment of questions of musical style and genre) emerges as an institutional and economic mediation as much as a performative genealogy. Here, again, we see Born's fourth order of social mediation at work: the large-scale social, cultural, economic, and political forces that provide for music's production, reproduction, or transformation are reproduced in discourse. We could argue that live coding's ontological politics oscillates between two levels, then: one, an explicit politics of technology and performance, where the black-boxed, obscurantist laptop music of the early part of the twenty-first century is directly challenged; and two, a more subtle politics of art and popular, where the hegemony of the former can be satirized and lampooned, but not *too* loudly. For live coding's effective institutionalization is ultimately contingent upon its being bracketed—historically, theoretically, and aesthetically—within those very same genealogies.

31.5 CONCLUSION

This chapter has offered a social perspective on algorithmic music. Drawing on theories of mediation, I have argued for an approach to algorithmic music's socialities that doesn't attempt a project of demystification and exposure relative to the social but that instead installs itself within the very ecology of these fields. To that end, I have used Actor-Network Theory to study the contribution of 'nonhuman actors' to the social world, via a case study of the network music pioneers The Hub. The example of The Hub drew forward the question of technological change, and the necessity of theorizing these external forces as part of technologized music's social ecology. Through this project, we discovered weaknesses in the ANT approach, to do with power and hierarchy, which led us to turn to Georgina Born's theory of musical mediation and the hierarchical notion of distinct 'orders' of social mediation. The second study centred on live coding. Using digital methods, I charted the development of the TOPLAP manifesto in order to illustrate how, far from being a technological determination, 'true' computer music was an ongoing social negotiation that continues to the present. The final section used the Issuecrawler software to analyse networks of association within live coding online. I argue that Born's fourth order of social mediation—the large-scale social, cultural, economic, and political forces that provide for music's production—bears strongly on the aesthetic and conceptual terrain of live coding, particularly in regard to the scene's careful negotiation of art and popular electronic musics.

NOTES

1. 'In accordance with this anti-labour aesthetic, the typical laptopper releases recordings of ice floes, radio interference or earthquakes. Laptop cool is about avoiding "the turgid, complex, actual, dirty 'thing'"—i.e. earning a living under capitalism—and instead losing oneself in the contemplation of unsullied nature. This is actually no more advanced in ideological terms than hanging a framed reproduction of a painting of a glade of silver birches on the wall of an urban living room' (Watson 2006, 8).
2. Even though it is there in the name, writers associated with ANT tend to eschew the term 'theory'. Properly speaking, ANT is a theory about how to study the social; as such, it is closer to ethnomethodology. It is the idea that ANT can be 'applied' so as to understand a given social phenomenon that the protagonists reject.
3. Autoethnography reverses ethnography's typical focus on another group's culture by focusing instead on the ethnographer's own subjective experience of her interaction with that culture.
4. This is also the ideology behind software companies' promises that their products are transparent conduits of the individual ideas of their users, a characteristic example of which can be found in Richard Boulanger's introduction to *The Csound Book*, where he writes, 'in the software synthesis world of Csound, there are no such limitations. In fact, the only limitations are the size of your hard disk, the amount of RAM in your PC, the speed of your CPU—and of course, the limits of your imagination' (Boulanger 2000, xxxvii).

5. '[T]he human-nonhuman pair does not refer us to a distribution of the beings of the pluriverse, but to an uncertainty, to a profound doubt about the nature of action, to a whole gamut of positions regarding the trials that make it possible to define an actor' (Latour 2004, 73).

6. Piekut 2014, 193.

7. Scot Gresham-Lancaster's statement that 'music is, at its core, a means of communication. Computers offer ways of enhancing interconnection' (1998, 39) shows the influences of writers like Gregory Bateson and Norbert Weiner on the musical thinking of The Hub.

8. Summarizing ANT, Feenberg draws on H. G. Wells's version of the myth of the 'sorceror's apprentice', where two early bioengineers invent a miracle food that causes animals and plants to grow to eight times their normal size. 'Sloppy experiments conducted on a farm near London result in the birth of giant wasps, rats, and even people. ... In Latour's terms, the delegation of the original program to sacks, walls, and guardians broke down as rats got at the food, and the network was unexpectedly prolonged (in its syntagmatic dimension) through its nonhuman rather than its human members. Of course from the standpoint of the preexisting experimental program the network was supposed to serve, this amounts to chaos, but if one views the matter objectively, i.e. not from the standpoint of the two scientists and their failed strategy, the network can be seen to grow. And this makes it possible for new actors to pursue new programs' (Feenberg 1999, 116).

9. In 2004, the TOPLAP acronym—a play on 'laptop'—was published on the web as standing for '(Temporary|Transnational|Terrestrial) Organisation for the (Promotion|Proliferation |Permanence) of Live (Audio|Art|Artistic) Programming'.

10. http://www.livecodenetwork.org/iclc2015/.

11. Here I am paraphrasing Adorno's definition of art from *Aesthetic Theory*: 'The defintion of art is art is at every point indicated by what art was, but it is legitimated only by what art became with respect to what it wants to and, perhaps can, become' (Adorno 2004, 3).

12. In 1994, John Major's Conservative government introduced the Criminal Justice and Public Order Act 1994, a sweeping bill that included within its many clauses a direct attack on the free party movement. Section 63 effectively gave police the powers to remove 'a gathering on land in the open air of 20 or more persons ... at which amplified music is played'. It included a clarificatory subclause referencing 'sounds wholly or predominantly characterized by the emission of a succession of repetitive beats.' Autechre's Anti-EP satirized the pending bill, bearing a black sticker on the front that read: 'Warning. Lost and Djarum contain repetitive beats. We advise you not to play these tracks if the Criminal Justice Bill becomes law. Flutter has been programmed in such a way that no bars contain identical beats and can therefore be played under the proposed new law. However, we advise DJs to have a lawyer and a musicologist present at all times to confirm the non-repetitive nature of the music in the event of police harassment' (Pattison 2014).

13. https://web.archive.org/web/20150310090045/; http://www.mappingcontroversies.net/.

14. Locating an issue network requires a list of starting URLs; key actors that together provide an overview of the issue at hand. Given this list ('seeds'), the Issuecrawler will crawl through the associated webpages and store in a database ('harvest') any hyperlinks that direct the user to another destination on the web ('outlinks'). The software then analyses the outlinks and stores only those that appear two or more times in the results ('co-link analysis'). The latter two stages of the analysis can be repeated for 'deeper' crawls; in this case, outlinks from the first set of results would also be harvested and a second co-link analysis would be performed on them, a process that dramatically increases the size of the

harvest. The process can be completed up to three times. The results are then plotted in a 2D network displaying inlink and outlink patterns amongst the key nodes (webpages), with the *x–y* position of the nodes on the map indicating their relatedness, i.e. how frequently links are exchanged between them. Node size corresponds either to the number of inlinks the associated site receives or to a mixture of inlinks received and outlinks made. In network analysis jargon, these two features are seen to represent the amount of 'authority' and 'knowledge' respectively a node contains. In other words, a node that receives a great number of inlinks is deemed an authoritative source of information, whereas one that makes a lot of outlinks is deemed to know where the 'debate' is happening (provided they appear in the network in the first place, receiving inlinks themselves). Further analysis is afforded by the domain name suffix associated with a website, with different colours being assigned to different namespaces (.org, .net, .com, and so on)

BIBLIOGRAPHY

Adorno, T. W. *Aesthetic Theory*. London: A. & C. Black, 2004.

Banks, M., and Morphy, H., eds. *Rethinking Visual Anthropology*. New edition. New Haven, CT: Yale University Press, 1999.

Becker, H. S. *Art Worlds*. Berkeley: University of California Press, 1984.

Born, G. *Rationalizing Culture: IRCAM, Boulez, and the Institutionalization of the Musical Avant-Garde*. Berkeley: University of California Press, 1995.

Born, G. *Western Music and Its Others: Difference, Representation and Appropriation in Music*. Berkeley: University of California Press, 2000.

Born, G. 'On Musical Mediation: Ontology, Technology and Creativity'. *Twentieth-Century Music* 2, no. 1 (2005): 7–36.

Born, G. 'The Social and the Aesthetic: For a Post-Bourdieuian Theory of Cultural Production'. *Cultural Sociology* 4, no. 2 (2010): 171–208.

Born, G. ' "Drifting, Merging and Bifurcating": Institutional and Ontological Politics in Digital Art Musics in the UK'. Paper presented at the conference Music, Digitisation, Mediation: Towards Interdisciplinary Music Studies, University of Oxford, Oxford, 11–13 July 2013.

Boulanger, R. C. *The Csound Book: Perspectives in Software Synthesis, Sound Design, Signal Processing, and Programming*. Cambridge, MA: MIT Press, 2000.

Collins, N. 'Electronica'. In *The Oxford Handbook of Computer Music*, edited by R. T. Dean, 334–353. Oxford: Oxford University Press, 2009.

Collins, N., and McLean, A. 'Algorave: A Survey of the History, Aesthetics and Technology of Live Performance of Algorithmic Electronic Dance Music'. In *Proceedings of the International Conference on New Interfaces for Musical Expression*, 355–358. Goldsmiths, University of London, 2014. http://www.nime.org/proceedings/2014/nime2014_426.pdf.

Collins, N., McLean, A., Rohrhuber, J., and Ward, A. 'Live Coding in Laptop Performance'. *Organised sound* 8, no. 3 (2003): 321–330.

DeNora, T. *Music in Everyday Life*. Cambridge: Cambridge University Press, 2000.

Di Scipio, A. 'Inseparable Models of Materials and of Musical Design in Electroacoustic and Computer Music'. *Journal of New Music Research* 24, no. 1 (1995): 34–50.

Drott, E. 'The End(s) of Genre'. *Journal of Music Theory* 57 (2013): 1–45.

Emmerson, S. 'From Dance! To "Dance": Distance and Digits'. *Computer Music Journal* 25, no. 1 (2001): 13–20.

Feenberg, A. *Questioning Technology*. Abingdon: Routledge, 1999.

Gresham-Lancaster, S. 'The Aesthetics and History of the Hub: The Effects of Changing Technology on Network Computer Music'. *Leonardo Music Journal* 8 (1998): 39–44.

Hennion, A. 'Music and Mediation: Toward a New Sociology of Music'. In *The Cultural Study of Music: A Critical Introduction*, edited by M. Clayton, T. Herbert, and R. Middleton, 80–91. London: Routledge, 2003.

Hennion, A. 'Loving Music: From a Sociology of Mediation to a Pragmatics of Taste / Gustos musicales: De una sociología de la mediación a una pragmática del gusto'. *Comunicar* 17, no. 34 (2010): 25.

Inglis, D. 'The Sociology of Art: Between Cynicism and Reflexivity'. In *The Sociology of Art: Ways of Seeing*, edited by D. Inglis and J. Hughson, 98–111. Basingstoke and New York: Palgrave Macmillan, 2005.

Law, J. *The Actor Network Resource*. April 2000. http://www.lancaster.ac.uk/fass/centres/css/ant/ant.htm.

Latour, B. 'The Enlightenment without the Critique: A Word on Michel Serres' Philosophy'. In *Contemporary French Philosophy*, edited by A. P. Griffiths, 83–98. Cambridge: Cambridge University Press, 1998.

Latour, B. *Reassembling the Social*. Oxford: Oxford University Press, 2005.

Latour, B. *Politics of Nature: How to Bring the Sciences into Democracy*. Cambridge, MA: Harvard University Press, 2004.

Latour, B., Jensen, P., Venturini, T., Grauwin, S., and Boullier, D. ' "The Whole Is Always Smaller than Its Parts": A Digital Test of Gabriel Tardes' Monads'. *British Journal of Sociology* 63, no. 4 (2012): 590–615.

Magnusson, T. 'Confessions of a Live Coder'. In *Proceedings of International Computer Music Conference*, 609–616. University of Huddersfield, 2011.

Magnusson, T. 'Herding Cats: Observing Live Coding in the Wild'. *Computer Music Journal* 38, no. 1 (2014): 8–16.

McLean, A. *Algorave*. www.algorave.com. Accessed 30 June 2017.

Mol, A. 'Ontological Politics: A Word and Some Questions'. In *Actor Network Theory and After*, edited by J. Law and J. Hassard, 74–89. Hoboken, NJ: Wiley, 1999.

Nilson, C. 'Acousmatics Anonymous', For Beer, Op-us 11000001. *Composer Programmer*, February 2013. http://composerprogrammer.com/music/ForBEER.pdf.

Pattison, L. 'How the Political Warning of Autechre's Anti EP Made It a Warp Records Classic'. *VICE Magazine*, 21 July 2014. https://thump.vice.com/en_us/article/how-the-political-warning-of-autechres-anti-ep-made-it-a-warp-records-classic.

Piekut, B. *Experimentalism Otherwise: The New York Avant-garde and Its Limits*. Berkeley: University of California Press, 2011.

Piekut, B. 'Actor-Networks in Music History: Clarifications and Critiques'. *Twentieth-Century Music* 11, no. 2 (2014): 191–215.

Prior, N. 'Putting a Glitch in the Field: Bourdieu, Actor Network Theory and Contemporary Music'. *Cultural Sociology* 2, no. 1 (2008): 301–319.

Prior, N. 'Critique and Renewal in the Sociology of Music: Bourdieu and Beyond'. *Cultural Sociology* 5, no. 1 (2011): 121–138.

Rogers, R. 'The Issue Crawler: The Makings of Live Social Science on the Web'. *EASST Review* 21, nos. 3–4 (2002): 8–11.

Sayes, E. "Actor-Network Theory and Methodology: Just What Does It Mean to Say that Nonhumans Have Agency?' *Social Studies of Science* 44, no. 1 (2014): 134–149.

Taylor, T. D. *Strange Sounds: Music, Technology and Culture*. London: Routledge, 2014.

Ward, A., Rohrhuber, J., Olofsson, F., McLean, A., Griffiths, D., Collins, C., and Alexander, A. 'Live Algorithm Programming and a Temporary Organisation for its Promotion'. In *read_ me - Software Art and Cultures*, edited by O. Goriunova and A. Shulgin, 242–261. University of Aarhus, 2005.

Waters, S. 'Performance Ecosystems: Ecological Approaches to Musical Interaction'. In *EMS: The 'Languages' of Electroacoustic Music: Proceedings of Electroacoustic Music Studies Network Conference*. Leicester, 2007. http://www.ems-network.org/IMG/pdf_ WatersEMS07.pdf.

Watson, B. 'Technology as Style, Product as Negation: the Aesthetics of the Esemplasm'. Keynote speech given to Sonorities Festival of Contemporary Music in Belfast, 29 April 2006. http://www.militantesthetix.co.uk/polemix/belfast.htm.

Williams, D. 'Sell Sibelius!' Change.org Petition. 2012. https://www.change.org/p/chris- gahagan-senior-vice-president-of-products-sell-sibelius?utm_campaign=petition_cre- ated_email&utm_medium=email&utm_source=guides. Accessed 30 June 2017.

Wright, M. 'Open Sound Control: An Enabling Technology for Musical Networking'. *Organised Sound* 10, no. 3 (2005): 193–200.

Xenakis, I. *Formalized Music: Thought and Mathematics in Composition*. 2nd ed. Hillsdale, NJ: Pendragon, 1992.

CHAPTER 32

....

ALGORITHMS AND COMPUTATION IN MUSIC EDUCATION

....

ANDREW BROWN

32.1 INTRODUCTION

MUSIC is often described as 'organized sound', and improving students' understanding about that organization, as part of developing their musicianship, is a typical educational goal. An algorithmic description of musical processes can contribute to such development by requiring students to externalize and formalize their understanding. Programming and playing back the results of those algorithms provide rapid feedback, allowing reflection on and refinement of ideas. Further, the design of musical algorithms serves both to demonstrate understanding and to provide a conduit for creativity.

This chapter explores ways in which algorithms and coding skills can be useful intellectual tools that assist in the development of musical intelligence and computational thinking. There are clear lessons for computer science educators here too; however, the emphasis in this chapter is on music education, given that elsewhere the role of music, and audiovisual media more generally, in enhancing the study of computer programming is quite well covered. The interested reader is directed to works such as diSessa (2000), Guzdial and Ericson (2010), Manaris and Brown (2014), Papert (1980), and Resnick (1994).

Music education is concerned with all aspects of music, including: listening, composition, performance, analysis, critique, recording, distribution, and cultural awareness. Algorithms and their computational implementation have the potential to be applied to many of these areas of study, from models of music perception, through analytical techniques of empirical musicology and music information retrieval, to computer-assisted composition and interactive performance systems.

Taube (2012) distinguishes between three levels of computational representations of music. First, the *acoustic* level is identified, where sound waves, synthesis, and physical properties of sound, space, and instrumentation are described and manipulated.

Second, the *performance* level involves score interpretation and physical sound control gestures. Third, the *compositional* level is concerned with structural organization and musical elements and events. At the second and third levels, stylistic, cultural, and equipmental conventions and constraints must be accounted for. At the third level, music is often represented as symbolic notation and structural relations. Taube emphasizes the significance of abstraction between these levels and the general role of abstraction in digital representations of music that are subject to computational algorithms. Describing abstractions as 'metalevels', he writes, 'the metalevel representation[s] are active representations of the processes, methods, algorithms, and techniques that a composer develops to craft the sounds of his or her compositions' (Taube 2004, 3).

Abstractions, and their intellectual potential, have been clearly articulated as key to computer science education at least since Papert (1980). In their landmark programming text, Abelson and Sussman begin by emphasizing that programming is about 'building abstractions and procedures' (1996, 1).

The understanding of musical patterns as abstractions of cultural practices (e.g. musical form) or as characteristics of cross-cultural sonic expression (e.g. auditory streaming) has been a mainstay of music education—although patterns (e.g. arpeggios) or processes (e.g. serialism) are not typically framed as algorithms, but more commonly as theories, models, or techniques. Nevertheless, the connections are not difficult to see. Composers were using algorithms well before the advent of computers to assist in exploring musical space to generate novel material and ideas (Nierhaus 2010), and with the advent of computing this has only expanded, from Xenakis to Eno. There is much to be gained by training musicians in the applications of algorithmic process, not only to help them understand existing musical processes, but also to prepare them to push the boundaries of musical possibility and help transform musical experiences. These are important objectives for music education (Dillon 2007; Jorgensen 2003).

Just like music educational practices in general, the application of algorithmic processes are often focussed on music creation and presentation. This will become evident in the examples presented later in the chapter. However, algorithmic representation and design emphasize a particular analytical engagement with music. As the music educator Keith Swanwick points out; 'drawing attention to certain features of music is inevitably a form of analysis. . . . It is simply a way of picking out patterns from an overall impression, for instance by focussing on such things as melodic development, harmony or instrumentation' (1994, 12–13). The description of music in algorithmic terms is an effective way of focussing attention on the organization of music.

This chapter explores the implications of an algorithmic approach to music learning. It then discusses a number of the challenges and opportunities that such an approach presents to existing educational practices. Finally, a series of examples reveals the variety of approaches that innovative educators have employed to introduce algorithmic music as part of their programs.

32.2 AUTOMATION AND AGENCY

Ever since Pythagoras there have been links between music and mathematics. As a result, the systematic description of processes that lead to sonic and musical structures is deeply embedded in our culture. The algorithmic description of these processes has long been applied to technological music making in the form of instrument design and to various mechanical music devices such as the music box and player piano (Collins, this volume, chapter 4; Levenson 1994). Computational descriptions of algorithmic processes are just the most recent, and also most powerful, application of this link between music and mathematics.

Programmability makes computers outstanding automators of musical processes and the autonomy that results provides computers with an unprecedented degree of agency in the music making that ensues. In these ways, algorithmic musical processes provide rich opportunities for enhancing music learning through automated support, through the articulation and reflection of musical structures, and via the codification of innate capabilities and cultural musical conventions and behaviours.

32.3 SCAFFOLDING AND ACCESS

Generative music systems rely on algorithmic descriptions for the real-time creation of music. The resulting outcomes often include variations at each generation, depending upon the amount of indeterminism in the processes. This provides an interesting parallel to the variability of interpretations that human performers provide, which is differentiated from fixed recordings, which were the dominant mode of music delivery in the past 100 years (see Levtov, this volume, chapter 34). Educating students about these new generative and interactive music methods is increasingly important. But generative systems can also provide support for developing traditional musical skills.

Because an algorithmic system can generate music autonomously, it can be a useful scaffold for beginner musicians; they can either play along with it, or 'direct' it through parametric control. As a result, students can be part of a musical outcome much richer than they alone might be able to produce—with obvious benefits to levels of motivation and self-esteem. The use of technologies in scaffolded learning has been well documented as a useful pedagogical strategy (Luckin 2008). Systems such as the authors' Jam2jam tool (see Figure 32.1) have allowed novice musicians to engage with musical concepts well beyond what would be accessible with their limited acoustic instrumental skills.

Interactive music systems, such as the numerous apps for mobile platforms or the various music games provide ready access to musical interactions because of their low skill requirements. From a learning perspective this can provide a shallow 'on ramp' to

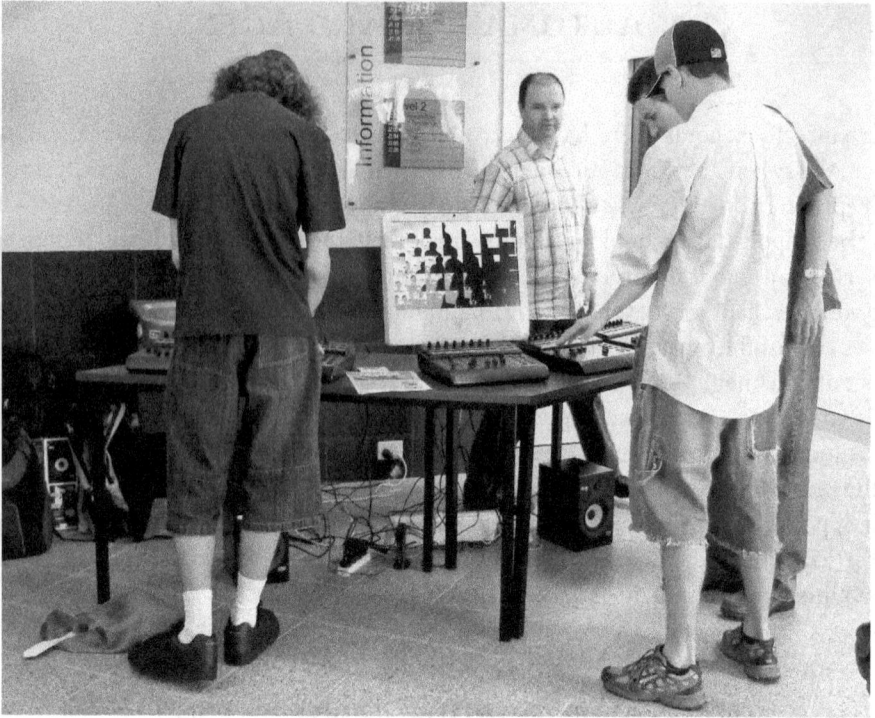

FIGURE 32.1 The Jam2jam AV interactive audio-visual system employs generative music algorithms.

engagement with music and, it is hoped, spark an ongoing interest in music and the developing of further musical skills. The same accessibility also makes algorithmic music systems ideal for those with special needs (Adkins et al. 2012).

32.3.1 Description and Notation

Beyond the *use* of algorithmic music systems by students is their involvement in system *design and development*. It has long been recognized that more can be learned by teaching a topic than by studying it. So, 'teaching' a computer to make music, by programming it to follow an algorithm, has a similar benefit.

Typically, algorithmic design and deployment require one to describe the process and then to codify it in a form the computer will understand. The initial design is often in a human readable form such as a diagram, a description written in prose, or a list of steps similar to a recipe. Prototypes can be created for manual testing, or steps might be performed in a digital audio workstation to explore procedures before coding. Programming is the task of articulating the algorithm in code; it requires the description to be rendered in a notation for the computer to interpret—usually a programming language. Like other musical notations (typically staff notation), the code description can be considered as a musical score; in the case of code, a score for the computer to follow.

Following Taube's three levels of computational representation of music, algorithms can be created to manage musical arrangements, compositions, performance renderings, or sound design; these tasks fit neatly into existing music education curricula. Coding can be considered an alternative notation just as graphic scores are. There are many compositional precedents in the use of alternative notations, including the instruction-based scores used by John Cage and other composers in the twentieth century. Algorithmic musical processes can be contextualized as part of the evolution of musical processes and have a logical place in a holistic view of notational literacy, as they are used in music texts like David Cope's *New Directions in Music* (2001).

32.3.2 Biological and Cultural Connections

Interacting with and designing algorithmic music systems involves immersion in sonic and musical conventions—even if it is to countervail them. These conventions are often cultural and they vary, in either dramatic or subtle ways, between cultures and within subcultures. At times, constraints are imposed on music making by our biological condition; for example, we have two ears and particular perceptual capabilities, and we embody certain motor skills and capabilities. At times our music algorithms need to mimic these constraints to fit in with musical conventions, or they might extend the musical opportunities available beyond those boundaries.

Learning music through engagement with algorithmic processes allows students to undertake activities such as simulating the cyclic patterns of Indonesian gamelan (see Matthews, this volume, chapter 5), exploring the limits of drum kit performance using four 'limbs', and using subtle variations in timbre, space, and temporal alignment to investigate the perceptual boundaries of polyphony and musical textures. Along the way, students learn to understand the rituals, tuning systems, and other aspects of these musical practices. For example, gamelan performance, or how typical drumming patterns are modulated by part coordination, or how psychoacoustic phenomena such as audio streaming and spectral masking affect the reception of music.

32.4 CHALLENGES AND SOLUTIONS

Given the many opportunities for algorithmic music to play a positive role in music education, as outlined in the previous section, it might seem odd that it is not already a standard part of all music education courses. Clearly, there are challenges involved in the adoption of algorithmic music activities for learning. These include the fact that often teachers are not experienced in algorithmic music, and that sometimes algorithmic methods are just a tedious route to mediocre musical outcomes, and therefore uncompetitive with traditional approaches. Examples and activities must be chosen with care to ensure that the relevance and the value of algorithmic approaches are maintained.

Another disincentive is that established disciplinary boundaries, often reinforced by educational institutions, mean that coding is perceived as a computing skill rather than a musical one, and therefore outside the responsibility (or even legitimacy) of a music education. As we will see in many of the examples described later in this chapter, a common approach to overcoming this division of responsibility is interdisciplinary collaboration, where music becomes a motivating context for learning programming and/or programming becomes an avenue to enhanced music making.

32.4.1 Fear of the Unknown

For this author, and probably most authors in this volume, algorithmic music brings together multiple passions across the creative arts, technology, and more. That same interdisciplinary proclivity, however, might serve as a barrier to entry for many others. Often musicians love music because it is an escape from maths and science; conversely, many computer scientists and engineers might feel uncomfortable with the ambiguities inherent in the creative arts, let alone the immateriality of music as an art form. Fear of the unknown and lack of self-confidence outside an established domain of understanding can be a significant barrier to educators' and students' serious engagement with algorithmic music.

Apprehension about the unfamiliar is not a new phenomenon in education and many approaches have been tried to overcome these barriers. Amongst them is the use of very structured tasks that lead people step by step through what might be unfamiliar territory. Brief excursions (small tasks) can also provide stepping stones to more in-depth engagement—for example, the use of exercises that require only short fragments of code, or tasks that do not rely on an advanced music theory background. Group projects allow people to share the journey and provide peer support when the inevitable challenges arise; they also allow for ensemble experiences, which are an important aspect of musical training. Mentoring and exposure to stories from experienced travellers help to show that the journey to understanding and using algorithmic music is possible and worthwhile. Exemplars can demystify the creative and/or technical process as one requiring only persistent and iterative steps rather than a mystical leap of understanding.

Educators, and certainly their students, are well aware that they live in a technological society and that the role of technology in music is significant. The music industry, after all, has been at the leading edge of disruption from digital technologies, in both the way music is made and how it is consumed. What seems less well appreciated, however, is the role of programming in driving that technological society and the opportunities for musicians who can code to increase control over their technological destiny. Two educational movements that have tried to help people learn to work with algorithms are courses in computational thinking and in creative coding.

32.4.2 Computational Thinking and Music

Courses in computational thinking introduce students to methods of description and design that follow algorithmic processes without immediately engaging in the representation of these as programming notations. Rather, algorithmic solutions to tasks are approached through familiar and tangible media such as writing, drawing, block building, and so on. When code is introduced, tasks often maintain a media orientation through computer graphics, sensor-based electronics, robotics, and sound. This approach to understand computing before attempting to code is designed for a nonspecialist audience (such as musicians), and is deliberately different from what might be considered appropriate for computing professionals (Guzdial 2008).

The main conceptual building blocks explored in computational thinking courses are abstraction and automation (Wing 2008). In music, abstractions include structural patterns such as canons, counterpoint, phasing, and so on, and the automation of these can be manual, mechanical, or computational. As Edwards notes: 'That algorithmic thinking is present in Western composition for at least 1,000 years has been established. That such thinking should lend itself to formalization in software algorithms was inevitable' (2011, 67). Similarly, algorithmic thinking about music is a logical development of existing theoretical and analytical approaches to music and, despite some reticence about the value of computational thinking curricula (Denning 2009), it seems reasonable to expect that approaches used for the teaching of computational thinking might be applied to thinking algorithmically about music (Ruthmann et al. 2010).

32.4.3 Creative Coding Movements

The application of computational processes in the creative arts has gathered momentum since at least the 1990s, when new media art or multimedia courses championed the production of digital creative works; most often these were screen-based works or animated installations. In more recent years, with the popularity of dedicated programming environments such as Processing (Reas and Fry 2003) and the increasing ubiquity of JavaScript automation in websites, artistic computational works have become popular under the banner of *creative coding*.

A leading early figure in this movement was John Maeda, who led the Aesthetics and Computation Group at the MIT Media Lab around the turn of the century. In his book *Creative Code* he outlines the shift in emphasis for creative artists, in which computer technology 'is not a tool; it is a new material for expression' (2004, 101). It is this shift in emphasis from using digital technology as a tool to understanding digital media as a medium that underscores creative coding activities—a shift for creative artists from technology *use* to technology *craftsmanship*.

A direct outcome of Maeda's ideas was the Processing environment, created by two of his students, which sparked an expansion of creative coding courses in art and design

colleges around the world. It includes built-in, and third-party, music and audio libraries that allow for algorithmic music activities such as building interactive soundscapes,
or creating a step sequencer. Other accessible creative coding environments with a
strong music focus include Fluxus (Griffiths 2007) (http://www.pawfal.org/fluxus/),
Gibber (Roberts 2013), and those used in the examples discussed below.

32.5 LEARNING CONTEXTS

Appropriate settings for learning about algorithmic music are many and varied, covering formal and informal learning contexts (Green 2002). Informal settings include
community-run 'hacker' sessions, 'maker' workshops, and after-school computer sessions, including the Computer Clubhouse (Sylvan 2006) and Code Club (Code Club
International 2012–2017). Formal settings mainly include courses at schools, colleges,
and universities, but increasingly include offshoots of these available freely as online
courses (e.g. Dave Conservatoire 2016; Freeman 2017; Kapur 2017).

Algorithmic processes can be applied to many areas of music making, as can be seen
by the variety of perspectives in this volume. Any of these areas of activity can provide a
context in which students can hear about algorithmic music. The best-established context is the use of algorithmic processes in composition, where they are used to generate
material and/or where algorithmic processes are integrated as part of the realization of
the work itself. Related to this, algorithmic music and/or sound design can be used in art
installations and interactive media such as computer games.

Interactive music devices (instruments) that incorporate algorithmic processes have
become quite popular over the past decade. Many of these are featured at the successful New Interfaces for Music Expression (NIME) conference series (http://nime.org).
As mobile computing power has increased, algorithmic music has become a part of
live performance practices. A prominent example is the live coding community that
is rapidly expanding (see Roberts and Wakefield, this volume, chapter 16). Live coding practice includes solo and ensemble performances with music and/or audiovisual
outcomes. A number of educational programs have included live coding in their 'laptop orchestras', where students make music with code as an ensemble (see Ogborn,
this volume, chapter 20). A wide range of musical genres is represented in live coding practices, including experimental, electroacoustic, electronic dance music, and
neoclassical.

Real-time programming environments are necessary for live coding and are useful
for learning about algorithmic music. Real-time programming environments allow
code to be updated while it is running, enabling changes to be made on the fly with an
uninterrupted flow of musical output. This immediacy of audible feedback is typical of
interaction with acoustic instruments and sound in the physical world, but has until
recently been uncommon in computer programming workflows, where applications are
typically halted and restarted after editing.

32.6 Examples

In this section, there are descriptions of several examples in which algorithmic music has been used in educational contexts. These exemplars will show how the issues and concerns discussed so far are managed in the context of real-world learning situations.

32.6.1 TuneBlocks

Jeanne Bamberger is Professor Emeritus of Music and Urban Education at the Massachusetts Institute of Technology. She designed the TuneBlocks compositional environment as a result of research that combined music psychology and computers in education. The software allows for the creation of musical fragments (blocks) and their combination in series and hierarchies. TuneBlocks was designed to support analysis (through reconstruction of existing works) and creation (through elaboration or creation of new material). First developed in the 1980s as part of Bamberger's Impromptu software, a version of TuneBlocks is available today for the Apple iPad (Figure 32.2). While its algorithmic capabilities are quite limited by today's standards, it is an important landmark in the use of computers and algorithmic thinking in music education.

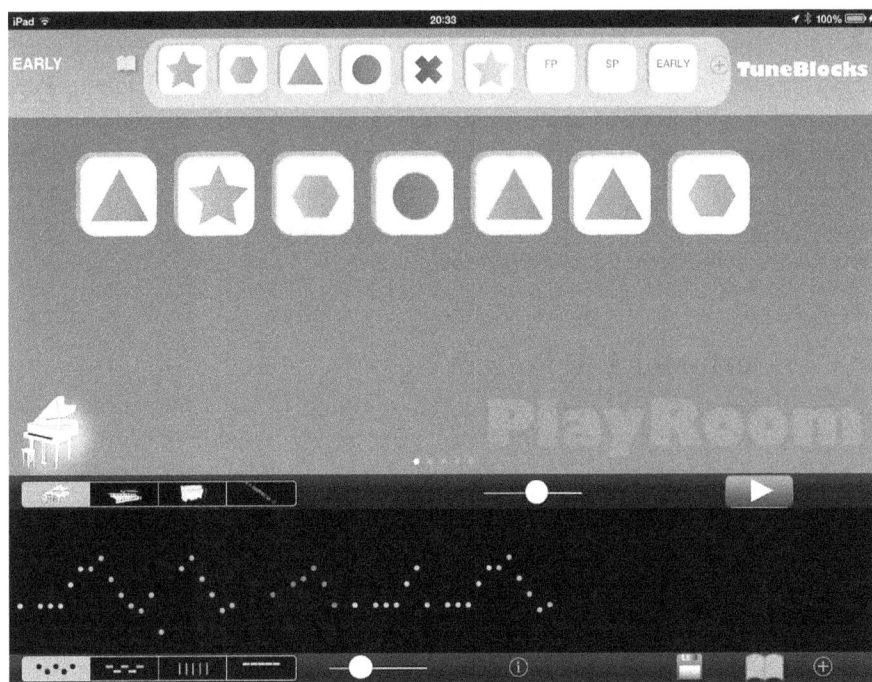

FIGURE 32.2 The playroom interface of TuneBlocks on the Apple iPad.

Individual tune blocks are visualized as square icons, each representing a musical motif (see Figure 32.2). This abstract representation is designed 'to focus the students' attention on *listening* rather than looking' (Bamberger 2003, 11). A common listening task with TuneBlocks is to have a melody divided amongst a set of blocks that a student must rearrange in the correct order. Extension activities can include arranging blocks in a new but effective order, and composing additional blocks to extend further the rearrangement and compositional possibilities. Through these recombinatorial processes, Bamberger's objective is to have students learn, through listening and experimentation, how to discern specific features of each block, how some are similar and others differentiated, what are the structural functions of each block (beginning, middle, end, and so on), why blocks combine well or not, and to appreciate how order and repetition matter. Students can also reflect on why they like or dislike particular blocks or combinations of blocks (Bamberger 2003).

While automation in the TuneBlocks application is limited, it is designed to support what is now called computational thinking. The simple interface was designed for children to use in an age when computing was not nearly as ubiquitous as it is today, and grew out of earlier experiences Bamberger had had with musical coding in the Logo language, in the 1970s, when she had collaborated with Seymour Papert and other early pioneers of teaching computer programming in schools (Bamberger 1979).

The notion, clearly evident in Bamberger's work, of developing understanding through practical activity is sympathetic with ideas of constructivist psychology. According to this theory, people develop and transform their understanding and ideas through experiences in the world, and in doing so they construct and internalize new knowledge. Constructivism is based on the developmental psychology theories that Jean Piaget established during the mid-1900s. Bamberger's colleague Seymour Papert was a student of Piaget. It was natural that these researchers would see the potential in computer programming and interface design for the articulation and externalization of processes and structures. Computers and computation, therefore, were taken up as vehicles for developing the kinds of systematic (algorithmic) thinking that are evident in many fields, including music composition.

32.6.2 PowerBooks UnPlugged

Laptop computer performance ensembles are increasingly common in educational settings, particularly at university level. One example is PowerBooks UnPlugged, based at the Institute of Time-Based Media at the Universität der Künste Berlin. As recently as 2017, the ensemble's website declared: 'Many have claimed that "The laptop is the new folk guitar"; if this is so, then PB_UP is the first acoustic computer music folk band: The laptop is their only instrument' (http://pbup.goto10.org/pbup.html). The ensemble, started by Alberto de Campo, Echo Ho, Hannes Hölzl, and Jankees van Kampen, with input from Renate Weiser and Julian Rohrhuber, gets its name from the practice of using the built-in speakers of the laptop as its playback system. Using this mobility to their advantage, the ensemble members typically distribute themselves amongst the audience during a concert, thus providing an inherently spatialized sound experience. Music is generated by live coding algorithms, which create the sound synthesis and execute them to improvise musical structures.

A software library called Republic was developed for PowerBooks UnPlugged, to enable collaborative and distributed code-based music performance over a wireless network; a practice they prefer to call 'just in time programming', where they (re)write programs while they are already running. As described by one of its developers:

> Republic is an extension library for the SuperCollider language designed to allow highly collaborative forms of live coding. In a symmetrical network of participants, everyone can write and evaluate scripts which generate musical processes that can distribute sounds across all machines. (Campo 2013, 22)

As well as sending music between machines, the system allows for chat communications that facilitate coordination amongst the ensemble. Performers are literally checking their email on stage! The environment is designed to be deeply collaborative. 'The implicit working model is as democratic and symmetrical as the spatial disposition of the music: everyone can make sounds on her own laptop as well as (simultaneously or sequentially) on everyone else's' (Rohrhuber et al. 2005).

The leaders of this ensemble are very clear that their focus is on 'improvising with algorithms', and their claim that this practice is a kind of 'folk' music for computers has echoes of the composer Iannis Xenakis's search for authentic characteristics of digital music in his algorithmic formalizations (Xenakis 1992). Writing about their ensemble, members suggest that

> a public improvisation with algorithms is no less plausible than experimenting with sounding objects on stage, and the numerous live coding approaches have led to an interesting variety of performances. Here, it is an ever changing dynamics of reprogrammed microcompositions that make up the improvisational situation, playing with the double time structure of processual change and change of the process. (Rohrhuber et al. 2007)

The educational affordances of laptop ensembles, such as PowerBooks UnPlugged, include sharing the workload and risk amongst the participants. That is because only a small fraction of the extraordinary sonic potential of each laptop needs to be harnessed by each performer, this allows performers to focus on manageable fragments of code and on the generation and manipulation of a restricted algorithmic process. This provides an achievable entry point for new musicians, and even in the case of a crash or error on one computer the distributed nature of the work means this has only a minor impact on the performance. The level of ensemble integration in a wirelessly shared computing environment, like Republic, can encourage a deep level of ensemble integration. 'Since instruments and control algorithms are shared, there's no real owner anymore; the creators are discrete musical entities only if they choose to be, ideas belong to everyone' (PowerBooks UnPlugged 2015). The inherently distributed nature of the Republic software used by PowerBooks UnPlugged enables performers to exploit these features maximally, as educational affordances, thus reinforcing ensemble performance skills.

32.6.3 Sound Thinking

The obviously interdisciplinary nature of computer music studies lends itself to collaborative courses between the arts and sciences. One such course is Sound Thinking, offered at the University of Massachusetts Lowell, which will be discussed in this section. Another is the course Computer Music on a Laptop: Composing, Performing, Interacting, taught at the College of Charleston, which is the topic of the next example. These courses exist though cooperation between teaching staff in music and computer science departments, and are particularly feasible in liberal arts educational contexts, where students are encouraged to take courses in both the arts and sciences—often computer music courses can count for credit in either or both areas.

The design and the teaching of the Sound Thinking course were led by Jesse Heines, Gena Greher, and Alex Ruthmann. It was designed as part of a teaching initiative called Performamatics that was devised to attract students to computing by tapping into their inherent interest in the performing arts (Ruthmann et al. 2010).

After some experimentation with a variety of technical platforms, the Sound Thinking course settled on the use of the Scratch environment (https://scratch.mit.edu/), developed at the MIT Media Lab. Scratch is a hybrid of text and visual programming paradigms, designed for young learners (see Figure 32.3). It has a strong focus on interaction and media outcomes such as animation and games. Its dynamic, media-rich environment suits the performative nature of musical activities. In the Sound Thinking course students develop various generative music algorithms and learn to manipulate these for variation during performances. The course also includes the integration of hardware controllers—the MaKey MaKey board, developed by Jay Silver and Eric Rosenbaum, and formerly the locally made IchiBoards—to facilitate real-time gestural control of musical parameters.

Algorithmic thinking is fostered by having students design musical flowcharts for various analytical tasks and compositional challenges. Flowcharts underscore the structural elements of musical compositions as a way of connecting algorithmic and musical designs. Algorithms studied in this course include random walk melodies, iteration through pitch and rhythm lists, and transposition through the use of offsets to MIDI note numbers.

In a book based on their experiences in these courses, Greher and Heines stress the course's focus on analytical skills and computational thinking which, they suggest, are just as important to music as they are to computer science. They also emphasize the benefits of the interdisciplinary mix of students that helps learners to break out of familiar habits of thinking and acting (Greher and Heines 2014).

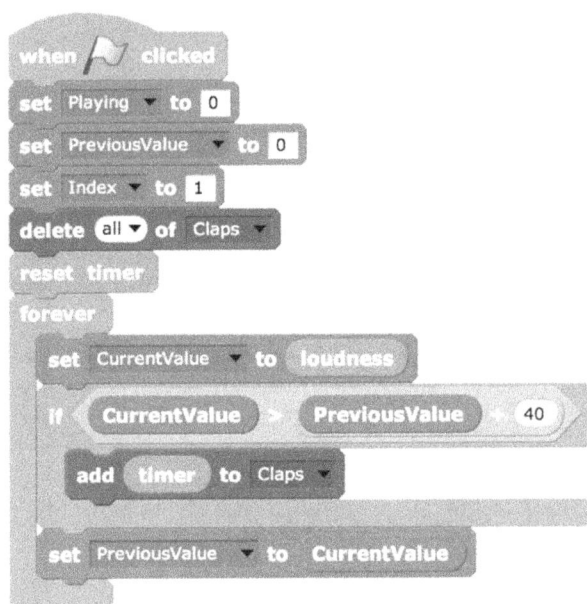

FIGURE 32.3 A code block from the Scratch environment.

32.6.4 JythonMusic

The College of Charleston, in South Carolina, is the site of another successful collaboration between music and computer science educators in the use of algorithmic music as the basis for interdisciplinary curriculum design. Courses in music fundamentals and introductory programming have been combined into new courses that include a variety of tasks including composition, interactive music systems, and ensemble performance. Several courses were designed and co-taught by Bill Manaris, Blake Stevens, and Yiorgos Vassilandonakis: Introduction to Computer Music and Aesthetics; Programming Music, Performing Computers and Computer Music on a Laptop; and Composing, Performing, Interacting. These courses were intended to 'synthesize creativity in the arts with the ability to model and automate processes in code' (Manaris, Brown, and Stevens 2016, 44). The introduction course has no prerequisites but students usually enter with some background in music performance or after having taken music classes at school; there is no expectation of prior computer programming experience. The Computer Music course is an honours-level offering and focusses on principles of music composition and computer programming for developing interactive computer music applications.

The JythonMusic programming environment was developed alongside the curriculum to support these courses (http://jythonmusic.org). It uses the Jython Environment for Music (JEM) editor for writing and evaluating code (see Figure 32.4). JythonMusic provides libraries for music data, audio playback, image manipulation, building graphical user interfaces (GUIs), and for connecting to external MIDI and OSC devices

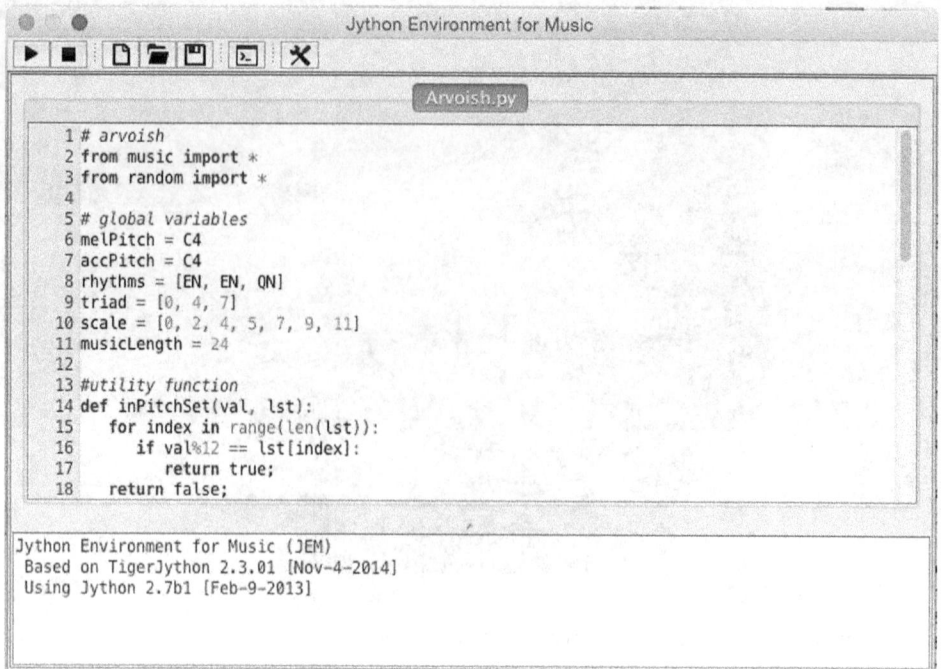

```
1 # arvoish
2 from music import *
3 from random import *
4
5 # global variables
6 melPitch = C4
7 accPitch = C4
8 rhythms = [EN, EN, QN]
9 triad = [0, 4, 7]
10 scale = [0, 2, 4, 5, 7, 9, 11]
11 musicLength = 24
12
13 #utility function
14 def inPitchSet(val, lst):
15     for index in range(len(lst)):
16         if val%12 == lst[index]:
17             return true;
18     return false;
```

```
Jython Environment for Music (JEM)
 Based on TigerJython 2.3.01 [Nov-4-2014]
 Using Jython 2.7b1 [Feb-9-2013]
```

FIGURE 32.4 The Jython Environment for Music (JEM) editor.

(Manaris and Brown 2014). A 'musical' data structure, inherited from jMusic, is used for representing musical scores. It includes classes for Note, Phrase, Part, and Score, as well as providing classes to represent audio material. Playback of score data is via internal or external virtual synthesizers. This capability fits well into, and supports, existing music curricula.

Activities in their curriculum include the study of temporal musical structures, algorithmic processes, soundscape design, graphical user interfaces, programming design patterns, data types, language syntax and semantics, musical terminology, and characteristics of musical style. Students are required to produce various musical artefacts, including standalone and interactive music software, solo and group compositions, and interactive performances. Project-based pedagogy is used as a vehicle to promote the integrated development of computing and musical skills and understandings.

Reflecting on their teaching of several iterations of these courses, staff say that a key benefit for students is that 'this coupling [of music and computing] leads to an increase in active and creative learning experiences, as each student gains proficiency in realizing and expressing musical ideas on a common instrument' (Manaris, Brown, and Stevens 2016). In addition to these participant observations, student surveys have been conducted. Results show that most are comfortable in regarding coding as a valuable medium for musical thinking and a legitimate form of musical composition. There is also strong support for music as an effective context for learning how to program. The surveyed cohort was less convinced that code-based music performances were capable of achieving musical outcomes comparable with more traditional methods of music making. As to their likelihood of continuing with music programming beyond the courses, the introductory class responded very positively, while students in the honours course were more polarized in their responses (Manaris, Brown, and Stevens 2016, 35).

Overall, this example reveals a comprehensive engagement with the use of musical algorithms as the basis for an interdisciplinary offering in computer music education. After a sustained effort over five years, the courses are more refined and the faculty involved have produced rich resources, including a development environment, a book full of examples, and educational research that reflects their experiences.

32.6.5 Sonic Pi

Sonic Pi (http://sonic-pi.net/) is an open-source music programming environment, created by Samuel Aaron and a team of voluntary developers. Sonic Pi's focus is on live coding and it is designed to be as easy as possible for beginners. From the outset there was a clear emphasis on learning pathways; and there are associated 'schemes of work' for music lessons. Sonic Pi has evolved through collaboration among musicians, academics, and educators interested in helping school children to learn programming by creating music. As with many of the previous examples, interdisciplinary teams have enriched the development of Sonic Pi, and experiences from workshops and classes have guided its evolution. The platform was developed alongside pedagogical strategies and teaching materials. It comes with a selection of example projects and lesson plans for teachers. The multifaceted motivations of those involved include: empowering children to access computing skills, promoting new musical habits and skills, and promoting creative partnerships between schools and communities (Aaron, Blackwell, and Bernard 2016).

The name Sonic Pi results from a close link with, and financial support from, the Raspberry Pi Foundation, which produces a low-cost computing platform on which Sonic Pi can run; there are also versions for macOS and Windows (see Figure 32.5). The partnership with the Raspberry Pi Foundation was facilitated by common connections at the University of Cambridge Computer Laboratory. The Sonic Pi environment includes an Interactive Development Environment (IDE; see Figure 32.5) and uses a customized version of the Ruby language as an interface to a SuperCollider server that provides the core audio capabilities.

Sonic Pi examples and materials focus on compositional structures described in its scripting code and make use of prebuilt instruments and audio samples for sound output. Algorithmic processes explored in the examples include: stochastic choice, repetition and iteration, data slicing and recombination, and isorhythms. The stylistic outcomes range from ambient to hip-hop.

In keeping with the Sonic Pi project's focus on teaching programming, it is not surprising that the teaching materials are organized around a series of coding topics such as syntax, debugging, data structures, and so on. For the music educator, the examples

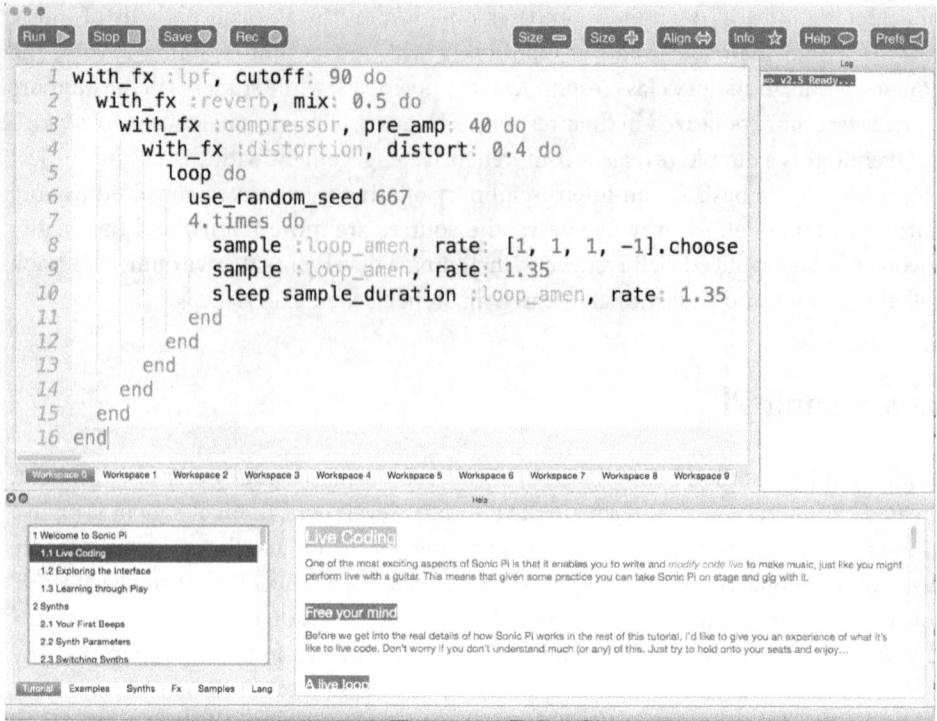

```
 1  with_fx :lpf, cutoff: 90 do
 2    with_fx :reverb, mix: 0.5 do
 3      with_fx :compressor, pre_amp: 40 do
 4        with_fx :distortion, distort: 0.4 do
 5          loop do
 6            use_random_seed 667
 7            4.times do
 8              sample :loop_amen, rate: [1, 1, 1, -1].choose
 9              sample :loop_amen, rate: 1.35
10              sleep sample_duration :loop_amen, rate: 1.35
11            end
12          end
13        end
14      end
15    end
16  end
```

FIGURE 32.5 The Sonic Pi interface on Mac OS.

provide a more convenient launch pad from which musical principles can be unpacked. Music educators might also be interested in exploring the Sonic Pi Live Coding Toolkit (http://www.sonicpiliveandcoding.com/), which provides support materials for music pedagogy with artistic examples in the form of 'Pop Pi' videos and a range of suggested activities focused on a music curriculum.

The pedagogical approaches developed during classes and workshops emphasize role-play and group work as ways of augmenting coding and listening. This is in keeping with teaching strategies employed in computational thinking courses. More aligned with arts pedagogy and constructivist approaches is the Sonic Pi project's emphasis on participatory culture, experimentation and open-ended creativity, rather than on individual work and tasks oriented to predetermined outcomes (Aaron, Blackwell, and Bernard 2016).

32.7 CONCLUSION

Algorithmic music processes have not been an everyday component of music education curricula. However, there are a growing number of tools and teaching materials that, in time, should bring about change in this direction. Musical techniques, theories, and methods have long been studied, but their formalization as algorithms and their articulation in computer programming languages are relatively new developments.

Algorithmic thinking privileges abstraction and generality—important concepts in both computer science and music composition—and therefore it has an important place in education in these fields. The leverage provided by computational automation means that algorithmic processes have a significant role to play in music production; their accessibility to educators and their prominence in music educational circles, however, are still developing.

This chapter has explored many of the issues involved in engaging with algorithmic representations of music for educational purposes. It has also provided an overview of several examples which show how the programming of music algorithms is being approached as a technique for assisting and motivating students to learn music and computer programming. Having identified some of the central opportunities and concerns of algorithmic music in music education, and after reviewing innovative examples, it seems clear that musical algorithms and coding skills are useful conceptual tools that can assist in the development of musical intelligence, but that they have yet to be fully embraced by the music education community.

Bibliography

Aaron, S., Blackwell, A. F., and Bernard, P. 'The Development of Sonic Pi and Its Use in Educational Partnerships: Co-Creating Pedagogies for Learning Computer Programming'. *Journal of Music, Technology and Education* 9, no. 1 (2016): 75–94.

Abelson, H., and Sussman, G. J. *Structure and Interpretation of Computer Programs*. 2nd ed. Cambridge, MA: MIT Press, 1996.

Adkins, B., Knox, M., Dillon, S., Brown, A. R., and Summerville, J. 'Digital Technologies and Musical Participation for People with Intellectual Disabilities'. *New Media and Society* 14, no. 7 (2012): 501–518.

Bamberger, J. *Logo Music Projects: Experiments in Musical Perception and Design*. Cambridge, MA: MIT Artificial Intelligence Laboratory, 1979.

Bamberger, J. 'The Development of Intuitive Musical Understanding: A Natural Experiment'. *Psychology of Music* 31, no. 1 (2003): 7–36.

Campo, A. de. 'Republic: Collaborative Live Coding 2003–2013'. In *Collaboration and Learning through Live Coding*. Dagstuhl Seminar 13382, *Dagstuhl Reports* 3, no. 9 (2014): 130–168.

Code Club International. 2012–2017. http://codeclubworld.org/, accessed 30 June 2017.

Cope, D. *New Directions in Music*. 7th ed. Prospect Heights, IL: Waveland, 2001.

Dave Conservatoire. *Introduction to Sonic Pi*. Last modified 2017. http://www.daveconservatoire.org/course/introduction-to-sonic-pi.

Denning, P. J. 'The Profession of IT beyond Computational Thinking'. *Communications of the ACM* 52, no. 6 (2009): 28–30.

Dillon, S. *Music, Meaning and Transformation: Meaningful Music Making for Life*. Cambridge: Cambridge Scholars, 2007.

diSessa, A. *Changing Minds: Computers, Learning, and Literacy*. Cambridge, MA: MIT Press, 2000.

Edwards, M. 'Algorithmic Composition: Computational Thinking in Music'. *Communications of the ACM* 54, no. 7 (2011): 58–67.

Freeman, J. A. 2017. *Survey of Music Technology*. Georgia Institute of Technology https://www.coursera.org/learn/music-technology, accessed 30 June 2017.

Green, L. *How Popular Musicians Learn: A Way Ahead for Music Education*. London: Ashgate, 2002.

Greher, G. R., and Heines, J. M. *Computational Thinking in Sound*. New York: Oxford University Press, 2014.

Griffiths, D. 2007. 'Fluxus'. In *Collaboration and Learning through Live Coding*. Dagstuhl Seminar 13382, *Dagstuhl Reports* 3, no. 9 (2014): 149–150.

Guzdial, M.. 'Education Paving the Way for Computational Thinking'. *Communications of the ACM* 51, no. 8 (2008): 25–27.

Guzdial, M., and Ericson, B. *Introduction to Computing and Programming in Python: A Multimedia Approach*. Upper Saddle River, NJ: Pearson, 2010.

Jorgensen, E. R. *Transforming Music Education*. Bloomington: Indiana University Press, 2003.

Kapur, A. 2017. *Introduction to Programming for Musicians and Digital Artists*. California Institute for the Arts. https://www.kadenze.com/courses/introduction-to-programming-for-musicians-and-digital-artists/info, accessed 30 June 2017.

Levenson, T. *Measure for Measure: A Musical History of Science*. New York: Touchstone, 1994.

Luckin, R. 'The Learner Centric Ecology of Resources: A Framework for Using Technology to Scaffold Learning'. *Computers and Education* 50, no. 2 (2008): 449–462.

Manaris, B., and Brown, A. R. *Making Music with Computers: Creative Programming in Python*. Boca Raton, FL: CRC Press, 2014.

Manaris, B., Brown, A. R., and Stevens, B. 'JythonMusic: An Environment for Teaching Algorithmic Music Composition, Dynamic Coding, and Musical Performativity'. *Journal of Music, Technology and Education* 9, no. 1 (2016): 33–56.

Maeda, J. *Creative Code*. London: Thames and Hudson, 2004.

Nierhaus, G. *Algorithmic Composition: Paradigms for Automated Music Generation*. New York: Springer, 2010.

Papert, S. *Mindstorms: Children, Computers, and Powerful Ideas*. New York: Basic Books1980.

Reas, C., and Fry, B. 'Processing: A Learning Environment for Creating Interactive Web Graphics'. In *International Conference on Computer Graphics and Interactive Techniques*, 1. New York: ACM 2003.

Resnick, M. *Turtles, Termites, and Traffic Jams*. Cambridge, MA: MIT Press1994.

Roberts, C. 2013. 'Gibber' (live coding platform). http://www.charlie-roberts.com/gibber/. Accessed April 2015.

Rohrhuber, J., Campo, A. de, and Wieser, R. 'Algorithms Today: Notes on Language Design for Just in Time Programming'. In *Proceedings of the International Computer Music Conference*, 455–458. Barcelona, 2005. http://quod.lib.umich.edu/i/icmc.

Rohrhuber, J, Campo, A. de, Wieser, R., van Kampen, J.-K., Ho, E., and Hölzl, H. 'Purloined Letters and Distributed Persons'. In *Proceedings of the Music in the Global Village Conference*. Mücsarnok, Budapest, 2007. http://www.wertlos.org/articles/Purloined_letters.pdf.

Ruthmann, A., Heines, J. M., Greher, G. R., Laidler, P., and Saulters, C. 'Teaching Computational Thinking through Musical Live Coding in Scratch'. In *Proceedings of the 41st ACM Technical Symposium on Computer Science Education*, 351–355. New York: ACM, 2010.

Sylvan, E. 'Who Knows Whom in a Virtual Learning Network? Applying Social Network Analysis to Communities of Learners at the Computer Clubhouse'. In *Proceedings of the 7th International Conference on Learning Sciences*, 758–764. Bloomington, IN: International Society of the Learning Sciences, 2006.

Swanwick, K. *Musical Knowledge: Intuition, Analysis, and Musical Education*. London: Routledge, 1994.

Taube, H. K.. 'Computation and Music'. In *Sound Musicianship: Understanding the Crafts of Music*, edited by A. R. Brown, 40–51. Newcastle upon Tyne: Cambridge Scholars, 2012.

Taube, H. K. *Notes from the Metalevel: Introduction to Algorithmic Music Composition*. London: Taylor and Francis, 2004.

Wing, J. M.. 'Computational Thinking and Thinking about Computing'. *Philosophical Transactions of the Royal Society A: Mathematical, Physical and Engineering Sciences* 366, no. 1881 (2008): 3717–3725.

Xenakis, I. *Formalized Music: Thought and Mathematics in Music*. Stuyvesant, NY: Pendragon, 1992.

CHAPTER 33

....

(MICRO)POLITICS OF
ALGORITHMIC MUSIC

Towards a Tactical Media Archaeology

....

GEOFF COX AND MORTEN RIIS

33.1 INTRODUCTION

ALGORITHMS have effects and play an increasingly determining role in our lives—from the inner workings of financial capitalism to decision-making processes in everyday circumstances. They are controlling the tiny microsounds of Curtis Roads and the minute generative drum structures from Autechre, but algorithms also provide structure to the 1000-year-long composition *Longplayer* by Jem Finer and John Cage's 639-year-long organ piece *As SLow aS Possible*.[1] More than simply instructions to be performed, algorithms modulate across different scales and temporalities; they are techniques that offer tactics to order and control materials between microtemporal processes and larger data sets. Based on mathematical models, algorithms process massive amounts of data, to predict gains and reduce risks, to exert control over increased complexity, but as can be seen from the financial crash of 2008, they are hardly infallible. Our reliance on them is a leap of faith and yet they are abstractions with powerful effects, as Andrew Goffey puts it: 'Algorithms act, but they do so as part of an ill-defined network of actions upon actions, part of a complex of power-knowledge relations, in which unintended consequences, like the side effects of a program's behavior, can become critically important' (2008, 19).

How algorithms act then becomes hugely significant for understanding how they operate within larger power–knowledge systems at different scales and for their creative potential. They do not merely manage and organize sound, as in the case of selection algorithms for Spotify or iTunes, but generate the deep rhythms and structures of what constitutes cultural and political life. This chapter unfolds the discussion in relation to algorithmic music to develop an argument for a shift of emphasis from the imposition of rules or instructions as a structuring device for new musical forms—as has been the tendency in electronic music—to a deeper techno-materialist understanding of what is happening at various scales and temporalities of operation. In this way our argument is that algorithms become epistemological tools to understand the increased dependency on scripts, scores, and programs as part of informational systems and wider socio-technical assemblages that shape our operational logic and decisions. This is especially important when algorithms like Google's PageRank or Facebook's EdgeRank make sense of big data in distorted ways to 'reify' knowledge and make sure that communication is linked to market forces.

In contrast to a perspective rooted in the political economy, this chapter develops an argument based upon ideas from media archaeology and the critical tradition of tactical media. These perspectives are significant as they shift attention to the ability of nonhuman entities to generate alternative forms of knowledge that are not easily perceptible to humans. The critical tradition of media archaeology is informed by the archaeological work of Michel Foucault and the media theory of Friedrich Kittler and Marshall McLuhan. Other significant writers that have influenced the development of media archaeology include Walter Benjamin, Siegfried Giedion, Ernst Robert Curtius, Dolf Sternberger, and Aby Warburg. In brief (as more detail is provided later in the chapter), the archaeological excavation becomes a mode of reverse-engineering of normative understandings, a transformation of what has already been written, and thereby counterhistory. Media archaeology builds upon these principles to assert that the material-technological dimension is not sufficiently developed in terms of accounting for the way that media produces knowledge and experience. In adding media to Foucault's 'archaeology of knowledge', the limits of the human sensory apparatus are exposed and other nonhuman forms of knowledge are made apparent through the operative use of technologies. By extension we can say that there are emergent kinds of music to discover that we simply would not be able to hear were it not for the use of certain algorithms.

The chapter is structured into four further sections. First, we provide an overview of our proposed shift of emphasis from the macro to micro scale of algorithmic music, in keeping with our understanding of political processes (section 33.2, 'From Macro to Micro Politics of Algorithmic Music'). Second, media archaeology is introduced in order to align it more closely to tactical media as methodological foundation for a critical practice in algorithmic music that is attentive to engineering (section 33.3, 'Tactics of Media Archaeology'). The next section (33.4, 'Microtemporality and Time-Criticality') leads to a more detailed discussion of microtemporality, with particular reference to the media archaeology of Wolfgang Ernst, and in turn to Gilles Deleuze and Félix Guattari's notion of micropolitics, and finally to Shintaro Miyazaki's concept of algo*rhythmics*. The final section (33.5, 'Tactical Algo*rhythmics*') aims to connect these concepts to form the main proposition of the chapter: that 'tactical media archaeology' offers an analytical method for developing alternative compositions of both algorithmic music and politics. Examples are provided throughout, but the argument is quite abstract, with the intention of emphasizing more speculative approaches and broader ecologies of practice (exemplified by the critical engineering of Martin Howse). Our claim is that it is only through exposing the way that algorithms operate as part of wider socio-technical assemblages that musical and political experimentation can really develop.

33.2 FROM MACRO TO MICRO POLITICS OF ALGORITHMIC MUSIC

It might be claimed that algorithmic music entails a specific aesthetic ideal or even genre—especially when taking into account the development of computer-based composition from the 1960s onwards. These compositional styles or strategies could, for instance, result in the implementation of bio-algorithms for musical organization (like the viral artificial life algorithms found in the works of Joseph Nechvatal [2011, 18, 40]), the creation of formalistic structures based on complex mathematics (Xenakis), or the more recent experiments of sound artist Florian Hecker in his *Untitled (F.A.N.N.)* (2013). With explicit references to David Tudor's work with neural networks (Riis 2013b), Hecker creates an ever-changing generative three-channel audio piece in which the artist starts an algorithmic process, and then steps back to let the computer execute a never-repeated autopoietic expression that stages technology as the creating force. It could be claimed that these examples have a specific 'sound' to them, but they also demonstrate a strategic quality—which comes out of the algorithmic approach to composition—to foreground computation and technology as a self-organizing creative force, while human agency fades into the background.

The politics of this is complex. The formal structure of algorithms offers an insight into the ways in which creative operations are organized and regulated, and a new kind of political rationality is generated that is based on control structures and accelerated automation. Algorithmic music is hardly new in this respect; automation and formalism within composition have traditions established long before the advent of computers at a macro level. Athanasius Kircher's *Arca Musarithma*, as described in his *Musurgia Universalis* from 1650 (Nierhaus 2009, 24–26), is an early example of an algorithmic compositional system that enabled untrained musicians to compose music, a system that recently became implemented within software (Bumgardner 2009). In the early twentieth century, Russian Formalism similarly sought to uncover the structural formations of language that might be applicable to wider social phenomena, founded on the idea that language consists of a set of rules that are executed according to command structures, rendering the creative subject—author, artist, musician, composer, programmer—a function of a discursive system or part of wider compositional assemblage. Taking the formalism of music into account is the notion of the music-in-itself, an autonomous entity that does not mean anything other than the sound it consists of, as exemplified by the iconic quote by Hanslick: 'Music has no subject beyond the combinations of notes we hear, for music speaks not only by means of sounds, it speaks nothing but sound' (Benestad 1977, 299). The autonomy of music is additionally expanded by incorporating algorithmic and serial techniques as seen in the works of Boulez, Stockhausen, and others. Collective works such as Arseny Avraamov's *The Symphony of the Factory Sirens* of 1922, in which trained musicians are replaced with workers combined with the machinery of industrial production, offers an example of how creative autonomy, authorship, and authority are undermined by wider human–nonhuman assemblages. In this massively orchestrated piece of public sound art, the ships' horns of the entire fleet, two batteries of artilleries, several full infantry regiments, trucks, seaplanes, twenty-five steam locomotives, an array of pitched whistles, and several massive choirs resonated together across the entire city of Baku.

Aside from the celebration of work, what about the politics of the composition? There are numerous examples of how the tactics of social hacking and sound cultures interact, including how sound can be utilized a tactical weapon (Goodman 2010). GlobalNoise is a more contemporary example that revives the traditional practice of making a loud noise by beating a pan or other kitchen equipment to gain attention to demonstrate unrest on a global scale. The algorithmic dimension is that the organizers, the international Occupy and Spanish Indignados movements, invite people to make noise on a certain day and at an agreed time (GlobalNoise 2012). Following our description of wider compositional assemblages, we would expand the notion of the political subject to not only include the demonstrators but also to the objects themselves as if they themselves exhibit political agency.[2] Political subjects and objects become 'entangled' and are activated by scripts and scores that operate under particular conditions that require technical, legal, and discursive detail to fully account for their transformative potential.

33.2.1 Composition as Political Reworking

More than simply an expression of formal logic, algorithms do things through their command structure to accomplish a decisive action with political effects. The way algorithms impose structure on material–discursive elements might be likened to compositional processes more broadly, and the socially engaged algorithmic practices of Cornelius Cardew provide an obvious example for the discussion of the political dimensions of algorithmic music. In the essay 'All Problems of Notation Will Be Solved by the Masses', Simon Yuill (2008) discusses the work of Cardew and the Scratch Orchestra in relation to other techno-social movements such as free-software development and hacker labs, as well as the practice of live coding as an instantiation of making source code available and modifiable in realtime. Also relevant here is ap's 'Life Coding' (involving Martin Howse), which plays with both software and hardware systems in realtime (Yuill 2008). Clearly, algorithms play an important part in this as they define how the music is generated but in the case of live coding, unlike much electronic music, the processes themselves are made somewhat apparent as changes to the code are made public at the time of the performance.

Although there is nothing particularly unusual about experimental and improvisation techniques such as this, it is the recognition of the political potential that has relevance for our argument. Emerging out of various creative energies of the late 1960s (such as the Anti-University of London), the Scratch Orchestra managed to develop a collective compositional form for the sharing of resources, self-organization, and peer critique.[3] The orchestra was open to all, regardless of musical training or ability, under the principles of free improvisation and experimentation. Notes, or 'scratches' as they were called, were performed and developed into larger collage forms, like the sharing of source code, distributed for further modification, and performed under 'copyleft' principles (reverse-engineering normative property relations). Works played with organizational forms and hierarchies, as in the instruction piece cited by Yuill, were likened to the procedures of generative automata: 'Each person entering the performance space receives a number in order. Anyone can give an order (imperatively obeyed) to a higher number, and must obey orders given him by a lower number' (Yuill 2008).[4]

The attention to algorithms highlights the implicit command and control structures of computation and the parallels that can be drawn between technical and social systems. Making explicit political allegiance to Marxism (Cardew was a founding member of the Revolutionary Communist Party in the United Kingdom), Cardew attacked the conservatism of musical notation and announced that 'all problems of notation will be solved by the masses' (Yuill 2008), in keeping with the perceived revolutionary potential of the worker to rewrite history. The utopianism of the Scratch Orchestra is explained by Yuill, as is its collapse as an ideological project through its overreliance on notation as a determining factor and the inherent contradiction that in legislating for nonconformity it operated its own tyranny (as, arguably, all algorithms do). Accordingly, we argue for something less totalizing for algorithmic music and more nuanced in recognition that the relations between subjects and objects are far more complex and that there are inherent conceptual problems with totalizations like 'history' or 'the masses' as the privileged agent of social change.

33.2.2 Microtemporal Rhythms

The conceptual distinction we wish to make with respect to algorithmic music is to shift attention from formalist instruction-based composition to an engagement with the temporal-sonic register of the algorithm itself. Algorithms do not simply define an ensuing event but also a movement and rhythm of signals that oscillate between various materialities and systems. This is what artist-researcher Shintaro Miyazaki has discussed through his concept of 'algorhythmics' (Miyazaki 2012, 2013a): on the one hand, it is a finite sequence of step-by-step instructions (algorithm), a procedure for solving a problem, and on the other a temporal ordering of infinite movement (rhythm). We use this concept centrally in this chapter to try to better understand the operation of algorithms and their inherent rhythmical structures and the musical potential of nonhuman objects. Rather than examine music as such, Miyazaki develops a method for recording the electromagnetic waves that fill the airwaves of our information society. When these recordings are played back and the speaker membrane begins to vibrate, a set of events unfold from a 'microtemporal' perspective (something discussed in more depth in the following section). In one recording, first there are rapid bursts of white noise, slowly transformed into more rhythmical structures accompanied by high-pitched melodic fluctuations hovering high over the low rumblings of metronomically accurate steps, forcing the speaker membrane to its utmost extreme positions. This close description of musical events taking place within the demodulated electromagnetic waves becomes an alternative way of analysing and conceptualizing the way in which algorithms function and somewhat control our experience and behaviour.

33.2.3 Micropolitical Tactics

The chapter develops this 'microtemporal' perspective in order to expose both the micropolitical lines within formalized and rule-based music and how algorithms increasingly mediate contemporary life and politics at multiple scales and temporalities. All politics is both *macro*political and *micro*political simultaneously, according to Deleuze and Guattari (2005), and we therefore ignore the small details at our peril. By connecting micropolitics and microtemporality more closely in this chapter, a more nuanced and subtle form of power can be seen to operate across scales and territories of social order and to address the increased importance of 'media'. This is also where 'tactical media' arguably becomes a useful critical tool for assessing the ability of oppositional movements to create meaningful change—building on Michel de Certeau's *The Practice of Everyday Life* (1984) and Hakim Bey's anarchist notion of 'Temporary Autonomous Zone' (1991). By 'reverse-engineering' aspects of the social order from within the very system itself, at a level of operation that is not directly apparent, tactical media offers contemporary forms of political action that lie somewhere between creative experimentation and a reflexive engagement with communicative forms and social change, using tactics that recognize shifting identifications, temporary alliances, and affinities according to relations, contingency, and context.

Derived from independent media activism and radical pragmatism, tactical media is also inspired by the ballistics of Dada and the Situationist notion of *détournement*, as well as other contemporary sources that draw together art, activism, and hacking, such as 'hacktivism' and 'artivism'. Much has been written on the topic of tactical media— a term that derives from 1996 and in particular the *Next 5 Minutes* festival, held in Amsterdam—especially by Geert Lovink and David Garcia, but also by Rita Raley in her book *Tactical Media* (Raley 2009) that describes some of the hacktivist practices of cultural producers such as Critical Art Ensemble, Electronic Disturbance Theater, and The Yes Men, amongst others. After the proliferation of social media and the Occupy social movement of 2011, the legacies of tactical media and its connections to the present have become the subject of ongoing critical examination, in keeping with its aim of adapting to historical conditions.[5]

Part of the difficulty of the use of tactical use is that it is prone to recuperation, effectively nullifying its critical effects. The noise and glitch music cultures make good examples in providing some useful critical tools for analysing informational flows and at the same time an aesthetic for commercial exploitation. The noise outbursts and screaming digital glitches of Merzbow and Oval in the early 1990s may have had a critical voice at that time, but today these explorations within the 'aesthetics of failure' (Cascone 2000) are subsumed within mainstream culture. Commercial sample packs containing 'the most original and inspiring glitch content possible' (Loopmasters 2012a), 'creating all manner of weird and unlikely sounds' (Loopmasters 2012b), and adding 'some serious twisted nastiness' (Loopmasters 2012c), are today a well-established part of the electronic music producer's sounding palette. Commercialization has transformed auditory glitches from a possible political critique into a well-established music genre, which Kim Cascone later recontextualizes as the 'failure of aesthetics' (Cascone 2010).

But whether tactical media works in the long term is the wrong question to ask. Instead, what should be asked is to what extent are its activities effective at any given moment in time. This describes the tactical as a temporary critical intervention that is highly contingent on the circumstances within which it finds itself, even to the point where its own contradiction can be embraced as a critical move. It is possible to speculate on how certain kinds of critical practice using algorithmic logic can also be pushed to their extremes—as an over-identification with financial capitalism. One challenging example, in the context of a critique of financial capitalism, is the Robin Hood Minor Asset Management project. Robin Hood uses financial technology upon itself, with a dynamic data-mining algorithm called Parasite. The algorithm analyses the big data of financial transactions to raise money and redistribute wealth. So rather than examine the broader political efficacy of these tactical interventions, this chapter aims to concentrate on those tactics that expose the relatively hidden layers of technological complexity through the inner workings of algorithmic machines. In this sense the chapter can be read as a tactical intervention in itself, to draw attention to algorithmic details in order to break down informational and networked flows of power—echoing Michel Foucault. This is why the micropolitical dimension of socio-technical assemblages needs greater elaboration and closer forensic analysis—as seen in the work of Kirschenbaum (2012)— in order to unfold practices that are sensitive to the mode of production and techno-materialist conditions.

33.3 Tactics of Media Archaeology

This section will briefly introduce and situate media archaeology, forming a methodological framing for our further investigation of the micropolitical aspects of algorithmic music. The term 'media archaeology' originates from Jacques Perriault's book *Mémoires de l'ombre et du son: Une archéologie de l'audio-visuel* [Memories of shadow and sound: An archaeology of audiovisuality] from 1981 (Huhtamo and Parikka 2011, 3), but the research field became more established during the 1990s, through scholars such as Erkki Huhtamo, Siegfried Zielinski, and Wolfgang Ernst, building on the media theory of Kittler and the counterhistory of Foucault. Foucault's notion of history opposes the traditional one, which is built upon drawing the great lines, building large-scale chronological tables that order series of continuous events and reflections (Foucault 1972, 3–4). Instead of lines of continuity, he insists on writing a history that focusses on discontinuity and rupture. The term 'archaeology' becomes central for Foucault in unfolding his counterhistory to investigate how various discourses within historical documents are often conflicting and contradictory. The archaeological excavation thus becomes a rewriting, a transformation of what has already been written, focusing greatly on the gaps or the forgotten details—meaning that the archaeological study always functions on many levels at the same time—comparing and opposing them (Foucault 1972, 157). The multiple and diverse discourses are not be regarded as obstacles or something to be overcome, but rather to be seen as what they exactly consist of in order to differentiate them and bring them into new formations of knowledge.

But despite the similarities of approach and connection to Foucauldian methodology, the media archaeological approach is far more complex and diverse. The media archaeological approach emphasizes cyclical rather than chronological development, which is in contrast to the customary way of thinking about technological culture in terms of a constant progress from one technological breakthrough to another, making earlier machines obsolete along the way. The aim of the media archaeological approach then is not to negate the 'reality' of technological development, but rather to balance it by placing it within a wider and more multifaceted social and cultural frame of reference (Huhtamo 1997, 223). This approach introduces a shift from the chronological and positivistic ordering of media technologies towards treating history as a multilayered dynamic system. In that perspective, media archaeology can be seen as a critique of media history in the narrative mode, where continuities are favoured instead of incorporating discontinuities.

The media archaeological focus on a materialistic temporality and processuality can be grounded in the work of Kittler, who redefined and modified Foucault's original concept of archaeology and gave it a more materialist approach. As Kittler claims (1999, 229), the factual conditions of any material object are no simple matter, and need to be examined and understood from a technological perspective, in order to be able to understand the development of technical media and its consequences for modern society. Furthermore, Kittler underlines the importance of the practitioners of cultural studies being able to thoroughly learn and understand mathematics in order for *Medienwissenschaft* to develop past its current media historical status. The fundamental technological logic, such as the procedures of data processing, must be studied from the engineer's point of view rather than merely evaluated 'from the point of view of their social usage' (xiv).

33.3.1 Antidisciplinary Tactics

Even though Foucauldian methodology echoes a distinct political agenda within the field of media archaeology, the approach that focuses on more fundamental materialistic layers of recording and archiving is often critiqued by the apparently nonpolitical nature of this special media ontology (Parikka 2011a, 257; 2011b, 54). In order to stress the inbound political aspects within the media archaeological research tradition, we connect the critique of linear history within Foucault and media archaeology to Certeau's conceptualization of tactics and strategics, which becomes useful in unfolding deeper perspectives of the political issues within technological apparatuses. In *The Practice of Everyday Life*, Certeau unfolds how the strategical can be understood as a force-relationship that occurs when a subject 'of will and power' can be isolated from an environment—executing what could be denoted as a long-time plan, which is a condition for, for example, political and economic models (Certeau 1988, xix). The tactic, or tactical, on the other hand, is described as not belonging to a spatial or institutional context. Tactics can be understood as dynamic processes that are dependent on flowing temporalities which constantly open up to new opportunities that must be seized on the wing. Thus tactics have no base at their disposal in which they can 'capitalise on its advantages, prepare its expansions, and secure independence with respect to circumstances' (xix).

In examining how autonomy might be reclaimed from the forces of commerce and politics, Certeau asserts that users operate opposing established rules in the most ordinary of circumstances. The concern is with the modes of operation, not human subjects as such but their actions, that together form a culture wherein models of action are characterized by users in ways that resist the idea of passive usage or consumption. According to Certeau, consumers negotiate discipline and power exerted on them by tactical forms and makeshift creativity; through what he calls 'antidiscipline'—making a direct reference to Foucault (Certeau 1988, xv). His examples are social practices like walking or cooking, but we might add playing or listening to music in the most ordinary of circumstances—everyday practices that constitute what has become known as popular culture in which social relations are reconstituted and thereby necessarily political. Everyday practices are potentially 'tactical in character', offering new ways of operating and doing politics (Certeau 1988, xix). For instance, the practice of hacking might be usefully described in these terms, as a tactical reuse of existing materials and structures, the modification of instructions and rule sets that we work and live by. Through these conceptualizations proposed by Certeau the media archaeological approach develops perspectives on the dynamic and fluid processes of technological apparatuses, thus unfolding what could be described as antidisciplinary tactics and antihistorical epistemologies.

33.3.2 The Media Archaeological Ear

Through media archaeology it is possible to discern new kinds of historical knowledge that exceed the visible and readable to include ways of listening to historical material: according to Ernst, it becomes possible to hear history. Following in the tradition of Kittler, Ernst takes media to be a blind spot in humanities research, and in the case of the phonograph, for instance, this opens up other sonic registers beyond music. His example is the way that the telephone or gramophone changes the way we understand the relations between writing and speech. According to Ernst, the human voice is frozen by technical media, by reducing it to symbolic code, but can be unfrozen by its replay with all its richness and variations of frequencies. In the case of a recording of a voice, the noise also becomes an important carrier of information—the frozen technical knowledge that is also made available. Ernst describes how technical repeatability generates 'an almost ahistorical functional reenactment' (2013, 175) and thus an experience of media time in contrast to the historicist notion of time. The tactic of 'reenactment' can operate as a time machine in this sense, not operating in the same time as such, but in the way knowledge is generated: 'the original experience is repeatable; the actual experiment allows for (com)munition across the temporal gap' (177).

The claim is that the human sensory apparatus is inadequate for certain tasks and that acoustic archaeology requires the media itself to uncover other hidden aspects of knowledge. This is what Ernst calls the 'media archaeological ear that listens to the sound of material tradition, in fact the technically mediated *sonic* processuality of what is otherwise called history' (2013, 181). Thus media archaeology comes into operation at the point where media (and not just humans) become active archaeologists of knowledge (Ernst 2011, 239), producing a conceptualization in which the media-archaeological ear is more like a microphone and the objects in question become less historical and more processual. Ernst's example is 'Fourier analysis', in which the machine performs a better cultural analysis than the human is capable of. For instance, in signal processing (of audio, radio waves, light waves, seismic waves, and images), Fourier analysis can isolate individual components of a compound waveform, concentrating them for easier detection or removal, in order to generate an expanded cultural understanding of how technology affects and intervenes in culture.[6] In this way, it becomes clear how media archaeology operates as an extension of a Foucauldian 'archaeology of knowledge', extending beyond the limitations of the human sensory apparatus to the nondiscursive realm of technical infrastructures and algorithmic processes.

Today there is an increasing tendency to historicize musical technology, and somehow try to connect a present-day technological reality to historical inventions. But in order for the media archaeological ear to be effective it is very important to distance oneself from the 'retro-maniac' (Reynolds 2011) tendencies within contemporary music technology consumption, and instead dig deeper into operational mechanisms. There are numerous examples of this problem. In 2014 Future Retro released the Zillion Algorithmic Sequencer (Futureretro 2014) a hardware MIDI sequencer that is heavily inspired by the Triadex Muse algorithmic event generator built by Edward Fredkin and Marvin Minsky at MIT in 1972 (Kendall, Haworth, and Cadiz 2014, 9; Sloane 2000, 192). The Triadex Muse was used extensively by composer Maryanne Amacher to create her famous 'eartone' music—a musical construction whose goal is to separate the musical stream from the subjective auditory ear—an aesthetic ideal that later was picked up by sound artist Florian Hecker on his 2012 tribute to Amacher on the album *Triadex Muse Treks* (Hecker 2012). What this contemporary example of reenacting 'vintage' algorithms is missing is that the algorithm cannot be reduced to its discursive symbolic meaning only and does not generate meaning or significance exclusively based on a historical dimension but needs a conceptualization that incorporates the knowledge

FIGURE 33.1 Section of the mechanical musical instrument and performance practice *Steam Machine Music* (2010) by Morten Riis. Courtesy of the artist.

gained from the nondiscursive realm of the technical apparatus itself (Ernst 2009b). So in order to achieve a more nuanced perspective of what constitutes an algorithm, we must shift our analysis towards the micro details of execution and temporality. The way algorithms are embedded within our lives is not just an operation on a symbolic level but an action with effects. Thus in order to understand how algorithms are acting (and controlling) we must go beyond the symbolic register of code and investigate the microtemporal events taking place through a techno-materialist understanding of what is happening at various scales, and across layers of operation, in complex assemblages of hardware and software.

Both *Steam Machine Music* (2010)[7] and *Opaque Sounding* (2014)[8] are examples of this critical media archaeological listening approach that through tactical reenactment operate as time machines and generate knowledge to highlight the materiality of the machine that runs its instructions. Both pieces use perforated paper strips (punch card systems) to store events which are then executed by either a steam engine or a mechanical construction in Meccano driven by a small 12V DC engine. With these programmable machines we are drawn to acknowledge that discursive structures are underwritten by the nondiscursive realm of technical infrastructures and algorithmic processes and that these both run and break down in particular ways. For this reason the combined conceptual and practical approach of media archaeology is successful in challenging the way that informational processes are understood and enacted. Against the grain of technical progress, such examples offer a challenge to media-historical narratives through 'epistemological reverse engineering' as Ernst puts it (2011, 239). From this perspective, the cultural lifespan of a technical object is not the same as its operational lifespan (e.g. a radio receiving an analogue signal), and there is a 'media-archaeological short circuit between otherwise historically clearly separated times' (240). By this, Ernst is suggesting that there is not necessarily a difference in its functional technical operation in the past and now. An algorithmic event undermines its own historicization.

33.4 MICROTEMPORALITY AND TIME-CRITICALITY

Under techno-epistemological conditions, it becomes clear that something other than a conventionally executed history of technical inventions is at stake. As previously explained, the media archaeological ear can short-circuit historical implications, and consequently it becomes necessary not only to understand time and cultural memory at a macro level, but to include the micro-level temporalities, or the time-critical elements which are crucial to both digital and mechanical technologies. 'Critical' in this context is not be understood as in, for example, critical theory, but as in criticality as the decisiveness of the temporal events that happen from the engineers' point of view (Parikka 2011b, 59), meaning that priority is given to the nonsemantic signals of technology; to some extent suspending the cultural or social implications. This media temporality or time-criticality is experienced by experimenting with the physical media themselves (Ernst 2009b). Thus the reenactment of the experiments conducted by Pythagoras on the monochord enables us, according to Ernst, to experience the relationship between integer numbers and harmonic musical intervals (2009b). Of course we are not in the same historical situation as Pythagoras, and even our mode of listening must be considered to be very different, but as Ernst claims, the monochord operates as a time machine: 'it lets us share, participate at the original discovery of musicological knowledge' (2009b).

Time-criticality or microtemporality is how modern technical media has manipulated the time axis since—for example—the Edison sound-reversing phonograph. It is the running of machines, code, and algorithms that is central to the time-critical understanding of digital media, defined by patterns of signals unfolding in time (Parikka 2011b, 59). Ernst focuses on the microtemporality, which combines technical memory with cultural memory as an active process and not just a stable permanent memory. For example, the television image is continuously being regenerated by the line-update frequency and is not just a stable image. The 'prepared televisions' of Nam June Paik, as for example his *Magnet TV* (1965), attest to the creative potential of this techno-materialist understanding, where magnetic fields interfere with the television's electronic signals, distorting the broadcast image into an abstract form. Similarly, the computer is not to be misinterpreted as a static machine with static memory but decidedly a temporal and mutable (sonic) machine.

33.4.1 Forensic Materiality

An understanding of this can be expanded through Mathew Kirschenbaum's notions of 'forensic materiality' and 'formal materiality'. Forensic materiality firmly rests upon its potential of an individualization inherent in matter (Kirschenbaum 2012, 10–11) and can be described as demonstrating how no two things in the physical world are alike; for instance the micron-sized nonconformity of the physical inscriptions on magnetic storage devices and the varying contours of the computer keyboard. Forensic materiality is positioned against the notion of formal materiality, which can be described as being how software sustains and propagates an illusion of immateriality (Kirschenbaum 2012, 11–12). This is an immateriality that is governed by the hyper-redundant error-checking routines within the hard drives' reading algorithms, which constantly minimize the possible reading errors of the physical inscriptions on the drives. Thus the formal can be understood as being concerned with restoring signals to near perfection (Kirschenbaum 2012, 133), which compensates for the reading errors of our rotating hard drives, whereas glitches in software remind us of forensic materiality (135). Forensic materiality and the physical motion and dynamics of the hard drive are demonstrated in numerous experimental artworks working with sound. *Harddisko* (2004) by Valentina Vuksic is an installation demonstrating the temporal qualities of the hard drive. Rhythmic noises are evolving from sixteen hard drives, which are orchestrated through simple power circuits. By cutting the hard disk's power in varying sequences and amplifying the peculiar sound characteristics of each drive, an unpredictable acoustic and visual interplay is taking place (Vuksic 2010). Patterns of sound are produced by the variances of manufacture, models, firmware versions, and the disk's history of usage. The physical properties of hard drives are further explored in *Analog HD1* (2011) and *Analog HD2* (2012) by Gijs Gieskes, as they are transformed into musical devices (Gieskes 2012).

33.4.2 Algorhythmics

In a similar explorative manner, a techno-materialist understanding of the algorithm allows for artistic and epistemological interventions in order to broaden our modes of listening to and analysing the inaccessible signals that occupy our experience of otherwise inaudible all-encompassing streams of communication data. Developed by Miyazaki in collaboration with artist Martin Howse, the concept of algo*rhythms* is an attempt to unfold how computation comprises symbolic *and* real physical structures, as exemplified when real matter becomes controlled by the symbolic and logical structures found in instructions and code. On the one hand, algorithms can be conceptualized as abstract symbolic step-by-step instructions, as seen in programming language, that through a compiler are translated into machine code. On the other hand, algorithms can transform data, but in the end everything can be conceptualized as microrhythmical structures,[9] a conceptualization that introduces the notion of algo*rhythm* (Miyazaki

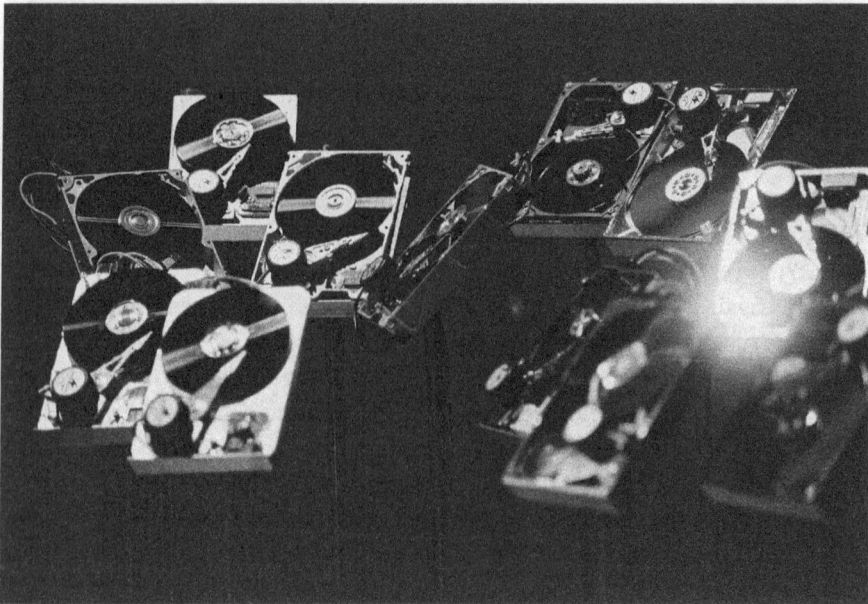

FIGURE 33.2 *Harddisko* by Valentina Vuksic, in performance at Kunsthalle, Vienna (2004). Courtesy of the artist.

2013a, 142). Through this understanding, algo*rhythms* make it possible to hear that our digital culture is not just immaterial, but instead something that performs and unfolds itself through rhythmical, performative, and sensual manifestations (Miyazaki 2013a, 135). This 'trans-sonic' (Miyazaki 2012) conceptualization opens up to a reality in which all signals that are inaudible to humans can be made audible through various media technological devices. The speeds within contemporary communication technologies rapidly exceed our human sensory apparatus, but nevertheless Miyazaki and Howse develop techniques that can be used to unfold these fast and elusive signals, as in the case of Detektor that scans the electromagnetic spectrum.

The Detektor is a small circuit developed by Miyazaki and Howse, an ongoing project that started in 2010. The fundamental design of the Detektor is based around a coil, amplifier, and headphone output, which makes it possible to listen to electromagnetic waves from 100MHz to about 5GHz. The key component in making the inaudible signals audible is the AD8313 Logarithmic Detector chip, which makes the demodulation of the electromagnetic signals possible (Miyazaki 2013b, 515). Through the circuit it becomes possible to listen to the signals of Wi-fi, Bluetooth, GPS, and the wireless signals that are distributed throughout the information ether. In addition, Miyazaki and Howse have conducted a series of workshops and alternative audio-walks in which participants are given the possibility to experience their environment through the conceptualization of deep algo*rhythmic* listening. Though these audio-walks the concept of algo*rhythm* not only provides a close reading of technical details within our communication technologies, but moreover it emphazises a need for aestheticization

(Miyazaki 2013b, 519) in order to propose alternative understandings of our digital and wirelessly distributed reality. The various transmission standards, for instance, the changes between second- and third-generation mobile phones, have profoundly different rhythmical structures, and thus expose changes in the microtemporal infrastructure that define our experience.

Making reference to a Pythagorean[10] understanding, Miyazaki's notion of rhythm proposes itself as an epistemological tool for understanding the world through musical categories (Miyazaki 2013a, 136), thus acting as a critical tool to reveal the hidden sonic structures of technology. Even though we as humans experience the rhythm from a mechanical metronome as being exactly uniform, there are in fact tiny differences which are introduced by irregularities in the wound-up spring within the mechanical metronome. There is also an inexactness present in the quartz crystal running within our computational devices, which is particularly vulnerable to temperature changes that can result in changes of the alleged stable tempo (139). As data become a stream of signals—for instance by plugging an ethernet cable into an audio mixer[11] (143)—a new operational process opens up. This becomes a process of analysing the microrhythmic Geiger-counter-like sounds, set against the rhythmical sequences of ultra-high pitches, combined with more or less melodic rhythmic noises that form continuously sound motifs (144–146).

This forensic approach further resonates with the work of Kittler, who argues for combining material conditions and epistemology, for example, in acknowledging the software used to write a particular chapter like this (Kittler 1995) to highlight how the

FIGURE 33.3 Detektors (2010–) by Shintaro Miyazaki and Martin Howse. Photograph by Samuel Hanselmann.

inscription tools are registered as part of the work itself. The deep listening techniques operating beyond human perceptual registers also offer epistemological challenge to what we know and how we know what we know, extending Foucault's *Archaeology of Knowledge* in seeking hitherto undiscovered knowledge, emergent knowledge forms, and their relation to forms of power. This is an important issue, as it helps to establish a better understanding of techno-material conditions and a politics of knowledge across human and nonhuman spheres.

33.4.3 Why Is Microtemporality Political?

That politics is necessarily related to a conception of historical time is clearly now a contestable position, as we have noted earlier, but the broader point we wish to make here is that the problem of temporality remains a political issue regardless of the position one takes or the scale of operation (Osborne 1995). As also established, the microtemporal aspect is important for its stress on relatively hidden knowledge as part of wider, complex socio-technical assemblages. In order to operate, algorithms exist as part of assemblages that include data, data structures, and bodies, together part of a process of automation that tries to reduce all aspects to the behaviours that can be controlled and determined. As a consequence, politics can be understood to no longer operate simply on a macro socio-economic level or in ideology alone but increasingly from mutations at a micro-scale or molecular level. In other words, there is a dynamic between micropolitics and politics in general, and algorithms modulate between the details of microtemporal processes and larger totalizing data sets like history or society.

For Deleuze and Guattari, the micromolecular offers 'lines of flight' from the over-coded machine organization of society, but can also develop into something which later becomes conceptualized within a macropolitical framing. An example of this could be the student revolution in France in 1968, a situation that started as a molecular oper-ation, and developed into a macropolitical overcoding machine. Deleuze and Guattari outline a conceptualization of events in which micropolitics is not defined by the small-ness of its elements, but instead by the nature of its mass—understood as the difference between a quantum flow and a segmented line. Through this perspective, the quantum flow always implies that something eludes or escapes the overcoding machine, because the segmented line implies a substitution of faltering perspectives in favour of totali-tarian overcoming lines (Deleuze and Guattari 2005, 217–219). The micropolitical thus becomes a shift of perspective away from high-level totalitarian decision makers, and instead provides focus on how power is exercised at the minute level of individual sub-jects and through 'everyday techniques that form perception, desires and judgments of individuals' (Scherer 2007, 564).

The execution and distribution of power within the microtemporal structures of algo*rhythmic* listening can be exemplified by focusing on how different parts of the signals impose a stronger audible fingerprint than other parts, as for example with the relationship between ethernet clock frequency, encoded bit streams through digital base-band modulation, and the self-clocking line code used in the 10Base-T protocol called the Manchester Code (Miyazaki 2013a, 143). These three layers of signals transmitted through ethernet cables provide a different notion of temporality, which goes against a totalizing and unifying understanding of what data and algorithms are ontologically. By conducting microtemporal analysis it becomes evident that the otherwise stable clock frequencies that constitute our information society in fact come attached with a multitude of inaccuracies due to changes of temperature, humidity and the physical materiality of cabling.

That the micropolitical quantum flow always escapes totalitarian lines further resonates with the idea of counterhistory proposed by Foucault and his description of forms of power as plural and decentralized. Accordingly, contingency and power relations need to be understood at all scales, as well as how algorithms execute a logic of command and control over these processes. These shifts of scale suggest that action can no longer be attributed to individual agents but to distributed action throughout more complex assemblages that indicate how algorithms need to be understood as 'relational, contingent, contextual in nature' (Kitchin 2017, 18). Microtemporality reminds us that politics is not simply human-centred but also involves nonhuman entities such as algorithms that express not only new forms of power that dictate how certain events unfold but also new lines of flight.

33.5 TACTICAL ALGORHYTHMICS

Under informational capitalism, automation of industrial production is accelerated by hidden algorithms to the extent that processes of control and command are hard to identify, let alone to exert control over. Importantly although, like contemporary forms of power, algorithms are everywhere and nowhere—anywhere perhaps—and they operate as part of larger assemblages that include data and infrastructures, software and hardware, programmers and other living entities, forms of knowledge, behaviours, and actions at micro spatio-temporal levels of operation.

As mentioned previously, tactics are required to adapt to any concrete situation, as opposed to strategy, which involves long-time planning often distanced from its actual execution. In this way tactical media's tactics are somewhat like the experimentation of Cardew, amongst others, inasmuch as they play with the spatio-temporal determinants of compositional assemblages as emergent forms of political organization. Algorithms are already tactical in this sense with respect to the way they operate relationally and contingently. Such an approach offers a conceptualization that corresponds very well to a microtemporal understanding of the algorithm, as we cannot reduce the algorithm to a set of predetermined instructions or rules, but we instead need to include an understanding that incorporates the wider apparatus and its execution in time. Through this tactical perspective the algorithm becomes very different from its otherwise prevailing conceptualization as something that governs data and processes our lives. Instead it shifts attention to something that in fact is dislocated from these strategic overcoded perspectives—because of its temporality and its constant and ever-changing embodiment within physical technology and its wider social context. This conceptualization resonates well with Deleuze and Guattari's understanding of the abstract machine—a machine that can be both an overcoding totalitarian machine and at the same time a decoding detotalitarian machine that flows and emits new quanta, something that is constantly changing on a molecular level (Deleuze and Guattari 2005, 223–224).

'Tactical' in this context might also be coupled with Walter Benjamin's notion of 'technique' in his essay *The Author as Producer* (1934), proposing and recasting the difference between overcoded political action and a more tactical politics that operates on the ground at the level of production. He writes: 'An author who has carefully thought about the conditions of production today ... will never be concerned with the products alone, but always, at the same time, with the means of production. In other words, his [or her] products must possess an organizing function besides and before their character as finished works' (Benjamin 1998, 98). Thus, Benjamin recommends that the cultural producer intervene in the production process, in the manner of an engineer. Again the point is that this is where social relations are operative and where techniques can be developed in order to change the apparatus. The spirit of this is captured more recently in the *Critical Engineering Manifesto* that similarly foregrounds the tactical potential of engineering: 'The Critical Engineer notes that written code expands into social and psychological realms, regulating behaviour between people and the machines they interact with. By understanding this, the Critical Engineer seeks to reconstruct user-constraints and social action through means of digital excavation.'[12]

33.5.1 Tactical Media Archaeology

'Tactical media archaeology' becomes operative at the point where media and more specifically the nondiscursive realm of technical infrastructures and algorithmic processes become active archaeologists of knowledge (Ernst 2011, 239). The work of Miyazaki and Howse exemplifies this approach as introduced above, but we would like to end this chapter with another work by Howse to stress our argument here through epistemic practice. Dark Interpreter (2015) is a noise instrument that takes its name and inspiration from the writings of Thomas de Quincey: 'The truth I heard often in sleep from the lips of the Dark Interpreter. Who is he? He is a shadow, reader, but a shadow with whom you must suffer me to make you acquainted' (Quincey 1891, 7). Dark Interpreter comes in three incarnations: Mater Lachrymarum, Our Lady of Tears, for complex granular sampling, is particularly suited for vocal reprocessing; Mater Suspiriorum, Our Lady of Sighs, has skin control and code entry to generate harsh noise with relinquished control; Mater Tenebrarum, Our Lady of Darkness, is an algorithmically designed, gold-plated skin breakout board and skin/EEG amplifier worn on the head or placed in the earth to produce harsh noise or granular processing with divined control.[13]

The overall project is an attempt to reveal the hidden dark forces of the electromagnetic spectrum that remain imperceptible to the human sensory apparatus. Howse describes it thus: 'The Dark Interpreter is modelled as a set of leaky, overlapping medieval village spaces within which various plague simulations run, and through which a motley of villagers (grains) wander, steered by electric fingers and touch' (2015a, unpaginated) It interprets the contemporary noise instrument as a 'dark symbolic mirror', placing control of parameters on the skin, and strapping a 'psyche/plague village interface' to the head. Howse describes the technique as 'modern live coding with no screen, no keyboard and little conscious control'. With the *Dark Interpreter* you are interfacing directly with the circuit, and by touch, your skin functions as an integrated resistor within the mechanism, thus altering resistor-capacitor time constants (Earl, 1977, 61–62), which introduces changes in the instrument's frequency response. A resistor-capacitor time constant is calculated by the time it takes a capacitor to charge and discharge its levels of voltage through a resistor. By changing the capacitor and/or resistor values it is possible to change the frequency response of a system. It takes time to charge or discharge a capacitor through a resistor. At low frequencies, there is plenty of time for the capacitor to charge up to practically the same voltage as the input voltage. At high frequencies, the capacitor has time to charge up only a small amount before the input switches direction.

FIGURE 33.4 Dark Interpreter: Mater Tenebrarum, Our Lady of Darkness (2015), by Martin Howse. Photograph courtesy of the artist.

The output goes up and down only a small fraction of the amount the input goes up and down. At double the frequency, there's time for it to charge up only half the amount—which then gives us the cut-off frequency determined by the resistor-capacitor time constant.

By focusing on these microtemporal changes modulated by human skin,[14] an awareness is raised to the dramatic effect that even the tiniest delays have on the timbral expressions within our sounding technology. Through this conceptualization, the tactical media archaeological algorithm not only develops time-critical perspectives on the discursive computational processes but also draws attention to the nondiscursive operations and their techno-epistemological conditions. Through sound, these techno-epistemological forces are cast in terms of medieval belief systems with contemporary environmental significance, in which the effects of algorithms can be understood as part of broader ecologies. Algorithms in this sense manage to operate across both discursive and nondiscursive registers and offer the potential to compose alternative musical and political epistemologies—beyond conventional anthropological notions of what constitutes music. Furthermore, the relation between sound and music does not make ontological sense either as an algorithm does not recognize the distinction.

The term 'dark' in Dark Interpreter could also be referencing the 'object-oriented philosophy'[15] of Timothy Morton and his notion of 'Dark Ecology'[16] (Morton 2007), which redefines the notion of ecology to become a way of collapsing the subject-object division, giving rise to a sense of coexistence and connection with other objects. The role of this dark ecologically aware art then becomes a way of 'attuning' to the inconsistency within and between objects, as a process that is always slightly out of phase, and which recognizes its fragility and thereby also its own uncanny strangeness (Morton 2013b, 177). Morton's key concepts of inconsistency, fragility, and strangeness additionally extends the tactical media archaeological perspectives on the macro and micro levels of computational culture. Thus the algorithm becomes something that is impossible to be reduced to consistent overcoding semantics, and instead develops into a fragile inconsistent 'hyperobject' (Morton 2013a), massively distributed in time and space as it transcends spatiotemporal specificity. Morton argues that in the age of the anthropocene (Morton 2011, 154; Whitehead 2014), the dark ecological perspective is not only important within analysis, but also in the creation of sounding pieces. This is because art is forced to relate to the current state of affairs concerning the human-made ecological crisis (Morton 2013b, 20, 22).

Connecting the aesthetic to the causal dimension through dark ecological art forces us to coexist with a vast plenum of nonhuman objects, such as algorithms, and helps us explore our own fragility and sense of contingency. Doing so collapses the belief that we can distance ourselves from the world through formal procedures, just as algorithmic music cannot separate itself from the world around us. Our claim is that algorithms need to be understood as part of temporal, relational, and contingent operations that are sensitive to their conditions and future trajectories. Only in this way can algorithmic music begin to make sense—politically at least.

NOTES

1. Jem Finer's *Longplayer* began playing at midnight on 31 December 1999 (see http://longplayer.org/). John Cage's *As SLow aS Possible* was first written in 1987 (see https://en.wikipedia.org/wiki/As_Slow_as_Possible).

2. The exhibition 'Disobedient Objects' made a similar claim by examining the powerful role of objects in movements for social change (Victoria and Albert Museum, London, 26 July 2014 to 1 February 2015). See http://www.vam.ac.uk/content/exhibitions/disobedient-objects/.

3. For more on the Anti-University, see Jakob Jakobsen's research from 2012, http://antihistory.org/.

4. Another example of this antitotalitarian mode of organization within musical composition is Cardew's *The Great Learning—Paragraph 7* (Cox and Warner 2004, 228).

5. Tactical Media Files is a research project tracing the legacies of Tactical Media (Kluitenberg and Garcia 2008–). A book project, *Legacies of Tactical Media: The Tactics of Occupation*, was initiated in 2011 by the Institute of Network Cultures.

6. The algorithm Fast Fourier Transform converts time (or space) to frequency (or wavenumber) and vice versa. For more on this, see Roads 1996, 1075–1112.

7. For video documentation of *Steam Machine Music*, see Riis 2010. Also see Riis 2013a for an expanded exposition of the piece within a media archaeological framing.

8. For video documentation of *Opaque Sounding*, see Riis 2014.

9. Here referencing Kittler (1995) and his conceptualization that software always is reducible to the movement of electrical current in registers.

10. Media archaeology becomes a means of listening to computation through technology as sonic events—opposed to musical theory, which in the occidental tradition continues from Pythagorean epistemology of harmonic calculation, entailing that sound is not perceived as a sonic event but instead becomes a phenomenon of mathematics (Ernst 2014).

11. This way of using our auditory sense as an epistemological tool for understanding computational processes can be contextualized as a reenactment of the way debugging and error checking was done on early computers in the 1940s and early 1950s. Like all other first-generation computers from this time, the Australian CSIRAC for instance, had a built-in loudspeaker which was used for warnings and debugging, and to signify the end of a programme (Doornbusch 2004, 12). The need to use auditive feedback was due to the fact the visual feedback devices such as display monitors had not been invented at this time. The sound reproduced by the loudspeaker was the raw bit pulse from the data bus, an on/off switching corresponding to the stream of bits, alterations of electronic current sonified by the movement of the speaker membrane.

12. *Critical Engineering Manifesto*, written by Julian Oliver, Gordan Savičić, and Danja Vasiliev (2011–2017), http://criticalengineering.org/.

13. See Howse 2015a for more details. Video documentation of the Dark Interpreter 'Mater Tenebrarum' can be found at Howse 2015b.

14. Or by the moisture in soil, as when the Dark Interpreter is being used in Howse's 'Earth Coding'—a project that explores alternative links between contemporary technology and the Earth, raising the question of whether the Earth as a process can be tempted to compose software (Medosch 2014).

15. Within object-oriented philosophy, Graham Harman additionally develops the notions of 'over-mining' and 'under-mining', a construction that parallels the macro (totalization)

and micro (details) levels of Deleuze and Guattari but that develops into an understanding in which an object is not reducible to its parts (undermining) but that also implies that an object cannot be reduced to its whole (overmining) (Harman 2011, 7–18; Morton 2011, 150; 2013b, 44). Following this line of thinking, the World—including other holistic concepts such as Nature and the Environment—also ceases to exist as a neutral background or stage for humans to occupy.

16. Martin Howse and Timothy Morton are often associated together (Parikka 2015; SonicActs 2015; Tuned City 2013), as both are trying to develop alternative conceptualizations of ecology through different notions of human and nonhuman interconnectivity.

Bibliography

Benestad, F. *Musikk og tanke: Hovedretninger i musikkestetikkens historie fra antikken til vår egen tid*. 2nd ed. Oslo: Aschehoug, 1977.

Benjamin, W. 'The Author as Producer'. In *Understanding Brecht*, translated by A. Bostock, 85–103. London: Verso, 1998.

Bey, H. *T.A.Z.: The Temporary Autonomous Zone*. New York: Autonomedia, 2011.

Bumgardner, J. 'Kircher's Mechanical Composer: A Software Implementation'. Paper presented at 'Bridges 2009: Mathematics, Music, Art, Architecture, Culture'. Banff, AB, 2009.

Cascone, K. 'The Aesthetics of Failure: "Post-Digital" Tendencies in Contemporary Computer Music'. *Computer Music Journal* 24, no. 4 (2000): 12–18.

Cascone, K. 'The Failure of Aesthetics'. Paper presented at the Share Festival, Torino, Italy, 2010. http://vimeo.com/17082963.

Certeau, M. de. *The Practice of Everyday Life*. Translated by S. Rendall. Berkeley: University of California Press, 1988.

Cox, C., and Warner, D.. *Audio Culture: Readings in Modern Music*. New York: Continuum, 2004.

Deleuze, G., and Guattari, F. *A Thousand Plateaus: Capitalism and Schizophrenia*. Translated by B. Massumi. Minneapolis: University of Minnesota Press, 2005.

Doornbusch, P. 'Computer Sound Synthesis in 1951: The Music of CSIRAC'. *Computer Music Journal* 28, no. 1 (2004): 10–25.

Earl, J. *Cassette Tape Recorders*. Watford: Fountain, 1977.

Ernst, W. '"… Else Loop Forever": The Untimeliness of Media'. Extended draft paper for 'Il Senso della Fine', conference. Universita degli Studi di Urbino, 2009a.

Ernst, W. 'Experimenting Media-Temporality: Pythagoras, Hertz, Turning'. Experimentation as event. Lecture manuscript. University of Lancaster, 2009b.

Ernst, W. 'Media Archaeography: Method and Machine versus History and Narrative of Media'. In *Media Archaeology: Approaches, Applications, and Implications*, edited by E. Huhtamo and J. Parikka, 239–255. Berkeley: University of California Press, 2011.

Ernst, W. *Digital Memory and the Archive*. Edited by J. Parikka. Minneapolis: University of Minnesota Press, 2013.

Ernst, W. 'The Sonic Time Machine: Explicit Sound and Implicit Sonicity in Terms of Media-Epistemological Knowledge'. Lecture given at Aarhus University, 22 May 2014.

Foucault, M. *The Archaeology of Knowledge*. Translated by A. M. Sheridan Smith. London: Tavistock, 1972.

Foucault, M. 'Nietzsche, Genealogy, History'. In *Language, Counter-Memory, Practice: Selected Essays and Interviews*, edited and translation by D. F. Bouchard and S. Simon, 139–164. Ithaca, NY: Cornell University Press, 1977.

Futureretro. 'Zillion Sequencer. 2014. http://www.future-retro.com/products.html#!/Zillion-Sequencer/p/50667567/category=0, accessed 30 June 2017.

Gieskes, G. 'Analog HD'. 2012. http://gieskes.nl/instruments/?file=analog-HD.

GlobalNoise. 'GlobalNoise'. 23 September 2012. http://www.tacticalmediafiles.net/campaigns/6446/GlobalNoise;jsessionid=FC75EB4798A76597DB4EB918D2C22E41.

Goffey, A. 'Algorithm'. In *Software Studies: A Lexicon*, edited by M. Fuller, 15–20. Cambridge, MA: MIT Press, 2008.

Goodman, S. *Sonic Warfare: Sound, Affect, and the Ecology of Fear*. Cambridge, MA: MIT Press, 2010.

Harman, G. *The Quadruple Object*. Winchester: Zero Books, 2011.

Hecker, F. *Triadex Muse Treks*. Berlin: Neuer Berliner Kunstverein, 2012.

Howse, M. 'The Dark Interpreter'. *The Dark Interpreter*, 6 September 2015a. http://www.1010.co.uk/org/darkint.html.

Howse, M. 'Dark Interpreter Simple Setup/Tutorial 001 No-input'. *Matrixsynth*, 16 March 2015a. http://www.matrixsynth.com/2015/03/dark-interpreter-simple-setuptutorial.html.

Huhtamo, E. 'From Kaleidoscomaniac to Cybernerd: Notes toward an Archaeology of the Media'. *Leonardo*, 30, no. 3 (1997): 221–224.

Huhtamo, E., and Parikka, J. *Media Archaeology: Approaches, Applications, and Implications*. Berkeley: University of California Press, 2011.

Kendall, G. S., Haworth, C., and Cadiz, R. F. 'Sound Synthesis with Auditory Distortion Products'. *Computer Music Journal* 38, no. 4 (2014): 5–23.

Kirschenbaum, M. G. *Mechanisms: New Media and the Forensic Imagination*. Cambridge, MA: MIT Press, 2012.

Kitchin, R. 'Thinking Critically about and Researching Algorithms'. *Information, Communication and Society* 20, no. 1 (2017): 14–29

Kittler, F. A. 'There is No Software'. *CTheory.net*. 18 October 1995. http://www.ctheory.net/articles.aspx?id=74.

Kittler, F. A. *Gramophone, Film, Typewriter*. Translated by G. Winthrop-Young and M. Wutz. Stanford, CA: Stanford University Press, 1999.

Kluitenberg, E., and Garcia, D. *Tactical Media Files*. 2008–. http://www.tacticalmediafiles.net/.

Loopmasters. 'Ambient Glitch Vol. 1'. 2012a. http://www.loopmasters.com/product/details/539/Ambient-Glitch-Vol-1.

Loopmasters. 'Circuit Bent Sounds Vol. 1'. 2012b. http://www.loopmasters.com/product/details/311/Circuit-Bent-Sounds-Vol1.

Loopmasters. 'Glitch Hop'. 2012c. http://www.loopmasters.com/product/details/1276/Glitch-Hop.

Medosch, A. 'Kunst Fields: Wild Ontologies in the Service of Activating History'. *Kunstradio—Radiokunst*, 30 November 2014. http://kunstradio.at/2014B/30_11_14en.html.

Miyazaki, S. 'Algorhythmics: Understanding Micro-Temporality in Computational Cultures'. *Computational Culture* 2 (2012). http://computationalculture.net/article/algorhythmics-understanding-micro-temporality-in-computational-cultures.

Miyazaki, S. 'AlgoRHYTHMS Everywhere: A Heuristic Approach to Everyday Technologies'. In *Off Beat: Pluralizing Rhythm*, edited by J. H. Hoogstad and B. Stougaard Pedersen, 135–148. Amsterdam and New York: Rodopi, 2013a.

Miyazaki, S. 'Urban Sounds Unheard-of: A Media Archaeology of Ubiquitous Infospheres'. *Continuum: Journal of Media and Cultural Studies* 27, no. 4 (2013b): 514–522.

Morton, T. *Ecology without Nature: Rethinking Environmental Aesthetics*. Cambridge, MA: Harvard University Press, 2007.

Morton, T. 'Objects as Temporary Autonomous Zones'. *Continent* 1, no. 3 (2011): 149–155.

Morton, T. *Hyperobjects: Philosophy and Ecology after the End of the World.* Minneapolis: University of Minnesota Press, 2013a.

Morton, T. *Realist Magic: Objects, Ontology, Causality.* Ann Arbor, MI: Open Humanities Press, 2013b.

Nechvatal, J. *Immersion into Noise.* Ann Arbor, MI: Open Humanitites Press, 2011.

Nierhaus, G. *Algorithmic Composition: Paradigms of Automated Music Generation.* Vienna: Springer, 2009.

Osborne, P. *The Politics of Time: Modernity and Avant-Garde.* London: Verso, 1995.

Parikka, J. 'Mapping Noise: Techniques and Tactics of Irregularities, Interception, and Disturbance'. In *Media Archaeology: Approaches, Applications, and Implications,* edited by E. Huhtamo and J. Parikka, 256–277. Berkeley: University of California Press, 2011a.

Parikka, J. 'Operative Media Archaeology: Wolfgang Ernst's Materialist Media Diagrammatics'. *Theory, Culture and Society* 28, no. 5 (2011b): 52–74. doi: 10.1177/0263276411411496.

Parikka, J. 'Mutating Media Ecologies'. *Continent* 4, no. 2 (2015): 24–32.

Quincey, T. P. de. *The Posthumous Works of Thomas De Quincey.* Edited by A. H. Japp. London: William Heinemann, 1891. http://www.gutenberg.org/files/23788/23788-h/23788-h.htm.

Raley, R. *Tactical Media.* Minneapolis: University of Minnesota Press, 2009.

Reynolds, S. *Retromania: Pop Culture's Addiction to Its Own Past.* New York: Farrar, Straus and Giroux, 2011.

Riis, M. 'Steam Machine Music'. *Vimeo,* 2010. http://vimeo.com/16995143.

Riis, M. 'The Media Archaeological Repairman'. *Organised Sound* 18, no. 3 (2013a): 255–256.

Riis, M. 'Systemics #2: As We May Think (Or, The Next World Library)'. 21 November 2013b. http://seismograf.org/systemic-2-as-we-may-think-or-the-next-world-library-0.

Riis, M. 'Opaque Sounding'. *YouTube,* 9 August 2014. http://www.youtube.com/watch?v=nzybYC5nqJM.

Roads, C. *The Computer Music Tutorial.* Cambridge, MA: MIT Press, 1996.

Robin Hood Asset Management Cooperative. 2017. http://www.robinhoodcoop.org/.

Scherer, M. *Micropolitics: Encyclopedia of Governance.* Thousand Oaks, CA: Sage, 2007.

Sloane, S. *Digital Fictions: Storytelling in a Material World.* Stamford, CT: Ablex, 2000.

SonicActs. 'Sonic Acts: The Geologic Imagination'. 26 February to 1 March 2015. http://www.sonicacts.com/2015.

Tuned City. 'Documentation'. *Tuned City,* 2013. http://www.tunedcity.net/?page_id=4594.

Vuksic, V. 'Harddisko'. 2004–2010. http://harddisko.ch/, accessed 30 June 2017.

Whitehead, M. *Environmental Transformations: A Geography of the Anthropocene.* New York: Routledge, 2014.

Yuill, S. 'All Problems of Notation Will Be Solved by the Masses'. *Mute,* 23 May 2008. http://www.metamute.org/editorial/articles/all-problems-notation-will-be-solved-masses.

CHAPTER 34

ALGORITHMIC MUSIC FOR MASS CONSUMPTION AND UNIVERSAL PRODUCTION

YULI LEVTOV

34.1 INTRODUCTION

HISTORICALLY, an individual performance of music has always been experiential and transient in form, with the performer and listener occupying the same physical space at the same time. Musicians were intrinsically connected to their surroundings, both physically and emotionally, making each rendition of a piece of music unique and, in some way, responsive to its immediate environment. If considered on this wider time-line, it is actually only recently (the last 100 years or so, compared to 39,000 years since the existence of the first musical instruments[1]) that we have been able to mechanically record and reproduce sound, and therefore capture the essence of an individual performance. It is even more recent that these recording methods have been developed and formalized to provide us with digital formats such as the MP3—currently the most common format for music (Sterne 2014). As such, it is remarkable that every music listener currently alive in the developed world will have grown up with these recorded formats as their default for music consumption, whether it was via radio, vinyl, CD, digital download, or streaming. These 'static' music distribution media have given birth to incredible advances in the art form as a whole, resulting in a worldwide industry, and making possible countless careers such as session musician, recording engineer, and promoter, not to mention the new musical genres that would simply not exist were it not for recorded, reproducible sound. However, the widespread formats of today have one particular defining characteristic that is fundamentally at odds with the experiential, transient nature of live music or indeed algorithmic music; once their sonic content is defined, they are designed to sound identical at every hearing. External factors that make a live performance unique, such as the physical characteristics of the venue, the size or physical proximity of the audience, the emotional mood of the performer, among countless others, have all been sacrificed during the development of a modern music format that can be distributed and consumed en masse. It is unsurprising, therefore, that any form of algorithmic music, where variation and process can define the end musical result in different ways in realtime, is not well accommodated by these standard formats, causing the speed at which the field can progress to be much slower than, say, pop music, which is wholly reliant on the predictability and perfect reproduction of sound.

As many authors in this book demonstrate, the wider creative opportunities for music composition and performance afforded by algorithms and computers are myriad, and have been explored and developed by an ever-growing population of musicians dating back to early pioneers of the field, such as Stockhausen and Xenakis, through to any modern-day producer who has used a procedural or algorithmic technique such as an arpeggiator, LFO, or randomizer.[2] However, the ways in which listeners can actually experience algorithmic music *in an algorithmic format*, that is, a format which presents the variation of the algorithms in realtime as opposed to just repeatedly presenting a frozen snapshot of one variation, are actually very rare.

While simply *writing* music using algorithmic techniques presents musicians with an incredible array of creative possibilities, actually *distributing* compositions in a way that allows listeners to experience the music algorithmically takes these principles into an entirely new creative realm. This chapter explores methods that aim to do just that, including examples of artists and albums that have exploited them. Such new artistic possibilities often require new skills and expertise from creators and new equipment and habits from listeners, so, on a more practical level, we also explore the challenges faced by creators, distributors, and listeners alike in making algorithmic music more mainstream, and how a new creative role is being forged in the process of distributing algorithmic music.

34.1.1 Types of Algorithmic Music for Mass Distribution

In the context of mass distribution, it is useful to define three distinct types of algorithmic music, which differ solely in the amount of listener interaction involved. Because all forms of mass-distributed algorithmic music effectively live as pieces of computer software, it is often helpful to borrow terms from this world.

34.1.1.1 *Generative*

The defining factor of generative music is that there is no way for any factor outside of the music system (software) to interact with or influence the algorithms and processes determining the end musical result. The software may include methods of random (or nonrandom) initialization for the sound banks, sequences, or other such musical factors, and therefore introduce the variety which defines algorithmic music, but beyond that, the listener is completely noninteractive with the system, other than that they possibly start or stop the software.

34.1.1.2 *Reactive*

Reactive music shares perhaps 90 percent of its DNA with generative music, but the key difference is that instead of the musical parameters and variations being defined purely by random or nonrandom number generators, some of the parameters can be influenced by external factors that are specific to the listener's environment. The scope of the environmental factors that can be used for this purpose is incredibly wide, and could range from something directly linked to an individual listener, such as their walking pace or heart rate, to something more indirect found in the wider world, such as the weather at a particular location and time. The defining characteristic of these external factors is that the listener does not try to actively manipulate them in order to influence the music—they are passive interactions which the music software uses to influence the musical results.

34.1.1.3 *Interactive*

Like reactive music, interactive music features ways for the outside world to influence the musical results of the composition; however, instead of the external factors being passive to the listener, the music software features some kind of interface that the listener is actively encouraged to use to interact with and deliberately take a direct role in the musical composition. This interface could be a touchscreen on a mobile phone, an accelerometer in a wearable device, or indeed any method of turning listener gestures into useable data for the music software.

Any one work, be it a mobile app, installation, and so on, can fit neatly into one of these categories or indeed involve a mixture of techniques. For example, an installation could be largely generative in its sound creation processes, yet involve some elements of interactivity with the listeners. Or a piece of music in a mobile app could be mostly reactive in that the listener's walking pace is dictating the overall energy level of the music, but there could also be some buttons on the touchscreen interface that allow the listener to switch between certain states within the music at will, thus adding a layer of interactivity.

34.2 DISTRIBUTED FORMATS FOR ALGORITHMIC MUSIC

Because algorithmic compositions rely on real-time execution of processes to generate sound, they cannot be listened to on standard digital music playback equipment such as home hi-fi systems or MP3 players—only on some kind of computer with a CPU powerful enough to execute the required processes, and with storage for any required audio samples. Furthermore, there are many different software options for how one could compose algorithmic music, ranging from writing code in environments such as SuperCollider[3] or ChucK,[4] to using dedicated software packages such as Noatikl[5] or

Open Music.[6] Although there is a wide range of options for composition, none offer a truly seamless path from composition to mass distribution to listeners. For decades this fragmentation meant that algorithmic music could be experienced in an algorithmic format only in live media such as installations, multimedia performances, and live-coding gigs. Despite this limitation, there are many examples of algorithmic music being distributed to mass audiences, but in nonalgorithmic formats.

34.2.1 Static Snapshots

The most poignant example of this disconnect between the production and consumption of algorithmic music is undoubtedly Brian Eno's seminal ambient work *Music For Airports* (Eno 1978). *Music For Airports* was created using a series of semi-unpredictable processes involving tape machines, synthesizers, recordings of vocals and instruments, and so on that would sound different every time one listened to it. The ways in which sounds blended, loops overlapped, and musical phrases combined to create new melodies and polyrhythms were results of a process which was never absolutely defined by Eno, and provided seemingly infinite variations of what was still considered one body of work. However, when it came to distributing this work to the masses, the only method available at the time was to press vinyls—a distinctly static medium. As such, although *Music For Airports* is the result of a process which can be exploited to deliver different results time after time, ad infinitum, the actual master recording of the album available today is merely a forty-eight-minute snapshot of one particular variation.

> I worked on things like [algorithmic music] for a while: *Music for Airports* and *Discreet Music* were examples, but what they represented were recordings of these processes in action. What I really wanted to do was to be able to sell the process to somebody, not just my output of it. (Brian Eno, in Dredge 2012)

This process of algorithmic composition coupled with static distribution was developed further in 2012 by UK electronic duo Icarus, with their album 'in 1,000 variations' *Fake Fish Distribution*. By using a combination of custom written software for the graphical programming language, Max MSP,[7] and the popular music creation tool Ableton Live,[8] *Fake Fish Distribution* was distributed as a limited edition of 1,000 digital downloads, each featuring the same source sounds and songs, but varying in more detailed sonic characteristics, making each album unique. Oliver Bown and Sam Britton also wrote a paper (2014) detailing the production methods used for *Fake Fish Distribution*, which goes on to explore the social and artistic implications of the practice.

34.2.2 Software Distribution

Early examples of any kind of algorithmic music being distributed to an audience that could listen to these compositions at their leisure started to occur only in the 1990s.

Koan, the first such generative music engine was released in 1992, with the authoring tool Koan Pro coming later in 1994, and Brian Eno released an album of generative pieces, *Generative Music 1*, for Koan Pro in 1996. Distributed as software on a floppy disc for PC, *Generative Music 1* still had an inconvenient system requirement, including a special soundcard, without which the music would not sound as intended. Also in 1996, *Let Us Play!*, the fourth studio album from UK electronica duo Coldcut, came with a CD-ROM featuring Playtime, an interactive music toy that allowed the user to remix and manipulate different parts of the overall musical composition in realtime. There has always been an active community of artists distributing their algorithmic works as source patches for programs like Pure Data and Max MSP, including well-known electronic artists such as Autechre and Aphex Twin (Holmes 2008). Max MSP also makes available a special version of its software dedicated to just playing back patches without being the user being able to edit them: Max Runtime.[9] Almost twenty years later, Yaxu (Alex McLean's performing name) released his *Peak Cut* EP (Yaxu 2015) as a USB stick containing static versions of his algorithmic compositions, the source code for all of the tracks, and a bootable operating system with his own programming language, Tidal, preinstalled.

34.2.3 Videogames

In as far as videogames are a well-established form of mass-distributed, interactive media, it is important to note their role in the development of algorithmic music, and this topic is covered in depth by David Kanaga in chapter 24 of this volume. Even the most basic of game titles will feature elements of algorithmic (usually interactive) music and industry-standard authoring tools for game sound designers and composers, such as Wwise[10] and FMOD,[11] are dedicated software packages for creating algorithmic music and sound, albeit solely in the context of games. The creative possibilities of generative, interactive, and reactive music have long been explored in this particular section of entertainment media, and this is possible because the format through which it is experienced—games console or computers—consists of interactive software packages, as opposed to static formats such as the MP3.

All of these systems rely on some kind of desktop computer or games console to generate music but, since the portable music revolution brought about by inventions like the Sony Walkman in the 1980s, music consumption patterns have become distinctly mobile and personal (Gopinath and Stanyek 2014). It was the proliferation of smartphones that gave algorithmic music composers a platform on which to easily distribute their compositions to a large audience in a truly dynamic format.

34.3 PROLIFERATION OF SMARTPHONES

From around 2007 (miniMIXA from Intermorphic), the smartphone revolution put a fully-fledged portable computer in people's pockets, and effectively created a whole new economy for developers of applications (apps) for these devices (Gopinath and Stanyek 2014). Dedicated devices such as MP3 players have become old-fashioned, and now people listen to their music on devices that are capable of making phone calls, sending emails, browsing the web, and much more. These handheld computers are entirely capable of executing the kind of processes required to generate algorithmic music in realtime, at the listener's whim. This was a significant milestone for the mass distribution of algorithmic music—millions of music listeners had adopted a platform that composers could use to distribute the actual software that generated their compositions, rather than just individual snapshots of what the software could do.

34.3.1 Generative Music Apps

In the world of generative music, the smartphone revolution meant that technology had caught up with the artists' original intentions, and provided artists such as Brian Eno with the platform that they had hoped for. It is therefore unsurprising that Eno released a series of apps in collaboration with software developer Peter Chilvers that not only revisited some of his previous works, but also expanded on them. *Air*[12] (for iPhone and iPod touch, 2009) is effectively a digital recreation of Eno's *Music For Airports*, where the listener can simply press 'Listen', and an infinite rendition of the generative composition will be produced. Eno's and Chilvers's other apps to date, *Bloom*[13] (iPhone and iPod touch, 2008), *Trope*[14] (iPhone and iPod touch, 2009), and *Scape*[15] (iPad, 2012), all continue to develop the notion of a generative music app as the progression of the album format. They can all be left to run endlessly, shifting through certain preprogrammed and generative states which sound 'always familiar, but never the same' (Eno 2012). They develop this principle further by taking advantage of the smartphone touchscreen displays and adding in layers of direct interactivity with the listener. Each app features a slightly different interaction method for the interactive modes. *Bloom* has a vertical keyboard system, prompting the listener to tap the screen to generate dulcet tones, whereas *Trope* has more of a swipe-based interaction.

Scape, the latest and most sophisticated of the apps, invites the listener to create their own compositions ('scapes') from scratch by using a simple drag and drop interface to position glyph-like objects and backgrounds on a square canvas. The various backgrounds and objects influence the resulting music in different ways, and can represent a particular musical phrase, an instrument, the 'mood' of the entire composition, and so on. The spatial positions of the objects, relative to the edges of the canvas and to each other, change their sonic properties further still, making the seemingly simple user interface surprisingly expressive. However, the exact function of each object and quite how their spatial positioning affects their properties are never explicitly shown to the listener, making *Scape* more about gradual experimentation than premeditated composition. In this way, *Scape* significantly blurs the line between composer and listener, and challenges the concept of a body of work created by an artist for the listener. It could easily be argued that because the listener is responsible for choosing and positioning the objects and backgrounds on the initially blank and silent canvas, they in fact become the composer. On the other hand, the fact that Eno and Chilvers have provided a limited number of sonic elements, each with predetermined behaviours and finite possible spatial positions, means that they retain their roles as the composers and make the listener more of an 'arranger'. Much like the conceptual shift that generative music caused for the idea of a finished 'composition', *Scape* actually makes the listener an active participant in the creation of the music itself, while the music still retains a distinctive sonic identity at all times. The app also comes with fifteen pre-made scapes from Eno and Chilvers, ten of which constitute the so-called album. These can be listened to in sequence or at random, with the app automatically transitioning from one scape to the next after a user-configurable period of time.

Eno's and Chilvers's apps effectively showcased the potential for mass-distributed algorithmic music, but solely within the ambient genre. In 2012, London-based pop-electronica musician Gwilym Gold created a version of his album *Tender Metal*[16] as an iOS app that uses Markov models and machine learning techniques to create new arrangements of the songs every time they are listened to.

34.3.2 Interactive Music Apps

Smartphones and tablets, such as the Apple iPad, brought upon a new wave of inter-active tools and apps for music creation. As demonstrated by Eno and Chilvers with their apps, these interfaces can turn music into a participatory art form, which includes and relies upon interaction with the listener. Indeed for this reason it is appropriate to label apps such as *Scape* as interactive, as well as generative.

In the parallel world of visual arts, San Francisco multimedia artist Scott Snibbe had been exploring interactive art using computer screens since the 1990s, with projects such as *Motion Phone* (1995),[17] *Bubble Harp* (1997),[18] and *Gravilux* (1998).[19] Much like Brian Eno with generative music, Snibbe did not revisit the medium in earnest until there was a widespread distribution platform available to him, and *Motion Phone, Bubble Harp*, and *Gravilux* were all later re-released as apps for various smartphone platforms shortly following the advent of the iPad. In a collaboration with Icelandic singer Björk, Snibbe created an interactive album app for Björk's 2012 album *Biophilia* (iPad, 2012; Android, 2013),[20] which features interactive versions of all the songs on the album. The interaction methods are different for each song, with some requiring more active participation than others, but all tie in to the narrative and subject matter of the album as a whole.

The practice of allowing the listener to create their own remixes of a song, as explored by Coldcut's CD-ROMs in the 1990s, was also repopularized by the smartphone platform, and it was indeed Matt Black from Coldcut who revisited their own concepts by releasing *NinjaJamm*[21] for iOS and Android (2013 and 2014 respectively). *NinjaJamm* featured a variety of different remixable tracks (called 'packs') from well-known artists on the NinjaTune label that the user could download and manipulate into their own reinterpretations of the track. There are also many other examples of remix apps, where existing music from an artist is broken into its constituent parts and presented via a touchscreen interface that allows the user to combine loops and add effects.

34.3.3 Reactive Music Apps

The fact that people now carry powerful computers in their pockets gives musicians a new platform on which to distribute algorithmic music. Beyond this, the advent of smartphones also presents exciting new opportunities and areas of exploration. Unlike the portable music players of previous generations, which were decidedly single-purpose, smartphones are powerful, flexible devices that inherently hold a huge amount of information about the listener. GPS chips transmit data about their current location, in-built accelerometers detect how they are moving, and their constant Internet connectivity can be used to gather real-time information from the world around them. Furthermore, many smartphone users use headphones with microphones built in to the cable that, while primarily intended for hands-free phone calls, may be used to detect the loudness of the listener's environment, or even record snippets of sound. This has brought about a logical step in algorithmic music where composers may use real-time information about the listener to directly influence the progression of the music, making every listening experience unique to its specific environment. The RjDj[22] software was first invented to explore this concept, as well as to act as a platform for distribution of generative, reactive, and interactive music on smartphones.

RjDj was truly pioneering in this field of reactive music, and the original iPhone app (2008) came with a collection of algorithmic music pieces ('scenes') from various artists. These ranged from small interactive sound-toys that the listener interacted with by pressing onscreen buttons and tilting the device, to fully-fledged music compositions which were generative, as well as reactive and interactive with the listeners sonic environment. This range of real-time information about the listener provides artists with a playground of new creative possibilities.

34.3.3.1 *Microphone Input*

The ability to 'listen in' on the acoustic environment of the listener provides the reactive music composer with a completely new, infinitely unique sonic palette to use in their compositions. In a range of scenes for the RjDj, Kids On DSP (artist name of in-house team at the company behind RjDj, Reality Jockey, and arguably the world's first widely distributed reactive music producers) exploited the possibilities of accessing the listener's microphone in two main ways; using it to determine the overall loudness of the listener's environment, and using it to actually record snippets of sound. For example, in *Doppelganger* (2009), the loudness of the environment is used to determine the octave in which the repeating melody at the beginning of the composition is played. A quieter environment results in the notes being played in a lower octave, whereas a louder environment results in a higher octave. Many of Kids On DSP's other reactive music compositions, as well as the other early scenes on the RjDj, actually used the microphone to record snippets of audio from the listener's environment and use them composition-ally. *Dimensions* (Kids On DSP, 2009) featured the microphone input being constantly treated with a pitch-shifter, delays, and reverbs to create swirling dubby soundscapes that were perfectly in tune and in time with the track beneath. *Timecruising* (Kids On DSP, 2009) also featured this kind of microphone treatment, and would also play chunks of sound backwards as soon as they were recorded, creating a surreal effect that made it feel like your world was suddenly being experienced in reverse.

34.3.3.2 *Motion*

The in-built accelerometers of iPhone and Android smartphones also provide inter-esting possibilities for reactive and interactive music, as they may be used to deduce whether the listener is still or in motion, as well as the exact angle the device is being held at. This is put to good use in *Inception: The App* (Reality Jockey and Remote Control Productions, iPhone, 2011)[23]—the companion app for the Christopher Nolan movie *Inception* (Nolan et al. 2011). Certain pieces create ambient soundscapes when the lis-tener is still, but as they start to walk or run, drums and other rhythmical elements come in, matching the intensity of the listener's actions.

Knowing the exact angle of the listener's device also allows more direct interaction with them in certain pieces of music. Indeed RjDj featured a whole subcategory of scenes called *Moovz* devoted to this kind of interaction. *Moovz* scenes presented the lis-tener with a simple four-button interface, where holding down a button would activate an instrument or an effect and tilting the device in various directions would control the instrument's or effect's parameters. Mapping the tonal elements to the musical scale of the track, and syncing any timed effects such as delays or glitchers with the track's tempo would guarantee that the listener's actions were always in key and in time with the music beneath, making for a fun, compelling, and easy remixing experience.

34.3.3.3 *Location*

Enabled by the built-in GPS chips in smartphones, Reality Jockey also explored the pos-sibilities of location-aware music in many of their apps. *Inception: The App* and their follow-up app, *Dimensions*, both featured scenes that used the listener's location to influ-ence the music in various ways. *Inception: The App* uses GPS to determine the listener's travelling speed and bases the music played back on this information. Location-aware albums such as *The National Mall*[24] (iPhone, 2011), by Washington-based electronica band Bluebrain, use the listener's position in Washington's Mall to determine the mix of various geo-tagged musical elements. The listener's path through The Mall, walking towards and away from various monuments and buildings, directly controls the volume of corresponding mini-compositions, all made and implemented to fit together in vari-ous combinations.

Many of these relationships are never made explicit to the listener, and are more about creating a natural synergy between the input of passive environmental information and the output of music that is tailored to it, effectively creating a soundtrack to the listen-er's life, as it is being lived. This was well demonstrated in a project that I worked on in

2013 in collaboration with UK dance duo Underworld and sound designer Nick Ryan, *Play the Road*.[25] By using a smartphone mounted to the dashboard of a car, *Play the Road* used GPS and accelerometer data to detect the driver's position on a racecourse, their travelling speed, rate of acceleration or deceleration, and angular velocity (i.e., how sharply they were turning left or right). All of this information was used to create a tailor-made soundtrack to the driver's tour around the racetrack.

34.3.4 Production Difficulties

Clearly the potential of smartphones as a distribution platform for algorithmic music was great, and provided a huge step forward in the possibilities for distributing music as dynamic software rather than static audio files. However, as is often true with advances in technology, this new platform came with its own set of challenges, and in the case of the production of apps for the smartphone market this meant needing to know how to design and program the user-interfaces for these apps as well as compose the algorithmic music. This additional barrier to production was significant, and meant that actually taking advantage of smartphones was possible only for those algorithmic musicians who were also programmers or who had access to programmers via collaboration or money. Ideally, composers would be able to harness the power of the numerous software packages available to them already but distribute their compositions via smartphones, without any need to do any programming of the smartphone itself.

Intermorphic, the company behind Koan,[26] already had widely adopted generative music software for Mac and PC in the form of Noatikl[27] and Mixtikl,[28] and later released versions of both these applications for mobile platforms. This allows generative music composers to create their music on a desktop and distribute it to users of the mobile apps without needing to know any additional programming at all.

Another significant development that followed the introduction of smartphones was the creation of libpd[29]—a library version of the popular graphical programming language, Pure Data (Pd)[30]—which allows any Pd patch to run on any hardware system capable of executing native code. Although there are many platforms other than smartphones that libpd can run on, it was the proliferation of iOS and Android devices that expedited its development. In particular, it was the work of Peter Brinkmann (2012) and programmers at Reality Jockey that was instrumental in providing users of Pd with a simple way of getting their work running on a mobile device without the need to program code for iOS or Android.

34.3.5 Platform Apps

Mixtikl and RjDj are early examples of platform apps, which provide an additional framework for algorithmic music that allows musicians to distribute their compositions easily. RjDj also developed this principle further by acting as a storefront for algorithmic music, with compositions being available to purchase in the app as downloadable content.

NinjaJamm also uses libpd as its audio engine, but makes the process even easier for artists by requiring them to simply provide a certain number of static audio files in a particular format—usually the constituent parts of a track as individual instrument stems, separated by song sections. The app then applies the required interface and interactivity, thus not requiring the composer to have any knowledge of Pd in order to create and distribute algorithmic music. The *Reactable* also has a mobile version of the node-based algorithmic music system (*Reactable Mobile,* iOS, 2010, Android, 2011) and allows composers to share their compositions (called 'tables') with other users of the app.

34.3.6 Limitations of the Smartphone Platform

The smartphone revolution undoubtedly created a large new audience for the possibilities of algorithmic music in an algorithmic format, but there are still many barriers to both production and consumption that limit the pace at which dynamic formats of music are adopted.

34.3.6.1 *Cost*

While the platform apps mentioned above can save algorithmic music composers the cost and hassle of creating a whole new interface and app architecture, the more interesting examples of distributed dynamic music do feature custom-built interfaces that are tailored to the individual compositions, such as *Inception: The App,* or the interactive concepts, such as *Scape.* These kinds of user interface cannot be accommodated by any platform app and therefore, developing a truly immersive smartphone-based experience still involves developing app code from scratch—a specialized and expensive process.

34.3.6.2 *Fragmentation*

Although all modern smartphones are technically capable of generating algorithmic music in realtime, the two most widely adopted operating systems require the user-interfaces for their apps to be written in different programming languages (Objective-C for iOS and Java for Android),[31] increasing development time and cost further. Furthermore, the Android hardware market is significantly fragmented within itself, as the hardware that Android (and therefore the Android app) can run on ranges from the latest, most powerful devices to the most basic ones that have a fraction of the processing power of their more expensive counterparts (Panzarino 2012). This makes developing consistent experiences across the entire range of mobile operating systems incredibly difficult. This is particularly true when attempting to use advanced sensor data such as the accelerometer, as all different hardware models have different accelerometer sensitivities.

34.3.6.3 *Lack of Integration with Static Music Formats*

The distribution methods for static music formats are as widespread and familiar as the formats themselves; physical media such as vinyl and CDs are distributed in bricks-and-mortar record shops or via mail-order, while digital formats are either downloaded from online stores or streamed. Listeners to these formats are comfortable with where they can buy or listen to music in the medium of their choice. Algorithmic music compositions in the form of apps, however, are not available through these existing channels, but instead must be purchased through the app stores of the smartphone platforms, such as the App Store for iOS, and Play Store for Android, making algorithmic music appear separate from the static formats. Furthermore, once an algorithmic music app is downloaded, it is stored in a different part of the smartphone operating system from the static music, often on a home screen along with all other apps, as opposed to an organized library that can be browsed by artist, album, or playlist. For even the most open-minded of listeners, these differences in procurement and procedure create a psychological disconnect between algorithmic music and static music, which limits the rate of its adoption. A future reality where algorithmic music is as common as static music will surely feature a listener experience where the various music formats sit together and on equal footing in the user interface.

34.4 WEB AUDIO

The web browser is the most ubiquitous interactive multimedia platform in existence, with over three billion people connected to the Internet at the time of writing,[32] and it has benefitted from active development in the visual domain since the days of Flash, with contemporary developments including SVG, WebGL, and Three.js. This has resulted in numerous innovative, visual web-based experiences, such as interactive music videos, beautifully rendered 3D website graphics, and live data visualizations. Since the introduction of the WebAudio API[33] in 2011, various groups of programmers have been exploring and expanding the capabilities of the modern web browser to create interactive audio experiences, with the first Web Audio Conference being held at IRCAM in January 2015.[34] The introduction of this standard has widened the potential for distributing algorithmic music beyond any other channel, including that of smartphones (considering that compliant WebAudio code is also largely supported on mobile web-browsers). There are distinct advantages and disadvantages of using the web as a platform for algorithmic audio experiences compared to using smartphones.

34.4.1 Easier Interface Design

Because the web and its associated technologies pre-date smartphones by many years, web developers and designers are significantly more common than app developers. Furthermore, the development times for websites are significantly shorter than those of apps, meaning that the overall cost of production is much lower on the web than it is on smartphones. The range of web-based tools and libraries available to interface designers is also mature and diverse and continues to be actively developed.

34.4.2 Instant Deployment

Websites can be deployed and updated instantly, without the need to submit builds of one's app to any platform-related approval boards, such as with Apple's App Store. In the same way that e-publishing has provided authors with a convenient way of updating their work after the initial publication, distributing algorithmic music via the web does away with the need to ever call a piece of music really 'finished'. A browser-based piece of algorithmic music could instead be viewed as an ongoing body of work, where the artist adds sounds and updates the code as and when desired.

34.4.3 Better Access

The concept of listening to music via dedicated mobile apps is unfamiliar for listeners, whereas websites like YouTube and SoundCloud have made music consumption on computers and the web commonplace. Furthermore, the act of distributing any piece of browser-based content is as simple as sharing a URL, with no need for intermediary app stores, thus making discovery more seamless for the listener and marketing much easier for the artist.

Despite these distinct advantages, Web Audio represents yet another innovation that is founded on technologies that are unfamiliar to algorithmic music composers, this time the JavaScript[35] programming language. While JavaScript is popular and widespread among web developers, it had never been used in the production of interactive audio until the introduction of the Web Audio API. In this sense, composers who would like to get their algorithmic music running in a web browser are still in need of significant expertise in JavaScript or at least of collaboration with a programmer who can translate their ideas. As such, although Web Audio represents a further widening of the potential audience for algorithmic music, it also comes with a significant narrowing of the group of potential composers capable of exploiting it until new tools which help bridge this gap are adopted.

In the case of both Web Audio and smartphones, algorithmic composers are still in need of a set of sophisticated authoring tools that can bridge the gap between mass production and mass consumption.

34.5 CRITICAL REACTION TO DYNAMIC MUSIC FORMATS

All of the ongoing developments with distribution platforms and creation tools could eventually lead to a place where composers and listeners can create and consume algorithmic music freely, but what has been the critical response to the examples of the limited amount of algorithmic music in existence today? Do listeners actually enjoy dynamic music more than its static counterpart? Do artists crave the new creative possibilities afforded by algorithmic music or do they see the current formats sufficient for expressing their musical intention? After all, there is no shortage of inventiveness in static music formats, with new genres constantly appearing from all corners of the world, some rejecting traditions and others subverting them (Harper 2014). Reviews of Gwilym Gold's *Tender Metal* generative work were equally positive and negative about its 'shapelessness' with some claiming that the format enhances the impact of the music (Adams 2012) and others lamenting that it makes it less enjoyable (Nicholson 2012). The interactivity of Björk's *Biophilia* has caused it to be the first app to appear in New York's Museum of Modern Art, yet this same interactivity has been criticized as being a mere distraction from the beautifully crafted songs (Diver 2011; Petridis 2011a).

On a more philosophical level, in reviewing *Tender Metal*, Nicholson (2012) touches upon a key feature of musical cognition that is lost in a format that strives to never sound the same twice; the deep-rooted necessity for repetition in the intelligibility of music (Margulis 2013; Schoenberg 1967). As Margulis (2014) points out in her various experiments, to recall and be able to mentally complete a musical phrase, as a result of having heard it before, is a form of active involvement in the music listening experience. Algorithmic music formats allow the potential for truly unrepeatable music, where notes, rhythms, and timbres constantly change, never settling on something that a listener could tap along to or hum,[36] but this strips away a large element of what makes music enjoyable and, ironically, engaging. Here, algorithmic music can be seen to negatively impact a piece's efficacy.

At its worst, algorithmic music of the most basic kind can be seen as a complete gimmick or marketing ploy, and this was indeed the reception given to early CD-interactive albums, such as Peter Gabriel's *XPLORA1* (1993) and *Jump: The David Bowie Interactive CD-ROM* (1994) (Talbot 2000), although it would seem unfair to compare these extremely primitive uses of multimedia, where listeners could do little more than change the volume level of certain parts of the track, with today's highly sophisticated offerings. Nevertheless, listeners, artists, and eventually the manufacturers (the Philips CDi format died out entirely in 1998) seemed to reach a consensus that this very basic form of interaction held little artistic interest, or at least any that was commercially viable at the time.

Critical responses to algorithmic music often feature feelings of being overwhelmed by the format itself, such that its technological novelty 'overshadows the music' (Adams 2012). In fact, many of the reviews of these algorithmic works have focussed on the technology in use, rather than the aesthetic possibilities that it unlocks (Aguilar 2012; Dredge 2013; Kincaid 2008), making it difficult to assess whether these formats add anything of true value to the musical expression. It is perhaps telling that Björk decided to release her *Biophilia* album via static media as well as in app form, treating the app as an optional accompaniment to the traditional album, rather than an essential embodiment of the music, much to the relief of her critics.

Many features of static music formats play perfectly into the hands of what makes music enjoyable to humans in the first place, and elements of algorithmic composition can be seen to go against a listener's direct desires and expectations from a piece of music. Reactive music, explicitly connected to one's real-time environment and thus able to provide a complementary multisensory experience, can be rather undesirable to those who listen to their favourite song in order to mentally escape their current situation. Interactive music can offer the listener great degrees of control over an artist's work, but a widespread understanding of modern music listening is that it is a passive endeavour that people do not want to actively engage with, but rather effortlessly consume (Petridis 2011).

From an artist's perspective, and on a practical level, the real-time cost of producing algorithmic music, especially to those who have never made it previously, is much greater than that of static music, causing artists to rightfully question the value of these new possibilities. And although few artists would say that the true motivation behind their work is simply to earn money, the fact that the potential audience for algorithmic music is likely to be a fraction of their traditional one makes the prospect less appealing still. On an aesthetic level, the wealth of sonic options in the creation of static music is already well into the realms of the paralysis of choice. Readily available software that is the staple of much music creation already overflows with more timbral, tonal, and rhythmic possibilities than any one artist could possibly claim to ever fully command, so to add the vast multiplying factors of generative, reactive, and interactive composition and distribution into the mix is understandably daunting, and perhaps seen as a distraction from a mastery of the current 'standard' techniques.

34.6 Conclusion

Recorded sound has brought about fundamental conceptual reevaluations of what music is, how it is experienced, and even why it exists in the first place, leading to new audiences, genres, and forms of artistic expression. Algorithmic music continues this trend of innovation, but has remained relatively niche due to a heavy reliance on new technology and conceptual practices for both production and consumption. These barriers to entry for artists and listeners result in a cycle that is common to any new content platform that seeks mainstream adoption; there is no content because there is no consuming market, and there is no consuming market because there is no content.

Despite its challenges, algorithmic music practices continue to grow, and even in the face of the deeply embedded position of static music, entirely new musical forms such as reactive music on smartphones come to light, and platforms such as Web Audio give algorithmic music composers an enormous audience to whom to distribute their work. It is yet to be seen what innovative musical forms will come as a result of the Web Audio API, but if they are to be compared to smartphones in scope, they are likely to be another significant milestone for algorithmic music.

These technology platforms seem to offer a satisfactory solution to the lack of mass distribution opportunities for algorithmic music, but the actual production of such music remains as specialized as ever, with no dedicated tool in existence that enables a composer to create and distribute to a mass audience without requiring any extra collaboration or skills. Again, this innovation is caught in an endless cycle, where the lack of software options is due to a lack of market (this time referring to the composers themselves), and the lack of market is, in part, due to lack of software options.

The role of a music programmer as a collaborator, who can take an artist's vision and directly translate it into code that can run as an app or a web-based experience, is currently of vital importance to this ecosystem, and will likely remain so until the market reaches a critical mass of interest from artists, listeners, and producers alike.

Exploiting the advantages of algorithmic music formats is not simply a case of using new techniques just because one can, and the responsibility of choosing when and with what it is best to experiment to create meaningful art lies, as ever, with the artist. Just as with making music for a piano, one can make 'good' or 'bad' algorithmic music, but this does not reflect inherently on the medium. It could be said that we are still novices in this new art form, with virtuosity yet to come. However, the influx of innovations in both consumer electronics and computer software have demonstrated that this area of musical creativity is being as actively explored as ever, and the new mass audiences afforded by smartphones and web-browser technologies offer some of the most exciting artistic opportunities since recorded sound itself.

NOTES

1. J. N. Wilford, 'Flute's Revised Age Dates the Sound of Music Earlier', *New York Times*, 29 May 2012, http://www.nytimes.com/2012/05/29/science/oldest-musical-instruments-are-even-older-than-first-thought.html?_r=0.
2. Many of these algorithmic techniques and principles such as arpeggiators, low frequency oscillators (LFOs), randomizers, and so on, can even be commonly found in popular music software designed for creating traditional static music.
3. http://supercollider.sourceforge.net/.
4. http://chuck.cs.princeton.edu/doc/develop/.
5. http://intermorphic.com/noatikl.
6. http://forumnet.ircam.fr/product/openmusic/.
7. https://cycling74.com/.
8. http://ableton.com/.
9. https://cycling74.com.
10. https://www.audiokinetic.com/products/wwise/.
11. http://www.fmod.org/.
12. Developed by Peter Chilvers featuring the voice of Irish vocalist Sandra O'Neill.
13. http://www.generativemusic.com/bloom.html.
14. http://www.generativemusic.com/trope.html.
15. http://www.generativemusic.com/scape.html.
16. http://bronzeformat.com/.

17. http://www.snibbe.com/projects/interactive/motionphone.
18. http://www.snibbe.com/projects/interactive/bubbleharp.
19. http://www.snibbe.com/projects/interactive/gravilux.
20. http://www.snibbe.com/blog/2011/10/11/biophilia/.
21. http://www.ninjajamm.com/.
22. http://rjdj.me/.
23. https://itunes.apple.com/in/app/inception-the-app/id405235483?mt=8.
24. http://bluebrainmusic.blogspot.co.uk/2011/03/national-mall.html.
25. http://www.volkswagen.co.uk/playtheroad.
26. http://intermorphic.com/sseyo/koan.
27. http://intermorphic.com/noatikl/.
28. http://intermorphic.com/mixtikl/.
29. https://github.com/libpd.
30. https://puredata.info/.
31. Code libraries that allow app developers to write cross-platform code once and deploy on multiple platforms are gaining popularity, but are far from ubiquitous at time of writing.
32. http://www.internetworldstats.com/stats.htm.
33. http://webaudio.github.io/web-audio-api/.
34. http://wac.ircam.fr/.
35. http://www.w3schools.com/js/.
36. Indeed there are many pieces of static music, particularly in the genre of so-called Intelligent Dance Music (IDM) and noise music, where repetition is rare if not nonexistent, but the static music formats would at least allow the listener to repeat the entire piece of music.

Bibliography

Adams, S. 'Album Review: Gwilym Gold, *Tender Metal*'. *Drowned In Sound*. 21 September 2012. http://drownedinsound.com/releases/17250/reviews/4145520.
Aguilar, M. 'Brian Eno's Bonkers New iPad App Thinks for Itself'. *Gizmodo*, 28 September 2012. http://gizmodo.com/5947115/brian-enos-most-bonkers-music-ipad-app-yet-thinks-for-itself.
Bown, O., and Britton, S. 'An Album in 1,000 Variations: Notes on the Composition and Distribution of a Parametric Musical Work'. *Leonardo* 47, no. 5 (2014): 437–441. doi:10.1162/LEON_a_00866.
Brinkmann, P. *Making Musical Apps: Real-Time Audio Synthesis on Android and iOS*. Sebastopol, CA: O'Reilly Media, 2012.
Collins, K. 'An Introduction to Procedural Music in Video Games'. *Contemporary Music Review* 28, no. 1 (2009): 5–15. doi:10.1080/07494460802663983.
Diver, M. 'Review of Björk, *Biophilia*'. *BBC*. 2011. http://www.bbc.co.uk/music/reviews/5zq4.
Dredge, S. 'Brian Eno and Peter Chilvers Talk Scape, iPad Apps and Generative Music'. *The Guardian*, 26 September 2012. http://www.theguardian.com/music/appsblog/2012/sep/26/brian-eno-scape-ipad-apps.
Eno, Bian. 'Air'. *Generative Music*. 2012. http://www.generativemusic.com/air.html, accessed 30 June 2017.
Eno, B. *Music for Airports*. Polydor Records, 1978. CD.

Gopinath, S., and Stanyek, J. 'Anytime, Anywhere? An Introduction to the Devices, Markets, and Theories of Mobile Music'. *The Oxford Handbook of Mobile Music Studies*, edited by S. Gopinath and J. Stanyek, 1:1–36. Oxford: Oxford University Press, 2014.

Gould, G. 'The Prospects of Recording'. In *Audio Culture: Readings in Modern Music*, edited by C. Cox and D. Warner, 115–126. New York: Continuum, 2004.

'Gwilym Gold Has a Totally New Idea'. *Dummy Mag*, 25 May 2011. http://www.dummymag.com/new-music/gwilym-gold-bronze.

Harper, A. 'Indie Goes Hi-Tech: The End of Analogue Warmth and Cosy Nostalgia'. Lecture, Berlin Music Week, Berlin, 5 September 2014.

Holmes, T. *Electronic and Experimental Music: Technology, Music, and Culture*. 3rd ed. New York: Routledge, 2008.

Kincaid, J. 'RjDj Continues To Be The Most Trippy App On The iPhone (And I Love It)'. *TechCrunch*, 29 December 2008. http://techcrunch.com/2008/12/29/rjdj-continues-to-be-the-trippiest-app-on-the-iphone-and-i-love-it/.

Margulis, E. H. *On Repeat: How Music Plays the Mind*. New York: Oxford University Press, 2014.

Margulis, E. H. 'One More Time'. *Aeon*. 2014. http://aeon.co/magazine/culture/why-we-love-repetition-in-music/.

Nicholson, R. 'Gwilym Gold and the Album That Will Never Sound the Same Way Twice'. *The Guardian*, 31 August 2012. http://www.theguardian.com/music/2012/sep/01/gwilym-gold-tender-metal-bronze.

Inception. Directed by C. Nolan. Burbank, CA: Warner Home Video, 2010.

Panzarino, M. 'The Shocking Toll of Hardware and Software Fragmentation on Android Development'. *The Next Web*, 30 May 2012. http://thenextweb.com/mobile/2012/03/30/the-shocking-toll-of-hardware-and-software-fragmentation-on-android-development/.

Petridis, A. 'Björk: Biophilia—Review'. *The Guardian*, 6 October 2011a. http://www.theguardian.com/music/2011/oct/06/bjork-biophilia-cd-review.

Petridis, A. 'Music Weekly Podcast: Battles and Gwilym Gold'. *The Guardian*, 3 June 2011b. http://www.theguardian.com/music/musicblog/audio/2011/jun/03/music-weekly-battles-gwilym-gold-audio.

Schoenberg, A. *Fundamentals of Musical Composition*. Edited and translated by G. Strang and L. Stein. London: Faber and Faber, 1967.

Sterne, J. 'How the MP3 Became Ubiquitous'. In *The Oxford Handbook of Mobile Music Studies*, edited by S. Gopinath and J. Stanyek, 1:37–54. Oxford: Oxford University Press, 2014.

Talbot, M., ed. *The Musical Work: Reality or Invention?* Liverpool: Liverpool University Press, 2000.

Yaxu. *Peak Cut*. Computer Club. USB card, 2015. https://computerclub.bandcamp.com/album/peak-cut.

Perspectives on Practice D

CHAPTER 35

ALGORITHMIC TRAJECTORIES

ALEX MCLEAN AND ROGER T. DEAN

WE jointly designed and edited this volume because of our complementary, overlapping, yet highly contrasting backgrounds (we have performed together and met first in the context of music research). The contrast between us stems both from our differing time frames of involvement and from the fact that Alex McLean (AM) makes music primarily (usually solely) via a computer and in realtime whereas Roger T. Dean (RTD) is an acoustic instrumentalist (particularly keyboards, often with computers) and a composer (offline), as well as an improviser (realtime). While AM was using computers from an early age, and began serious programming around 1986 (aged eleven), RTD first used a (desktop) computer in around 1982 (already aged more than thirty).

So in this final Perspective on Practice, we will discuss our own experiences and the development of our current enthusiasms. We hope that brief consideration of these trajectories will have some interest for readers seeking to engage with the breadth of our field of algorithmic music. We drafted our own sections, and then jointly edited the chapter, providing a brief conclusion; we also took advantage of helpful suggestions from external reviewers. See note below for information on CD and other sources of the music mentioned in the authors' sections that follow.[1]

35.1 ROGER DEAN (AND MY ENSEMBLE australYSIS)

My algorithmic music experiences centre on real-time use of computation in jazz and freer improvisation, but they also contribute to my compositional work. If we consider that one thing that can distinguish an algorithm from a traditional sequencer is real-time mutability of timbre, pitch, and event timing, then my first experiences of algorithmic music making were with the DX7 Yamaha FM Synthesizer, released in 1984. I still occasionally use this pioneering instrument, but subsequently it was complemented in my performing armoury by a succession of sampler instruments (EPS, Kurzweil) with parallel capacities. The DX7 was one of the first affordable digital instruments to permit real-time timbral and pitch manipulation, and its sequencer functions could be transformed while running. I found the timbral transformation very stimulating, and used it both within improvised solos in jazz (e.g. with Graham Collier Music on the 1985 album *Something British*) and with my own European group LYSIS. My piece *Alela* (1986) was an algorithmic 11/8 rhythmic and harmonic accompaniment with which instrumentalists could improvise. The 11/8 pattern was transformed or rotated in ways akin to *Clapping Music* or *Piano Phase* of Steve Reich, and tempo and timbre (pitched versus unpitched) might change in performance. I found this broad potential for variability one of the most appealing aspects of algorithmic music; at the time I saw less point in using absolutely deterministic algorithms. This is not a surprising stance for an improviser. Indeed, between 1970 and 1980 I was (temporarily) opposed to the idea of making fully notated compositions.

In the 1980s I experimented with some of the early live algorithms for improvisers, such as Laurie Spiegel's Music Mouse and the program M. But my usage became more serious after Max was released, when soon after this I was made aware of it by my colleague the late keyboardist and researcher Jeff Pressing (see, for example, Pressing 1988, 1994). My usage was subsequently aided by collaborator Greg White, a member of austraLYSIS, into which I converted LYSIS when I migrated to Australia in 1988. Max is now a well-known graphic object-oriented programming language for music, initially note-based, later (with Max/MSP signal processing) adding sound-based music and image (with Jitter). It remains my core platform, together with several others.

As implied already, I found that my distinct perspectives as instrumentalist, improviser, and composer all found application in algorithmic music making. Much of what I have written as algorithms constitutes performing interfaces or hyperinstruments, that I can use while also playing the piano, or that run fairly automatically with little intervention. I also write algorithms which require full-time attention, and since 2014 I have also performed live coding with Gibber.

Algorithms in principle can be applied to any aspect of music. For me, an early appeal of algorithmic music was to make a range of 'rhythm machines' (see Dean 2003), which I used interactively while playing the piano. Some operated on isochronic rhythmic units (making pulsed rhythms of various complexities), while others treated rhythm as a continuous variable (creating greater irregularity, sometimes essentially without patterns). In Max, fewer than twenty objects are needed to produce metrical patterns, or repeating patterns of any diversity: each unit can be an integer multiple of an isochronic pulse, each can have a different duration, or some may repeat. The same patch can allow the pattern to be varied every time it repeats, when required maintaining an initial pattern of shorter (s) or longer (l) events, such as sslls, even if every s and l duration changes each time the pattern is enunciated. This work complemented my 1980s acoustic work on multiplicity of rhythms within an improvising group and on competitive metrical and pulse shifts, somewhat akin to Elliott Carter's metric modulation or the earlier complexity of Conlon Nancarrow. I also made a transformed (perhaps contorted) drum 'n' bass generator algorithm, which is functional both as an autonomous automaton and under user input (Dean 2003). Most recently Andy Milne and I have developed MeanTimes (now part of the freely available software Xronomorph), a Max implementation of research ideas on the nature of well-formed rhythms that go well beyond isochronic pulse assemblies and metrical concepts of well-formedness, and may contain several disparate pulse rates and components, even to an extreme where there is no shared pulse amongst any of the metrical or hierarchical levels (Milne and Dean 2016).

Postminimalism (spearheaded and coined by William Duckworth) is a compositional approach for which algorithms are ideal. It encourages slow melodic transformation, within and without tonality, alongside the other features. I used such algorithmic approaches to make *The Egg The Cart The Horse The Chicken* (2001) and *Mutase* (2008), which has also been used in empirical studies of music perception. In *MultiMutase* (2008) and *Ligating the Rhizome* (2006) I used chain annealing techniques to allow various simple forms of 'cross-over' between melodies.

I emphasize variability, but this does not preclude algorithmic approaches which from the perspective of music theory are highly constrained. For example, key features of serial composition using the techniques pioneered by the Second Viennese School (Schoenberg and his colleagues) can readily be coded, and I made such an algorithm (the Serial Collaborator; Dean 2014), which operates either autonomously or interactively, to create multi-handed piano music. It can even be applied to tonal motives. I also used related algorithms for pitch, dynamics, and rhythms in some sections of my orchestral piece *SonoPetal* (1995; commissioned and premiered by the Australian Chamber Orchestra). Simple forms of machine listening contribute to many of my algorithmic patches (see also Dean 2003).

My most recent work on algorithmic music has focused on using time-series analysis models of note-based (and currently sound-based) music as generative devices. In my BANG (beat and note generator) algorithm (Dean 2017), time series models of an ongoing performance are regularly constructed (and cumulated). I have also embarked on deep-learning neural net approaches to modelling music in performance, again for use in BANG. My currently developing TIMETRANS (timbre event transformer) will apply similar processes within continuous timbral flux. Music generation in BANG then uses the models to simulate the embodied process, and the output may be subject to other transformation before being sounded. BANG is somewhat like Pachet's Continuator (Pachet 2003; Pachet 2004), except that BANG is based on time series analysis or deep neural nets, rather than the Markov chain approach, and mounting evidence shows that these provide distinct representations of music, which may be complementary (Gingras et al. 2016). The purpose of BANG is to encourage more posttonal melody, harmony, and free rhythm (continuous temporal durations) in the output, which will always diverge substantially from the input. BANG presently uses a multivariate approach with four music parameters: pitch, key velocity (a surrogate of loudness), note and chord duration, and event inter-onset interval. These are jointly modelled in the BANG algorithm, giving a model which simultaneously predicts all four outputs and which allows for their mutual influences. Memory functions within BANG permit a wide range of hybridization of different periods of the current (and earlier) performances, according to the choice of the instrumental co-performer. It may also operate autonomously.

Algorithmic manipulation of timbre and its spatialization (see Schacher, chapter 25 in this volume) is used in my solo MultiPiano performances, involving piano and computer producing multiple sonic strand: I use at least a 4.1 diffusion system whenever possible. Similarly, Jitter permits what I call algorithmic synaesthesia (Dean et al. 2006; Sagiv, Bailes, and Dean 2009), in which algorithmic processing and/or input data are shared between sound and image. This is analogous to the impressive laser, sound, and video works of Robin Fox and others.

Within austraLYSIS, my ongoing creative ensemble, several artists contribute major inputs (Dean and Smith, 2003). Figure 35.1 illustrates the opening of one of our intermedia works (Smith, Luers, and Dean 2016). Hazel Smith, our poet and performer, collaborates on numerous works involving spoken, performed, and/or displayed text, and since 2001 we have pursued algorithmic text manipulation within our Text Transformation Toolkit, developed in Python. Published works such as *Instabilities 2* are amongst the results. David Worrall collaborated in developing this Python toolkit, and is well known for his pioneering work on computer music in Australia and on (algorithmic) sonification. Sonification is the sonic analogue of visualization, processes for (hopefully

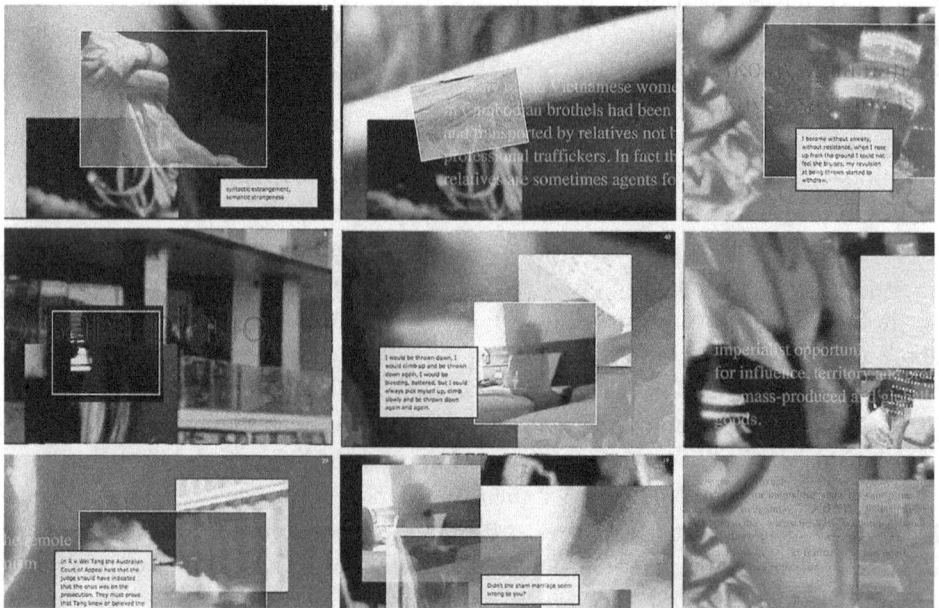

FIGURE 35.1 The landing page of *motions* (2013), an intermedia work by Hazel Smith (text), Will Luers (image, programming), and Roger Dean (sound, programming) (Smith, Luers, and Dean 2016).

accessible) representation of data content in sound and image. With David and colleague Greg White, I made a 22:4 spatialized algorithmic sonification of multichannel brain EEG recordings (2002; http://www.icad.org). Greg also contributed technical guidance in the early stages of austraLYSIS's application of Max/MSP, Flash, and Director, for example, with our *Walking the Faultlines* (1996–1997, published by ICMA) and *Wordstuffs*, a 1997 1Mb (!) floppy disk work occasioned by the Australian Film Commission and still available on the Australian Broadcasting Corporation (ABC) web site (http://www.abc.net. au/arts/stuff-art/stuff-art99/stuff98/wordstuf/index.htm).

White's recent thesis (2015) describes the trajectory of his work, often in our context. Alongside using Fibonacci series for pitch and rhythm in an austraLYSIS Electroband piece, Greg's notable early contributions (*Silence of Eyes, Glass Bead Game*, and *Scrolls of Time*, from 1995–2002) involved live performer interaction with the computational system and continuously generated score material which the performer then realized, either in text (*Silence of Eyes*) or musically. For *Glass Bead Game*, the felicity or precision of fulfilment of the 'comprovisation' score (i.e. improvising in ways which are consistent with the score) was assessed in realtime, with consequent changes in the next score output. With *Scrolls of Time*, the challenge was to perform in realtime more exactly what the score progressively indicated.

Finally, I return to my own priorities, aside from those of austraLYSIS. In the temporal domain I distinguish influence and interaction from the performer to the algorithm, in the sense defined by Edmonds (Boden and Edmonds 2009): where influence is a delayed consequence of a performer action, while interaction involves an algorithmic response which is almost instantaneous, such that it may well be detectable by the performer and perhaps by the audience. As developed also by others such as Laurence Casserley (Dean 2003), I do timbral manipulation by DSP with various time delays. I use a continuously updating three-minute 'buffer' of recent sound. Live algorithms remain a primary performance vehicle for me, but I have also written JavaScript code exploiting the audio API for in-browser algorithmic sonic manipulation by users in installations and across the web (for example, in *Long Time No See*, a 2013 work by a team lead by Keith Armstrong).

Overall, I see that the early stages of my own algorithmic music, and I think also that of many others, were strongly driven by the appearance of new technologies at affordable prices. Ideally, one might spend the 10,000 or so hours considered necessary for 'expert' performance in any field on aspects of algorithmic performance in order to gain such 'professional'-level facility. Since the advent of Max/MSP, the plethora of possibilities has been such that choices and enthusiasms can drive what I do; again, I think this is a shared experience. I repeatedly engage a research-led practice, practice-led research cyclic web (Smith and Dean 2009). This is not to say that technologies, or artistic innovations by others, have no impact. It is to say that apart from computational speed, most of the challenges one would like to erect in making algorithmic music can in principle be overcome with sufficient coding effort. Algorithmic music has come of age in this respect, as in many others, which is one of the reasons I hope the present book is timely and will be useful to our colleagues and our listeners.

ALEX MCLEAN

Many algorithmic music practitioners, including Roger, as described above, have come to algorithms looking for ways to extend their music practice. For me it was the other way around; I was a young computer programmer obsessed with the world of code but yearning to connect more with the 'outside world', and finding music to be an ideal outlet. While working during the first Internet boom in the 1990s, programmers were emerging from an antisocial stereotype, but still programming was quite far from being seen as creative in the public mind. Even discussion within the algorithmic art field was obsessed with questions around the authorship and autonomy of computers, with little heed paid to the creativity of the programmer. The difference was stark between this wider view of code as technical and mundane, and my own experience of it as an expressive, revelatory, and humbling medium, its very use requiring rigorous self-reflection of ideas and intentions.

My first experience with algorithmic music was encouraged by software artist Adrian Ward. Ade had already produced generative artworks parodying mainstream software (which would later lead into the release of the award-winning Auto-Illustrator) and was starting to move from making Autechre-influenced techno in old-school tracker software into generating music with code. We started writing software to make music together, and collaborating on events as VXSLAB; on 17 June 2000 we advertised an event with 'Pure Perl code generating "dance music" and ambience—some of this Perl will be written live'. Collective memory is unsure what actually happened there, but our early motivation clearly centred on bringing programming into the public realm. Adrian and I began performing as Slub later that year, and we were joined by long-time collaborator Dave Griffiths in 2005. We became known for projecting our screens, exposing our handmade interfaces.

As young, idealistic digital artists, Ade and I wrote 'The Generative Manifesto', partly inspired by Ade's interest in the Fluxus movement. This was presented at the Institute for Contemporary Arts in London on 23 August 2000, with Ade shouting out the words while I (being shy) ran one of our music-generating Perl scripts to underline each point:

Attention to detail, that only handmade generative music can allow (code allows you to go deeper into creative structures).

Real-time output and compositional control: we hate to wait (it is inconceivable to expect non-real-time systems to exhibit signs of life).

Construct and explore new sonic environments with echoes from our own (art reflects human narrative, code reflects human activity).

Open process, open minds: we have nothing to hide. (Code is unambiguous, it can never hide behind obscurity. We seek to abolish obscurity in the arts.)

Only use software applications written by ourselves: software dictates output, we dictate software. (Authorship cannot be granted to those who have not authored!)

Our emphasis on human authorship in algorithmic music, on celebrating creativity in computer programming, and on a live and open approach, is clear. Our aim to have people dance to our code was fulfilled in late 2001 at the legendary Paradiso club in Amsterdam during the Sonic Acts festival. Ade and I controlled our generative music scripts from the back of the room, with VJ (video jockey) Pfadfinderei controlling visuals from the front. Being able to watch the crowd go crazy, without their knowing where the music was coming from, was a great moment, which has fed into my ongoing ambivalence about projecting our screens. With rare exceptions, however, we always have, although we have often obscured the code by projecting our screens on top of one another. Balancing the need to be open and forthright about our methods while keeping focus on the resulting music has always been difficult.

As a band, Slub developed outside academia, through regular performances at digital arts festivals, and two dearly missed venues in London: the KLF's Hoxton Foundry and Siraj Izhar's Public Life. Nonetheless, significant crossover between the fringes of electronic music and academia led me to submit papers to academic venues such as 'Music without Walls' in De Montfort University, Leicester, in 2001. Here I was amazed to discover that many academics in computer music looked upon dance music with derision, dismissing it as popular, and rejecting the notion of repetition. One exception was Nick Collins, who presented a seminal paper on breakbeat science (Collins 2001). (Nick Collins and I later became co-founders of Chordpunch algorithmic music record label and Algorave event promoters, hopefully helping improve the landscape of dance-oriented computer music.)

A major turning point came by an invitation to the 'Changing Grammars' 2004 symposium, organized by Julian Rohrhuber and Renate Wieser (both are authors in this book, as is Nick Collins, who was also present). This symposium brought together people interested in what was variously called just-in-time programming, on-the-fly programming, or live programming, and acknowledged that there are important nuances in these terms, now generally referred to as live coding, with good coverage in this book (including chapter 16 by Roberts and Wakefield). At this moment, Ade had made his MAP/MSG live coding system (a working parody of Max/MSP), Ge Wang was releasing ChucK, and users of SuperCollider 3 were beginning to explore the possibilities of live interpretation of code in music making. Interactive programming has always been possible in computing, but there was certainly something in the air; with all this activity, live coding was ready to come together into a community of practice. In a smoky club at the close of the event, the Temporary Organisation for the Promotion of Live Algorithm Programming (TOPLAP) was born, with a manifesto hastily drafted on the plane home (see Haworth, chapter 31 in this volume, for some revealing ethnographical work on this document). Many of those present had been through grad school, myself included, and are now increasingly senior academics, enjoying seeing where the next generation of young students is taking things.

Following our involvement with the founding of the TOPLAP live coding collective, Adrian and I decided to delete our old Slub system in 2004, continuing only with live coded performances from then on. I developed feedback.pl, a Perl live coding environment, so named because the live coded process was able to modify and reinterpret its own code. In practice this allowed the code to become a dynamic user interface much in the same way as a Max/MSP or PureData patcher language does, the code that updates data represented within the code as it is continually updated. The shift to live coding meant I was able to explore collaborations in jazz improvisation, starting with percussionist Alex Garacotche, who pushed me towards the ideal of live coding from scratch, where live coding ideas are formed and implemented during a performance. This addressed the free improvisation ethic well, but was impractical when coding in Perl; it just took too long to start making sound, or to make changes in response to those of an instrumental co-performer. When I finally succumbed to the call of academia and became a research student with Professor Geraint Wiggins in 2006, it gave me the time and resources to explore functional programming and alternative models of computation that lead to the TidalCycles system described with Thor Magnusson in in chapter 14 of the present volume (and also discussed by Wiggins and Forth in chapter 15). Finally, with TidalCycles I feel I am able to improvise with instrumentalists as equals, including through collaboration with percussionists Paul Hession and Matthew Yee-King.

I have not said a great deal about the actual algorithms I have explored in my music. This is in part left to the chapter on pattern in this volume (chapter 14, with Thor Magnusson), but I will say a few words here. I have not instigated an archaeology of my old code bases, which I fear may not be accessible on the stack of deteriorating hard drives in my studio, but I can pick out some memories. I had an early interest in simple agent-based automata, such as flower.pl, exhibited at the Sonar Festival, where agents travelled around the petals of an abstract flower, triggering sounds. Slight interaction between agents at junctions was enough for the agents to fall into a repeating pattern. Another obsession was with fork bombs, where processes split in two, rapidly taking over a host computer until it crashed; my performances often concluded with sonified fork bombs. In general, 'glitch' was embraced in Slub performances, the handmade nature of our software leaving it prone to audio drop-outs and synthesis artefacts, although because we knew the software intimately, only very rarely causing unintended crashes. Fredrik Olofsson appreciated this aesthetic, and even implemented a SuperCollider plugin, called Slub, to purposefully overload the host computer to induce glitching, still available in the official SuperCollider library of extensions. My main obsession throughout has been on different aspects of patterning, such as polymetric sequences, reflective and rotational symmetries, and the endlessly fascinating interference configurations that emerge from combining patterns in different ways. Engaging with pure functional representations has taken this obsession to a new level in TidalCycles, and I am now happily surprised by what others produce with it almost on a daily basis.

In recent years, my collaborations (see Figure 35.2 for an example) have diversified and become increasingly important to the development of my practice. Work with choreographer Kate Sicchio, live artist Hester Reeve, performance artist Susanne Palzer, and audiovisual noisemaker xname, and many one-off musical improvisations

FIGURE 35.2 Live coding performance Idle Chatter Manchester, October 2016. Pictured: Alex McLean with TidalCycles code (left) and visuals with superimposed Cyril code by Dan Hett (right). Photo Vitalija Glovackyte.

have allowed me to consider the bridge between my abstract algorithmic patterns and corporeal views of performance. A particular influence has been working with mathematician, philosopher, and textile artist Ellen Harlizius-Klück, weaver and live coder Dave Griffiths, and interlocutor Emma Cocker, looking at the correspondence between patterns in weaves, computation, and music in our 'Weaving Codes, Coding Weaves' project. A common thread in all this work is looking for ways to ground computer programming in corporeal experience, through dance, movement, or cloth.

In summary, my motivations in algorithmic music have in large part arisen from a need to express myself and connect with people through a medium. Perhaps a greater motivation is the direct experience of algorithmic music, particularly of exploring and enjoying music beyond my imagination. Computer programming is often used to model external phenomena, but in my musical practice its usefulness is in-and-of-itself. When writing code to make music, I am able to describe an experience of music that I have not yet encountered and cannot imagine until I hear it. For this reason, I feel the creativity of writing algorithmic music is just as much in listening as it is in coding. Writing code feels like series of a syntactical twists and combinations, and the physical and cognitive result can be known only by experiencing the output. This is of course not unique to algorithmic music; indeed with reference to Paul Klee, Timothy Ingold (2010) argues that all making is about *following* material, rather than the hylomorphic imposition of ideas onto the world.

35.3 OUTLOOK

Hidden under the surface of many of our comments, and emerging more strongly against this concept of hylomorphism (which has both philosophical and computational connotations) is the idea of recursion: a computational process generates a result which is immediately fed back into a repetition of that process, continuing until some predefined terminating condition is reached, a user intervenes, or a machine stops working. Recursion is also close to, or at least applicable to, many of the ideas discussed in other chapters.

Perhaps algorithmic music at large can also be seen as many simultaneous large-scale recursive processes, where outputs are not only used by their initiator, but by numerous other people. Internet exchange, web crawling or scraping, database exploration, and many other current computational processes facilitate this. In this image of algorithmic music it may be a long time before one result returns, however transmuted, to the original creator and their code (no doubt already significantly changed itself); but this may happen.

Perhaps this is merely a microcosm of creative processes across the arts and sciences at large. Or perhaps it is something in which algorithmic steps are necessary to ensure that the eventual range of outcomes is as wide-ranging as it can conceivably be. This book encourages hope that creative opportunities continue to expand and diversify. We certainly hope that the reader who has reached this point in our book will feel inclined to explore and contribute further, as we ourselves continue to do.

NOTE

1. The following gives information on CDs and other sources of the music mentioned and pointers to selected published recordings of music by the editors.

    Dean: Yamaha DX7 work with Graham Collier is on *Something British* (1985; Mosaic LP GCM 871, reissued on CD). With austraLYSIS: *Moving the Landscapes* (Tall Poppies TP007, 1992) and six subsequent CDs on the Tall Poppies label include algorithmic work, particularly *MultiPiano* (2012; TP 225) and *History Goes Everywhere* (2015). Algorithmic work with text performance (with Hazel Smith) and involving voice manipulation, is on two CDs on Rufus, and several intermedia works for radio, web, and CD-ROM. Noise sculpting is on the CD-ROM with Dean (2003), as well as on several web works. Algorithmic image as well as sound manipulation is in numerous collaborative web pieces, installation works, and an algorithmic digital novel, with Smith and Luers, called *novelling* (Cork: New Binary Press, 2016), comprising text for reading, sound, and video. Besides Hazel Smith and austraLYSIS, key collaborators include Keith Armstrong and Will Luers. Links to the intermedia work, mostly freely accessible, and details of other releases, can be found at www.australysis.com.

    McLean: Many recordings of Slub live performances are freely available online (http://slub.org/), including one released via ChordPunch (CP0X08, 2011), and two via Fals.ch (6374,

6444, both in 2002). The main published solo work is under the name Yaxu; the six-track *Peak Cut* EP on Computer Club (DISK02, 2015), released as a limited edition USB stick, containing a bootable operating system running the TidalCycles software and including the source code for the music. I also released a two-track single called *Broken*, again on Chordpunch (CP0X0D, 2013). The TidalCycles software that I initiated is available as free, open-source software online (http://tidalcycles.org/).

BIBLIOGRAPHY

Boden, M. A., and Edmonds, E. A. 'What Is Generative Art?' *Digital Creativity* 20, nos. 1–2 (2009): 21–46.

Collins, N. 'Algorithmic composition methods for breakbeat science.' *Proceedings of Music Without Walls* (2001): 21–23.

Dean, R. T. *Hyperimprovisation: Computer Interactive Sound Improvisation*. Madison, WI: A-R Editions, 2003.

Dean, R. T. 'The Serial Collaborator: A Meta-Pianist for Real-Time Tonal and Non-Tonal Music Generation.' *Leonardo* 47, no. 3 (2014): 260–261.

Dean, R. T. 'Generative Live Music-Making Using Autoregressive Times Series Models: Melodies and Beats'. *Journal of Creative Music Systems* 1, no. 2 (2017). http://jcms.org.uk/issues/Vol1Issue2/generative-live-music-making/generative-live-music-making.html.

Dean, R. T., and Smith, H. 'Sonic Narratives: Intermedia Ttransformations in the Work of austraLYSIS'. *Australasian Music Research* 8 (2003): 91–105.

Dean, R. T., and Smith, H. 'The Evolving Technology of Performance in the Work of austraLYSIS, and the Politics of Co-operativity'. *Sounds Australian* 65 (2005): 16–21.

Dean, R. T., Whitelaw, M. Smith, H. and Worrall, D. 'The Mirage of Algorithmic Synaesthesia: Some Compositional Mechanisms and Research Agendas in Computer Music and Sonification'. *Contemporary Music Review* 25, no. 4 (2006): 311–327.

Gingras, B., Pearce, M. T., Goodchild, M., Dean, R. T., Wiggins, G., and McAdams, S. 'Linking Melodic Expectation to Expressive Performance Timing and Perceived Musical Tension'. *Journal of Experimental Psychology: Human Perception and Performance* 42, no. 4 (2016): 594–609.

Ingold, T. 'The Textility of Making'. *Cambridge Journal of Economics* 34, no. 1 (2010): 91–102.

McLean, A. 'Reflections on Live Coding Collaboration'. In *Proceedings of the Third Conference on Computation, Communication, Aesthetics and X*. Porto: Universidade do Porto, 2015.

Milne, A. J., and Dean, R. T. 'Computational Creation and Morphing of Multi-Level Rhythms by Control of Evenness'. *Computer Music Journal* 40, 1 (2016): 35–53.

Pachet, F. 'The Continuator: Musical Interaction with Style'. *Journal of New Music Research* 32, no. 3 (2003): 333–341.

Pressing, J. 'Improvisation: Methods and Models'. In *Generative Processes in Music: The Psychology of Performance, Improvisation and Composition*, edited by J. A. Sloboda, 129–178. Oxford: Clarendon Press, 1988.

Pressing, J. *Compositions for Improvisers: An Australian Perspective*. Melbourne: La Trobe University Press, 1994.

Sagiv, N., Bailes, F. and Dean, R. T. 'Algorithmic Synaesthesia'. In *The Oxford Handbook of Computer Music*, edited by R. T. Dean, 294–311. Oxford: Oxford University Press, 2009.

Smith, H., and Dean, R. T. 'Introduction: Practice-Led Research, Research-Led Practice—Towards the Iterative Cyclic Web'. In *Practice-Led Research, Research-Led Practice in the Creative Arts*, edited by H. Smith and R. T. Dean, 1–38. Edinburgh: Edinburgh University Press, 2009.

Smith, H., Luers, W., and Dean, R. T. 'Motions' (2014). In *Electronic Literature Collection*, vol. 3. Cambridge, MA: Electronic Literature Organization, 2016. http://collection.eliterature.org/3/work.html?work=motions.

White, G. *Towards Maximal Convergence: The Relationship between Composition, Performance, and Production in Realtime Software Environments*. PhD dissertation, University of Newcastle, Australia, 2015.

INDEX

Page numbers followed by *t* and *f* indicate tables and figures, respectively. Numbers followed by n indicate notes.

CPSIA information can be obtained
at www.ICGtesting.com
Printed in the USA
BVHW072253050122
624973BV00003B/4